CONTENTS

		Page
	Introduction	v
Chapter 1:	How to use this book	vi

Part A: UK Regulation of Pesticides

Chapter 2:	The Legislation	1
Chapter 2.1:	The Food and Environment Protection Act 1985 (FEPA)	1
	The Control of Pesticides Regulations 1986, (as amended) (COPR)	
	Codes Of Practice	4
Chapter 2.2:	The Plant Protection Products Directive (91/414/EEC)	6
	The Plant Protection Products Regulations 1995, (as amended) (PPPR)	8
	The Plant Protection Products (Basic Conditions) Regulations 1997 (BCR)	8
	The Pesticides (Maximum Residue Levels In Crops Food and Feeding Stuffs) Regulations 1994 (as amended).	10
Chapter 2.3:	The Control of Substances Hazardous to Health Regulations 1999 (COSHH)	10
Chapter 2.4:	The Biocidal Products Directive (98/8/EC)	12
Chapter 2.5:	The Campaign Against Illegal Poisoning	14
Chapter 3:	The Approvals Process	16
	Annex A: Schedules to the Control Of Pesticides Regulations (as amended) 1986 given by Ministers.	21
	Annex B: Products approved for use by aerial application	29
	Annex C: Off-label arrangements	41
	Annex D: Commodity substance approvals	48
	Annex E: Banned and non-authorised pesticides in the UK	75
	Annex F: Active Substances subject to the Poisons Law	79
	Annex G: List of ACP Published Evaluations	80

Annex H: products approved for use in or near water (other than for public hygiene or antifouling use)   83

PART B: PSD Registered Products

Section 1: Professional Products   89
            1.1 Herbicides   90
            1.2 Fungicides   162
            1.3 Insecticides   207
            1.4 Vertebrate Control Products   221
            1.5 Biological Pesticides   223
            1.6 Miscellaneous   224
Section 2: Amateur Products   229
Section 3: Product Trade Name Index   283
Section 4: Active Substance Index   313

PART C: HSE Registered Products

Section 1:   Antifouling Products   327
Section 2:   Biocidal Paints   375
Section 3:   Insect Repellents   381
Section 4:   Insecticides (including Insecticidal Paints)   385
Section 5:   Rodenticides   437
Section 6:   Surface Biocides   443
Section 7:   Wood Preservatives   469
Section 8:   Wood Treatments   553
Section 9:   Product Name Index   557
Section 10:   Active Ingredient Index   599

PESTICIDES SAFETY DIRECTORATE
An Executive Agency of the Ministry of Agriculture, Fisheries and Food
HEALTH AND SAFETY EXECUTIVE

Pesticides 2001

Pesticides approved under:
The Control of Pesticides Regulations 1986

and

The Plant Protection Products Regulations 1995

London: The Stationery Office

First published 2001

ISBN 0 11 243060 0

INTRODUCTION

1. Pesticides 2001 contains lists of agricultural and non-agricultural pesticide products whose uses held either full or provisional approval in the UK as at **13 October 2000** for PSD products and **27 October 2000** for HSE products. The exception are:

– PSD products for which approvals were known on **15 October 2000** to be due to expire on or before **31 December 2000** and which therefore do not appear

– HSE products for which approvals were known on **26 October 2000** to be due to expire on or before **31 December 2000** and which therefore appear with the appropriate symbol (see page vi).

Unless otherwise specified products listed in this book may be sold, supplied, stored, used or advertised *subject to the requirements of the pesticides legislation*, as explained in Chapter 2, and the conditions of their approvals, which are explained in Chapter 3.

2. Additional information on the products listed is available on product labels and literature; or from manufacturers and suppliers. You should always read the product label and use pesticides safely. *You have a duty to check that you know how to use a product safely in accordance with the statutory conditions of approval.* These approval conditions are set out on the product label and in the Schedules to the Control Of Pesticides Regulations (as amended) 1986 given by Ministers.

3. Further information on the approval status of pesticides may be obtained, in the case of PSD registered products listed in Part B, from the Pesticides Safety Directorate, External Relations Branch, Mallard House, Kings Pool, 3 Peasholme Green, York, YO1 7PX (Telephone No. 01904 455775) (e-mail: p.s.d.information@psd.maff.gsi.gov.uk). In the case of HSE registered products listed in Part C, from the Health and Safety Executive, Biocides & Pesticides Assessment Unit, Magdalen House, Stanley Precinct, Bootle, Merseyside, L20 3QZ (Telephone No. 0151 951 3535 fax: 0151 951 3317, e-mail: biocides@hse.gsi.gov.uk).

1. HOW TO USE THIS BOOK

PART A

The introductory chapters give general guidance on the use of pesticides and the relevant legislative controls. The remainder of the text is divided into lists of approved products, which are indexed.

PART B

This lists products registered by the Pesticides Safety Directorate (PSD), an Executive Agency of the Ministry of Agriculture, Fisheries and Food (MAFF). It is divided into four sections:

Section 1: Lists products approved for professional use in agricultural, horticultural, forestry and other pest control areas. This is sub-divided into the following classification groups: herbicides, fungicides, insecticides and vertebrate control products, biological pesticides and miscellaneous.

Within each of these sub-sections, approved products are listed by trade name under the appropriate alphabetically listed active substance along with details of the marketing company, registration number and (where approvals for the product are being revoked) the expiry date of the product. This is the final date for approval for advertisement, sale, supply, storage and use of the product.

For products with two or more active substances, the entry is repeated in respect of each, e.g.:
   2,4-D+Mecoprop and
   Mecoprop+2,4-D

Where there are three or more active substances, the secondary active substance in each entry are in alphabetical order, e.g.:
   Bentazone+Dichlorprop+Isoproturon
   Dichlorprop+Bentazone+Isoproturon
   Isoproturon+Bentazone+Dichlorprop

Products which have more than one area of use e.g. herbicide and fungicide, appear under all the appropriate activity headings.

Section 2: Products approved for amateur use feature in this section, although these products may also be used by professionals. The areas of use are detailed as part of the active substance heading.

Section 3: An alphabetical index of product trade names with a cross reference to the first entry of the product in each section by means of a code number.

Section 4: An alphabetical index of all active substance combinations with a cross-reference to the relevant part of each section by means of a code number.

Key to symbols

C – some or all uses of this product are "approved for agricultural use" as defined by the Schedules to the Control Of Pesticides Regulations (as amended) 1986 given by Ministers under Regulation 6 of the Control of Pesticides Regulations 1986, (for further information see Chapter 2.1, paragraph 10).

A – these products are approved for use in or near water as defined by the Schedules to the Control Of Pesticides Regulations (as amended) 1986 given by Ministers under Regulation 6 of the Control of Pesticides Regulations 1986, (for further information see Chapter 3 paragraph 17).

Expiry Dates – In any cases this expiry date will be preceded by a period in which approvals for the product are wound down. The terms of a wind-down can vary but, most commonly, will see the approval holder and their agents no longer benefiting from approval for advertisement, sale, supply and use from a specific date; whilst approvals for advertisement, sale, supply, storage and use by other persons, and for storage by the approval holder and their agents, remain for a further period (up to two years). It is this later date which is listed.

PART C

This covers products registered with the Biocides & Pesticides Assessment Unit of the Health and Safety Executive (HSE) as antifouling products, biocidal paints, insect repellents, insecticides (including insecticidal paints), rodenticides, surface biocides, wood preservatives and wood treatments.

Within sections 1–8, relating to their area of use, approved products are listed below their active ingredients (which are given a code number and are listed alphabetically). In addition details of the marketing company, who can use the product and the HSE registration number are provided.

For products with two or more active ingredients, the entry is repeated for each active ingredient eg;

(1) 2-Phenylphenol and Cypermethrin
(2) Cypermethrin and 2-Phenylphenol

Products which have more than one area of use, e.g. wood preservation and surface biocide use, appear under each of the headings.

Who can use the product is given by the terms amateur, professional and industrial, which mean:

Amateur use – means that the product can be used by the general public.

Professional use – means that the product can only be used by people who are required to use pesticides as part of their work and who have received appropriate information, instruction and training.

Industrial use – means that the product is a wood preservative/treatment that can only be used for industrial pre-treatment of wood by operatives who have received appropriate information, instruction and training.

Key to symbols

W these products have either been voluntarily withdrawn for commercial reasons or revoked and hold approval for storage, supply and use but not for advertisement and sale under The Control of Pesticides Regulations 1986.

D these products have either been voluntarily withdrawn for commercial reasons or revoked and hold approval for the purposes of disposal only. HSE should be contacted for further information on individual products.

Active Ingredient Names

A number of active ingredients have several possible names. In order to compile the lists for sections 1–8, it has been necessary to standardise these names. Thus the active ingredient under which a product is listed may not be the same as that which appears on its label. These changes have been made for the purpose of Pesticides 2001 only and do not affect the name that should appear on the product label, which should be in accordance with the notice and schedule of approval.

The names that have been standardised are as follows:

Alkyldimethylbenzyl Ammonium Chloride appears as **Benzalkonium chloride**

Ammonium Alkyldimethylbenzyl Chloride appears as **Benzalkonium chloride**

Chromic Acetate appears as **Chromium acetate**

Copper (I) appears as **Cuprous**

Copper (II) appears as **Copper**

Didecyldimethylammonium Chloride appears as **Dialkyldimethyl ammonium Chloride**

Diethoxy Cetostearyl Alcohol appears as **Ceto-stearyl diethoxylate**

Dioctyldimethylammonium Chloride appears as **Dialkyldimethyl ammonium Chloride**

Disodium Tetraborate appears as **Sodium tetraborate**

Gamma-HCH appears as **Lindane**

Mono-ethoxy Oleic Acid appears as **Oleyl monoethoxylate**

Paradichlorobenzene appears as **1, 4-Dichlorobenzene**

Pyrethrum/Pyrethrins/Pyrethrum extract appears as **Pyrethrins**

Tributyltinmethacrylate Copolymer appears as **Tributyltin methacrylate**

Hydrates – Hydrated and anhydrous substances are listed as the anhydrous form.

Part C is divided into 10 sections;

SECTION 1:   Antifouling Products

SECTION 2:   Biocidal Paints

SECTION 3:   Insect Repellents

SECTION 4:   Insecticides (including Insecticidal Paints)

SECTION 5:   Rodenticides

SECTION 6:   Surface Biocides

SECTION 7:   Wood Preservatives

SECTION 8:   Wood Treatments

SECTION 9:   An alphabetical index of product names, cross-referenced to the relevant active ingredient(s) listed in parts 1–8 by code numbers.

SECTION 10: An alphabetical index of all active ingredients, cross-referenced to the relevant products listed in parts 1–8 by code numbers.

PART A

UK Regulation of Pesticides

2. THE LEGISLATION

2.1 Food and Environment Protection Act 1985 (FEPA) and The Control of Pesticides Regulations 1986 (as amended) (COPR)

Aims

1. Statutory powers to control pesticides are contained within Part III of The Food and Environment Protection Act 1985 (FEPA). Section 16 of the Act describes the aims of the controls as being to:

   i.   protect the health of human beings, creatures and plants;

   ii. safeguard the environment;

   iii. secure safe, efficient and humane methods of controlling pests; and

   iv. make information about pesticides available to the public.

2. The mechanism by which these aims are to be achieved is set out in regulations made under the Act. The Control of Pesticides Regulations 1986 (SI 1986/1510) (as amended) (COPR) define in detail those types of pesticides which are subject to control and those which are excluded; prescribe the approvals required before any pesticide may be sold, stored, supplied, used or advertised; and allow for general conditions on sale, supply, storage, advertisement, and use, including aerial application, of pesticides. The 1986 Regulations were updated by the Control of Pesticides (Amendment) Regulations 1997 (SI 1997/188) and the text of the Schedules that set out these general conditions are at Annex A. Similar provisions apply in respect of the Plant Protection Product Regulations 1995 (as amended) (PPPR) (see chapter 2.2). Similar legislation exists in Northern Ireland and the majority of products approved for use in Great Britain are subsequently approved for use in Northern Ireland.

Definition of pesticides

3. Under FEPA, a pesticide is any substance, preparation or organism prepared or used, among other uses, to protect plants or wood or other plant products from harmful organisms; to regulate the growth of plants; to give protection against harmful creatures; or to render such creatures harmless.

The term pesticides therefore has a very broad definition which embraces herbicides, fungicides, insecticides, rodenticides, soil-sterilants, wood preservatives and surface biocides among others. A more complete definition and details of pesticides which fall outside the scope of the legislation is given in Regulation 3 of COPR.

What does COPR require?

4. The law requires that only pesticides approved by Ministers in the responsible Departments shall be sold, supplied, used, stored, or advertised. Post-devolution, the Minister of Agriculture, Fisheries and Food and the UK Environmental Minister, the UK Minister for Health and Ministers for the Scottish Executive, the National Assembly for Wales and (for N.I. approvals) the Department of Agriculture and Rural Development in Northern Ireland are together responsible for the approval of all pesticides. There are three categories of approvals;

   i. an experimental approval, to enable testing and development to be carried out with a view to providing the Ministers with safety and other data;

   ii. a provisional approval, for a stipulated period; or

   iii. a full approval, for an unstipulated period.

5. Products granted only an experimental approval cannot be advertised or sold. Such products do not appear in this book.

6. A data submission deadline, as the name suggests, is the deadline by which the Registration Authorities must be in receipt of any requested data required to secure continued approval. Failure to meet this deadline will result in the revocation of approval for products concerned. For PSD products, approval for advertisement, sale, supply and use by the approval holder and their agents is immediate, whilst approval for advertisement, sale, supply, storage and use by other parties, and storage by the approval holder and their agents remains for a limited period of up to two years.

7. Phased revocation of approval also occurs where the approval holder requests that a product be commercially withdrawn. This allows set periods of time to allow existing stocks to be marketed and used. For PSD products this allows up to two years for any persons other than the approval holder or their agents to market and use existing stocks. For HSE products this allows up to

three years for the advertisement and sale of the product, and up to five years for the supply, storage and use of the product.

8. Products granted provisional or full approval are normally allowed to be advertised, sold, supplied, stored and used.

9. New approvals are set out in "The Pesticides Register" available monthly from The Stationery Office book shops. For more details see on the back cover.

10. Products listed in Part B and approved for agricultural use (as defined in the Schedules to the Control Of Pesticides Regulations 1986)(as amended) are indicated by the symbol "C". This formal classification is important for deciding whether the person storing, selling, supplying or using the pesticide needs to hold a recognised certificate of competence. The certification requirements are set out in Schedules 2 and 3 to the Control Of Pesticides Regulations 1986 (as amended) reproduced at Annex A at the end of chapter 3.

11. Products approved for amateur use (as listed in Part B, Section 2) and those registered with the HSE (as listed in Part C) do not attract storekeeper, salesperson or operator certification under COPR.

12. The supply of methyl bromide and chloropicrin is restricted to those who can prove that they are competent and have the necessary experience to use them safely.

The British Pest Control Association (BPCA) provide training and their own certification which is available to both members and non members. Competence can be proved by the possession of the relevant BPCA Certificate.

For further guidance on the use of methyl bromide, chloropicrin or other fumigants please refer to L86 "The Control of Substances Hazardous to Health in Fumigation Operations".

13. Controls over the use of adjuvants and the tank mixing of pesticides are set out in the Schedules to the Regulations, which should be consulted. The list of authorised adjuvants is published annually as a supplement in The Pesticides Register. Updates are published in subsequent monthly editions of the Register.

14. Anyone who advertises, sells, supplies, stores or uses a pesticide is affected by the legislation, including people who use pesticides in their own homes, gardens and allotments.

Access to Information on Pesticides

15. A wide range of information about pesticides is published or made available by the Government. The Pesticides Act 1998 will widen public access to information on pesticides by enabling Ministers to provide the public with information on all pesticides, by way of new regulations. Previously, Ministers were able to provide information to the public only on those chemicals approved or reviewed since the introduction of statutory controls in 1986. PSD plan to have the regulations in place by early 2001.

The independent ACP publishes an annual report on its activities. The agenda and minutes of each meeting are published on the website (www.pesticides.gov.uk/committees/ACP/acp.htm). Detailed evaluations of new or reviewed active ingredients are published, and a list of those available as at 30 November 2000 is at Annex G. The data underlying published evaluations is available for inspection at PSD York, and HSE Bootle on application (for addresses see page v).

Full results of pesticide residues monitoring in food stuffs and water are now published annually, as are usage surveys and reports of suspected incidents of human and wildlife poisoning. News of policy and procedural matters are published in The Pesticides Monitor (incorporating The Pesticides Register) and product approvals and revocations are published monthly in "The Pesticides Register". Details of how to order "The Pesticides Register" can be found on the back cover.

Codes of Practice

16. Government Departments have produced guidance to help those working with pesticides meet the requirements of the legislation. These are:

   i. The MAFF "Code of Practice for Suppliers of Pesticides to Agriculture, Horticulture and Forestry" (PB3259), gives practical guidance on the storage and transport of pesticides and the obligations on those who sell, supply and store for sale and supply. Copies are available free from: MAFF Publications, ADMAIL 6000, London SW1A 2XX Email: maff@sr-comms.co.uk and;

ii. the MAFF "Code of Practice for the Safe Use of Pesticides on Farms and Holdings" (PB3528) which promotes the safe use of pesticides. It covers the requirements of the Control of Pesticides Regulations 1986 (COPR) and the Control of Substances Hazardous to Health Regulations 1994 (COSHH). Copies are available free from MAFF Publications ADMAIL 6000, London SW1A 2XX Email: maff@sr-comms.co.uk

iii. the HSC "The Safe Use of Pesticides for Non-Agricultural Purposes: Control of Substances Hazardous to Health Regulations 1994. Approved Code of Practice" which provides advice to those in the non-agricultural sector on compliance with COSHH priced £6.95 (L9, ISBN 0-7176-0542-6)'.

iv. the HSC "Recommendations for Training Users of Non-Agricultural Pesticides" priced £2.00 (ISBN 0-11-885548-4).

v. the HSC "Approved Code of Practice for the Control of Substances Hazardous to Health in Fumigation Operations" provides practical guidance on the COSHH Regulations 1999 priced at £8.50 (L86, ISBN 0-7176-1195-7).

vi. the HSC "General COSHH (Control of Substances Hazardous to Health) ACoP, Carcinogens (Control of carcinogenic substances) ACoP and Biological Agents (Control of biological agents) ACoP" which provides guidance on the COSHH Regulations 1999 priced at £8.50 (L5, ISBN 0-7176-1670-3).

The above HSC and HSE publications are available by mail order from:

HSE Books
PO Box 1999
SUDBURY
Suffolk
CO10 2WA
Tel. (01787 881165 Fax. (01787) 313995
Email. Hsebooks@prologistics.co.uk
website: http:/www.hsebooks.co.uk

vii. the HSE/DOE "Remedial Timber Treatment in Buildings A Guide to Good Practice and the Safe Use of Wood Preservatives" priced at £4.00 (ISBN 0-11-885-9870). Copies can be ordered from The Publications Centre (mail, telephone and fax orders only) PO Box 29, Norwich NR3 1GN Telephone orders/General enquiries 0870 600 5522 Fax orders 0870 600 5533 Website: www.tso-online.co.uk

They can also be purchased from Dillons Bookstores, and ordered from Ryman the Stationer and Ryman Computer Centres telephone 020 - 856 93000.

2.2 The Plant Protection Products Directive (91/414/EEC), The Plant Protection Products Regulations 1995 (as amended)(PPPR); The Plant Protection Products (Basic Conditions) Regulations 1997 (BCR)

The Directive

1. Council Directive 91/414/EEC aims to harmonise the arrangements for authorisation of plant protection products within the Community, although product authorisation remains the responsibility of individual Member States. Annex I to the Directive is a positive list of active substances which have been successfully evaluated at EC level and considered to meet the requirements of 91/414. This list will be added to over time as and when active substances, both existing and new are evaluated. The data requirements which must be met in order to ensure that the active substance, and the products on which that active substance are based, are set out in Annex II and Annex III to the Directive. This will be built up over a period of time as existing active substances are reviewed and new ones authorised.

2. The requirements of this Directive have been implemented in Great Britain under the Plant Protection Products Regulations 1995 (PPPR) (see Sections 10 – 14 below). The Pesticides Safety Directorate is the lead authority in Great Britain for the authorisation of products under this Directive.

Review Programme

3. As well as the authorisation of plant protection products, Directive 91/414 provides for the review of active substances on the market on or before 25 July 1993. The first round of the review programme was established by Commission Regulation 3600/92 (as amended). Member states were allocated active substances from a list of 90 to evaluate on behalf of the Community.

4. Following the evaluation of the submitted data, Member States are required to send their report to the Commission with a recommendation as to whether the active substance should be listed in Annex I of the Directive, refused Annex I listing, suspended from the market pending the provision of further

data, or whether a decision on listing should be postponed pending the provision of further data.

5. The second and future rounds of the review programme were established by Commission Regulation 451/2000. This Regulation sets out the list of substances to be evaluated in the second round and establishes a procedure for dealing with all other active substances used in plant protection products.

New Active Substances

6. Companies with active substances new to the Community apply to a Member State of their choice for assessment on behalf of the Commission. The Member State initially ensures that the application dossier is compliant with the requirements of the Directive and then, following detailed evaluation of the data, submits a review report to the Commission with recommendations regarding whether or not the active substance should be included in Annex I of the Directive.

7. The recommendation of the Member States in relation to the Annex I inclusion of new and existing active substances will be considered by all Member States in the framework of the Standing Committee on Plant Health and a decision taken by the Commission. The Commission has agreed the inclusion of a number of active substances (new and existing) on Annex I.

Uniform Principles, Transitional arrangements, Mutual Recognition

8. On completion of an EC review of an active substance Member States will review each of the products that contain it. "Uniform Principles" establishing common criteria for evaluating products at a national level were agreed in September 1997 to enable harmonisation of the evaluation of applications for pesticide products by Member States.

9. Under transitional arrangements Member States may, prior to the listing of active substances in Annex I:

a. Provisionally authorise products containing active substances not on the market by 25 July 1993 for periods of up to three years in advance of Annex I listing;

b. maintain existing national approvals (in the case of Great Britain under the Control of Pesticides Regulations) until 25 July 2003, or until completion of the review of the active substance, whichever is earlier; and

c. authorise new products containing active substances on the Community market on 25 July 1993 under existing national rules.

10. The Directive provides for "mutual recognition" for plant protection products with Annex I listed active substances which are authorised by a Member State in accordance with the provisions of the Directive. Subject to certain conditions, such plant protection products must be authorised by other Member States. Authorisation may be subject to conditions set to take account of differences such as climate and agricultural practice.

11. Detailed consideration of the impact of the Directive on approval procedures in Great Britain continues. Revised procedures are published in the "The Pesticides Monitor (incorporating The Pesticides Register)" and in the Registration Handbook.

The Plant Protection Products Regulations 1995 (as amended); The Plant Protection Products (Basic Conditions) Regulations 1997

Aim

12. The Plant Protection Products Regulations 1995 (SI 1995/887) (as amended) (PPPR) implement the Plant Protection Products Directive, 91/414/EEC, in Great Britain. They define those products which are subject to regulation and prescribe the approvals required before any plant protection product (mainly agricultural pesticides) may be placed on the market or used. The Plant Protection Products (Basic Conditions) Regulations 1997 (SI 1997/189) (BCR) apply control and enforcement provisions similar to those of COPR to plant protection products, and thus ensure that enforcement powers for products authorised under PPPR are equivalent to those under COPR. The general conditions relating to the sale, supply, storage, advertisement and use of plant protection products are set out in Schedules 1 to 4 of BCR. As with COPR, similar legislation exists in Northern Ireland.

Definition of a plant protection product

13. A plant protection product is an active substance or a preparation containing one or more active substances, put up in the form in which it is supplied to the user, intended to:

a. protect plants or plant products against all harmful organisms or prevent the action of such organisms;

b. influence the life processes of plants, other than as a nutrient (for example, as a growth regulator);

c. preserve plant products, in so far as such substances or products are not subject to provisions of Community law on preservatives;

d. destroy undesired plants; or

e. destroy parts of plants or check or prevent the undesired growth of plants.

14. As active substances are placed on Annex I of the Directive, plant protection products containing only those active substances are required to be regulated under PPPR. Products based on active substances on the market after 25th July 1993 (and not yet included in Annex I) are also required to be regulated, on a provisional basis, under PPPR.

What does PPPR require?

15. As with COPR, the Regulations require that only plant protection products approved by Ministers in the responsible Departments (see Section 2.1, para 4) shall be used or placed on the market. There are three main categories of approvals;

i. Standard approval, to place on the market and use any plant protection product for a period not exceeding ten years. A standard approval is only appropriate where the active ingredient is on Annex I to Directive 91/414/EEC.

ii. Provisional approval, to place on the market and use any plant protection product for a period not exceeding three years. A provisional approval is appropriate where the Commission decision on the Annex I listing of the active substance is awaited.

iii. Approval for research and development, to enable experiments or tests for research and development to be carried out where such tests involve the release into the environment of a product which has not been otherwise approved.

16. Approvals may be renewed by Ministers for standard approvals (if requirements for approval continue to be satisfied) and for provisional approvals where the Commission's decision on the Annex I listing of the active substance continues to be awaited.

The Pesticides (Maximum Residue Levels In Crops, Food and Feeding Stuffs) (England and Wales) Regulations 1999 (as amended).

17. Residues of pesticides in food and feeding stuffs are controlled by a system of statutory maximum residue levels (MRLs). To date these have been set for certain key pesticides used on the more important components of the average national diet. MRLs are intended to check that pesticides are being used correctly, and also to provide additional protection for the consumer.

2.3 The Control of Substances Hazardous to Health Regulations 1999 (COSHH)

1. Pesticides should only be used when necessary, and if the benefit from using them significantly outweighs the risks to human health and the environment. If a decision is reached to use a pesticide, the product selected should be one that poses the least risk to people, livestock and the environment yet is still effective in controlling the pest, disease or weed problem that has been identified.

2. The COSHH Regulations apply to a wide range of pesticides used at work. They lay down essential requirements and a step by step approach for the assessment and control of exposure to hazardous substances. Failure to comply with COSHH, in addition to possibly exposing people to risk, constitutes an offence and is subject to penalties under the Health and Safety at Work etc. Act 1974.

3. Substances (including pesticides and other chemicals) that are "hazardous to health" include those labelled very toxic, toxic, harmful, irritant or corrosive, substances with occupational exposure limits, harmful micro-organisms, or substantial quantities of dust. They also include fumes, gases and other materials which could cause comparable harm to health.

4. Employers and the self-employed are required by the COSHH Regulations 1999 to carry out an assessment of the risks to human health. The assessment may confirm that the pesticide chosen is the most appropriate, but if it becomes apparent that using another pesticide may involve less risk, then the choice will need to be reconsidered. Always make sure that the product is approved for the intended use and situation.

Making a COSHH Assessment

5. Before carrying out an assessment it should be remembered that the harm from pesticides depends mainly on the active substance, its form, e.g. granules, liquid etc. and how the product is to be applied. Hazardous substances can get into the body through the skin, the mouth or by inhalation. The greatest risk of exposure is often through skin absorption.

6. Information supplied by the manufacturer will provide the basis for an assessment but don't forget other sources of advice, such as Codes of Practice and industry guidance.

7. The COSHH assessment needs to be suitable and sufficient. It should include the following steps:

Step 1 – Identify the hazards

- Read the pesticide label and any other literature available from the manufacturer.
- What form is the pesticide in?

Step 2 – Who might be harmed and how?

- Who might be exposed – workers, members of the public?
- Consider both normal and accidental exposure

Step 3 – Decide what needs to be done.

- Consider all aspects of the job, e.g. storage, transport, application, disposal, cleaning and maintenance of both application equipment and personal protective equipment (PPE).
- Apply the hierarchy of control:–

    1. Could the product be substituted for a less hazardous pesticide

    2. Can engineering/technical controls reduce exposure, e.g. induction hoppers, closed transfer systems and in-cab controls?

    3. Are operational controls in use, e.g. warning signs and keeping people out of treated areas, notification of adjacent occupiers and others who need to know?

    4. PPE should only be used where substitution, engineering/technical controls and operational controls are not feasible. The PPE must be suitable for the job. If there is any doubt, the manufacturer or supplier of the pesticide should be consulted.

- Are those involved adequately trained and informed about items such as the importance of equipment calibration, weather conditions, timing of application and harvest intervals?
- Has the most appropriate application equipment been selected?
- Is health or medical surveillance required?

Step 4 – Record the findings if it is useful to do so.

Step 5 – Review the assessment from time to time.

2.4 The Biocidal Products Directive

1. The Biocidal Products Directive (98/8/EC) will be implemented in the UK by the Biocidal Products Regulations. Due to be implemented in all Member States by 14 May 2000, the Directive harmonises trade in biocidal products by introducing a community wide scheme for authorising the placing of biocidal products on the market. Biocidal products include: all products currently regulated under CoPR where HSE takes the lead and, in addition, preservatives, disinfectants, water biocides, certain cleaners claiming biocidal properties and other specialist products.

The Directive will not apply to certain products already subject to European legislation including plant protection products, human medicines, veterinary medicines (such as sheep dips), medical devices or cosmetics. The legislation also excludes products that act by physical means and the non-biocidal uses of products and active substances.

The authorisation scheme will be similar to that introduced by the Plant Protection Products Directive, described in 2.2 above. A two-stage process will be introduced whereby active substances are assessed at community level for possible inclusion on Annex I of the Directive and then products containing Annex I-listed active substances are authorised by Member States. Biocidal products containing an active substance which was not on the market before 14 May 2000 cannot be placed on the market until they are authorised following Annex I listing of the active substance.

2. Biocidal products containing active substances already on the market before 14 May 2000 will be able to stay on the European market, subject to national legislation, if any, and the Transition Arrangements. The Directive envisages a 10-year programme for the review of existing active substances

for possible inclusion on Annex I. Implementation of this programme will be by a series of Review Regulations. The first of these was published in September 2000 (No. 1896/2000).

This first Review Regulation requires that Industry Identifies existing active substances and Notifies those which they wish to have evaluated for possible inclusion on Annex I. The deadline for identification and notification is 27 March 2002. This first Regulation also gives the deadline, 27 March 2004, for submission of Dossiers supporting the Annex I listing application for active substances used in Product types (PT) 8 (Wood preservatives) and 14 (Rodenticides).

Active substances which are not Identified or Notified and biocidal products containing them, will have to be withdrawn from the EU market. The deadlines for these withdrawals will be given in a subsequent Review Regulation. Additionally, active substances used in PT's 8 or 14 will have to be withdrawn from the market if not Identified or Notified or if the dossiers are not submitted according to the deadlines.

HSE, as lead authority in Great Britain for BPD, has produced guidance on the first Review Regulation, available from the Biocides & Pesticides Assessment Unit (ref 1/2000/RP).

3. The Directive will be supported by a number of EU-wide Technical Notes for Guidance (TNG's). The first of these, on Data Requirements, is currently available on the European Chemical Bureaus website (http://ecb.ei.jrc.it/biocides/) as a draft, to be finalised following experience in use. Guidance on the preparation of Dossiers is expected to be similarly available in draft form soon. Those on procedures for listing active substances on Annex I and for the authorisation of products will be finalised later.

The European Commission, together with the Member States and Industry continue to work hard to finalise other aspects of the implementation of the Directive. Discussions on the scope of the Directive and boundaries with other Directives are close to completion and work is progressing in other areas (e.g. data protection, implementation of Art 25 on charging).

4. HSE has been developing guidance for operating the new Regulations as soon as they are in force and work is required. Guidance for suppliers, importers and users of biocidal products accompanying the Regulations will be available from HSE Books once the Regulations are published.

Procedures for receiving and processing applications for Annex I listing and authorisation of products have been developed. An 'Applicant's Handbook' giving guidance to applicants will be available from BPAU once the Regulations are published. As it draws heavily on the EU guidance, there will be gaps which can only be filled when they become a priority at community level. Work to establish the independent Biocides Consultative Committee has commenced and the Committee should meet for the first time in 2001.

5. The Directive requires that Member States charge for the work done under the Directive. Work on the charging regime is being finalised, including, where appropriate, the level of individual fees.

2.5 The Campaign Against Illegal Poisoning

The Campaign Against Illegal Poisoning of Wildlife, aimed at protecting some of Britain's rarest birds of prey and wildlife whilst also safeguarding domestic animals, was launched in March 1991 by MAFF and the Department of the Environment, Transport and the Regions. It is strongly supported by a range of organisations associated with animal welfare, nature preservation, field sports and game keeping including the RSPB, English Nature, the British Field Sports Society and the Game Conservancy Trust.

The three objectives of the Campaign are;
- To advise farmers, gamekeepers and other land managers on legal ways of controlling pests;
- To advise the public on how to report illegal poisoning incidents and to respect the need for legal alternatives;
- To investigate incidents and prosecute offenders.

All these contribute to the main objective which is to deter those who may be considering using pesticides illegally from doing so.

Since the campaign started, much work has been done to achieve these objectives. A freephone number (0800 321 600) has been established to make it easier for the public to report incidents and numerous leaflets, posters, postcards and stickers have been created and distributed in order to publicise the existence of the Campaign. A video has also been produced and this is used to illustrate the many talks, demonstrations and exhibitions which are regularly presented by ADAS Consulting Ltd. on behalf of MAFF in England and Wales.

The Campaign arose from the results of a MAFF scheme for the investigation of possible cases of illegal poisoning, the Wildlife Incident Investigation Scheme. Under this scheme, all reported cases are considered and thoroughly investigated where appropriate. Enforcement action is taken wherever sufficient evidence of an offence can be obtained. Since the launch of the Campaign there have been over 45 prosecutions.

The message is getting through that this dangerous and illegal poisoning will not be tolerated. Your help in reporting possible incidents and spreading the word will be much appreciated.

Further information about the Campaign is available from:

Pesticides Safety Directorate
Research Co-ordination and Environmental Policy Branch
Room 317
Mallard House
Kings Pool
3 Peasholme Green
York
YO1 7PX

Web-site address is: www.pesticides.gov.uk

3. THE APPROVALS PROCESS

1. Approvals are granted by Ministers in response to an application from a manufacturer, formulator, importer or distributor (or in certain circumstances a user) supported by the necessary data on safety, efficacy and, where relevant, humaneness. Approvals are normally granted only in relation to individual products and only for specified uses, and remain subject to immediate revocation, suspension or amendment at any time if safety considerations so demand.

Statutory Conditions of Use

2. It is an offence not to follow the statutory conditions of use of a pesticide: these are detailed in the relevant notice of approval. Details of approvals are published in "The Pesticides Register". Additionally copies of approvals may be obtained from PSD, Information Section, Mallard House, Kings Pool, 3 Peasholme Green, York, Y01 7PX (Telephone No. 01904 455775). Additionally COPR require that the product label be consistent with the statutory conditions of use. For "off-label" approvals and commodity substances – see below. Statutory conditions of use will differ from product to product and use to use: they cover:

(i)   Field of use restrictions, e.g. agriculture, wood preservative.

(ii)   User restrictions

(iii)   The crop or situation which may be treated.

(iv)   Maximum individual dose/application rate.

(v)   Maximum number of treatments or maximum total dose.

(vi)   Maximum area or quantity which may be treated;

(vii)   Latest time of application, harvest or re-entry interval.

(viii) Operator protection or training requirements.

(ix)   Environmental protection requirements.

(x)   Any other specific restrictions which Ministers may require.

It is therefore important to *read the label (including any accompanying leaflet) carefully, if possible before purchase and certainly before use.*

3. In order to highlight the statutory conditions to the user the following steps have been taken:

Since 31 December 1994 all approved pesticides must display the statutory conditions of use in a designated area on the label known as the "statutory box".

Approvals for (Extensions of Use) Off-Label Uses

4. Users of agricultural pesticides may apply to have the approval of specific pesticides extended to cover uses additional to those approved and shown on the manufacturer's product label. Any such specific off-label approvals granted may have additional conditions of use attached to them. Use in these cases is undertaken at the user's choosing, and the commercial risk is entirely theirs. Users are required to be in possession of the relevant notice of approval. Approvals are published in the "Pesticides Register" and are available from ADAS or NFU offices. For further details contact ADAS Terrington, Terrington St Clements, Kings Lynn, Norfolk, PE34 4PW Tel (01553) 828621 Fax (01553) 827229 EMAIL sarah.eastwood@adas.co.uk. Additionally electronic copies of off-label approvals can be obtained from the PSD website: www.pesticides.gov.uk (follow the links through "Farmers and Growers" and "search the off-label database"). For further information contact PSDs Approvals Support Group (01904 455714, a.s.g@psd.maff.gsi.gov.uk)

5. Since 1 January 1990 arrangements have been in place which permit many pesticide products to be used for additional specific minor uses, subject to adherence to various conditions. These arrangements were updated in December 1999 to current standards and additional guidance notes provided to improve clarity. Full details are given in Annex C at the end of this section. These arrangements are valid until 31 December 2005. Use in these cases is undertaken at the user's choosing, and the commercial risk is entirely theirs.

Approval of Commodity Substances

6. Commodity substances are chemicals which have a variety of non-pesticidal uses and also have minor uses as pesticides. If such a substance is to be used as a pesticide, it requires approval under COPR, and is granted approval for use only.

7. *Sale, supply, storage and advertisement* of a commodity substance as a pesticide is an offence unless specific approval has been granted for the substance to be marketed as a pesticide under an approved label. Anyone wishing to sell, supply, store or advertise any of these substances specifically as pesticides therefore has to seek specific approval under COPR.

8. The commodity substances that can be used as a pesticide are listed below. For some substances separate approval is given for different fields of use. The approvals are listed at Annex D and detail the conditions which a user must follow when the commodity substance is used as a pesticide;

4-chloro-*m*-cresol
Camphor
Carbon dioxide
Disodium octoborate
Ethanol
Ethyl acetate
Formaldehyde
Isopropanol
Methyl bromide
Liquid nitrogen
Paraffin oil
Sodium chloride
Sodium hypochlorite
Strychnine hydrochloride
Sulphuric acid
Tetrachloroethylene
Thymol
Urea
White spirit

List of Products Approved for Aerial Application

9. Products approved for use by aerial application, together with details of the crops on which they are approved for use are listed in Annex B at the end of this chapter. Only products approved for aerial application may be used for that purpose, and regular returns by any person who undertakes the aerial application will have to be made to the Pesticides Usage Survey Group, MAFF, Central Science Laboratory, Sand Hutton, York, YO4 1LZ.

Amendment, Expiry, Suspension and Revocation of Approvals

10. Pesticide approvals may at any time be subject to review, amendment, suspension or revocation. Revocation of approval may occur for a number of reasons, for example the identification of safety concerns or an approval holder's failure to meet a data submission deadline. Where possible a "phased revocation" will be implemented, but when safety considerations make it necessary immediate revocation may take effect. On expiry or revocation of approvals it becomes unlawful to advertise, sell, supply, store or use the products for the uses concerned.

11. Suspension of approval may take place when it is anticipated that the action is temporary, e.g. where approval might be reinstated by the provision of required and satisfactory data.

12. A list of pesticide active ingredients banned or severely restricted in the UK can be found at Annex E.

Authorised Adjuvants

Adjuvants

13. An adjuvant is a substance other than water which is not in itself a pesticide but which enhances or is intended to enhance the effectiveness of the pesticide with which it is used. Adjuvants for use with agricultural pesticides have been categorised as extenders, wetting agents, sticking agents and fogging agents.

14. Schedule 3, paragraph 5(1) of COPR (as amended) permits the use of an adjuvant with a pesticide only where:
    i.   the use is in accordance with the conditions of approval of the pesticide;
    ii. the adjuvant appears on a list of adjuvants published by Ministers; and
    iii. the use is in accordance with any conditions to which use of that adjuvant with that pesticide is subject.

15. The List of adjuvants includes details of all the conditions to which use of particular adjuvants are subject. The List is published annually in full as a supplement to The Pesticides Register. Updates are published in subsequent monthly editions of the Register. Copies of the List or of individual list entries are available on request from PSD and the list can also be viewed on the PSD website www.pesticides.gov.uk.

List of Products Approved for Use in or Near Water

16. Pesticides must not be used in or near water unless the conditions of approval, specifically allows such use. Such pesticides may be subject to additional statutory conditions which will be detailed on the product label. Revised "Guidelines for the use of Herbicides on Weeds in or near Watercourses and Lakes" (PB 2289) were published in September 1995 and advise on the precautions that should be taken especially where fresh water may be abstracted for the irrigation of crops, watering of livestock, fisheries or in fish farming. Situations defined for such treatment include drainage channels, streams, rivers, ponds, lakes, reservoirs, canals and dry ditches. These guidelines also cover the control of vegetation growing on the banks, or areas immediately adjacent to such water bodies, but not the control of vegetation growing on nearby cropped or amenity land. Copies can be obtained free from MAFF Publications, ADMAIL 6000, London SW1A 2XX (Telephone No. 0645 556000).

17. Products approved for use in or near water (other than for public hygiene or antifouling use) are listed in Annex H at the end of this chapter.

ANNEX A

FOOD AND ENVIRONMENT PROTECTION ACT 1985
PART III

SCHEDULES TO THE CONTROL OF PESTICIDES REGULATIONS (AS AMENDED) 1986 GIVEN BY MINISTERS

The advertisement, sale, supply, storage and use of pesticides is prohibited unless the conditions of consent set out in the appropriate Schedules to the Control Of Pesticides Regulations 1986 (as amended) are met.

Ministers gave their consent to the advertisment, sale, supply and storage and use of pesticides, subject to the conditions set out in the Schedules to COPR, on 31 January 1997 (these were published, on that date, in the London and Edinburgh Gazettes.) The schedules to COPR may be changed at any time following Parliamentary approval. The schedules to COPR are published in the "Pesticides Register" and notified to interested parties.

SCHEDULE 1 – CONDITIONS RELATING TO CONSENT TO THE ADVERTISMENT OF PESTICIDES

1.–(1) An advertisement of a pesticide shall relate only to such conditions as are permitted by the approval given in relation to that pesticide.

(2) No advertisement of a pesticide shall contain any claim for safety in relation to that pesticide which is not permitted by the approval given in relation to that pesticide to be on the label for the pesticide.

2. –(1) Any advertisement of a pesticide, other than a notice at the point of sale which is intended to draw attention solely to product name and price, shall include –

(a) a statement of each active ingredient of each pesticide mentioned in the advertisement, such statement being the name by which each active ingredient is identified in the approval given in relation to the pesticide in which it is contained;

(b) a general warning as follows:

"Always read the label. Use pesticides safely"; and

(c) where required by a condition of the approval given in relation to a pesticide mentioned in the advertisement, a statement of any special degree of risk to human beings, creatures, plants or the environment.

(2) Notwithstanding sub-paragraph (1)(a) above –

(a) any price list consisting only of an indication of product availability and price need not state the active ingredient of each pesticide;

(b) any advertisement of a range of pesticides need only state the active ingredients of those individual products which are identified by name.

(3) Any statement or warning given under this paragraph shall be –

(a) in the case of a printed or pictorial advertisement, clearly presented separately from any other text; and

(b) in the case of an advertisement which is broadcast or recorded or is stored or transmitted by electronic means, clearly spoken or shown separately.

3. In this Schedule "advertisement" means any printed, pictorial, broadcast or recorded advertisement and includes any advertisement which is stored or transmitted by electronic means

SCHEDULE 2 – CONDITIONS RELATING TO CONSENT TO THE SALE, SUPPLY AND STORAGE OF PESTICIDES

1. It shall be the duty of all employers to ensure that persons in their employment who may be required during the course of their employment to sell, supply or store pesticides are provided with such instruction, training and guidance as is necessary to enable those persons to comply with any requirements provided in and under these Regulations.

2.–(1) Any person who sells, supplies or stores a pesticide shall –

(a) take all reasonable precautions, particularly with regard to storage and transport, to protect the health of human beings, creatures and plants, safeguard the environment and in particular avoid the pollution of water; and

(b) be competent for the duties which that person is called upon to perform.

(2) In this paragraph "water" means –

(a) any surface water;

(b) any ground water.

3. No person shall sell, supply or otherwise market to the end-user an approved pesticide other than in the container which has been supplied for that purpose by the holder of the approval of that pesticide and labelled in a manner consistent with the approval.

4. No person shall store for the purpose of sale or supply a pesticide approved for agricultural use in a quantity in excess of, at any one time, 200 kg or 200 litres, or a similar mixed quantity, unless that person –

(a) has obtained a certificate of competence recognised by the Ministers, or

(b) stores that pesticide under the direct supervision of a person who holds such a certificate.

5. No person shall sell, supply or otherwise market to the end-user a pesticide approved for agricultural use unless that person –

(a) has obtained a certificate of competence recognised by the Ministers, or

(b) sells or supplies that pesticide under the direct supervision of a person who holds such a certificate.

6.–(1) In paragraphs 4 and 5 above –

"pesticide approved for agricultural use" means a pesticide (other than a pesticide with methyl bromide or chloropicrin as one of its active ingredients) approved for one or more of the following uses –

(a) agriculture and horticulture (including amenity horticulture);

(b) forestry;

(c) in or near water other than for amateur, public hygiene or anti-fouling uses;

(d) industrial herbicides, including weed-killers for use on land not intended for the production of any crop.

(2) In this paragraph "water" means any surface water.

SCHEDULE 3 – CONDITIONS RELATING TO CONSENT TO THE USE OF PESTICIDES

1. It shall be the duty of all employers to ensure that persons in their employment who may be required during the course of their employment to use pesticides are provided with such instruction, training and guidance as is necessary to enable those persons to comply with any requirements provided in and under these Regulations.

2.–(1) Any person who uses a pesticide shall take all reasonable precautions to protect the health of human beings, creatures and plants, safeguard the environment and in particular avoid the pollution of water.

(2) In this paragraph "water" means –

    (a) any surface water;

    (b) any ground water.

3. No person in the course of a business or employment shall use a pesticide, or give an instruction to others on the use of a pesticide, unless that person –

    (a) has received adequate instruction, training and guidance in the safe, efficient and humane use of pesticides, and

    (b) is competent for the duties which that person is called upon to perform.

4. Any person who uses a pesticide shall confine the application of that pesticide to the land, crop, structure, material or other area intended to be treated.

5.–(1) Subject to sub-paragraph (4) below, no person shall use a pesticide in conjunction with an adjuvant in any manner unless –

    (a) that adjuvant has been specified, upon application by any person (in this paragraph 5 referred to as "the applicant") to the Ministers, in a list of adjuvants published by the Ministers from time to time (in this paragraph 5 referred to as "the list"); and

    (b) the use of that pesticide with that adjuvant in that manner is in accordance with –

        (i) the conditions of the approval given in relation to that pesticide; and

        (ii) any requirements to which the use of that adjuvant with that pesticide is subject, as determined or amended under sub-paragraph (2)(a)(ii) or (iii) below.

(2) In the application of this paragraph –

    (a) the Ministers may, in relation to any adjuvant specified in the list, at any time –

        (i) determine data requirements (concerning human safety or environmental protection) to which the specification of that adjuvant in the list shall be subject;

        (ii) determine requirements to which the use of that adjuvant with approved pesticides shall be subject;

        (iii) for reasons of human safety or environmental protection, or with the consent of the applicant, amend any requirement which has been determined under sub-paragraph (ii) above;

    (b) the Ministers shall, in relation to any adjuvant specified in the list, also specify in that list any requirements which they have determined or amended under paragraph (a)(ii) or (iii) above.

(3) In the application of this paragraph –

    (a) the Ministers may, in relation to any adjuvant specified in the list, remove that adjuvant from the list –

(i) if it appears to them that the applicant has failed to comply with any data requirement which has been determined in relation to that adjuvant under sub-paragraph (2)(a)(i) above;

(ii) if it appears to them that any relevant literature relating to the adjuvant is not in accordance with any requirement to which the use of that adjuvant is subject, as determined or amended under sub-paragraph (2)(a)(ii) or (iii) above;

(iii) if it appears to them that –

(aa) any relevant literature relating to the adjuvant refers to a pesticide, and

(bb) the use of that adjuvant with that pesticide is not in accordance with the conditions of the approval given in relation to that pesticide;

(iv) for reasons of human safety or environmental protection;

(v) at the request of the applicant;

(b) the Ministers shall, upon a decision to remove an adjuvant from the list specify in the list –

(i) that decision, and

(ii) the date on which, and any conditions in accordance with which, the removal is to take effect;

(c) "relevant literature", in relation to any adjuvant, means –

(i) the labelling of the packaging in which the adjuvant is contained;

(ii) any leaflet accompanying that package;

(iii) any other literature produced by, or on behalf of, the applicant describing the adjuvant.

(4) This paragraph shall not apply where the use of an adjuvant with an approved pesticide is for the purpose of research or development and is carried out under the direct control of the person intending to place the adjuvant on the market.

(5) In this paragraph "adjuvant" means a substance other than water, without significant pesticidal properties, which enhances or is intended to enhance the effectiveness of a pesticide when it has been added to that pesticide.

6.–(1) No person shall combine or mix for use two or more pesticides which are anticholinesterase compounds unless such a mixture is expressly permitted by the conditions of the approval given in relation to at least one of those pesticides or by the labelling of the container in which at least one of those pesticides has been sold, supplied or otherwise marketed to that person.

(2) No person shall combine or mix for use two or more pesticides unless –

(a) all of the conditions of approval given in relation to each of those pesticides, and

(b) the labelling of the container in which each of those pesticides has been sold, supplied or otherwise marketed to that person,

can be complied with.

7.–(1)No person in the course of a commercial service shall use a pesticide approved for agricultural use unless that person –

(a) has obtained a certificate of competence recognised by the Ministers; or

(b) uses that pesticide under the direct and personal supervision of a person who holds such a certificate; or

(c) uses it in accordance with an approval, if any, for one or more of the following uses –

(i) home garden (amateur gardening);

(ii) animal husbandry;

    (iii) food storage practice;

    (iv) vertebrate control (including rodenticides and repellents);

    (v)   domestic use;

    (vi) wood preservation;

    (vii) as a surface biocide;

    (viii) public hygiene or prevention of public nuisance;

    (ix) other industrial biocides;

    (x)   as an anti-fouling product;

    (xi) "other" (as may be defined by the Ministers).

(2) In this paragraph "commercial service" means the application of a pesticide by a person –

  (a) to crops, land, produce, materials, buildings or the contents of buildings not in the ownership or occupation of that person or that person's employer;

  (b) to seed other than seed intended solely for use by that person or that person's employer.

8. No person who was born later than 31 December 1964 shall use a pesticide approved for agricultural use unless that person –

  (a) has obtained a certificate of competence recognised by the Ministers; or

  (b) uses that pesticide under the direct and personal supervision of a person who holds such a certificate; or

  (c) uses it in accordance with an approval, if any, for one of the uses specified in paragraph 7(1)(c) above.

9.–(1) In paragraphs 7 and 8 above "pesticide approved for agricultural use" means a pesticide (other than a pesticide with methyl bromide or chloropicrin as one of its active ingredients) approved for one or more of the following uses –

  (a) agriculture and horticulture (including amenity horticulture);

  (b) forestry;

  (c) in or near water, other than for amateur, public hygiene or anti-fouling uses;

  (d) industrial herbicides, including weed-killers for use on land not intended for the production of any crop.

(2) In this paragraph "water" means any surface water.

SCHEDULE 4 – CONDITIONS RELATING TO CONSENT TO THE USE OF PESTICIDES BY AERIAL APPLICATION

1. No person shall undertake an aerial application of a pesticide unless –

    (a) an aerial application certificate granted under article 42(2) of the Air Navigation Order 1985 is held by that person, that person's employer or the main contractor undertaking the aerial application, and

    (b) the pesticide to be used has been approved for the intended aerial application.

2.–(1) No person shall undertake an aerial application of a pesticide unless that person, or a person specifically designated in writing on that person's behalf, has –

    (a) not less than 72 hours before the commencement of the aerial application consulted the relevant authority if any part of land which is a Local Nature Reserve, a Marine Nature Reserve, National Nature Reserve or Site of Special Scientific Interest lies within 1500 metres of any part of the land to which that pesticide is to be applied;

    (b) not less than 72 hours before the commencement of the aerial application consulted the appropriate area office of the Environment Agency (if the area in which the intended aerial application is to take place is in England and Wales) or the appropriate area office of the Scottish Environment Protection Agency (if such area is in Scotland) if the land to which that pesticide is to be applied is adjacent to, or within 250 metres of, water;

    (c) obtained the consent of such office if that pesticide is to be applied for the purpose of controlling aquatic weeds or weeds on the banks of watercourses or lakes;

    (d) not less than 24 hours and (so far as is practicable) not more than 48 hours before the commencement of the aerial application, given notice of the intended aerial application to the Chief Environmental Health Officer for the district in which the intended aerial application is to take place;

    (e) not less than 24 hours and (so far as is reasonably practicable) not more than 48 hours before the commencement of the aerial application given notice of the intended aerial application to the occupants or their agents of all property within 25 metres of the boundary of the land to which that pesticide is to be applied;

    (f) not less than 24 hours and (so far as is practicable) not more than 48 hours before the commencement of the aerial application, given notice of the intended aerial application to the person in charge of any hospital, school or other institution any part of the curtilage of which lies within 150 metres of any flight path intended to be used for the aerial application; and

    (g) not less than 48 hours before the commencement of the aerial application, given notice of the intended aerial application to the appropriate reporting point of the local beekeepers' spray warning scheme operating within the district in which the intended aerial application is to take place.

(2) A notice of an intended aerial application given under paragraph (e) or (f) of sub-paragraph (1) above shall be in writing and include details of–

    (a) the name and address, and telephone number (if any), of the person intending to carry out the aerial application;

    (b) the name of the pesticide to be applied and its active ingredient and approval registration number;

    (c) the intended time and date of application; and

    (d) an indication that the same details have been served on the Chief Environmental Health Officer for the district in which the intended aerial application is to take place.

3. No person shall undertake an aerial application of a pesticide unless –

(a) the wind velocity at the height of application at the place of intended aerial application does not exceed 10 knots, except where the approval given in relation to that pesticide permits aerial application when such wind velocity exceeds 10 knots;

(b) not less than 24 hours before the aerial application, that person has provided and put in place within 60 metres of the land to which that pesticide is to be applied signs, of adequate robustness and legibility, to warn pedestrians and drivers of vehicles of the time and place of the intended aerial application; and

(c) before the aerial application that person has provided ground markers in all circumstances where a ground marker will assist the pilot to comply with the provisions of paragraph 5 below.

4. Any person who undertakes the aerial application of a pesticide shall –

(a) keep and retain for not less than 3 years after each application records of –

    (i)   the nature, place and date of that application;

    (ii) the registration number of the aircraft used;

    (iii) the name and permanent address of the pilot of that aircraft;

    (iv) the name and quantity of the pesticide applied;

    (v) the dilution and volume of application of the pesticide applied;

    (vi) the type and specification of application system (which may include nozzle type and size);

    (vii) the method of application;

    (viii) the flight times of the aerial application;

    (ix) the speed and direction of the wind during the application; and

    (x) any unusual occurrences which affected the application;

(b) provide the Ministers with summaries of the records required by sub-paragraph (a) above, in any manner which they may require under section 16(11) of the 1985 Act, within 30 days after the end of the calendar month to which those records relate.

5. The pilot of an aircraft engaged in an aerial application shall –

(a) maintain the aircraft at a height of not less than 200 feet from ground level when flying over an occupied building or its curtilage;

(b) maintain the aircraft at a horizontal distance from any occupied building and its curtilage, children's playground, sports ground or building containing livestock of –

    (i) not less than 30 metres, if the pilot has the written consent of the occupier; and

    (ii) not less than 60 metres, in any other case;

(c) maintain the aircraft at a height of not less than 250 feet from ground level over any motorway, or of not less than 100 feet from ground level over any other public highway, unless that motorway or public highway has been closed to traffic during the course of the application.

6. For the purposes of this Schedule –

"appropriate nature conservation agency" means English Nature, Scottish Natural Heritage and the Countryside Council for Wales;

"curtilage", in relation to any building, means the land attached to, and forming one enclosure with, that building;

"ground marker" includes a person who is instructed by a person intending to carry out an aerial application to be present on or near to the land to which the pesticide is to be applied so that that person is able to communicate with the pilot of the aircraft engaged in the aerial application for the purpose of ensuring the safe application of that pesticide;

"local beekeepers' spray warning scheme" means any scheme for the advance notification of the application of pesticides, organised by local beekeepers and notified to the Minister of Agriculture, Fisheries and Food, the Secretary of State for Scotland or the Secretary of State for Wales (being the Secretaries of State respectively concerned with agriculture in Scotland and Wales);

"Local Nature Reserve" means a nature reserve established by a local authority under section 21 of the National Parks and Access to the Countryside Act 1949 and "the relevant authority" in regard to such a reserve shall be the local authority which is providing or securing the provision of the reserve;

"Marine Nature Reserve" means an area designated as such by the Secretary of State under section 36 of the Wildlife and Countryside Act 1981 and the "relevant authority" in regard to such an area shall be the appropriate nature conservation agency;

"National Nature Reserve" means any land declared as such by the appropriate nature conservation agency under section 19 of the National Parks and Access to the Countryside Act 1949, or under section 35 of the Wildlife and Countryside Act 1981, and "the relevant authority" in regard to such land shall be the appropriate nature conservation agency;

"Site of Special Scientific Interest" means any area designated as such by the appropriate nature conservation agency under section 28 of the Wildlife and Countryside Act 1981, or in respect of which the Secretary of State has made an Order under section 29 of the Wildlife and Countryside Act 1981, and "the relevant authority" in regard to such an area shall be the appropriate nature conservation agency;
"water" means any surface water.

ANNEX B
PRODUCTS APPROVED FOR USE BY AERIAL APPLICATION

Products approved for use by aerial application, together with details of the crops on which they are approved for use are listed below. Only products approved for aerial application may be used for that purpose and regular returns have to be made to the Pesticides Usage Survey Group, MAFF, Central Science Laboratory, Sand Hutton Lane, Sand Hutton, York, YO4 1LW.

Note reference must be made to the label for conditions of use:

Product (Reg No.)	Expiry Date	Crops/Uses
Asulam		
Asulox (06124)		Agricultural Grassland, Amenity Grassland, Forestry, Rough Upland Intended For Grazing
Asulox (09969)		Agricultural Grassland, Amenity Grassland, Forestry, Rough Upland Intended For Grazing
IT Asulam (10186)		Agricultural Grassland, Amenity Vegetation, Forestry, Rough Upland Intended For Grazing
Benalaxyl + Mancozeb		
Galben M (07220)		Potato (Early), Potato (Main), Potato (Seed)
Tairel (07767)		Potato (Early), Potato (Main), Potato (Seed)
Carbendazim		
Ashlade Carbendazim Flowable (06213)		Oilseed Rape, Wheat (Winter)
Delsene 50 Flo (09469)		Oilseed Rape, Wheat (Winter)
HY-CARB (05933)		Wheat (Winter)
MSS Mircarb (08788)		Oilseed Rape, Wheat (Winter)
Quadrangle Hinge (08070)		Oilseed Rape, Wheat (Winter)
Top Farm Carbendazim - 435 (05307)		Oilseed Rape, Wheat (Winter)
Tripart Defensor FL (02752)		Oilseed Rape, Wheat (Winter)
Chlormequat		
3C Chlormequat 460 (03916)	28/02/2001	Barley (Winter), Oats, Wheat
3C Chlormequat 600 (04079)	28/02/2001	Barley (Winter), Oats, Wheat
Agriguard 5C Chlormequat 460 (09851)		Oats (Spring), Oats (Winter), Wheat (Spring), Wheat (Winter)
Agriguard Chlormequat 700 (09282)		Oats (Spring), Oats (Winter), Wheat (Spring), Wheat (Winter)
Agriguard Chlormequat 700 (09782)		Oats (Spring), Oats (Winter), Wheat (Spring), Wheat (Winter)

Product (Reg No.)	Expiry Date	Crops/Uses
Allied Colloids Chlormequat 460 (07859)	30/06/2002	Oats (Spring), Oats (Winter), Wheat (Spring), Wheat (Winter)
Allied Colloids Chlormequat 460:320 (07861)	30/06/2002	Oats (Spring), Oats (Winter), Wheat (Spring), Wheat (Winter)
Allied Colloids Chlormequat 730 (07860)	30/06/2002	Oats (Spring), Oats (Winter), Wheat (Spring), Wheat (Winter)
Ashlade 460 CCC (06474)		Oats (Spring), Oats (Winter), Wheat (Spring), Wheat (Winter)
Ashlade 700 5C (07046)		Oats (Spring), Oats (Winter), Wheat (Spring), Wheat (Winter)
Ashlade 700 CCC (06473)		Oats (Spring), Oats (Winter), Wheat (Spring), Wheat (Winter)
Ashlade Brevis (08119)		Oats (Spring), Oats (Winter), Wheat (Spring), Wheat (Winter)
Atlas 3C:645 Chlormequat (05710)		Oats, Wheat (Spring), Wheat (Winter)
Atlas 3C:645 Chlormequat (07700)		Oats, Wheat (Spring), Wheat (Winter)
Atlas 5C Chlormequat (03084)		Oats (Spring), Oats (Winter), Wheat (Spring), Wheat (Winter)
Atlas 5C Chlormequat (07701)		Oats (Spring), Oats (Winter), Wheat (Spring), Wheat (Winter)
Atlas Chlormequat 46 (07704)		Oats (Spring), Oats (Winter), Wheat (Spring), Wheat (Winter)
Atlas Chlormequat 460:46 (06258)		Oats (Spring), Oats (Winter), Wheat (Spring), Wheat (Winter)
Atlas Chlormequat 460:46 (07705)		Oats (Spring), Oats (Winter), Wheat (Spring), Wheat (Winter)
Atlas Chlormequat 700 (03402)		Oats (Spring), Oats (Winter), Wheat (Spring), Wheat (Winter)
Atlas Chlormequat 700 (07708)		Oats (Spring), Oats (Winter), Wheat (Spring), Wheat (Winter)
Atlas Quintacel (07706)		Oats (Spring), Oats (Winter), Wheat (Winter)
Atlas Terbine (06523)		Oats (Spring), Oats (Winter), Wheat (Spring), Wheat (Winter)
Atlas Terbine (07709)		Oats (Spring), Oats (Winter), Wheat (Spring), Wheat (Winter)
Atlas Tricol (07190)		Oats (Spring), Oats (Winter), Wheat (Spring), Wheat (Winter)
Atlas Tricol (07707)		Oats (Spring), Oats (Winter), Wheat (Spring), Wheat (Winter)
Barclay Holdup (06799)		Barley (Winter), Oats (Spring), Oats (Winter), Wheat (Spring), Wheat (Winter)
Barclay Holdup 600 (08794)		Barley (Winter), Oats (Spring), Oats (Winter), Wheat (Spring), Wheat (Winter)
Barclay Holdup 640 (08795)		Barley (Winter), Oats (Spring), Oats (Winter), Wheat (Spring), Wheat (Winter)

Product (Reg No.)	Expiry Date	Crops/Uses
Barclay Liffey (09856)		Barley (Winter), Oats (Spring), Oats (Winter), Wheat (Spring), Wheat (Winter)
Barclay Lucan (09855)		Barley (Winter), Oats (Spring), Oats (Winter), Wheat (Spring), Wheat (Winter)
Barclay Take 5 (08524)		Barley (Winter), Oats (Spring), Oats (Winter), Rye, Triticale, Wheat (Spring), Wheat (Winter)
BASF 3C Chlormequat 600 (04077)		Barley (Winter), Oats, Wheat
BASF 3C Chlormequat 720 (06514)		Barley (Winter), Oats (Spring), Oats (Winter), Rye, Triticale, Wheat (Spring), Wheat (Winter)
BASF 3C Chlormequat 750 (06878)		Barley (Winter), Oats (Spring), Oats (Winter), Rye, Triticale, Wheat (Spring), Wheat (Winter)
Ciba Chlormequat 460 (09525)		Oats (Spring), Oats (Winter), Wheat (Spring), Wheat (Winter)
Ciba Chlormequat 5C 460:320 (09527)		Oats (Spring), Oats (Winter), Wheat (Spring), Wheat (Winter)
Ciba Chlormequat 730 (09526)		Oats (Spring), Oats (Winter), Wheat (Spring), Wheat (Winter)
Clayton Standup (08771)		Oats (Spring), Oats (Winter), Wheat (Spring), Wheat (Winter)
Greencrop Carna (09403)		Barley (Winter), Oats (Spring), Oats (Winter), Wheat (Spring), Wheat (Winter)
Greencrop Coolfin (09449)		Oats (Spring), Oats (Winter), Wheat (Spring), Wheat (Winter)
Hyquat 70 (03364)		Barley (Winter), Oats (Spring), Oats (Winter), Wheat (Spring), Wheat (Winter)
Intracrop Balance (08037)		Barley (Winter), Oats (Winter), Wheat (Spring), Wheat (Winter)
Larke (10045)		Barley (Winter), Oats, Wheat (Spring), Wheat (Winter)
Mandops Barleyquat B (06001)		Barley (Winter)
Mandops Bettaquat B (06004)		Oats, Wheat, Wheat (Spring)
Mandops Chlormequat 700 (06002)		Oats, Wheat (Spring), Wheat (Winter)
MSS Mircell (06939)		Oats (Spring), Oats (Winter), Wheat (Spring), Wheat (Winter)
MSS Mirquat (08166)		Oats (Spring), Oats (Winter), Wheat (Spring), Wheat (Winter)
New 5C Cycocel (01482)		Barley (Winter), Oats (Spring), Oats (Winter), Rye, Triticale, Wheat (Spring), Wheat (Spring) (Autumn Drilled), Wheat (Winter)
New 5C Cycocel (01483)		Oats, Rye, Triticale, Wheat (Spring), Wheat (Winter)

Product (Reg No.)	Expiry Date	Crops/Uses
Sigma PCT (08663)		Barley (Winter), Wheat (Spring), Wheat (Winter)
Stabilan 5C (08144)		Oats (Spring), Oats (Winter), Wheat (Spring), Wheat (Winter)
Stabilan 750 (08004)	31/10/2001	Barley (Winter), Oats (Winter), Wheat (Spring), Wheat (Winter)
Stabilan 750 (09303)		Barley (Spring), Oats (Winter), Wheat (Spring), Wheat (Winter)
Supaquat (09381)		Barley (Winter), Oats (Spring), Oats (Winter), Rye, Triticale, Wheat (Spring), Wheat (Winter)
Trio (08883)	31/12/2001	Barley (Winter), Oats (Spring), Oats (Winter), Rye (Winter), Triticale, Wheat (Spring), Wheat (Winter)
Tripart Brevis (03754)		Oats (Spring), Oats (Winter), Wheat (Spring), Wheat (Winter)
Tripart Chlormequat 460 (03685)		Oats (Spring), Oats (Winter), Wheat (Spring), Wheat (Winter)
Uplift (07527)		Barley (Winter), Oats (Spring), Oats (Winter), Wheat (Spring), Wheat (Winter)
Whyte Chlormequat 700 (09641)		Oats (Spring), Oats (Winter), Wheat (Spring), Wheat (Winter)

2-Chloroethylphosphonic acid

Aventis Cerone (09748)		Barley (Winter)
Barclay Coolmore (07917)		Barley (Winter)
Cerone (06185)		Barley (Winter)
Cerone (09985)		Barley (Winter)
Charger (08827)		Barley (Winter)
Charger (09986)		Barley (Winter)
CleanCrop Fonic (09868)		Barley (Winter)
EXP03149D (08828)		Barley (Winter)
EXP03149D (09989)		Barley (Winter)
Stantion (06205)	31/01/2001	Barley (Winter)
Unistar Ethephon 480 (06282)	20/04/2001	Barley (Winter)

Chlorothalonil

Agriguard Chlorothalonil (09390)		Potato
Barclay Corrib 500 (08981)		Potato
Baton 500 SC (09850)		Potato
BB Chlorothalonil (03320)		Potato
Bombardier (02675)		Potato
Bombardier FL (07910)		Potato
Bravo 500 (05637)		Potato
Bravo 500 (05638)	31/05/2001	Potato
Bravo 500 (09059)		Potato
Bravo 720 (09104)		Potato
Clayton Turret (09400)		Potato

Product (Reg No.)	Expiry Date	Crops/Uses
CleanCrop Chlorothalonil 720 (10102)		Potato
Clortosip (06126)		Potato
Clortosip 500 (09320)		Potato
Flute (08953)		Potato
Greencrop Orchid (09566)		Potato
ISK 375 (07455)	30/06/2002	Potato
ISK 375 (09103)		Potato
Jupital (05554)	30/06/2002	Potato
Jupital (09109)		Potato
Jupital DG (09181)		Potato
Mainstay (05625)		Potato
Miros DF (04966)		Potato
Repulse (07641)		Potato
Sipcam Echo 75 (08302)		Potato
Sipcam UK Rover (09848)		Potato
Sipcam UK Rover 500 (04165)	30/06/2001	Potato
Strada (08824)	31/12/2002	Potato
Strada 500 (09849)		Potato
Tripart Faber (04549)		Potato
Ultrafaber (05627)		Potato
Visclor 500 SC (09404)		Potato

Chlorothalonil + Cymoxanil

Ashlade Cyclops (04857)	30/06/2001	Potato

Chlorotoluron

Lentipur CL 500 (08743)		Barley (Winter), Durum Wheat, Triticale, Wheat (Winter)
Luxan Chlorotoluron 500 Flowable (09165)		Barley (Winter), Durum Wheat, Triticale, Wheat (Winter)
NWA CTU 500 (10173)		Barley (Winter), Durum Wheat, Triticale, Wheat (Winter)
Talisman (03109)		Barley (Winter), Durum Wheat, Triticale, Wheat (Winter)

Copper oxychloride

Cuprokylt (00604)		Potato

Diflubenzuron

Dimilin Flo (08769)		Forestry
Dimilin Flo (08985)		Forestry

Dimethoate

Barclay Dimethosect (08538)	31/07/2002	Barley (Winter), Oats (Winter), Pea, Potato (Ware), Rye (Winter), Sugar Beet, Triticale, Wheat (Winter)
BASF Dimethoate 40 (00199)		Barley (Spring), Barley (Winter), Oats (Spring), Oats (Winter), Pea, Potato (Ware), Rye, Sugar Beet, Triticale, Wheat (Spring), Wheat (Winter)

Product (Reg No.)	Expiry Date	Crops/Uses
Danadim Dimethoate 40 (07351)	31/05/2002	Barley (Winter), Oats (Winter), Pea, Potato (Ware), Rye (Winter), Sugar Beet, Triticale, Wheat (Winter)
PA Dimethoate 40 (01527)		Cereal, Fodder Beet, Mangel, Pea, Potato (Chitting House), Potato (Seed Crop), Potato (Ware), Red Beet, Sugar Beet
Rogor L40 (07611)		Barley (Spring), Barley (Winter), Oats (Spring), Oats (Winter), Pea, Potato (Ware), Rye, Sugar Beet, Triticale, Wheat (Spring), Wheat (Winter)
Sector (08882)		Barley (Winter), Oats (Winter), Pea, Potato (Ware), Rye (Winter), Sugar Beet, Triticale, Wheat (Winter)

Disulfoton

Disyston P-10 (00715)	20/04/2001	Brussels Sprout, Cabbage, Carrot, Cauliflower, Sugar Beet

Fenitrothion

Dicofen (00693)	20/04/2001	Barley (Spring), Barley (Winter), Pea, Pine (Lodgepole), Wheat (Spring), Wheat (Winter)
Dicofen (09598)	20/04/2001	Barley (Spring), Barley (Winter), Pea, Wheat (Spring), Wheat (Winter)
Unicrop Fenitrothion 50 (02267)	20/04/2001	Barley (Spring), Barley (Winter), Pea, Wheat (Spring), Wheat (Winter)

Flamprop-M-isopropyl

Commando (07005)		Barley (Spring), Barley (Winter), Durum Wheat, Rye, Triticale, Wheat (Spring), Wheat (Winter)

Mancozeb

Agrichem Mancozeb 80 (06354)		Potato
Ashlade Mancozeb FL (06226)		Potato
Barclay Manzeb 455 (07990)		Potato
Deny WP (09899)		Potato
Dequiman MZ (06870)		Potato
Dithane 945 (00719)		Potato
Dithane 945 (04017)	31/12/2001	Potato
Dithane 945 (09889)		Potato
Dithane 945 (09897)		Potato
Dithane Dry Flowable (04251)	30/09/2002	Potato
Dithane Dry Flowable (04255)	31/12/2001	Potato
Dithane Dry Flowable Newtec (09754)		Potato
Dithane Dry Flowable Newtec (09892)		Potato
Dithane NT Dry Flowable (09898)		Potato
Dithane Superflo (06290)	31/05/2002	Potato
Headland Zebra Flo (07442)	31/03/2002	Potato

Product (Reg No.)	Expiry Date	Crops/Uses
Headland Zebra WP (07441)		Potato
Helm 75 WG Newtec (09757)		Potato
Helm 75WG (08309)	30/09/2002	Potato
Kor DF (08979)	30/09/2002	Potato
Kor DF Newtec (09758)		Potato
Kor Flo (08019)		Potato
Kor Flo (09895)		Potato
Kor NT Dry Flowable (09893)		Potato
Landgold Mancozeb 80 W (06507)	25/07/2003	Potato
Luxan Mancozeb Flowable (06812)	30/09/2002	Potato
Manconex (09555)		Potato
Mancozeb 80 (09896)		Potato
Mandate 80 WP (09080)		Potato
Manex II (07637)		Potato
Manzate 200 PI (07209)	28/02/2002	Potato
Manzate 200 PI (09480)		Potato
Micene 80 (09112)		Potato
Micene DF (09957)		Potato
Mortar Flo (09592)		Potato
Opie 80 WP (08301)		Potato
Opie 80 WP (09890)		Potato
Penncozeb (07820)		Potato
Penncozeb WDG (07833)	31/07/2002	Potato
Penncozeb WDG (09690)		Potato
Quell Flo (08317)		Potato
Quell Flo (09894)		Potato
Restraint DF (09499)	30/09/2002	Potato
Restraint DF Newtec (09755)		Potato
Stefes Deny (08932)		Potato
Stefes Restraint (08945)	31/03/2001	Potato
Tariff 75 WG Newtec (09756)		Potato
Tariff 75 WG Newtec (09891)		Potato
Tariff 75WG (08308)	30/09/2002	Potato
Tridex (07922)	31/12/2001	Potato
Trimanzone (09278)	31/05/2002	Potato
Trimanzone (09584)		Potato
Unicrop Mancozeb (05467)		Potato
Unicrop Mancozeb 80 (07451)	31/03/2002	Potato

Mancozeb + Metalaxyl

Fubol 75 WP (08409)		Potato
Osprey 58 WP (08428)		Potato (Early), Potato (Main)
Ridomil MZ 75 WP (08438)		Potato

Mancozeb + Oxadixyl

Recoil (08483)		Potato

Maneb

Agrichem Maneb 80 (05474)		Potato
Ashlade Maneb Flowable (06477)		Potato
Headland Spirit (04548)		Potato

Product (Reg No.)	Expiry Date	Crops/Uses
Maneb 80 (01276)		Potato
RH Maneb 80 (01796)		Potato
Trimangol 80 (06070)		Potato
Trimangol 80 (06871)		Potato
Trimangol WDG (06992)		Potato
Unicrop Maneb 80 (06926)		Potato
X-Spor SC (08077)		Potato

Metaldehyde

Product (Reg No.)	Expiry Date	Crops/Uses
Aristo (09622)	30/11/2002	Around Edible Crop (Outdoor), Around Edible Crop (Protected), Around Non-Edible Crop (Outdoor), Around Non-Edible Crop (Protected), Bare Soil
Aristo (09824)		Bare Soil, Edible Crop (Around Outdoor), Edible Crop (Around Protected), Non-Edible Crop (Around Protected), Non-Edible Crop (Outdoor) (Around)
Brits (10116)		Around Edible Crop (Outdoor), Around Edible Crop (Protected), Around Non-Edible Crop (Outdoor), Around Non-Edible Crop (Protected), Bare Soil
Clartex (09213)		All Edible Crop, All Non-Edible Crop
Dixie 6 (10052)		Around Edible Crop (Outdoor), Around Edible Crop (Protected), Around Non-Edible Crop (Outdoor), Around Non-Edible Crop (Protected), Bare Soil
Doff Agricultural Slug Killer with Animal Repellent (06058)	31/07/2001	Edible Crop (Around Outdoor), Edible Crop (Around Protected), Non-Edible Crop (Around Protected), Non-Edible Crop (Outdoor) (Around), Soil (Bare)
Doff Horticultural Slug Killer Blue Mini Pellets (05688)	30/06/2002	Edible Crop, Non-Edible Crop, Soil (Bare)
Doff Horticultural Slug Killer Blue Mini Pellets (09666)		All Edible Crop, All Non-Edible Crop, Bare Soil
Doff New Formula Metaldehyde Slug Killer Mini Pellets (09338)	31/10/2002	Around Edible Crop (Outdoor), Around Edible Crop (Protected), Around Non-Edible Crop (Outdoor), Around Non-Edible Crop (Protected), Bare Soil
Doff New Formula Metaldehyde Slug Killer Mini Pellets (09772)	31/05/2003	Around Edible Crop (Outdoor), Around Edible Crop (Protected), Around Non-Edible Crop (Outdoor), Around Non-Edible Crop (Protected), Bare Soil
EM 1617/01 (09344)		All Edible Crop, All Non-Edible Crop, Bare Soil
ESP (09428)		All Edible Crop, All Non-Edible Crop
FP 107 (06666)	31/05/2001	Bare Soil, Edible Crop (Outdoor), Edible Plant (Around), Non-Edible Crop (Outdoor), Non-Edible Plant (Around)

Product (Reg No.)	Expiry Date	Crops/Uses
FP 107 (09060)		Around Edible Crop (Outdoor), Around Non-Edible Crop (Outdoor), Bare Soil
Lynx (09137)	31/10/2002	Around Edible Crop (Outdoor), Around Edible Crop (Protected), Around Non-Edible Crop (Outdoor)
Lynx (09770)		Around Edible Crop (Outdoor), Around Edible Crop (Protected), Around Non-Edible Crop (Outdoor), Around Non-Edible Crop (Protected), Bare Soil
Optimol (06688)	31/05/2001	Bare Soil, Edible Plant (Around), Non-Edible Plant (Around)
Optimol (09061)		Around Edible Crop (Outdoor), Around Non-Edible Crop (Outdoor), Bare Soil
pbi Slug Pellets (01558)	25/07/2003	All Edible Crop (Outdoor), All Non-Edible Crop (Outdoor), Bare Soil
pbi Slug Pellets (09607)	25/07/2003	All Edible Crop (Outdoor), All Non-Edible Crop (Outdoor), Bare Soil
Pesta (10070)		Bare Soil, Edible Crop (Around Protected), Non-Edible Crop (Around Protected), Non-Edible Crop (Outdoor) (Around)
Slug Pellets (01558)	28/02/2001	All Edible Crop (Outdoor), All Non-Edible Crop (Outdoor),Bare Soil
Slug Pellets (10150)		All Edible Crop (Outdoor), All Non-Edible Crop (Outdoor),Bare Soil
Super-flor 6% Metaldehyde Slug Killer Mini Pellets (05453)	31/10/2002	Around Edible Crop (Outdoor), Around Edible Crop (Protected), Around Non-Edible Crop (Outdoor), Around Non-Edible Crop (Protected), Bare Soil
Super-Flor 6% Metaldehyde Slug Killer Mini Pellets (09773)		Around Edible Crop (Outdoor), Around Edible Crop (Protected), Around Non-Edible Crop (Outdoor), Around Non-Edible Crop (Protected), Bare Soil
Unicrop 6% Mini Slug Pellets (02275)	31/10/2002	Around Edible Crop (Outdoor), Around Edible Crop (Protected), Around Non-Edible Crop (Outdoor), Around Non-Edible Crop (Protected), Bare Soil
Unicrop 6% Mini Slug Pellets (09771)		Around Edible Crop (Outdoor), Around Edible Crop (Protected), Around Non-Edible Crop (Outdoor), Around Non-Edible Crop (Protected), Bare Soil
Yeoman (09623)	30/11/2002	Around Edible Crop (Outdoor), Around Edible Crop (Protected), Around Non-Edible Crop (Outdoor), Around Non-Edible Crop (Protected), Bare Soil
Yeoman (09825)	31/05/2003	Bare Soil, Edible Crop (Around Protected), Non-Edible Crop (Around Protected), Non-Edible Crop (Outdoor) (Around)

Product (Reg No.)	Expiry Date	Crops/Uses

Methabenzthiazuron

Tribunil (02169) — Barley (Spring), Barley (Winter), Durum Wheat, Oats (Winter), Rye (Winter), Ryegrass (Perennial), Triticale, Wheat (Spring) (Autumn Sown), Wheat (Winter)

Methiocarb

Barclay Poacher (09031) — All Edible Crop (Outdoor), All Non-Edible Crop (Outdoor), Bare Soil

Bayer UK 808 (09513) — Bare Soil, Cereal Seed (Admixture), Edible Crop (Outdoor), Non-Edible Crop (Outdoor), Ryegrass (Seed) (Admixture)

Bayer UK 809 (09514) — Bare Soil, Cereal Seed (Admixture), Edible Crop (Outdoor), Non-Edible Crop (Outdoor), Ryegrass (Seed) (Admixture)

Bayer UK 892 (09540) — Bare Soil, Edible Crop (Outdoor), Non-Edible Crop (Outdoor)

Bayer UK 935 (09541) — Bare Soil, Edible Crop (Outdoor), Non-Edible Crop (Outdoor)

Club (07176) — Bare Soil, Edible Crop (Outdoor), Non-Edible Crop (Outdoor)

Decoy (06535) — Bare Soil, Edible Crop (Outdoor), Non-Edible Crop (Outdoor)

Decoy Plus (07615) — Bare Soil, Edible Crop (Outdoor), Non-Edible Crop (Outdoor)

Decoy Wetex (09707) — Bare Soil, Cereal Seed (Admixture), Edible Crop (Outdoor), Non-Edible (Outdoor), Ryegrass (Seed) (Admixture)

Draza (00765) — Bare Soil, Edible Crop (Outdoor), Non-Edible Crop (Outdoor)

Draza 2 (04748) — Bare Soil, Edible Crop (Outdoor), Non-Edible Crop (Outdoor)

Draza Plus (06553) — Bare Soil, Edible Crop (Outdoor), Non-Edible Crop (Outdoor)

Draza Wetex (09704) — Bare Soil, Cereal Seed (Admixture), Edible Crop (Outdoor), Non-Edible Crop (Outdoor), Ryegrass (Seed) (Admixture)

Elvitox (06738) — Bare Soil, Edible Crop (Outdoor), Non-Edible Crop (Outdoor)

Epox (06737) — Bare Soil, Edible Crop (Outdoor), Non-Edible Crop (Outdoor)

Exit (07632) — Bare Soil, Edible Crop (Outdoor), Non-Edible Crop

Product (Reg No.)	Expiry Date	Crops/Uses
Exit Wetex (10149)		Bare Soil, Edible Crop (Outdoor), Non-Edible Crop (Outdoor)
Huron (10148)		Bare Soil, Edible Crop (Outdoor), Non-Edible Crop (Outdoor)
Karan (09637)		All Edible Crop (Outdoor), All Non-Edible Crop (Outdoor), Bare Soil, Cereal Seed (Admixture), Ryegrass (Seed) (Admixture)
Lupus (09638)		All Edible Crop (Outdoor), All Non-Edible Crop (Outdoor),Bare Soil, Cereal Seed (Admixture), Ryegrass (Seed) (Admixture)
Rescur (07942)		Bare Soil, Edible Crop (Outdoor), Non-Edible Crop
(Outdoor)		
Rivet (09512)		Bare Soil, Cereal Seed (Admixture), Edible Crop (Outdoor), Non-Edible Crop (Outdoor), Ryegrass (Seed) (Admixture)

Monolinuron

Product (Reg No.)	Expiry Date	Crops/Uses
Arresin (07303)	09/09/2001	Dwarf French Bean, Potato

Phosalone

Product (Reg No.)	Expiry Date	Crops/Uses
Zolone Liquid (06173)	20/04/2001	Brassica (Seed Crop), Cereal, Oilseed Rape

Pirimicarb

Product (Reg No.)	Expiry Date	Crops/Uses
Agriguard Pirimicarb (09620)		Barley, Durum Wheat, Oats, Rye, Triticale, Wheat
Aphox (06633)		Barley, Durum Wheat, Oats, Rye, Triticale, Wheat
Barclay Pirimisect (09057)		Barley (Spring), Barley (Winter), Oats (Spring), Oats (Winter), Wheat (Spring), Wheat (Winter)
Clayton Pirimicarb 50SG (09221)		Barley, Durum Wheat, Oats, Rye, Triticale, Wheat
Greencrop Glenroe (09903)		Barley, Durum Wheat, Oats, Rye, Triticale, Wheat
Helocarb Granule 500 (08157)	20/04/2001	Barley, Durum Wheat, Oats, Rye, Triticale, Wheat
Phantom (04519)		Barley, Durum Wheat, Oats, Rye, Triticale, Wheat
Pirimate (09568)		Barley, Oats, Wheat
Unistar Pirimicarb 500 (06975)	20/04/2001	Barley, Durum Wheat, Oats, Rye, Triticale, Wheat

Propiconazole

Product (Reg No.)	Expiry Date	Crops/Uses
Barclay Bolt (08341)		Barley (Spring), Barley (Winter), Oats (Spring), Oats (Winter), Rye, Wheat (Spring), Wheat (Winter)

Product (Reg No.)	Expiry Date	Crops/Uses
Radar (09168)		Barley (Spring), Barley (Winter), Oats, Rye, Wheat (Spring), Wheat (Winter)
Standon Propiconazole (07037)	25/07/2003	Barley (Spring), Barley (Winter), Oats, Rye, Wheat (Spring), Wheat (Winter)

Sulphur
Stoller Flowable Sulphur (03760)   Barley (Spring), Barley (Winter), Sugar Beet, Wheat (Spring), Wheat (Winter)

Tri-allate
Avadex 15 G (10167)   Barley (Spring), Barley (Winter), Field Bean (Winter), Pea (Dried), Pea (Seed), Pea (Vining), Wheat (Winter)

Avadex BW Granular (00174)   Barley (Spring), Barley (Winter), Field Bean (Winter), Pea (Dried), Pea (Seed), Pea (Vining), Wheat (Winter)

Avadex Excel 15G (07117)   Barley, Barley (Spring), Barley (Winter), Field Bean (Winter), Pea (Dried), Pea (Seed), Pea (Vining), Wheat (Winter)

Triadimefon
100-Plus (05112)   31/12/2001   Barley (Spring), Barley (Winter), Oats (Spring), Oats (Winter), Rye (Spring), Rye (Winter), Wheat (Spring),

Wheat (Winter)
Bayleton (00221)   Barley (Spring), Barley (Winter), Oats (Spring), Oats (Winter), Rye, Rye (Spring), Rye (Winter), Wheat, Wheat (Spring), Wheat (Winter)

Zineb
Unicrop Zineb (02279)   Potato

Zineb-ethylene thiuram disulphide adduct
Polyram DF (08234)   Potato

ANNEX C

THE LONG TERM ARRANGEMENTS FOR EXTENSION OF USE (2000)

PLEASE NOTE THAT THESE EXTENSIONS OF USE ARE AT ALL TIMES DONE AT THE USER'S CHOOSING, AND THE COMMERCIAL RISK IS ENTIRELY THEIRS.

SPECIFIC RESTRICTIONS FOR EXTENSION OF USE UNDER THESE ARRANGEMENTS

To ensure that the extension of use does not increase the risk to the operator, the consumer or the environment, **the following conditions MUST be followed** when applying pesticides under the terms of this scheme:

GENERAL RESTRICTIONS

1. These arrangements apply to label and specific off-label recommendations for use of ONLY products approved for use as Agricultural/Horticultural pesticides.

2. All safety precautions and statutory conditions relating to use (which are clearly identified in the statutory box on product labels) MUST be observed. If extrapolation from a specific off-label is to be used then in addition to all safety precautions and statutory conditions relating to use specified on the product's label, all conditions relating to use specified on the Notice of Approval for the specific off-label use MUST be observed.

3. Pesticides MUST only be used in the same situation (outdoor or protected) as that specified on the product label/specific off-label Notice of Approval for the use on which the extrapolation is to be based, specifically:

Pesticides must not be used on protected crops, i.e. crops grown in glasshouses, poly tunnels, cloches or polythene covers or in any other building, unless the product label specifically allows use under protection on the crop on which the extrapolation is to be based. Similarly, pesticides approved only for use in protected situations must not be applied outdoors.

PLEASE NOTE: Unless specifically restricted to outdoor crops only, pesticides approved for use on tomatoes, cucumbers, lettuce, chrysanthemum and mushrooms are assumed to be approved for use under protection. **For all other uses, if the label/specific off-label Notice of Approval does not specify a situation, then only extrapolation to an outdoor use is permitted**.

APPLICATION METHOD RESTRICTIONS

4. The method of application must be as stated on the pesticide label and in accordance with the relevant codes of practice and requirements under COSHH 1994 (Control of Substances Hazardous to Health).

5. When planning to use hand held equipment to apply a pesticide under these arrangements, users MUST ensure that hand held use is appropriate for the current on-label recommendations/ specific off-label Notice of Approval conditions. **Note:** unless otherwise stated spray applications to protected crops include hand held uses.

Where hand held use is not appropriate for the use on which the extrapolation is to be based, hand held application should NOT be made if the pesticide label/specific off-label Notice of Approval.

    (a) prohibits hand held use;

    (b) requires the use of personal protective clothing when using the pesticide diluted to the minimum volume rate recommended on the label/specific off-label Notice of Approval for the dose required;

(c) is classified with one of the following hazard warnings:

"Corrosive", "Very toxic", "Toxic", or "Risk of serious damage to eyes".

In other cases hand held application is permitted provided that:

(i) the concentration of the spray volume for the extension of use is no greater than the maximum concentration recommended on the pesticide label;

(ii) spray quality is at least as coarse as the British Crop Protection Council medium or coarse spray;

(iii) operators wear at least a coverall, gloves and rubber boots when applying pesticides below waist level. Use of a faceshield is also required for applications which are above waist height.

(iv) where there are label precautions with regards to buffer zone restrictions for vehicle mounted use, then users must observe a buffer zone distance of 1 m from the top of the bank of any static or flowing water body when applying by hand held equipment.

ENVIRONMENTAL RESTRICTIONS

6. When planning to apply a pesticide under these arrangements by broadcast air-assisted sprayer (any equipment which broadcasts spray droplets by means of fan assistance which carry outwards and upwards from the source of the spray), only pesticides with specific on-label/off-label recommendations for such use on the crop on which the extrapolation is to be based (e.g. on hops, bush, cane or top fruit) can be used. Any associated buffer zone restrictions must also be observed.

7. Pesticides classified as Harmful, Dangerous, Extremely Dangerous or High Risk to bees must not be used during flowering of any crop (i.e. from first flower to complete petal fall) unless otherwise permitted. Applications of such pesticides must also not be made when flowering weeds are present or where bees are actively foraging.

8. If there is an aquatic buffer zone restriction set for the on-label/off-label use, then where appropriate, users are also obliged to conduct a Local Environmental Risk Assessment for Pesticides (LERAP) for the extension of use.

9. All reasonable precautions MUST be taken to safeguard wildlife and the environment.

EXCLUSIONS

10. The following uses are NOT PERMITTED under these arrangements.

(a) Aerial applications

(b) Use in or near water (in or near water includes drainage channels, streams, rivers, ponds, lakes, reservoirs, canals, dry ditches, areas designated for water storage).

(c) Use in or near coastal waters.

(d) Use of rodenticides and other vertebrate control agents.

(e) Use on land not intended for cropping, land not intended to bear vegetation, amenity grassland, managed amenity turf and amenity vegetation (this includes areas such as paths, pavements, roads, ground around buildings, motorway verges, railway embankments, public parks, turf, sports fields, upland areas, moorland areas, nature reserves, etc.).

EXTENSIONS OF USE

I. NON-EDIBLE CROPS AND PLANTS

(a) Subject to the SPECIFIC RESTRICTIONS FOR EXTENSION OF USE set out above, pesticides approved for use on any growing crop may be used on commercial agricultural and horticultural holdings and in forest nurseries on the following crops and plants:

  (i) hardy ornamental nursery stock, ornamental plants, ornamental bulbs and flowers and ornamental crops grown for seed where neither the seed nor any part of the plant is to be consumed by humans or animals;

  (ii) forest nursery crops prior to final planting out.

(b) Subject to the SPECIFIC RESTRICTIONS FOR EXTENSION OF USE set out above, pesticides approved for use on any growing edible crop may be used on commercial agricultural and horticultural holdings on non-ornamental crops grown for seed where neither the seed nor any part of the plant is to be consumed by humans or animals. This extrapolation EXCLUDES use on potatoes, cereals, oilseeds, peas, beans and other pulses grown for seed. Seed treatments themselves are NOT included in this extension of use.

(c) Subject to the SPECIFIC RESTRICTIONS FOR EXTENSION OF USE set out above, pesticides approved for use on oilseed rape may be used on commercial agricultural and horticultural holdings on hemp grown for fibre. Seed treatments are NOT included in this extension of use.

(d) Subject to the SPECIFIC RESTRICTIONS FOR EXTENSION OF USE set out above, herbicides approved for use on cereals, grass and maize may be used on commercial agricultural and horticultural holdings on *Miscanthus spp* (Elephant grass). Applications must NOT be made after the crop has reached 1 metre in height. The crop or products of the crop must NOT be used for food or feed.

PLEASE NOTE:

For a – d above, all on-label/off-label conditions of use must be observed, including any harvest interval i.e. any interval between application and harvest/exposure to the public specified on the label for the use on which the extrapolation is to be based, must be observed.

Before making hand held applications see paragraph 5 of the SPECIFIC RESTRICTIONS FOR EXTENSION OF USE.

II. FARM FORESTRY AND ROTATIONAL COPPICING

Subject to the SPECIFIC RESTRICTIONS FOR EXTENSION OF USE set out above, herbicides approved for use on:

(a) cereals may be used in the first five years of establishment in farm forestry (including short rotation coppicing) on land previously under arable cultivation or improved grassland (as defined by the Farm Woodland Scheme) and reclaimed brownfield sites;

(b) cereals, oilseed rape, sugar beet, potatoes, peas and beans may be used in the first year of re-growth after cutting in coppices (short term, rotational, intensive wood production e.g. poplar or willow biofuel production) established on land previously under arable cultivation or improved grassland (as defined by the Farm Woodland Scheme) and reclaimed brownfield sites;

III NURSERY FRUIT CROPS

Subject to the SPECIFIC RESTRICTIONS FOR EXTENSION OF USE set out above, pesticides approved for use on any crop for human or animal consumption may be used on commercial agricultural and horticultural holdings on nursery fruit trees, nursery grape vines prior to final planting out, bushes, canes and non-fruiting strawberry plants provided any fruit harvested within 12 months of treatment is destroyed. Applications must NOT be made where there are fruit present.

If hand held or broadcast air assisted use is required see paragraphs 5 and 6 respectively of the SPECIFIC RESTRICTIONS FOR EXTENSION OF USE.

IV HOPS *(Humulus spp.)*

Subject to the SPECIFIC RESTRICTIONS FOR EXTENSION OF USE set out above, pesticides may be used on commercial agricultural and horticultural holdings on the following hop plants grown in the circumstance below:

(a) Mature stock or mother plants which are kept specifically for the supply of propagation material.

(b) Propagation of hop planting material- propagules prior to final planting out.

(c) "Nursery hops". First year plants not taken to harvest that year, in their final planting out position.

PLEASE NOTE:

For a – c above, treated hops must NOT be harvested for human or animal consumption (including idling) within 12 months of treatment.

If hand held or broadcast air assisted application is required, users must comply with paragraphs 5 and 6 respectively of the SPECIFIC RESTRICTIONS FOR EXTENSION OF USE.

V. CROPS USED PARTLY OR WHOLLY FOR HUMAN OR ANIMAL CONSUMPTION.

Subject to the SPECIFIC RESTRICTIONS FOR EXTENSION OF USE set out above, pesticides may be used on commercial agricultural or horticultural holdings on the crops listed in TABLE ONE and TWO below in the first column if they have been approved for use on the crop(s) listed opposite them in the second column.

HOWEVER, BEFORE USING ANY OF THE FOLLOWING EXTRAPOLATIONS (TABLES ONE AND TWO), THE USER MUST FIRST OBSERVE THE FOLLOWING:

(a) **It is the responsibility of the user to ensure that the proposed use does not result in any statutory UK Maximum Residue Levels (MRLs) being exceeded. MRLs are set out in statutory instrument No. 1985 of 1994: >The Pesticides (Maximum Residue Levels in Crops, Food and Feeding Stuffs) Regulations 1994= (The Stationery Office, ISBN 0-11-044985-1) and any subsequent updates.**

(b) **These extrapolations DO NOT APPLY in the following situations:**

(i) **Where the MRL for the crop in column 1 is lower than the MRL for the crop in column 2.**

(ii) **Where no MRL is set for the crop in Column 2, but a MRL has been established for the crops in column 1.**

In any of the above circumstances use on the crop in column 1 is NOT PERMITTED.

TABLE ONE

Column 1: Minor use	Column 2: Crops on which use is approved	Additional special conditions
A. ARABLE CROPS		
Poppy (grown for oilseed), Sesame	Sunflower	
Mustard, Linseed, Honesty, Evening primrose	Oilseed rape	
Borage (grown for oilseed) Canary flower e.g. *Echium vulgare/Echium plantaginium* (grown for oilseed)	Oilseed rape	Seed treatments are not permitted
Rye, Triticale	Barley	Treatments applied before first spikelet of inflorescence just visible
Rye, Triticale	Wheat	
Grass seed crop	Grass for grazing or fodder	
Grass seed crop	Wheat, barley, oats, rye, triticale	Treated crops must not be grazed or cut for fodder until 90 days after treatment. Seed treatments are not permitted Use of chlormequat-containing products is not permitted
Lupins	Combining peas or field beans	
B. FRUIT CROPS		
Almond, Chestnut, Walnut, Hazelnut	Apple or cherry or plum	For herbicides used on the orchard floor ONLY
Almond, Chestnut, Walnut, Hazelnut	Products approved for use on two of the following: almond, chestnut, hazelnut and walnut	
Quince, Crab apple	Apple or pear	
Nectarine, Apricot	Peach	
Blackberry, Dewberry Rubus species (e.g. tayberry, loganberry)	Raspberry	
Whitecurrant, Bilberry, Cranberry	Blackcurrant or redcurrant	
Redcurrant	Blackcurrant	
C. VEGETABLE CROPS		
Parsley root	Carrot or radish	
Fodder beet, Mangel	Sugar beet	

Column 1: Minor use	Column 2: Crops on which use is approved	Additional special conditions
Horseradish	Carrot or radish	
Parsnip	Carrot	
Salsify	Carrot or celeriac	
Swede	Turnip	
Turnip	Swede	
Garlic, Shallot	Bulb onion	
Aubergine	Tomato	
Squash, Pumpkin, Marrow, Watermelon	Melon	
Broccoli	Calabrese	
Calabrese	Broccoli	
Roscoff cauliflower	Cauliflower	
Collards	Kale	
Lamb's lettuce, frisee/frise, radicchio, cress, scarole	Lettuce	
Leaf herbs and edible flowers*	Lettuce or spinach or parsley or sage or mint or tarragon	
Beet leaves, Red chard, White chard, Yellow chard	Spinach	
Edible podded peas (e.g. mange-tout, sugar snap)	Edible podded beans	
Runner beans	Dwarf French beans	
Rhubarb, Cardoon	Celery	
Edible fungi other than mushroom (e.g. oyster mushroom)	Mushroom	

*This extension of use applies to the following leaf herbs and edible flowers: angelica, balm, basil, bay, borage, burnet (salad), caraway, camomile, chervil, chives, clary, coriander, dill, fennel, fenugreek, feverfew, hyssop, land cress, lovage, marjoram, marigold, mint, nasturtium, nettle, oregano, parsley, rocket, rosemary, rue, sage, savory, sorrel, tarragon, thyme, verbena (lemon), woodruff.

For applications in store on crops PARTLY OR WHOLLY FOR HUMAN OR ANIMAL CONSUMPTION, the following extensions of use apply:

TABLE TWO

Column 1: Minor Use	Column 2: Crops on which use is approved.
Rye, Barley, Oats, Buckwheat, Millet, Sorghum, Triticale	Wheat
Dried pea	Dried bean
Dried bean	Dried pea
Mustard, Sunflower, Honesty, Sesame, Linseed, Evening primrose, Poppy (grown for oilseed), Borage (grown for oilseed) Canary flower e.g. *Echium vulgare/Echium plantaginium* (grown for oilseed)	Oilseed rape

VI. CLARIFICATIONS:

Under these arrangement the following crops are considered to be synonymous or equivalent and as such, uses on crops in Column 1 can be read across to uses in Column 2.

Column 1:	Column 2: equivalent
Hazelnut	Cobnuts, Filberts
French bean	Navy bean
Vining pea	Picking pea, Shelling pea, Non-edible podded pea
Linseed	Linola, Flax
Wheat	Durum wheat

ANNEX D

COMMODITY SUBSTANCE APPROVALS

HEALTH AND SAFETY EXECUTIVE

Food and Environment Protection Act 1985

Schedule: COMMODITY SUBSTANCE: **4-CHLORO-*m*-CRESOL**

Date of issue: 18 February 1993

Date of expiry: 28 February 2011

This approval is subject to the following conditions:

  1 *FIELD OF USE*:   ONLY AS A FUNGICIDE

  2 *PEST AND USAGE AREA*: FOR THE CONTROL OF FUNGI ON INSECT SPECIMENS

  3 *APPLICATION METHOD*:   4-CHLORO-*m*-CRESOL CRYSTALS IN A COLLECTING BOX

Operator protection:

(1) A written COSHH assessment must be made before using 4-chloro-*m*-cresol.

(2) Engineering control of operator exposure must be used where reasonably practicable in addition to the following items of personal protective equipment.

Operators must wear suitable protective clothing, including protective gloves and eye protection and a dust mask, when handling or applying the material.

(3) However engineering controls may replace personal protective equipment if a COSHH assessment shows they provide an equal or higher standard of protection.

Other specific restrictions:

(1) Operators should be provided with adequate information about the hazards of the substance and the precautions necessary for safe use. Sources of information include the supplier's Safety Data Sheet.

(2) Unprotected persons and animals must be excluded from any areas where treatment is taking place, and such areas should be ventilated after treatment.

(3) Must be used only by operators who are suitably trained and competent to carry out this work.

HEALTH AND SAFETY EXECUTIVE

Food and Environment Protection Act 1985

Schedule: COMMODITY SUBSTANCE: **CAMPHOR**

Date of issue: 18 February 1993

Date of expiry: 28 February 2011

This approval is subject to the following conditions:

1 *FIELD OF USE*: ONLY AS AN INSECT REPELLENT IN MUSEUMS AND BUILDINGS OF CULTURAL, ARTISTIC AND HISTORICAL INTEREST.

2 *PEST AND USAGE AREA*: FOR THE CONTROL OF FLYING AND CRAWLING INSECTS

3 *APPLICATION METHOD*: CRYSTALS OF CAMPHOR IN A SEALED SPECIMEN CASE

Operator protection:

(1) A written COSHH assessment must be made before using camphor. Operators should also observe the OES set out in HSE guidance note EH40/99 or subsequent issues.

(2) Engineering control of operator exposure must be used where reasonably practicable in addition to the following items of personal protective equipment.

Operators must wear suitable protective clothing, including protective gloves and eye protection, when handling or applying the material.

(3) However engineering controls may replace personal protective equipment if a COSHH assessment shows they provide an equal or higher standard of protection.

Other specific restrictions:

(1) Operators should be provided with adequate information about the hazards of the substance and the precautions necessary for safe use. Sources of information include the supplier's Safety Data Sheet.

(2) Unprotected persons and animals must be excluded from any areas where treatment is taking place, and such areas should be ventilated after treatment.

(3) Must be used only by operators who are suitably trained and competent to carry out this work.

Food and Environment Protection Act 1985
Control of Pesticides Regulations 1986 (SI 1986 No. 1510) : APPROVAL

In exercise of the powers conferred by regulation 5 of the Control of Pesticides Regulations 1986 (SI 1986/1510) and of all other powers enabling them in that behalf, the Minister of Agriculture, Fisheries and Food and the Secretary of State (acting jointly), and the Scottish Ministers (as regards Scotland) and the National Assembly for Wales and the Minister of Agriculture, Fisheries and Food (acting jointly as regards Wales), hereby jointly give full approval for the use of:

Commodity substance: being 99.9% v/v **CARBON DIOXIDE** subject to the conditions set out below:

Date of issue: 8 October 1993

Use:

Field of use: **Only as a rodenticide**

Situations: Trapped rodents.

Operator protection:

(1) Engineering control of operator exposure must be used where reasonably practicable in addition to the following personal protective equipment:

Operators must wear self-contained breathing apparatus when CO_2 levels are greater than 0.5% v/v.

(2) However, engineering controls may replace personal protective equipment if a COSHH assessment shows they provide an equal or higher standard of protection.

Other specific restrictions:

(1) Unprotected persons and non-target animals must be excluded from the treatment enclosures and from the area surrounding the treatment enclosures unless CO_2 levels are below 0.5% v/v.

(2) This substance must only be used by operators who are suitably trained and competent to carry out this work.

ADVISORY NOTE

This approval allows the use of CO_2 to destroy trapped rodent pests.

Food and Environment Protection Act 1985
Control of Pesticides Regulations 1986 (SI 1986 No. 1510) : APPROVAL

In exercise of the powers conferred by regulation 5 of the Control of Pesticides Regulations 1986 (SI 1986/1510) and of all other powers enabling them in that behalf, the Minister of Agriculture, Fisheries and Food and the Secretary of State (acting jointly), and the Scottish Ministers (as regards Scotland) and the National Assembly for Wales and the Minister of Agriculture, Fisheries and Food (acting jointly as regards Wales), hereby jointly give full approval for the use of:

Commodity substance: being 99.9% v/v **CARBON DIOXIDE** subject to the conditions set out below:

Date of issue: 8 October 1993

Use:

Field of use: **Only in vertebrate control**

Situations: Birds covered by general licences issued by the Agriculture and Environment Departments under Section 16(1) of the Wildlife and Countryside Act (1981) for the control of opportunistic bird species, where birds have been trapped or stupefied with alphachloralose/seconal.

Operator protection:

(1) Engineering control of operator exposure must be used where reasonably practicable in addition to the following personal protective equipment:

Operators must wear self-contained breathing apparatus when CO_2 levels are greater than 0.5% v/v.

(2) However, engineering controls may replace personal protective equipment if a COSHH assessment shows they provide an equal or higher standard of protection.

Other specific restrictions:

(1) Unprotected persons and non-target animals must be excluded from the treatment enclosures and from the area surrounding the treatment enclosures unless CO_2 levels are below 0.5% v/v.

(2) This substance must only be used by operators who are suitably trained and competent to carry out this work.

(3) Only to be used where a licence has been issued in accordance with Section 16(1) of the Wildlife and Countryside Act 1981 to permit the use of a substance otherwise prohibited under Section 5 of the Wildlife and Countryside Act 1981.

Food and Environment Protection Act 1985
Control of Pesticides Regulations 1986 (SI 1986 No. 1510) : APPROVAL

In exercise of the powers conferred by regulation 5 of the Control of Pesticides Regulations 1986 (SI 1986/1510) and of all other powers enabling them in that behalf, the Minister of Agriculture, Fisheries and Food and the Secretary of State (acting jointly), and the Scottish Ministers (as regards Scotland) and the National Assembly for Wales and the Minister of Agriculture, Fisheries and Food (acting jointly as regards Wales), hereby jointly give full approval for the use of:

Commodity substance: being 99.9% v/v **CARBON DIOXIDE** subject to the conditions set out below:

Date of issue: 8 October 1993

Use:

Field of use: **Only as an insecticide, acaricide and rodenticide in food storage practice**

Situations: Raw and processed food commodities.

Operator protection:

(1) Engineering control of operator exposure must be used where reasonably practicable in addition to the following personal protective equipment:

  Operators must wear self-contained breathing apparatus when CO_2 levels are greater than 0.5% v/v.

(2) However, engineering controls may replace personal protective equipment if a COSHH assessment shows they provide an equal or higher standard of protection.

Other specific restrictions:

(1) Unprotected persons and non-target animals must be excluded from the treatment enclosures and from the area surrounding the treatment enclosures unless CO_2 levels are below 0.5% v/v.

(2) This substance must only be used by operators who are suitably trained and competent to carry out this work.

ADVISORY NOTE

Ensure adequate ventilation of premises during all treatment and venting operations.

Food and Environment Protection Act 1985

Control of Pesticide Regulations 1986 (SI 1986 No. 1510): APPROVAL

Notice is hereby given that in exercise of the powers conferred by Regulation 5 of the Control of Pesticides Regulations 1986 (SI 1986/1510) (as amended) and of all other powers enabling them in that behalf, the Minister of Agriculture, Fisheries and Food and the Secretary of State (acting jointly), and the Scottish Ministers (as regards Scotland) and the National Assembly for Wales and the Minister of Agriculture, Fisheries and Food (acting jointly as regards Wales), hereby jointly give full approval for the use of the following:

Commodity Substance: being **DISODIUM OCTABORATE TETRAHYDRATE** subject to the conditions set out below:

Date of issue: 4 December 2000

Use:

Field of Use: **Only as a forestry fungicide**

Crop/situation	Maximum individual dose	Maximum number of treatments	Latest time of application
Coniferous tree stumps	1 litre of a 10% w/v aqueous solution to 1 sq m of conifer stumps, with a total application of boron not to exceed 2 kg boron/ha, measured as the element.	One per stump per year	Immediately after felling

Operator protection:

(1) A written COSHH assessment must be made before using disodium octaborate tetrahydrate (DOT).

(2) Engineering control of operator exposure must be used where reasonably practicable in addition to the following personal protective equipment.

Operators must wear suitable protective clothing as listed below:

Staff involved in mixing products and preparing solutions.	Boots, coveralls, particle-filtering dust-masks, and waterproof gloves
Operators and service personnel during mechanical application.	Gloves, coveralls and eye protection while repairing or maintaining spray equipment.
Operators and service personnel during hand-held application.	Gloves, coveralls and faceshield.

(3) However, engineering controls may replace personal protective equipment if a COSHH assessment shows they provide an equal or higher standard of protection.

Other specific restrictions:

(1) Must be used only by professional operators who must be suitably trained and competent to carry out this work.

(2) Operators must be supplied with a Section 6 (HSW) safety data sheet before commencing work.

(3) Unprotected persons and animals must be kept out of areas undergoing treatment until tree-felling operations are completed.

ADVISORY NOTES

1. This field of use covers pesticides used for the control of harmful organisms, chiefly fungi, detrimental to plant health.

2. Product must be applied as soon as possible after felling, in order to protect wildlife that may subsequently colonise the stump area.

3. To support use at a rate of 4 kg boron/ha, further data would be required to address its effects in the aquatic environment; protocols should be discussed with PSD.

HEALTH AND SAFETY EXECUTIVE

Food and Environment Protection Act 1985

Schedule: COMMODITY SUBSTANCE: **ETHANOL**

Date of issue: 18 February 1993

Date of expiry: 28 February 2011

This approval is subject to the following conditions:

1 *FIELD OF USE:*	i) AS AN INSECTICIDE IN MUSEUMS AND BUILDINGS OF CULTURAL, ARTISTIC AND HISTORICAL INTEREST.
	ii) AS A PRESERVATIVE IN MUSEUMS, AND BUILDINGS OF CULTURAL ARTISTIC AND HISTORICAL INTEREST
2 *PEST AND USAGE AREA:*	FOR THE CONTROL OF FLYING AND CRAWLING INSECTS AND FUNGI
3 *APPLICATION METHOD:*	i) IMMERSION IN A TANK ENCLOSED IN A FUME CUPBOARD
	ii) STORAGE OF SPECIMENS IN MATERIAL.

Operator protection:

(1) A written COSHH assessment must be made before using ethanol. Operators should also observe the OES set out in HSE guidance note EH40/99 or subsequent issues.

(2) Engineering control of operator exposure must be used where reasonably practicable in addition to the following items of personal protective equipment.

Operators must wear suitable protective clothing, including protective gloves and eye protection, when handling or applying the material.

(3) However engineering controls may replace personal protective equipment if a COSHH assessment shows they provide an equal or higher standard of protection.

Other specific restrictions:

(1) Operators should be provided with adequate information about the hazards of the substance and the precautions necessary for safe use. Sources of information include the supplier's Safety Data Sheet.

(2) Unprotected persons and animals must be excluded from any areas where treatment is taking place, and such areas should be ventilated after treatment.

(3) Must be used only by operators who are suitably trained and competent to carry out this work.

HEALTH AND SAFETY EXECUTIVE

Food and Environment Protection Act l985

Schedule: COMMODITY SUBSTANCE: **ETHYL ACETATE**

Date of issue: 18 February 1993

Date of expiry: 28 February 2011

This approval is subject to the following conditions;

  1 *FIELD OF USE:* ONLY AS AN INSECTICIDE IN MUSEUMS AND BUILDINGS OF CULTURAL, ARTISTIC AND HISTORICAL INTEREST.

  2 *PEST AND USAGE AREA:* FOR THE CONTROL OF FLYING AND CRAWLING INSECTS.

  3 *APPLICATION METHOD:* TREATMENT IN SEALED CONTAINERS.

Operation protection:

(1) A written COSHH assessment must be made before using ethyl acetate. Operators should also observe the OES set out in HSE guidance note EH40/99 or subsequent issues.

(2) Engineering control of operator exposure must be used where reasonably practicable in addition to the following items of personal protective equipment.

Operators must wear suitable protective clothing, including protective gloves and eye protection, when handling or applying the substance.

(3) However engineering controls may replace personal protective equipment if a COSHH assessment shows they provide an equal or higher standard of protection.

Other specific restrictions:

(1) Operators should be provided with adequate information about the hazards of the substance and the precautions necessary for safe use. Sources of information include the supplier's Safety Data Sheet.

(2) Unprotected persons and animals must be excluded from any areas where treatment is taking place, and such areas should be ventilated after treatment.

(3) Must be used only by operators who are suitably trained and competent to carry out this work.

Food and Environment Protection Act 1985
Control of Pesticides Regulations 1986 (SI 1986 NO 1510): APPROVAL

In exercise of the powers conferred by regulation 5 of the Control of Pesticides Regulations 1986 (SI 1986/1510) and of all other powers enabling them in that behalf, the Minister of Agriculture, Fisheries and Food and the Secretary of State (acting jointly), and the Scottish Ministers (as regards Scotland) and the National Assembly for Wales and the Minister of Agriculture, Fisheries and Food (acting jointly as regards Wales), hereby jointly give full approval for the use of:

Commodity Substance: being **FORMALDEHYDE** (formalin 38–40% aqueous solution) subject to the conditions set out below:

Date of issue: 1 March 1991

Date of expiry: 28 February 2001 (see advisory note 3)

Use:

Field of Use: **Only as an agricultural/horticultural fungicide and sterilant**

Crops/Situations	Maximum individual dose	Other specific restrictions (1) Maximum concentration
Soil and compost sterilant, indoors and outdoors	As a drench: 0.5 litre formalin/m^2	1:4 parts water
Bulb dip	—	As a dip: 1:200 parts water
Mushroom houses	As a spray: 0.5 litre formalin/m^2	1:50 parts water
	As a fumigant: 400 ml formalin/$100m^3$ or 100 g of potassium permanganate added to 500 ml formalin/$100m^3$	
Greenhouse hygiene, boxes, pots etc.	As a fumigant: 400 ml formalin/$100m^3$ or 100 g of potassium permanganate added to 500 ml of formalin/$100m^3$	As a spray or dip: 1:50 parts water

Operator protection:

(1) A written COSHH assessment must be made before using formaldehyde. Operators should observe the Maximum Exposure Limit set out in HSE guidance note EH40/93 or subsequent issues and COP 30 "The control of substances hazardous to health in fumigation operations".

(2) Engineering control of operator exposure must be used where reasonably practicable in addition to the following personal protective equipment:

   (a) Operators must wear suitable respiratory equipment and other suitable protective equipment when handling or applying the fumigant.

   (b) Operators must wear suitable protective clothing and gloves when handling the concentrate.

(3) However, engineering controls may replace personal protective equipment if a COSHH assessment shows they provide an equal or higher standard of protection.

Other specific restrictions:

(1) Maximum concentration of formalin in water: see table.

(2) Operators must be supplied with a Section 6 (HSW) Safety Data Sheet before commencing work.

ADVISORY NOTES

1 Use as a disinfectant for equipment, greenhouse and public hygiene purposes are outside the Regulations.

2 Use of this substance for sterilising hatching eggs are outside the Regulations.

3 Whilst data submitted by interested parties is being considered by PSD for Continuation of approval, the date of expiry shall be extended to 25 July 2003.

Food and Environment Protection Act 1985
Control of Pesticides Regulations 1986 (SI 1986 NO 1510): APPROVAL

In exercise of the powers conferred by regulation 5 of the Control of Pesticides Regulations 1986 (SI 1986/1510) and of all other powers enabling them in that behalf, the Minister of Agriculture, Fisheries and Food and the Secretary of State (acting jointly), and the Scottish Ministers (as regards Scotland) and the National Assembly for Wales and the Minister of Agriculture, Fisheries and Food (acting jointly as regards Wales), hereby jointly give full approval for the use of:

Commodity Substance: being **FORMALDEHYDE** (paraformaldehyde) subject to the conditions set out below:

Date of issue: 1 March 1991

Date of expiry: 28 February 2001 (see advisory note)

Use:

Field of Use: **Only as an animal husbandry fungicide**

Situations: Animal houses

Maximum individual dose: As a fumigant: 5g paraformaldehyde/m^3 or 20 g potassium permanganate added to 40 ml of formalin/m^3.

Operator protection:

(1) A written COSHH assessment must be made before using formaldehyde. Operators should observe the Maximum Exposure Limit set out in the HSE guidance note EH40/93 or subsequent issues and COP 30 "The control of substances hazardous to health in fumigation operations".

(2) Engineering control of operator exposure must be used where reasonably practicable in addition to the following personal protective equipment:

Operators must wear suitable respiratory equipment and other suitable protective equipment when handling or applying the fumigant.

(3) However, engineering controls may replace personal protective equipment if a COSHH assessment shows they provide an equal or higher standard of protection.

Other specific restrictions:

(1) Operators must remove livestock, feed, exposed milk and water and collect eggs before application.

(2) Operators must be supplied with a Section 6 (HSW) safety data sheet before commencing work.

ADVISORY NOTE

Whilst data submitted by interested parties is being considered by PSD for Continuation of approval, the date of expiry shall be extended to 25 July 2003.

HEALTH AND SAFETY EXECUTIVE

Food and Environment Protection Act 1985

Schedule:　　　COMMODITY SUBSTANCE: **FORMALDEHYDE**

Date of issue:　18 February 1993

Date of expiry:　28 February 2011

This approval is subject to the following conditions:

1 *FIELD OF USE:*	i) AS AN INSECTICIDE IN MUSEUMS AND BUILDINGS OF CULTURAL, ARTISTIC AND HISTORICAL INTEREST ii) AS A PRESERVATIVE IN MUSEUMS AND BUILDINGS OF CULTURAL, ARTISTIC AND HISTORICAL INTEREST
2 *PEST AND USAGE AREA:*	FOR THE CONTROL OF FLYING AND CRAWLING INSECTS AND FUNGI
3 *APPLICATION METHOD:*	i) TREATMENT IN SEALED CONTAINERS ii) STORAGE IN 5-10% AQUEOUS SOLUTION

Operator protection:

(1) A written COSHH assessment must be made before using formaldehyde. Operators should also observe the MEL set out in HSE guidance note EH40/99 or subsequent issues.

(2) Engineering control of operator exposure must be used where reasonably practicable in addition to the following items of personal protective equipment.

Operators must wear suitable protective clothing, including protective gloves, eye protection and suitable respiratory protective equipment, when handling or applying the substance.

(3) However engineering controls may replace personal protective equipment if a COSHH assessment shows they provide an equal or higher standard of protection.

Other specific restrictions:

(1) Operators should be provided with adequate information about the hazards of the substance and the precautions necessary for safe use. Sources of information include the supplier's Safety Data Sheet.

(2) Unprotected persons and animals must be excluded from any areas where treatment is taking place, and such areas should be ventilated after treatment.

(3) Must be used only by operators who are suitably trained and competent to carry out this work.

HEALTH AND SAFETY EXECUTIVE

Food and Environment Protection Act 1985

Schedule: COMMODITY SUBSTANCE: **ISOPROPANOL**

Date of issue: 18 February 1993

Date of expiry: 28 February 2011

This approval is subject to the following conditions:

1 *FIELD OF USE:*	ONLY AS A PRESERVATIVE IN MUSEUMS AND BUILDINGS OF CULTURAL, ARTISTIC AND HISTORICAL INTEREST.
2 *PEST AND USAGE AREA:*	FOR THE CONTROL OF FUNGI ON SPECIMENS
3 *APPLICATION METHOD:*	STORAGE OF SPECIMENS IN 50-60% AQUEOUS SOLUTION

Operator protection:

(1) A written COSHH assessment must be made before using isopropanol. Operators should also observe the OES set out in HSE guidance note EH40/99 or subsequent issues.

(2) Engineering control of operator exposure must be used where reasonably practicable in addition to the following items of personal protective equipment.

   Operators must wear suitable protective clothing, including protective gloves and eye protection, when handling or applying the material.

(3) However engineering controls may replace personal protective equipment if a COSHH assessment shows they provide an equal or higher standard of protection.

Other specific restrictions:

(1) Operators should be provided with adequate information about the hazards of the substance and the precautions necessary for safe use. Sources of information include the supplier's Safety Data Sheet.

(2) Unprotected persons and animals must be excluded from any areas where treatment is taking place, and such areas should be ventilated after treatment.

(3) Must be used only by operators who are suitably trained and competent to carry out this work.

HEALTH AND SAFETY EXECUTIVE

Food and Environment Protection Act 1985

Schedule: COMMODITY SUBSTANCE: **METHYL BROMIDE**

Date of Issue: 20 November 2000

Date of Expiry: 28 February 2005

This approval is subject to the following conditions:

1. *FIELD OF USE:*	ONLY AS A FUMIGANT IN PUBLIC HYGIENE AND VERTEBRATE CONTROL
2. *PEST AND USAGE AREA*	FOR THE CONTROL OF VERTEBRATES AND FLYING AND CRAWLING INSECTS. FOR USE IN UNOCCUPIED AIRCRAFT, SHIPS, ENCLOSED STRUCTURES (E.G. WAREHOUSES, FACTORIES, MILLS AND OTHER BUILDINGS, CONTAINERS, CHAMBERS AND OTHER TRANSPORT UNITS) AND STACKED COMMODITIES.
3. *APPLICATION METHOD*	USERS SHOULD FOLLOW HSE CS22 GUIDANCE NOTES "FUMIGATION WITH METHYL BROMIDE" AND THE HSE APPROVED CODE OF PRACTICE L86 "THE CONTROL OF SUBSTANCES HAZARDOUS TO HEALTH IN FUMIGATION OPERATIONS".

Operator protection:

(1) A written COSHH assessment must be made before using methyl bromide.

(2) Engineering control of operator exposure must be used were reasonably practicable in addition to the following items of personal protection equipment:

Operators must wear suitable approved respiratory equipment and other suitable protective equipment when handling or applying fumigant.

(3) However, engineering controls may replace personal protective equipment if a COSHH assessment show that provide an equal or higher standard of protection.

Other specific restrictions:

(1) Must be used only by professional operators of servicing companies, local authorities and Government Departments who must be suitably trained and competent to carry out this work.

(2) Operators must be supplied with a Section 6 (HSW) safety data sheet before commencing work.

(3) Unprotected persons and animals must be kept out of premises under fumigation or being aired following fumigation until any exposure levels are below the Occupational Exposure Standard.

Advisory notes

(1) Approved products containing methyl bromide are available for agricultural, horticultural, food storage and space spray use.

(2) This field of use covers pesticides used for the control of harmful organisms, chiefly insects and vertebrates detrimental to public health.

HEALTH AND SAFETY EXECUTIVE

Food and Environment Protection Act 1985

Schedule: COMMODITY SUBSTANCE: **LIQUID NITROGEN**

Date of Issue: 18 February 1993

Date of expiry: 28 February 2011

This approval is subject to the following conditions:

  1 *FIELD OF USE:*   ONLY AS AN ACARICIDE IN DOMESTIC AND PUBLIC HYGIENE SITUATIONS.

  2 *PEST AND USAGE AREA:* FOR THE CONTROL OF DUST MITES ON FURNISHINGS.

  3 *APPLICATION METHOD:* SUITABLE LIQUID NITROGEN APPLICATOR.

Operator protection:

(1) An assessment under The Management of Health and Safety at Work Regulations 1999, or subsequent issues.

(2) Operators must wear suitable protective clothing, including protective gloves, eye protection a dust mask and an oxygen monitor, when handling or applying the material.

Other specific restrictions:

(1) Operators should be provided with adequate information about the hazards of the substance and the precautions necessary for safe use. Sources of information include the supplier's Safety Data Sheet.

(2) Safe working procedures must be specified where it is possible that an oxygen deficient atmosphere could develop. Guidance on such procedures is given in HSE Approved Code of Practice L101 "Safe Work in Confined Spaces".

(3) Unprotected persons and animals must be excluded from any areas where treatment is taking place, and such areas should be ventilated after treatment.

(4) Must be used only by operators who are suitably trained and competent to carry out this work.

Food and Environment Protection Act 1985
Control of Pesticides Regulations 1986 (SI 1986 No. 1510) : APPROVAL

In exercise of the powers conferred by regulation 5 of the Control of Pesticides Regulations 1986 (SI 1986/1510) and of all other powers enabling them in that behalf, the Minister of Agriculture, Fisheries and Food and the Secretary of State (acting jointly), and the Scottish Ministers (as regards Scotland) and the National Assembly for Wales and the Minister of Agriculture, Fisheries and Food (acting jointly as regards Wales), hereby jointly give full approval for the use of:

Commodity substance: being **PARAFFIN OIL** subject to the conditions set out below:

Date of issue: 25 April 1995

Use*:*

Field of use: **Only in vertebrate control**

Situations: Eggs of birds covered by licences issued by the Agriculture and Environment Departments under Section 16 (1) of the Wildlife and Countryside Act (1981)

Operator protection:

(1) Engineering control of operator exposure must be used where reasonably practicable in addition to the following personal protective equipment:

Operators must wear suitable protective gloves and faceshield when handling and applying the substance.

(2) However, engineering controls may replace personal protective equipment if a COSHH assessment shows they provide an equal or higher standard of protection.

Other specific restrictions:

Only to be used where a licence has been approved in accordance with Section 16 (1) of the Wildlife and Countryside Act 1981 to permit the use of a substance otherwise prohibited under Section 5 of the Wildlife and Countryside Act 1981.

ADVISORY NOTES

1. Egg treatment should be undertaken as soon as a clutch is complete.

2. Eggs should be treated once by complete immersion in paraffin oil.

Food and Enviroment Protection Act 1985
Control of Pesticides Regulations 1986 (SI 1986 No. 1510): Commodity Substance Approval

In exercise of the powers conferred by regulation 5 of the Control of Pesticides Regulations 1986 (SI 1986/1510) and of all other powers enabling them in that behalf, the Minister of Agriculture, Fisheries and Food and the Secretary of State (acting jointly), and the Scottish Ministers (as regards Scotland) and the National Assembly for Wales and the Minister of Agriculture, Fisheries and Food (acting jointly as regards Wales), hereby jointly give full approval for the use of:

Commodity substance: being **SODIUM CHLORIDE** subject to the conditions set out below:

Date of issue: 11 September 1996

Use:

Field of Use: **Only as an agricultural herbicide**

Crop: Sugar beet

ADVISORY NOTES:

1. Typical treatment consists of a single application of 1000 litres of saturated sodium chloride (salt) per hectare (or split applications) containing a non-ionic wetter (0.1% w/w) to the crop from after the emergence of three pairs of true leaves growth stage to the end of July for the control of volunteer potatoes and activity against other weeds.

2. Crop phytotoxicity may occur after treatments.

Food and Environment Protection Act 1985
Control of Pesticides Regulations 1986 (SI 1986 No. 1510): Approval

In exercise of the powers conferred by regulation 5 of the Control of Pesticides Regulations 1986 (SI 1986/1510) and of all other powers enabling them in that behalf, the Minister of Agriculture, Fisheries and Food and the Secretary of State (acting jointly), and the Scottish Ministers (as regards Scotland) and the National Assembly for Wales and the Minister of Agriculture, Fisheries and Food (acting jointly as regards Wales), hereby jointly give full approval for the use of:

Commodity substance: being **SODIUM HYPOCHLORITE** subject to the conditions set out below:

Date of issue:	5 December 1996

Use:

Field of Use:	**Only as an horticultural bactericide**
Crop:	Mushroom
Maximum individual dose:	See 'Other specific restrictions'
Latest time of application:	One day before harvest
Operator protection:	(1) Engineering control of operator exposure must be used where reasonably practicable in addition to the following personal protective equipment: Operators must wear suitable protective clothing (coveralls), suitable protective gloves and face protection (faceshield) when handling the concentrate.
	(2) However, engineering controls may replace personal protective equipment if a COSHH assessment shows they provide an equal or higher standard of protection.
Environmental protection:	Since this substance is harmful to fish or other aquatic life surface waters or ditches must not be contaminated with chemical or used container.
Other specific restrictions:	(1) This substance must only be used by operators who are suitably trained and competent to carry out this work.
	(2) The maximum concentration must not exceed 315 mg sodium hypochlorite per litre of water (equivalent to 150 mg available chlorine per litre of water).

ADVISORY NOTE:

Mixing and loading must only take place in a ventilated area.

Food and Environment Protection Act 1985
Control of Pesticides Regulations 1986 (SI 1986 No. 1510): APPROVAL

In exercise of the powers conferred by regulation 5 of the Control of Pesticides Regulations 1986 (SI 1986/1510) and of all other powers enabling them in that behalf, the Minister of Agriculture, Fisheries and Food and the Secretary of State (acting jointly), and the Scottish Ministers (as regards Scotland) and the National Assembly for Wales and the Minister of Agriculture, Fisheries and Food (acting jointly as regards Wales), hereby jointly give full approval for the use of:

Commodity substance: being **STRYCHNINE HYDROCHLORIDE** subject to the conditions set out below:

Date of issue: 19 June 1997

Use:

Field of Use: **Only as a vertebrate control agent for the destruction of moles underground**

Crops/Situations: Commercial agricultural/horticultural land where public access is restricted; grassland associated with aircraft landing strips, horse paddocks, gallops and race courses; golf courses and other areas specifically approved by Agricultural Departments.

Sale and Supply:

Label: Substance to be supplied with a label in accordance in the Poisons Act 1972

Container: Substance to be supplied only in the original sealed packaging of the manufacturer and only in units of up to 2 g.

Other   (a) Must only be supplied to the holders of an Authority to Purchase issued by the appropriate Agriculture Department (in England – the Ministry of Agriculture, Fisheries and Food; in Wales – the Welsh Office Agriculture Department; and in Scotland – the Department of Agriculture and Fisheries for Scotland).

(b) Quantities of more than 8 g must only be supplied to providers of a commercial service.

Storage: (a) Providers of a commercial service must not hold more strychnine than the amount specified by the authorising Agriculture Department.

(b) Substance must be stored in the original container under lock and key and only on the premises and under the control of the holder of an authority to purchase or a named individual who satisfies the appropriate Agriculture Department.

Advertisement: The substance must not be advertised except that individual pharmacists may provide details of availability and price to a person authorised to purchase the substance.

Operator protection:

(a) A written Control of Substances Hazardous to Health (COSHH) assessment must be made before using strychnine (see Advisory Note 1).

(b) Gloves must be worn when preparing and laying bait and when handling contaminated utensils.

Environmental protection:

(a) The substance must be prepared for applications with great care so that there is no contamination of the surface of the ground.

(b) Any prepared bait remaining at the end of the day must be buried (see Advisory Note 2).

Other specific restrictions:

(a) Authorities to purchase may only be issued to persons who satisfy the appropriate Agriculture Department that they are trained and competent in its use and can be entrusted with it (see Advisory Note 3).

(b) Access to strychnine must be restricted to those who hold the authority to purchase and to named individuals who satisfy the appropriate Agriculture Department.

(c) The substance must be used as and where directed by the appropriate Agriculture Department.

(d) Operators must be supplied with a CHIP Safety Data Sheet before commencing work (see Advisory Note 4).

(e) Providers of a commercial service must use suitable dedicated utensils capable of being washed clean. Such equipment must be cleaned after every treatment and washings disposed of in a mole run. The equipment must be stored securely between treatments.

(f) Non commercial users must use suitable disposal utensils which must be disposed of by burial on the land where the treatment takes place. They must not be retained for re-use.

(g) Providers of a commercial service must advise the local office of the appropriate Agriculture Department of the treatments applied in the previous year, report the quantity of the substance held in store and the arrangements for secure storage i) when applying for further quantities, or ii) within 12 months of last using it, whichever is the earlier.

ADVISORY NOTES:

1. Guidance on how to carry out a COSHH assessment may be obtained from HSE

2. Burial of contaminated materials should be in accordance with advice given in the Code of Practice for the Safe Use of Pesticides on Farms and Holdings.

3. When deciding whether a person is fit to be entrusted with the substance, account will be taken of character and expertise.

4. The Chemicals (Hazard Information and Packaging for Supply) Regulations 1994 as amended (CHIP) data sheet is obtainable from the supplier of strychnine eg the pharmacist.

Food and Enviroment Protection Act 1985
Control of Pesticides Regulations 1986 (SI 1986 No 1510): Approval

In exercise of the powers conferred by regulation 5 of the Control of Pesticides Regulations 1986 (SI 1986/1510) and of all other powers enabling them in that behalf, the Minister of Agriculture, Fisheries and Food and the Secretary of State (acting jointly), and the Scottish Ministers (as regards Scotland) and the National Assembly for Wales and the Minister of Agriculture, Fisheries and Food (acting jointly as regards Wales), hereby jointly give full approval for the use of:

Commodity Substance: being 77% clean **SULPHURIC ACID** subject to the conditions set out below:

Date of issue: 23 November 1995

Date of expiry: 25 July 2003

Use:

Field of Use: **Only as an agricultural desiccant**

Crops	*Maximum individual dose*	*Maximum number of treatments*	*Times of application*
Potato (grown for canning)	800 litres substance/hectare	3 per crop (See 'Other specific restrictions')	1 May – 15 November
Potato (Other)	340 litres substance/hectare	3 per crop (See 'Other specific restrictions')	1 May – 15 November
Bulbs and Corms	280 litres substance/hectare	1 per year	1 May – 15 November
Peas	220 litres substance/hectare	1 per crop	1 May – 15 November

Operator protection:

(1) A written COSHH assessment must be made before using sulphuric acid. Operators should observe the Occupational Exposure Standard set out in HSE guidance note EH40/90 or subsequent issues.

(2) Engineering control of operator exposure must be used where reasonably practicable in addition to the following personal protective equipment:

Operators must wear suitable protective clothing as listed below:

(a) When filling sprayer; spraying carrying out adjustment to application equipment or cleaning equipment; re-entering the sprayed area within 24 hours.

Face shield of acid resistant type; acid resistant coveralls either single or combination garment (or for persons operating bulk installations, acid proof); gauntlet gloves either natural rubber or PVC material, rubber boots, acid proof apron and suitable respiratory protective equipment.

(b) When re-entering the sprayed area between 24 and 96 hours.   Acid resistant coveralls, gloves and boots.

(3) However, engineering controls may replace personal protective equipment if a COSHH assessment shows they provide an equal or higher standard of protection.

(4) Operators must have liquid suitable for eye irrigation immediately available at all times throughout the operation.

Other specific restrictions:

(1) Must only be used by operators holding a relevant recognised certificate of competence in the use of the equipment for the application of sulphuric acid.

(2) Must not be applied using pedestrian controlled applicators or hand held equipment.

(3) All equipment must be constructed of materials suitable for use with or exposure to sulphuric acid.

(4) Application must be confined to the land intended to be treated.

(5) Spray must not be deposited within one metre of public footpaths.

(6) At least 24 hours written notice of the intended operation and the possibility of a hazard must be given to occupants of any premises and to the owner, or his agent, of any livestock or crops within 25 metres of any boundary of the land intended to be treated.

(7) Before the spraying takes place, readable notices must be posted on adjacent roads and paths warning passers by and drivers of vehicles of the time and place of the intended application and possibility of hazard. Notices to be kept in place for 96 hours following treatment

(8) Unprotected persons must be kept out of treated areas for at least 96 hours following treatment.

(9) The maximum quantity to be applied to potatoes must not exceed 800 litres per hectare per crop.

(10) Do not apply to crops in which bees are actively foraging. Do not apply when flowering weeds are present.

(11) Only unused "sulphur burnt" sulphuric acid to be used.

HEALTH AND SAFETY EXECUTIVE
Food and Environment Protection Act 1985

Schedule: COMMODITY SUBSTANCE: **TETRACHLOROETHYLENE**

Date of issue: 18 February 1993

Date of expiry: 28 February 2001

This approval is subject to the following conditions:

1 *FIELD OF USE:*   ONLY AS AN INSECTICIDE IN MUSEUMS AND BUILDINGS
                              OF CULTURAL, ARTISTIC AND HISTORICAL INTEREST

2 *PEST AND USAGE AREA:*   FOR THE CONTROL OF FLYING AND CRAWLING INSECTS
                              ON TEXTILES.

3 *APPLICATION METHOD:*   IMMERSION IN A TANK ENCLOSED IN A FUME CUPBOARD

Operator protection:

(1) A written COSHH assessment must be made before using tetrachloroethylene. Operators
    should also observe the OES set out in HSE guidance note EH40/93 or subsequent issues.

(2) Engineering controls of operator exposure must be used where reasonably practicable in
    addition to the following items of personal protective equipment.

    Operators must wear suitable protective clothing, including protective gloves and eye
    protection, when handling and using the material.

(3) However engineering controls may replace personal protective equipment if a COSHH
    assessment shows they provide an equal or higher standard of protection.

Other specific restrictions

(1) Operators should be provided with adequate information about the hazards of the substance
    and the precautions necessary for safe use. Sources of information include the supplier's
    Safety Data Sheet.

(2) Unprotected persons and animals must be excluded from any areas where treatment is taking
    place, and such areas should be ventilated after treatment.

(3) Must be used only by operators who are suitably trained and competent to carry out this work.

HEALTH AND SAFETY EXECUTIVE
Food and Environment Protection Act 1985

Schedule: COMMODITY SUBSTANCE: **THYMOL**

Date of issue: 18 February 1993

Date of expiry: 28 February 2011

This approval is subject to the following conditions:

1 *FIELD OF USE:*   ONLY AS A FUNGICIDE IN MUSEUMS AND BUILDINGS OF
                           CULTURAL, ARTISTIC AND HISTORICAL INTEREST.

2 *PEST AND USAGE AREA:*   FOR THE CONTROL OF FUNGI.

3 *APPLICATION METHOD:*   TREATMENT IN SEALED TREATMENT CABINETS

Operator protection:

(1) A written COSHH assessment must be made before using thymol.

(2) Engineering control of operator exposure must be used where reasonably practicable in
    addition to the following items of personal protective equipment.

    Operators must wear suitable protective clothing, including protective gloves and eye
    protection, when handling or applying the material.

(3) However engineering controls may replace personal protective equipment if a COSHH
    assessment shows they provide an equal or higher standard of protection.

Other specific restrictions:

(1) Operators should be provided with adequate information about the hazards of the substance
    and the precautions necessary for safe use. Sources of information include the supplier's
    Safety Data Sheet.

(2) Unprotected persons and animals must be excluded from any areas where treatment is taking
    place, and such areas should be ventilated after treatment.

(3) Must be used only by operators who are suitably trained and competent to carry out this work.

Food and Environment Protection Act 1985
Control of Pestices Regulations 1986 (SI 1986 NO 1510): APPROVAL

In exercise of the powers conferred by regulation 5 of the Control of Pesticides Regulations 1986 (SI 1986/1510) and of all other powers enabling them in that behalf, the Minister of Agriculture, Fisheries and Food and the Secretary of State (acting jointly), and the Scottish Ministers (as regards Scotland) and the National Assembly for Wales and the Minister of Agriculture, Fisheries and Food (acting jointly as regards Wales), hereby jointly give full approval for the use of:

Commodity Substance: being **UREA** subject to the conditions set out below:

Date of issue: 1 March 1991

Date of expiry: 28 February 2001 (see advisory note 4)

Use:

Field of Use: **Only as a home garden fungicide**

Crop/ Situation	*Maximum individual dose*	*Maximum number of treatments*	*Latest time of application*
Cut stumps of trees	1 litre of 37% w/v aqueous solution/m^2 of cut stump	One per cut stump	At felling
Apple and pear trees	0.07 litres of 7% w/v solution/m^2	One per tree per year	Post harvest, pre leaf-fall

Operator protection:

When used at work, the following must be observed:

(1) Engineering control of operator exposure must be used where reasonably practicable in addition to the following personal protective equipment:

Operators must wear suitable protective gloves when mixing urea solution.

(2) However, engineering controls may replace personal protective equipment if a COSHH assessment shows they provide an equal or higher standard of protection.

ADVISORY NOTES

1 Pesticides approved for amateur use may be used by professional operators without user certification.

2 For the treatment of cut stumps of trees the following dyes may be used at 0.04% w/v concentration when mixed with solution of urea:

(i) Kenacid Turquoise V5898 (CI Acid Blue 42045)

(ii) Denacid Turquoise AN 200 } (CI Acid Blue 9 42090)
Duasyn Acid Blue AE-20   }

3 Operators should wear rubber or other chemical-proof gloves, a nuisance dust mask and a cotton/terylene overall when handling dye powder.

4 Whilst data submitted by interested parties is being considered by PSD for Continuation of approval, the date of expiry shall be extended to 25 July 2003.

HEALTH AND SAFETY EXECUTIVE
Food and Environment Protection Act 1985

Schedule: COMMODITY SUBSTANCE: **WHITE SPIRIT**

Date of issue: 18 February 1993

Date of expiry: 28 February 2011

This approval is subject to the following conditions:

1 *FIELD OF USE:* ONLY AS AN INSECTICIDE IN MUSEUMS AND BUILDINGS OF CULTURAL, ARTISTIC AND HISTORICAL INTEREST

2 *PEST AND USAGE AREA:* FOR THE CONTROL OF FLYING AND CRAWLING INSECTS ON TEXTILES

3 *APPLICATION METHOD:* IMMERSION IN A TANK ENCLOSED IN A FUME CUPBOARD

Operator protection:

(1) A written COSHH assessment must be made before using white spirit. Operators should also observe the OES set out in HSE guidance note EH40/99 or subsequent issues.

(2) Engineering controls of operator exposure must be used where reasonably practicable in addition to the following items of personal protective equipment.

Operators must wear suitable protective clothing, including protective gloves and eye protection, when handling and using the material.

(3) However engineering controls may replace personal protective equipment if a COSHH assessment shows they provide an equal or higher standard of protection.

Other specific restrictions:

(1) Operators must be provided with adequate information on hazards and precautions. Sources of information may include the supplier's Safety Data Sheet.

(2) Unprotected persons and animals must be excluded from any areas where treatment is taking place, and such areas should be ventilated after treatment.

(3) Must be used only by operators who are suitably trained and competent to carry out this work.

ANNEX E

BANNED AND NON-AUTHORISED PESTICIDES IN THE UNITED KINGDOM

ACTIVE SUBSTANCES BANNED IN THE EUROPEAN UNION UNDER COUNCIL DIRECTIVE 79/117/EEC*

Council Directive 79/117/EEC dated 21 December 1978 prohibits the placing on the market and the use of plant protection products containing certain active substances which, even if applied in an approved manner, could give rise to harmful effects on human health or the environment.

ACTIVE SUBSTANCE	EFFECTIVE DATE OF BAN
Mercury Compounds	
Mercuric oxide (mercury oxide)	1992
Mercurous chloride (calomel)	1992
Other inorganic mercury compounds	1981
Alkyl mercury compounds	1991
Alkoxyalkyl and aryl mercury compounds	1992
Persistent Organo-chlorine Compounds	
Aldrin	1991
Chlordane	1981
Dieldrin	1981
DDT	1986
Endrin	1991
HCH containing less than 99% of gamma isomer	1981
Heptachlor	1984
Camphechlor	1984
Hexachlorobenzene	1981
Other Compounds	
Ethylene oxide	1991
Nitrofen	1988
1,2 Dibromoethane (Ethylene dibromide)	1988
1,2 Dichloroethane (Ethylene dichloride)	1989
Dinoseb, its acetate and salts	1991
Binapacryl	1991
Captafol	1991
Dicofol containing less than 78% of pp[1] – dicofol or more than 1g/kg DDT and DDT related compounds.	1991

(a) Maleic hydrazide and its salts, other than its choline, potassium and sodium salts	1991
(b) Choline, potassium and sodium salts of maleic hydrazide containing more than 1g/kg of free hydrazine expressed on the basis of the acid equivalent.	1991
Quintozene containing more than 1g/kg of HCB or more than 10g/kg pentachlorobenzene	1991

Note: Some of these active substances had already been banned in the UK prior to inclusion in Directive 79/117/EEC. Further details are available from PSD.

* Directive 79/117/EEC was published in the Official Journal of the European Communities L33 of 8 February 1979. It has been amended by Council Directives 83/131/EEC dated 14 March 1983, 85/298/EEC dated 22 May 1985, 86/214/EEC dated 26 May 1986, 86/355/EEC dated 21 July 1986, 87/181/EEC dated 9 March 1987, 87/477/EEC dated 9 September 1987, 89/365/EEC dated 30 May 1989, 90/335/EEC dated 7 June 1990, 90/533/EEC dated 15 October 1990, 91/188/EEC dated 19 March 1991. Copies of all these Directives are available from The Stationery Office, Publications Centre, PO Box 29, Norwich, NR3 1GN; telephone orders 0870 600 5522 or Fax 0870 600 5533.

ACTIVE SUBSTANCES BANNED IN THE UK

In addition to the substances listed above, the UK has also banned the following substances:

ACTIVE SUBSTANCE	EFFECTIVE DATE OF BAN	REASON FOR BAN
Antu (thiourea)	1966	Evidence of carcinogenicity
Azobenzene	1975	Evidence of carcinogenicity
Cadmium compounds	1965	Evidence of carcinogenicity
Calcium arsenate	1968	High acute toxicity; persistence in soil; evidence of carcinogenicity
Chlordecone	1977	Evidence of carcinogenicity
Cyhexatin	1988	Evidence of teratogenicity
Methyl mercury	1971	Environmental hazard (accumulation in the food chain)
Phenylmercury salicylate	1972	Acute toxicity; accumulation in the environment
Potassium arsenite	1961	Acute toxicity to wildlife and livestock
Selenium compounds eg sodium selenate	1962	Acute toxicity to humans and livestock
Sodium arsenite	1961	Acute toxicity to wildlife and livestock
1,1,2,2-tetrachloroethane	1969	Acute and chronic toxicity to humans

Note: In addition to the above, various other active substances were effectively banned pre-1986 by being withdrawn under the non-statutory scheme that was in place before COPR. Further details of these, and other substances which are no longer approved in the UK, are available from PSD.

ACTIVE SUBSTANCES NOT INCLUDED IN ANNEX I OF COUNCIL DIRECTIVE 91/414/EEC (NON-AUTHORISED PESTICIDES)

Council Directive 91/414/EEC provided for a review of all active substances on the market prior to 25 July 1993. This review is an ongoing process and below is a list of the active substances reviewed and not included in Annex I of the Directive as at 1 December 2000.

Non-inclusion means that authorisations for plant protection products containing these substances must be withdrawn within a period specified in the Commission Decision.

ACTIVE SUBSTANCE DECISION	COMMISSION REFERENCE	EC OFFICIAL JOURNAL
Azinphos ethyl[1]	95/276/EC, dated 13/7/95	L 170, 20 July 1995, pps 22-23
Chlozolinate	2000/626/EC, dated 13/7/00	L 263, 18 October 2000, pps 32-33
Cyhalothrin[1]	94/643/EC, dated 12/9/94	L 249, 24 September 1994, p 18
Dinoterb	98/269/EC, dated 7/4/98	L 117, 21 April 1998, pps 13-14
DNOC	1999/164/EC, dated 17/2/99	L 54, 2 March 1999, pps 21-22
Fenvalerate[1]	98/270/EC, dated 7/4/98	L 117, 21 April 1998, p 15
Ferbam[1]	95/276/EC, dated 13/7/95	L 170, 20 July 1995, pps 22-23
Lindane	2000/801/EC, dated 20/12/00	L 324, 21 December 2000, pps 42-43
Monolinuron	2000/234/EC, dated 9/3/00	L 73, 22 March 2000, pps 18-19
Permethrin	2000/817/EC, dated 27/12/00	L 332, 28 December 2000, pps 114-115
Propham[1]	96/586/EC, dated 9/4/96	L 257, 10 October 2000, pps 41-42
Pyrazaphos	2000/233/EC, dated 9/3/00	L 73, 22 March 2000, pps 16-17
Quintozene	2000/816/EC, dated 27/12/00	L 332, 28 December 2000, pps 112-113
Tecnazene	2000/725/EC, dated 2/11/00	L 292, 21 November 2000, pps 30-31

[1] Non-inclusion does not preclude the possibility that these substances could be considered again at a future date under the terms of Article 6 of Directive 91/414/EEC.

ANNEX F

ACTIVE INGREDIENTS SUBJECT TO THE POISONS LAW

Certain products in this book are subject to the provisions of the Poisons Act 1972, the Poisons List Order 1982, the Poisons Rules 1982 and amending Orders made to these statutory instruments (copies of all these are obtainable from The Stationery Office). These Rules include general and specific provisions for the storage and sale and supply of listed non-medicine poisons.

The Active Ingredients approved for use in the UK and included in this book are specified under Parts I and II of the Poisons List as follows:

Part I (sale restricted to registered retail pharmacists)

aluminium phosphide	methyl bromide
chloropicrin	sodium cyanide
magnesium phosphide	strychnine

Part II (sale restricted to registered retail pharmacists and listed sellers registered with local authority)

aldicarb	endosulfan	paraquat (d)
alphachloralose	fentin acetate	phorate (a)
ammonium bifluoride	fentin hydroxide	sodium fluoride
carbofuran (a)	formaldehyde	sulphuric acid
chlorfenvinphos (a,b)	mephosfolan	thiometon
demeton-s-methyl	methomyl (f)	triazaphos (a)
dichlorvos (a,e)	nicotine (c)	zinc phosphide
disulfuton (a)	oxamyl (a)	

(a) Granular formulations which do not contain more than 12% w/w of this, or a combination of similarly flagged poisons, are exempt

(b) Treatments on seeds are exempt

(c) Formulations containing not more than 7.5% of nicotine are exempt

(d) Pellets containing not more than 5% paraquat ion are exempt

(e) Preparations in aerosol dispensers containing not more than 1% w/w active ingredient are exempt. Materials impregnated with dichlorvos for slow release are exempt

(f) Solid substances containing not more than 1% w/w ai are exempt.

ANNEX G
PUBLISHED EVALUATIONS LIST

PUBLISHED EVALUATIONS, NOVEMBER 2000

Title Price		Document Number
A		
Abamectin	£13.50	(no.60)
AEF107892	£24.50	(no.186)
Alachlor (1)	£ 8.50	(no.22)
Alachlor (2)	£ 7.00	(no.41)
Aldicarb	£ 9.50	(no.109)
Amidosulfuron	£21.00	(no.91)
Amorphous Silicon Dioxide	£ 5.50	(no.129)
Anilazine	£19.50	(no.93)
2-Aminobutane	£ 4.00	(no.29)
Assessment of Humaneness of Vertebrate Control Agents	£ 6.00	(no.171)
Atrazine (1)	£ 9.50	(no.51)
Atrazine (2)	£20.00	(no.71)
Azaconazole	£ 3.00	(no.08)
B		
Benfuracarb	£ 4.00	(no.05)
Benomyl	£14.00	(no.57)
Benzyl Benzoate	£ 6.00	(no.159)
Bifenthrin	£ 3.50	(no.12)
Bitertanol	£11.50	(no.92)
Bromoxynil. Review of Agricultural and Horticultural Uses	£10.00	(no.124)
Bromuconazole	£25.00	(no.160)
Bti (1)	£ 3.50	(no.101)
Bti (2)	£ 4.00	(no.105)
Buprofezin	£19.00	(no.81)
C		
Carbaryl - (Maff Approved Uses)	£11.50	(no.155)
Carbaryl - (Review of its Uses in Public Hygiene and Amateur Insecticides)	£ 5.50	(no.156)
Carbendazim	£18.00	(no.58)
Carbetamide	£ 5.50	(no.125)
Chlorfenvinphos	£25.00	(no.107
Chlorothalonil	£22.50	(no.112)

Title Price		Document Number
Chlorpropham	£ 5.50	(no.87)
Chlorsulfuron	£ 4.50	(no.37)
Clodinafop-Propargyl and Cloquintocet-Mexyl	£25.00	(no.121)
Commodity Substances	£ 6.00	(no.103)
Copper Compounds	£18.00	(no.183)
Cycloxydim	£12.50	(no.31)
Cyfluthrin	£ 4.00	(no.03)
Cyfluthrin: Use in Wood Preservation	£ 8.00	(no.195)
Cyhalothrin	£ 4.00	(no.21)
Cyprodinil	£25.00	(no.169)
Cyromazine (1)	£11.00	(no.89)
Cyromazine (2)	£ 6.00	(no.116)
D		
2,4-D	£25.00	(no.68)
Daminozide	£ 5.00	(no.14)
Demeton-S-Methyl	£ 4.00	(no.77)
Desmedipham	£17.50	(no.75)
Diazinon (1)	£25.00	(no.35)
Diazinon (2)	£ 3.00	(no.43)
Dichlorvos	£17.50	(no.120)
Diclofop-Methyl (1)	£25.00	(no.46)
Diclofop-Methyl (2)	£ 4.50	(no.117)
Diclofop-Methyl (3)	£ 5.00	(no.194)
Dicofol (Review of)	£13.50	(no.157)
Difenoconazole (1)	£14.00	(no.106)
Difenoconazole-Ecotoxicity (2)	£ 3.50	(no.152)
Diflufenican	£ 9.50	(no.137)
Dimefuron	£11.50	(no.79)
Dimethoate (1)	£ 3.50	(no.30)
Dimethoate (2)	£21.50	(no.86)
Dimethomorph	£21.50	(no.99)
Dinocap	£ 9.00	(no.32)
Dinocap (Review of)	£13.00	(no.175)
DPX M6316	£ 3.50	(no.07)
E		
Epoxiconazole (1)	£25.00	(no.108)
Epoxiconazole (2)	£ 5.50	(no.189)
Esfenvalerate	£25.00	(no.55)
Ethoprophos	£ 5.50	(no.04)
Ethylene Bisdithiocarbamates (1)	£10.00	(no.16)
Ethylene Bisdithiocarbamates (2)	£25.00	(no.36)

Title	Price	Document Number
F		
Fenazaquin	£25.00	(no.150)
Fenbuconazole	£25.00	(no.128)
Fenbutatin Oxide	£ 4.00	(no.26)
Fenoxaprop Ethyl	£10.50	(no.18)
Fenoxaprop-P-Ethyl	£ 8.00	(no.17)
Fenoxycarb	£25.00	(no.161)
Fenpiclonil	£18.50	(no.78)
Fenpropathrin	£ 9.50	(no.23)
Fenpropidin	£17.00	(no.67)
Fenpyroximate	£22.00	(no.130)
Fentin Acetate	£ 3.00	(no.27)
Fentin Hydroxide	£ 4.00	(no.25)
Flocoumafen	£ 3.00	(no.01)
Fluazifop-P-Butyl	£ 5.50	(no.10)
Fluazinam	£17.00	(no.100)
Fludioxonil	£20.00	(no.126)
Flufenoxuron	£11.00	(no.143)
Flufenoxuron 2 (Use as a Wood Preservative)	£14.50	(no.180)
Fluoroglycofen-Ethyl	£25.00	(no.50)
Flurtamone	£23.00	(no.196)
Flusilazole	£10.50	(no.11)
Flutriafol	£11.50	(no.158)
Fluquinconazole	£25.00	(no.184)
Fipronil: Use as a public Hygiene insecticide	£16.00	(no.187)
Fomesafen	£15.00	(no.118)
G		
Gamma-HCH (Lindane 1)	£ 4.00	(no.47)
Gamma-HCH (Lindane 2)	£18.50	(no.64)
Gamma-HCH (Lindane 3) (Agricultural Uses)	£25.00	(no.151)
Glufosinate-ammonium	£21.50	(no.33)
Grain Protectants in the UK Review of the use of.	£ 5.00	(no.59)
Guazatine	£ 9.50	(no.53)
H		
HOE 070542 Triazole Coformulant	£ 7.50	(no.19)
Hydramethylnon	£ 6.50	(no.102)
Hydroprene (1)	£ 5.50	(no.44)
Hydroprene (2)	£ 7.00	(no.145)
I		
Imazaquin	£17.00	(no.66)
Imazethapyr	£10.00	(no.149)

Title	Price	Document Number
Imidacloprid	£23.00	(no.73)
Ioxynil - Review of the Agricultural and Horticultural Uses.	£11.00	(no.123)
IPBC (1)	£ 4.50	(no.13)
IPBC (2)	£ 8.50	(no. 115)
Iprodione	£ 4.00	(no.28)
Isoproturon	£23.00	(no.140)
K		
Kathon 886 (1)	£ 9.00	(no.90)
Kathon 886 (2)	£ 6.00	(no.176)
Kresoxim-Methyl	£25.00	(no.163)
L		
Lambda-Cyhalothrin	£ 3.00	(no.104)
Lindane - Reproductive Toxicity Effects in Dogs	£ 6.50	(no.164)
Lindane- The Review of	£13.50	(no.191)
Lindane 1 (Gamma-HCH)	£ 4.00	(no.47)
Lindane 2 (Gamma-HCH)	£18.50	(no.64)
Lindane 3 (Gamma-HCH) (Agricultural Uses)	£25.00	(no.151)
Linuron	£18.00	(no.132)
M		
Malathion	£19.50	(no.135)
MBC Fungicides - Benomyl and Carbendazim	£ 6.00	(no.170)
MBC Fungicide - Thiophanate Methyl	£ 4.50	(no. 181)
Mecoprop	£17.00	(no.95)
Mecoprop-P	£17.00	(no.96)
Metaldehyde	£19.50	(no.153)
Metconazole	£25.00	(no.193)
Methiocarb (Review of)	£25.00	(no.179)
Methyl Bromide	£12.00	(no.63)
Metosulam	£22.00	(no.148)
Metsulfuron Methyl	£ 7.50	(no.38)
Metsulfuron-Methyl and Thifensulfuron-Methyl (Review of)	£ 5.50	(no.119)
Review of Environmental Persistence Monolinuron	£ 9.00	(no.133)
O		
Omethoate	£13.50	(no.83)
Oxine Copper	£ 3.50	(no.09)
Oxydemeton-Methyl	£11.00	(no.76)

Title	Price	Document Number
P		
Paclobutrazol	£11.00	(no.142)
Pentachlorophenol	£17.50	(no.114)
2-Phenyl Phenol	£ 8.00	(no.82)
Phlebiopsis Gigantea	£ 8.50	(no173)
Phorate	£14.00	(no.98)
Pirimicarb	£24.00	(no.134)
Pirimiphos-Methyl	£22.50	(no.167)
PP321 (Lambda-Cyhalothrin)	£ 5.00	(no.20)
Prallethrin	£ 6.50	(no.131)
Propamocarb Hydrochloride	£21.00	(no.62)
Propaquizafop	£23.00	(no.94)
Propiconazole	£ 8.00	(no.80)
Propyzamide	£ 3.50	(no.154)
Pyrimethanil	£18.00	(no.138)
Pyriproxyfen	£11.50	(no.147)
Q		
Quinmerac	£25.00	(no.177)
Quizalofop-Ethyl	£ 3.50	(no.02)
R		
RH 3866	£ 4.00	(no.06)
Rimsulfuron	£25.00	(no.146)
S		
SAN 619F (Cyproconazole)	£25.00	(no.45)
Simazine (1)	£ 9.50	(no.52)
Simazine (2)	£18.50	(no.72)
S-Methoprene (1)	£ 5.50	(no.85)
S-Methoprene (2)	£ 8.00	(no.166)
Sodium Cyanide	£11.00	(no.144)
Sodium Cyanide (UK Review of)	£ 6.50	(no.192)
Strychnine Hydrochloride	£ 6.00	(no.168)
Sulphuric Acid	£ 4.00	(no174)
T		
Tau - Fluvalinate	£25.00	(no.162)
Tebuconazole (1)	£25.00	(no.65)

Title	Price	Document Number
Tebuconazole (2)	£ 7.00	(no.88)
Tebufenpyrad	£25.00	(no.122)
Tecnazene (Review of)	£20.00	(no.127)
Teflubenzuron	£ 1.50	(no.40)
Tefluthrin	£13.00	(no.42)
Tetraconazole	£25.00	(no.185)
Thiabendazole	£10.00	(no.54)
Thifensulfuron-Methyl	£ 5.00	(no.39)
Thiodicarb	£25.00	(no.49)
Thiophanate-Methyl	£11.00	(no.56)
Tolclofos-Methyl	£16.00	(no.69)
Tolclofos-Methyl in the Products 'Rizolex'	£ 8.50	(no.178)
Tolylfluanid	£19.50	(no.136)
Tralkoxydim (PP604)	£25.00	(no.70)
Transfluthrin (1)	£10.00	(no.165)
Transfluthrin (2)	£ 7.50	(no.188)
Triasulfuron	£25.00	(no.48)
Triazamate	£25.00	(no.172)
Triazophos	£24.00	(no.84)
Triazoxide	£15.00	(no.97)
Tribenuron Methyl	£15.00	(no.61)
Tributyltin Naphthenate (1)	£ 3.00	(no.15)
Tributyltin Naphthenate (2)	£ 4.50	(no.110)
Tributyltin Oxide (1)	£ 7.00	(no.24)
Tributyltin Oxide (2)	£ 4.50	(no.74)
Tridemorph	£ 8.00	(no.190)
Triflusulfuron-Methyl	£25.00	(no.139)
Trinexapac Ethyl	£18.00	(no.141)
Triorganotin Compounds (1)	£16.50	(no.111)
Triorganotin Compounds (2)	£ 8.50	(no.182)
V		
Vinclozolin (1)	£12.00	(no.34)
Vinclozolin(2)	£12.50	(no.113)

Evaluations are available on application in writing to Mrs D Wells, Finance and Corporate Services Unit, Pesticides Safety Directorate, Room 313, Mallard House, Kings Pool, 3 Peasholme Green, York, YO1 7PX. Tel. 01904 455728.

ANNEX H

PRODUCTS APPROVED FOR USE IN OR NEAR WATER
(other than for Public Hygiene or Antifouling Use)

"Products approved for use in or near water as at 31 October 2000, are listed below. Before you use any product in or near water you should first consult the appropriate water regulatory body (Environment Agency/Scottish Environment Protection Agency). Always read the label before use."

Product name	Marketing company	Reg No.
Acrion	Aventis Environmental Science	09965
Agricorn 2,4-D	Farmers Crop Chemicals Ltd	07349
Atlas 2,4-D	Atlas Crop Protection Ltd	07699
Azural	Monsanto Plc	09582
Barclay Barbarian	Barclay Chemicals Manufacturing Ltd	09865
Barclay Gallup	360 Barclay Plant Protection	09127
Barclay Gallup Amenity	Barclay Chemicals Manufacturing Ltd	06753
Barclay Gallup Biograde 360	Barclay Chemicals Manufacturing Ltd	09840
Barclay Gallup Biograde Amenity	Barclay Chemicals Manufacturing Ltd	10203
Barclay Garryowen	Barclay Chemicals (R & D) Ltd	09869
Biactive 270	Monsanto Plc	10031
Bioglyce	Applyworld Limited	10054
Buggy Sg	Sipcam UK Ltd	08573
Cardel Egret	Cardel	09703
Cardel Egret	Cardel Agro Sas	10252
Cardel Glyphosate	Cardel	09581
Cardel Glyphosate	Cardel Agro Sas	10253
Casoron G	Miracle Professional	07926
Casoron G	Zeneca Crop Protection	08065
Casoron G	Uniroyal Chemical Ltd	09022
Casoron G	Nomix-Chipman Ltd	09023
Casoron G	Rigby Taylor Ltd	09326
Casoron Gsr	Imperial Chemical Industries Plc	00451
Casoron Gsr	Miracle Professional	07925
Clarosan	Novartis Crop Protection UK Ltd	08396
Clarosan	The Scotts Company (UK) Limited	09394
Clarosan 1fg	Ciba Agriculture	03859
Clayton Swath	Clayton Plant Protection (UK) Ltd	06715
Cleancrop Diquat	United Agri Products	09687
Clinic	Nufarm UK Ltd	08579
Clinic	Nufarm UK Ltd	09378
Danagri Glyphosate 360	Danagri Aps	06955
Do-Away	Nch Europe Ltd	10112
Dormone	Rhone-Poulenc Amenity	05412
Dormone	Aventis Environmental Science	09932
Glyfos	Cheminova Agro (UK) Ltd	07109
Glyfos 480	Cheminova Agro (UK) Ltd	08014
Glyfos Proactive	Nomix-Chipman Ltd	07800

Product name	Marketing company	Reg No.
Glyper	Pan Britannica Industries Ltd	07968
Glyphogan	Makhteshim-Agan (UK) Ltd	05784
Glyphosate 360	Danagri Aps	08568
Glyphosate 360	Applyworld Limited	09151
Glyphosate 360	Danagri Aps	09233
Glyphosate 360 C	Monsanto Plc	09846
Glyphosate 360 D	Monsanto Plc	09847
Glyphosate 360a	Danagri Aps	09234
Glyphosate 360b	Danagri Aps	09235
Glyphosate Biactive	Monsanto Plc	08307
Helosate	Helm AG	06499
Krenite	Du Pont (UK) Ltd	01165
Levi	P.S.I. Phoenix Scientific Innovation UK Ltd	07845
Luxan Dichlobenil Granules	Luxan (UK) Ltd	09250
Marnoch Glyphosate	Marnoch Ventures Limited	09744
Midstream	Imperial Chemical Industries Plc	01348
Midstream	Zeneca Professional Products	06824
Midstream	Miracle Professional	07739
Midstream	The Scotts Company (UK) Limited	09267
Mon 44068 Pro	Monsanto Plc	06815
Mon 52276	Monsanto Plc	06949
Mon 77158a	Monsanto Plc	09904
Monty	Quadrangle Agrochemicals	07796
Mss 2,4-D Amine	Mirfield Sales Services Ltd	01391
Mss 2,4-D Amine	Nufarm Whyte Agriculture Ltd	10183
Mss Glyfield	Mirfield Sales Services Ltd	08009
Nufosate	Nufarm UK Ltd	10096
Ragox	Nufarm Whyte Agriculture Ltd	09876
Reglone	Zeneca Crop Protection	06703
Reglone	Zeneca Crop Protection	09646
Regulox K	Rhone-Poulenc Amenity	05405
Regulox K	Aventis Environmental Science	09937
Reliance	Monsanto Plc	09954
Rival	Monsanto Plc	09220
Roundup	Monsanto Plc	01828
Roundup A	Monsanto Plc	08375
Roundup Amenity	Monsanto Plc	08721
Roundup Biactive	Monsanto Plc	06941
Roundup Biactive Dry	Monsanto Plc	06942
Roundup Pro	Monsanto Plc	04146
Roundup Pro Biactive	Monsanto Plc	06954
Roundup Ultra St	Monsanto (UK) Ltd	10199
Samurai	Monsanto Plc	09952
Scorpion	Monsanto Plc	09953
Sierraron G	The Scotts Company (UK) Limited	09263
Sierraron G	The Scotts Company (UK) Limited	09675
Spasor	Rhone-Poulenc Amenity	07211
Spasor	Aventis Environmental Science	09945

Product name	Marketing company	Reg No.
Spasor Biactive	Rhone-Poulenc Amenity	07651
Spasor Biactive	Aventis Environmental Science	09940
Stetson	Monsanto Plc	06956
Typhoon 360	Feinchemie Schwebda Gmbh	09322
Typhoon 360	Feinchemie (UK) Ltd	09792

PART B

PSD Registered Products

1

PROFESSIONAL PRODUCTS

Product Name	Marketing Company	Reg. No.	Expiry Date

1.1 Herbicides

including growth regulators, defoliants, rooting agents and desiccants

1 Amidosulfuron

C	Aventis Eagle	Aventis CropScience UK Limited	09765
C	Barclay Cleave	Barclay Chemicals Manufacturing Ltd	09489
C	Druid	Aventis CropScience UK Limited	08714
C	Eagle	Aventis CropScience UK Limited	07318
C	Landgold Amidosulfuron	Landgold & Co Ltd	09021
C	Pursuit	Aventis CropScience UK Limited	07333
C	Pursuit 50	Aventis CropScience UK Limited	08716
C	Squire	Aventis CropScience UK Limited	08715

2 Amidosulfuron + Metribuzin

C	Bayer UK 590	Bayer plc	08149	30/06/2001
C	Galis	Bayer plc	08039	30/06/2001

3 Amitrole

C	Loft	A H Marks & Co Ltd	06030
C	MSS Aminotriazole Technical	Mirfield Sales Services Ltd	04645
C	Weedazol-TL	A H Marks & Co Ltd	02349
C	Weedazol-TL	Bayer plc	02979

4 Amitrole + Bromacil + Diuron

C	BR Destral	Rhone-Poulenc Amenity	05184	31/12/2001

5 Amitrole + 2,4-D + Diuron

C	Trik	Mirfield Sales Services Ltd	07853

6 Amitrole + Simazine

C	Alpha Simazol	Makhteshim-Agan (UK) Ltd	04799
C	Alpha Simazol T	Makhteshim-Agan (UK) Ltd	04874

7 Ammonium sulphamate

C	Amcide	Battle Hayward & Bower Ltd	04246

8 Anthracene oils

C	Sterilite Hop Defoliant	Coventry Chemicals Ltd	05060

C   These products are "approved for agricultural use". For further details refer to page vii.
A   These products are approved for use in or near water. For further details refer to page vii.

	Product Name	Marketing Company	Reg. No.	Expiry Date

9 Asulam

C	Asulox	Aventis CropScience UK Limited	09969	
C	Asulox	Rhone-Poulenc Agriculture	06124	
C	IT Asulam	IT Agro Ltd	10186	

10 Atrazine

C	Alpha Atrazine 50 SC	Makhteshim-Agan (UK) Ltd	04877	
C	Alpha Atrazine 50 WP	Makhteshim-Agan (UK) Ltd	04793	
C	Atlas Atrazine	Atlas Crop Protection Ltd	07702	
C	Atlas Atrazine	Atlas Interlates Ltd	03097	30/06/2002
C	Atrazine 90WG	Sipcam UK Ltd	09310	
C	Atrazol	Sipcam UK Ltd	07598	
C	Dapt Atrazine 50 SC	DAPT Agrochemicals Ltd	08031	
C	Gesaprim	Novartis Crop Protection UK Ltd	08411	
C	MSS Atrazine 50 FL	Mirfield Sales Services Ltd	01398	
C	MSS Atrazine 80 WP	Mirfield Sales Services Ltd	04360	
C	Unicrop Atrazine 50	Universal Crop Protection Ltd	02645	
C	Unicrop Atrazine FL	Universal Crop Protection Ltd	08045	
C	Unicrop Flowable Atrazine	Universal Crop Protection Ltd	05446	

11 Aziprotryne

C	Brasoran	Novartis Crop Protection UK Ltd	08394	

12 Benazolin

C	Aventis Galtak 50 SC	Aventis CropScience UK Limited	09711	
C	Galtak 50 SC	Aventis CropScience UK Limited	07258	

13 Benazolin + Bromoxynil + Ioxynil

C	Asset	Aventis CropScience UK Limited	07243	

14 Benazolin + Clopyralid

C	Benazalox	Aventis CropScience UK Limited	07246	

15 Benazolin + 2,4-DB + MCPA

C	Legumex Extra	AgrEvo UK Ltd	07901	31/05/2001
C	Legumex Extra	Aventis CropScience UK Limited	08676	
C	Master Sward	Stefes Plant Protection Ltd	07883	31/03/2002
C	Setter 33	Aventis CropScience UK Limited	07282	
C	Setter 33	Dow AgroSciences Ltd	05623	
C	Stefes Clover Ley	Stefes Plant Protection Ltd	07842	31/03/2002
C	Stefes Legumex Extra	Aventis CropScience UK Limited	10085	
C	Stefes Legumex Extra	Stefes Plant Protection Ltd	07841	

C   These products are "approved for agricultural use". For further details refer to page vii.
A   These products are approved for use in or near water. For further details refer to page vii.

Product Name	Marketing Company	Reg. No.	Expiry Date
16 Bentazone			
C Basagran	BASF plc	00188	
C Basagran SG	BASF plc	08360	
C IT Bentazone	IT Agro Ltd	08677	
C IT Bentazone 48	IT Agro Ltd	09283	
C Standon Bentazone	Standon Chemicals Ltd	09204	31/10/2002
C Standon Bentazone S	Standon Chemicals Ltd	10124	
17 Bentazone + Dichlorprop-P			
C Quitt SL	BASF plc	08108	
18 Bentazone + MCPA + MCPB			
C Acumen	BASF plc	00028	
C Headland Archer	Headland Agrochemicals Ltd	08814	
19 Bentazone + MCPB			
C Pulsar	BASF plc	04002	
20 Bentazone + Pendimethalin			
C Impuls	BASF plc	09720	
21 Bifenox + Chlorotoluron			
C Dicurane Duo 446 SC	Novartis Crop Protection UK Ltd	08404	
22 Bifenox + Isoproturon			
C Banco	Portman Agrochemicals Ltd	09027	
C RP 4169	Rhone-Poulenc Agriculture	05801	
23 Bifenox + MCPA + Mecoprop-P			
C Sirocco	Aventis Environmental Science	09939	
C Sirocco	Rhone-Poulenc Amenity	09645	
24 Bromacil			
C Alpha Bromacil 80 WP	Makhteshim-Agan (UK) Ltd	04802	
C Hyvar X	Du Pont (UK) Ltd	01105	
25 Bromacil + Amitrole + Diuron			
C BR Destral	Rhone-Poulenc Amenity	05184	31/12/2001
26 Bromacil + Diuron			
C Borocil K	Aventis Environmental Science	09924	
C Borocil K	Rhone-Poulenc Amenity	05183	

C These products are "approved for agricultural use". For further details refer to page vii.
A These products are approved for use in or near water. For further details refer to page vii.

Product Name	Marketing Company	Reg. No.	Expiry Date

27 Bromacil + Picloram

C	Hydon	Chipman Ltd	01088	

28 Bromoxynil

C	Alpha Bromolin 225 EC	Makhteshim-Agan (UK) Ltd	08255	
C	Alpha Bromotril P	Makhteshim-Agan (UK) Ltd	07099	
C	Barclay Mutiny	Barclay Chemicals Manufacturing Ltd	08933	31/12/2002
C	Flagon 400 EC	Makhteshim-Agan (UK) Ltd	08875	
C	Greencrop Tassle	Greencrop Technology Ltd	09659	

29 Bromoxynil + Benazolin + Ioxynil

C	Asset	Aventis CropScience UK Limited	07243	

30 Bromoxynil + Clopyralid

C	Vindex	Dow AgroSciences Ltd	05470	

31 Bromoxynil + Clopyralid + Fluroxypyr + Ioxynil

C	Crusader S	Dow AgroSciences Ltd	05174	

32 Bromoxynil + Dichlorprop + Ioxynil + MCPA

C	Atlas Minerva	Atlas Interlates Ltd	03046	

33 Bromoxynil + Diflufenican + Ioxynil

C	Capture	Aventis CropScience UK Limited	09982	
C	Capture	Rhone-Poulenc Agriculture	06881	25/07/2003

34 Bromoxynil + Ethofumesate + Ioxynil

C	Leyclene	Aventis CropScience UK Limited	07263	
C	Stefes Leyclene	Aventis CropScience UK Limited	10086	
C	Stefes Leyclene	Stefes Plant Protection Ltd	08173	

35 Bromoxynil + Fluroxypyr

C	Sickle	Dow AgroSciences Ltd	05187	
C	Tomahawk Plus	Makhteshim-Agan (UK) Ltd	09836	

36 Bromoxynil + Fluroxypyr + Ioxynil

C	Advance	Dow AgroSciences Ltd	05173	

37 Bromoxynil + Ioxynil

C	Alpha Briotril	Makhteshim-Agan (UK) Ltd	04876	
C	Alpha Briotril Plus 19/19	Makhteshim-Agan (UK) Ltd	04740	
C	Deloxil	Aventis CropScience UK Limited	09987	

C These products are "approved for agricultural use". For further details refer to page vii.
A These products are approved for use in or near water. For further details refer to page vii.

Product Name	Marketing Company	Reg. No.	Expiry Date

37 Bromoxynil + Ioxynil—continued

C	Deloxil	Aventis CropScience UK Limited	07313
C	Deloxil	Rhone-Poulenc Agriculture	07405
C	Mextrol-Biox	Nufarm UK Ltd	09470
C	Oxytril CM	Aventis CropScience UK Limited	10005
C	Oxytril CM	Rhone-Poulenc Agriculture	08667
C	Percept	MTM Agrochemicals Ltd	05481
C	Status	Aventis CropScience UK Limited	10019
C	Status	Rhone-Poulenc Agriculture	08668
C	Stellox 380 EC	Novartis Crop Protection UK Ltd	08451

38 Bromoxynil + Ioxynil + Mecoprop-P

C	Swipe P	Novartis Crop Protection UK Ltd	08452

39 Bromoxynil + Ioxynil + Triasulfuron

C	Teal	Novartis Crop Protection UK Ltd	08453
C	Teal-M	Novartis Crop Protection UK Ltd	08454

40 Bromoxynil + Ioxynil + Trifluralin

C	Masterspray	Pan Britannica Industries Ltd	02971
C	Masterspray	SumiAgro (UK) Ltd	09603

41 Bromoxynil + Prosulfuron

C	Jester	Novartis Crop Protection UK Ltd	08681

42 Bromoxynil + Terbuthylazine

C	Alpha Bromotril PT	Makhteshim-Agan (UK) Ltd	09435
C	Templar	Makhteshim-Agan (UK) Ltd	10254

43 Carbendazim + Tecnazene

C	Hickstor 6 Plus MBC	Hickson&Welch Ltd	04176	31/01/2002
C	Hortag Tecnacarb	Avon Packers Ltd	02929	31/01/2002
C	New Arena Plus	Hickson & Welch	04598	31/01/2002
C	New Hickstor 6 Plus MBC	Hickson & Welch Ltd	04599	31/01/2002
C	Tripart Arena Plus	Hickson & Welch	05602	31/01/2002

44 Carbetamide

C	Carbetamex	Aventis CropScience UK Limited	09983	30/09/2003
C	Carbetamex	Feinchemie (UK) Ltd	10265	
C	Carbetamex	Rhone-Poulenc Agriculture	06186	30/09/2003

45 Carbetamide + Diflufenican + Oxadiazon

C	Helmsman	Aventis Environmental Science	09934

C   These products are "approved for agricultural use". For further details refer to page vii.
A   These products are approved for use in or near water. For further details refer to page vii.

Product Name	Marketing Company	Reg. No.	Expiry Date

45 Carbetamide + Diflufenican + Oxadiazon—continued

C	Helmsman	Rhone-Poulenc Amenity	09696	25/07/2003

46 Carfentrazone-ethyl

C	Aurora	FMC Corporation (UK) Ltd	09807
C	Platform	FMC Corporation (UK) Ltd	09661

47 Carfentrazone-ethyl + Flupyrsulfuron-methyl

C	Lexus Class	Du Pont (UK) Ltd	08636
C	Lexus Class WSB	Du Pont (UK) Ltd	08637

48 Carfentrazone-ethyl + Isoproturon

C	Affinity	FMC Corporation (UK) Ltd	09745

49 Carfentrazone-ethyl + Mecoprop-P

C	Platform S	FMC Corporation (UK) Ltd	09706	
C	Platform S	FMC Europe NV	08638	30/11/2002

50 Carfentrazone-ethyl + Metsulfuron-methyl

C	Ally Express	Du Pont (UK) Ltd	08640

51 Carfentrazone-ethyl + Thifensulfuron-methyl

C	Harmony Express	Du Pont (UK) Ltd	09467

52 Chloridazon

C	Better DF	Sipcam UK Ltd	06250
C	Better Flowable	Sipcam UK Ltd	04924
C	Burex 430 SC	Agricola Ltd	09494
C	Gladiator	Tripart Farm Chemicals Ltd	00986
C	Gladiator DF	Tripart Farm Chemicals Ltd	06342
C	Luxan Chloridazon	Luxan (UK) Ltd	06304
C	Portman Weedmaster	Portman Agrochemicals Ltd	06018
C	Pyramin DF	BASF plc	03438
C	Pyramin FL	BASF plc	01661
C	Questar	BASF plc	07955
C	Sculptor	Sipcam UK Ltd	08836
C	Starter Flowable	Truchem Ltd	03421
C	Stefes Chloridazon	Aventis CropScience UK Limited	10110
C	Stefes Chloridazon	Stefes UK Ltd	07678
C	Takron	BASF plc	06237
C	Tripart Gladiator 2	Tripart Farm Chemicals Ltd	06618
C	Weed master SC	Portman Agrochemicals Ltd	08793

C   These products are "approved for agricultural use". For further details refer to page vii.
A   These products are approved for use in or near water. For further details refer to page vii.

Product Name	Marketing Company	Reg. No.	Expiry Date

53 Chloridazon + Ethofumesate

C	Gremlin	Sipcam UK Ltd	09468	
C	Magnum	BASF plc	08635	
C	Spectron	Aventis CropScience UK Limited	07284	
C	Stefes Spectron	Stefes UK Ltd	09337	

54 Chloridazon + Lenacil

C	Advizor	Du Pont (UK) Ltd	06571	
C	Advizor S	Du Pont (UK) Ltd	06960	31/10/2002
C	Advizor S	Griffin (Europe) Marketing NV (UK Branch)	09799	
C	Varmint	Pan Britannica Industries Ltd	01561	31/12/2001

55 Chloridazon + Metamitron

C	Volcan Combi	Sipcam UK Ltd	10256	

56 Chloridazon + Propachlor

C	Ashlade CP	Ashlade Formulations Ltd	06481	

57 Chloridazon + Quinmerac

C	Fiesta T	BASF plc	10260	

58 Chlormequat

C	3C Chlormequat 460	Pennine Chemical Services Ltd	03916	28/02/2001
C	3C Chlormequat 600	Pennine Chemical Services Ltd	04079	28/02/2001
C	Adjust	Mandops (UK) Ltd	05589	
C	Agriguard 5C Chlormequat 460	Agriguard Ltd	09851	
C	Agriguard Chlormequat 700	Agriguard Ltd	09782	
C	Agriguard Chlormequat 700	Agriguard Ltd	09282	
C	Agriguard Chlormequat 720	Agriguard Ltd	09919	
C	Allied Colloids Chlormequat 460	Ciba Speciality Chemicals, Water Treatments Ltd	07859	30/06/2002
C	Allied Colloids Chlormequat 460:320	Ciba Speciality Chemicals, Water Treatments Ltd	07861	30/06/2002
C	Allied Colloids Chlormequat 730	Ciba Speciality Chemicals, Water Treatments Ltd	07860	30/06/2002
C	Alpha Chlormequat 460	Makhteshim-Agan (UK) Ltd	04804	
C	Alpha Pentagan	Makhteshim-Agan (UK) Ltd	04794	
C	Alpha Pentagan Extra	Makhteshim-Agan (UK) Ltd	04796	
C	Ashlade 460 CCC	Ashlade Formulations Ltd	06474	
C	Ashlade 5C	Ashlade Formulations Ltd	06227	
C	Ashlade 700 5C	Ashlade Formulations Ltd	07046	
C	Ashlade 700 CCC	Ashlade Formulations Ltd	06473	
C	Ashlade Brevis	Ashlade Formulations Ltd	08119	

C   These products are "approved for agricultural use". For further details refer to page vii.
A   These products are approved for use in or near water. For further details refer to page vii.

Product Name	Marketing Company	Reg. No.	Expiry Date

58 Chlormequat—continued

C	Atlas 3C:645 Chlormequat	Atlas Crop Protection Ltd	07700
C	Atlas 3C:645 Chlormequat	Atlas Interlates Ltd	05710
C	Atlas 5C Chlormequat	Atlas Crop Protection Ltd	07701
C	Atlas 5C Chlormequat	Atlas Interlates Ltd	03084
C	Atlas Chlormequat 46	Atlas Crop Protection Ltd	07704
C	Atlas Chlormequat 460:46	Atlas Crop Protection Ltd	07705
C	Atlas Chlormequat 460:46	Atlas Interlates Ltd	06258
C	Atlas Chlormequat 700	Atlas Crop Protection Ltd	07708
C	Atlas Chlormequat 700	Atlas Interlates Ltd	03402
C	Atlas Quintacel	Atlas Crop Protection Ltd	07706
C	Atlas Terbine	Atlas Crop Protection Ltd	07709
C	Atlas Terbine	Atlas Interlates Ltd	06523
C	Atlas Tricol	Atlas Crop Protection Ltd	07707
C	Atlas Tricol	Atlas Interlates Ltd	07190
C	Barclay Holdup	Barclay Chemicals Manufacturing Ltd	06799
C	Barclay Holdup 600	Barclay Chemicals Manufacturing Ltd	08794
C	Barclay Holdup 640	Barclay Chemicals Manufacturing Ltd	08795
C	Barclay Liffey	Barclay Chemicals Manufacturing Ltd	09856
C	Barclay Lucan	Barclay Chemicals Manufacturing Ltd	09855
C	Barclay Take 5	Barclay Chemicals Manufacturing Ltd	08524
C	Barleyquat B	Mandops (UK) Ltd	07051
C	BASF 3C Chlormequat 600	BASF plc	04077
C	BASF 3C Chlormequat 720	BASF plc	06514
C	BASF 3C Chlormequat 750	BASF plc	06878
C	Belcocel	UCB Chemicals	08248
C	Bettaquat B	Mandops (UK) Ltd	07050
C	Ciba Chlormequat 460	Ciba Speciality Chemicals, Water Treatments Ltd	09525
C	Ciba Chlormequat 5C 460:320	Ciba Speciality Chemicals, Water Treatments Ltd	09527
C	Ciba Chlormequat 730	Ciba Speciality Chemicals, Water Treatments Ltd	09526
C	Clayton CCC 750	Clayton Plant Protection (UK) Ltd	07952
C	Clayton Manquat	Clayton Plant Protection (UK) Ltd	09916
C	Clayton Standup	Clayton Plant Protection Ltd	08771
C	Cleancrop Chlormequat 700	United Agri Products	10143
C	Cropsafe 5C Chlormequat	Hortichem Ltd	07897
C	Fargro Chlormequat	Fargro Ltd	02600

C   These products are "approved for agricultural use". For further details refer to page vii.
A   These products are approved for use in or near water. For further details refer to page vii.

Product Name	Marketing Company	Reg. No.	Expiry Date

58 Chlormequat—continued

	Product Name	Marketing Company	Reg. No.	Expiry Date
C	Greencrop Carna	Greencrop Technology Ltd	09403	
C	Greencrop Cong 750	Greencrop Technology Ltd	09383	
C	Greencrop Coolfin	Greencrop Technology Ltd	09449	
C	Hyquat 70	Agrichem (International) Ltd	03364	
C	Intracrop Balance	Intracrop	08037	
C	Intracrop MCCC	Intracrop	08506	
C	Landgold CC 720	Landgold & Co Ltd	08527	
C	Larke	Atlas Crop Protection Ltd	10045	
C	Mandops Barleyquat B	Mandops (UK) Ltd	06001	
C	Mandops Bettaquat B	Mandops (UK) Ltd	06004	
C	Mandops Chlormequat 460	Mandops (UK) Ltd	06090	
C	Mandops Chlormequat 700	Mandops (UK) Ltd	06002	
C	Manipulator	Mandops (UK) Ltd	05871	
C	MSS Chlormequat 40	Mirfield Sales Services Ltd	01401	
C	MSS Chlormequat 460	Mirfield Sales Services Ltd	03935	
C	MSS Chlormequat 60	Mirfield Sales Services Ltd	03936	
C	MSS Chlormequat 70	Mirfield Sales Services Ltd	03937	
C	MSS Mircell	Mirfield Sales Services Ltd	06939	
C	MSS Mirquat	Mirfield Sales Services Ltd	08166	
C	New 5C Cycocel	BASF plc	01482	
C	New 5C Cycocel	Cyanamid Agriculture Ltd	01483	
C	PA Chlormequat 400	Portman Agrochemicals Ltd	01523	
C	PA Chlormequat 460	Portman Agrochemicals Ltd	02549	
C	Podquat	Mandops (UK) Ltd	03003	
C	Portman Chlormequat 700	Portman Agrochemicals Ltd	03465	
C	Portman Supaquat	Portman Agrochemicals Ltd	03466	
C	Quadrangle Chlormequat 700	Quadrangle Agrochemicals	03401	
C	Renown	Aventis CropScience UK Limited	10103	
C	Renown	Stefes UK Ltd	09058	
C	Sigma PCT	Atlas Crop Protection Ltd	08663	
C	Stabilan 460	Ashlade Formulations Ltd	08005	31/10/2001
C	Stabilan 460	Nufarm UK Ltd	09304	
C	Stabilan 5C	Ashlade Formulations Ltd	08144	
C	Stabilan 640	Nufarm UK Ltd	09401	
C	Stabilan 670	Nufarm UK Ltd	09402	
C	Stabilan 700	Ashlade Formulations Ltd	08006	31/10/2001
C	Stabilan 700	Nufarm UK Ltd	09302	
C	Stabilan 750	Ashlade Formulations Ltd	08004	31/10/2001
C	Stabilan 750	Nufarm UK Ltd	09303	
C	Stay Up	Nufarm UK Ltd	09444	
C	Stefes CCC	Aventis CropScience UK Limited	10104	
C	Stefes CCC	Stefes UK Ltd	05959	
C	Stefes CCC 640	Aventis CropScience UK Limited	10111	
C	Stefes CCC 640	Stefes UK Ltd	06993	

C   These products are "approved for agricultural use". For further details refer to page vii.
A   These products are approved for use in or near water. For further details refer to page vii.

Product Name	Marketing Company	Reg. No.	Expiry Date

58 Chlormequat—continued

C	Stefes CCC 700	Aventis CropScience UK Limited	10105	
C	Stefes CCC 700	Stefes UK Ltd	07116	
C	Stefes CCC 720	Aventis CropScience UK Limited	10108	
C	Stefes CCC 720	Stefes UK Ltd	05834	
C	Stefes K2	Stefes Plant Protection Ltd	07054	31/03/2002
C	Supaquat	Portman Agrochemicals Ltd	09381	
C	Trio	Portman Agrochemicals Ltd	08883	31/12/2001
C	Tripart 5C	Tripart Farm Chemicals Ltd	04726	
C	Tripart Brevis	Tripart Farm Chemicals Ltd	03754	
C	Tripart Brevis 2	Tripart Farm Chemicals Ltd	06612	
C	Tripart Chlormequat 460	Tripart Farm Chemicals Ltd	03685	
C	Uplift	United Phosphorus Ltd	07527	
C	Whyte Chlormequat 700	Nufarm Whyte Agriculture Ltd	09641	

59 Chlormequat + 2-Chloroethylphosphonic acid

C	Barclay Banshee XL	Barclay Chemicals Manufacturing Ltd	09201	
C	Greencrop Tycoon	Greencrop Technology Ltd	09571	
C	Nomad	Aventis CropScience UK Limited	10004	
C	Nomad	Rhone-Poulenc Agriculture	07888	31/03/2003
C	Strate	Aventis CropScience UK Limited	10020	
C	Strate	Rhone-Poulenc Agriculture	07430	31/03/2003
C	Sypex	BASF plc	04650	
C	Terpal C	BASF plc	07062	
C	Upgrade	Aventis CropScience UK Limited	10029	
C	Upgrade	Rhone-Poulenc Agriculture	06177	31/03/2003

60 Chlormequat + 2-Chloroethylphosphonic acid + Imazaquin

C	Ice	Cyanamid Agriculture Ltd	08970
C	Satellite	Cyanamid Agriculture Ltd	08969

61 Chlormequat + 2-Chloroethylphosphonic acid + Mepiquat

C	Cyclade	BASF plc	08958

62 Chlormequat + Imazaquin

C	Meteor	Cyanamid Agriculture Ltd	06505
C	Standon Imazaquin 5C	Standon Chemicals Ltd	08813
C	Upright	Cyanamid Agriculture Ltd	08290

63 Chlormequat + Mepiquat

C	Cyter	BASF plc	09133
C	Stronghold	BASF plc	09134

C   These products are "approved for agricultural use". For further details refer to page vii.
A   These products are approved for use in or near water. For further details refer to page vii.

Product Name	Marketing Company	Reg. No.	Expiry Date

64 2-Chloroethylphosphonic acid

C Aventis Cerone	Aventis CropScience UK Limited	09972	
C Aventis Cerone	Rhone-Poulenc Agriculture	09748	
C Barclay Coolmore	Barclay Chemicals Manufacturing Ltd	07917	
C Cerone	A H Marks & Co Ltd	00462	20/04/2001
C Cerone	Aventis CropScience UK Limited	09985	
C Cerone	Rhone-Poulenc Agriculture	06185	
C Charger	Aventis CropScience UK Limited	09986	
C Charger	Rhone-Poulenc Agriculture	08827	
C CleanCrop Fonic	United Agri Products	09868	
C Ethrel C	Hortichem Ltd	06995	
C EXP03149D	Aventis CropScience UK Limited	09989	
C EXP03149D	Rhone-Poulenc Agriculture	08828	
C Stantion	Rhone-Poulenc Agriculture	06205	31/01/2001
C Unistar Ethephon 480	Unistar Ltd	06282	20/04/2001

65 2-Chloroethylphosphonic acid + Chlormequat

C Barclay Banshee XL	Barclay Chemicals Manufacturing Ltd	09201	
C Greencrop Tycoon	Greencrop Technology Ltd	09571	
C Nomad	Aventis CropScience UK Limited	10004	
C Nomad	Rhone-Poulenc Agriculture	07888	31/03/2003
C Strate	Aventis CropScience UK Limited	10020	
C Strate	Rhone-Poulenc Agriculture	07430	31/03/2003
C Sypex	BASF plc	04650	
C Terpal C	BASF plc	07062	
C Upgrade	Aventis CropScience UK Limited	10029	
C Upgrade	Rhone-Poulenc Agriculture	06177	31/03/2003

66 2-Chloroethylphosphonic acid + Chlormequat + Imazaquin

C Ice	Cyanamid Agriculture Ltd	08970	
C Satellite	Cyanamid Agriculture Ltd	08969	

67 2-Chloroethylphosphonic acid + Chlormequat + Mepiquat

C Cyclade	BASF plc	08958	

68 2-Chloroethylphosphonic acid + Mepiquat

C Barclay Banshee	Barclay Chemicals Manufacturing Ltd	08175	
C CleanCrop Fonic M	United Agri Products	09553	
C Marnoch Phonet	Marnoch Ventures Limited	09979	
C Standon Mepiquat Plus	Standon Chemicals Ltd	09373	
C Stefes Mepiquat	Stefes Plant Protection Ltd	06970	20/04/2001

C   These products are "approved for agricultural use". For further details refer to page vii.
A   These products are approved for use in or near water. For further details refer to page vii.

Product Name	Marketing Company	Reg. No.	Expiry Date

68 2-Chloroethylphosphonic acid + Mepiquat—continued

C	Sypex M	BASF plc	06810	30/06/2001
C	Terpal	BASF plc	02103	
C	Terpal	Clayton Plant Protection (UK) Ltd	07626	
C	Terpitz	Me^2 Crop Protection Ltd	09634	

69 Chlorotoluron

C	Alpha Chlorotoluron 500	Makhteshim-Agan (UK) Ltd	04848	
C	Atol	Ashlade Formulations Ltd	07347	25/07/2003
C	Atol	Nufarm Whyte Agriculture Ltd	10235	
C	Chlortoluron 500	Pan Britannica Industries Ltd	03686	31/12/2001
C	Clayton Chloron	Clayton Plant Protection (UK) Ltd	08148	
C	Dicurane	Novartis Crop Protection UK Ltd	08403	
C	Lentipur CL 500	Nufarm UK Ltd	08743	
C	Luxan Chlorotoluron 500 Flowable	Luxan (UK) Ltd	09165	
C	MSS Chlortoluron 500	Mirfield Sales Services Ltd	07871	
C	NWA CTU 500	Nufarm Whyte Agriculture Ltd	10173	
C	Portman Chlortoluron	Portman Agrochemicals Ltd	03068	
C	Stefes Toluron	Aventis CropScience UK Limited	10106	
C	Stefes Toluron	Stefes Plant Protection Ltd	05779	
C	Talisman	Farmers Crop Chemicals Ltd	03109	
C	Tolugan 700	Makhteshim-Agan (UK) Ltd	08064	
C	Top Farm Toluron 500	Top Farm Formulations Ltd	05986	
C	Tripart Culmus	Tripart Farm Chemicals Ltd	06619	

70 Chlorotoluron + Bifenox

C	Dicurane Duo 446 SC	Novartis Crop Protection UK Ltd	08404

71 Chlorotoluron + Isoproturon

C	Tolugan Extra	Makhteshim-Agan (UK) Ltd	09393

72 Chlorotoluron + Pendimethalin

C	Totem	Cyanamid Agriculture Ltd	04670

73 Chlorphonium

C	Phosfleur 1.5	Perifleur Products Ltd	05750
C	Phosfleur 10% Liquid	Perifleur Products Ltd	01587

74 Chlorpropham

C	Atlas CIPC 40	Atlas Crop Protection Ltd	07710
C	Atlas Herbon Pabrac	Atlas Crop Protection Ltd	07714
C	Atlas Herbon Pabrac	Atlas Interlates Ltd	03997
	BL500	Wheatley Chemical Co Ltd	00279

C These products are "approved for agricultural use". For further details refer to page vii.
A These products are approved for use in or near water. For further details refer to page vii.

Product Name	Marketing Company	Reg. No.	Expiry Date

74 Chlorpropham—continued

C Comrade	United Phosphorus Ltd	10181	
C Croptex Pewter	Hortichem Ltd	02507	
Luxan Gro-Stop 300 EC	Luxan (UK) Ltd	08602	
Luxan Gro-Stop Basis	Luxan (UK) Ltd	08601	
Luxan Gro-Stop Fog	Luxan (UK) Ltd	09388	
Luxan Gro-Stop HN	Luxan (UK) Ltd	07689	
C MSS CIPC 30M	Whyte Agrochemicals Division	10064	
C MSS CIPC 40 EC	Mirfield Sales Services Ltd	01403	
MSS CIPC 50 LF	Mirfield Sales Services Ltd	03285	
MSS CIPC 50M	Mirfield Sales Services Ltd	01404	
MSS CIPC 5G	Mirfield Sales Services Ltd	01402	
C MTM CIPC 40	MTM Agrochemicals Ltd	05895	30/06/2003
Standon CIPC 300 HN	Standon Chemicals Ltd	09187	
C Triherbicide CIPC	Atochem Agri BV	06426	
C Triherbicide CIPC	Elf Atochem Agri SA	06874	
Warefog 25	Mirfield Sales Services Ltd	06776	

75 Chlorpropham + Fenuron

C Atlas Red	Atlas Crop Protection Ltd	07724	
C Atlas Red	Atlas Interlates Ltd	03091	
C Croptex Chrome	Hortichem Ltd	02415	

76 Chlorpropham + Linuron

C Profalon	AgrEvo UK Crop Protection Ltd	07331	31/01/2001
C Profalon	Hortichem Ltd	08900	

77 Chlorpropham + Pentanochlor

C Atlas Brown	Atlas Crop Protection Ltd	07703	

78 Chlorthal-dimethyl

C Dacthal W75	AMVAC Chemicals UK Ltd	10289	
C Dacthal W75	Hortichem Ltd	05500	
C Dacthal W75	ISK Biosciences Ltd	05556	

79 Chlorthal-dimethyl + Propachlor

C Decimate	ISK Biosciences Ltd	05626	28/02/2002

80 Cinidon-ethyl

C Lotus	BASF plc	09231	

81 Citronella oil

C Barrier H	Barrier Biotech Ltd	10136	

C These products are "approved for agricultural use". For further details refer to page vii.
A These products are approved for use in or near water. For further details refer to page vii.

Product Name	Marketing Company	Reg. No.	Expiry Date

82 Clodinafop-propargyl

C Greencrop Boulevard	Greencrop Technology Ltd	09960	
C Landgold Clodinafop	Landgold & Co Ltd	10172	
C Landgold Clodinafop	Landgold & Co Ltd	09017	31/05/2002
C Marathon	Me^2 Crop Protection Limited	09959	
C Standon Clodinafop 240	Standon Chemicals Ltd	10174	
C Standon Clodinafop 240	Standon Chemicals Ltd	09343	31/05/2002
C Topik	Ciba Agriculture	07763	28/02/2001
C Topik	Novartis Crop Protection UK Ltd	08461	
C Viscount	Ciba Agriculture	08127	28/02/2001

83 Clodinafop-propargyl + Diflufenican

C Amazon	Aventis CropScience UK Limited	10266	
C Amazon	Novartis Crop Protection UK Ltd	08128	
C Lucifer	Aventis CropScience UK Limited	10001	
C Lucifer	Rhone-Poulenc Agriculture	08129	30/06/2003

84 Clodinafop-propargyl + Diflufenican + Isoproturon

C SL 520	Ciba Agriculture	07687	28/02/2001
C SL 520	Novartis Crop Protection UK Ltd	08447	30/10/2001
C Unite	Rhone-Poulenc Agriculture	07686	30/10/2001

85 Clodinafop-propargyl + Trifluralin

C Hawk	Novartis Crop Protection UK Ltd	08417	
C Reserve	Ciba Agriculture	08169	28/02/2001

86 Clopyralid

C Barclay Karaoke	Barclay Chemicals Manufacturing Ltd	08185	
C Cliophar	Chimac-Agriphar SA	09430	
C Dow Shield	Dow AgroSciences Ltd	05578	
C I T Clopyralid	IT Agro Ltd	09808	
C Lontrel 100	Dow AgroSciences Ltd	05737	

87 Clopyralid + Benazolin

C Benazalox	Aventis CropScience UK Limited	07246	

88 Clopyralid + Bromoxynil

C Vindex	Dow AgroSciences Ltd	05470	

89 Clopyralid + Bromoxynil + Fluroxypyr + Ioxynil

C Crusader S	Dow AgroSciences Ltd	05174	

C These products are "approved for agricultural use". For further details refer to page vii.
A These products are approved for use in or near water. For further details refer to page vii.

Product Name	Marketing Company	Reg. No.	Expiry Date

90 Clopyralid + 2,4-D + MCPA

C Lonpar	Dow AgroSciences Ltd	08686	

91 Clopyralid + Dichlorprop + MCPA

C Lontrel Plus	Dow AgroSciences Ltd	05269	

92 Clopyralid + Diflufenican + MCPA

C Spearhead	Aventis Environmental Science	09941	
C Spearhead	Rhone-Poulenc Amenity	07342	

93 Clopyralid + Fluroxypyr + Ioxynil

C Hotspur	Dow AgroSciences Ltd	05185	

94 Clopyralid + Fluroxypyr + MCPA

C Greenor	Rigby Taylor Ltd	07848	

94a Clopyralid + Fluroxypyr + Triclopyr

C Pastor	DowAgroSciences Ltd	07440	

95 Clopyralid + Ioxynil

C Escort	Dow AgroSciences Ltd	05466	

96 Clopyralid + Propyzamide

C Matrikerb	Pan Britannica Industries Ltd	01308	
C Matrikerb	Rohm & Haas (UK) Ltd	02443	
C Matrikerb	SumiAgro (UK) Ltd	09604	

97 Clopyralid + Triclopyr

C Grazon 90	Dow AgroSciences Ltd	05456	

98 Copper hydroxide

C Spin-Out	Fargro Ltd	07610	

99 Cyanazine

C Barclay Canter	Barclay Chemicals Manufacturing Ltd	07530	
C Clayton Gazette	Clayton Plant Protection (UK) Ltd	10141	
C Fortrol	BASF plc	10222	
C Fortrol	Cyanamid Agriculture Ltd	07009	
C I.T. Cyanazine	IT Agro Ltd	10179	
C Reply	Cyanamid Agriculture Ltd	07084	31/12/2001
C Standon Cyanazine 50	Standon Chemicals Ltd	09230	

C These products are "approved for agricultural use". For further details refer to page vii.
A These products are approved for use in or near water. For further details refer to page vii.

Product Name	Marketing Company	Reg. No.	Expiry Date

100 Cyanazine + Pendimethalin

C	Activus	BASF plc	10207	
C	Activus	Cyanamid Agriculture Ltd	09174	
C	Bullet	BASF plc	10212	
C	Bullet	Cyanamid Agriculture Ltd	08049	

101 Cyanazine + Terbuthylazine

C	Angle	Novartis Crop Protection UK Ltd	08385	31/10/2001

102 Cycloxydim

C	Greencrop Valentia	Greencrop Technology Ltd	10197
C	Landgold Cycloxydim	Landgold & Co Ltd	06269
C	Laser	BASF plc	05251
C	Marnoch Clodim	Marnoch Ventures Limited	09852
C	Standon Cycloxydim	Standon Chemicals Ltd	08830
C	Stratos	BASF plc	06891

103 Cypermethrin

C	MCC 25 EC	Chiltern Farm Chemicals Ltd	09115

104 2,4-D

C		Agrichem 2,4-D	Agrichem Ltd	04098	
C	A	Agricorn 2,4-D	Farmers Crop Chemicals Ltd	07349	
C		Agricorn D	Farmers Crop Chemicals Ltd	00056	
C		Agricorn D11	Nufarm UK Ltd	09415	
C	A	Atlas 2,4-D	Atlas Crop Protection Ltd	07699	
C		Barclay Haybob II	Barclay Plant Protection	08532	
C		Depitox	Nufarm UK Ltd	08000	
C		Dicotox Extra	Aventis Environmental Science	09930	
C		Dicotox Extra	Rhone-Poulenc Amenity	05330	
C		Dioweed 50	United Phosphorus Ltd	08050	
C	A	Dormone	Aventis Environmental Science	09932	
C	A	Dormone	Rhone-Poulenc Amenity	05412	
C		Easel	Nufarm Whyte Agriculture Ltd	09878	
C		For-ester	Synchemicals Ltd	00914	
C		GroWell 2,4-D Amine	GroWell Ltd	08750	
C		Headland Staff	Headland Agrochemicals Ltd	07189	
C		Herboxone	A H Marks&Co Ltd	09692	28/02/2003
C		Herboxone	Headland Agrochemicals Ltd	10032	
C		Herboxone 60	A H Marks&Co Ltd	09693	
C		HY-D	Agrichem Ltd	06278	
C		Luxan 2,4-D	Luxan (UK) Ltd	09379	
C		Marks 2,4-D-A	A H Marks&Co Ltd	01282	
C	A	MSS 2,4-D Amine	Mirfield Sales Services Ltd	01391	25/07/2003

C   These products are "approved for agricultural use". For further details refer to page vii.
A   These products are approved for use in or near water. For further details refer to page vii.

Product Name	Marketing Company	Reg. No.	Expiry Date

104 2,4-D—continued

C A	MSS 2,4-D Amine	Nufarm Whyte Agriculture Ltd	10183	
C	MSS 2,4-D Ester	Mirfield Sales Services Ltd	01393	
C	Nomix 2,4-D Herbicide	Nomix-Chipman Ltd	06394	
C	Palormone D	Universal Crop Protection Ltd	01534	
C A	Ragox	Nufarm Whyte Agriculture Ltd	09876	
C	Ritefeed 2,4-D Amine	Ritefeed Ltd	05309	
C	Silvapron D	BP Chemicals Ltd	01935	
C	Syford	Synchemicals Ltd	02062	

105 2,4-D + Amitrole + Diuron

C	Trik	Mirfield Sales Services Ltd	07853

106 2,4-D + Clopyralid + MCPA

C	Lonpar	Dow AgroSciences Ltd	08686

107 2,4-D + Dicamba

C	Lawn Builder Plus Weed Control	Scotts UK Professional	08499
C	New Estermone	Vitax Ltd	06336

108 2,4-D + Dicamba + Triclopyr

C	Broadsword	United Phosphorus Ltd	09140
C	Nufarm Nu-Shot	Nufarm UK Ltd	09139
C	Terbel Triple	Chimac-Agriphar SA	09138

109 2,4-D + Dichlorprop

C	Marks Polytox-M	A H Marks & Co Ltd	01301
C	Polymone X	Universal Crop Protection Ltd	01613

110 2,4-D + Dichlorprop + MCPA + Mecoprop

C	Camppex	United Phosphorus Ltd	08266	31/10/2001

111 2,4-D + Dichlorprop + MCPA + Mecoprop-P

C	UPL Camppex	United Phosphorus Ltd	09121

112 2,4-D + MCPA

C	Agroxone Combi	Headland Agrochemicals Ltd	10277
C	Headland Polo	Headland Agrochemicals Ltd	10283

113 2,4-D + Mecoprop

C	Supertox 30	Rhone-Poulenc Amenity	05340	30/06/2001

C These products are "approved for agricultural use". For further details refer to page vii.
A These products are approved for use in or near water. For further details refer to page vii.

Product Name	Marketing Company	Reg. No.	Expiry Date
114 2,4-D + Mecoprop-P			
C Mascot Selective-P	Rigby Taylor Ltd	06105	
C Supertox 30	Aventis Environmental Science	09946	
C Supertox 30	Rhone-Poulenc Amenity	09102	28/02/2003
C Sydex	Vitax Ltd	06412	
115 2,4-D + Picloram			
C Atladox HI	Nomix-Chipman Ltd	05559	
C Tordon 101	Dow AgroSciences Ltd	05816	
116 Daminozide			
C B-Nine	Hortichem Ltd	07844	
C B-Nine	Uniroyal Chemical Ltd	04468	
C Dazide	Fine Agrochemicals Ltd	02691	
117 2,4-DB			
C DB Straight	United Phosphorus Ltd	07523	
C Marks 2,4-DB	A H Marks & Co Ltd	01283	
118 2,4-DB + Benazolin + MCPA			
C Legumex Extra	AgrEvo UK Ltd	07901	31/05/2001
C Legumex Extra	Aventis CropScience UK Limited	08676	
C Master Sward	Stefes Plant Protection Ltd	07883	31/03/2002
C Setter 33	Aventis CropScience UK Limited	07282	
C Setter 33	Dow AgroSciences Ltd	05623	
C Stefes Clover Ley	Stefes Plant Protection Ltd	07842	31/03/2002
C Stefes Legumex Extra	Aventis CropScience UK Limited	10085	
C Stefes Legumex Extra	Stefes Plant Protection Ltd	07841	
119 2,4-DB + Linuron + MCPA			
C Alistell	Zeneca Crop Protection	06515	
120 2,4-DB + MCPA			
C Agrichem DB Plus	Agrichem Ltd	00044	
C Headland Cedar	Headland Agrochemicals Ltd	09948	
C Marks 2,4-DB Extra	A H Marks&Co Ltd	01284	
C MSS 2,4-DB + MCPA	Mirfield Sales Services Ltd	01392	
C Redlegor	United Phosphorus Ltd	07519	
121 DE-570			
C Boxer	Dow AgroSciences Ltd	09819	
C Primus 25 SC	Dow AgroSciences Ltd	10175	

C   These products are "approved for agricultural use". For further details refer to page vii.
A   These products are approved for use in or near water. For further details refer to page vii.

Product Name	Marketing Company	Reg. No.	Expiry Date

122 Desmedipham + Ethofumesate + Phenmedipham

C	Aventis Betanal Progress OF	Aventis CropScience UK Limited	09722	
C	Betanal Congress	AgrEvo UK Ltd	07534	31/05/2001
C	Betanal Progress	AgrEvo UK Ltd	07533	31/05/2001
C	Betanal Progress OF	Aventis CropScience UK Limited	07629	
C	Betanal Ultima	AgrEvo UK Ltd	07535	31/05/2001

123 Desmedipham + Phenmedipham

C	Betanal Compact	Aventis CropScience UK Limited	07247	
C	Betanal Quorum	AgrEvo UK Ltd	07252	31/05/2001
C	Betanal Rostrum	AgrEvo UK Ltd	07253	31/05/2001

124 Desmetryn

C	Semeron	Novartis Crop Protection UK Ltd	08441	

125 Dicamba

C	Cadence	Barclay Chemicals Manufacturing Ltd	09578	
C	Cadence	Novartis Crop Protection UK Ltd	08796	
C	San 845H	Novartis Crop Protection UK Ltd	08797	

126 Dicamba + 2,4-D

C	Lawn Builder Plus Weed Control	Scotts UK Professional	08499	
C	New Estermone	Vitax Ltd	06336	

127 Dicamba + 2,4-D + Triclopyr

C	Broadsword	United Phosphorus Ltd	09140	
C	Nufarm Nu-Shot	Nufarm UK Ltd	09139	
C	Terbel Triple	Chimac-Agriphar SA	09138	

128 Dicamba + Dichlorprop + Ferrous sulphate + MCPA

C	Longlife Renovator 2	Miracle Professional	08379	
C	Renovator 2	The Scotts Company (UK) Limited	09272	

129 Dicamba + Dichlorprop + MCPA

C	Cleanrun 2	The Scotts Company (UK) Limited	09271	
C	Intrepid	Miracle Professional	07819	
C	Intrepid	The Scotts Company (UK) Limited	09266	
C	Longlife Clearun 2	Miracle Professional	07851	30/06/2003

130 Dicamba + Maleic hydrazide + MCPA

C	Mazide Selective	Vitax Ltd	05753	

C   These products are "approved for agricultural use". For further details refer to page vii.

A   These products are approved for use in or near water. For further details refer to page vii.

Product Name	Marketing Company	Reg. No.	Expiry Date

131 Dicamba + MCPA

C	Banvel M	Novartis Crop Protection UK Ltd	08469	

132 Dicamba + MCPA + Mecoprop

C	Barclay Hat-Trick	Barclay Chemicals Manufacturing Ltd	07579	28/02/2002
C	Hyprone	Agrichem (International) Ltd	01093	30/06/2001
C	Hysward	Agrichem (International) Ltd	01096	31/05/2001
C	Premier Triple Selective	Amenity Land Services Limited	08529	31/05/2001

133 Dicamba + MCPA + Mecoprop-P

C	ALS Premier Selective Plus	Amenity Land Services Limited	08940	
C	Banlene Super	AgrEvo UK Crop Protection Ltd	07168	30/04/2003
C	Banlene Super	Aventis CropScience UK Limited	10053	
C	Field Marshal	MTM Agrochemicals Ltd	06077	31/03/2001
C	Field Marshal	United Phosphorus Ltd	08956	
C	Headland Relay P	Headland Agrochemicals Ltd	08580	
C	Headland Relay Turf	Headland Agrochemicals Ltd	08935	
C	Herrisol New	Bayer plc	07166	
C	Hycamba Plus	Agrichem (International) Ltd	10180	
C	Hyprone P	Agrichem (International) Ltd	09125	
C	Hysward P	Agrichem (International) Ltd	09052	
C	Mascot Super Selective-P	Rigby Taylor Ltd	06106	
C	Mircam Plus	Nufarm Whyte Agriculture Ltd	09884	
C	MSS Mircam Plus	Mirfield Sales Services Ltd	01416	
C	MTM Grassland Herbicide	MTM Agrochemicals Ltd	06089	28/02/2001
C	Pasturol D	Farmers Crop Chemicals Ltd	07033	
C	Pasturol Plus	Farmers Crop Chemicals Ltd	08581	30/09/2003
C	Pasturol Plus	Headland Agrochemicals Ltd	10278	
C	Pasturol-P	Farmers Crop Chemicals Ltd	09099	
C	Premier Amenity Selective	Amenity Land Services Limited	10098	
C	Premier Triple Selective P	Amenity Land Services Limited	09053	
C	Quadrangle Quadban	Quadrangle Agrochemicals	07090	
C	Stefes Banlene Super	Aventis CropScience UK Limited	10089	
C	Stefes Banlene Super	Stefes Plant Protection Ltd	07691	
C	Stefes Docklene Super	AgrEvo UK Crop Protection Ltd	09873	
C	Stefes Docklene Super	Stefes Plant Protection Ltd	07696	31/01/2003
C	Stefes Swelland Super	Stefes Plant Protection Ltd	08189	31/03/2002
C	Tribute	Nomix-Chipman Ltd	06921	
C	Tribute Plus	Nomix-Chipman Ltd	09493	
C	Tritox	The Scotts Company (UK) Limited + Levington Horticulture Ltd	07764	
C	UPL Grassland Herbicide	United Phosphorus Ltd	08934	

C These products are "approved for agricultural use". For further details refer to page vii.
A These products are approved for use in or near water. For further details refer to page vii.

Product Name	Marketing Company	Reg. No.	Expiry Date

134 Dicamba + Mecoprop

C	Condox	Zeneca Crop Protection	06519	31/05/2002
C	Hyban	Agrichem Ltd	01084	30/06/2001
C	Hygrass	Agrichem (International) Ltd	01090	30/06/2001

135 Dicamba + Mecoprop-P

C	Camber	Headland Agrochemicals Ltd	09901
C	Condox	Zeneca Crop Protection	09429
C	Di-Farmon R	Headland Agrochemicals Ltd	08472
C	Dockmaster	Nufarm Whyte Agriculture Ltd	10033
C	Foundation	Novartis Crop Protection UK Ltd	08475
C	Hyban P	Agrichem (International) Ltd	09129
C	Hygrass P	Agrichem (International) Ltd	09130
C	Mircam	Nufarm Whyte Agriculture Ltd	09888
C	MSS Mircam	Mirfield Sales Services Ltd	01415
C	Optica Forte	A H Marks & Co Ltd	07432

136 Dicamba + Paclobutrazol

C	Holdfast D	Imperial Chemical Industries Plc	05056	30/10/2001
C	Holdfast D	Miracle Professional	07864	31/03/2001

137 Dicamba + Triasulfuron

C	Banvel T	Novartis Crop Protection UK Ltd	08470
C	Framolene	Novartis Crop Protection UK Ltd	08408

138 Dichlobenil

C A	Casoron G	Miracle Professional	07926	
C A	Casoron G	Nomix-Chipman Ltd	09023	
C A	Casoron G	Rigby Taylor Ltd	09326	
C A	Casoron G	Uniroyal Chemical Ltd	09022	
C A	Casoron G	Zeneca Crop Protection	08065	
C	Casoron G4	ICI Agrochemicals	02406	31/05/2001
C	Casoron G4	Miracle Professional	07927	
C	Casoron G4	Uniroyal Chemical Ltd	09215	
C A	Casoron GSR	Imperial Chemical Industries Plc	00451	31/05/2001
C A	Casoron GSR	Miracle Professional	07925	
C A	Luxan Dichlobenil Granules	Luxan (UK) Ltd	09250	
C	Prefix D	Cyanamid Agriculture Ltd	07013	30/06/2001
C A	Sierraron G	The Scotts Company (UK) Limited	09675	
C A	Sierraron G	The Scotts Company (UK) Limited	09263	
C	Standon Dichlobenil 6G	Standon Chemicals Ltd	08874	

139 Dichlorophen

C	50-50 Liquid Mosskiller	Vitax Ltd	07191

C These products are "approved for agricultural use". For further details refer to page vii.
A These products are approved for use in or near water. For further details refer to page vii.

Product Name	Marketing Company	Reg. No.	Expiry Date
139 Dichlorophen—continued			
C Enforcer	Miracle Professional	07866	30/06/2002
C Enforcer	The Scotts Company (UK) Limited	09288	
C Enforcer	Zeneca Professional Products	07079	31/05/2001
C Mascot Mosskiller	Rigby Taylor Ltd	02439	
C Mossicide	Pan Britannica Industries Ltd	08183	
C Mossicide	SumiAgro (UK) Ltd	09606	
C Nomix-Chipman Mosskiller	Nomix-Chipman Ltd	06271	
C Panacide M	Coalite Chemicals	05611	
C Panacide Technical Solution	Coalite Chemicals	05612	
C Ritefeed Dichlorophen	Ritefeed Ltd	05265	
C Super Mosstox	Rhone-Poulenc Amenity	05339	
C SuperMosstox	Aventis Environmental Science	09942	
140 Dichlorophen + Ferrous sulphate			
C Aitkens Lawn Sand Plus	RCC Agro	04542	
C SHL Lawn Sand Plus	Sinclair Horticulture & Leisure Ltd	04439	
141 Dichlorprop			
C Marks Polytox-K	A H Marks & Co Ltd	01300	
C MSS 2,4-DP	Mirfield Sales Services Ltd	01394	
C Redipon	United Phosphorus Ltd	07524	
142 Dichlorprop + Bromoxynil + Ioxynil + MCPA			
C Atlas Minerva	Atlas Interlates Ltd	03046	
143 Dichlorprop + Clopyralid + MCPA			
C Lontrel Plus	Dow AgroSciences Ltd	05269	
144 Dichlorprop + 2,4-D			
C Marks Polytox-M	A H Marks & Co Ltd	01301	
C Polymone X	Universal Crop Protection Ltd	01613	
145 Dichlorprop + 2,4-D + MCPA + Mecoprop			
C Camppex	United Phosphorus Ltd	08266	31/10/2001
146 Dichlorprop + 2,4-D + MCPA + Mecoprop-P			
C UPL Camppex	United Phosphorus Ltd	09121	
147 Dichlorprop + Dicamba + Ferrous sulphate + MCPA			
C Longlife Renovator 2	Miracle Professional	08379	
C Renovator 2	The Scotts Company (UK) Limited	09272	

C   These products are "approved for agricultural use". For further details refer to page vii.
A   These products are approved for use in or near water. For further details refer to page vii.

Product Name	Marketing Company	Reg. No.	Expiry Date

148 Dichlorprop + Dicamba + MCPA

C	Cleanrun 2	The Scotts Company (UK) Limited	09271	
C	Intrepid	Miracle Professional	07819	
C	Intrepid	The Scotts Company (UK) Limited	09266	
C	Longlife Clearun 2	Miracle Professional	07851	30/06/2003

149 Dichlorprop + Ferrous sulphate + MCPA

C	SHL Granular Feed, Weed & Mosskiller	William Sinclair Horticulture Ltd	09925
C	SHL Turf Feed and Weed and Mosskiller	William Sinclair Horticulture Ltd	04438
C	Vitagrow Mini-Granular Feed, Weed & Mosskiller	William Sinclair Horticulture Ltd	10295

150 Dichlorprop + MCPA

C	MSS 2,4-DP + MCPA	Mirfield Sales Services Ltd	01396
C	Redipon Extra	United Phosphorus Ltd	07518
C	Seritox 50	Aventis CropScience UK Limited	10016
C	Seritox 50	Rhone-Poulenc Agriculture	06170
C	Seritox Turf	Rhone-Poulenc Environmental Products	06802
C	SHL Granular Feed&Weed	William Sinclair Horticulture Ltd	09926
C	SHL Turf Feed&Weed	William Sinclair Horticulture Ltd	04437

151 Dichlorprop-P

C	Optica DP	A H Marks & Co Ltd	07818

152 Dichlorprop-P + Bentazone

C	Quitt SL	BASF plc	08108

153 Dichlorprop-P + MCPA + Mecoprop-P

C	Hymec Trio	Agrichem (International) Ltd	09764	28/02/2003
C	Hymec Triple	Agrichem (International) Ltd	09949	
C	Optica Trio	A H Marks & Co Ltd	09747	

154 Diclofop-methyl

C	Hoegrass	AgrEvo UK Ltd	07323	31/07/2001

155 Diclofop-methyl + Fenoxaprop-P-ethyl

C	Aventis Tigress Ultra	Aventis CropScience UK Limited	09725
C	Corniche	Aventis CropScience UK Limited	08947
C	Tigress Ultra	Aventis CropScience UK Limited	08946

C   These products are "approved for agricultural use". For further details refer to page vii.
A   These products are approved for use in or near water. For further details refer to page vii.

Product Name	Marketing Company	Reg. No.	Expiry Date

156 Difenzoquat

C Avenge 2	BASF plc	10210	
C Avenge 2	Cyanamid Agriculture Ltd	03241	

157 Diflufenican + Bromoxynil + Ioxynil

C Capture	Aventis CropScience UK Limited	09982	
C Capture	Rhone-Poulenc Agriculture	06881	25/07/2003

158 Diflufenican + Carbetamide + Oxadiazon

C Helmsman	Aventis Environmental Science	09934	
C Helmsman	Rhone-Poulenc Amenity	09696	25/07/2003

159 Diflufenican + Clodinafop-propargyl

C Amazon	Aventis CropScience UK Limited	10266	
C Amazon	Novartis Crop Protection UK Ltd	08128	
C Lucifer	Aventis CropScience UK Limited	10001	
C Lucifer	Rhone-Poulenc Agriculture	08129	30/06/2003

160 Diflufenican + Clodinafop-propargyl + Isoproturon

C SL 520	Ciba Agriculture	07687	28/02/2001
C SL 520	Novartis Crop Protection UK Ltd	08447	30/10/2001
C Unite	Rhone-Poulenc Agriculture	07686	30/10/2001

161 Diflufenican + Clopyralid + MCPA

C Spearhead	Aventis Environmental Science	09941	
C Spearhead	Rhone-Poulenc Amenity	07342	

162 Diflufenican + Flurtamone

C Bacara	Aventis CropScience UK Limited	09976	
C Bacara	Rhône-Poulenc Agriculture	08323	30/04/2003

163 Diflufenican + Flurtamone + Isoproturon

C EXP 31165	Aventis CropScience UK Limited	09990	30/04/2002
C EXP 31165	Rhone-Poulenc Agriculture	08611	30/04/2002
C EXP 31502A	Aventis CropScience UK Limited	09991	
C EXP 31502A	Rhône-Poulenc Agriculture	09838	30/04/2003
C Ingot	Aventis CropScience UK Limited	09997	
C Ingot	Rhone-Poulenc Agriculture	08610	30/04/2003

164 Diflufenican + Isoproturon

C Barclay Slogan	Barclay Chemicals Manufacturing Ltd	09016	25/07/2003
C Clayton Fenican 625	Clayton Plant Protection (UK) Ltd	09395	31/07/2003

C These products are "approved for agricultural use". For further details refer to page vii.
A These products are approved for use in or near water. For further details refer to page vii.

Product Name	Marketing Company	Reg. No.	Expiry Date

164 Diflufenican + Isoproturon—continued

C	Cougar	Rhone-Poulenc Ireland Ltd	06611	30/09/2001
C	DFF + IPU WDG	Aventis CropScience UK Limited	10204	
C	Gavel	Aventis CropScience UK Limited	09992	
C	Gavel	Rhone-Poulenc Agriculture	08135	25/07/2003
C	Grenadier	Aventis CropScience UK Limited	09996	
C	Grenadier	Rhone-Poulenc Agriculture	08136	25/07/2003
C	Javelin	Aventis CropScience UK Limited	09998	
C	Javelin	Rhone-Poulenc Agriculture	06192	25/07/2003
C	Javelin Gold	Aventis CropScience UK Limited	09999	
C	Javelin Gold	Rhone-Poulenc Agriculture	06200	25/07/2003
C	Landgold DFF 625	Landgold & Co Ltd	06274	31/07/2003
C	Panther	Aventis CropScience UK Limited	10008	
C	Panther	Rhone-Poulenc Agriculture	06491	25/07/2003
C	Shire	Aventis CropScience UK Limited	10017	
C	Shire	Rhone-Poulenc Agriculture	08117	25/07/2003
C	Standon DFF-IPU	Standon Chemicals Ltd	08730	31/07/2001
C	Standon Diflufenican IPU	Standon Chemicals Ltd	09175	31/10/2003
C	Tolkan Turbo	Aventis CropScience UK Limited	10024	
C	Tolkan Turbo	Rhone-Poulenc Agriculture	06795	25/07/2003

165 Diflufenican + Terbuthylazine

C	Bolero	Ciba Agriculture	07436	31/07/2001
C	Bolero	Novartis Crop Protection UK Ltd	08392	

166 Diflufenican + Trifluralin

C	Ardent	Aventis CropScience UK Limited	09968	
C	Ardent	Rhone-Poulenc Agriculture	06203	25/07/2003

167 Diquat

C	Barclay Desiquat	Barclay Chemicals Manufacturing Ltd	09063	31/07/2002
C	Clayton Diquat	Clayton Plant Protection (UK) Ltd	07650	31/07/2002
C	Clayton Quatrow	Clayton Plant Protection (UK) Ltd	07175	25/07/2003
C A	CleanCrop Diquat	United Agri Products	09687	
C	Greencrop Boomerang	Greencrop Technology Ltd	09563	31/07/2002
C	Landgold Diquat	Landgold & Co Ltd	09020	31/07/2002
C A	Levi	P.S.I. Phoenix Scientific Innovation UK Ltd	07845	31/07/2002
C A	Midstream	Imperial Chemical Industries Plc	01348	31/05/2001
C A	Midstream	Miracle Professional	07739	
C A	Midstream	The Scotts Company (UK) Limited	09267	
C A	Midstream	Zeneca Professional Products	06824	31/05/2001
C A	Reglone	Zeneca Crop Protection	09646	

C These products are "approved for agricultural use". For further details refer to page vii.
A These products are approved for use in or near water. For further details refer to page vii.

Product Name	Marketing Company	Reg. No.	Expiry Date

167 Diquat—continued

C A	Reglone	Zeneca Crop Protection	06703	30/06/2002
C	Standon Diquat	Standon Chemicals Ltd	05587	31/07/2002

168 Diquat + Paraquat

C	PDQ	Zeneca Crop Protection	10241	
C	PDQ	Zeneca Crop Protection	06518	25/07/2003
C	Speedway 2	Miracle Professional	08164	31/10/2001
C	Speedway 2	The Scotts Company (UK) Limited	09273	

169 Diquat + Paraquat + Simazine

C	Pathclear S	Miracle Garden Care Ltd	08991	
C	Pathclear S	The Scotts Company (UK) Limited	09269	

170 Diuron

C	Atlas Diuron	Atlas Crop Protection Ltd	08214	
C	Chipko Diuron 80	Chipman Ltd	00497	
C	Chipman Diuron 80	Nomix-Chipman Ltd	08054	
C	Chipman Diuron Flowable	Nomix-Chipman Ltd	05701	
C	Diuron 50 FL	Staveley Chemicals Ltd	02814	
C	Diuron 80 WG	Rhone-Poulenc Amenity	08509	31/05/2002
C	Diuron 80 WP	Rhone-Poulenc Amenity	05199	
C	Diuron 80% WP	Staveley Chemicals Ltd	00730	
C	Diuron 80WP	Aventis Environmental Science	09931	
C	Epsilon	Rhone-Poulenc Amenity	08508	31/05/2002
C	Freeway	Aventis Environmental Science	09933	
C	Freeway	Rhone-Poulenc Amenity	06047	
C	Karmex	Du Pont (UK) Ltd	01128	28/02/2002
C	Karmex	Griffin (Europe) SA	09475	
C	MSS Diuron 500 FL	Mirfield Sales Services Ltd	08171	
C	MSS Diuron 50FL	Mirfield Sales Services Ltd	07160	
C	MSS Diuron 80 WP	Mirfield Sales Services Ltd	09787	
C	Rescind	Rhone-Poulenc Amenity	08036	
C	Sanuron	Dow AgroSciences Ltd	09236	
C	Sanuron	Sanachem International Ltd	08213	30/09/2001
C	Unicrop Flowable Diuron	Universal Crop Protection Ltd	02270	
C	UPL Diuron 80	United Phosphorus Ltd	07619	

171 Diuron + Amitrole + Bromacil

C	BR Destral	Rhone-Poulenc Amenity	05184	31/12/2001

172 Diuron + Amitrole + 2,4-D

C	Trik	Mirfield Sales Services Ltd	07853	

C   These products are "approved for agricultural use". For further details refer to page vii.
A   These products are approved for use in or near water. For further details refer to page vii.

Product Name	Marketing Company	Reg. No.	Expiry Date

173 Diuron + Bromacil

C	Borocil K	Aventis Environmental Science	09924
C	Borocil K	Rhone-Poulenc Amenity	05183

174 Diuron + Glyphosate

C	NX 3083	Nomix-Chipman Ltd	08712	
C	Total	Nomix-Chipman Ltd	07932	
C	Touche	Nomix-Chipman Ltd	07913	
C	Touche	Nomix-Chipman Ltd	06797	
C	Xanadu	Aventis Environmental Science	09943	
C	Xanadu	Rhone-Poulenc Amenity	09228	31/03/2003

175 Diuron + Paraquat

C	Dexuron	Nomix-Chipman Ltd	07169

176 Ethofumesate

C	Agriguard Ethofumesate Flo	Tronsan Ltd	09478	
C	Atlas Thor	Atlas Crop Protection Ltd	07732	
C	Atlas Thor	Atlas Interlates Ltd	06966	
C	Barclay Keeper	Barclay Chemicals Manufacturing Ltd	08835	
C	Barclay Keeper 200	Barclay Chemicals Manufacturing Ltd	05266	25/07/2003
C	Barclay Keeper 500 FL	Barclay Chemicals Manufacturing Ltd	09438	
C	Ethosan	KVK Agro A/S	08555	30/06/2001
C	Griffin Ethofumesate 200	Griffin (Bermuda) SA	08818	28/02/2002
C	Kubist	Griffin (Europe) Marketing NV (UK Branch)	09454	
C	Kubist 2	Griffin (Europe) Marketing NV (UK Branch)	09880	
C	Landgold Ethofumesate 200	Landgold & Co Ltd	08980	
C	MSS Primasan	KVK Agro A/S	08556	30/06/2001
C	MSS Thor	Mirfield Sales Services Ltd	08817	
C	Nortron	Aventis CropScience UK Limited	07266	
C	Nortron Flo	Aventis CropScience UK Limited	08154	
C	Salute	United Phosphorus Ltd	07660	
C	Standon Ethofumesate 200	Standon Chemicals Ltd	09360	
C	Standon Ethofumesate 200	Standon Chemicals Ltd	08726	
C	Stefes Fumat 2	Aventis CropScience UK Limited	10084	
C	Stefes Fumat 2	Stefes Plant Protection Ltd	07856	

177 Ethofumesate + Bromoxynil + Ioxynil

C	Leyclene	Aventis CropScience UK Limited	07263

C These products are "approved for agricultural use". For further details refer to page vii.
A These products are approved for use in or near water. For further details refer to page vii.

Product Name	Marketing Company	Reg. No.	Expiry Date

177 Ethofumesate + Bromoxynil + Ioxynil—continued

C	Stefes Leyclene	Aventis CropScience UK Limited	10086	
C	Stefes Leyclene	Stefes Plant Protection Ltd	08173	

178 Ethofumesate + Chloridazon

C	Gremlin	Sipcam UK Ltd	09468	
C	Magnum	BASF plc	08635	
C	Spectron	Aventis CropScience UK Limited	07284	
C	Stefes Spectron	Stefes UK Ltd	09337	

179 Ethofumesate + Desmedipham + Phenmedipham

C	Aventis Betanal Progress OF	Aventis CropScience UK Limited	09722	
C	Betanal Congress	AgrEvo UK Ltd	07534	31/05/2001
C	Betanal Progress	AgrEvo UK Ltd	07533	31/05/2001
C	Betanal Progress OF	Aventis CropScience UK Limited	07629	
C	Betanal Ultima	AgrEvo UK Ltd	07535	31/05/2001

180 Ethofumesate + Metamitron

C	Stefes Duplex	Stefes UK Ltd	09252	31/12/2001
C	Stefes ST/STE	Aventis CropScience UK Limited	10109	
C	Stefes ST/STE	Stefes Plant Protection Ltd	09111	

181 Ethofumesate + Metamitron + Phenmedipham

C	Bayer UK 540	Bayer plc	07399	
C	Betanal Trio OF	AgrEvo UK Crop Protection Ltd	08680	31/05/2001
C	Betanal Trio WG	Aventis CropScience UK Limited	07537	

182 Ethofumesate + Phenmedipham

C	Barclay Goalpost	Barclay Chemicals Manufacturing Ltd	09497	
C	Barclay Goalpost	Barclay Chemicals Manufacturing Ltd	09192	
C	Barclay Goldpost	Barclay Chem. Manufacturing Ltd	08964	31/03/2002
C	Betanal Montage	AgrEvo UK Ltd	07250	28/02/2001
C	Betanal Tandem	Aventis CropScience UK Limited	07254	
C	Betosip Combi	Sipcam UK Ltd	08630	
C	Stefes Medimat	Stefes Plant Protection Ltd	07886	30/04/2001
C	Stefes Medimat 2	Stefes Plant Protection Ltd	07577	31/03/2002
C	Stefes Tandem	Aventis CropScience UK Limited	10088	
C	Stefes Tandem	Stefes Plant Protection Ltd	08906	30/04/2003
C	Twin	Feinchemie (UK) Ltd	09875	
C	Twin	Feinchemie (UK) Ltd	09784	31/01/2003
C	Twin	Feinchemie Schwebda GmbH	07374	31/10/2002

C   These products are "approved for agricultural use". For further details refer to page vii.
A   These products are approved for use in or near water. For further details refer to page vii.

Product Name	Marketing Company	Reg. No.	Expiry Date

183 Fenoxaprop-P-ethyl

C	Aventis Cheetah Super	Aventis CropScience UK Limited	09730	
C	Cheetah Super	Aventis CropScience UK Limited	08723	
C	FPE 120	AgrEvo UK Ltd	08742	31/05/2001
C	FPE 55	AgrEvo UK Ltd	08725	31/05/2001
C	Gazelle	AgrEvo UK Ltd	08741	31/05/2001
C	Triumph	Aventis CropScience UK Limited	08740	
C	Wildcat	AgrEvo UK Ltd	08724	31/05/2001

184 Fenoxaprop-P-ethyl + Diclofop-methyl

C	Aventis Tigress Ultra	Aventis CropScience UK Limited	09725
C	Corniche	Aventis CropScience UK Limited	08947
C	Tigress Ultra	Aventis CropScience UK Limited	08946

185 Fenoxaprop-P-ethyl + Isoproturon

C	Pinion	Griffin (Europe) Marketing NV (UK Branch)	10287	
C	Puma X	AgrEvo UK Ltd	07332	31/12/2001
C	Puma X	Aventis CropScience UK Limited	08779	
C	Whip	AgrEvo UK Ltd	08780	31/05/2001
C	Whip X	AgrEvo UK Ltd	07339	31/12/2001

186 Fenuron + Chlorpropham

C	Atlas Red	Atlas Crop Protection Ltd	07724
C	Atlas Red	Atlas Interlates Ltd	03091
C	Croptex Chrome	Hortichem Ltd	02415

187 Ferrous sulphate

C	Aitkens Lawn Sand	RCC Agro	05253
C	Elliott's Lawn Sand	Thomas Elliott Ltd	04860
C	Elliott's Mosskiller	Thomas Elliott Ltd	04909
C	Fisons Greenmaster Autumn	Fisons Plc	03211
C	Fisons Greenmaster Mosskiller	Fisons Plc	00881
C	Greenmaster Autumn	The Scotts Company (UK) Limited + Levington Horticulture Ltd	07508
C	Greenmaster Mosskiller	The Scotts Company (UK) Limited + Levington Horticulture Ltd	07509
C	Maxicrop Mosskiller and Conditioner	Maxicrop (UK) Ltd	04635
C	Moss Control Plus Lawn Fertilizer	Miracle Garden Care Ltd	07912
C	SHL Lawn Sand	Sinclair Horticulture & Leisure Ltd	05254
C	Taylor's Lawn Sand	Rigby Taylor Ltd	04451
C	Vitagrow Lawn Sand	Vitagrow (Fertilisers) Ltd	05097

C   These products are "approved for agricultural use". For further details refer to page vii.
A   These products are approved for use in or near water. For further details refer to page vii.

Product Name	Marketing Company	Reg. No.	Expiry Date

187 Ferrous sulphate—continued

C Vitax Microgran 2	Vitax Ltd	04541	
C Vitax Turf Tonic	Vitax Ltd	04354	

188 Ferrous sulphate + Dicamba + Dichlorprop + MCPA

C Longlife Renovator 2	Miracle Professional	08379	
C Renovator 2	The Scotts Company (UK) Limited	09272	

189 Ferrous sulphate + Dichlorophen

C Aitkens Lawn Sand Plus	RCC Agro	04542	
C SHL Lawn Sand Plus	Sinclair Horticulture & Leisure Ltd	04439	

190 Ferrous sulphate + Dichlorprop + MCPA

C SHL Granular Feed, Weed & Mosskiller	William Sinclair Horticulture Ltd	09925	
C SHL Turf Feed and Weed and Mosskiller	William Sinclair Horticulture Ltd	04438	
C Vitagrow Mini-Granular Feed, Weed & Mosskiller	William Sinclair Horticulture Ltd	10295	

191 Flamprop-M-isopropyl

C Commando	BASF plc	10215	
C Commando	Cyanamid Agriculture Ltd	07005	

192 Fluazifop-P-butyl

C Barclay Winner	Barclay Chemicals Manufacturing Ltd	06986	25/07/2003
C Citadel	Zeneca Crop Protection	06762	30/09/2002
C Corral	Zeneca Crop Protection	06647	30/09/2002
C Fusilade 250 EW	Zeneca Crop Protection	06531	
C Fusilade 5	Zeneca Crop Protection	06669	30/09/2002
C PP 007	Zeneca Crop Protection	06533	
C Wizzard	Zeneca Crop Protection	06521	

193 Fluoroglycofen-ethyl + Isoproturon

C Compete Forte	AgrEvo UK Crop Protection Ltd	08708	
C Competitor	Aventis CropScience UK Limited	07310	
C Effect	AgrEvo UK Ltd	07319	31/05/2001
C Stefes Competitor	Stefes Plant Protection Ltd	08753	31/03/2002

194 Flupyrsulfuron-methyl

C Clayton Tranche	Clayton Plant Protection (UK) Ltd	10274	
C DPX-KE459 WSB	Du Pont (UK) Ltd	08540	
C KE 459 DF	Du Pont (UK) Ltd	09835	

C These products are "approved for agricultural use". For further details refer to page vii.
A These products are approved for use in or near water. For further details refer to page vii.

119

Product Name	Marketing Company	Reg. No.	Expiry Date

194 Flupyrsulfuron-methyl—continued

| C | Lexus 50 DF | Du Pont (UK) Ltd | 09026 | |

195 Flupyrsulfuron-methyl + Carfentrazone-ethyl

| C | Lexus Class | Du Pont (UK) Ltd | 08636 | |
| C | Lexus Class WSB | Du Pont (UK) Ltd | 08637 | |

196 Flupyrsulfuron-methyl + Metsulfuron-methyl

C	Lexus XPE	Du Pont (UK) Ltd	08541	
C	Lexus XPE WSB	Du Pont (UK) Ltd	08542	
C	Standon Flupyrsulfuron MM	Standon Chemicals Ltd	09098	

197 Flupyrsulfuron-methyl + Thifensulfuron-methyl

| C | Lexus Millenium | Du Pont (UK) Ltd | 09206 | |
| C | Lexus Millenium WSB | Du Pont (UK) Ltd | 09207 | |

198 Fluquinconazole + Prochloraz

| C | Jockey Plus | Aventis CropScience UK Limited | 10075 | |

199 Fluroxypyr

C	Agriguard Fluroxypyr	Agriguard Ltd	09298	
C	Barclay Hurler	Barclay Chemicals Manufacturing Ltd	08791	
C	Barclay Hurler 200	Barclay Chemicals Manufacturing Ltd	09917	
C	Clayton Fluroxypyr	Clayton Plant Protection (UK) Ltd	06356	
C	Gala	Dow AgroSciences Ltd	09793	
C	Greencrop Reaper	Greencrop Technology Ltd	09359	
C	Greencrop Reaper 2	Greencrop Technology Ltd	10060	
C	Standon Fluroxypyr	Standon Chemicals Ltd	08923	
C	Starane 2	Dow AgroSciences Ltd	09405	31/12/2002
C	Starane 2	Dow AgroSciences Ltd	05496	
C	Tomahawk	Makhteshim-Agan (UK) Ltd	09249	
C	Tomahawk 2000	Makhteshim-Agan (UK) Ltd	09307	

200 Fluroxypyr + Bromoxynil

| C | Sickle | Dow AgroSciences Ltd | 05187 | |
| C | Tomahawk Plus | Makhteshim-Agan (UK) Ltd | 09836 | |

201 Fluroxypyr + Bromoxynil + Clopyralid + Ioxynil

| C | Crusader S | Dow AgroSciences Ltd | 05174 | |

C   These products are "approved for agricultural use". For further details refer to page vii.
A   These products are approved for use in or near water. For further details refer to page vii.

Product Name	Marketing Company	Reg. No.	Expiry Date

202 Fluroxypyr + Bromoxynil + Ioxynil

| C | Advance | Dow AgroSciences Ltd | 05173 | |

203 Fluroxypyr + Clopyralid + Ioxynil

| C | Hotspur | Dow AgroSciences Ltd | 05185 | |

204 Fluroxypyr + Clopyralid + MCPA

| C | Greenor | Rigby Taylor Ltd | 07848 | |

204a Fluroxypyr + Clopyralid + Triclopyr

| C | Pastor | Dow AgroSciences Ltd | 07440 | |

205 Fluroxypyr + Mecoprop-P

| C | Bastion T | Rigby Taylor Ltd | 06011 | |

206 Fluroxypyr + Metosulam

| C | EF 1166 | Dow AgroSciences Ltd | 07966 | |

207 Fluroxypyr + Thifensulfuron-methyl + Tribenuron-methyl

C	DP 353	Du Pont (UK) Ltd	09626	
C	GEM 353	Griffin (Europe) Marketing NV (UK Branch)	09826	25/07/2003
C	GEX 353	Griffin (Europe) Marketing NV (UK Branch)	10185	
C	Starane Super	Dow AgroSciences Ltd	09625	

208 Fluroxypyr + Triclopyr

| C | Doxstar | Dow AgroSciences Ltd | 06050 | |
| C | Evade | Dow AgroSciences Ltd | 08071 | 31/05/2003 |

209 Flurtamone + Diflufenican

| C | Bacara | Aventis CropScience UK Limited | 09976 | |
| C | Bacara | Rhône-Poulenc Agriculture | 08323 | 30/04/2003 |

210 Flurtamone + Diflufenican + Isoproturon

C	EXP 31165	Aventis CropScience UK Limited	09990	30/04/2002
C	EXP 31165	Rhône-Poulenc Agriculture	08611	30/04/2002
C	EXP 31502A	Aventis CropScience UK Limited	09991	
C	EXP 31502A	Rhône-Poulenc Agriculture	09838	30/04/2003
C	Ingot	Aventis CropScience UK Limited	09997	
C	Ingot	Rhone-Poulenc Agriculture	08610	30/04/2003

C   These products are "approved for agricultural use". For further details refer to page vii.
A   These products are approved for use in or near water. For further details refer to page vii.

Product Name	Marketing Company	Reg. No.	Expiry Date

211 Fomesafen

C   Flex   Novartis Crop Protection UK Ltd   08885

212 Fomesafen + Terbutryn

C   Reflex T   Novartis Crop Protection UK Ltd   08884

213 Fosamine-ammonium

C A   Krenite   Du Pont (UK) Ltd   01165

214 Gibberellins

C	Berelex	Whyte Chemicals Ltd	08903	
C	Berelex	Zeneca Crop Protection	06637	31/01/2001
C	Novagib	Fine Agrochemicals Ltd	08954	
C	Regulex	Sumitomo Chemical Agro Europe SA	10147	
C	Regulex	Zeneca Crop Protection	07997	31/05/2002

215 Glufosinate-ammonium

C	Aventis Challenge	Aventis CropScience UK Limited	09721
C	Aventis Harvest	Aventis CropScience UK Limited	09810
C	Challenge	Aventis CropScience UK Limited	07306
C	Challenge 2	Aventis CropScience UK Limited	07307
C	Challenge 60	Aventis CropScience UK Limited	08236
C	Dash	Nomix-Chipman Ltd	05177
C	Finale	Aventis Environmental Science	10092
C	Harvest	Aventis CropScience UK Limited	07321
C	Headland Sword	Headland Agrochemicals Ltd	07676
C	Nomix Touchweed	Nomix-Chipman Ltd	09596

216 Glyphosate

C A	Acrion	Aventis Environmental Science	09965	
C	Agriguard Glyphosate 180	Agriguard Ltd	09806	
C	Agriguard Glyphosate 360	Agriguard Ltd	09184	
C	Amega Pro TMF	Nufarm UK Ltd	09630	
C	Apache	Zeneca Crop Protection	06748	
C A	Azural	Monsanto Plc	09582	
C	Barbarian	Barclay Chemicals Manufacturing Ltd	07625	
C	Barbarian	Barclay Chemicals Manufacturing Ltd	07980	31/01/2002
C A	Barclay Barbarian	Barclay Chemicals Manufacturing Ltd	09865	
C	Barclay Cleanup	Barclay Plant Protection	09179	

C   These products are "approved for agricultural use". For further details refer to page vii.
A   These products are approved for use in or near water. For further details refer to page vii.

	Product Name	Marketing Company	Reg. No.	Expiry Date
	216 Glyphosate—continued			
C	Barclay Dart	Barclay Chemicals Manufacturing Ltd	05129	
C	Barclay Gallup	Barclay Chemicals Manufacturing Ltd	05161	
C A	Barclay Gallup 360	Barclay Plant Protection	09127	
C A	Barclay Gallup Amenity	Barclay Chemicals Manufacturing Ltd	06753	
C A	Barclay Gallup Biograde 360	Barclay Chemicals Manufacturing Ltd	09840	
C A	Barclay Gallup Biograde Amenity	Barclay Chemicals Manufacturing Ltd	10203	
C	Barclay Garryowen	Barclay Chemicals (R & D) Ltd	09869	
C	Barclay Garryowen	Barclay Chemicals Manufacturing Ltd	08599	31/01/2002
C	Barclay Trustee	Barclay Plant Protection	09177	
C A	Biactive 270	Monsanto Plc	10031	
C	Bioglyce	Applyworld Limited	10054	
C A	Buggy SG	Sipcam UK Ltd	08573	
C A	Cardel Egret	Cardel	09703	
C A	Cardel Egret	Cardel Agro SAS	10252	
C A	Cardel Glyphosate	Cardel	09581	
C A	Cardel Glyphosate	Cardel Agro SAS	10253	
C	CDA Vanquish	Aventis Environmental Science	09927	
C	CDA Vanquish	Rhone-Poulenc Amenity	08577	
C	Clarion	Zeneca Crop Protection	08272	
C	Clayton Glyphosate	Clayton Plant Protection (UK) Ltd	06608	
C	Clayton Rhizeup	Clayton Plant Protection Ltd	08920	
C A	Clayton Swath	Clayton Plant Protection (UK) Ltd	06715	
C	Clean-Up-360	Top Farm Formulations Ltd	05076	
C A	Clinic	Nufarm UK Ltd	09378	
C A	Clinic	Nufarm UK Ltd	08579	31/10/2001
C	Clinic Pro TMF	Nufarm UK Ltd	09443	30/06/2002
C	Do-Away	NCH Europe Ltd	10112	
C	Economix	Nomix-Chipman Ltd	08008	
C	Gly_480	Powaspray CDA Ltd	09837	
C	Glyfonex 480	Danagri ApS	07464	
C A	Glyfos	Cheminova Agro (UK) Ltd	07109	
C A	Glyfos 480	Cheminova Agro (UK) Ltd	08014	
C A	Glyfos Proactive	Nomix-Chipman Ltd	07800	
C	Glyfosate-360	Top Farm Formulations Ltd	05319	
C A	Glyper	Pan Britannica Industries Ltd	07968	
C A	Glyphogan	Makhteshim-Agan (UK) Ltd	05784	
C	Glyphosate 120B	Monsanto Plc	08292	
C A	Glyphosate 360	Applyworld Limited	09151	

C   These products are "approved for agricultural use". For further details refer to page vii.
A   These products are approved for use in or near water. For further details refer to page vii.

Product Name	Marketing Company	Reg. No.	Expiry Date

216 Glyphosate—continued

C A	Glyphosate 360	Danagri ApS	09233	
C A	Glyphosate 360	Danagri ApS	08568	31/07/2001
C A	Glyphosate 360A	Danagri ApS	09234	
C A	Glyphosate 360B	Danagri ApS	09235	
C A	Glyphosate Biactive	Monsanto Plc	08307	
C	Gorgon	Portman Agrochemicals Ltd	09182	
C	Greencrop Gypsy	Greencrop Technology Ltd	09432	
C A	Helosate	Helm AG	06499	
C	Hilite	Nomix-Chipman Ltd	06261	
C	IT Glyphosate	IT Agro Ltd	07212	
C	Landgold Glyphosate 360	Landgold & Co Ltd	05929	
C A	Marnoch Glyphosate	Marnoch Ventures Limited	09744	
C	Mogul	Monsanto Plc	07076	
C	Mon 240	Monsanto Plc	04538	
C A	MON 44068 Pro	Monsanto Plc	06815	
C A	MON 52276	Monsanto Plc	06949	
C A	MON 77158A	Monsanto Plc	09904	31/03/2003
C A	Monty	Quadrangle Agrochemicals	07796	
C A	MSS Glyfield	Mirfield Sales Services Ltd	08009	
C	Muster	Zeneca Crop Protection	06685	
C	Muster LA	Zeneca Professional Products	05762	
C	Nomix G	Nomix-Chipman Ltd	08781	
C	Nomix Nova	Nomix-Chipman Ltd	08647	
C	Nomix Revenge	Nomix-Chipman Ltd	09483	
C	Nomix Supernova	Nomix-Chipman Ltd	09473	
C A	Nufosate	Nufarm UK Ltd	10096	
C	NX 8049	Nomix-Chipman Ltd	10099	
C	PMG	Agriguard Ltd	10153	
C	Poise	Universal Crop Protection Ltd	08276	
C	Portman Glider	Portman Agrochemicals Ltd	04695	
C	Portman Glyphosate	Portman Agrochemicals Ltd	05891	
C	Portman Glyphosate 360	Portman Agrochemicals Ltd	04699	
C	Portman Glyphosate 480	Portman Agrochemicals Ltd	07194	31/08/2001
C A	Reliance	Monsanto Plc	09954	
C A	Rival	Monsanto Plc	09220	
C A	Roundup	Monsanto Plc	01828	
C	Roundup 2000	Monsanto Plc	08069	
C A	Roundup A	Monsanto Plc	08375	31/08/2001
C	Roundup Accuply	Accuply Ltd	09544	
C A	Roundup Amenity	Monsanto Plc	08721	
C A	Roundup Biactive	Monsanto Plc	06941	
C A	Roundup Biactive Dry	Monsanto Plc	06942	
C	Roundup CF	Monsanto Plc	09547	
C	Roundup Four 80	Monsanto Plc	03176	

C   These products are "approved for agricultural use". For further details refer to page vii.
A   These products are approved for use in or near water. For further details refer to page vii.

Product Name	Marketing Company	Reg. No.	Expiry Date

216 Glyphosate—continued

C	Roundup GT	Monsanto Plc	08068	
C	Roundup MX	Monsanto Plc	09631	
C A	Roundup Pro	Monsanto Plc	04146	
C A	Roundup Pro Biactive	Monsanto Plc	06954	
C	Roundup Rapide	Monsanto Plc	08067	
C A	Roundup Ultra ST	Monsanto (UK) Ltd	10199	
C A	Samurai	Monsanto Plc	09952	
C A	Scorpion	Monsanto Plc	09953	
C	Smart 360	Mastra Europe Limited	09476	
C A	Spasor	Aventis Environmental Science	09945	
C A	Spasor	Rhone-Poulenc Amenity	07211	
C A	Spasor Biactive	Aventis Environmental Science	09940	
C A	Spasor Biactive	Rhone-Poulenc Amenity	07651	30/04/2003
C	Stacato	Sipcam UK Ltd	05892	
C	Stampede	Zeneca Crop Protection	06327	
C	Standon Glyphosate 360	Standon Chemicals Ltd	05582	
C	Stefes Glyphosate	Stefes Plant Protection Ltd	05819	31/03/2002
C	Stefes Kickdown	Stefes Plant Protection Ltd	06329	31/03/2002
C	Stefes Kickdown 2	Stefes Plant Protection Ltd	06548	31/03/2002
C	Sting	Monsanto Plc	02789	
C	Sting CT	Monsanto Plc	04754	
C	Sting Eco	Monsanto Plc	08291	
C	Stirrup	Nomix-Chipman Ltd	06132	
C	Stride	Zeneca Crop Protection	08282	
C	Touchdown	Zeneca Crop Protection	06326	
C	Touchdown LA	Miracle Professional	07747	30/09/2001
C	Touchdown LA	The Scotts Company (UK) Limited	09270	
C A	Typhoon 360	Feinchemie (UK) Ltd	09792	
C A	Typhoon 360	Feinchemie Schwebda GmbH	09322	31/10/2002
C	Unistar Glyfosate 360	Unistar Ltd	05928	
C	Unistar Glyphosate 360	Unistar Ltd	06332	

217 Glyphosate + Diuron

C	NX 3083	Nomix-Chipman Ltd	08712	
C	Total	Nomix-Chipman Ltd	07932	
C	Touche	Nomix-Chipman Ltd	07913	
C	Touche	Nomix-Chipman Ltd	06797	
C	Xanadu	Aventis Environmental Science	09943	
C	Xanadu	Rhone-Poulenc Amenity	09228	31/03/2003

218 Glyphosate + Oxadiazon

C	Zapper	Aventis Environmental Science	09944	
C	Zapper	Rhone-Poulenc Amenity	08605	31/12/2003

C   These products are "approved for agricultural use". For further details refer to page vii.
A   These products are approved for use in or near water. For further details refer to page vii.

Product Name	Marketing Company	Reg. No.	Expiry Date

219 Imazamethabenz-methyl

C Assert	BASF plc	10209	
C Assert	Cyanamid Agriculture Ltd	08656	
C Dagger	BASF plc	10218	
C Dagger	Cyanamid Agriculture Ltd	03737	

220 Imazapyr

C Arsenal	BASF plc	10208	
C Arsenal	Cyanamid Agriculture Ltd	04064	
C Arsenal	Nomix-Chipman Ltd	05537	
C Arsenal 50	Cyanamid Agriculture Ltd	04070	31/03/2002
C Arsenal 50	Nomix-Chipman Ltd	05567	

221 Imazaquin + Chlormequat

C Meteor	Cyanamid Agriculture Ltd	06505	
C Standon Imazaquin 5C	Standon Chemicals Ltd	08813	
C Upright	Cyanamid Agriculture Ltd	08290	

222 Imazaquin + 2-Chloroethylphosphonic acid + Chlormequat

C Ice	Cyanamid Agriculture Ltd	08970	
C Satellite	Cyanamid Agriculture Ltd	08969	

223 Indol-3-ylacetic acid

C Rhizopon A Powder	Fargro Ltd	07131	
C Rhizopon A Powder	Rhizopon UK Limited	09087	
C Rhizopon A Tablets	Fargro Ltd	07132	
C Rhizopon A Tablets	Rhizopon UK Limited	09088	

224 4-Indol-3-ylbutric acid

C Chryzoplus Grey 0.8%	Fargro Ltd	07984	
C Chryzoplus Grey 0.8%	Rhizopon UK Limited	09094	
C Chryzopon Rose 0.1%	Fargro Ltd	07982	
C Chryzopon Rose 0.1%	Rhizopon UK Limited	09092	
C Chryzosan White 0.6%	Fargro Ltd	07983	
C Chryzosan White 0.6%	Rhizopon UK Limited	09093	
C Chryzotek Beige	Fargro Ltd	07125	
C Chryzotek Beige	Rhizopon UK Limited	09081	
C Chryzotop Green	Fargro Ltd	07129	
C Chryzotop Green	Rhizopon UK Limited	09085	
C Rhizopon AA Powder (0.5%)	Fargro Ltd	07126	
C Rhizopon AA Powder (0.5%)	Rhizopon UK Limited	09082	
C Rhizopon AA Powder (1 %)	Fargro Ltd	07127	
C Rhizopon AA Powder (1 %)	Rhizopon UK Limited	09084	
C Rhizopon AA Powder (2%)	Fargro Ltd	07128	

C These products are "approved for agricultural use". For further details refer to page vii.
A These products are approved for use in or near water. For further details refer to page vii.

Product Name	Marketing Company	Reg. No.	Expiry Date

224 4-Indol-3-ylbutric acid—continued

C	Rhizopon AA Powder 2%	Rhizopon UK Limited	09083
C	Rhizopon AA Tablets	Fargro Ltd	07130
C	Rhizopon AA Tablets	Rhizopon UK Limited	09086
C	Seradix 1	Rhone-Poulenc Agriculture	06191
C	Seradix 2	Rhone-Poulenc Agriculture	06193
C	Seradix 3	Rhone-Poulenc Agriculture	06194

225 4-Indol-3-ylbutyric acid + 1-Naphthylacetic acid

C	Synergol	Hortichem Ltd	07386

226 Ioxynil

C	Actrilawn 10	Rhone-Poulenc Amenity	05247
C	Totril	Aventis CropScience UK Limited	10026
C	Totril	Rhone-Poulenc Agriculture	06116

227 Ioxynil + Benazolin + Bromoxynil

C	Asset	Aventis CropScience UK Limited	07243

228 Ioxynil + Bromoxynil

C	Alpha Briotril	Makhteshim-Agan (UK) Ltd	04876
C	Alpha Briotril Plus 19/19	Makhteshim-Agan (UK) Ltd	04740
C	Deloxil	Aventis CropScience UK Limited	09987
C	Deloxil	Aventis CropScience UK Limited	07313
C	Deloxil	Rhone-Poulenc Agriculture	07405
C	Mextrol-Biox	Nufarm UK Ltd	09470
C	Oxytril CM	Aventis CropScience UK Limited	10005
C	Oxytril CM	Rhone-Poulenc Agriculture	08667
C	Percept	MTM Agrochemicals Ltd	05481
C	Status	Aventis CropScience UK Limited	10019
C	Status	Rhone-Poulenc Agriculture	08668
C	Stellox 380 EC	Novartis Crop Protection UK Ltd	08451

229 Ioxynil + Bromoxynil + Clopyralid + Fluroxypyr

C	Crusader S	Dow AgroSciences Ltd	05174

230 Ioxynil + Bromoxynil + Dichlorprop + MCPA

C	Atlas Minerva	Atlas Interlates Ltd	03046

231 Ioxynil + Bromoxynil + Diflufenican

C	Capture	Aventis CropScience UK Limited	09982	
C	Capture	Rhone-Poulenc Agriculture	06881	25/07/2003

C These products are "approved for agricultural use". For further details refer to page vii.
A These products are approved for use in or near water. For further details refer to page vii.

Product Name	Marketing Company	Reg. No.	Expiry Date

232 Ioxynil + Bromoxynil + Ethofumesate

C	Leyclene	Aventis CropScience UK Limited	07263
C	Stefes Leyclene	Aventis CropScience UK Limited	10086
C	Stefes Leyclene	Stefes Plant Protection Ltd	08173

233 Ioxynil + Bromoxynil + Fluroxypyr

C	Advance	Dow AgroSciences Ltd	05173

234 Ioxynil + Bromoxynil + Mecoprop-P

C	Swipe P	Novartis Crop Protection UK Ltd	08452

235 Ioxynil + Bromoxynil + Triasulfuron

C	Teal	Novartis Crop Protection UK Ltd	08453
C	Teal-M	Novartis Crop Protection UK Ltd	08454

236 Ioxynil + Bromoxynil + Trifluralin

C	Masterspray	Pan Britannica Industries Ltd	02971
C	Masterspray	SumiAgro (UK) Ltd	09603

237 Ioxynil + Clopyralid

C	Escort	Dow AgroSciences Ltd	05466

238 Ioxynil + Clopyralid + Fluroxypyr

C	Hotspur	Dow AgroSciences Ltd	05185

239 Isoproturon

C	Agriguard Isoproturon	Agriguard Ltd	09769
C	Aligran WDG	Portman Agrochemicals Ltd	08931
C	Alpha Isoproturon 500	Makhteshim-Agan (UK) Ltd	05882
C	Alpha Isoproturon 650	Makhteshim-Agan (UK) Ltd	07034
C	Arelon	AgrEvo UK Crop Protection Ltd	06716	31/05/2001
C	Arelon 2	Griffin (Europe) Marketing NV (UK Branch)	10276
C	Arelon 500	Aventis CropScience UK Limited	08100
C	Arelon 500	Griffin (Europe) Marketing NV (UK Branch)	10259
C	Atlas Fieldguard	Atlas Crop Protection Ltd	08582
C	Atum	Aventis CropScience UK Limited	09970
C	Atum	Rhone-Poulenc Agriculture	07379
C	Atum WDG	Aventis CropScience UK Limited	09971
C	Atum WDG	Rhone-Poulenc Agriculture	07778
C	Auger	Rhone-Poulenc Agriculture	06581
C	Auger WDG	Rhone-Poulenc Agriculture	07777	31/03/2001

C   These products are "approved for agricultural use". For further details refer to page vii.
A   These products are approved for use in or near water. For further details refer to page vii.

Product Name	Marketing Company	Reg. No.	Expiry Date

239 Isoproturon—continued

	Product Name	Marketing Company	Reg. No.	Expiry Date
C	Barclay Guideline 500	Barclay Chemicals Manufacturing Ltd	09257	30/09/2003
C	Bison	Gharda Chemicals Ltd	08699	
C	Bison 83 WG	Nufarm Whyte Agriculture Ltd	10063	
C	Bison 83WG	Gharda Chemicals Ltd	10062	
C	Clayton IPU	Clayton Plant Protection (UK) Ltd	07661	31/08/2001
C	Clayton Siptu 50 FL	Clayton Plant Protection (UK) Ltd	08751	
C	Clayton Siptu 500	Clayton Plant Protection (UK) Ltd	08735	30/11/2003
C	Cordelia	Griffin (Europe) Marketing NV (UK Branch)	09780	
C	Cordelia 2	Griffin (Europe) Marketing NV (UK Branch)	10238	
C	DAPT Isoproturon 500 FL	DAPT Agrochemicals Ltd	08092	
C	Hytane 500 SC	Novartis Crop Protection UK Ltd	08419	
C	I.P.U Sun	Phytorus UK Ltd	08798	
C	IPU 500	Rhone-Poulenc Agriculture	07380	
C	IPU 500 WDG	Rhone-Poulenc Agriculture	07780	31/03/2001
C	ISO 500	Quality Assured Marketing Ltd	08775	31/08/2001
C	Isoguard	Chiltern Farm Chemicals Ltd	08497	
C	Isoguard 83WG	Gharda Chemicals Ltd	10061	
C	Isoguard SVR	Cyanamid Agriculture Ltd	08696	
C	Isoproturon 500	Aventis CropScience UK Limited	06718	25/07/2003
C	Isoproturon 500	Griffin (Europe) Marketing NV (UK Branch)	10261	
C	Isoproturon 553	AgrEvo UK Crop Protection Ltd	06719	31/05/2001
C	Isotop SC	Portman Agrochemicals Ltd	07663	
C	Landgold Isoproturon	Landgold & Co Ltd	06012	
C	Luxan Isoproturon 500 Flowable	Luxan (UK) Ltd	07437	
C	MSS Iprofile	Mirfield Sales Services Ltd	06341	31/12/2001
C	Mysen	Portman Agrochemicals Ltd	07695	
C	Mysen WDG	Portman Agrochemicals Ltd	09141	
C	Portman Isotop	Portman Agrochemicals Ltd	03434	
C	Protugan	Makhteshim-Agan (UK) Ltd	10043	
C	Quintil 500	Phytorus UK Ltd	06800	
C	RP 500	Rhone-Poulenc Agriculture	07378	
C	RP 500 WDG	Rhone-Poulenc Agriculture	07779	31/03/2001
C	Sabre	AgrEvo UK Crop Protection Ltd	06717	31/05/2001
C	Sabre 500	AgrEvo UK Ltd	08101	31/05/2001
C	Standon IPU	Standon Chemicals Ltd	08671	
C	Stefes IPU	Aventis CropScience UK Limited	10107	
C	Stefes IPU	Stefes Plant Protection Ltd	05776	
C	Stefes IPU 500	Aventis CropScience UK Limited	10082	25/07/2003
C	Stefes IPU 500	Stefes UK Ltd	08102	25/07/2003

C   These products are "approved for agricultural use". For further details refer to page vii.
A   These products are approved for use in or near water. For further details refer to page vii.

Product Name	Marketing Company	Reg. No.	Expiry Date

239 Isoproturon—continued

C Stefes IPU 700	Aventis CropScience UK Limited	10152	
C Tolkan	Aventis CropScience UK Limited	10022	
C Tolkan	Rhone-Poulenc Agriculture	07365	
C Tolkan Liquid	Aventis CropScience UK Limited	10023	
C Tolkan Liquid	Rhone-Poulenc Agriculture	06172	25/07/2003
C Tolkan WDG	Rhone-Poulenc Agriculture	07776	31/03/2001
C Tripart Pugil 5	Tripart Farm Chemicals Ltd	08606	

240 Isoproturon + Bifenox

C Banco	Portman Agrochemicals Ltd	09027	
C RP 4169	Rhone-Poulenc Agriculture	05801	

241 Isoproturon + Carfentrazone-ethyl

C Affinity	FMC Corporation (UK) Ltd	09745	

242 Isoproturon + Chlorotoluron

C Tolugan Extra	Makhteshim-Agan (UK) Ltd	09393	

243 Isoproturon + Clodinafop-propargyl + Diflufenican

C SL 520	Ciba Agriculture	07687	28/02/2001
C SL 520	Novartis Crop Protection UK Ltd	08447	30/10/2001
C Unite	Rhone-Poulenc Agriculture	07686	30/10/2001

244 Isoproturon + Diflufenican

C Barclay Slogan	Barclay Chemicals Manufacturing Ltd	09016	25/07/2003
C Clayton Fenican 625	Clayton Plant Protection (UK) Ltd	09395	31/07/2003
C Cougar	Rhone-Poulenc Ireland Ltd	06611	30/09/2001
C DFF + IPU WDG	Aventis CropScience UK Limited	10204	
C Gavel	Aventis CropScience UK Limited	09992	
C Gavel	Rhone-Poulenc Agriculture	08135	25/07/2003
C Grenadier	Aventis CropScience UK Limited	09996	
C Grenadier	Rhone-Poulenc Agriculture	08136	25/07/2003
C Javelin	Aventis CropScience UK Limited	09998	
C Javelin	Rhone-Poulenc Agriculture	06192	25/07/2003
C Javelin Gold	Aventis CropScience UK Limited	09999	
C Javelin Gold	Rhone-Poulenc Agriculture	06200	25/07/2003
C Landgold DFF 625	Landgold & Co Ltd	06274	31/07/2003
C Panther	Aventis CropScience UK Limited	10008	
C Panther	Rhone-Poulenc Agriculture	06491	25/07/2003
C Shire	Aventis CropScience UK Limited	10017	
C Shire	Rhone-Poulenc Agriculture	08117	25/07/2003
C Standon DFF-IPU	Standon Chemicals Ltd	08730	31/07/2001

C These products are "approved for agricultural use". For further details refer to page vii.
A These products are approved for use in or near water. For further details refer to page vii.

Product Name	Marketing Company	Reg. No.	Expiry Date

244 Isoproturon + Diflufenican—continued

C	Standon Diflufenican IPU	Standon Chemicals Ltd	09175	31/10/2003
C	Tolkan Turbo	Aventis CropScience UK Limited	10024	
C	Tolkan Turbo	Rhone-Poulenc Agriculture	06795	25/07/2003

245 Isoproturon + Diflufenican + Flurtamone

C	EXP 31165	Aventis CropScience UK Limited	09990	30/04/2002
C	EXP 31165	Rhone-Poulenc Agriculture	08611	30/04/2002
C	EXP 31502A	Aventis CropScience UK Limited	09991	
C	EXP 31502A	Rhône-Poulenc Agriculture	09838	30/04/2003
C	Ingot	Aventis CropScience UK Limited	09997	
C	Ingot	Rhone-Poulenc Agriculture	08610	30/04/2003

246 Isoproturon + Fenoxaprop-P-ethyl

C	Pinion	Griffin (Europe) Marketing NV (UK Branch)	10287	
C	Puma X	AgrEvo UK Ltd	07332	31/12/2001
C	Puma X	Aventis CropScience UK Limited	08779	
C	Whip	AgrEvo UK Ltd	08780	31/05/2001
C	Whip X	AgrEvo UK Ltd	07339	31/12/2001

247 Isoproturon + Fluoroglycofen-ethyl

C	Compete Forte	AgrEvo UK Crop Protection Ltd	08708	
C	Competitor	Aventis CropScience UK Limited	07310	
C	Effect	AgrEvo UK Ltd	07319	31/05/2001
C	Stefes Competitor	Stefes Plant Protection Ltd	08753	31/03/2002

248 Isoproturon + Isoxaben

C	IPSO	Dow AgroSciences Ltd	05736

249 Isoproturon + Pendimethalin

C	Artillery	Cyanamid Agriculture Ltd	07768
C	Artillery SC	Cyanamid Agriculture Ltd	08504
C	Clayton Pendalin-IPU 37	Clayton Plant Protection (UK) Ltd	09154
C	Clayton Pendalin-IPU 47	Clayton Plant Protection (UK) Ltd	09170
C	Encore	Cyanamid Agriculture Ltd	04737
C	Encore SC	Cyanamid Agriculture Ltd	08502
C	Jolt	Cyanamid Agriculture Ltd	05488
C	Jolt SC	Cyanamid Agriculture Ltd	08503
C	Orient	Cyanamid Agriculture Ltd	07798
C	Orient SC	Cyanamid Agriculture Ltd	08501
C	Trump	Cyanamid Agriculture Ltd	03687
C	Trump SC	Cyanamid Agriculture Ltd	08500

C   These products are "approved for agricultural use". For further details refer to page vii.
A   These products are approved for use in or near water. For further details refer to page vii.

131

Product Name	Marketing Company	Reg. No.	Expiry Date

250 Isoproturon + Simazine
C	Alpha Protugan Plus	Makhteshim-Agan (UK) Ltd	08799	
C	Harlequin 500 SC	Makhteshim-Agan (UK) Ltd	09779	
C	Harlequin 500 SC	Novartis Crop Protection UK Ltd	08416	30/09/2002

251 Isoproturon + Trifluralin
C	Autumn Kite	Aventis CropScience UK Limited	07119	
C	Stefes Union	Stefes Plant Protection Ltd	08146	31/03/2002

252 Isoxaben
C	Flexidor	Dow AgroSciences Ltd	05121	
C	Flexidor 125	Dow AgroSciences Ltd	05104	
C	Gallery 125	Rigby Taylor Ltd	06889	
C	Knot Out	Vitax Ltd	05163	

253 Isoxaben + Isoproturon
C	IPSO	Dow AgroSciences Ltd	05736	

254 Isoxaben + Methabenzthiazuron
C	Glytex	Bayer plc	04230	

255 Isoxaben + Terbuthylazine
C	Skirmish	Novartis Crop Protection UK Ltd	08444	

256 Isoxaben + Trifluralin
C	Axit GR	Dow AgroSciences Ltd	08892	
C	Premiere Granules	Dow AgroSciences Ltd	07987	

257 Lenacil
C	Agricola Lenacil FL	Agricola Ltd	09481	
C	Clayton Lenacil 80W	Clayton Plant Protection (UK) Ltd	09488	
C	Cleancrop Lenflow	United Agri Products Limited	10059	
C	Venzar 80 WP	Du Pont (UK) Ltd	09981	
C	Venzar Flowable	Du Pont (UK) Ltd	06907	
C	Vizor	Du Pont (UK) Ltd	06572	

258 Lenacil + Chloridazon
C	Advizor	Du Pont (UK) Ltd	06571	
C	Advizor S	Du Pont (UK) Ltd	06960	31/10/2002
C	Advizor S	Griffin (Europe) Marketing NV (UK Branch)	09799	
C	Varmint	Pan Britannica Industries Ltd	01561	31/12/2001

C   These products are "approved for agricultural use". For further details refer to page vii.
A   These products are approved for use in or near water. For further details refer to page vii.

Product Name	Marketing Company	Reg. No.	Expiry Date

259 Lenacil + Linuron

C	Lanslide	Pan Britannica Industries Ltd	01184	31/12/2001

260 Lenacil + Phenmedipham

C	Agricola Lens	Agricola Lens	10257	
C	DUK-880	Du Pont (UK) Ltd	04121	
C	Stefes 880	Stefes Plant Protection Ltd	08914	31/03/2002
C	Stefes Excel	Stefes Plant Protection Ltd	08913	31/03/2002

261 Linuron

C	Afalon	AgrEvo UK Crop Protection Ltd	07299	30/04/2002
C	Afalon	Aventis CropScience UK Limited	08186	
C	Alpha Linuron 50 SC	Makhteshim-Agan (UK) Ltd	06967	
C	Alpha Linuron 50 WP	Makhteshim-Agan (UK) Ltd	04870	
C	Ashlade Linuron FL	Ashlade Formulations Ltd	06221	
C	DAPT Linuron 50 SC	DAPT Agrochemicals Ltd	08058	
C	Linurex 50SC	Makhteshim-Agan (UK) Ltd	07950	
C	Linuron Flowable	Pan Britannica Industries Ltd	02965	
C	Linuron Flowable	SumiAgro (UK) Ltd	09602	
C	Liquid Linuron	Pan Britannica Industries Ltd	01556	31/12/2001
C	MSS Linuron 500	Mirfield Sales Services Ltd	08893	
C	Stefes Linuron	Stefes Plant Protection Ltd	08187	31/03/2002
C	Stefes Linuron	Stefes Plant Protection Ltd	07902	31/03/2002
C	UPL Linuron 45% Flowable	United Phosphorus Ltd	07435	

262 Linuron + Chlorpropham

C	Profalon	AgrEvo UK Crop Protection Ltd	07331	31/01/2001
C	Profalon	Hortichem Ltd	08900	

263 Linuron + 2,4-DB + MCPA

C	Alistell	Zeneca Crop Protection	06515	

264 Linuron + Lenacil

C	Lanslide	Pan Britannica Industries Ltd	01184	31/12/2001

265 Linuron + Trifluralin

C	Chandor	Dow AgroSciences Ltd	05631	
C	Linnet	Pan Britannica Industries Ltd	01555	
C	Linnet	SumiAgro (UK) Ltd	09601	
C	Neminfest	Montedison UK Ltd	02546	
C	Neminfest	Sipcam UK Ltd	07219	

266 Maleic hydrazide

C	Bos MH 180	Uniroyal Chemical Ltd	06502	

C   These products are "approved for agricultural use". For further details refer to page vii.
A   These products are approved for use in or near water. For further details refer to page vii.

Product Name	Marketing Company	Reg. No.	Expiry Date

266 Maleic hydrazide—continued

C	Fazor	DowElanco Ltd	05558	
C	Fazor	Uniroyal Chemical Ltd	05461	
C	Mazide 25	Synchemicals Ltd	02067	
C A	Regulox K	Aventis Environmental Science	09937	
C A	Regulox K	Rhone-Poulenc Amenity	05405	
C	Royal MH 180	Mirfield Sales Services Ltd	07840	
C	Royal MH180	Uniroyal Chemical Ltd	07043	
C	Source II	Chiltern Farm Chemicals Ltd	08314	

267 Maleic hydrazide + Dicamba + MCPA

C	Mazide Selective	Vitax Ltd	05753	

268 MCPA

C	Agrichem MCPA 50	Agrichem Ltd	04097	
C	Agricorn 500	Farmers Crop Chemicals Ltd	00055	
C	Agricorn 500 II	Farmers Crop Chemicals Ltd	09155	
C	Agritox 50	Nufarm UK Ltd	07400	
C	Agritox D	Nufarm UK Ltd	07061	31/03/2002
C	Agroxone	Headland Agrochemicals Ltd	09947	
C	Agroxone 40	A H Marks & Co Ltd	10264	
C	AGROXONE 50	Mirfield Sales Services Ltd	08345	
C	Agroxone 75	A H Marks & Co Ltd	09208	
C	Atlas MCPA	Atlas Crop Protection Ltd	07717	
C	Atlas MCPA	Atlas Interlates Ltd	03055	
C	Barclay Meadowman	Barclay Chemicals Manufacturing Ltd	07639	
C	Barclay Meadowman II	Barclay Plant Protection	08525	
C	BASF MCPA Amine 50	BASF plc	00209	
C	BH MCPA 75	Rhone-Poulenc Amenity	05395	
C	Campbell's MCPA 50	J D Campbell & Sons Ltd	00381	
C	Circium II	Nufarm Whyte Agriculture Ltd	09866	
C	Cirsium	Mirfield Sales Services Ltd	08729	31/10/2002
C	Empal	Universal Crop Protection Ltd	00795	
C	FCC Agricorn 50M	Farmers Crop Chemicals Ltd	09032	
C	Headland Spear	Headland Agrochemicals Ltd	07115	
C	HY-MCPA	Agrichem Ltd	06293	
C	Luxan MCPA 500	Luxan (UK) Ltd	07470	
C	Marks MCPA 50A	A H Marks & Co Ltd	02710	28/02/2003
C	Marks MCPA P30	A H Marks & Co Ltd	01292	
C	Marks MCPA S25	A H Marks & Co Ltd	01294	
C	Marks MCPA SP	A H Marks & Co Ltd	01293	
C	MCPA 500	Nufarm UK Ltd	08655	
C	MSS MCPA 50	Mirfield Sales Services Ltd	01412	25/07/2003
C	MSS MCPA 50	Nufarm Whyte Agriculture Ltd	10240	

C These products are "approved for agricultural use". For further details refer to page vii.
A These products are approved for use in or near water. For further details refer to page vii.

Product Name	Marketing Company	Reg. No.	Expiry Date

268 MCPA—continued

C	Nufarm MCPA DMA 480	Nufarm UK Ltd	09686	
C	Nufarm MCPA DMA 500	Nufarm UK Ltd	09685	
C	Quadrangle MCPA 50	Quadrangle Agrochemicals	06935	
C	Sanaphen M	Dow AgroSciences Ltd	09238	31/10/2002
C	Sanaphen M	Sanachem International Ltd	08728	30/09/2001
C	Stefes Phenoxylene 50	Stefes Plant Protection Ltd	07612	
C	Tripart MCPA 50	Tripart Farm Chemicals Ltd	02206	
C	Vitax MCPA Ester	Vitax Ltd	07353	31/05/2001

269 MCPA + Bentazone + MCPB

C	Acumen	BASF plc	00028
C	Headland Archer	Headland Agrochemicals Ltd	08814

270 MCPA + Bifenox + Mecoprop-P

C	Sirocco	Aventis Environmental Science	09939
C	Sirocco	Rhone-Poulenc Amenity	09645

271 MCPA + Bromoxynil + Dichlorprop + Ioxynil

C	Atlas Minerva	Atlas Interlates Ltd	03046

272 MCPA + Clopyralid + Dichlorprop

C	Lontrel Plus	Dow AgroSciences Ltd	05269

273 MCPA + Clopyralid + Diflufenican

C	Spearhead	Aventis Environmental Science	09941
C	Spearhead	Rhone-Poulenc Amenity	07342

274 MCPA + Clopyralid + Fluroxypyr

C	Greenor	Rigby Taylor Ltd	07848

275 MCPA + 2,4-D

C	Agroxone Combi	Headland Agrochemicals Ltd	10277
C	Headland Polo	Headland Agrochemicals Ltd	10283

276 MCPA + 2,4-D + Clopyralid

C	Lonpar	Dow AgroSciences Ltd	08686

277 MCPA + 2,4-D + Dichlorprop + Mecoprop

C	Camppex	United Phosphorus Ltd	08266	31/10/2001

278 MCPA + 2,4-D + Dichlorprop + Mecoprop-P

C	UPL Camppex	United Phosphorus Ltd	09121

C These products are "approved for agricultural use". For further details refer to page vii.
A These products are approved for use in or near water. For further details refer to page vii.

Product Name	Marketing Company	Reg. No.	Expiry Date

279 MCPA + 2,4-DB

C	Agrichem DB Plus	Agrichem Ltd	00044	
C	Headland Cedar	Headland Agrochemicals Ltd	09948	
C	Marks 2,4-DB Extra	A H Marks & Co Ltd	01284	
C	MSS 2,4-DB + MCPA	Mirfield Sales Services Ltd	01392	
C	Redlegor	United Phosphorus Ltd	07519	

280 MCPA + 2,4-DB + Benazolin

C	Legumex Extra	AgrEvo UK Ltd	07901	31/05/2001
C	Legumex Extra	Aventis CropScience UK Limited	08676	
C	Master Sward	Stefes Plant Protection Ltd	07883	31/03/2002
C	Setter 33	Aventis CropScience UK Limited	07282	
C	Setter 33	Dow AgroSciences Ltd	05623	
C	Stefes Clover Ley	Stefes Plant Protection Ltd	07842	31/03/2002
C	Stefes Legumex Extra	Aventis CropScience UK Limited	10085	
C	Stefes Legumex Extra	Stefes Plant Protection Ltd	07841	

281 MCPA + 2,4-DB + Linuron

C	Alistell	Zeneca Crop Protection	06515	

282 MCPA + Dicamba

C	Banvel M	Novartis Crop Protection UK Ltd	08469	

283 MCPA + Dicamba + Dichlorprop

C	Cleanrun 2	The Scotts Company (UK) Limited	09271	
C	Intrepid	Miracle Professional	07819	
C	Intrepid	The Scotts Company (UK) Limited	09266	
C	Longlife Clearun 2	Miracle Professional	07851	30/06/2003

284 MCPA + Dicamba + Dichlorprop + Ferrous sulphate

C	Longlife Renovator 2	Miracle Professional	08379	
C	Renovator 2	The Scotts Company (UK) Limited	09272	

285 MCPA + Dicamba + Maleic hydrazide

C	Mazide Selective	Vitax Ltd	05753	

286 MCPA + Dicamba + Mecoprop

C	Barclay Hat-Trick	Barclay Chemicals Manufacturing Ltd	07579	28/02/2002
C	Hyprone	Agrichem (International) Ltd	01093	30/06/2001
C	Hysward	Agrichem (International) Ltd	01096	31/05/2001
C	Premier Triple Selective	Amenity Land Services Limited	08529	31/05/2001

C These products are "approved for agricultural use". For further details refer to page vii.
A These products are approved for use in or near water. For further details refer to page vii.

Product Name	Marketing Company	Reg. No.	Expiry Date

287 MCPA + Dicamba + Mecoprop-P

C	ALS Premier Selective Plus	Amenity Land Services Limited	08940	
C	Banlene Super	AgrEvo UK Crop Protection Ltd	07168	30/04/2003
C	Banlene Super	Aventis CropScience UK Limited	10053	
C	Field Marshal	MTM Agrochemicals Ltd	06077	31/03/2001
C	Field Marshal	United Phosphorus Ltd	08956	
C	Headland Relay P	Headland Agrochemicals Ltd	08580	
C	Headland Relay Turf	Headland Agrochemicals Ltd	08935	
C	Herrisol New	Bayer plc	07166	
C	Hycamba Plus	Agrichem (International) Ltd	10180	
C	Hyprone P	Agrichem (International) Ltd	09125	
C	Hysward P	Agrichem (International) Ltd	09052	
C	Mascot Super Selective-P	Rigby Taylor Ltd	06106	
C	Mircam Plus	Nufarm Whyte Agriculture Ltd	09884	
C	MSS Mircam Plus	Mirfield Sales Services Ltd	01416	
C	MTM Grassland Herbicide	MTM Agrochemicals Ltd	06089	28/02/2001
C	Pasturol D	Farmers Crop Chemicals Ltd	07033	
C	Pasturol Plus	Farmers Crop Chemicals Ltd	08581	30/09/2003
C	Pasturol Plus	Headland Agrochemicals Ltd	10278	
C	Pasturol-P	Farmers Crop Chemicals Ltd	09099	
C	Premier Amenity Selective	Amenity Land Services Limited	10098	
C	Premier Triple Selective P	Amenity Land Services Limited	09053	
C	Quadrangle Quadban	Quadrangle Agrochemicals	07090	
C	Stefes Banlene Super	Aventis CropScience UK Limited	10089	
C	Stefes Banlene Super	Stefes Plant Protection Ltd	07691	
C	Stefes Docklene Super	AgrEvo UK Crop Protection Ltd	09873	
C	Stefes Docklene Super	Stefes Plant Protection Ltd	07696	31/01/2003
C	Stefes Swelland Super	Stefes Plant Protection Ltd	08189	31/03/2002
C	Tribute	Nomix-Chipman Ltd	06921	
C	Tribute Plus	Nomix-Chipman Ltd	09493	
C	Tritox	The Scotts Company (UK) Limited + Levington Horticulture Ltd	07764	
C	UPL Grassland Herbicide	United Phosphorus Ltd	08934	

288 MCPA + Dichlorprop

C	MSS 2,4-DP + MCPA	Mirfield Sales Services Ltd	01396	
C	Redipon Extra	United Phosphorus Ltd	07518	
C	Seritox 50	Aventis CropScience UK Limited	10016	
C	Seritox 50	Rhone-Poulenc Agriculture	06170	
C	Seritox Turf	Rhone-Poulenc Environmental Products	06802	
C	SHL Granular Feed & Weed	William Sinclair Horticulture Ltd	09926	
C	SHL Turf Feed & Weed	William Sinclair Horticulture Ltd	04437	

C These products are "approved for agricultural use". For further details refer to page vii.
A These products are approved for use in or near water. For further details refer to page vii.

Product Name	Marketing Company	Reg. No.	Expiry Date

289 MCPA + Dichlorprop + Ferrous sulphate

C	SHL Granular Feed, Weed & Mosskiller	William Sinclair Horticulture Ltd	09925
C	SHL Turf Feed and Weed and Mosskiller	William Sinclair Horticulture Ltd	04438
C	Vitagrow Mini-Granular Feed, Weed & Mosskiller	William Sinclair Horticulture Ltd	10295

290 MCPA + Dichlorprop-P + Mecoprop-P

C	Hymec Trio	Agrichem (International) Ltd	09764	28/02/2003
C	Hymec Triple	Agrichem (International) Ltd	09949
C	Optica Trio	A H Marks & Co Ltd	09747

291 MCPA + MCPB

C	Bellmac Plus	United Phosphorus Ltd	07521
C	MSS MCPB + MCPA	Mirfield Sales Services Ltd	01413
C	Trifolex-Tra	Cyanamid Agriculture Ltd	07147
C	Tropotox Plus	Aventis CropScience UK Limited	10028
C	Tropotox Plus	Rhone-Poulenc Agriculture	06156

292 MCPA + Mecoprop-P

C	Greenmaster Extra	The Scotts Company (UK) Limited + Levington Horticulture Ltd	07594
C	Optica Combi	A H Marks & Co Ltd	10118

293 MCPB

C	Bellmac Straight	United Phosphorus Ltd	07522
C	Marks MCPB	A H Marks & Co Ltd	03497
C	Tropotox	Aventis CropScience UK Limited	10027
C	Tropotox	Rhone-Poulenc Agriculture	06179

294 MCPB + Bentazone

C	Pulsar	BASF plc	04002

295 MCPB + Bentazone + MCPA

C	Acumen	BASF plc	00028
C	Headland Archer	Headland Agrochemicals Ltd	08814

296 MCPB + MCPA

C	Bellmac Plus	United Phosphorus Ltd	07521
C	MSS MCPB + MCPA	Mirfield Sales Services Ltd	01413
C	Trifolex-Tra	Cyanamid Agriculture Ltd	07147
C	Tropotox Plus	Aventis CropScience UK Limited	10028
C	Tropotox Plus	Rhone-Poulenc Agriculture	06156

C These products are "approved for agricultural use". For further details refer to page vii.
A These products are approved for use in or near water. For further details refer to page vii.

Product Name	Marketing Company	Reg. No.	Expiry Date

297 Mecoprop

C	Barclay Mecrop	Barclay Chemicals Manufacturing Ltd	07578	28/02/2002
C	Campbell's CMPP	J D Campbell & Sons Ltd	02918	31/03/2001
C	Clenecorn	Farmers Crop Chemicals Ltd	00542	31/03/2001
C	Hymec	Agrichem Ltd	01091	30/06/2001

298 Mecoprop + 2,4-D

C	Supertox 30	Rhone-Poulenc Amenity	05340	30/06/2001

299 Mecoprop + 2,4-D + Dichlorprop + MCPA

C	Camppex	United Phosphorus Ltd	08266	31/10/2001

300 Mecoprop + Dicamba

C	Condox	Zeneca Crop Protection	06519	31/05/2002
C	Hyban	Agrichem Ltd	01084	30/06/2001
C	Hygrass	Agrichem (International) Ltd	01090	30/06/2001

301 Mecoprop + Dicamba + MCPA

C	Barclay Hat-Trick	Barclay Chemicals Manufacturing Ltd	07579	28/02/2002
C	Hyprone	Agrichem (International) Ltd	01093	30/06/2001
C	Hysward	Agrichem (International) Ltd	01096	31/05/2001
C	Premier Triple Selective	Amenity Land Services Limited	08529	31/05/2001

302 Mecoprop-P

C	Astix	Aventis CropScience UK Limited	10038	
C	Astix	Rhône-Poulenc Agriculture	06174	
C	Astix K	Aventis CropScience UK Limited	10039	
C	Astix K	Rhône-Poulenc Agriculture	06904	
C	Barclay Melody	Barclay Chemicals Manufacturing Ltd	08790	31/03/2002
C	BAS 03729H	Mirfield Sales Services Ltd	05912	
C	Clenecorn Super	Farmers Crop Chemicals Ltd	09818	
C	Clenecorn Super	Farmers Crop Chemicals Ltd	09382	31/03/2002
C	Clovotox	Aventis Environmental Science	09928	
C	Clovotox	Rhone-Poulenc Amenity	05354	
C	Compitox Extra P	Nufarm UK Ltd	09558	30/04/2003
C	Compitox Plus	Nufarm UK Ltd	10077	
C	Compitox Plus	Nufarm UK Ltd	07930	31/03/2002
C	Compitox Plus	Nufarm UK Ltd	08997	31/12/2001
C	Duplosan	BASF plc	05889	28/02/2002
C	Duplosan 500	BASF plc	07889	
C	Duplosan KV	BASF plc	09431	

C   These products are "approved for agricultural use". For further details refer to page vii.
A   These products are approved for use in or near water. For further details refer to page vii.

Product Name	Marketing Company	Reg. No.	Expiry Date

302 Mecoprop-P—continued

C	Duplosan KV 500	BASF plc	08027	
C	Duplosan New System CMPP	BASF plc	04481	
C	Isomec	Nufarm Whyte Agriculture Ltd	09881	
C	Landgold Mecoprop-P	Landgold & Co Ltd	06052	
C	Landgold Mecoprop-P 600	Landgold & Co Ltd	09014	
C	MSS Mirprop	Mirfield Sales Services Ltd	05911	
C	MSS Optica	Mirfield Sales Services Ltd	04973	
C	Optica	A H Marks & Co Ltd	05814	
C	Optica	Bayer UK Ltd	04922	31/12/2001
C	Optica	Headland Agrochemicals Ltd	09963	
C	Optica 50	A H Marks & Co Ltd	08283	

303 Mecoprop-P + Bifenox + MCPA

C	Sirocco	Aventis Environmental Science	09939
C	Sirocco	Rhone-Poulenc Amenity	09645

304 Mecoprop-P + Bromoxynil + Ioxynil

C	Swipe P	Novartis Crop Protection UK Ltd	08452

305 Mecoprop-P + Carfentrazone-ethyl

C	Platform S	FMC Corporation (UK) Ltd	09706	
C	Platform S	FMC Europe NV	08638	30/11/2002

306 Mecoprop-P + 2,4-D

C	Mascot Selective-P	Rigby Taylor Ltd	06105	
C	Supertox 30	Aventis Environmental Science	09946	
C	Supertox 30	Rhone-Poulenc Amenity	09102	28/02/2003
C	Sydex	Vitax Ltd	06412	

307 Mecoprop-P + 2,4-D + Dichlorprop + MCPA

C	UPL Camppex	United Phosphorus Ltd	09121

308 Mecoprop-P + Dicamba

C	Camber	Headland Agrochemicals Ltd	09901
C	Condox	Zeneca Crop Protection	09429
C	Di-Farmon R	Headland Agrochemicals Ltd	08472
C	Dockmaster	Nufarm Whyte Agriculture Ltd	10033
C	Foundation	Novartis Crop Protection UK Ltd	08475
C	Hyban P	Agrichem (International) Ltd	09129
C	Hygrass P	Agrichem (International) Ltd	09130
C	Mircam	Nufarm Whyte Agriculture Ltd	09888
C	MSS Mircam	Mirfield Sales Services Ltd	01415
C	Optica Forte	A H Marks & Co Ltd	07432

C   These products are "approved for agricultural use". For further details refer to page vii.
A   These products are approved for use in or near water. For further details refer to page vii.

Product Name	Marketing Company	Reg. No.	Expiry Date

309 Mecoprop-P + Dicamba + MCPA

	Product Name	Marketing Company	Reg. No.	Expiry Date
C	ALS Premier Selective Plus	Amenity Land Services Limited	08940	
C	Banlene Super	AgrEvo UK Crop Protection Ltd	07168	30/04/2003
C	Banlene Super	Aventis CropScience UK Limited	10053	
C	Field Marshal	MTM Agrochemicals Ltd	06077	31/03/2001
C	Field Marshal	United Phosphorus Ltd	08956	
C	Headland Relay P	Headland Agrochemicals Ltd	08580	
C	Headland Relay Turf	Headland Agrochemicals Ltd	08935	
C	Herrisol New	Bayer plc	07166	
C	Hycamba Plus	Agrichem (International) Ltd	10180	
C	Hyprone P	Agrichem (International) Ltd	09125	
C	Hysward P	Agrichem (International) Ltd	09052	
C	Mascot Super Selective-P	Rigby Taylor Ltd	06106	
C	Mircam Plus	Nufarm Whyte Agriculture Ltd	09884	
C	MSS Mircam Plus	Mirfield Sales Services Ltd	01416	
C	MTM Grassland Herbicide	MTM Agrochemicals Ltd	06089	28/02/2001
C	Pasturol D	Farmers Crop Chemicals Ltd	07033	
C	Pasturol Plus	Farmers Crop Chemicals Ltd	08581	30/09/2003
C	Pasturol Plus	Headland Agrochemicals Ltd	10278	
C	Pasturol-P	Farmers Crop Chemicals Ltd	09099	
C	Premier Amenity Selective	Amenity Land Services Limited	10098	
C	Premier Triple Selective P	Amenity Land Services Limited	09053	
C	Quadrangle Quadban	Quadrangle Agrochemicals	07090	
C	Stefes Banlene Super	Aventis CropScience UK Limited	10089	
C	Stefes Banlene Super	Stefes Plant Protection Ltd	07691	
C	Stefes Docklene Super	AgrEvo UK Crop Protection Ltd	09873	
C	Stefes Docklene Super	Stefes Plant Protection Ltd	07696	31/01/2003
C	Stefes Swelland Super	Stefes Plant Protection Ltd	08189	31/03/2002
C	Tribute	Nomix-Chipman Ltd	06921	
C	Tribute Plus	Nomix-Chipman Ltd	09493	
C	Tritox	The Scotts Company (UK) Limited + Levington Horticulture Ltd	07764	
C	UPL Grassland Herbicide	United Phosphorus Ltd	08934	

310 Mecoprop-P + Dichlorprop-P + MCPA

	Product Name	Marketing Company	Reg. No.	Expiry Date
C	Hymec Trio	Agrichem (International) Ltd	09764	28/02/2003
C	Hymec Triple	Agrichem (International) Ltd	09949	
C	Optica Trio	A H Marks & Co Ltd	09747	

311 Mecoprop-P + Fluroxypyr

	Product Name	Marketing Company	Reg. No.	Expiry Date
C	Bastion T	Rigby Taylor Ltd	06011	

C These products are "approved for agricultural use". For further details refer to page vii.
A These products are approved for use in or near water. For further details refer to page vii.

Product Name	Marketing Company	Reg. No.	Expiry Date

312 Mecoprop-P + MCPA

C	Greenmaster Extra	The Scotts Company (UK) Limited + Levington Horticulture Ltd	07594	
C	Optica Combi	A H Marks & Co Ltd	10118	

313 Mecoprop-P + Metribuzin

C	Bayer UK 683	Bayer plc	08113	31/12/2001
C	Centra	Bayer plc	08112	
C	Optica Plus	A H Marks & Co Ltd	09325	

314 Mecoprop-P + Metsulfuron-methyl

C	Headland Neptune	Headland Agrochemicals Ltd	10230	

315 Mecoprop-P + Triasulfuron

C	Raven	Novartis Crop Protection UK Ltd	08434	

316 Mefluidide

C	Check Turf II	Certified Laboratories Ltd	04463	
C	Embark	Gordon International Corp	04810	
C	Embark Lite	Intracrop	08749	
C	Gro-Tard II	Chemsearch	04462	

317 Mepiquat + Chlormequat

C	Cyter	BASF plc	09133	
C	Stronghold	BASF plc	09134	

318 Mepiquat + 2-Chloroethylphosphonic acid

C	Barclay Banshee	Barclay Chemicals Manufacturing Ltd	08175	
C	CleanCrop Fonic M	United Agri Products	09553	
C	Marnoch Phonet	Marnoch Ventures Limited	09979	
C	Standon Mepiquat Plus	Standon Chemicals Ltd	09373	
C	Stefes Mepiquat	Stefes Plant Protection Ltd	06970	20/04/2001
C	Sypex M	BASF plc	06810	30/06/2001
C	Terpal	BASF plc	02103	
C	Terpal	Clayton Plant Protection (UK) Ltd	07626	
C	Terpitz	Me^2 Crop Protection Ltd	09634	

319 Mepiquat + 2-Chloroethylphosphonic acid + Chlormequat

C	Cyclade	BASF plc	08958	

320 Metamitron

C	Agriguard Metamitron	Agriguard Ltd	09859	

C These products are "approved for agricultural use". For further details refer to page vii.
A These products are approved for use in or near water. For further details refer to page vii.

Product Name	Marketing Company	Reg. No.	Expiry Date

320 Metamitron—continued

C	Barclay Seismic	Barclay Chemicals Manufacturing Ltd	09323	
C	Barclay Seismic XL	Barclay Chemicals Manufacturing Ltd	09328	
C	Goltix 90	Bayer plc	08654	
C	Goltix Flowable	Bayer plc	08986	
C	Goltix WG	Bayer plc	02430	
C	Homer	Griffin (Europe) Marketing NV (UK Branch)	09834	
C	Landgold Metamitron	Landgold & Co Ltd	06287	
C	Lektan	Bayer plc	06111	
C	Marquise	Makhteshim-Agan (UK) Ltd	08738	
C	Mitron WG	Hermoo Belgium NV	10171	
C	MM70	Gharda Chemicals Ltd	09490	
C	Quartz	Quadrangle Agrochemicals	05167	
C	Skater	Feinchemie (UK) Ltd	09790	
C	Skater	Feinchemie Schwebda GmbH	08623	31/10/2002
C	Standon Metamitron	Standon Chemicals Ltd	07885	
C	Stefes 7G	Stefes Plant Protection Ltd	06350	31/03/2002
C	Stefes Metamitron	Stefes Plant Protection Ltd	05821	31/03/2002
C	Tripart Accendo	Tripart Farm Chemicals Ltd	06110	
C	Valiant	Stefes UK Ltd	09251	30/06/2002
C	Volcan	Sipcam UK Ltd	09295	

321 Metamitron + Chloridazon

C	Volcan Combi	Sipcam UK Ltd	10256

322 Metamitron + Ethofumesate

C	Stefes Duplex	Stefes UK Ltd	09252	31/12/2001
C	Stefes ST/STE	Aventis CropScience UK Limited	10109	
C	Stefes ST/STE	Stefes Plant Protection Ltd	09111	

323 Metamitron + Ethofumesate + Phenmedipham

C	Bayer UK 540	Bayer plc	07399	
C	Betanal Trio OF	AgrEvo UK Crop Protection Ltd	08680	31/05/2001
C	Betanal Trio WG	Aventis CropScience UK Limited	07537	

324 Metazachlor

C	Barclay Metaza	Barclay Chemicals Manufacturing Ltd	09169
C	Butisan S	BASF plc	00357
C	Clayton Metazachlor	Clayton Plant Protection (UK) Ltd	09688
C	CleanCrop MTZ 500	United Agri Products	09222

C These products are "approved for agricultural use". For further details refer to page vii.
A These products are approved for use in or near water. For further details refer to page vii.

Product Name	Marketing Company	Reg. No.	Expiry Date

324 Metazachlor—continued

C Landgold Metazachlor 50	Landgold & Co Ltd	09726	
C Marnoch Metazachlor	Marnoch Ventures Limited	09761	

325 Metazachlor + Quinmerac

C Katamaran	BASF plc	09049	
C Standon Metazachlor-Q	Standon Chemicals Ltd	09676	

326 Methabenzthiazuron

C Clayton Benson	Clayton Plant Protection (UK) Ltd	08687	
C Tribunil	Bayer plc	02169	
C Tribunil WG	Bayer plc	03260	

327 Methabenzthiazuron + Isoxaben

C Glytex	Bayer plc	04230	

328 Metosulam

C EF 1077	Dow AgroSciences Ltd	07965	

329 Metosulam + Fluroxypyr

C EF 1166	Dow AgroSciences Ltd	07966	

330 Metoxuron

C Deftor	Novartis Crop Protection UK Ltd	08471	
C Dosaflo	Novartis Crop Protection UK Ltd	09351	
C Dosaflo	Novartis Crop Protection UK Ltd	08473	28/02/2002
C Endspray II	Pan Britannica Industries Ltd	05079	31/12/2001

331 Metribuzin

C Agriguard Metribuzin	Agriguard Ltd	09853	
C Bayer UK 093	Bayer plc	07903	
C Citation	IT Agro Ltd	08685	30/11/2001
C Citation 70	United Phosphorus Ltd	09370	
C CleanCrop Metribuzin	United Agri Products	09621	
C Inter-Metribuzin WG	Agriguard Ltd	09336	31/10/2002
C Inter-Metribuzin WG	IT Agro Ltd	09801	
C Lexone 70 DF	Du Pont (UK) Ltd	04991	
C Python	Feinchemie (UK) Ltd	09791	
C Python	Headland Agrochemicals Ltd	09243	31/10/2002
C Sencorex WG	Bayer plc	03755	

332 Metribuzin + Amidosulfuron

C Bayer UK 590	Bayer plc	08149	30/06/2001

C These products are "approved for agricultural use". For further details refer to page vii.
A These products are approved for use in or near water. For further details refer to page vii.

Product Name	Marketing Company	Reg. No.	Expiry Date

332 Metribuzin + Amidosulfuron—continued

C	Galis	Bayer plc	08039	30/06/2001

333 Metribuzin + Mecoprop-P

C	Bayer UK 683	Bayer plc	08113	31/12/2001
C	Centra	Bayer plc	08112	
C	Optica Plus	A H Marks&Co Ltd	09325	

334 Metsulfuron-methyl

C	Agriguard Metsulfuron	Tronsan Ltd	09569
C	Ally	Du Pont (UK) Ltd	02977
C	Ally WSB	Du Pont (UK) Ltd	06588
C	Barclay Flumen	Barclay Chemicals Manufacturing Ltd	08752
C	Clayton Metsulfuron	Clayton Plant Protection (UK) Ltd	06734
C	Gem 690	Griffin (Europe) HQ NV	09436
C	Jubilee	Du Pont (UK) Ltd	06082
C	Jubilee 20DF	Du Pont (UK) Ltd	06136
C	Landgold Metsulfuron	Landgold&Co Ltd	06280
C	Lorate 20DF	Du Pont (UK) Ltd	06135
C	Simba 20 DF	Griffin (Europe) Marketing NV (UK Branch)	09437
C	Standon Metsulfuron	Standon Chemicals Ltd	05670

335 Metsulfuron-methyl + Carfentrazone-ethyl

C	Ally Express	Du Pont (UK) Ltd	08640

336 Metsulfuron-methyl + Flupyrsulfuron-methyl

C	Lexus XPE	Du Pont (UK) Ltd	08541
C	Lexus XPE WSB	Du Pont (UK) Ltd	08542
C	Standon Flupyrsulfuron MM	Standon Chemicals Ltd	09098

337 Metsulfuron-methyl + Mecoprop-P

C	Headland Neptune	Headland Agrochemicals Ltd	10230

338 Metsulfuron-methyl + Thifensulfuron-methyl

C	DP 928	Du Pont (UK) Ltd	09632
C	Harmony M	Du Pont (UK) Ltd	03990

339 Metsulfuron-methyl + Tribenuron-methyl

C	DP 911 WSB	Du Pont (UK) Ltd	09867

C   These products are "approved for agricultural use". For further details refer to page vii.
A   These products are approved for use in or near water. For further details refer to page vii.

Product Name	Marketing Company	Reg. No.	Expiry Date

340 Monolinuron

C	Arresin	Aventis CropScience UK Limited	07303	09/09/2001

341 1-Naphthylacetic acid

C	Rhizopon B Powder (0.1%)	Fargro Ltd	07133	
C	Rhizopon B Powder (0.1%)	Rhizopon UK Limited	09089	
C	Rhizopon B Powder (0.2 %)	Rhizopon UK Limited	09090	
C	Rhizopon B Powder (0.2%)	Fargro Ltd	07134	
C	Rhizopon B Tablets	Fargro Ltd	07135	
C	Rhizopon B Tablets	Rhizopon UK Limited	09091	
C	Tipoff	Universal Crop Protection Ltd	05878	

342 1-Naphthylacetic acid + 4-Indol-3-ylbutyric acid

C	Synergol	Hortichem Ltd	07386	

343 2-Naphthyloxyacetic Acid

C	Betapal Concentrate	Synchemicals Ltd	00234	

344 Napropamide

C	Agriguard Napropamide	Agriguard Ltd	09223	30/04/2002
C	Banweed	United Phosphorus Ltd	09376	
C	Banweed	Zeneca Crop Protection	06672	31/12/2001
C	Devrinol	Rhone-Poulenc Agriculture	06195	31/12/2001
C	Devrinol	United Phosphorus Ltd	09375	
C	Devrinol	United Phosphorus Ltd	09374	
C	Devrinol	Zeneca Crop Protection	06653	31/12/2001

345 Oxadiazon

C	Ronstar 2G	Aventis CropScience UK Limited	10011	
C	Ronstar 2G	Rhone-Poulenc Agriculture	06492	31/03/2003
C	Ronstar Liquid	Hortichem Ltd	08974	
C	Ronstar Liquid	Rhone-Poulenc Agriculture	06766	30/09/2001

346 Oxadiazon + Carbetamide + Diflufenican

C	Helmsman	Aventis Environmental Science	09934	
C	Helmsman	Rhone-Poulenc Amenity	09696	25/07/2003

347 Oxadiazon + Glyphosate

C	Zapper	Aventis Environmental Science	09944	
C	Zapper	Rhone-Poulenc Amenity	08605	31/12/2003

348 Paclobutrazol

C	Bonzi	Zeneca Crop Protection	06640	

C   These products are "approved for agricultural use". For further details refer to page vii.
A   These products are approved for use in or near water. For further details refer to page vii.

Product Name	Marketing Company	Reg. No.	Expiry Date

348 Paclobutrazol—continued

C Cultar	Zeneca Crop Protection	06649	

349 Paclobutrazol + Dicamba

C Holdfast D	Imperial Chemical Industries Plc	05056	30/10/2001
C Holdfast D	Miracle Professional	07864	31/03/2001

350 Paraquat

C Agriguard Paraquat	Agriguard Ltd	09951	
C Barclay Total	Barclay Chemicals Manufacturing Ltd	08822	
C Dextrone X	Chipman Ltd	00687	
C Gramoxone 100	Zeneca Crop Protection	06674	
C Scythe LC	Cyanamid Agriculture Ltd	05877	30/06/2001
C Speedway	Miracle Professional	07743	30/10/2001
C Speedway Liquid	Imperial Chemical Industries Plc	04365	31/05/2001
C Speedway Liquid	Miracle Professional	07744	
C Speedway Liquid	Zeneca Professional Products	06825	31/05/2001

351 Paraquat + Diquat

C PDQ	Zeneca Crop Protection	10241	
C PDQ	Zeneca Crop Protection	06518	25/07/2003
C Speedway 2	Miracle Professional	08164	31/10/2001
C Speedway 2	The Scotts Company (UK) Limited	09273	

352 Paraquat + Diquat + Simazine

C Pathclear S	Miracle Garden Care Ltd	08991	
C Pathclear S	The Scotts Company (UK) Limited	09269	

353 Paraquat + Diuron

C Dexuron	Nomix-Chipman Ltd	07169	

354 Pendimethalin

C Barclay Tremor	Barclay Chemicals Manufacturing Ltd	09188	
C Blazer	Cheminova Agro (UK) Ltd	10258	
C Claymore	BASF plc	10214	
C Claymore	Cyanamid Agriculture Ltd	09232	
C Clayton Pendalin	Clayton Plant Protection (UK) Ltd	09708	
C Inter-Pendimethalin	IT Agro Ltd	09841	
C Ipimethalin	pbi Agrochemicals Ltd	09701	
C Landgold Pendimethalin 400	Landgold & Co Ltd	10288	
C Plinth	pbi Agrochemicals Ltd	09702	
C Sovereign	BASF plc	10228	

C These products are "approved for agricultural use". For further details refer to page vii.
A These products are approved for use in or near water. For further details refer to page vii.

Product Name	Marketing Company	Reg. No.	Expiry Date

354 Pendimethalin—continued

C	Sovereign	Cyanamid Agriculture Ltd	08151
C	Sovereign	Novartis Crop Protection UK Ltd	08533
C	Stomp 400 SC	BASF plc	10229
C	Stomp 400 SC	Cyanamid Agriculture Ltd	04183

355 Pendimethalin + Bentazone

C	Impuls	BASF plc	09720

356 Pendimethalin + Chlorotoluron

C	Totem	Cyanamid Agriculture Ltd	04670

357 Pendimethalin + Cyanazine

C	Activus	BASF plc	10207
C	Activus	Cyanamid Agriculture Ltd	09174
C	Bullet	BASF plc	10212
C	Bullet	Cyanamid Agriculture Ltd	08049

358 Pendimethalin + Isoproturon

C	Artillery	Cyanamid Agriculture Ltd	07768
C	Artillery SC	Cyanamid Agriculture Ltd	08504
C	Clayton Pendalin-IPU 37	Clayton Plant Protection (UK) Ltd	09154
C	Clayton Pendalin-IPU 47	Clayton Plant Protection (UK) Ltd	09170
C	Encore	Cyanamid Agriculture Ltd	04737
C	Encore SC	Cyanamid Agriculture Ltd	08502
C	Jolt	Cyanamid Agriculture Ltd	05488
C	Jolt SC	Cyanamid Agriculture Ltd	08503
C	Orient	Cyanamid Agriculture Ltd	07798
C	Orient SC	Cyanamid Agriculture Ltd	08501
C	Trump	Cyanamid Agriculture Ltd	03687
C	Trump SC	Cyanamid Agriculture Ltd	08500

359 Pendimethalin + Simazine

C	Deuce	Cyanamid Agriculture Ltd	06746
C	Merit	Cyanamid Agriculture Ltd	04976

360 Pentanochlor

C	Atlas Solan 40	Atlas Crop Protection Ltd	07726
C	Croptex Bronze	Hortichem Ltd	04087

361 Pentanochlor + Chlorpropham

C	Atlas Brown	Atlas Crop Protection Ltd	07703

C   These products are "approved for agricultural use". For further details refer to page vii.
A   These products are approved for use in or near water. For further details refer to page vii.

Product Name	Marketing Company	Reg. No.	Expiry Date

362 Phenmedipham

C Atlas Protrum K	Atlas Crop Protection Ltd	07723	
C Atlas Protrum K	Atlas Interlates Ltd	03089	
C Barclay Punter XL	Barclay Chemicals Manufacturing Ltd	08047	
C Beetomax	Fine Agrochemicals Ltd	03129	30/11/2001
C Beetup	United Phosphorus Ltd	07520	
C Betanal E	Aventis CropScience UK Limited	07248	
C Betanal Flo	Aventis CropScience UK Limited	08898	
C Betosip	Sipcam UK Ltd	06787	
C Betosip 114	Sipcam UK Ltd	05910	
C Cirrus	Quadrangle Ltd	06367	31/05/2001
C CleanCrop Phenmedipham	United Agri Products	09495	
C Dancer	Sipcam UK Ltd	10048	
C Griffin Phenmedipham 114	Griffin (Bermuda) SA	08343	28/02/2002
C Herbasan	KVK Agro A/S	08553	
C Hickson Phenmedipham	Griffin (Europe) Marketing NV (UK Branch)	09961	
C Hickson Phenmedipham	Hickson & Welch Ltd	02825	28/02/2003
C Kemifam E	AgrEvo UK Ltd	08978	
C Kemifam E	Kemira Cropcare Ltd	06104	
C Kemifam Flow	AgrEvo UK Ltd	07991	31/05/2001
C Luxan Phenmedipham	Luxan (UK) Ltd	06933	
C Mandolin	Griffin (Europe) Marketing NV (UK Branch)	09456	
C MSS Betaren Flow	KVK Agro A/S	08554	
C MSS Protrum G	Mirfield Sales Services Ltd	08342	
C Rubenal Flow	Aventis CropScience UK Limited	09227	
C Rubenal Flow	Pan Britannica Industries Ltd	07647	30/09/2001
C Stefes Forte	Stefes Plant Protection Ltd	06427	31/03/2002
C Stefes Forte 2	Aventis CropScience UK Limited	10083	
C Stefes Forte 2	Stefes Plant Protection Ltd	08204	
C Stefes Medipham 2	Aventis CropScience UK Limited	10087	
C Stefes Medipham 2	Stefes Plant Protection Ltd	08203	
C Tripart Beta	Tripart Farm Chemicals Ltd	03111	
C Tripart Beta S	Tripart Farm Chemicals Ltd	08202	30/06/2001

363 Phenmedipham + Desmedipham

C Betanal Compact	Aventis CropScience UK Limited	07247	
C Betanal Quorum	AgrEvo UK Ltd	07252	31/05/2001
C Betanal Rostrum	AgrEvo UK Ltd	07253	31/05/2001

364 Phenmedipham + Desmedipham + Ethofumesate

C Aventis Betanal Progress OF	Aventis CropScience UK Limited	09722	
C Betanal Congress	AgrEvo UK Ltd	07534	31/05/2001

C   These products are "approved for agricultural use". For further details refer to page vii.
A   These products are approved for use in or near water. For further details refer to page vii.

Product Name	Marketing Company	Reg. No.	Expiry Date

364 Phenmedipham + Desmedipham + Ethofumesate—continued

C	Betanal Progress	AgrEvo UK Ltd	07533	31/05/2001
C	Betanal Progress OF	Aventis CropScience UK Limited	07629	
C	Betanal Ultima	AgrEvo UK Ltd	07535	31/05/2001

365 Phenmedipham + Ethofumesate

C	Barclay Goalpost	Barclay Chemicals Manufacturing Ltd	09497	
C	Barclay Goalpost	Barclay Chemicals Manufacturing Ltd	09192	
C	Barclay Goldpost	Barclay Chem. Manufacturing Ltd	08964	31/03/2002
C	Betanal Montage	AgrEvo UK Ltd	07250	28/02/2001
C	Betanal Tandem	Aventis CropScience UK Limited	07254	
C	Betosip Combi	Sipcam UK Ltd	08630	
C	Stefes Medimat	Stefes Plant Protection Ltd	07886	30/04/2001
C	Stefes Medimat 2	Stefes Plant Protection Ltd	07577	31/03/2002
C	Stefes Tandem	Aventis CropScience UK Limited	10088	
C	Stefes Tandem	Stefes Plant Protection Ltd	08906	30/04/2003
C	Twin	Feinchemie (UK) Ltd	09875	
C	Twin	Feinchemie (UK) Ltd	09784	31/01/2003
C	Twin	Feinchemie Schwebda GmbH	07374	31/10/2002

366 Phenmedipham + Ethofumesate + Metamitron

C	Bayer UK 540	Bayer plc	07399	
C	Betanal Trio OF	AgrEvo UK Crop Protection Ltd	08680	31/05/2001
C	Betanal Trio WG	Aventis CropScience UK Limited	07537	

367 Phenmedipham + Lenacil

C	Agricola Lens	Agricola Lens	10257	
C	DUK-880	Du Pont (UK) Ltd	04121	
C	Stefes 880	Stefes Plant Protection Ltd	08914	31/03/2002
C	Stefes Excel	Stefes Plant Protection Ltd	08913	31/03/2002

368 Picloram

C	Tordon 22K	Dow AgroSciences Ltd	05083	
C	Tordon 22K	Nomix-Chipman Ltd	05790	

369 Picloram + Bromacil

C	Hydon	Chipman Ltd	01088	

370 Picloram + 2,4-D

C	Atladox HI	Nomix-Chipman Ltd	05559	
C	Tordon 101	Dow AgroSciences Ltd	05816	

C   These products are "approved for agricultural use". For further details refer to page vii.
A   These products are approved for use in or near water. For further details refer to page vii.

Product Name	Marketing Company	Reg. No.	Expiry Date

371 Prochloraz + Fluquinconazole

C	Jockey Plus	Aventis CropScience UK Limited	10075

372 Prometryn

C	Alpha Prometryne 50 WP	Makhteshim-Agan (UK) Ltd	04871
C	Alpha Prometryne 80 WP	Makhteshim-Agan (UK) Ltd	04795
C	Atlas Prometryn 50 WP	Atlas Interlates Ltd	03502
C	Gesagard	Novartis Crop Protection UK Ltd	08410

373 Prometryn + Terbutryn

C	P-Weed	Pan Britannica Industries Ltd	08574
C	P-Weed	SumiAgro (UK) Ltd	09608

374 Propachlor

C	Alpha Propachlor 50 SC	Makhteshim-Agan (UK) Ltd	04873
C	Alpha Propachlor 65 WP	Makhteshim-Agan (UK) Ltd	04807
C	Atlas Orange	Atlas Interlates Ltd	03096
C	Atlas Propachlor	Atlas Interlates Ltd	06462
C	Portman Brasson	Portman Agrochemicals Ltd	08158
C	Portman Propachlor 50 FL	Portman Agrochemicals Ltd	06892
C	Portman Propachlor SC	Portman Agrochemicals Ltd	08159
C	Ramrod 20 Granular	Hortichem Ltd	05806
C	Ramrod 20 Granular	Monsanto Plc	01687
C	Ramrod Flowable	Monsanto Plc	01688
C	Sentinel 2	Tripart Farm Chemicals Ltd	05140
C	Tripart Sentinel	Tripart Farm Chemicals Ltd	03250

375 Propachlor + Chloridazon

C	Ashlade CP	Ashlade Formulations Ltd	06481

376 Propachlor + Chlorthal-dimethyl

C	Decimate	ISK Biosciences Ltd	05626	28/02/2002

377 Propaquizafop

C	Barclay Rebel II	Barclay Chemicals Manufacturing Ltd	08897	31/03/2003
C	CleanCrop PropaQ	United Agri Products	09297	31/03/2003
C	Falcon	Cyanamid Agriculture Ltd	08288	31/03/2003
C	Falcon	Novartis Crop Protection UK Ltd	09384	
C	Landgold PQF 100	Landgold & Co Ltd	08976	31/03/2003
C	Shogun	Cyanamid Agriculture Ltd	08251	31/03/2003
C	Shogun	Novartis Crop Protection UK Ltd	08443	
C	Standon Propaquizafop	Standon Chemicals Ltd	09120	31/03/2003

C   These products are "approved for agricultural use". For further details refer to page vii.
A   These products are approved for use in or near water. For further details refer to page vii.

Product Name	Marketing Company	Reg. No.	Expiry Date

378 Propyzamide

	Product Name	Marketing Company	Reg. No.	Expiry Date
C	Agriguard Proyzamide 50 WP	Tronsan Ltd	10187	
C	Barclay Piza 400 FL	Barclay Chemicals Manufacturing Ltd	09633	30/09/2003
C	Base 50W	Interfarm (UK) Limited	10202	
C	Bulwark Flo	Headland Agrochemicals Ltd	09135	25/07/2003
C	Bulwark Flo	Interfarm (UK) Limited	10161	
C	Clayton Propel	Clayton Plant Protection (UK) Ltd	09783	
C	Greencrop Saffron FL	Greencrop Technology Ltd	10244	
C	Headland Judo	Headland Agrochemicals Ltd	08339	25/07/2003
C	Headland Judo	Headland Agrochemicals Ltd	07387	30/09/2003
C	Interfarm Base	Interfarm (UK) Limited	09733	30/09/2003
C	Interfarm Quaver	Interfarm (UK) Limited	09728	30/09/2003
C	Kerb 50 W	Interfarm (UK) Limited	10200	
C	Kerb 50 W	Pan Britannica Industries Ltd	01133	
C	Kerb 50 W	Rohm & Haas (UK) Ltd	02986	25/07/2003
C	Kerb 50 W	SumiAgro (UK) Ltd	10166	
C	Kerb 80 EDF	Interfarm (UK) Limited	10201	
C	Kerb 80 EDF	Pan Britannica Industries Ltd	09256	25/07/2003
C	Kerb 80 EDF	SumiAgro (UK) Ltd	10163	
C	Kerb Flo	Interfarm (UK) Limited	10160	
C	Kerb Flo	Pan Britannica Industries Ltd	04521	
C	Kerb Flo	Pan Britannica Industries Ltd	02759	25/07/2003
C	Kerb Flo	SumiAgro (UK) Ltd	10157	
C	Kerb Granules	Rohm&Haas (UK) Ltd	01136	
C	Kerb Granules	SumiAgro (UK) Ltd	08917	
C	Kerb Pro Flo	Pan Britannica Industries Ltd	08679	25/07/2003
C	Kerb Pro Flo	SumiAgro (UK) Ltd	10158	
C	Kerb Pro Granules	Pan Britannica Industries Ltd	08698	
C	Kerb Pro Granules	SumiAgro (UK) Ltd	09600	
C	Landgold Propyzamide 50	Landgold&Co Ltd	09753	30/09/2003
C	Menace 80 EDF	Headland Agrochemicals Ltd	09255	25/07/2003
C	Menace 80 EDF	Interfarm (UK) Limited	10165	
C	Mithras 80 EDF	Headland Agrochemicals Ltd	09219	25/07/2003
C	Mithras 80 EDF	Interfarm (UK) Limited	10164	
C	Precis	Pan Britannica Industries Ltd	08678	25/07/2003
C	Precis	SumiAgro (UK) Ltd	10159	
C	Precis Flo	Pan Britannica Industries Ltd	08530	
C	Quaver Flo	Interfarm (UK) Limited	10162	
C	Rapier	MTM Agrochemicals Ltd	05314	
C	Redeem Flo	Headland Agrochemicals Ltd	08563	25/07/2003
C	Redeem Flo	Interfarm (UK) Limited	10236	
C	Standon Propyzamide 400 SC	Standon Chemicals Ltd	10255	
C	Standon Propyzamide 50	Standon Chemicals Ltd	09054	30/09/2003
C	Stefes Pride	Stefes Plant Protection Ltd	08899	

C   These products are "approved for agricultural use". For further details refer to page vii.
A   These products are approved for use in or near water. For further details refer to page vii.

Product Name	Marketing Company	Reg. No.	Expiry Date

378 Propyzamide—continued

C	Stefes Pride Flo	Stefes Plant Protection Ltd	07812	31/03/2002
C	Top Farm Propyzamide 500	Top Farm Formulations Ltd	05484	

379 Propyzamide + Clopyralid

C	Matrikerb	Pan Britannica Industries Ltd	01308
C	Matrikerb	Rohm & Haas (UK) Ltd	02443
C	Matrikerb	SumiAgro (UK) Ltd	09604

380 Prosulfuron

C	SL 600	Novartis Crop Protection UK Ltd	08633

381 Prosulfuron + Bromoxynil

C	Jester	Novartis Crop Protection UK Ltd	08681

382 Pyridate

C	Lentagran EC	Novartis Crop Protection UK Ltd	08987
C	Lentagran WP	Novartis Crop Protection UK Ltd	08478

383 Quinmerac + Chloridazon

C	Fiesta T	BASF plc	10260

384 Quinmerac + Metazachlor

C	Katamaran	BASF plc	09049
C	Standon Metazachlor-Q	Standon Chemicals Ltd	09676

385 Quizalofop-ethyl

C	Everest	AgrEvo UK Ltd	07341	30/04/2001
C	Mission	AgrEvo UK Ltd	07264	30/04/2001
C	Pilot	AgrEvo UK Ltd	07268	31/07/2001

386 Quizalofop-P-ethyl

C	Alias	AgrEvo UK Ltd	08044	31/05/2001
C	CoPilot	Aventis CropScience UK Limited	08042	
C	Pilot D	Aventis CropScience UK Limited	08041	
C	Sceptre	Aventis CropScience UK Limited	08043	

387 Rimsulfuron

C	Landgold Rimsulfuron	Landgold & Co Ltd	08959
C	Me[2] Aducksbackside	Me[2] Crop Protection Limited	10065
C	Standon Rimsulfuron	Standon Chemicals Ltd	09955
C	Tarot	Du Pont (UK) Ltd	07909
C	Titus	Du Pont (UK) Ltd	07908

C These products are "approved for agricultural use". For further details refer to page vii.
A These products are approved for use in or near water. For further details refer to page vii.

Product Name	Marketing Company	Reg. No.	Expiry Date

388 Sethoxydim

C	Checkmate	Hortichem Ltd	09122	
C	Checkmate	Rhone-Poulenc Agriculture	06129	30/06/2001

389 Simazine

C	Alpha Simazine 50 SC	Makhteshim-Agan (UK) Ltd	04801
C	Alpha Simazine 50 WP	Makhteshim-Agan (UK) Ltd	04879
C	Alpha Simazine 80 WP	Makhteshim-Agan (UK) Ltd	04800
C	Atlas Simazine	Atlas Crop Protection Ltd	07725
C	Atlas Simazine	Atlas Interlates Ltd	05610
C	Gesatop	Novartis Crop Protection UK Ltd	08412
C	MSS Simazine 50 FL	Mirfield Sales Services Ltd	01418
C	Sipcam Simazine Flowable	Sipcam UK Ltd	07622
C	Unicrop Flowable Simazine	Universal Crop Protection Ltd	05447
C	Unicrop Simazine 50	Universal Crop Protection Ltd	02646
C	Unicrop Simazine FL	Universal Crop Protection Ltd	08032

390 Simazine + Amitrole

C	Alpha Simazol	Makhteshim-Agan (UK) Ltd	04799
C	Alpha Simazol T	Makhteshim-Agan (UK) Ltd	04874

391 Simazine + Diquat + Paraquat

C	Pathclear S	Miracle Garden Care Ltd	08991
C	Pathclear S	The Scotts Company (UK) Limited	09269

392 Simazine + Isoproturon

C	Alpha Protugan Plus	Makhteshim-Agan (UK) Ltd	08799	
C	Harlequin 500 SC	Makhteshim-Agan (UK) Ltd	09779	
C	Harlequin 500 SC	Novartis Crop Protection UK Ltd	08416	30/09/2002

393 Simazine + Pendimethalin

C	Deuce	Cyanamid Agriculture Ltd	06746
C	Merit	Cyanamid Agriculture Ltd	04976

394 Simazine + Trietazine

C	Remtal SC	Aventis CropScience UK Limited	07270

395 Sodium chlorate

C	Ace-Sodium chlorate (Fire Suppressed) Weedkiller	Ace Chemicals Ltd	06413
C	Atlacide Soluble Powder Weedkiller	Chipman Ltd	00125

C   These products are "approved for agricultural use". For further details refer to page vii.
A   These products are approved for use in or near water. For further details refer to page vii.

Product Name	Marketing Company	Reg. No.	Expiry Date

395 Sodium chlorate—continued

C	Cooke's Professional Sodium chlorate Weedkiller with Fire Depressant	Cooke's Chemicals (Sales) Ltd	06796	
C	Cooke's Weedclear	Cooke's Chemicals (Sales) Ltd	06512	
C	Deosan Chlorate Weedkiller	DiverseyLever Ltd	08521	
C	Doff Sodium chlorate Weedkiller	Doff Portland Ltd	06049	
C	Gem Sodium chlorate Weedkiller	Joseph Metcalf Ltd	04276	
C	Morgan's Sodium chlorate Weedkiller	David Morgan (Nottingham) Ltd	08927	
C	Sodium chlorate	Marlow Chemical Co Ltd	06294	
C	Strathclyde Sodium chlorate Weedclear	Strathclyde Chemical Co Ltd	07420	
C	TWK-Total Weedkiller	Yule Catto Consumer Chemicals Ltd	06393	

396 Sodium chloroacetate

C	Atlas Somon	Atlas Crop Protection Ltd	07727	
C	Atlas Somon	Atlas Interlates Ltd	03045	

397 Sodium monochloroacetate

C	Croptex Steel	Hortichem Ltd	02418	

398 Sodium silver thiosulphate

C	Argylene	Fargro Ltd	03386	28/02/2002

399 Tebutam

C	Comodor 600	Agrichem (International) Ltd	08398	

400 Tecnazene

C	Atlas Tecgran 100	Atlas Crop Protection Ltd	07730	31/01/2002
C	Atlas Tecgran 100	Atlas Interlates Ltd	05574	31/01/2001
C	Atlas Tecnazene 6% dust	Atlas Crop Protection Ltd	07731	31/01/2002
C	Atlas Tecnazene 6% Dust	Atlas Interlates Ltd	06351	31/01/2002
C	Bygran F	Wheatley Chemical Co Ltd	00365	31/01/2002
C	Fusarex 10G	Hickson & Welch Ltd	09727	31/01/2002
C	Hickstor 10	Hickson & Welch Ltd	03121	31/01/2002
	Hickstor 3	Hickson & Welch Ltd	03105	31/01/2002
C	Hickstor 5	Hickson & Welch Ltd	03180	31/01/2002
	Hickstor 6	Hickson & Welch Ltd	03106	31/01/2002
C	Hystore 10	Agrichem (International) Ltd	03581	31/01/2002
	Hytec	Agrichem Ltd	01099	31/01/2002

C These products are "approved for agricultural use". For further details refer to page vii.
A These products are approved for use in or near water. For further details refer to page vii.

Product Name	Marketing Company	Reg. No.	Expiry Date

400 Tecnazene—continued

C	Hytec 6	Agrichem Ltd	03580	31/01/2002
C	New Hickstor 6	Hickson & Welch Ltd	04221	31/01/2002
C	Tripart Arena 10G	Tripart Farm Chemicals Ltd	05603	31/01/2002
	Tripart Arena 3	Tripart Farm Chemicals Ltd	05605	31/01/2002
C	Tripart Arena 5G	Tripart Farm Chemicals Ltd	05604	31/01/2002
C	Tripart New Arena 6	Tripart Farm Chemicals Ltd	05813	31/01/2002

401 Tecnazene + Carbendazim

C	Hickstor 6 Plus MBC	Hickson & Welch Ltd	04176	31/01/2002
C	Hortag Tecnacarb	Avon Packers Ltd	02929	31/01/2002
C	New Arena Plus	Hickson & Welch	04598	31/01/2002
C	New Hickstor 6 Plus MBC	Hickson & Welch Ltd	04599	31/01/2002
C	Tripart Arena Plus	Hickson & Welch	05602	31/01/2002

402 Tecnazene + Thiabendazole

C	Hytec Super	Agrichem (International) Ltd	01100	31/01/2002
	New Arena TBZ 6	Tripart Farm Chemicals Ltd	05606	31/01/2002

403 Tepraloxydim

C	Aramo	BASF plc	10280	

404 Terbacil

C	Sinbar	Du Pont (UK) Ltd	01956	

405 Terbuthylazine + Bromoxynil

C	Alpha Bromotril PT	Makhteshim-Agan (UK) Ltd	09435	
C	Templar	Makhteshim-Agan (UK) Ltd	10254	

406 Terbuthylazine + Cyanazine

C	Angle	Novartis Crop Protection UK Ltd	08385	31/10/2001

407 Terbuthylazine + Diflufenican

C	Bolero	Ciba Agriculture	07436	31/07/2001
C	Bolero	Novartis Crop Protection UK Ltd	08392	

408 Terbuthylazine + Isoxaben

C	Skirmish	Novartis Crop Protection UK Ltd	08444	

409 Terbuthylazine + Terbutryn

C	Batallion	Makhteshim-Agan (UK) Ltd	08305	
C	Opogard	Novartis Crop Protection UK Ltd	08427	

C   These products are "approved for agricultural use". For further details refer to page vii.
A   These products are approved for use in or near water. For further details refer to page vii.

Product Name	Marketing Company	Reg. No.	Expiry Date

410 Terbutryn

C Alpha Terbutryne 50 SC	Makhteshim-Agan (UK) Ltd	04809	
C Alpha Terbutryne 50 WP	Makhteshim-Agan (UK) Ltd	04875	
C A Clarosan	Novartis Crop Protection UK Ltd	08396	31/08/2001
C A Clarosan	The Scotts Company (UK) Limited	09394	
C Prebane	Novartis Crop Protection UK Ltd	08432	

411 Terbutryn + Fomesafen

C Reflex T	Novartis Crop Protection UK Ltd	08884	

412 Terbutryn + Prometryn

C P-Weed	Pan Britannica Industries Ltd	08574	
C P-Weed	SumiAgro (UK) Ltd	09608	

413 Terbutryn + Terbuthylazine

C Batallion	Makhteshim-Agan (UK) Ltd	08305	
C Opogard	Novartis Crop Protection UK Ltd	08427	

414 Terbutryn + Trietazine

C Senate	Aventis CropScience UK Limited	07279	31/12/2002

415 Terbutryn + Trifluralin

C Alpha Terbalin 35 SC	Makhteshim-Agan (UK) Ltd	04792	
C Ashlade Summit	Ashlade Formulations Ltd	06214	

416 Thiabendazole + Tecnazene

C Hytec Super	Agrichem (International) Ltd	01100	31/01/2002
New Arena TBZ 6	Tripart Farm Chemicals Ltd	05606	31/01/2002

417 Thifensulfuron-methyl

C DUK 118	Du Pont (UK) Ltd	04596	
C Landgold Thifensulfuron	Landgold & Co Ltd	08523	25/07/2003
C Prospect	Du Pont (UK) Ltd	06541	31/12/2001
C Prospect	Nufarm UK Ltd	09389	

418 Thifensulfuron-methyl + Carfentrazone-ethyl

C Harmony Express	Du Pont (UK) Ltd	09467	

419 Thifensulfuron-methyl + Flupyrsulfuron-methyl

C Lexus Millenium	Du Pont (UK) Ltd	09206	
C Lexus Millenium WSB	Du Pont (UK) Ltd	09207	

C These products are "approved for agricultural use". For further details refer to page vii.
A These products are approved for use in or near water. For further details refer to page vii.

Product Name	Marketing Company	Reg. No.	Expiry Date

420 Thifensulfuron-methyl + Fluroxypyr + Tribenuron-methyl

C	DP 353	Du Pont (UK) Ltd	09626	
C	GEM 353	Griffin (Europe) Marketing NV (UK Branch)	09826	25/07/2003
C	GEX 353	Griffin (Europe) Marketing NV (UK Branch)	10185	
C	Starane Super	Dow AgroSciences Ltd	09625	

421 Thifensulfuron-methyl + Metsulfuron-methyl

C	DP 928	Du Pont (UK) Ltd	09632
C	Harmony M	Du Pont (UK) Ltd	03990

422 Thifensulfuron-methyl + Tribenuron-methyl

C	Calibre	Du Pont (UK) Ltd	07795	
C	DUK 110	Du Pont (UK) Ltd	09189	
C	DUK 110	Du Pont (UK) Ltd	06266	31/08/2001

423 Tralkoxydim

C	Grasp	Zeneca Crop Protection	06675
C	Greencrop Gweedore	Greencrop Technology Ltd	09882
C	Landgold Tralkoxydim	Landgold & Co Ltd	08604
C	Standon Tralkoxydim	Standon Chemicals Ltd	09579

424 Tri-allate

C	Avadex 15 G	Monsanto Plc	10167	
C	Avadex BW	Monsanto Plc	00173	31/07/2001
C	Avadex BW 480	Monsanto Plc	04742	31/07/2001
C	Avadex BW Granular	Monsanto Plc	00174	
C	Avadex Excel 15G	Monsanto Plc	07117	
C	Landgold Triallate 480	Landgold & Co Ltd	08505	31/07/2001

425 Triasulfuron

C	Lo-Gran	Novartis Crop Protection UK Ltd	08421

426 Triasulfuron + Bromoxynil + Ioxynil

C	Teal	Novartis Crop Protection UK Ltd	08453
C	Teal-M	Novartis Crop Protection UK Ltd	08454

427 Triasulfuron + Dicamba

C	Banvel T	Novartis Crop Protection UK Ltd	08470
C	Framolene	Novartis Crop Protection UK Ltd	08408

C   These products are "approved for agricultural use". For further details refer to page vii.
A   These products are approved for use in or near water. For further details refer to page vii.

Product Name	Marketing Company	Reg. No.	Expiry Date

428 Triasulfuron + Mecoprop-P

C	Raven	Novartis Crop Protection UK Ltd	08434	

429 Tribenuron-methyl

C	Quantum	Du Pont (UK) Ltd	06270	
C	Quantum 75 DF	Du Pont (UK) Ltd	09340	

430 Tribenuron-methyl + Fluroxypyr + Thifensulfuron-methyl

C	DP 353	Du Pont (UK) Ltd	09626	
C	GEM 353	Griffin (Europe) Marketing NV (UK Branch)	09826	25/07/2003
C	GEX 353	Griffin (Europe) Marketing NV (UK Branch)	10185	
C	Starane Super	Dow AgroSciences Ltd	09625	

431 Tribenuron-methyl + Metsulfuron-methyl

C	DP 911 WSB	Du Pont (UK) Ltd	09867	

432 Tribenuron-methyl + Thifensulfuron-methyl

C	Calibre	Du Pont (UK) Ltd	07795	
C	DUK 110	Du Pont (UK) Ltd	09189	
C	DUK 110	Du Pont (UK) Ltd	06266	31/08/2001

433 Triclopyr

C	Chipman Garlon 4	Nomix-Chipman Ltd	06016	
C	Convoy	Nufarm UK Ltd	09185	
C	Garlon 2	Dow AgroSciences Ltd	05682	
C	Garlon 2	Zeneca Crop Protection	06616	
C	Garlon 4	Dow AgroSciences Ltd	05090	
C	Timbrel	Dow AgroSciences Ltd	05815	
C	Tribel 480	United Phosphorus Ltd	08977	31/10/2001
C	Triptic 48 EC	United Phosphorus Ltd	09294	

434 Triclopyr + Clopyralid

C	Grazon 90	Dow AgroSciences Ltd	05456	

434a Triclopyr + Clopyralid + Fluroxypyr

C	Pastor	Dow AgroSciences Ltd	07440	

435 Triclopyr + 2,4-D + Dicamba

C	Broadsword	United Phosphorus Ltd	09140	
C	Nufarm Nu-Shot	Nufarm UK Ltd	09139	
C	Terbel Triple	Chimac-Agriphar SA	09138	

C   These products are "approved for agricultural use". For further details refer to page vii.
A   These products are approved for use in or near water. For further details refer to page vii.

Product Name	Marketing Company	Reg. No.	Expiry Date
436 Triclopyr + Fluroxypyr			
C Doxstar	Dow AgroSciences Ltd	06050	
C Evade	Dow AgroSciences Ltd	08071	31/05/2003
437 Trietazine + Simazine			
C Remtal SC	Aventis CropScience UK Limited	07270	
438 Trietazine + Terbutryn			
C Senate	Aventis CropScience UK Limited	07279	31/12/2002
439 Trifluralin			
C Alpha Trifluralin 48 EC	Makhteshim-Agan (UK) Ltd	07406	
C Ashlade Trifluralin	Ashlade Formulations Ltd	08303	
C Ashlade Trimaran	Ashlade Formulations Ltd	06228	
C Atlas Trifluralin	Atlas Crop Protection Ltd	08498	
C DAPT Trifluralin 48 EC	DAPT Agrochemicals Ltd	07906	
C Digermin	Montedison UK Ltd	00701	
C Digermin	Sipcam UK Ltd	07221	
C FCC Trigard	Farmers Crop Chemicals Ltd	09030	
C Ipifluor	I Pi Ci	04692	
C MSS Trifluralin 48 EC	Mirfield Sales Services Ltd	07753	
C MTM Trifluralin	MTM Agrochemicals Ltd	05313	30/04/2001
C Nufarm Triflur	Nufarm UK Ltd	08311	31/08/2002
C Portman Trifluralin	Portman Agrochemicals Ltd	05751	
C Treflan	Dow AgroSciences Ltd	05817	
C Triflur	Nufarm UK Ltd	09670	
C Triflurex 48EC	Makhteshim-Agan (UK) Ltd	07947	
C Trifsan	Dow AgroSciences Ltd	09237	
C Trifsan	Sanachem International Ltd	08233	30/09/2001
C Trigard	Farmers Crop Chemicals Ltd	02178	31/08/2002
C Trigard	pbi Agrochemicals Ltd	09699	
C Trilogy	United Phosphorus Ltd	08996	
C Tripart Trifluralin 48 EC	Tripart Farm Chemicals Ltd	02215	
C Triplen	Sipcam UK Ltd	05897	
C Tristar	Pan Britannica Industries Ltd	02219	
C Tristar	SumiAgro (UK) Ltd	09612	
C Whyte Trifluralin	Nufarm Whyte Agriculture Ltd	09286	
440 Trifluralin + Bromoxynil + Ioxynil			
C Masterspray	Pan Britannica Industries Ltd	02971	
C Masterspray	SumiAgro (UK) Ltd	09603	
441 Trifluralin + Clodinafop-propargyl			
C Hawk	Novartis Crop Protection UK Ltd	08417	

C These products are "approved for agricultural use". For further details refer to page vii.
A These products are approved for use in or near water. For further details refer to page vii.

Product Name	Marketing Company	Reg. No.	Expiry Date

441 Trifluralin + Clodinafop-propargyl—continued

C Reserve	Ciba Agriculture	08169	28/02/2001

442 Trifluralin + Diflufenican

C Ardent	Aventis CropScience UK Limited	09968	
C Ardent	Rhone-Poulenc Agriculture	06203	25/07/2003

443 Trifluralin + Isoproturon

C Autumn Kite	Aventis CropScience UK Limited	07119	
C Stefes Union	Stefes Plant Protection Ltd	08146	31/03/2002

444 Trifluralin + Isoxaben

C Axit GR	Dow AgroSciences Ltd	08892	
C Premiere Granules	Dow AgroSciences Ltd	07987	

445 Trifluralin + Linuron

C Chandor	Dow AgroSciences Ltd	05631	
C Linnet	Pan Britannica Industries Ltd	01555	
C Linnet	SumiAgro (UK) Ltd	09601	
C Neminfest	Montedison UK Ltd	02546	
C Neminfest	Sipcam UK Ltd	07219	

446 Trifluralin + Terbutryn

C Alpha Terbalin 35 SC	Makhteshim-Agan (UK) Ltd	04792	
C Ashlade Summit	Ashlade Formulations Ltd	06214	

447 Triflusulfuron-methyl

C Debut	Du Pont (UK) Ltd	07804	
C Debut WSB	Du Pont (UK) Ltd	07809	
C DUK 440	Du Pont (UK) Ltd	07811	
C DUK 550	Du Pont (UK) Ltd	07810	
C Landgold TFS 50	Landgold & Co Ltd	08941	
C Standon Triflusulfuron	Standon Chemicals Ltd	09487	

448 Trinexapac-ethyl

C Moddus	Novartis Crop Protection UK Ltd	08801	
C Moddus	Novartis Crop Protection UK Ltd	09035	31/03/2001

C These products are "approved for agricultural use". For further details refer to page vii.
A These products are approved for use in or near water. For further details refer to page vii.

Product Name	Marketing Company	Reg. No.	Expiry Date

1.2 Fungicides
including bactericides

449 Acibenzolar-S-Methyl

C	Bion	Novartis Crop Protection UK Ltd	09803

450 Aminobutane-2

C	Hortichem 2-Aminobutane	Hortichem Ltd	06147

451 Azaconazole + Imazalil

C	Nectec Paste	Hortichem Ltd	08510

452 Azoxystrobin

C	Amistar	Zeneca Crop Protection	08517
C	Barclay ZX	Barclay Chemicals Manufacturing Ltd	09570
C	Clayton Stobik	Clayton Plant Protection (UK) Ltd	09440
C	Landgold Strobilurin 250	Landgold & Co Ltd	09595
C	Me^2 Azoxystrobin	Me^2 Crop Protection Ltd	09654
C	Olympus	Zeneca Crop Protection	08520
C	Ortiva	Zeneca Crop Protection	09843
C	Priori	Zeneca Crop Protection	08516
C	Standon Azoxystrobin	Standon Chemicals Ltd	09515

453 Azoxystrobin + Flutriafol

C	Amigo	Zeneca Crop Protection	08834
C	Amistar Gem	Zeneca Crop Protection	08833

454 Benalaxyl + Mancozeb

C	Galben M	Sipcam UK Ltd	07220
C	Tairel	Sipcam UK Ltd	07767
C	Trecatol	Rohm & Haas (UK) Ltd	09823

455 Benodanil

C	Calirus	BASF plc	00368

456 Benomyl

C	Benlate Fungicide	Du Pont (UK) Ltd	00229

C These products are "approved for agricultural use". For further details refer to page vii.
A These products are approved for use in or near water. For further details refer to page vii.

Product Name	Marketing Company	Reg. No.	Expiry Date

457 Bitertanol + Fuberidazole

C Sibutol	Bayer plc	07238	
C Sibutol CF	Bayer plc	08174	
C Sibutol LS	Bayer plc	08109	31/12/2001
C Sibutol New Formula	Bayer plc	08270	

458 Bitertanol + Fuberidazole + Imidacloprid + Triadimenol

C Cereline Secur	Bayer plc	09511	

459 Bitertanol + Fuberidazole + Triadimenol

C Cereline	Bayer plc	07239	

460 Bromuconazole

C Granit	Aventis CropScience UK Limited	09995	
C Granit	Rhone-Poulenc Agriculture	08268	

461 Bromuconazole + Fenpropimorph

C Granit M	Rhone-Poulenc Agriculture	08269	31/03/2002

462 Bupirimate

C Nimrod	Zeneca Crop Protection	06686	

463 Bupirimate + Triforine

C Nimrod T	The Scotts Company (UK) Limited	09268	
C Nimrod T	Miracle Professional	07865	
C Nimrod T	Zeneca Professional Products	06859	31/01/2001
C Nimrod T	ICI Agrochemicals	01499	31/05/2001

464 Captan

C Alpha Captan 50 WP	Makhteshim-Agan (UK) Ltd	04797	
C Alpha Captan 80 WDG	Makhteshim-Agan (UK) Ltd	07096	
C Alpha Captan 83 WP	Makhteshim-Agan (UK) Ltd	04806	
C PP Captan 80 WG	Tomen (UK) Plc	08971	
C PP Captan 80 WG	Zeneca Crop Protection	06696	31/03/2001
C PP Captan 83	Tomen (UK) Plc	08768	

465 Captan + Penconazole

C Topas C 50 WP	Novartis Crop Protection UK Ltd	08459	

466 Carbendazim

C Ashlade Carbendazim Flowable	Ashlade Formulations Ltd	06213	

C These products are "approved for agricultural use". For further details refer to page vii.
A These products are approved for use in or near water. For further details refer to page vii.

Product Name	Marketing Company	Reg. No.	Expiry Date

466 Carbendazim—continued

	Product Name	Marketing Company	Reg. No.	Expiry Date
C	Barclay Shelter	Barclay Chemicals Manufacturing Ltd	09000	
C	BASF Turf Systemic Fungicide	BASF plc	05774	
C	Bavistin	BASF plc	00217	
C	Bavistin DF	BASF plc	03848	
C	Bavistin FL	BASF plc	00218	
C	Carbate Flowable	SumiAgro (UK) Ltd	08957	
C	Carbate Flowable	Pan Britannica Industries Ltd	03341	31/03/2001
C	Clayton Chizm	Clayton Plant Protection (UK) Ltd	09124	25/07/2003
C	Delsene 50 Flo	Griffin (Europe) Marketing NV (UK	09469	
C	Derosal Liquid	Aventis CropScience UK Limited	07315	
C	Derosal WDG	Aventis CropScience UK Limited	07316	
C	Focal Liquid	AgrEvo UK Crop Protection Ltd	08763	31/05/2001
C	Focal WDG	AgrEvo UK Crop Protection Ltd	08762	31/05/2001
C	Greencrop Mooncoin	Greencrop Technology Ltd	09392	25/07/2003
C	Headland Addstem	Headland Agrochemicals Ltd	06755	
C	Headland Addstem DF	Headland Agrochemicals Ltd	08904	
C	Headland Regain	Headland Agrochemicals Ltd	08675	
C	HY-CARB	Agrichem Ltd	05933	
C	Mascot Systemic	The Scotts Company (UK) Ltd	09132	
C	Mascot Systemic	Rigby Taylor Ltd	08776	
C	Mascot Systemic	Rigby Taylor Ltd	07654	30/11/2001
C	MSS Mircarb	Mirfield Sales Services Ltd	08788	
C	pbi Turf Fungicide	Pan Britannica Industries Ltd	08194	30/11/2001
C	pbi Turf Systemic Fungicide	pbi Agrochemicals Ltd	09349	
C	Quadrangle Hinge	Quadrangle Agrochemicals	08070	
C	Stefes C-Flo 2	Stefes Plant Protection Ltd	08059	31/03/2002
C	Stefes Carbendazim Flo	Stefes Plant Protection Ltd	05677	31/03/2002
C	Stefes Derosal Liquid	Aventis CropScience UK Limited	10079	
C	Stefes Derosal Liquid	Stefes Plant Protection Ltd	07649	31/05/2003
C	Stefes Derosal WDG	Aventis CropScience UK Limited	10078	
C	Stefes Derosal WDG	Stefes Plant Protection Ltd	07658	31/05/2003
C	Supercarb	SumiAgro (UK) Ltd	09610	
C	Supercarb	Pan Britannica Industries Ltd	01560	
C	Top Farm Carbendazim - 435	Top Farm Formulations Ltd	05307	
C	Tripart Defensor FL	Tripart Farm Chemicals Ltd	02752	
C	Tripart Defensor Liq	Tripart Farm Chemicals Ltd	07857	30/06/2001
C	Tripart Defensor WDG	Tripart Farm Chemicals Ltd	07855	30/06/2001
C	Turfclear	The Scotts Company (UK) Limited	07506	
C	Turfclear WDG	The Scotts Company (UK) Limited	07490	
C	Twincarb	Vitax Ltd	08777	
C	UPL Carbendazim 500 FL	United Phosphorus Ltd	07472	30/09/2001

C   These products are "approved for agricultural use". For further details refer to page vii.
A   These products are approved for use in or near water. For further details refer to page vii.

Product Name	Marketing Company	Reg. No.	Expiry Date

467 Carbendazim + Chlorothalonil

C	Bravocarb	Zeneca Crop Protection	09105	
C	Bravocarb	ISK Biosciences Ltd	05119	28/02/2003
C	Greenshield	The Scotts Company (UK) Limited + Miracle Professional	07988	

468 Carbendazim + Chlorothalonil + Maneb

C	Ashlade Mancarb Plus	Ashlade Formulations Ltd	08160	31/12/2001
C	Tripart Victor	Tripart Farm Chemicals Ltd	08161	31/12/2001

469 Carbendazim + Cymoxanil + Oxadixyl + Thiram

C	Apron Elite	Novartis Crop Protection UK Ltd	08770	

470 Carbendazim + Cyproconazole

C	Alto Combi	Novartis Crop Protection UK Ltd	08465	30/04/2001

471 Carbendazim + Epoxiconazole

C	Capricorn	Aventis Environmental Science	10035	

472 Carbendazim + Flusilazole

C	Contrast	Du Pont (UK) Ltd	06150	
C	Landgold Flusilazole MBC	Landgold & Co Ltd	08528	
C	Punch C	Du Pont (UK) Ltd	06801	
C	Standon Flusilazole Plus	Standon Chemicals Ltd	07403	

473 Carbendazim + Flutriafol

C	Early Impact	Zeneca Crop Protection	06659	31/12/2001
C	Pacer	Zeneca Crop Protection	06690	31/12/2001
C	Palette	Zeneca Crop Protection	06691	31/12/2001

474 Carbendazim + Iprodione

C	Calidan	Aventis CropScience UK Limited	09980	
C	Calidan	Rhone-Poulenc Agriculture	06536	
C	Vitesse	Aventis CropScience UK Limited	10042	
C	Vitesse	Rhone-Poulenc Amenity	06537	

475 Carbendazim + Mancozeb

C	Headland Maple	Headland Agrochemicals Ltd	09044	
C	Kombat WDG	Rohm & Haas (UK) Ltd	05509	

476 Carbendazim + Maneb

C	Ashlade Mancarb FL	Ashlade Formulations Ltd	07977	
C	Headland Dual	Headland Agrochemicals Ltd	03782	

C   These products are "approved for agricultural use". For further details refer to page vii.
A   These products are approved for use in or near water. For further details refer to page vii.

Product Name	Marketing Company	Reg. No.	Expiry Date

476 Carbendazim + Maneb—continued

C	MC Flowable	United Phosphorus Ltd	08198	
C	Multi-W FL	SumiAgro (UK) Ltd	09447	
C	Multi-W FL	Pan Britannica Industries Ltd	04131	28/02/2002
C	Protector	Procam Group Ltd	09448	
C	Protector	Procam Group Ltd	07075	28/02/2002
C	Tripart 147	Tripart Farm Chemicals Ltd	07978	

477 Carbendazim + Maneb + Sulphur

C	Bolda FL	Atlas Crop Protection Ltd	07653	
C	Legion S	Ashlade Formulations Ltd	08121	30/06/2002

478 Carbendazim + Maneb + Tridemorph

C	Cosmic FL	BASF plc	03473	28/02/2002

479 Carbendazim + Metalaxyl

C	Ridomil MBC 60 WP	Novartis Crop Protection UK Ltd	08437	

480 Carbendazim + Prochloraz

C	Novak	Aventis CropScience UK Limited	08020	
C	Sportak Alpha	AgrEvo UK Ltd	07222	31/05/2001
C	Sportak Alpha HF	Aventis CropScience UK Limited	07225	

481 Carbendazim + Propiconazole

C	Hispor 45 WP	Novartis Crop Protection UK Ltd	08418	
C	Sparkle 45 WP	Novartis Crop Protection UK Ltd	08450	

482 Carbendazim + Tebuconazole

C	Bayer UK413	Bayer plc	08277	
C	Tricur	Bayer plc	10281	

483 Carbendazim + Tecnazene

C	Hickstor 6 Plus MBC	Hickson & Welch Ltd	04176	31/01/2002
C	Hortag Tecnacarb	Avon Packers Ltd	02929	31/01/2002
C	New Arena Plus	Hickson & Welch	04598	31/01/2002
C	New Hickstor 6 Plus MBC	Hickson & Welch Ltd	04599	31/01/2002
C	Tripart Arena Plus	Hickson & Welch	05602	31/01/2002

484 Carbendazim + Vinclozolin

C	Konker	BASF plc	03988	

485 Carboxin + Imazalil + Thiabendazole

C	Vitaflo Extra	Uniroyal Chemical Ltd	07048	

C   These products are "approved for agricultural use". For further details refer to page vii.
A   These products are approved for use in or near water. For further details refer to page vii.

Product Name	Marketing Company	Reg. No.	Expiry Date
486 Carboxin + Prochloraz			
C Prelude Universal LS	AgrEvo UK Crop Protection Ltd	08650	31/05/2001
C Provax	Uniroyal Chemical Company	08651	31/05/2001
487 Carboxin + Thiram			
C Anchor	Uniroyal Chemical Ltd	08684	
488 Chlorothalonil			
C Agriguard Chlorothalonil	Tronsan Ltd	09390	
C Atlas Cropguard	Mirfield Sales Services Ltd	09123	
C Barclay Corrib 500	Barclay Chemicals Manufacturing Ltd	08981	
C Baton 500 SC	Sipcam UK Ltd	09850	
C Baton SC	Bayer plc	07945	31/12/2002
C Baton WG	Bayer plc	07944	31/03/2002
C BB Chlorothalonil	Brown Butlin Group	03320	
C Bombardier	Universal Crop Protection Ltd	02675	
C Bombardier FL	Universal Crop Protection Ltd	07910	
C Bravo 500	Zeneca Crop Protection	09059	
C Bravo 500	ISK Biosciences Ltd	05638	31/05/2001
C Bravo 500	BASF plc	05637	
C Bravo 720	Zeneca Crop Protection	09104	
C Bravo 720	ISK Biosciences Ltd	05544	30/06/2002
C Clayton Turret	Clayton Plant Protection (UK) Ltd	09400	
C CleanCrop Chlorothalonil 720	United Agri Products Limited	10102	
C Clortosip	Sipcam UK Ltd	06126	
C Clortosip 500	Sipcam UK Ltd	09320	
C Contact 75	ISK Biosciences Ltd	05563	28/02/2002
C Daconil Turf	The Scotts Company (UK) Limited	09265	
C Daconil Turf	Miracle Professional	07929	28/02/2002
C Flute	Sipcam UK Ltd	08953	
C Fusonil Turf	Rigby Taylor Ltd	09695	
C Greencrop Orchid	Greencrop Technology Ltd	09566	
C ISK 375	Zeneca Crop Protection	09103	
C ISK 375	ISK Biosciences Ltd	07455	30/06/2002
C Jupital	Zeneca Crop Protection	09109	
C Jupital	ISK Biosciences Ltd	05554	30/06/2002
C Jupital DG	Zeneca Crop Protection	09181	
C Mainstay	Quadrangle Agrochemicals	05625	
C Marnoch Chlorothalonil	Marnoch Ventures Limited	09763	
C Miros DF	Sipcam UK Ltd	04966	
C MSS Chlorothalonil	Mirfield Sales Services Ltd	08366	30/06/2001
C Mycoguard	Chiltern Farm Chemicals Ltd	08115	
C Repulse	Hortichem Ltd	07641	
C Repulse	Zeneca Crop Protection	06705	

C These products are "approved for agricultural use". For further details refer to page vii.
A These products are approved for use in or near water. For further details refer to page vii.

Product Name	Marketing Company	Reg. No.	Expiry Date

488 Chlorothalonil—continued

C	Rover DF	Sipcam UK Ltd	06151	31/03/2002
C	Sipcam Echo 75	Sipcam UK Ltd	08302	
C	Sipcam UK Rover	Sipcam UK Ltd	09848	
C	Sipcam UK Rover 500	Sipcam UK Ltd	04165	30/06/2001
C	Standon Chlorothalonil 500	Standon Chemicals Ltd	08597	
C	Strada	Sipcam UK Ltd	08824	31/12/2002
C	Strada 500	Sipcam UK Ltd	09849	
C	Tripart Faber	Tripart Farm Chemicals Ltd	04549	
C	Ultrafaber	Tripart Farm Chemicals Ltd	05627	
C	Visclor 500 SC	Sipcam UK Ltd	09404	
C	Visclor 75 DF	Vischim SRL	09361	

489 Chlorothalonil + Carbendazim

C	Bravocarb	Zeneca Crop Protection	09105	
C	Bravocarb	ISK Biosciences Ltd	05119	28/02/2003
C	Greenshield	The Scotts Company (UK) Limited + Miracle Professional	07988	

490 Chlorothalonil + Carbendazim + Maneb

C	Ashlade Mancarb Plus	Ashlade Formulations Ltd	08160	31/12/2001
C	Tripart Victor	Tripart Farm Chemicals Ltd	08161	31/12/2001

491 Chlorothalonil + Cymoxanil

C	Ashlade Cyclops	ICI Agrochemicals	04857	30/06/2001
C	DUK 44	Du Pont (UK) Ltd	07475	30/11/2002
C	GEM 44	Griffin (Europe) Marketing NV (UK	09813	25/07/2003
C	GEX 44	Griffin (Europe) Marketing NV (UK	10168	

492 Chlorothalonil + Cyproconazole

C	Alto Elite	Novartis Crop Protection UK Ltd	08467	
C	CleanCrop Cyprothal	United Agri Products	09580	
C	Octolan	Novartis Crop Protection UK Ltd	08480	
C	SAN 703	Novartis Crop Protection UK Ltd	08487	

493 Chlorothalonil + Fenpropimorph

C	BAS 438	BASF plc	03451	31/03/2002
C	Corbel CL	BASF plc	04196	31/03/2002

494 Chlorothalonil + Fluquinconazole

C	Trident	Aventis CropScience UK Limited	09369	
C	Vista CT	Aventis CropScience UK Limited	09368	

C   These products are "approved for agricultural use". For further details refer to page vii.
A   These products are approved for use in or near water. For further details refer to page vii.

	Product Name	Marketing Company	Reg. No.	Expiry Date

495 Chlorothalonil + Flutriafol

C	Halo	Zeneca Crop Protection	06520	31/12/2001
C	Impact Excel	Zeneca Crop Protection	06680	31/12/2001

496 Chlorothalonil + Mancozeb

C	Adagio	Rohm & Haas (UK) Ltd	10057	
C	Adagio	pbi Agrochemicals Ltd	09597	30/04/2003
C	Adagio	Pan Britannica Industries Ltd	07832	30/04/2003
C	Adagio C	Rohm & Haas (UK) Ltd	10056	
C	Adagio C	pbi Agrochemicals Ltd	09842	
C	Adagio C	pbi Agrochemicals Ltd	09309	30/04/2003
C	Delphi	Sipcam UK Ltd	09860	
C	Dreadnought Flo	pbi Agrochemicals Ltd	09599	
C	Dreadnought Flo	Rohm & Haas (UK) Ltd	07600	
C	Dreadnought Flo	Pan Britannica Industries Ltd	07599	
C	Sipcam Flo	Sipcam UK Ltd	07601	

497 Chlorothalonil + Metalaxyl

C	Folio	Novartis Crop Protection UK Ltd	08547	

498 Chlorothalonil + Propamocarb Hydrochloride

C	Aventis Merlin	Aventis CropScience UK Limited	09719	
C	Merlin	Aventis CropScience UK Limited	07943	
C	Tattoo C	Aventis CropScience UK Limited	07623	

499 Chlorothalonil + Propiconazole

C	Sambarin 312.5 SC	Novartis Crop Protection UK Ltd	08439	

500 Chlorothalonil + Tetraconazole

C	TC Elite	Monsanto Plc	09412	
C	Voodoo	Sipcam UK Ltd	09414	

501 Chlorothalonil + Vinclozolin

C	Curalan CL	BASF plc	07174	

502 Copper Ammonium Carbonate

C	Croptex Fungex	Hortichem Ltd	02888	

503 Copper Complex - Bordeaux

C	Wetcol 3 Copper Fungicide	Ford Smith & Co Ltd	02360	

504 Copper Oxychloride

C	Cuprokylt	Universal Crop Protection Ltd	00604	

C These products are "approved for agricultural use". For further details refer to page vii.
A These products are approved for use in or near water. For further details refer to page vii.

Product Name	Marketing Company	Reg. No.	Expiry Date

504 Copper Oxychloride—continued

C	Cuprokylt FL	Universal Crop Protection Ltd	08299
C	Cuprosana H	Universal Crop Protection Ltd	00605
C	Headland Inorganic Liquid Copper	Headland Agrochemicals Ltd	07799

505 Copper Oxychloride + Maneb + Sulphur

C	Ashlade SMC Flowable	Ashlade Formulations Ltd	06494

506 Copper Oxychloride + Metalaxyl

C	Ridomil Plus	Novartis Crop Protection UK Ltd	08353	
C	Ridomil Plus 50 WP	Ciba Agriculture	01803	31/03/2001

507 Cymoxanil + Carbendazim + Oxadixyl + Thiram

C	Apron Elite	Novartis Crop Protection UK Ltd	08770

508 Cymoxanil + Chlorothalonil

C	Ashlade Cyclops	ICI Agrochemicals	04857	30/06/2001
C	DUK 44	Du Pont (UK) Ltd	07475	30/11/2002
C	GEM 44	Griffin (Europe) Marketing NV (UK	09813	25/07/2003
C	GEX 44	Griffin (Europe) Marketing NV (UK	10168	

509 Cymoxanil + Mancozeb

C	Ashlade Solace	Ashlade Formulations Ltd	08087	
C	Besiege	Du Pont (UK) Ltd	08086	
C	Besiege WSB	Du Pont (UK) Ltd	08075	
C	Clayton Krypton	Clayton Plant Protection (UK) Ltd	09398	
C	CleanCrop Xanilite	United Agri Products	10050	
C	Curzate M68	Du Pont (UK) Ltd	08072	
C	Curzate M68 WSB	Du Pont (UK) Ltd	08073	
C	Fytospore 68	Du Pont (UK) Ltd	09827	
C	Fytospore 68	Du Pont (UK) Ltd	08649	30/11/2002
C	Marnoch Mancym	Marnoch Ventures Limited	10130	
C	Me^2 Cymoxeb	Me^2 Crop Protection Ltd	09486	
C	Rhythm	Interfarm (UK) Limited	09909	
C	Rhythm	Interfarm (UK) Limited	09636	
C	Standon Cymoxanil Extra	Standon Chemicals Ltd	09442	
C	Systol M	Quadrangle Agrochemicals	08085	
C	Systol M	Quadrangle Agrochemicals	03480	
C	Systol M WSB	Quadrangle Agrochemicals	08074	

510 Cymoxanil + Mancozeb + Oxadixyl

C	Ripost Pepite	Bayer Ltd	08895
C	Ripost Pepite	Novartis Crop Protection UK Ltd	08485

C These products are "approved for agricultural use". For further details refer to page vii.
A These products are approved for use in or near water. For further details refer to page vii.

Product Name	Marketing Company	Reg. No.	Expiry Date

510 Cymoxanil + Mancozeb + Oxadixyl—continued

C Ripost Pepite	Sandoz Agro Ltd	06485	31/03/2001
C Trustan WDG	Du Pont (UK) Ltd	05050	

511 Cyproconazole

C Agriguard Cyproconazole	Tronsan Ltd	09406	25/07/2003
C Alto 100 SL	Novartis Crop Protection UK Ltd	08350	30/04/2001
C Alto 240 EC	Novartis Crop Protection UK Ltd	08354	
C Aplan	Novartis Crop Protection UK Ltd	08351	31/03/2001
C Aplan 240 EC	Novartis Crop Protection UK Ltd	08355	31/03/2002
C Barclay Shandon	Barclay Chemicals Manufacturing Ltd	06464	30/04/2002
C Clayton Cyprocon	Clayton Plant Protection (UK) Ltd	07668	30/04/2002
C Greencrop Gentian	Greencrop Technology Ltd	09380	30/04/2002
C Landgold Cyproconazole 100	Landgold & Co Ltd	06463	25/07/2003
C SAN 619F 240 EC	Novartis Crop Protection UK Ltd	08357	31/03/2002
C Standon Cyproconazole	Standon Chemicals Ltd	07751	30/04/2002
C Star Cyproconazole	Star Agrochem Ltd	09163	30/04/2002

512 Cyproconazole + Carbendazim

C Alto Combi	Novartis Crop Protection UK Ltd	08465	30/04/2001

513 Cyproconazole + Chlorothalonil

C Alto Elite	Novartis Crop Protection UK Ltd	08467	
C CleanCrop Cyprothal	United Agri Products	09580	
C Octolan	Novartis Crop Protection UK Ltd	08480	
C SAN 703	Novartis Crop Protection UK Ltd	08487	

514 Cyproconazole + Cyprodinil

C Radius	Novartis Crop Protection UK Ltd	09387	

515 Cyproconazole + Mancozeb

C Alto Eco	Novartis Crop Protection UK Ltd	08466	

516 Cyproconazole + Prochloraz

C Profile	Aventis CropScience UK Limited	08134	
C SAN 710	Novartis Crop Protection UK Ltd	08488	31/05/2001
C SAN 710 HF	Novartis Crop Protection UK Ltd	08489	
C Sportak Delta 460	AgrEvo UK Crop Protection Ltd	07224	31/05/2001
C Sportak Delta 460 HF	Aventis CropScience UK Limited	07431	
C Tiptor	Novartis Crop Protection UK Ltd	08495	
C Tiptor	AgrEvo UK Ltd	07295	31/05/2001

C These products are "approved for agricultural use". For further details refer to page vii.
A These products are approved for use in or near water. For further details refer to page vii.

Product Name	Marketing Company	Reg. No.	Expiry Date

517 Cyproconazole + Propiconazole

C	Menara	Novartis Crop Protection UK Ltd	09321	

518 Cyproconazole + Quinoxyfen

C	Divora	Novartis Crop Protection UK Ltd	08960	
C	DOE 1762	Dow AgroSciences Ltd	08962	
C	NOV 1390	Novartis Crop Protection UK Ltd	08961	

519 Cyproconazole + Tridemorph

C	Alto Major	Novartis Crop Protection UK Ltd	08468	30/09/2001
C	Alto Major	Sandoz Agro Ltd	06979	30/09/2001
C	Moot	Novartis Crop Protection UK Ltd	08479	30/09/2001
C	Moot	Sandoz Agro Ltd	06990	30/09/2001
C	SAN 735	Novartis Crop Protection UK Ltd	08490	30/09/2001
C	San 735	Sandoz Agro Ltd	06991	30/09/2001

520 Cyproconazole + Trifloxystrobin

C	Sphere	Novartis Crop Protection UK Ltd	10263	

521 Cyprodinil

C	Barclay Amtrak	Barclay Chemicals Manufacturing Ltd	09562	
C	Chieftain	Novartis Crop Protection UK Ltd	09301	
C	CleanCrop Cyprodinil	United Agri Products	09668	
C	Helmet	Novartis Crop Protection UK Ltd	08765	
C	NOV 219	Novartis Crop Protection UK Ltd	08766	
C	Skua	Novartis Crop Protection UK Ltd	08767	
C	Standon Cyprodinil	Standon Chemicals Ltd	09345	
C	Unix	Novartis Crop Protection UK Ltd	08764	

522 Cyprodinil + Cyproconazole

C	Radius	Novartis Crop Protection UK Ltd	09387	

523 Dichlofluanid

C	Elvaron WG	Bayer plc	04855	

524 Dichlorophen

C	Halophen RE 49	McMillan Technical Services Ltd	04636	28/02/2002
C	Super Mosstox	Rhone-Poulenc Amenity	05339	
C	SuperMosstox	Aventis Environmental Science	09942	

525 Dicloran

C	Fumite Dicloran Smoke	Octavius Hunt Ltd	09291	

C   These products are "approved for agricultural use". For further details refer to page vii.
A   These products are approved for use in or near water. For further details refer to page vii.

Product Name	Marketing Company	Reg. No.	Expiry Date

525 Dicloran—continued

C	Fumite Dicloran Smoke	Octavius Hunt Ltd	00930	31/10/2001

526 Difenoconazole

C	Landgold Difenoconazole	Landgold & Co Ltd	09964
C	Plover	Novartis Crop Protection UK Ltd	08429

527 Difenzoquat

C	Match	BASF plc	10225
C	Match	Cyanamid Agriculture Ltd	07186

528 Dimethomorph + Mancozeb

C	Invader	Cyanamid Agriculture Ltd	06989
C	Invader WDG	Cyanamid Agriculture Ltd	09556
C	Saracen	Cyanamid Agriculture Ltd	07872
C	Saracen WDG	Cyanamid Agriculture Ltd	09557

529 Dinocap

C	Karathane Liquid	Landseer Limited	09262
C	Karathane Liquid	Rohm & Haas (UK) Ltd	07198	30/09/2001

530 Dithianon

C	Barclay Cluster	Barclay Chemicals Manufacturing Ltd	08792
C	Dithianon Flowable	BASF plc	10219
C	Dithianon Flowable	Cyanamid Agriculture Ltd	07007

531 Dithianon + Penconazole

C	Topas D275 SC	Novartis Crop Protection UK Ltd	08460

532 Dodemorph

C	F238	BASF plc	00206

533 Dodine

C	Barclay Dodex	Barclay Chemicals Manufacturing Ltd	09055
C	Radspor FL	Truchem Ltd	01685

534 Epoxiconazole

C	Agriguard Epoxiconazole	Tronsan Ltd	09407
C	Epic	BASF plc	08320
C	Landgold Epoxiconazole	Landgold & Co Ltd	09821
C	Opus	BASF plc	08319

C These products are "approved for agricultural use". For further details refer to page vii.
A These products are approved for use in or near water. For further details refer to page vii.

Product Name	Marketing Company	Reg. No.	Expiry Date

534 Epoxiconazole—continued

C	Standon Epoxiconazole	Standon Chemicals Ltd	09517	

535 Epoxiconazole + Carbendazim

C	Capricorn	Aventis Environmental Science	10035	

536 Epoxiconazole + Fenpropimorph

C	Barclay Riverdance	Barclay Chemicals Manufacturing Ltd	09658	
C	Eclipse	BASF plc	07361	
C	Greencrop Galore	Greencrop Technology Ltd	09561	
C	Landgold Epoxiconazole FM	Landgold & Co Ltd	08806	
C	Opus Team	BASF plc	07362	
C	Standon Epoxifen	Standon Chemicals Ltd	08972	

537 Epoxiconazole + Fenpropimorph + Kresoxim-Methyl

C	CleanCrop Kresoxazole Plus	United Agri Products	09742	31/03/2002
C	Mantra	BASF plc	08886	
C	Standon Kresoxim Super	Standon Chemicals Ltd	09794	

538 Epoxiconazole + Kresoxim-Methyl

C	Barclay Avalon	Barclay Chemicals Manufacturing Ltd	09466	
C	Clayton Gantry	Clayton Plant Protection (UK) Ltd	09482	
C	CleanCrop Kresoxazole	United Agri Products	09698	
C	Landgold Strobilurin KE	Landgold & Co Ltd	09908	
C	Landmark	BASF plc	08889	
C	Marnoch Kempo	Marnoch Ventures Limited	09918	
C	Me^2 KME	Me^2 Crop Protection Ltd	09594	
C	Standon Kresoxim-Epoxiconazole	Standon Chemicals Ltd	09281	

539 Epoxiconazole + Tridemorph

C	Opus Plus	BASF plc	07363	28/02/2002

540 Ethirimol + Flutriafol + Thiabendazole

C	Ferrax	Zeneca Crop Protection	06662	28/02/2002
C	Ferrax	Bayer plc	05284	28/02/2002

541 Ethirimol + Fuberidazole + Triadimenol

C	Bay UK 292	Bayer plc	04335	

C   These products are "approved for agricultural use". For further details refer to page vii.
A   These products are approved for use in or near water. For further details refer to page vii.

Product Name	Marketing Company	Reg. No.	Expiry Date

542 Etridiazole

C AAterra WP	Zeneca Crop Protection	06625	
C Standon Etridiazole 35	Standon Chemicals Ltd	08778	
C Terrazole 35 WP	Uniroyal Chemical Ltd	09800	

543 Fenarimol

C Rimidin	Dow AgroSciences Ltd	07938	
C Rimidin	Rigby Taylor Ltd	05907	
C Rubigan	Dow AgroSciences Ltd	05489	

544 Fenbuconazole

C Indar 5EC	Rohm&Haas (UK) Ltd	07581	31/03/2002
C Indar 5EW	T P Whelehan Son & Co Ltd	09644	
C Indar 5EW	Landseer Limited	09518	
C Indar 5EW	Rohm & Haas (UK) Ltd	08011	31/03/2002
C Kruga 5 EC	SumiAgro (UK) Ltd	09874	
C Kruga 5EC	Interfarm (UK) Limited	09863	
C Kruga 5EC	Headland Agrochemicals Ltd	08756	31/03/2002
C Reward 5EC	Interfarm (UK) Limited	09862	
C Reward 5EC	Headland Agrochemicals Ltd	08757	31/03/2002
C Scarab 5EW	Headland Agrochemicals Ltd	08012	31/03/2002
C Surpass 5EC	Interfarm (UK) Limited	09861	
C Surpass 5EC	Headland Agrochemicals Ltd	08758	31/03/2002

545 Fenbuconazole + Fenpropidin

C Accolade	SumiAgro (UK) Ltd	09950	

546 Fenbuconazole + Prochloraz

C Mirage Extra	Stefes UK Ltd	07958	
C Stefes Fortune	Stefes Plant Protection Ltd	08537	31/01/2001
C Stefes Inception	Stefes Plant Protection Ltd	08829	

547 Fenbuconazole + Propiconazole

C Indar CG	Novartis Crop Protection UK Ltd	08578	
C Graphic	Rohm & Haas (UK) Ltd	09907	

548 Fenbuconazole + Tridemorph

C Unison	Pan Britannica Industries Ltd	08318	28/02/2002

549 Fenhexamid

C Lattice	Bayer plc	09198	
C Teldor	Bayer plc	08955	

C These products are "approved for agricultural use". For further details refer to page vii.
A These products are approved for use in or near water. For further details refer to page vii.

Product Name	Marketing Company	Reg. No.	Expiry Date

550 Fenhexamid + Tolylfluanid

C	Talat	Bayer plc	09655

551 Fenpiclonil

C	Gambit	Novartis Crop Protection UK Ltd	08535	28/02/2002

552 Fenpropidin

C	Landgold Fenpropidin 750	Landgold & Co Ltd	08973
C	Mallard	Novartis Crop Protection UK Ltd	08662
C	Patrol	Novartis Crop Protection UK Ltd	08661
C	Tern	Novartis Crop Protection UK Ltd	08660

553 Fenpropidin + Fenbuconazole

C	Accolade	SumiAgro (UK) Ltd	09950

554 Fenpropidin + Fenpropimorph

C	Agrys	Novartis Crop Protection UK Ltd	08382	30/09/2002
C	Boscor	Novartis Crop Protection UK Ltd	08682	30/09/2002

555 Fenpropidin + Prochloraz

C	SL 552A	Novartis Crop Protection UK Ltd	08673
C	Sponsor	Aventis CropScience UK Limited	08674

556 Fenpropidin + Propiconazole

C	Prophet	Novartis Crop Protection UK Ltd	08433
C	Sheen	Novartis Crop Protection UK Ltd	08442
C	Zulu	Novartis Crop Protection UK Ltd	08464

557 Fenpropidin + Propiconazole + Tebuconazole

C	Bayer UK 593	Bayer plc	08285
C	Gladio	Novartis Crop Protection UK Ltd	08413

558 Fenpropidin + Tebuconazole

C	Monicle	Bayer plc	07375
C	SL-556 500 EC	Novartis Crop Protection UK Ltd	08449

559 Fenpropimorph

C	Aura	Novartis Crop Protection UK Ltd	08388
C	BAS 421F	BASF plc	06127
C	CleanCrop Fenpro	United Agri Products Limited	09885
C	CleanCrop Fenpropimorph	United Agri Products	09445
C	Corbel	BASF plc	00578
C	Keetak	BASF plc	06950

C These products are "approved for agricultural use". For further details refer to page vii.
A These products are approved for use in or near water. For further details refer to page vii.

Product Name	Marketing Company	Reg. No.	Expiry Date

559 Fenpropimorph—continued

C Mistral	Novartis Crop Protection UK Ltd	08425	
C Standon Fenpropimorph 750	Standon Chemicals Ltd	08965	
C Widgeon	Novartis Crop Protection UK Ltd	08463	

560 Fenpropimorph + Bromuconazole

C Granit M	Rhone-Poulenc Agriculture	08269	31/03/2002

561 Fenpropimorph + Chlorothalonil

C BAS 438	BASF plc	03451	31/03/2002
C Corbel CL	BASF plc	04196	31/03/2002

562 Fenpropimorph + Epoxiconazole

C Barclay Riverdance	Barclay Chemicals Manufacturing Ltd	09658	
C Eclipse	BASF plc	07361	
C Greencrop Galore	Greencrop Technology Ltd	09561	
C Landgold Epoxiconazole FM	Landgold & Co Ltd	08806	
C Opus Team	BASF plc	07362	
C Standon Epoxifen	Standon Chemicals Ltd	08972	

563 Fenpropimorph + Epoxiconazole + Kresoxim-Methyl

C CleanCrop Kresoxazole Plus	United Agri Products	09742	31/03/2002
C Mantra	BASF plc	08886	
C Standon Kresoxim Super	Standon Chemicals Ltd	09794	

564 Fenpropimorph + Fenpropidin

C Agrys	Novartis Crop Protection UK Ltd	08382	30/09/2002
C Boscor	Novartis Crop Protection UK Ltd	08682	30/09/2002

565 Fenpropimorph + Flusilazole

C BAS 48500F	BASF plc	06784	
C Colstar	Du Pont (UK) Ltd	06783	
C DUK 7876	Du Pont (UK) Ltd	09588	

566 Fenpropimorph + Flusilazole + Tridemorph

C Bingo	BASF plc	06920	28/02/2002
C DUK 51	Du Pont (UK) Ltd	06764	30/09/2001
C Justice	Du Pont (UK) Ltd	07963	30/09/2001

567 Fenpropimorph + Kresoxim-Methyl

C Ensign	BASF plc	08362	
C Greencrop Monsoon	Greencrop Technology Ltd	09573	

C   These products are "approved for agricultural use". For further details refer to page vii.
A   These products are approved for use in or near water. For further details refer to page vii.

Product Name	Marketing Company	Reg. No.	Expiry Date

567 Fenpropimorph + Kresoxim-Methyl—continued

C	Landgold Strobilurin KF	Landgold&Co Ltd	09196
C	Standon Kresoxim FM	Standon Chemicals Ltd	08922

568 Fenpropimorph + Prochloraz

C	SL571A	Novartis Crop Protection UK Ltd	08420	31/03/2002
C	Sprint	AgrEvo UK Ltd	07291	31/05/2001
C	Sprint HF	AgrEvo UK Ltd	07292	31/03/2002

569 Fenpropimorph + Propiconazole

C	Belvedere	Makhteshim-Agan (UK) Ltd	08084
C	Decade	Novartis Crop Protection UK Ltd	08402
C	Glint	Novartis Crop Protection UK Ltd	08414
C	Mantle	Novartis Crop Protection UK Ltd	08424

570 Fenpropimorph + Quinoxyfen

C	EF-1288	Dow AgroSciences Ltd	08241
C	Orka	Dow AgroSciences Ltd	08879

571 Fenpropimorph + Tebuconazole

C	Bayer UK506	Bayer plc	07540	31/03/2001
C	BUK 84000F	BASF plc	07541	31/03/2001

572 Fenpropimorph + Tridemorph

C	BAS 46402F	BASF plc	03313	28/02/2002
C	Gemini	BASF plc	05684	28/02/2002

573 Fentin Acetate + Maneb

C	Brestan 60 SP	Aventis CropScience UK Limited	07305

574 Fentin Hydroxide

C	Ashlade Flotin 2	Ashlade Formulations Ltd	06224	
C	Barclay Fentin Flow	Barclay Chemicals Manufacturing Ltd	07914	
C	Barclay Fentin Flow 532	Barclay Chemicals Manufacturing Ltd	09434	31/01/2003
C	Farmatin 560	Aventis CropScience UK Limited	09877	
C	Farmatin 560	AgrEvo UK Ltd	07320	31/01/2003
C	Greencrop Rosette	Greencrop Technology Ltd	09648	
C	Keytin	Chiltern Farm Chemicals Ltd	08894	31/01/2003
C	MSS Flotin 480	Mirfield Sales Services Ltd	07616	
C	Stefes Blytin	Stefes UK Ltd	07907	31/03/2002
C	Super-Tin 4L	Nufarm Whyte Agriculture Ltd	09028	
C	Super-Tin 4L	Chiltern Farm Chemicals Ltd	02995	

C   These products are "approved for agricultural use". For further details refer to page vii.
A   These products are approved for use in or near water. For further details refer to page vii.

Product Name	Marketing Company	Reg. No.	Expiry Date

574 Fentin Hydroxide—continued

C	Super-Tin 80WP	Chiltern Farm Chemicals Ltd	07606	31/05/2001
C	Super-Tin 80WP	Pan Britannica Industries Ltd	07605	
C	Supertin 4L	Griffin (Europe) Marketing NV (UK	09559	
C	Supertin 80 WP	Griffin (Europe) Marketing NV (UK	09560	

575 Fentin Hydroxide + Glufosinate-Ammonium

C	Safran	AgrEvo UK Crop Protection Ltd	08559	31/05/2001

576 Fluazinam

C	Barclay Cobbler	Barclay Chemicals Manufacturing Ltd	08349	
C	Landgold Fluazinam	Landgold & Co Ltd	08060	
C	Legacy	ISK Biosciences Europe SA	09966	
C	Legacy	ISK Biosciences Ltd	07401	31/03/2003
C	Salvo	Zeneca Crop Protection	07092	
C	Shirlan	Zeneca Crop Protection	07091	
C	Shirlan Programme	Zeneca Crop Protection	08761	
C	Standon Fluazinam 500	Standon Chemicals Ltd	08670	
C	Top Farm Fluazinam	Top Farm Formulations Ltd	07683	

577 Fludioxonil

C	Beret Gold	Novartis Crop Protection UK Ltd	08390	
C	Celest	Novartis Crop Protection UK Ltd	08617	

578 Fluquinconazole

C	Aventis Flamenco	Aventis CropScience UK Limited	09915	
C	Aventis Flamenco	AgrEvo UK Ltd	09729	28/02/2003
C	Diablo	Aventis CropScience UK Limited	09914	
C	Diablo	AgrEvo UK Ltd	09386	28/02/2003
C	Flamenco	Aventis CropScience UK Limited	09913	
C	Flamenco	AgrEvo UK Ltd	09385	28/02/2003
C	Jockey F	Aventis CropScience UK Limited	10074	
C	Jockey Flexi	Aventis CropScience UK Limited	10073	

579 Fluquinconazole + Chlorothalonil

C	Trident	Aventis CropScience UK Limited	09369	
C	Vista CT	Aventis CropScience UK Limited	09368	

580 Fluquinconazole + Prochloraz

C	Aventis Foil	Aventis CropScience UK Limited	09709	
C	Baron	Aventis CropScience UK Limited	09364	
C	Flamenco Plus	Aventis CropScience UK Limited	09362	
C	Foil	Aventis CropScience UK Limited	09363	

C   These products are "approved for agricultural use". For further details refer to page vii.
A   These products are approved for use in or near water. For further details refer to page vii.

Product Name	Marketing Company	Reg. No.	Expiry Date

580 Fluquinconazole + Prochloraz—continued

| C | Jockey | Aventis CropScience UK Limited | 10076 | |
| C | Jockey Plus | Aventis CropScience UK Limited | 10075 | |

581 Flusilazole

C	DUK 747	Du Pont (UK) Ltd	08239	
C	Genie	Du Pont (UK) Ltd	08238	
C	Genie 25	Du Pont (UK) Ltd	10285	
C	Lyric	Du Pont (UK) Ltd	08252	
C	Sanction	Du Pont (UK) Ltd	08237	
C	Sanction 25	Du Pont (UK) Ltd	10284	

582 Flusilazole + Carbendazim

C	Contrast	Du Pont (UK) Ltd	06150	
C	Landgold Flusilazole MBC	Landgold & Co Ltd	08528	
C	Punch C	Du Pont (UK) Ltd	06801	
C	Standon Flusilazole Plus	Standon Chemicals Ltd	07403	

583 Flusilazole + Fenpropimorph

C	BAS 48500F	BASF plc	06784	
C	Colstar	Du Pont (UK) Ltd	06783	
C	DUK 7876	Du Pont (UK) Ltd	09588	

584 Flusilazole + Fenpropimorph + Tridemorph

C	Bingo	BASF plc	06920	28/02/2002
C	DUK 51	Du Pont (UK) Ltd	06764	30/09/2001
C	Justice	Du Pont (UK) Ltd	07963	30/09/2001

585 Flusilazole + Tridemorph

C	Fusion	Du Pont (UK) Ltd	04908	30/09/2001
C	Meld	BASF plc	04914	28/02/2002
C	Option	Du Pont (UK) Ltd	07951	30/09/2001

586 Flutriafol

C	Barclay Rascal	Barclay Chemicals Manufacturing Ltd	08921	31/12/2002
C	Impact	Zeneca Crop Protection	06679	31/12/2001
C	Pointer	Zeneca Crop Protection	06695	31/12/2001
C	PP 450	Zeneca Crop Protection	06700	31/12/2001

587 Flutriafol + Azoxystrobin

| C | Amigo | Zeneca Crop Protection | 08834 | |
| C | Amistar Gem | Zeneca Crop Protection | 08833 | |

C   These products are "approved for agricultural use". For further details refer to page vii.
A   These products are approved for use in or near water. For further details refer to page vii.

Product Name	Marketing Company	Reg. No.	Expiry Date

588 Flutriafol + Carbendazim

C	Early Impact	Zeneca Crop Protection	06659	31/12/2001
C	Pacer	Zeneca Crop Protection	06690	31/12/2001
C	Palette	Zeneca Crop Protection	06691	31/12/2001

589 Flutriafol + Chlorothalonil

| C | Halo | Zeneca Crop Protection | 06520 | 31/12/2001 |
| C | Impact Excel | Zeneca Crop Protection | 06680 | 31/12/2001 |

590 Flutriafol + Ethirimol + Thiabendazole

| C | Ferrax | Zeneca Crop Protection | 06662 | 28/02/2002 |
| C | Ferrax | Bayer plc | 05284 | 28/02/2002 |

591 Fosetyl-Aluminium

C	Aliette	Aventis CropScience UK Limited	10036	
C	Aliette	Rhone-Poulenc Agriculture	05648	
C	Aliette	Hortichem Ltd	02484	
C	Aliette 80 WG	Hortichem Ltd	09156	

592 Fuberidazole + Bitertanol

C	Sibutol	Bayer plc	07238	
C	Sibutol CF	Bayer plc	08174	
C	Sibutol LS	Bayer plc	08109	31/12/2001
C	Sibutol New Formula	Bayer plc	08270	

593 Fuberidazole + Bitertanol + Imidacloprid + Triadimenol

| C | Cereline Secur | Bayer plc | 09511 | |

594 Fuberidazole + Bitertanol + Triadimenol

| C | Cereline | Bayer plc | 07239 | |

595 Fuberidazole + Ethirimol + Triadimenol

| C | Bay UK 292 | Bayer plc | 04335 | |

596 Fuberidazole + Imazalil + Triadimenol

| C | Baytan IM | Bayer plc | 00226 | |

597 Fuberidazole + Imidacloprid + Triadimenol

| C | Baytan Secur | Bayer plc | 09510 | |

598 Fuberidazole + Triadimenol

| C | Baytan | Bayer plc | 00225 | |
| C | Baytan CF | Bayer plc | 08193 | |

C These products are "approved for agricultural use". For further details refer to page vii.
A These products are approved for use in or near water. For further details refer to page vii.

Product Name	Marketing Company	Reg. No.	Expiry Date

598 Fuberidazole + Triadimenol—continued

C	Baytan Flowable	Bayer plc	02593	

599 Furalaxyl

C	Fongarid	Novartis Crop Protection UK Ltd	08407	

600 Glufosinate-Ammonium + Fentin Hydroxide

C	Safran	AgrEvo UK Crop Protection Ltd	08559	31/05/2001

601 Guazatine

C	Panoctine	Aventis CropScience UK Limited	10094	
C	Panoctine	Aventis CropScience UK Limited	10006	31/05/2003
C	Panoctine	Rhone-Poulenc Agriculture	06207	31/05/2003
C	Ravine	Aventis CropScience UK Limited	10095	
C	Ravine	Aventis CropScience UK Limited	10009	31/05/2003
C	Ravine	Rhone-Poulenc Agriculture	07193	31/05/2003

602 Guazatine + Imazalil

C	Panoctine Plus	Aventis CropScience UK Limited	10007	31/08/2002
C	Panoctine Plus	Rhone-Poulenc Agriculture	06208	31/08/2002
C	Ravine Plus	Aventis CropScience UK Limited	10010	31/08/2002
C	Ravine Plus	Rhone-Poulenc Agriculture	07343	31/08/2002

603 Guazatine + Triticonazole

C	Premis	Aventis CropScience UK Limited	10177	

604 Hymexazol

C	Tachigaren 70 WP	Sumitomo Corporation (UK) Plc	02649	

605 Imazalil

C	Fungaflor	Brinkman UK Ltd	05968	
C	Fungaflor	Hortichem Ltd	05967	
C	Fungaflor Smoke	Brinkman UK Ltd	06009	
C	Fungaflor Smoke	Hortichem Ltd	05969	
C	Fungazil 100 SL	Aventis CropScience UK Limited	10270	
C	Fungazil 100 SL	Rhone-Poulenc Agriculture	06202	30/09/2003
C	Sphinx	Rhone-Poulenc Agriculture	07607	
C	Stryper	Uniroyal Chemical Ltd	08310	

606 Imazalil + Azaconazole

C	Nectec Paste	Hortichem Ltd	08510	

C   These products are "approved for agricultural use". For further details refer to page vii.
A   These products are approved for use in or near water. For further details refer to page vii.

Product Name	Marketing Company	Reg. No.	Expiry Date

607 Imazalil + Carboxin + Thiabendazole

C	Vitaflo Extra	Uniroyal Chemical Ltd	07048

608 Imazalil + Fuberidazole + Triadimenol

C	Baytan IM	Bayer plc	00226

609 Imazalil + Guazatine

C	Panoctine Plus	Aventis CropScience UK Limited	10007	31/08/2002
C	Panoctine Plus	Rhone-Poulenc Agriculture	06208	31/08/2002
C	Ravine Plus	Aventis CropScience UK Limited	10010	31/08/2002
C	Ravine Plus	Rhone-Poulenc Agriculture	07343	31/08/2002

610 Imazalil + Pencycuron

C	Monceren IM	Bayer plc	06259
C	Monceren IM Flowable	Bayer plc	06731

611 Imazalil + Thiabendazole

C	Extratect Flowable	Seedcote Systems Ltd	09157
C	Extratect Flowable	Novartis Crop Protection UK Ltd	08704

612 Imazalil + Triticonazole

C	Robust	Aventis CropScience UK Limited	10176

613 Imidacloprid + Bitertanol + Fuberidazole + Triadimenol

C	Cereline Secur	Bayer plc	09511

614 Imidacloprid + Fuberidazole + Triadimenol

C	Baytan Secur	Bayer plc	09510

615 Imidacloprid + Tebuconazole + Triazoxide

C	Raxil Secur	Bayer plc	08966

616 Iprodione

C	Aventis Rovral Flo	Aventis CropScience UK Limited	09974	
C	Aventis Rovral Flo	Rhone-Poulenc Agriculture	09767	31/05/2003
C	CDA Rovral	Rhone-Poulenc Amenity	04679	
C	IT Iprodione	IT Agro Ltd	08267	
C	Landgold Iprodione 250	Landgold & Co Ltd	06465	31/05/2003
C	Rovral Flo	Aventis CropScience UK Limited	10013	
C	Rovral Flo	Rhone-Poulenc Agriculture	06328	31/05/2003
C	Rovral Green	Aventis Environmental Science	09938	
C	Rovral Green	Rhone-Poulenc Amenity	05702	
C	Rovral Liquid FS	Aventis CropScience UK Limited	10014	

C   These products are "approved for agricultural use". For further details refer to page vii.
A   These products are approved for use in or near water. For further details refer to page vii.

183

Product Name	Marketing Company	Reg. No.	Expiry Date

616 Iprodione—continued

C Rovral Liquid FS	Rhone-Poulenc Agriculture	09776	
C Rovral Liquid FS	Rhone-Poulenc Agriculture	06366	31/10/2002
C Rovral WP	Aventis CropScience UK Limited	10015	
C Rovral WP	Rhone-Poulenc Agriculture	06091	
C Turbair Rovral	Pan Britannica Industries Ltd	02248	31/12/2001

617 Iprodione + Carbendazim

C Calidan	Aventis CropScience UK Limited	09980	
C Calidan	Rhone-Poulenc Agriculture	06536	
C Vitesse	Aventis CropScience UK Limited	10042	
C Vitesse	Rhone-Poulenc Amenity	06537	

618 Iprodione + Thiophanate-Methyl

C Aventis Compass	Aventis CropScience UK Limited	10040	
C Aventis Compass	Rhone-Poulenc Agriculture	09760	31/10/2003
C Compass	Aventis CropScience UK Limited	10041	
C Compass	Rhone-Poulenc Agriculture	06190	31/10/2003
C Snooker	Aventis CropScience UK Limited	10018	
C Snooker	Rhone-Poulenc Agriculture	07940	31/10/2003

619 Kresoxim-Methyl

C Stroby WG	BASF plc	08653	

620 Kresoxim-Methyl + Epoxiconazole

C Barclay Avalon	Barclay Chemicals Manufacturing Ltd	09466	
C Clayton Gantry	Clayton Plant Protection (UK) Ltd	09482	
C CleanCrop Kresoxazole	United Agri Products	09698	
C Landgold Strobilurin KE	Landgold & Co Ltd	09908	
C Landmark	BASF plc	08889	
C Marnoch Kempo	Marnoch Ventures Limited	09918	
C Me^2 KME	Me^2 Crop Protection Ltd	09594	
C Standon Kresoxim-Epoxiconazole	Standon Chemicals Ltd	09281	

621 Kresoxim-Methyl + Epoxiconazole + Fenpropimorph

C CleanCrop Kresoxazole Plus	United Agri Products	09742	31/03/2002
C Mantra	BASF plc	08886	
C Standon Kresoxim Super	Standon Chemicals Ltd	09794	

622 Kresoxim-Methyl + Fenpropimorph

C Ensign	BASF plc	08362	
C Greencrop Monsoon	Greencrop Technology Ltd	09573	

C These products are "approved for agricultural use". For further details refer to page vii.
A These products are approved for use in or near water. For further details refer to page vii.

Product Name	Marketing Company	Reg. No.	Expiry Date

622 Kresoxim-Methyl + Fenpropimorph—continued

C	Landgold Strobilurin KF	Landgold & Co Ltd	09196
C	Standon Kresoxim FM	Standon Chemicals Ltd	08922

623 Mancozeb

C	Agrichem Mancozeb 80	Agrichem (International) Ltd	06354	
C	Ashlade Mancozeb FL	Ashlade Formulations Ltd	06226	
C	Barclay Manzeb 455	Barclay Chemicals Manufacturing Ltd	07990	
C	Deny WP	Interfarm (UK) Limited	09899	
C	Dequiman MZ	Elf Atochem Agri SA	06870	
C	Dithane 945	Interfarm (UK) Limited	09897	
C	Dithane 945	SumiAgro (UK) Ltd	09889	
C	Dithane 945	Pan Britannica Industries Ltd	04017	31/12/2001
C	Dithane 945	Pan Britannica Industries Ltd	00719	
C	Dithane Dry Flowable	Pan Britannica Industries Ltd	04255	31/12/2001
C	Dithane Dry Flowable	Pan Britannica Industries Ltd	04251	30/09/2002
C	Dithane Dry Flowable Newtec	SumiAgro (UK) Ltd	09892	
C	Dithane Dry Flowable Newtec	pbi Agrochemicals Ltd	09754	
C	Dithane NT Dry Flowable	Interfarm (UK) Limited	09898	
C	Dithane Superflo	Pan Britannica Industries Ltd	06593	31/12/2001
C	Dithane Superflo	Pan Britannica Industries Ltd	06290	31/05/2002
C	Headland Zebra Flo	Headland Agrochemicals Ltd	07442	31/03/2002
C	Headland Zebra WP	Headland Agrochemicals Ltd	07441	
C	Helm 75 WG Newtec	pbi Agrochemicals Ltd	09757	
C	Helm 75WG	Pan Britannica Industries Ltd	08309	30/09/2002
C	Karamate Dry Flo	Landseer Limited	09259	
C	Karamate Dry Flo	Rohm & Haas (UK) Ltd	04250	30/09/2001
C	Karamate Dry Flo Newtec	Landseer Limited	09759	
C	Karamate N	Rohm&Haas (UK) Ltd	01125	
C	Kor DF	Headland Agrochemicals Ltd	08979	30/09/2002
C	Kor DF Newtec	Headland Agrochemicals Ltd	09758	
C	Kor Flo	Interfarm (UK) Limited	09895	
C	Kor Flo	Headland Agrochemicals Ltd	08019	
C	Kor NT Dry Flowable	Interfarm (UK) Limited	09893	
C	Landgold Mancozeb 80 W	Landgold & Co Ltd	06507	25/07/2003
C	Luxan Mancozeb Flowable	Luxan (UK) Ltd	06812	30/09/2002
C	Manconex	Griffin (Europe) Marketing NV (UK	09555	
C	Mancozeb 80	Rohm & Haas (UK) Ltd	09896	
C	Mandate 75 WDG	Portman Agrochemicals Ltd	09051	31/12/2002
C	Mandate 80 WP	Portman Agrochemicals Ltd	09080	
C	Manex II	Agrichem (International) Ltd	07637	
C	Manzate 200 DF	Griffin (Europe) Marketing NV (UK	10081	
C	Manzate 200 DF	Du Pont (UK) Ltd	06010	31/03/2003
C	Manzate 200 PI	Griffin (Europe) SA	09480	

C   These products are "approved for agricultural use". For further details refer to page vii.
A   These products are approved for use in or near water. For further details refer to page vii.

Product Name	Marketing Company	Reg. No.	Expiry Date

623 Mancozeb—continued

	Product Name	Marketing Company	Reg. No.	Expiry Date
C	Manzate 200 PI	Du Pont (UK) Ltd	07209	28/02/2002
C	Micene 80	Sipcam UK Ltd	09112	
C	Micene 80	Sipcam UK Ltd	08560	
C	Micene DF	Sipcam UK Ltd	09957	
C	Mortar Flo	Griffin (Europe) Marketing NV (UK	09592	
C	Nemispor	Sipcam UK Ltd	07348	
C	Opie 80 WP	SumiAgro (UK) Ltd	09890	
C	Opie 80 WP	Pan Britannica Industries Ltd	08301	
C	Penncozeb	Elf Atochem Agri SA	07820	
C	Penncozeb WDG	Nufarm Whyte Agriculture Ltd	09690	
C	Penncozeb WDG	Elf Atochem Agri SA	07833	31/07/2002
C	Quell Flo	Interfarm (UK) Limited	09894	
C	Quell Flo	Headland Agrochemicals Ltd	08317	
C	Restraint DF	Griffin (Europe) Marketing NV (UK	09499	30/09/2002
C	Restraint DF Newtec	Griffin (Europe) Marketing NV (UK	09755	
C	Stefes Deny	Stefes Plant Protection Ltd	08932	
C	Stefes Restraint	Stefes Plant Protection Ltd	08945	31/03/2001
C	Tariff 75 WG Newtec	SumiAgro (UK) Ltd	09891	
C	Tariff 75 WG Newtec	pbi Agrochemicals Ltd	09756	
C	Tariff 75WG	Pan Britannica Industries Ltd	08308	30/09/2002
C	Tridex	Elf Atochem Agri SA	07922	31/12/2001
C	Trimanzone	Intracrop	09584	
C	Trimanzone	Brian Lewis Agriculture Ltd	09278	31/05/2002
C	Unicrop Mancozeb	Universal Crop Protection Ltd	05467	
C	Unicrop Mancozeb 80	Universal Crop Protection Ltd	07451	31/03/2002

624 Mancozeb + Benalaxyl

	Product Name	Marketing Company	Reg. No.	Expiry Date
C	Galben M	Sipcam UK Ltd	07220	
C	Tairel	Sipcam UK Ltd	07767	
C	Trecatol	Rohm & Haas (UK) Ltd	09823	

625 Mancozeb + Carbendazim

	Product Name	Marketing Company	Reg. No.	Expiry Date
C	Headland Maple	Headland Agrochemicals Ltd	09044	
C	Kombat WDG	Rohm & Haas (UK) Ltd	05509	

626 Mancozeb + Chlorothalonil

	Product Name	Marketing Company	Reg. No.	Expiry Date
C	Adagio	Rohm & Haas (UK) Ltd	10057	
C	Adagio	pbi Agrochemicals Ltd	09597	30/04/2003
C	Adagio	Pan Britannica Industries Ltd	07832	30/04/2003
C	Adagio C	Rohm & Haas (UK) Ltd	10056	
C	Adagio C	pbi Agrochemicals Ltd	09842	
C	Adagio C	pbi Agrochemicals Ltd	09309	30/04/2003
C	Delphi	Sipcam UK Ltd	09860	

C These products are "approved for agricultural use". For further details refer to page vii.
A These products are approved for use in or near water. For further details refer to page vii.

Product Name	Marketing Company	Reg. No.	Expiry Date

626 Mancozeb + Chlorothalonil—continued

C	Dreadnought Flo	pbi Agrochemicals Ltd	09599
C	Dreadnought Flo	Rohm & Haas (UK) Ltd	07600
C	Dreadnought Flo	Pan Britannica Industries Ltd	07599
C	Sipcam Flo	Sipcam UK Ltd	07601

627 Mancozeb + Cymoxanil

C	Ashlade Solace	Ashlade Formulations Ltd	08087
C	Besiege	Du Pont (UK) Ltd	08086
C	Besiege WSB	Du Pont (UK) Ltd	08075
C	Clayton Krypton	Clayton Plant Protection (UK) Ltd	09398
C	CleanCrop Xanilite	United Agri Products	10050
C	Curzate M68	Du Pont (UK) Ltd	08072
C	Curzate M68 WSB	Du Pont (UK) Ltd	08073
C	Fytospore 68	Du Pont (UK) Ltd	09827
C	Fytospore 68	Du Pont (UK) Ltd	08649	30/11/2002
C	Marnoch Mancym	Marnoch Ventures Limited	10130
C	Me^2 Cymoxeb	Me^2 Crop Protection Ltd	09486
C	Rhythm	Interfarm (UK) Limited	09909
C	Rhythm	Interfarm (UK) Limited	09636
C	Standon Cymoxanil Extra	Standon Chemicals Ltd	09442
C	Systol M	Quadrangle Agrochemicals	08085
C	Systol M	Quadrangle Agrochemicals	03480
C	Systol M WSB	Quadrangle Agrochemicals	08074

628 Mancozeb + Cymoxanil + Oxadixyl

C	Ripost Pepite	Bayer Ltd	08895
C	Ripost Pepite	Novartis Crop Protection UK Ltd	08485
C	Ripost Pepite	Sandoz Agro Ltd	06485	31/03/2001
C	Trustan WDG	Du Pont (UK) Ltd	05050

629 Mancozeb + Cyproconazole

C	Alto Eco	Novartis Crop Protection UK Ltd	08466

630 Mancozeb + Dimethomorph

C	Invader	Cyanamid Agriculture Ltd	06989
C	Invader WDG	Cyanamid Agriculture Ltd	09556
C	Saracen	Cyanamid Agriculture Ltd	07872
C	Saracen WDG	Cyanamid Agriculture Ltd	09557

631 Mancozeb + Maneb

C	Barclay Manzeb Flow	Barclay Chemicals Manufacturing Ltd	05872

C   These products are "approved for agricultural use". For further details refer to page vii.
A   These products are approved for use in or near water. For further details refer to page vii.

Product Name	Marketing Company	Reg. No.	Expiry Date

632 Mancozeb + Metalaxyl

C	Fubol 58 WP	Novartis Crop Protection UK Ltd	08534
C	Fubol 75 WP	Novartis Crop Protection UK Ltd	08409
C	Osprey 58 WP	Novartis Crop Protection UK Ltd	08428
C	Ridomil MZ 75 WP	Novartis Crop Protection UK Ltd	08438

633 Mancozeb + Metalaxyl-M

C	Fubol Gold	Novartis Crop Protection UK Ltd	08812
C	Fubol Gold WG	Novartis Crop Protection UK Ltd	10184

634 Mancozeb + Ofurace

C	Patafol	Mirfield Sales Services Ltd	07397

635 Mancozeb + Oxadixyl

C	Recoil	Novartis Crop Protection UK Ltd	08483

636 Mancozeb + Propamocarb Hydrochloride

C	Aventis Tattoo	Aventis CropScience UK Limited	09811
C	Tattoo	Aventis CropScience UK Limited	07293

637 Maneb

C	Agrichem Maneb 80	Agrichem (International) Ltd	05474
C	Ashlade Maneb Flowable	Ashlade Formulations Ltd	06477
C	Headland Spirit	Headland Agrochemicals Ltd	04548	30/06/2002
C	Luxan Maneb 80	Luxan (UK) Ltd	06570
C	Maneb 80	Rohm & Haas (UK) Ltd	01276
C	Manex	Griffin (Europe) Marketing NV (UK	09554
C	Manex	Agrichem (International) Ltd	07935
C	Manex	Chiltern Farm Chemicals Ltd	05731
C	RH Maneb 80	Rohm & Haas (UK) Ltd	01796
C	Trimangol 80	Elf Atochem Agri SA	06871
C	Trimangol 80	Atochem Agri BV	06070
C	Trimangol WDG	Elf Atochem Agri SA	06992
C	Unicrop Maneb 80	Universal Crop Protection Ltd	06926
C	Unicrop Maneb FL	Universal Crop Protection Ltd	08025
C	X-Spor SC	United Phosphorus Ltd	08077

638 Maneb + Carbendazim

C	Ashlade Mancarb FL	Ashlade Formulations Ltd	07977
C	Headland Dual	Headland Agrochemicals Ltd	03782
C	MC Flowable	United Phosphorus Ltd	08198
C	Multi-W FL	SumiAgro (UK) Ltd	09447
C	Multi-W FL	Pan Britannica Industries Ltd	04131	28/02/2002
C	Protector	Procam Group Ltd	09448

C These products are "approved for agricultural use". For further details refer to page vii.
A These products are approved for use in or near water. For further details refer to page vii.

Product Name	Marketing Company	Reg. No.	Expiry Date

638 Maneb + Carbendazim—continued

C	Protector	Procam Group Ltd	07075	28/02/2002
C	Tripart 147	Tripart Farm Chemicals Ltd	07978	

639 Maneb + Carbendazim + Chlorothalonil

C	Ashlade Mancarb Plus	Ashlade Formulations Ltd	08160	31/12/2001
C	Tripart Victor	Tripart Farm Chemicals Ltd	08161	31/12/2001

640 Maneb + Carbendazim + Sulphur

C	Bolda FL	Atlas Crop Protection Ltd	07653	
C	Legion S	Ashlade Formulations Ltd	08121	30/06/2002

641 Maneb + Carbendazim + Tridemorph

C	Cosmic FL	BASF plc	03473	28/02/2002

642 Maneb + Copper Oxychloride + Sulphur

C	Ashlade SMC Flowable	Ashlade Formulations Ltd	06494	

643 Maneb + Fentin Acetate

C	Brestan 60 SP	Aventis CropScience UK Limited	07305	

644 Maneb + Mancozeb

C	Barclay Manzeb Flow	Barclay Chemicals Manufacturing Ltd	05872	

645 Metalaxyl

C	Polycote Universal	Seedcote Systems Ltd	08431	

646 Metalaxyl + Carbendazim

C	Ridomil MBC 60 WP	Novartis Crop Protection UK Ltd	08437	

647 Metalaxyl + Chlorothalonil

C	Folio	Novartis Crop Protection UK Ltd	08547	

648 Metalaxyl + Copper Oxychloride

C	Ridomil Plus	Novartis Crop Protection UK Ltd	08353	
C	Ridomil Plus 50 WP	Ciba Agriculture	01803	31/03/2001

649 Metalaxyl + Mancozeb

C	Fubol 58 WP	Novartis Crop Protection UK Ltd	08534	
C	Fubol 75 WP	Novartis Crop Protection UK Ltd	08409	
C	Osprey 58 WP	Novartis Crop Protection UK Ltd	08428	
C	Ridomil MZ 75 WP	Novartis Crop Protection UK Ltd	08438	

C   These products are "approved for agricultural use". For further details refer to page vii.
A   These products are approved for use in or near water. For further details refer to page vii.

Product Name	Marketing Company	Reg. No.	Expiry Date

650 Metalaxyl + Thiabendazole

C Apron T69 WS	Novartis Crop Protection UK Ltd	08387	
C Polycote Select	Germains UK Ltd	09718	
C Polycote Select	Seedcote Systems Ltd	08430	

651 Metalaxyl + Thiabendazole + Thiram

C Apron Combi FS	Novartis Crop Protection UK Ltd	08386	

652 Metalaxyl + Thiram

C Favour 600 SC	Novartis Crop Protection UK Ltd	08405	

653 Metalaxyl-M

C SL 567A	Novartis Crop Protection UK Ltd	08811	

654 Metalaxyl-M + Mancozeb

C Fubol Gold	Novartis Crop Protection UK Ltd	08812	
C Fubol Gold WG	Novartis Crop Protection UK Ltd	10184	

655 Metconazole

C Caramba	BASF plc	10213	
C Caramba	Cyanamid Agriculture Ltd	09864	

656 Myclobutanil

C Systhane 20 EW	T P Whelehan Son & Co Ltd	09397	
C Systhane 20 EW	Landseer Limited	09396	
C Systhane 6 Flo	AgrEvo UK Ltd	07334	
C Systhane 6 Flo	T P Whelehan Son & Co Ltd	06551	
C Systhane 6W	SumiAgro (UK) Ltd	09611	
C Systhane 6W	Pan Britannica Industries Ltd	04571	
C Systhane 6W	Pan Britannica Industries Ltd	04570	
C Systhane ST	Uniroyal Chemical Ltd	08034	30/10/2001

657 Nuarimol

C Chemtech Nuarimol	Chemtech (Crop Protection) Ltd	04517	
C Triminol	Dow AgroSciences Ltd	05818	

658 Octhilinone

C Pancil T	Landseer Limited	09261	
C Pancil T	Rohm & Haas (UK) Ltd	01540	30/09/2001

659 Ofurace + Mancozeb

C Patafol	Mirfield Sales Services Ltd	07397	

C   These products are "approved for agricultural use". For further details refer to page vii.
A   These products are approved for use in or near water. For further details refer to page vii.

660 Oxadixyl + Carbendazim + Cymoxanil + Thiram

C	Apron Elite	Novartis Crop Protection UK Ltd	08770	

661 Oxadixyl + Cymoxanil + Mancozeb

C	Ripost Pepite	Bayer Ltd	08895	
C	Ripost Pepite	Novartis Crop Protection UK Ltd	08485	
C	Ripost Pepite	Sandoz Agro Ltd	06485	31/03/2001
C	Trustan WDG	Du Pont (UK) Ltd	05050	

662 Oxadixyl + Mancozeb

C	Recoil	Novartis Crop Protection UK Ltd	08483

663 Oxadixyl + Thiabendazole + Thiram

C	Leap	Uniroyal Chemical Ltd	08477	
C	Leap	Uniroyal Chemical Ltd	07884	31/12/2001

664 Oxycarboxin

C	Plantvax 20	Uniroyal Chemical Ltd	01600	31/05/2001
C	Plantvax 75	Fargro Ltd	01601	
C	Ringmaster	Rhone-Poulenc Amenity	05334	30/09/2001

665 Parahydroxyphenylsalicylamide

	Fumispore	Laboratoire de Chimie et Biologie	08300

666 Penconazole

C	Topas	Novartis Crop Protection UK Ltd	09717	
C	Topas	Novartis Crop Protection UK Ltd	08458	31/10/2002

667 Penconazole + Captan

C	Topas C 50 WP	Novartis Crop Protection UK Ltd	08459

668 Penconazole + Dithianon

C	Topas D275 SC	Novartis Crop Protection UK Ltd	08460

669 Pencycuron

C	Marnoch Penor D	Marnoch Ventures Limited	09871
C	Monceren DS	Bayer plc	04160
C	Monceren FS	Bayer plc	04907
C	Standon Pencycuron DP	Standon Chemicals Ltd	08774

670 Pencycuron + Imazalil

C	Monceren IM	Bayer plc	06259
C	Monceren IM Flowable	Bayer plc	06731

C   These products are "approved for agricultural use". For further details refer to page vii.
A   These products are approved for use in or near water. For further details refer to page vii.

Product Name	Marketing Company	Reg. No.	Expiry Date

671 Permethrin + Thiram

C	Combinex	Pan Britannica Industries Ltd	00562	31/10/2001

672 Prochloraz

C	Admiral	AgrEvo UK Ltd	08162	31/05/2001
C	Barclay Eyetak	Barclay Chemicals Manufacturing Ltd	06813	30/06/2001
C	Barclay Eyetak 40	Barclay Chemicals Manufacturing Ltd	07843	
C	Barclay Eyetak 450	Barclay Chemicals Manufacturing Ltd	09484	
C	Levington Octave	Levington Horticulture Ltd	07505	
C	Mirage 40 EC	Makhteshim-Agan (UK) Ltd	06770	
C	Octave	Aventis CropScience UK Limited	07267	
C	Prelude 20 LF	Aventis CropScience UK Limited	07269	
C	Prelude 20 LF	Agrichem (International) Ltd	04371	
C	Scotts Octave	The Scotts Company (UK) Limited	09275	
C	Sporgon 50 WP	Sylvan Spawn Ltd	08802	
C	Sportak 45 EW	Aventis CropScience UK Limited	07996	
C	Sportak 45 HF	Aventis CropScience UK Limited	07287	
C	Sportak Focus EW	AgrEvo UK Ltd	08002	30/09/2001
C	Sportak Sierra EW	AgrEvo UK Ltd	08003	31/05/2001
C	Sportak Sierra HF	AgrEvo UK Ltd	07290	31/05/2001
C	Stefes Poraz	Aventis CropScience UK Limited	10090	
C	Stefes Poraz	Stefes UK Ltd	07528	

673 Prochloraz + Carbendazim

C	Novak	Aventis CropScience UK Limited	08020	
C	Sportak Alpha	AgrEvo UK Ltd	07222	31/05/2001
C	Sportak Alpha HF	Aventis CropScience UK Limited	07225	

674 Prochloraz + Carboxin

C	Prelude Universal LS	AgrEvo UK Crop Protection Ltd	08650	31/05/2001
C	Provax	Uniroyal Chemical Company	08651	31/05/2001

675 Prochloraz + Cyproconazole

C	Profile	Aventis CropScience UK Limited	08134	
C	SAN 710	Novartis Crop Protection UK Ltd	08488	31/05/2001
C	SAN 710 HF	Novartis Crop Protection UK Ltd	08489	
C	Sportak Delta 460	AgrEvo UK Crop Protection Ltd	07224	31/05/2001
C	Sportak Delta 460 HF	Aventis CropScience UK Limited	07431	
C	Tiptor	Novartis Crop Protection UK Ltd	08495	
C	Tiptor	AgrEvo UK Ltd	07295	31/05/2001

C   These products are "approved for agricultural use". For further details refer to page vii.
A   These products are approved for use in or near water. For further details refer to page vii.

Product Name	Marketing Company	Reg. No.	Expiry Date

676 Prochloraz + Fenbuconazole

C Mirage Extra	Stefes UK Ltd	07958	
C Stefes Fortune	Stefes Plant Protection Ltd	08537	31/01/2001
C Stefes Inception	Stefes Plant Protection Ltd	08829	

677 Prochloraz + Fenpropidin

C SL 552A	Novartis Crop Protection UK Ltd	08673	
C Sponsor	Aventis CropScience UK Limited	08674	

678 Prochloraz + Fenpropimorph

C SL571A	Novartis Crop Protection UK Ltd	08420	31/03/2002
C Sprint	AgrEvo UK Ltd	07291	31/05/2001
C Sprint HF	AgrEvo UK Ltd	07292	31/03/2002

679 Prochloraz + Fluquinconazole

C Aventis Foil	Aventis CropScience UK Limited	09709	
C Baron	Aventis CropScience UK Limited	09364	
C Flamenco Plus	Aventis CropScience UK Limited	09362	
C Foil	Aventis CropScience UK Limited	09363	
C Jockey	Aventis CropScience UK Limited	10076	
C Jockey Plus	Aventis CropScience UK Limited	10075	

680 Prochloraz + Propiconazole

C Bumper P	Makhteshim-Agan (UK) Ltd	08548	
C Greencrop Twinstar	Greencrop Technology Ltd	09516	

681 Prochloraz + Tebuconazole

C Agate	Bayer plc	08826	

682 Propamocarb Hydrochloride

C Filex	The Scotts Company (UK) Limited	07631	
C Previcur N	Aventis CropScience UK Limited	08575	
C Proplant	Chimac-Agriphar SA	08572	

683 Propamocarb Hydrochloride + Chlorothalonil

C Aventis Merlin	Aventis CropScience UK Limited	09719	
C Merlin	Aventis CropScience UK Limited	07943	
C Tattoo C	Aventis CropScience UK Limited	07623	

684 Propamocarb Hydrochloride + Mancozeb

C Aventis Tattoo	Aventis CropScience UK Limited	09811	
C Tattoo	Aventis CropScience UK Limited	07293	

C   These products are "approved for agricultural use". For further details refer to page vii.
A   These products are approved for use in or near water. For further details refer to page vii.

Product Name	Marketing Company	Reg. No.	Expiry Date

685 Propiconazole

C Barclay Bolt	Barclay Chemicals Manufacturing Ltd	08341	
C Bumper 250 EC	Makhteshim-Agan (UK) Ltd	09039	
C Controller Fungicide	Novartis Crop Protection UK Ltd	08399	
C Landgold Propiconazole	Landgold & Co Ltd	06291	25/07/2003
C Mantis	Novartis Crop Protection UK Ltd	08423	
C Radar	Zeneca Crop Protection	09168	
C Standon Propiconazole	Standon Chemicals Ltd	07037	25/07/2003
C Tilt	Novartis Crop Protection UK Ltd	08456	

686 Propiconazole + Carbendazim

C Hispor 45 WP	Novartis Crop Protection UK Ltd	08418	
C Sparkle 45 WP	Novartis Crop Protection UK Ltd	08450	

687 Propiconazole + Chlorothalonil

C Sambarin 312.5 SC	Novartis Crop Protection UK Ltd	08439	

688 Propiconazole + Cyproconazole

C Menara	Novartis Crop Protection UK Ltd	09321	

689 Propiconazole + Fenbuconazole

C Graphic	Rohm & Haas (UK) Ltd	09907	
C Indar CG	Novartis Crop Protection UK Ltd	08578	

690 Propiconazole + Fenpropidin

C Prophet	Novartis Crop Protection UK Ltd	08433	
C Sheen	Novartis Crop Protection UK Ltd	08442	
C Zulu	Novartis Crop Protection UK Ltd	08464	

691 Propiconazole + Fenpropidin + Tebuconazole

C Bayer UK 593	Bayer plc	08285	
C Gladio	Novartis Crop Protection UK Ltd	08413	

692 Propiconazole + Fenpropimorph

C Belvedere	Makhteshim-Agan (UK) Ltd	08084	
C Decade	Novartis Crop Protection UK Ltd	08402	
C Glint	Novartis Crop Protection UK Ltd	08414	
C Mantle	Novartis Crop Protection UK Ltd	08424	

693 Propiconazole + Prochloraz

C Bumper P	Makhteshim-Agan (UK) Ltd	08548	
C Greencrop Twinstar	Greencrop Technology Ltd	09516	

C These products are "approved for agricultural use". For further details refer to page vii.
A These products are approved for use in or near water. For further details refer to page vii.

Product Name	Marketing Company	Reg. No.	Expiry Date

694 Propiconazole + Tebuconazole

C	Cogito	Novartis Crop Protection UK Ltd	08397	
C	Endeavour	Bayer plc	07385	

695 Propiconazole + Tridemorph

C	Joust	Universal Crop Protection Ltd	08122	28/02/2002
C	Tilt Turbo 475 EC	Novartis Crop Protection UK Ltd	08457	30/09/2001
C	Tilt Turbo 475 EC	Ciba Agriculture	03476	30/09/2001

696 Propiconazole + Trifloxystrobin

C	Rombus	Novartis Crop Protection UK Ltd	10262	

697 Propineb

C	Antracol	Bayer plc	00104	

698 Propoxur

C	Fumite Propoxur Smoke	Octavius Hunt Ltd	09290	31/12/2001

699 Pyrazophos

C	Afugan	AgrEvo UK Ltd	07301	20/04/2001

700 Pyrifenox

C	Dorado	Zeneca Crop Protection	06657	
C	SL 471 200 EC	Novartis Crop Protection UK Ltd	08446	31/07/2001

701 Pyrimethanil

C	Scala	Aventis CropScience UK Limited	07806	

702 Quinomethionate

C	Morestan	Bayer plc	01376	

703 Quinoxyfen

C	Apres	Dow AgroSciences Ltd	08881	
C	Erysto	Dow AgroSciences Ltd	08697	
C	Fortress	Dow AgroSciences Ltd	08279	

704 Quinoxyfen + Cyproconazole

C	Divora	Novartis Crop Protection UK Ltd	08960	
C	DOE 1762	Dow AgroSciences Ltd	08962	
C	NOV 1390	Novartis Crop Protection UK Ltd	08961	

705 Quinoxyfen + Fenpropimorph

C	EF-1288	Dow AgroSciences Ltd	08241	

C   These products are "approved for agricultural use". For further details refer to page vii.
A   These products are approved for use in or near water. For further details refer to page vii.

Product Name	Marketing Company	Reg. No.	Expiry Date

705 Quinoxyfen + Fenpropimorph—continued

C	Orka	Dow AgroSciences Ltd	08879

706 Quintozene

C	Quintozene Wettable Powder	Rhone-Poulenc Amenity	05404
C	Quintozene WP	Aventis Environmental Science	09936
C	Terraclor 20D	Uniroyal Chemical Ltd	06578
C	Terraclor Flo	Hortichem Ltd	08666
C	Terraclor Flo	Uniroyal Chemical Ltd	08665

707 Spiroxamine

C	Accrue	Bayer plc	08335
C	Bayer UK 477	Bayer plc	08330
C	Neon	Bayer plc	08337
C	Talvin	Bayer plc	09253
C	Torch	Bayer plc	08336
C	Zenon	Bayer plc	09193

708 Spiroxamine + Tebuconazole

C	Array	Bayer plc	09352
C	Bayer UK 552	Bayer plc	08331
C	Beam	Bayer plc	08332
C	Bronze	Bayer plc	08333
C	Draco	Bayer plc	09353
C	Sage	Bayer plc	08334

709 Sulphur

C	Ashlade Sulphur FL	Ashlade Formulations Ltd	06478	
C	Headland Sulphur	Headland Agrochemicals Ltd	03714	
C	Headland Venus	Headland Agrochemicals Ltd	09572	
C	Kumulus DF	BASF plc	04707	
C	Kumulus S	BASF plc	01170	
C	Luxan Micro-Sulphur	Luxan (UK) Ltd	06565	
C	Microsul Flowable Sulphur	Stoller Chemical Ltd	03907	31/05/2002
C	Microthiol Special	Elf Atochem Agri SA	06268	
C	MSS Sulphur 80	Mirfield Sales Services Ltd	05752	
C	MSS Sulphur 80 WP	Mirfield Sales Services Ltd	03225	
C	Solfa	Atlas Interlates Ltd	06959	
C	Stoller Flowable Sulphur	Stoller Chemical Ltd	03760	31/05/2002
C	Sulphur Flowable	United Phosphorus Ltd	07526	
C	Thiovit	Novartis Crop Protection UK Ltd	08493	
C	Tripart Imber	Tripart Farm Chemicals Ltd	04050	

C These products are "approved for agricultural use". For further details refer to page vii.
A These products are approved for use in or near water. For further details refer to page vii.

Product Name	Marketing Company	Reg. No.	Expiry Date
710 Sulphur + Carbendazim + Maneb			
C Bolda FL	Atlas Crop Protection Ltd	07653	
C Legion S	Ashlade Formulations Ltd	08121	30/06/2002
711 Sulphur + Copper Oxychloride + Maneb			
C Ashlade SMC Flowable	Ashlade Formulations Ltd	06494	
712 Tar acids			
C Bray's Emulsion	Fargro Ltd	08316	28/02/2002
713 Tar Oils			
C Sterilite Tar Oil Winter Wash 60% Stock Emulsion	Coventry Chemicals Ltd	05061	
C Sterilite Tar Oil Winter Wash 80% Miscible Quality	Coventry Chemicals Ltd	05062	
714 Tebuconazole			
C Barclay Busker	Barclay Chemicals Manufacturing Ltd	08994	
C Bayer UK226	Bayer plc	09504	
C Clayton Tebucon	Clayton Plant Protection (UK) Ltd	08707	
C Clayton Tebucon	Clayton Plant Protection (UK) Ltd	07045	25/07/2003
C Folicur	Bayer plc	08691	
C Gainer	Bayer plc	08692	
C Halt	Bayer plc	08693	
C Me^2 Tebuconazole	Me^2 Crop Protection Ltd	09751	
C Raxil	Bayer plc	06460	
C Standon Tebuconazole	Standon Chemicals Ltd	09056	
C UK 200	Bayer plc	08246	
715 Tebuconazole + Carbendazim			
C Bayer UK413	Bayer plc	08277	
C Tricur	Bayer plc	10281	
716 Tebuconazole + Fenpropidin			
C Monicle	Bayer plc	07375	
C SL-556 500 EC	Novartis Crop Protection UK Ltd	08449	
717 Tebuconazole + Fenpropidin + Propiconazole			
C Bayer UK 593	Bayer plc	08285	
C Gladio	Novartis Crop Protection UK Ltd	08413	

C These products are "approved for agricultural use". For further details refer to page vii.
A These products are approved for use in or near water. For further details refer to page vii.

	Product Name	Marketing Company	Reg. No.	Expiry Date

718 Tebuconazole + Fenpropimorph

C	Bayer UK506	Bayer plc	07540	31/03/2001
C	BUK 84000F	BASF plc	07541	31/03/2001

719 Tebuconazole + Imidacloprid + Triazoxide

C	Raxil Secur	Bayer plc	08966	

720 Tebuconazole + Prochloraz

C	Agate	Bayer plc	08826	

721 Tebuconazole + Propiconazole

C	Cogito	Novartis Crop Protection UK Ltd	08397	
C	Endeavour	Bayer plc	07385	

722 Tebuconazole + Spiroxamine

C	Array	Bayer plc	09352	
C	Bayer UK 552	Bayer plc	08331	
C	Beam	Bayer plc	08332	
C	Bronze	Bayer plc	08333	
C	Draco	Bayer plc	09353	
C	Sage	Bayer plc	08334	

723 Tebuconazole + Triadimenol

C	Garnet	Bayer plc	06391	
C	Ruby	Bayer plc	06389	
C	Silvacur	Bayer plc	06387	
C	Silvacur 300	Bayer plc	08056	
C	Veto F	Bayer plc	08057	

724 Tebuconazole + Triazoxide

C	Raxil S	Bayer plc	06974	
C	Raxil S CF	Bayer plc	08192	

725 Tebuconazole + Tridemorph

C	Allicur	Bayer plc	06468	30/09/2001
C	BAS 91580F	BASF plc	06469	31/12/2001

726 Tecnazene

C	Atlas Tecgran 100	Atlas Crop Protection Ltd	07730	31/01/2002
C	Atlas Tecgran 100	Atlas Interlates Ltd	05574	31/01/2001
C	Atlas Tecnazene 6% dust	Atlas Crop Protection Ltd	07731	31/01/2002
C	Atlas Tecnazene 6% Dust	Atlas Interlates Ltd	06351	31/01/2002
C	Bygran F	Wheatley Chemical Co Ltd	00365	31/01/2002

C   These products are "approved for agricultural use". For further details refer to page vii.
A   These products are approved for use in or near water. For further details refer to page vii.

	Product Name	Marketing Company	Reg. No.	Expiry Date

726 Tecnazene—continued

	Bygran S	Wheatley Chemical Co Ltd	00366	31/01/2002
	Fusarex 10% Granules	Zeneca Crop Protection	06668	18/02/2001
C	Fusarex 10G	Hickson & Welch Ltd	09727	31/01/2002
	Fusarex 6% Dust	Zeneca Crop Protection	06667	18/02/2001
C	Hickstor 10	Hickson & Welch Ltd	03121	31/01/2002
	Hickstor 3	Hickson & Welch Ltd	03105	31/01/2002
C	Hickstor 5	Hickson & Welch Ltd	03180	31/01/2002
	Hickstor 6	Hickson & Welch Ltd	03106	31/01/2002
C	Hortag Tecnazene 10% Granules	Hortag Chemicals Ltd	03966	31/01/2002
	Hortag Tecnazene Double Dust	Hortag Chemicals Ltd	01072	31/01/2002
	Hortag Tecnazene Potato Dust	Hortag Chemicals Ltd	01074	31/01/2002
	Hortag Tecnazene Potato Granules	Hortag Chemicals Ltd	01075	31/01/2002
C	Hystore 10	Agrichem (International) Ltd	03581	31/01/2002
C	Hystore 10G	Agrichem (International) Ltd	06899	31/01/2002
	Hytec	Agrichem Ltd	01099	31/01/2002
C	Hytec 6	Agrichem Ltd	03580	31/01/2002
C	Nebulin	Wheatley Chemical Co Ltd	01469	31/03/2001
C	New Hickstor 6	Hickson & Welch Ltd	04221	31/01/2002
	New Hystore	Agrichem Ltd	01485	31/01/2002
C	Tripart Arena 10G	Tripart Farm Chemicals Ltd	05603	31/01/2002
	Tripart Arena 3	Tripart Farm Chemicals Ltd	05605	31/01/2002
C	Tripart Arena 5G	Tripart Farm Chemicals Ltd	05604	31/01/2002
C	Tripart New Arena 6	Tripart Farm Chemicals Ltd	05813	31/01/2002

727 Tecnazene + Carbendazim

C	Hickstor 6 Plus MBC	Hickson&Welch Ltd	04176	31/01/2002
C	Hortag Tecnacarb	Avon Packers Ltd	02929	31/01/2002
C	New Arena Plus	Hickson & Welch	04598	31/01/2002
C	New Hickstor 6 Plus MBC	Hickson & Welch Ltd	04599	31/01/2002
C	Tripart Arena Plus	Hickson & Welch	05602	31/01/2002

728 Tecnazene + Thiabendazole

C	Hytec Super	Agrichem (International) Ltd	01100	31/01/2002
	New Arena TBZ 6	Tripart Farm Chemicals Ltd	05606	31/01/2002

729 Tetraconazole

C	Digit	Monsanto Plc	09409	
C	Eminent	Monsanto Plc	09410	
C	Juggler	Sipcam UK Ltd	09391	
C	MON 10EC	Monsanto Plc	09347	

C These products are "approved for agricultural use". For further details refer to page vii.
A These products are approved for use in or near water. For further details refer to page vii.

Product Name	Marketing Company	Reg. No.	Expiry Date

729 Tetraconazole—continued

C	Omen	Sipcam UK Ltd	09413	
C	Tetra 5634	Monsanto Plc	09411	

730 Tetraconazole + Chlorothalonil

C	TC Elite	Monsanto Plc	09412	
C	Voodoo	Sipcam UK Ltd	09414	

731 Thiabendazole

C	Hykeep	Agrichem (International) Ltd	06744	
C	Hymush	Agrichem (International) Ltd	07937	31/01/2002
C	Storite Clear Liquid	Novartis Crop Protection UK Ltd	09503	
C	Storite Clear Liquid	Seedcote Systems Ltd	08982	
C	Storite Clear Liquid	Novartis Crop Protection UK Ltd	08700	31/03/2001
C	Storite Clear Liquid	MSD Agvet	02032	31/03/2001
C	Storite Excel	Novartis Crop Protection UK Ltd	09542	
C	Storite Excel	Seedcote Systems Ltd	08993	
C	Storite Flowable	Novartis Crop Protection UK Ltd	08703	
C	Tecto Flowable Turf Fungicide	Vitax Ltd	06273	
C	Tecto Superflowable Turf Fungicide	Vitax Ltd	09258	

732 Thiabendazole + Carboxin + Imazalil

C	Vitaflo Extra	Uniroyal Chemical Ltd	07048	

733 Thiabendazole + Ethirimol + Flutriafol

C	Ferrax	Zeneca Crop Protection	06662	28/02/2002
C	Ferrax	Bayer plc	05284	28/02/2002

734 Thiabendazole + Imazalil

C	Extratect Flowable	Seedcote Systems Ltd	09157	
C	Extratect Flowable	Novartis Crop Protection UK Ltd	08704	

735 Thiabendazole + Metalaxyl

C	Apron T69 WS	Novartis Crop Protection UK Ltd	08387	
C	Polycote Select	Germains UK Ltd	09718	
C	Polycote Select	Seedcote Systems Ltd	08430	

736 Thiabendazole + Metalaxyl + Thiram

C	Apron Combi FS	Novartis Crop Protection UK Ltd	08386	

737 Thiabendazole + Oxadixyl + Thiram

C	Leap	Uniroyal Chemical Ltd	08477	

C These products are "approved for agricultural use". For further details refer to page vii.
A These products are approved for use in or near water. For further details refer to page vii.

Product Name	Marketing Company	Reg. No.	Expiry Date

737 Thiabendazole + Oxadixyl + Thiram—continued

C Leap	Uniroyal Chemical Ltd	07884	31/12/2001

738 Thiabendazole + Tecnazene

C Hytec Super	Agrichem (International) Ltd	01100	31/01/2002
New Arena TBZ 6	Tripart Farm Chemicals Ltd	05606	31/01/2002

739 Thiabendazole + Thiram

C HY-TL	Agrichem (International) Ltd	06246	
C HY-VIC	Agrichem (International) Ltd	06247	
C sHYlin	Agrichem (International) Ltd	10030	

740 Thiophanate-Methyl

C Cercobin Liquid	Aventis CropScience UK Limited	09984	
C Cercobin Liquid	Rhone-Poulenc Agriculture	06188	
C Mildothane Liquid	Aventis CropScience UK Limited	10002	
C Mildothane Liquid	Rhone-Poulenc Agriculture	06211	
C Mildothane Turf Liquid	Aventis Environmental Science	09935	
C Mildothane Turf Liquid	Rhone-Poulenc Amenity	05331	

741 Thiophanate-Methyl + Iprodione

C Aventis Compass	Aventis CropScience UK Limited	10040	
C Aventis Compass	Rhone-Poulenc Agriculture	09760	31/10/2003
C Compass	Aventis CropScience UK Limited	10041	
C Compass	Rhone-Poulenc Agriculture	06190	31/10/2003
C Snooker	Aventis CropScience UK Limited	10018	
C Snooker	Rhone-Poulenc Agriculture	07940	31/10/2003

742 Thiram

C Agrichem Flowable Thiram	Agrichem (International) Ltd	06245	
C Hy-Flo	Agrichem (International) Ltd	04637	
C Robinson's Thiram 60	Agrichem (International) Ltd	04638	
C Thiraflo	Uniroyal Chemical Ltd	09496	
C Unicrop Thianosan DG	Universal Crop Protection Ltd	05454	

743 Thiram + Carbendazim + Cymoxanil + Oxadixyl

C Apron Elite	Novartis Crop Protection UK Ltd	08770	

744 Thiram + Carboxin

C Anchor	Uniroyal Chemical Ltd	08684	

745 Thiram + Metalaxyl

C Favour 600 SC	Novartis Crop Protection UK Ltd	08405	

C   These products are "approved for agricultural use". For further details refer to page vii.
A   These products are approved for use in or near water. For further details refer to page vii.

Product Name	Marketing Company	Reg. No.	Expiry Date

746 Thiram + Metalaxyl + Thiabendazole

C Apron Combi FS	Novartis Crop Protection UK Ltd	08386	

747 Thiram + Oxadixyl + Thiabendazole

C Leap	Uniroyal Chemical Ltd	08477	
C Leap	Uniroyal Chemical Ltd	07884	31/12/2001

748 Thiram + Permethrin

C Combinex	Pan Britannica Industries Ltd	00562	31/10/2001

749 Thiram + Thiabendazole

C HY-TL	Agrichem (International) Ltd	06246	
C HY-VIC	Agrichem (International) Ltd	06247	
C sHYlin	Agrichem (International) Ltd	10030	

750 Tolclofos-Methyl

C Basilex	The Scotts Company (UK) Limited + Levington Horticulture Ltd	07494	
C Basilex Soluble Sachets	The Scotts Company (UK) Limited + Levington Horticulture Ltd	06958	31/07/2001
C Rizolex	Sumitomo Chemical Agro Europe SA	09673	
C Rizolex	AgrEvo UK Ltd	07271	31/07/2002
C Rizolex 50 WP	Aventis CropScience UK Limited	07272	
C Rizolex 50 WP in Soluble Sachets	AgrEvo UK Ltd	06957	20/04/2001
C Rizolex Flowable	Sumitomo Chemical Agro Europe SA	09358	
C Rizolex Flowable	AgrEvo UK Ltd	07273	31/12/2001

751 Tolylfluanid

C Bayer UK 456	Bayer plc	07995	
C Elvaron M	Bayer plc	07772	
C Elvaron Multi	Bayer plc	10080	

752 Tolylfluanid + Fenhexamid

C Talat	Bayer plc	09655	

753 Triadimefon

C 100-Plus	Dalgety Agriculture Ltd	05112	31/12/2001
C Bayleton	Bayer plc	00221	
C Bayleton 5	Bayer plc	00222	
C Landgold Triadimefon 25	Landgold & Co Ltd	06139	

C These products are "approved for agricultural use". For further details refer to page vii.
A These products are approved for use in or near water. For further details refer to page vii.

Product Name	Marketing Company	Reg. No.	Expiry Date

754 Triadimenol

C	Bayfidan	Bayer plc	02672
C	Hi-Shot	Cyanamid Agriculture Ltd	06508
C	Spinnaker	Cyanamid Agriculture Ltd	07023

755 Triadimenol + Bitertanol + Fuberidazole

C	Cereline	Bayer plc	07239

756 Triadimenol + Bitertanol + Fuberidazole + Imidacloprid

C	Cereline Secur	Bayer plc	09511

757 Triadimenol + Ethirimol + Fuberidazole

C	Bay UK 292	Bayer plc	04335

758 Triadimenol + Fuberidazole

C	Baytan	Bayer plc	00225
C	Baytan CF	Bayer plc	08193
C	Baytan Flowable	Bayer plc	02593

759 Triadimenol + Fuberidazole + Imazalil

C	Baytan IM	Bayer plc	00226

760 Triadimenol + Fuberidazole + Imidacloprid

C	Baytan Secur	Bayer plc	09510

761 Triadimenol + Tebuconazole

C	Garnet	Bayer plc	06391
C	Ruby	Bayer plc	06389
C	Silvacur	Bayer plc	06387
C	Silvacur 300	Bayer plc	08056
C	Veto F	Bayer plc	08057

762 Triadimenol + Tridemorph

C	Dorin	Bayer plc	08361	30/09/2001
C	Dorin	Bayer plc	03292	30/09/2001
C	Jasper	BASF plc	06044	28/02/2002

763 Triazoxide + Imidacloprid + Tebuconazole

C	Raxil Secur	Bayer plc	08966

764 Triazoxide + Tebuconazole

C	Raxil S	Bayer plc	06974
C	Raxil S CF	Bayer plc	08192

C   These products are "approved for agricultural use". For further details refer to page vii.
A   These products are approved for use in or near water. For further details refer to page vii.

Product Name	Marketing Company	Reg. No.	Expiry Date
765 Tridemorph			
C Calixin	BASF plc	00369	28/02/2002
C Standon Tridemorph 750	Standon Chemicals Ltd	05667	28/02/2002
766 Tridemorph + Carbendazim + Maneb			
C Cosmic FL	BASF plc	03473	28/02/2002
767 Tridemorph + Cyproconazole			
C Alto Major	Novartis Crop Protection UK Ltd	08468	30/09/2001
C Alto Major	Sandoz Agro Ltd	06979	30/09/2001
C Moot	Novartis Crop Protection UK Ltd	08479	30/09/2001
C Moot	Sandoz Agro Ltd	06990	30/09/2001
C SAN 735	Novartis Crop Protection UK Ltd	08490	30/09/2001
C San 735	Sandoz Agro Ltd	06991	30/09/2001
768 Tridemorph + Epoxiconazole			
C Opus Plus	BASF plc	07363	28/02/2002
769 Tridemorph + Fenbuconazole			
C Unison	Pan Britannica Industries Ltd	08318	28/02/2002
770 Tridemorph + Fenpropimorph			
C BAS 46402F	BASF plc	03313	28/02/2002
C Gemini	BASF plc	05684	28/02/2002
771 Tridemorph + Fenpropimorph + Flusilazole			
C Bingo	BASF plc	06920	28/02/2002
C DUK 51	Du Pont (UK) Ltd	06764	30/09/2001
C Justice	Du Pont (UK) Ltd	07963	30/09/2001
772 Tridemorph + Flusilazole			
C Fusion	Du Pont (UK) Ltd	04908	30/09/2001
C Meld	BASF plc	04914	28/02/2002
C Option	Du Pont (UK) Ltd	07951	30/09/2001
773 Tridemorph + Propiconazole			
C Joust	Universal Crop Protection Ltd	08122	28/02/2002
C Tilt Turbo 475 EC	Novartis Crop Protection UK Ltd	08457	30/09/2001
C Tilt Turbo 475 EC	Ciba Agriculture	03476	30/09/2001
774 Tridemorph + Tebuconazole			
C Allicur	Bayer plc	06468	30/09/2001
C BAS 91580F	BASF plc	06469	31/12/2001

C These products are "approved for agricultural use". For further details refer to page vii.
A These products are approved for use in or near water. For further details refer to page vii.

Product Name	Marketing Company	Reg. No.	Expiry Date

775 Tridemorph + Triadimenol

C	Dorin	Bayer plc	08361	30/09/2001
C	Dorin	Bayer plc	03292	30/09/2001
C	Jasper	BASF plc	06044	28/02/2002

776 Trifloxystrobin

C	Flint	Novartis Crop Protection UK Ltd	10126
C	Twist	Novartis Crop Protection UK Ltd	10125

777 Trifloxystrobin + Cyproconazole

C	Sphere	Novartis Crop Protection UK Ltd	10263

778 Trifloxystrobin + Propiconazole

C	Rombus	Novartis Crop Protection UK Ltd	10262

779 Triforine

C	Fairy Ring Destroyer	Vitax Ltd	05541
C	Saprol	BASF plc	10227
C	Saprol	Cyanamid Agriculture Ltd	07016

780 Triforine + Bupirimate

C	Nimrod T	The Scotts Company (UK) Limited	09268	
C	Nimrod T	Miracle Professional	07865	
C	Nimrod T	Zeneca Professional Products	06859	31/01/2001
C	Nimrod T	ICI Agrochemicals	01499	31/05/2001

781 Triticonazole + Guazatine

C	Premis	Aventis CropScience UK Limited	10177

782 Triticonazole + Imazalil

C	Robust	Aventis CropScience UK Limited	10176

783 Vinclozolin

C	Barclay Flotilla	Barclay Chemicals Manufacturing Ltd	07905
C	Landgold Vinclozolin DG	Landgold & Co Ltd	06500
C	Landgold Vinclozolin SC	Landgold & Co Ltd	06459
C	Marnoch Vinol	Marnoch Ventures Limited	09815
C	Ronilan DF	BASF plc	04456
C	Ronilan FL	BASF plc	02960
C	Standon Vinclozolin	Standon Chemicals Ltd	07836

C   These products are "approved for agricultural use". For further details refer to page vii.
A   These products are approved for use in or near water. For further details refer to page vii.

Product Name	Marketing Company	Reg. No.	Expiry Date

784 Vinclozolin + Carbendazim

C Konker | BASF plc | 03988

785 Vinclozolin + Chlorothalonil

C Curalan CL | BASF plc | 07174

786 Zineb

| C | Tritoftorol | Elf Atochem Agri SA | 06872 |
| C | Unicrop Zineb | Universal Crop Protection Ltd | 02279 |

787 Zineb-Ethylene Thiuram Disulphide Adduct

C Polyram DF | BASF plc | 08234

C   These products are "approved for agricultural use". For further details refer to page vii.
A   These products are approved for use in or near water. For further details refer to page vii.

Product Name	Marketing Company	Reg. No.	Expiry Date

1.3 Insecticides
including acaricides and nematicides

789 Abamectin

C	Dynamec	Novartis Crop Protection UK Ltd	08701	
C	Dynamec	MSD Agvet	06804	28/02/2001

790 Aldicarb

C	Aventis Temik 10G	Aventis CropScience UK Limited	09975	
C	Aventis Temik 10G	Rhone-Poulenc Agriculture	09749	
C	CleanCrop Aldicarb 10 G	United Agri Products	09694	25/07/2003
C	Landgold Aldicarb 10G	Landgold & Co Ltd	06036	25/07/2003
C	Me² Aldee	Me² Crop Protection Ltd	09781	25/07/2003
C	Standon Aldicarb 10G	Standon Chemicals Ltd	05915	25/07/2003
C	Temik 10G	Aventis CropScience UK Limited	10021	
C	Temik 10G	Rhone-Poulenc Agriculture	06210	25/07/2003

791 Alphacypermethrin

C	Acquit	Du Pont (UK) Ltd	07000	
C	Agriguard Alpha-Cyper	Agriguard Ltd	09920	
C	Alert	BASF plc	10272	
C	Alphathrin	Whyte Agrochemicals Division	10125	
C	Cleancrop scope	United Agri Products Limited	10273	
C	Contest	BASF plc	10216	
C	Contest	Cyanamid Agriculture Ltd	09024	
C	Contest Eco	BASF plc	10217	30/09/2003
C	Contest Eco	Cyanamid Agriculture Ltd	08995	30/09/2003
C	Fastac	BASF plc	10220	
C	Fastac	Cyanamid Agriculture Ltd	07008	
C	Fastac Dry	BASF plc	10221	
C	Fastac Dry	Cyanamid Agriculture Ltd	09025	
C	IT Alpha-cypermethrin	IT Agro Ltd	08274	
C	IT Alpha-Cypermethrin 100 EC	Nufarm Whyte Agriculture Ltd	09372	31/03/2003
C	Standon Alpha-C10	Standon Chemicals Ltd	08823	

792 Aluminium phosphide

C	Detia Gas Ex-P	Igrox Chemicals Limited	09802	
	Detia Gas Ex-T	Igrox Ltd	03792	
	Detia Gas-Ex-B	Igrox Ltd	06927	
	Fumitoxin	Igrox Ltd	04207	31/08/2002
	Phostek	Terminix Peter Cox	09029	

C These products are "approved for agricultural use". For further details refer to page vii.
A These products are approved for use in or near water. For further details refer to page vii.

Product Name	Marketing Company	Reg. No.	Expiry Date

792 Aluminium phosphide—continued

Phostek	Killgerm Chemicals Ltd	05115	
Phostoxin I	Rentokil Initial UK Ltd	09313	
Phostoxin I	Rentokil Initial UK Ltd	05694	

793 Amitraz

C	Bye Bye 20EC	Chimac-Agriphar SA	09346	
C	Mitac 20	AgrEvo UK Ltd	07265	31/05/2001
C	Mitac HF	Aventis CropScience UK Limited	07358	
C	Ovasyn	Aventis CropScience UK Limited	09190	
	Taktic Buildings Spray	Hoechst Roussel Vet Ltd	09136	
	Taktic Buildings Spray	Hoechst Animal Health Ltd	06504	31/07/2001

794 Azamethiphos

Alfacron 10 WP	Novartis Animal Health UK Ltd	08587	28/02/2002
Alfacron 10WP	Ciba Agriculture	02832	28/02/2002
Alfacron 50 WP	Novartis Animal Health UK Ltd	08588	20/04/2001
Farm Fly Spray 10 WP	Rentokil Initial UK Ltd	09314	20/04/2001
Farm Fly Spray 10 WP	Rentokil Ltd	05274	20/04/2001

795 Azamethiphos + (Z)-9-Tricosene

Alfacron Plus	Novartis Animal Health UK Ltd	09439

797 Bendiocarb

	Ficam ULV	AgrEvo UK Ltd	07863	20/04/2001
C	Garvox 3G	AgrEvo UK Ltd	07259	20/04/2001
C	Seedox SC	Uniroyal Chemical Ltd	07591	20/04/2001

798 Benfuracarb

C	Oncol 10G	Mirfield Sales Services Ltd	08249

799 Bifenthrin

C	Starion	FMC Europe NV	09795
C	Talstar	FMC Europe NV	06913

800 Bitertanol + Fuberidazole + Imidacloprid + Triadimenol

C	Cereline Secur	Bayer plc	09511

801 Buprofezin

C	Applaud	Zeneca Crop Protection	06900

802 Carbaryl

C	Cavalier	Rhone-Poulenc Amenity	07027	20/04/2001

C These products are "approved for agricultural use". For further details refer to page vii.
A These products are approved for use in or near water. For further details refer to page vii.

Product Name	Marketing Company	Reg. No.	Expiry Date

802 Carbaryl—continued

C	Thinsec	Zeneca Crop Protection	06710	31/07/2001

803 Carbofuran

C	Nex	Tripart Farm Chemicals Ltd	05165	31/12/2001
C	Rampart	Sipcam UK Ltd	05166	31/12/2001
C	Throttle	Quadrangle Agrochemicals Ltd	05204	31/12/2001
C	Yaltox	Bayer plc	02371	31/12/2001

804 Carbosulfan

C	Landgold Carbosulfan 10 G	Landgold & Co Ltd	09019	20/04/2001
C	Marshal 10G	Rhone-Poulenc Agriculture	06165	20/04/2001
C	Marshal Soil Insecticide suSCon CR granules	Fargro Ltd	06978	
C	Marshal 10G	FMC Corporation (UK) Ltd	09804	
C	Marshal 10G	FMC Europe NV	08873	31/10/2002
C	Posse 10 G	FMC Corporation (UK) Ltd	09805	

805 Chlorfenvinphos

C	Birlane 24	Cyanamid Agriculture Ltd	07002	31/12/2001

806 Chlorpyrifos

C	Alpha Chlorpyrifos 48EC	Makhteshim-Agan (UK) Ltd	04821	
C	Ballad	Headland Amenity Ltd	09775	
C	Barclay Clinch II	Barclay Chemicals Manufacturing Ltd	08596	
C	Barclay Clinch II	Barclay Chemicals Manufacturing Ltd	07815	31/07/2001
C	Choir	Nufarm Whyte Agriculture Ltd	09778	
C	Crossfire 480	Aventis Environmental Science	09929	
C	Crossfire 480	Rhone-Poulenc Amenity	08142	
C	Crossfire 480	Dow AgroSciences Ltd	08141	
C	CYREN	Cheminova Agro (UK) Ltd	08358	
C	Dispatch	Dow AgroSciences Ltd	08139	
C	Dursban 4	Dow AgroSciences Ltd	07815	
C	Dursban WG	Dow AgroSciences Ltd	09153	
C	Greencrop Pontoon	Greencrop Technology Ltd	09667	
C	Lorsban 480	Dow AgroSciences Ltd	08076	
C	Lorsban T	Rigby Taylor Ltd	07813	
C	Lorsban WG	Dow AgroSciences Ltd	10139	
C	Maraud	The Scotts Company (UK) Limited	09274	
C	Pyrinex 48EC	Makhteshim-Agan (UK) Ltd	08644	
C	Spannit	SumiAgro (UK) Ltd	08744	

C These products are "approved for agricultural use". For further details refer to page vii.
A These products are approved for use in or near water. For further details refer to page vii.

Product Name	Marketing Company	Reg. No.	Expiry Date

806 Chlorpyrifos—continued

C Spannit Granules	SumiAgro (UK) Ltd	08984	
C Spannit Granules	Pan Britannica Industries Ltd	04048	31/03/2001
C Standon Chlorpyrifos 48	Standon Chemicals Ltd	08286	
C Suscon Green	Scotts Europe BV	09226	30/06/2001
C Suscon Green Soil Insecticide	Fargro Ltd	06312	
C suSCon Indigo Soil Insecticide	Fargro Ltd	09902	
C Talon	Farmers Crop Chemicals Ltd	08746	
C Tripart Audax	Tripart Farm Chemicals Ltd	08745	

807 Chlorpyrifos + Disulfoton

C Twinspan	pbi Agrochemicals Ltd	08983	31/12/2001

808 Chlorpyrifos-methyl

C Reldan 22	Dow AgroSciences Ltd	08191	
C Reldan 50	Dow AgroSciences Ltd	05742	20/04/2001

809 Clofentezine

C Apollo 50 SC	Aventis CropScience UK Limited	07242	

810 Cyfluthrin

C Baythroid	Bayer plc	04273	

811 Cypermethrin

C Afrisect 10	Stefes UK Ltd	09114	
C Agriguard Cypermethrin	Agriguard Ltd	10093	
C Ashlade Cypermethrin 10 EC	Cyanamid Agriculture Ltd	07412	30/06/2001
C Chemtech Cypermethrin 10 EC	Cyanamid Agriculture Ltd	07413	30/06/2001
C Chemtech Cypermethrin 10 EC	Chemtech (Crop Protection) Ltd	04827	31/12/2001
C Cyperkill 10	Chiltern Farm Chemicals Ltd	04119	
C Cyperkill 25	Chiltern Farm Chemicals Ltd	09038	
C Cyperkill 25	Mitchell Cotts Chemicals Ltd	03741	31/05/2001
C Cyperkill 5	Mitchell Cotts Chemicals Ltd	00625	
C Cypertox	Cyanamid Agriculture Ltd	07414	30/06/2001
C Cypertox	Farmers Crop Chemicals Ltd	05122	
C Fernpath Banjo	Agriguard Ltd	10182	
C Luxan Cypermethrin 10	Luxan (UK) Ltd	06283	
C Permasect C	Nufarm Whyte Agriculture Ltd	09200	
C Permasect C	Mirfield Sales Services Ltd	07680	31/08/2001
C Quadrangle Cyper 10	Cyanamid Agriculture Ltd	07410	30/06/2001
C Quadrangle Cyper 10	Quadrangle Agrochemicals	03242	
C Ripcord	Cyanamid Agriculture Ltd	07014	30/06/2001

C These products are "approved for agricultural use". For further details refer to page vii.
A These products are approved for use in or near water. For further details refer to page vii.

Product Name	Marketing Company	Reg. No.	Expiry Date

811 Cypermethrin—continued

C	Stefes Cypermethrin 3	Stefes Plant Protection Ltd	09172	31/03/2002
C	Toppel 10	United Phosphorus Ltd	08772	
C	Toppel 10	Zeneca Crop Protection	06516	31/01/2001
C	Vassgro Cypermethrin Insecticide	Cyanamid Agriculture Ltd	07411	30/06/2001
C	Vassgro Cypermethrin Insecticide	L W Vass (Agricultural) Ltd	03240	

812 Cyromazine

	Neporex 2 SG	Novartis Animal Health UK Ltd	08589	
	Neporex 2 SG	Ciba Agriculture	06985	31/01/2001

813 Deltamethrin

C	Aventis Decis	Aventis CropScience UK Limited	09710
C	Decis	Aventis CropScience UK Limited	07172
C	Decis Micro	Aventis CropScience UK Limited	08618
C	Deleet	Rentokil Initial UK Ltd	09312
C	Deleet	Rentokil Ltd	08195
C	Landgold Deltaland	Landgold & Co Ltd	09906
C	Pearl Micro	Aventis CropScience UK Limited	08620
C	Thripstick	Aquaspersions Ltd	02134

814 Deltamethrin + Heptenophos

C	Decisquick	AgrEvo UK Ltd	07312	20/04/2001

815 Deltamethrin + Pirimicarb

C	ACP 105	Aventis CropScience UK Limited	08989
C	Best	Aventis CropScience UK Limited	08988
C	Evidence	Aventis CropScience UK Limited	06934
C	Patriot EC	Aventis CropScience UK Limited	08990

816 Dicamba + Dichlorprop + MCPA

C	Cleanrun 2	The Scotts Company (UK) Limited	09271

817 1,3-Dichloropropene

C	Telone 2000	Dow AgroSciences Ltd	05748
C	Telone II	Dow AgroSciences Ltd	05749

818 Dichlorprop + Dicamba + MCPA

C	Cleanrun 2	The Scotts Company (UK) Limited	09271

C   These products are "approved for agricultural use". For further details refer to page vii.
A   These products are approved for use in or near water. For further details refer to page vii.

Product Name	Marketing Company	Reg. No.	Expiry Date

819 Dichlorvos

C Luxan Dichlorvos 600	Luxan (UK) Ltd	08297	
C Luxan Dichlorvos Aerosol 15	Luxan (UK) Ltd	08298	
Nuvan 500 EC	Novartis Animal Health UK Ltd	08590	
Nuvan 500 EC	Ciba Agriculture	03861	31/12/2001

820 Dicofol

C Kelthane	Landseer Limited	09260	31/05/2002

821 Dicofol + Tetradifon

C Childion	Hortichem Ltd	03821	31/05/2002

822 Diflubenzuron

C Dimilin 25 WP	Uniroyal Chemical Ltd	08902	
C Dimilin Flo	Zeneca Crop Protection	08985	
C Dimilin Flo	Uniroyal Chemical Ltd	08769	
C Dimilin WP	Uniroyal Chemical Ltd	08870	
C Dimilin WP	Zeneca Crop Protection	06656	31/01/2001

823 Dimethoate

C Barclay Dimethosect	Barclay Chemicals Manufacturing Ltd	08538	31/07/2002
C BASF Dimethoate 40	BASF plc	00199	
C Danadim	Cheminova Agro (UK) Ltd	09583	
C Danadim Dimethoate 40	Cheminova Agro (UK) Ltd	07351	31/05/2002
C Greencrop Pelethon	Greencrop Technology Ltd	09477	31/07/2002
C PA Dimethoate 40	Portman Agrochemicals Ltd	01527	
C Rogor L40	Sipcam UK Ltd	07611	
C Sector	Portman Agrochemicals Ltd	08882	

824 Disulfoton

C Campbell's Disulfoton FE 10	J D Campbell & Sons Ltd	00402	20/04/2001
C Disulfoton P10	United Phosphorus Ltd	08023	31/12/2001
C Disyston FE-10	Bayer plc	00714	20/04/2001
C Disyston P-10	Bayer plc	00715	20/04/2001

825 Disulfoton + Chlorpyrifos

C Twinspan	pbi Agrochemicals Ltd	08983	31/12/2001

826 Endosulfan

C Thiodan 20 EC	Aventis CropScience UK Limited	07335	

C   These products are "approved for agricultural use". For further details refer to page vii.
A   These products are approved for use in or near water. For further details refer to page vii.

Product Name	Marketing Company	Reg. No.	Expiry Date

827 Esfenvalerate

C Sumi-Alpha	Cyanamid Agriculture Ltd	07207	

828 Ethiofencarb

C Croneton	Bayer plc	00593	20/04/2001

829 Ethoprophos

C Aventis Mocap 10G	Aventis CropScience UK Limited	09973	
C Aventis Mocap 10G	Rhone-Poulenc Agriculture	09750	
C Mocap 10G	Aventis CropScience UK Limited	10003	
C Mocap 10G	Rhone-Poulenc Agriculture	06773	

830 Etrimfos

Satisfar	Nickerson Seed Specialists Ltd	04180	31/12/2001
Satisfar Dust	Nickerson Seed Specialists Ltd	04085	31/12/2001

831 Fatty acids

C Safer's Insecticidal Soap	Safer Ltd	07197	
C Savona	Koppert (UK) Ltd	06057	
C Savona	Koppert (UK) Ltd	03137	

832 Fenazaquin

C Matador 200 SC	Dow AgroSciences Ltd	07960	

833 Fenbutatin oxide

C Torq	Fargro Ltd	08370	
C Torque	Cyanamid Agriculture Ltd	07148	

834 Fenitrothion

C Dicofen	pbi Agrochemicals Ltd	09598	20/04/2001
C Dicofen	Pan Britannica Industries Ltd	00693	20/04/2001
C Unicrop Fenitrothion 50	Universal Crop Protection Ltd	02267	20/04/2001

835 Fenitrothion + Permethrin + Resmethrin

Turbair Grain Store Insecticide	Pan Britannica Industries Ltd	02238	20/04/2001

836 Fenoxycarb

C Insegar	Novartis Crop Protection UK Ltd	08558	31/03/2002
C Insegar WG	Novartis Crop Protection UK Ltd	09789	

837 Fenpropathrin

C Meothrin	Cyanamid Agriculture Ltd	07206	

C   These products are "approved for agricultural use". For further details refer to page vii.
A   These products are approved for use in or near water. For further details refer to page vii.

Product Name	Marketing Company	Reg. No.	Expiry Date
838 Fenpyroximate			
C NNI-850 5SC	Nihon Nohyaku Co., Ltd	09887	
C Sequel	Hortichem Ltd	09886	
C Sequel	AgrEvo UK Ltd	07624	31/01/2003
839 Fosthiazate			
C Nemathorin 10G	Zeneca Crop Protection	08915	
840 Fuberidazole + Bitertanol + Imidacloprid + Triadimenol			
C Cereline Secur	Bayer plc	09511	
841 Fuberidazole + Imidacloprid + Triadimenol			
C Baytan Secur	Bayer plc	09510	
842 Heptenophos			
C Hostaquick	AgrEvo UK Ltd	07326	20/04/2001
843 Heptenophos + Deltamethrin			
C Decisquick	AgrEvo UK Ltd	07312	20/04/2001
844 Imidacloprid			
C Admire	Bayer plc	07481	
C Bayer UK 397	Bayer plc	08930	
C Bayer UK 479	Bayer plc	08584	
C Bayer UK368	Bayer plc	08125	
C Gaucho	Bayer plc	06590	
C Gaucho FS	Bayer plc	08496	
C Intercept 5GR	The Scotts Company (UK) Limited + Levington Horticulture Ltd	08126	
C Intercept 70 WG	The Scotts Company (UK) Limited + Levington Horticulture Ltd	08585	
C Levington Professional Plus Intercept	The Scotts Company (UK) Limited + Levington Horticulture Ltd	08569	
C Rentokil Desyst	Rentokil Initial UK Ltd	09446	
845 Imidacloprid + Bitertanol + Fuberidazole + Triadimenol			
C Cereline Secur	Bayer plc	09511	
846 Imidacloprid + Fuberidazole + Triadimenol			
C Baytan Secur	Bayer plc	09510	
847 Lambda-cyhalothrin			
C Agriguard Lambda-Cyhalo	Tronsan Ltd	10129	
C Hallmark	Zeneca Crop Protection	06434	31/05/2002

C   These products are "approved for agricultural use". For further details refer to page vii.
A   These products are approved for use in or near water. For further details refer to page vii.

Product Name	Marketing Company	Reg. No.	Expiry Date

847 Lambda-cyhalothrin—continued

C	Hallmark with Zeon Technology	Zeneca Crop Protection	09809	
C	Hero	Zeneca Crop Protection	07821	31/05/2002
C	Landgold Lambda-C	Landgold & Co Ltd	09205	31/05/2003
C	Marnoch Lamoc	Marnoch Ventures Limited	10051	31/05/2003
C	Me² Lambda	Me² Crop Protection Ltd	09712	31/05/2003
C	Stampout	Phytheron 2000 SA	09777	31/05/2003
C	Standon Lambda-C	Standon Chemicals Ltd	08831	31/05/2003

848 Lambda-cyhalothrin + Pirimicarb

C	Dovetail	Zeneca Crop Protection	07973	

849 Lindane

C	Atlas Steward	Atlas Crop Protection Ltd	07728	
C	Fumite Lindane 10	Octavius Hunt Ltd	00933	30/06/2002
C	Fumite Lindane 40	Octavius Hunt Ltd	00934	30/06/2002
C	Fumite Lindane Pellets	Octavius Hunt Ltd	00937	30/06/2002
C	Lindane Flowable	pbi Agrochemicals Ltd	09113	31/03/2002
C	Lindane Flowable	Pan Britannica Industries Ltd	02610	31/08/2001

850 Lindane + Thiophanate-methyl

C	Castaway Plus	Rhone-Poulenc Amenity	05327	31/07/2001

851 Magnesium Phosphide

	Degesch Plates	Rentokil Initial UK Ltd	07603	

852 Malathion

C	Malathion 60	United Phosphorus Ltd	08018	

853 MCPA + Dicamba + Dichlorprop

C	Cleanrun 2	The Scotts Company (UK) Limited	09271	

855 Mephosfolan

C	Cytro-Lane	Cyanamid Agriculture Ltd	00626	20/04/2001

856 Methiocarb

C	Bayer UK 808	Bayer plc	09513	
C	Bayer UK 809	Bayer plc	09514	
C	Bayer UK 892	Bayer plc	09540	
C	Bayer UK 935	Bayer plc	09541	
C	Club	Bayer plc	07176	
C	Decoy	Bayer plc	06535	

C   These products are "approved for agricultural use". For further details refer to page vii.
A   These products are approved for use in or near water. For further details refer to page vii.

215

Product Name	Marketing Company	Reg. No.	Expiry Date

856 Methiocarb—continued

C Decoy Plus	Bayer plc	07615	
C Decoy Wetex	Bayer plc	09707	
C Draza	Bayer plc	00765	
C Draza 2	Bayer plc	04748	
C Draza Plus	Bayer plc	06553	
C Draza Wetex	Bayer plc	09704	
C Elvitox	Bayer plc	06738	
C Epox	Bayer plc	06737	
C Exit	Bayer plc	07632	
C Exit Wetex	Bayer plc	10149	
C Huron	Bayer plc	10148	
C Karan	Bayer plc	09637	
C Lupus	Bayer plc	09638	
C Rivet	Bayer plc	09512	

857 Methomyl

Golden Malrin	Novartis Animal Health UK Ltd	08821	20/04/2001

858 Methoprene

C Apex 5E	Novartis Animal Health UK Ltd	08739	30/09/2001

859 Nicotine

C Nico Soap	United Phosphorus Ltd	07517	
C Nicotine 40% Shreds	Dow AgroSciences Ltd	05725	
C No-FID	Hortichem Ltd	07959	
C XL All Insecticide	Synchemicals Ltd	02369	
C XL All Nicotine 95%	Vitax Ltd	07402	

860 Oxamyl

C Fielder	Du Pont (UK) Ltd	05279	
C Vydate 10G	Du Pont (UK) Ltd	02322	

861 Permethrin

C Darmycel Agarifume Smoke Generator	Sylvan Spawn Ltd	09564	
C Darmycel Agarifume Smoke Generator	Darmycel	07904	30/04/2002
C Fumite Permethrin Smoke	Octavius Hunt Ltd	00940	
Geest Fumite MK2	Octavius Hunt Ltd	07939	
Geestline Fumite MK2	Octavius Hunt Ltd	07218	
C Permasect 10 EC	Mitchell Cotts Chemicals Ltd	03920	
C Permasect 25 EC	Mitchell Cotts Chemicals Ltd	01576	
C Permethrin 12 ED	Techneat Chemicals Limited	09956	

C These products are "approved for agricultural use". For further details refer to page vii.
A These products are approved for use in or near water. For further details refer to page vii.

Product Name	Marketing Company	Reg. No.	Expiry Date

861 Permethrin—continued

C	Permethrin 12 ED	Zeneca Crop Protection	09629	31/03/2002
C	Permit	SumiAgro (UK) Ltd	09609	
C	Permit	Pan Britannica Industries Ltd	01577	
C	Turbair Permethrin	pbi Agrochemicals Ltd	09616	28/02/2002
C	Turbair Permethrin	Pan Britannica Industries Ltd	02246	28/02/2002

862 Permethrin + Fenitrothion + Resmethrin

	Turbair Grain Store Insecticide	Pan Britannica Industries Ltd	02238	20/04/2001

863 Permethrin + Thiram

C	Combinex	Pan Britannica Industries Ltd	00562	31/10/2001

864 Phorate

C	Phorate 10G	United Phosphorus Ltd	08007	

865 Phosalone

C	Zolone Liquid	Rhone-Poulenc Agriculture	06173	20/04/2001

866 Pirimicarb

C	Agriguard Pirimicarb	Tronsan Ltd	09620	
C	Aphox	Zeneca Crop Protection	06633	
C	Barclay Pirimisect	Barclay Chemicals Manufacturing Ltd	09057	
C	Clayton Pirimicarb 50SG	Clayton Plant Protection (UK) Ltd	09221	
C	Greencrop Glenroe	Greencrop Technology Ltd	09903	
C	Helocarb Granule 500	Helm AG	08157	20/04/2001
C	Landgold Pirimicarb 50	Landgold & Co Ltd	09018	
C	Phantom	Bayer plc	04519	
C	Pirimate	Portman Agrochemicals Ltd	09568	
C	Pirimor	Zeneca Crop Protection	06694	20/04/2001
C	Standon Pirimicarb 50	Standon Chemicals Ltd	08878	
C	Standon Pirimicarb H	Standon Chemicals Ltd	05669	20/04/2001
C	Unistar Pirimicarb 500	Unistar Ltd	06975	20/04/2001

867 Pirimicarb + Deltamethrin

C	ACP 105	Aventis CropScience UK Limited	08989	
C	Best	Aventis CropScience UK Limited	08988	
C	Evidence	Aventis CropScience UK Limited	06934	
C	Patriot EC	Aventis CropScience UK Limited	08990	

868 Pirimicarb + Lambda-cyhalothrin

C	Dovetail	Zeneca Crop Protection	07973	

C   These products are "approved for agricultural use". For further details refer to page vii.
A   These products are approved for use in or near water. For further details refer to page vii.

Product Name	Marketing Company	Reg. No.	Expiry Date

869 Pirimiphos-methyl

	Actellic 2% Dust	Zeneca Crop Protection	06931	
	Actellic D	Zeneca Crop Protection	06930	
	Actellic Smoke Generator No 10	Zeneca Public Health	07900	
	Actellic Smoke Generator No 20	Zeneca Crop Protection	06627	
C	Actellifog	Hortichem Ltd	08078	31/03/2002
C	Actellifog	Zeneca Crop Protection	06628	20/04/2001
C	Blex	Zeneca Crop Protection	06639	20/04/2001
C	Dazzel	Zeneca Crop Protection	08024	20/04/2001
C	Fumite Pirimiphos methyl Smoke	Octavius Hunt Ltd	00941	

870 Propoxur

| C | Fumite Propoxur Smoke | Octavius Hunt Ltd | 00942 | 31/10/2001 |

871 Pymetrozine

| C | Chess | Novartis Crop Protection UK Ltd | 09817 | |
| C | Plenum | Novartis Crop Protection UK Ltd | 09816 | |

872 Pyrethrins

	Alfadex	Novartis Animal Health UK Ltd	08591	
	Alfadex	Ciba Agriculture	00074	
	Turbair Flydown	SumiAgro (UK) Ltd	09613	
	Turbair Flydown	Pan Britannica Industries Ltd	05482	
	Turbair Kilsect Short Life Grade	SumiAgro (UK) Ltd	09615	
	Turbair Kilsect Short Life Grade	Pan Britannica Industries Ltd	02240	
	Turbair Super Flydown	SumiAgro (UK) Ltd	09617	
	Turbair Super Flydown	Pan Britannica Industries Ltd	02249	

873 Pyrethrins + Resmethrin

| C | Pynosect 30 Water Miscible | Mitchell Cotts Chemicals Ltd | 01653 | |

874 Resmethrin

| C | Turbair Resmethrin Extra | SumiAgro (UK) Ltd | 02247 | |

875 Resmethrin + Fenitrothion + Permethrin

| | Turbair Grain Store Insecticide | Pan Britannica Industries Ltd | 02238 | 20/04/2001 |

876 Resmethrin + Pyrethrins

| C | Pynosect 30 Water Miscible | Mitchell Cotts Chemicals Ltd | 01653 | |

C These products are "approved for agricultural use". For further details refer to page vii.
A These products are approved for use in or near water. For further details refer to page vii.

Product Name	Marketing Company	Reg. No.	Expiry Date

877 Rotenone

C	Devcol Liquid Derris	Devcol Ltd	06063
C	Liquid Derris	Ford Smith & Co Ltd	01213

878 Sulphur

C	Ashlade Sulphur FL	Ashlade Formulations Ltd	06478
C	Kumulus DF	BASF plc	04707
C	MSS Sulphur 80	Mirfield Sales Services Ltd	05752

879 Tar oils

C	Sterilite Tar Oil Winter Wash 60% Stock Emulsion	Coventry Chemicals Ltd	05061
C	Sterilite Tar Oil Winter Wash 80% Miscible Quality	Coventry Chemicals Ltd	05062

880 Tau-fluvalinate

C	Mavrik	Novartis Crop Protection UK Ltd	09697	
C	Mavrik Aquaflow	Novartis Crop Protection UK Ltd	08347	31/08/2002

881 Tebufenpyrad

C	Masai	BASF plc	10223
C	Masai	Cyanamid Agriculture Ltd	07452
C	Masai G	BASF plc	10224
C	Masai G	Cyanamid Agriculture Ltd	07453

882 Teflubenzuron

C	Nemolt	BASF plc	10226
C	Nemolt	Cyanamid Agriculture Ltd	07012

883 Tefluthrin

C	Evict	Bayer plc	09150	
C	Evict	Zeneca Crop Protection	08731	31/07/2001
C	Force ST	Bayer plc	09713	
C	Force ST	Zeneca Crop Protection	06665	31/08/2002

884 Tetradifon

C	Tedion V-18 EC	Hortichem Ltd	03820

885 Tetradifon + Dicofol

C	Childion	Hortichem Ltd	03821	31/05/2002

886 Thiophanate-methyl + Lindane

C	Castaway Plus	Rhone-Poulenc Amenity	05327	31/07/2001

C These products are "approved for agricultural use". For further details refer to page vii.
A These products are approved for use in or near water. For further details refer to page vii.

Product Name	Marketing Company	Reg. No.	Expiry Date
887 Thiram + Permethrin			
C Combinex	Pan Britannica Industries Ltd	00562	31/10/2001
888 Triadimenol + Bitertanol + Fuberidazole + Imidacloprid			
C Cereline Secur	Bayer plc	09511	
889 Triadimenol + Fuberidazole + Imidacloprid			
C Baytan Secur	Bayer plc	09510	
890 Triazamate			
C Aztec	BASF plc	10211	
C Aztec	Cyanamid Agriculture Ltd	07817	
891 Trichlorfon			
C Dipterex 80	Bayer plc	00711	20/04/2001
892 Zetacypermethrin			
C Fury 10 EW	FMC Corporation (UK) Ltd	10268	
C Fury 10 EW	SumiAgro (UK) Ltd	09689	30/09/2003
C Fury 10 EW	Pan Britannica Industries Ltd	08153	31/07/2002
C Minuet EW	FMC Corporation (UK) Ltd	10269	
C Minuet EW	SumiAgro (UK) Ltd	09605	30/09/2003
C Minuet EW	Pan Britannica Industries Ltd	08820	31/07/2002

C These products are "approved for agricultural use". For further details refer to page vii.
A These products are approved for use in or near water. For further details refer to page vii.

Product Name	Marketing Company	Reg. No.	Expiry Date

1.4 Vertebrate Control Products
including rodenticides, mole killers and bird repellents

893 Alphachloralose

Alphabird	Killgerm Chemicals Ltd	10146	
Alphachloralose Technical	Rentokil Initial UK Ltd	09319	
Rentokil Alphachloralose	Rentokil Initial UK Ltd	01720	25/07/2003

894 Aluminium ammonium sulphate

Curb	Sphere Laboratories (London) Ltd	02480	
Guardsman B	Chiltern Farm Chemicals Ltd	05494	
Guardsman M	Chiltern Farm Chemicals Ltd	05495	
Guardsman STP	Sphere Laboratories (London) Ltd	03606	
Liquid Curb Crop Spray	Sphere Laboratories (London) Ltd	03164	
Rezist	Barrettine Environmental Health	08576	

895 Aluminium phosphide

Luxan Talunex	Luxan (UK) Ltd	06563	
Phostek	Killgerm Chemicals Ltd	07921	
Phostek	Killgerm Chemicals Ltd	05115	
Phostoxin	Rentokil Initial UK Ltd	09315	
Rentokil Phostoxin	Rentokil Initial UK Ltd	01775	

896 Bromadiolone

Bromabloc	Rentokil Initial UK Ltd	09854	31/05/2003
Deadline Place Packs	Rentokil Ltd	06624	31/05/2003

897 Chlorophacinone

Karate Ready to Use Rat and Mouse Bait	DiverseyLever Ltd	08658	30/06/2003
Rat Eyre Rat and Mouse Bait	Vermin Eradication Supplies	03952	31/05/2003
Rat Eyre Rat Bait	Vermin Eradication Supplies	03952	31/05/2003

898 Coumatetralyl

Racumin Master Mix	Bayer plc	01677	31/01/2002

899 Difenacoum

Difenard	Rentokil Initial UK Ltd	09012	30/06/2003
Killgerm Rat Rod	Killgerm Chemicals Ltd	05154	30/06/2003
Killgerm Wax Bait	Killgerm Chemicals Ltd	04096	30/06/2003

C   These products are "approved for agricultural use". For further details refer to page vii.
A   These products are approved for use in or near water. For further details refer to page vii.

Product Name	Marketing Company	Reg. No.	Expiry Date
899 Difenacoum—continued			
Rentokil Difenard	Rentokil Initial Plc	08124	28/02/2002
900 Diphacinone			
Ditrac All-Weather Bait Bar	Bell Laboratories Inc	07227	31/05/2001
Ditrac Rat and Mouse Bait	Bell Laboratories Inc	07170	31/05/2001
Tomcat All-Weather Rodent Bar	Antec International Ltd	07231	31/05/2001
Tomcat Rat and Mouse Bait	Antec International Ltd	07171	31/05/2001
Tomcat Super Blox All-Weather	Antec International Ltd	07446	31/05/2001
901 Seconal			
Seconal	Killgerm Chemicals Ltd	04715	31/07/2002
902 Sodium cyanide			
Cymag	Sorex Ltd	09308	
Cymag	Zeneca Crop Protection	06651	30/11/2001
903 Sulphonated cod liver oil			
Scuttle	Fine Agrochemicals Ltd	06232	
903a (Z)-9-Tricosene + Azamethiphos			
Alfacron Plus	Novartis Animal Health UK Ltd	09439	
904 Warfarin			
Grey Squirrel Liquid Concentrate	Killgerm Chemicals Ltd	06455	
Grey Squirrel Liquid Concentrate	Rodent Control Ltd	01009	31/03/2002
905 Zinc phosphide			
ZP Rodent Bait	Bell Laboratories Inc	02822	
ZP Rodent Pellets	Antec International Ltd	07814	
906 Ziram			
AAprotect	Universal Crop Protection Ltd	03784	

C   These products are "approved for agricultural use". For further details refer to page vii.
A   These products are approved for use in or near water. For further details refer to page vii.

Product Name	Marketing Company	Reg. No.	Expiry Date

1.5 Biological Pesticides

907 Bacillus thuringiensis

C	Bactura WP	Koppert (UK) Ltd	08732	30/09/2002
C	DiPel WP	English Woodlands Biocontrol	08634	30/09/2002
C	Novosol Flowable Concentrate	Ashlade Formulations Ltd	06566	31/03/2001
C	Thuricide HP	Novartis Crop Protection UK Ltd	08494	31/07/2002

907a Bacillus thuringiensis Berliner var kurstaki

C	Novosol FC	Whyte Agrochemcials Ltd	08819

908 Peniophora Gigantea

C	PG Suspension	Ecological Laboratories Ltd	08975

909 Verticillium Lecanii

C	Mycotal	Koppert (UK) Ltd	04782
C	Vertalec	Koppert (UK) Ltd	04781

C   These products are "approved for agricultural use". For further details refer to page vii.
A   These products are approved for use in or near water. For further details refer to page vii.

Product Name	Marketing Company	Reg. No.	Expiry Date

1.6 Miscellaneous

including molluscicides, lumbricides, soil sterilants, anti-oxidants and fumigants

910 Bitertanol + Fuberidazole + Imidacloprid

C Sibutol Secur	Bayer plc	09131	

911 Carbaryl

C Cavalier	Rhone-Poulenc Amenity	07027	20/04/2001
C Twister Flow	Rhone-Poulenc Amenity	05712	20/04/2001

912 Carbendazim

C Mascot Systemic	Rigby Taylor Ltd	07654	30/11/2001
C Turfclear WDG	The Scotts Company (UK) Limited	07490	
C Twincarb	Vitax Ltd	08777	

913 Chloropicrin

C Chloropicrin Fumigant	Dewco-Lloyd Ltd	04216	
C K & S Chlorofume	K & S Fumigation Services Ltd	08722	

914 Chlorpropham

Aceto Chlorpropham 50M	Aceto Agricultural Chemicals Corporation (UK) Ltd	08929	

915 Dazomet

C Basamid	Hortichem Ltd	07204	
C Basamid	BASF plc	00192	

916 1,2-Dichloropropane + 1,3-Dichloropropene

C Cyanamid DD Soil Fumigant	Cyanamid Agriculture Ltd	07017	31/03/2002

917 1,3-Dichloropropene + 1,2-Dichloropropane

C Cyanamid DD Soil Fumigant	Cyanamid Agriculture Ltd	07017	31/03/2002

918 Diphenylamine

C No Scald DPA 31	Elf Atochem Agri SA	08312	
Shield DPA 15%	United Agri Products Limited	09550	
Shield DPA 15%	Willmot Pertwee Ltd	08313	31/05/2002

C These products are "approved for agricultural use". For further details refer to page vii.
A These products are approved for use in or near water. For further details refer to page vii.

Product Name	Marketing Company	Reg. No.	Expiry Date

919 Fuberidazole + Bitertanol + Imidacloprid

C	Sibutol Secur	Bayer plc	09131	

920 Guazatine + Triticonazole

C	Premis	Aventis CropScience UK Limited	10177	

921 Imazalil + Triticonazole

C	Robust	Aventis CropScience UK Limited	10176	

922 Imidacloprid + Bitertanol + Fuberidazole

C	Sibutol Secur	Bayer plc	09131	

923 Lindane + Thiophanate-methyl

C	Castaway Plus	Rhone-Poulenc Amenity	05327	31/07/2001

924 Metaldehyde

C	Aristo	De Sangosse (UK) SA	09824	
C	Aristo	De Sangosse (UK) SA	09622	30/11/2002
C	Brits	Doff Portland Ltd	10116	
C	Chiltern Blues	Chiltern Farm Chemicals Ltd	10071	
C	Chiltern Hardy	Chiltern Farm Chemicals Ltd	06948	
C	Chiltern Hundreds	Chiltern Farm Chemicals Ltd	10072	
C	Clartex	pbi Agrochemicals Ltd	09213	
C	Clayton Musket	Clayton Plant Protection (UK) Ltd	10198	
C	Cookes 6% Metaldehyde Slug Killer	Nehra Cooke's Chemicals Ltd	09619	31/10/2001
C	Devcol Morgan 6% Metaldehyde Slug Killer	Devcol Morgan Ltd	07422	31/10/2001
C	Dixie 6	Greencrop Technology Ltd	10052	
C	Doff Agricultural Slug Killer with Animal Repellent	Doff Portland Ltd	06058	31/07/2001
C	Doff Horticultural Slug Killer Blue Mini Pellets	Doff Portland Ltd	09666	
C	Doff Horticultural Slug Killer Blue Mini Pellets	Doff Portland Ltd	05688	30/06/2002
C	Doff Metaldehyde Slug Killer Mini Pellets	Doff Portland Ltd	00741	30/11/2001
C	Doff New Formula Metaldehyde Slug Killer Mini Pellets	Doff Portland Ltd	09772	31/05/2003
C	Doff New Formula Metaldehyde Slug Killer Mini Pellets	Doff Portland Ltd	09338	31/10/2002
C	EM 1617/01	SumiAgro (UK) Ltd	09344	

C These products are "approved for agricultural use". For further details refer to page vii.
A These products are approved for use in or near water. For further details refer to page vii.

Product Name	Marketing Company	Reg. No.	Expiry Date

924 Metaldehyde—continued

	Product Name	Marketing Company	Reg. No.	Expiry Date
C	Escar-go 6	Chiltern Farm Chemicals Ltd	06076	
C	Escar-Go Z	Lonza Ltd	08754	
C	ESP	SumiAgro (UK) Ltd	09428	
C	Fisons Helarion	Fisons Plc	02520	31/05/2001
C	FP 107	SumiAgro (UK) Ltd	09060	
C	FP 107	Zeneca Crop Protection	06666	31/05/2001
C	Hardy Z	Lonza Ltd	08755	
C	Helimax	De Sangosse (UK) SA	07350	
C	Luxan 9363	Luxan (UK) Ltd	07359	
C	Luxan Metaldehyde	Luxan (UK) Ltd	06564	
C	Lynx	De Sangosse (UK) SA	09770	
C	Lynx	De Sangosse (UK) SA	09137	31/10/2002
C	Metarex	De Sangosse (UK) SA	07752	31/10/2001
C	Metarex Green	De Sangosse (UK) SA	10113	
C	Metarex Green	De Sangosse (UK) SA	08131	31/05/2002
C	Metarex RG	De Sangosse (UK) SA	06754	30/06/2002
C	Metarex RGS	De Sangosse (UK) SA	10115	
C	Metarex RGS	De Sangosse (UK) SA	09664	31/05/2002
C	Mifaslug	Farmers Crop Chemicals Ltd	10292	
C	Mifaslug	Farmers Crop Chemicals Ltd	01349	31/07/2002
C	Molotov	Chiltern Farm Chemicals Ltd	08295	
C	Optimol	SumiAgro (UK) Ltd	09061	
C	Optimol	Zeneca Crop Protection	06688	31/05/2001
C	pbi Slug Pellets	pbi Agrochemicals Ltd	09607	25/07/2003
C	pbi Slug Pellets	Pan Britannica Industries Ltd	01558	25/07/2003
C	Pesta	De Sangosse (UK) SA	10070	
C	Regel	De Sangosse (UK) SA	10114	
C	Regel	De Sangosse (UK) SA	08155	31/05/2002
C	Slug Pellets	SumiAgro (UK) Ltd	10150	
C	Slug Pellets	Pan Britannica Industries Ltd	01558	28/02/2001
C	Super-Flor 6% Metaldehyde Slug Killer Mini Pellets	CMI Limited	09773	
C	Super-flor 6% Metaldehyde Slug Killer Mini Pellets	CMI Limited	05453	31/10/2002
C	Unicrop 6% Mini Slug Pellets	Universal Crop Protection Ltd	09771	
C	Unicrop 6% Mini Slug Pellets	Universal Crop Protection Ltd	02275	31/10/2002
C	Yeoman	De Sangosse (UK) SA	09825	31/05/2003
C	Yeoman	De Sangosse (UK) SA	09623	30/11/2002

925 Metam-Sodium

	Product Name	Marketing Company	Reg. No.	Expiry Date
C	Fumetham	Chemical & Nutritional Supplies Ltd	10047	
C	Metam 510	UCB (Chem) Ltd	09796	
C	Metham Sodium 400	United Phosphorus Ltd	08051	

C   These products are "approved for agricultural use". For further details refer to page vii.
A   These products are approved for use in or near water. For further details refer to page vii.

Product Name	Marketing Company	Reg. No.	Expiry Date
925 Metam-Sodium—continued			
C Sistan	Universal Crop Protection Ltd	01957	
C Sistan 38	Universal Crop Protection Ltd	08646	
C Sistan 51	Universal Crop Protection Ltd	10046	
C Vapam	Willmot Pertwee Ltd	09194	
926 Methiocarb			
C Barclay Poacher	Barclay Chemicals Manufacturing Ltd	09031	
C Bayer UK 808	Bayer plc	09513	
C Bayer UK 809	Bayer plc	09514	
C Bayer UK 892	Bayer plc	09540	
C Bayer UK 935	Bayer plc	09541	
C Club	Bayer plc	07176	
C Decoy	Bayer plc	06535	
C Decoy Plus	Bayer plc	07615	
C Decoy Wetex	Bayer plc	09707	
C Draza	Bayer plc	00765	
C Draza 2	Bayer plc	04748	
C Draza Plus	Bayer plc	06553	
C Draza ST	Bayer plc	05315	
C Draza Wetex	Bayer plc	09704	
C Elvitox	Bayer plc	06738	
C Epox	Bayer plc	06737	
C Exit	Bayer plc	07632	
C Exit Wetex	Bayer plc	10149	
C Huron	Bayer plc	10148	
C Karan	Bayer plc	09637	
C Lupus	Bayer plc	09638	
C Rescur	Bayer plc	07942	
C Rivet	Bayer plc	09512	
927 Methyl bromide			
Brom-O-Gas	Great Lakes Chemical (Europe) Ltd	04508	31/01/2001
Bromomethane 100%	Brian Jones&Associates Ltd	09244	
Fumyl-O-Gas	Brian Jones&Associates Ltd	04833	
Mebrom 100	Mebrom NV	04869	
Mebrom 98	Mebrom NV	04779	31/07/2001
Methyl bromide	Rentokil Initial UK Ltd	09316	
Methyl bromide 100%	Bromine & Chemicals Ltd	01336	
Methyl bromide 98%	Bromine & Chemicals Ltd	01335	
Rentokil Methyl bromide	Rentokil Initial UK Ltd	05646	
Sobrom BM 100	Brian Jones & Associates Ltd	04381	
Sobrom BM 98	Brian Jones & Associates Ltd	04189	

C These products are "approved for agricultural use". For further details refer to page vii.
A These products are approved for use in or near water. For further details refer to page vii.

227

Product Name	Marketing Company	Reg. No.	Expiry Date
928 Thiodicarb			
C Genesis	Aventis CropScience UK Limited	09993	
C Genesis	Rhone-Poulenc Agriculture	06168	
C Genesis ST	Aventis CropScience UK Limited	09994	
C Genesis ST	Rhone-Poulenc Agriculture	08211	30/09/2002
C Judge	Aventis CropScience UK Limited	10000	
C Judge	Rhone-Poulenc Agriculture	08163	30/09/2002
C Me^2 Exodus	Me^2 Crop Protection Ltd	09786	
929 Thiophanate-methyl + Lindane			
C Castaway Plus	Rhone-Poulenc Amenity	05327	31/07/2001
930 Trinexapac-ethyl			
C Shortcut	The Scotts Company (UK) Limited	09254	
931 Triticonazole + Guazatine			
C Premis	Aventis CropScience UK Limited	10177	
932 Triticonazole + Imazalil			
C Robust	Aventis CropScience UK Limited	10176	

C   These products are "approved for agricultural use". For further details refer to page vii.
A   These products are approved for use in or near water. For further details refer to page vii.

2

AMATEUR PRODUCTS

Product Name	Marketing Company	Reg. No.	Expiry Date

Amateur Products

935 Aluminium ammonium sulphate (Vertebrate Control)

Bio Catapult	Pan Britannica Industries Ltd	07195	
Curb (Garden Pack)	Sphere Laboratories (London) Ltd	03983	
Scoot	William Sinclair Horticulture Ltd	07388	
Stay-Off	Synchemicals Ltd	02019	

936 Aluminium sulphate (Miscellaneous)

6X Slug Killer	Organic Concentrates Ltd	04702	
Fertosan Slug and Snail Killer	Growing Success Organics Ltd	10049	
Fertosan Slug and Snail Powder	Growing Success Organics Ltd	08507	30/04/2003
Growing Success Slug Killer	Growing Success Organics Ltd	04386	
Septico Slug Killer	Growing Success Organics Ltd	09724	
Septico Slug Killer	Greenco	07803	30/09/2002

937 Amitrole + Atrazine (Herbicide)

Atlazin D-Weed	Nomix-Chipman Ltd	08210	
Deeweed	Direct Mail Products Ltd	10145	
Deeweed	Arable & Bulb Chemicals Ltd	00659	31/05/2003
Do-It-All Path and Patio Weedkiller	Do-It-All Ltd	08156	
Doff Total Path Weedkiller	Doff Portland Ltd	04632	
Murphy Path Weedkiller	Fisons Plc	03630	
Wilko Path Weedkiller	Wilkinson Group of Companies	06976	
Woolworths Path Weedkiller	Woolworths Plc	07456	

938 Amitrole + 2,4-D + Diuron (Herbicide)

B & Q Path & Patio Weedkiller	B & Q Plc	09545	
Doff Long-Lasting Path Weedkiller	Doff Portland Ltd	09289	
Trik	Nufarm Whyte Agriculture Ltd	09788	

939 Amitrole + 2,4-D + Diuron + Simazine (Herbicide)

Hytrol	Agrichem (International) Ltd	04540	30/10/2001

940 Amitrole + Diquat + Paraquat + Simazine (Herbicide)

Pathclear	The Scotts Company (UK) Limited	09287	
Pathclear	Miracle Garden Care	07789	

C   These products are "approved for agricultural use". For further details refer to page vii.
A   These products are approved for use in or near water. For further details refer to page vii.

Product Name	Marketing Company	Reg. No.	Expiry Date

941 Amitrole + Diuron + Simazine (Herbicide)

Hytrol Total	Agrichem (International) Ltd	08296	

942 Amitrole + Simazine (Herbicide)

Homebase Path&Drive Weed Killer	Sainsbury's Homebase	07755	
Path and Drive Weedkiller	pbi Home & Garden Ltd	05958	

943 Ammonium sulphamate (Herbicide)

Amcide	Battle Hayward & Bower Ltd	00089	
Deep Root	Growing Success Organics Ltd	08368	
Deep Root Path&Patio Weedkiller	Growing Success Organics Ltd	09365	
Root-Out	Dax Products Ltd	03510	

944 Atrazine + Amitrole (Herbicide)

Atlazin D-Weed	Nomix-Chipman Ltd	08210	
Deeweed	Direct Mail Products Ltd	10145	
Deeweed	Arable & Bulb Chemicals Ltd	00659	31/05/2003
Do-It-All Path and Patio Weedkiller	Do-It-All Ltd	08156	
Doff Total Path Weedkiller	Doff Portland Ltd	04632	
Murphy Path Weedkiller	Fisons Plc	03630	
Wilko Path Weedkiller	Wilkinson Group of Companies	06976	
Woolworths Path Weedkiller	Woolworths Plc	07456	

945 Bacillus thuringiensis (Biological)

DiPel WP	English Woodlands Biocontrol	08733	30/09/2002

946 Bendiocarb (Insecticide)

B & Q Ant Killer	B & Q Plc	04880	
B & Q Woodlice Killer	B & Q Plc	07155	
Camco Insect Powder	AgrEvo UK Ltd	07576	
Delta Insect Powder	AgrEvo UK Ltd	07547	20/04/2001
Do It All Ant Killer	Do-It-All Ltd	04854	
Do-It-All Woodlice Killer	Do-It-All Ltd	08105	
Doff Ant Control Powder	Doff Portland Ltd	04881	
Doff Wasp Nest Killer	Doff Portland Ltd	06114	
Doff Woodlice Killer	Doff Portland Ltd	06081	
Ficam Insect Powder	AgrEvo UK Ltd	07546	20/04/2001
Focus Ant Killer	Focus DIY Ltd	06235	
Great Mills Ant Killer	Great Mills (Retail) Ltd	04852	
Great Mills Woodlice Killer	Great Mills (Retail) Ltd	07066	
Homebase Ant Killer	Homebase Ltd	06236	

C   These products are "approved for agricultural use". For further details refer to page vii.
A   These products are approved for use in or near water. For further details refer to page vii.

Product Name	Marketing Company	Reg. No.	Expiry Date

946 Bendiocarb (Insecticide)—continued

Product Name	Marketing Company	Reg. No.	Expiry Date
Homebase Wasp Nest Killer	Homebase Ltd	08281	
Homebase Woodlice Killer	Homebase Ltd	06994	
Portland Brand Ant Killer	Doff Portland Ltd	07068	
Premier Ant Killer	Premier Way Ltd	07644	
Secto Ant and Crawling Insect Powder	Sinclair Animal & Household Care Ltd	07895	
Secto Wasp Killer Powder	Sinclair Animal & Household Care Ltd	07896	
Wilko Ant Destroyer	Wilkinson Group of Companies	04853	
Wilko Woodlice Killer	Wilkinson Group of Companies	07156	
Woolworths Ant Killer	Woolworths Plc	07067	
Woolworths Woodlice Killer	Woolworths Plc	07057	

947 Benzalkonium chloride + Copper sulphate (Herbicide)

Product Name	Marketing Company	Reg. No.	Expiry Date
Algizin P	Waterlife Research Industries Ltd	08851	
Lotus Algicide	Waterlife Research Industries Ltd	08852	
Lotus Fungicide	Waterlife Research Industries Ltd	08853	

948 Bifenthrin (Insecticide)

Product Name	Marketing Company	Reg. No.	Expiry Date
Bio New Greenfly Killer Plus	pbi Home & Garden Ltd	09043	
Bio New Sprayday	pbi Home & Garden Ltd	09042	31/07/2002
Bio Sprayday Greenfly Killer Plus	pbi Home & Garden Ltd	09669	
Blitz Bug Gun	Miracle Garden Care Ltd	09040	28/02/2002
Bug Clear	The Scotts Company (UK) Limited	10128	
Bug Clear Gun!	The Scotts Company (UK) Limited	10127	
Bug Free Concentrate	pbi Home & Garden Ltd	10170	
Bug Gun! For Gardens	The Scotts Company (UK) Limited	09463	
Bug-Free Concentrate	Phostrogen Ltd	09762	25/07/2003
Bug-Free Ready To Use	pbi Home & Garden Ltd	10169	
Bug-Free Ready To Use	Phostrogen Ltd	09766	25/07/2003
Polysect Insecticide	The Scotts Company (UK) Limited	10195	
Polysect Insecticide	The Scotts Company (UK) Limited	09650	
Polysect Insecticide	Solaris, Garden Division of Monsanto	08625	25/07/2003
Polysect Insecticide	Solaris, Garden Division of Monsanto	08624	30/06/2002
Polysect Insecticide Ready To Use	The Scotts Company (UK) Limited	10196	
Polysect Insecticide Ready To Use	The Scotts Company (UK) Limited	09651	
Polysect Insecticide Ready To Use	Solaris, Garden Division of Monsanto	08627	25/07/2003

C   These products are "approved for agricultural use". For further details refer to page vii.
A   These products are approved for use in or near water. For further details refer to page vii.

Product Name	Marketing Company	Reg. No.	Expiry Date

948 Bifenthrin (Insecticide)—continued

Polysect Insecticide Ready To Use	Solaris, Garden Division of Monsanto	08626	30/06/2002
Sybol Extra	The Scotts Company (UK) Limited	09723	
Sybol Extra	Miracle Garden Care Ltd	09041	

949 Bifenthrin + Flutriafol (Fungicide) (Insecticide)

| Roseclear Gun ! | The Scotts Company (UK) Limited | 09923 | |

950 Bifenthrin + Myclobutanil (Fungicide) (Insecticide)

| Bug & Fungus Free | pbi Home & Garden Ltd | 10205 | |
| Multirose ready-to-use | pbi Home & Garden Ltd | 10206 | |

951 Bioallethrin + Permethrin (Insecticide)

Bio Spraydex Greenfly Killer	pbi Home & Garden Ltd	07404	31/01/2001
Floracid	Perycut Insectengun Ltd	06798	30/09/2002
Longer Lasting Bug Gun	Miracle Garden Care	08021	30/09/2002

952 Bone oil (Vertebrate Control)

| Renardine 72-2 | Roebuck-Eyot Ltd | 06769 | |
| Renardine 72/2 | Roebuck-Eyot Ltd | 06769 | 31/03/2002 |

953 Borax (Insecticide)

| Nippon Ant Killer Liquid | Vitax Ltd | 05270 | |

954 Bupirimate + Pirimicarb + Triforine (Fungicide) (Insecticide)

| 'Roseclear' 2 | The Scotts Company (UK) Limited | 09498 | |
| 'Roseclear' 2 | Miracle Garden Care Ltd | 08736 | |

955 Bupirimate + Triforine (Fungicide)

Nimrod T	The Scotts Company (UK) Limited	09502	
Nimrod T	Miracle Garden Care	07876	
Nimrod T	Zeneca Garden Care	06843	31/01/2001
Nimrod T	Imperial Chemical Industries Plc	03982	31/05/2001

956 Buprofezin (Insecticide)

| Whitefly Gun! | Miracle Garden Care | 08235 | 31/08/2002 |

957 Butoxycarboxim (Insecticide)

| Systemic Insecticide Pins | pbi Home & Garden Ltd | 10190 | 30/06/2003 |
| Systemic Insecticide Pins | The Scotts Company & Subsidiaries | 09831 | 30/06/2003 |

C   These products are "approved for agricultural use". For further details refer to page vii.
A   These products are approved for use in or near water. For further details refer to page vii.

Product Name	Marketing Company	Reg. No.	Expiry Date

957 Butoxycarboxim (Insecticide)—continued

Systemic Insecticide Pins	Monsanto Plc, Solaris - Garden Division	09330	
Systemic Insecticide Pins	Phostrogen Ltd	08209	30/11/2001

958 Captan + 1-Naphthylacetic acid (Fungicide) (Herbicide)

Doff Hormone Rooting Powder	Doff Portland Ltd	01065	
Homebase Hormone Rooting Powder	Homebase Ltd	09798	
Homebase Rooting Hormone Powder	Sainsbury's Homebase	07630	
Murphy Hormone Rooting Powder	The Scotts Company (UK) Limited + Levington Horticulture Ltd	07923	
Murphy Hormone Rooting Powder	Fisons Plc	03618	
New Strike	pbi Home & Garden Ltd	05956	31/07/2002
Rooting Powder	Vitax Ltd	06334	

959 Captan + 1-Naphthylacetic acid (Herbicide)

Bio Strike	pbi Home & Garden Ltd	09674	

960 Carbendazim (Fungicide)

AgrEvo Garden Systemic Fungicide	AgrEvo UK Ltd	08133	
B & Q Systemic Fungicide Concentrate	B & Q Plc	07417	
Do It All Plant Disease Control	Do-It-All Ltd	08132	
Doff Plant Disease Control	Doff Portland Ltd	07159	
Murphy Systemic Action Fungicide	Levington Horticulture Ltd	07558	
PBI Supercarb	pbi Home & Garden Ltd	03981	31/07/2001
Spotless	The Scotts Company (UK) Limited	09534	
Spotless	AgrEvo UK Ltd	08167	
Wilko Plant Disease Control	Wilkinson Group of Companies	08963	

961 Chlorpyrifos (Insecticide)

New Chlorophos	pbi Home & Garden Ltd	08773	30/06/2002

962 Cholecalciferol + Difenacoum (Vertebrate Control)

Sorexa CD Mouse Killer	Sorex Ltd	08208	25/07/2003

963 Citronella oil (Vertebrate Control)

The Ultimate Cat Repellent	Claytek Ltd	09822	

C   These products are "approved for agricultural use". For further details refer to page vii.
A   These products are approved for use in or near water. For further details refer to page vii.

Product Name	Marketing Company	Reg. No.	Expiry Date

964 Citronella oil + Methyl nonyl ketone (Vertebrate Control)

Secto Keep Off	Sinclair Animal & Household Care Ltd	07890	

965 Citronella oil + Tar oils (Fungicide) (Miscellaneous)

Bio Scat-A-Cat	pbi Home & Garden Ltd	10037	

966 Citrus Extract (Herbicide)

Algae Control No 3 Anti-Slime Algae	Interpet Ltd	09071	
Biotal Bio-Clear	Interpet Ltd	09072	
Blagdon Pondsafe Bio Clear	Interpet Ltd	09857	
Blagdon Pondsafe Bio Clear	Biotal Industrial Products Ltd	09073	31/12/2002
Duckweed Control	Interpet Ltd	09070	

967 Citrus Extract (Miscellaneous)

Bio-Active Algaway	Interpet Ltd	09076	

968 Clopyralid + Fluroxypyr + MCPA (Herbicide)

Weed-B-Gon	Solaris, Garden Division of Monsanto	08926	
Weed-B-Gon	Dow AgroSciences Ltd	08925	
Weed-B-Gon Ready To Use	Solaris, Garden Division of Monsanto	09451	
Weed-B-Gon Ready To Use	Dow AgroSciences Ltd	09450	

969 Copper (Fungicide)

Bordeaux Mixture	Vitax Ltd	07162	

970 Copper chloride + Metaldehyde (Miscellaneous)

Sera Snailpur	Sera GmbH	09587	
Sera Snailpur	Sera Werke Heimtierbedarf	09064	30/04/2001

971 Copper hydrate (Miscellaneous)

Algimycin PLL	Certikin International Ltd	09096	
Algimycin PLL	Primemix Ltd	08859	30/06/2001

972 Copper oxychloride (Fungicide)

Murphy Traditional Copper Fungicide	Levington Horticulture Ltd	07574	
Murphy Traditional Copper Fungicide	Fisons Plc	04585	

C   These products are "approved for agricultural use". For further details refer to page vii.
A   These products are approved for use in or near water. For further details refer to page vii.

Product Name	Marketing Company	Reg. No.	Expiry Date
973 Copper sulphate (Fungicide)			
B & Q Garden Fungicide Spray	B & Q Plc	06285	
PBI Cheshunt Compound	pbi Home & Garden Ltd	00485	
974 Copper sulphate (Herbicide)			
Algalit	Symbionics	09077	30/06/2002
Aqualife Pond Algicide	Rosewood Pet Products Ltd	08838	
Betta Algae Clear	J & K Aquatics	08843	
Green Algae Control	King British Aquarium Accessories	08855	
Hydroperfect Algalit HP-352	Symbionics Sarl	09519	
Hydroperfect Algalit HP-355	Symbionics Sarl	09520	
Hydroperfect Algalit HP-357	Symbionics Sarl	09521	
Hydroperfect Lemnalit HP-372	Symbionics Sarl	09528	
Hydroperfect Lemnalit HP-375	Symbionics Sarl	09529	
Hydroperfect Lemnalit HP-377	Symbionics Sarl	09530	
Lemnalit	Symbionics	09078	30/04/2002
Mister Green Algalit MG-352	Symbionics Sarl	09522	
Mister Green Algalit MG-355	Symbionics Sarl	09523	
Mister Green Algalit MG-357	Symbionics Sarl	09524	
Mister Green Lemnalit MG-372	Symbionics Sarl	09531	
Mister Green Lemnalit MG-375	Symbionics Sarl	09532	
Mister Green Lemnalit MG-377	Symbionics Sarl	09533	
Pond Pride Green Algae Control	Sinclair Animal & Household Care Ltd	09100	
Sera Algopur	Sera Werke Heimtierbedarf	09065	
Waterscene Coldwater - Algae Control	Waterscene Enterprises Co. Ltd	08869	
975 Copper sulphate (Herbicide) (Miscellaneous)			
Aqualife Algae and Snail Killer	Rosewood Pet Products Ltd	08839	
976 Copper sulphate (Miscellaneous)			
Betta Snail Clear	J & K Aquatics	08842	
BWG Anti Snails	Technical Aquatic Products	08847	
King British Formula KB7	King British Aquarium Accessories	08854	30/06/2001
Molluzin	Waterlife Research Industries Ltd	08849	
Snail Control	Sinclair Animal & Household Care Ltd	09101	
Snail Smasher	Technical Aquatic Products	08840	
977 Copper sulphate + Benzalkonium chloride (Herbicide)			
Algizin P	Waterlife Research Industries Ltd	08851	

C These products are "approved for agricultural use". For further details refer to page vii.
A These products are approved for use in or near water. For further details refer to page vii.

Product Name	Marketing Company	Reg. No.	Expiry Date

977 Copper sulphate + Benzalkonium chloride (Herbicide)—continued

Lotus Algicide	Waterlife Research Industries Ltd	08852	
Lotus Fungicide	Waterlife Research Industries Ltd	08853	

978 Copper sulphate + Ferrous sulphate (Miscellaneous)

Snailaway	Interpet Ltd	02457	

979 Copper sulphate + Ferrous sulphate + Magnesium sulphate (Herbicide) (Miscellaneous)

Anti-Snail	Interpet Ltd	10117	

980 Copper sulphate + Ferrous sulphate + Magnesium sulphate (Miscellaneous)

Snail Away	Interpet Ltd	08868	31/03/2003

981 Copper sulphate + Simazine (Herbicide)

PPI Clarity Excel	Pet Products International Ltd	09095	
Tap Algasan	Technical Aquatic Products	08845	

982 Coumatetralyl (Vertebrate Control)

PBI Racumin Mouse Bait	pbi Home & Garden Ltd	01678	31/01/2002

983 Cypermethrin + Propiconazole (Fungicide) (Insecticide)

Murphy Roseguard	Levington Horticulture Ltd	08564	

984 2,4-D + Amitrole + Diuron (Herbicide)

B & Q Path & Patio Weedkiller	B & Q Plc	09545	
Doff Long-Lasting Path Weedkiller	Doff Portland Ltd	09289	
Trik	Nufarm Whyte Agriculture Ltd	09788	

985 2,4-D + Amitrole + Diuron + Simazine (Herbicide)

Hytrol	Agrichem (International) Ltd	04540	30/10/2001

986 2,4-D + Dicamba (Herbicide)

B & Q Granular Weed and Feed	B & Q Plc	05294	
B & Q Lawnweed Spray	B & Q Plc	05804	
Betterware Spot Weedkiller Spray	Betterware UK Ltd	08377	
Bio Lawn Weedkiller	pbi Home & Garden Ltd	00268	
Bio Toplawn Feed and Weed	pbi Home & Garden Ltd	09958	
Elliott Easyweeder	Thomas Elliott Ltd	07693	31/07/2001
Elliott's Touchweeder	Thomas Elliott Ltd	09160	

C   These products are "approved for agricultural use". For further details refer to page vii.
A   These products are approved for use in or near water. For further details refer to page vii.

Product Name	Marketing Company	Reg. No.	Expiry Date

986 2,4-D + Dicamba (Herbicide)—continued

Green Up Lawn Feed and Weed	Vitax Ltd	06419	
Green Up Weedfree Lawn Weedkiller	Vitax Ltd	05322	
Green Up Weedfree Spot Weedkiller for Lawns	Vitax Ltd	06321	
Lawn Builder Plus Weed Control	Miracle Garden Care Ltd	07936	
Lawn Weed Gun	Zeneca Garden Care	06838	31/05/2001
Lawn Weed Gun	ICI Garden & Professional Products	04407	31/05/2001
Toplawn Feed with Weedkiller	Pan Britannica Industries Ltd	08289	28/02/2003
Verdone Gun!	The Scotts Company (UK) Limited	09643	
Verdone Gun!	Miracle Garden Care Ltd	07671	

987 2,4-D + Dicamba + Ferrous sulphate (Herbicide)

Bio Supergreen 123	pbi Home & Garden Ltd	07962	31/01/2001
Toplawn Feed with Weed & Mosskiller	pbi Home & Garden Ltd	08120	31/12/2001

988 2,4-D + Dicamba + Ferrous sulphate heptahydrate (Herbicide)

Toplawn Feed, Weed & Mosskiller	pbi Home & Garden Ltd	09350	

989 2,4-D + Dicamba + Mecoprop (Herbicide)

B & Q Tree Stump and Brushwood Killer	B & Q Plc	07645	

990 2,4-D + Dicamba + Mecoprop-P (Herbicide)

New Formulation SBK Brushwood Killer	Vitax Ltd	08737	

991 2,4-D + Dichlorprop (Herbicide)

AgrEvo Lawn Weed Killer	AgrEvo UK Ltd	08137	
AgrEvo Lawn Weedkiller Concentrate	AgrEvo UK Ltd	08912	
B & Q Lawn Feed and Weed	B & Q Plc	05487	
B & Q Lawn Spot Weeder	B & Q Plc	05486	
B & Q Lawn Weedkiller	B & Q Plc	05324	
B & Q Nettle Gun	B & Q Plc	06897	
Do It All Lawn Feed & Weed	Do-It-All Ltd	09311	
Do It All Wipeout! Lawn Weed Killer	Do-It-All Ltd	08197	

C   These products are "approved for agricultural use". For further details refer to page vii.
A   These products are approved for use in or near water. For further details refer to page vii.

Product Name	Marketing Company	Reg. No.	Expiry Date

991 2,4-D + Dichlorprop (Herbicide)—continued

Product Name	Marketing Company	Reg. No.	Expiry Date
Do-It-All Lawn Weedkiller	Do-It-All Ltd	08106	
Doff Lawn Feed and Weed	Doff Portland Ltd	05117	
Doff Lawn Spot Weeder	Doff Portland Ltd	03995	
Doff Nettle Gun	Doff Portland Ltd	07158	
Doff New Formula Lawn Weedkiller	Doff Portland Ltd	05666	
Fisons Ready-to-Use Lawn Weedkiller	Fisons Plc	05868	
Fisons Water-on Lawn Weedkiller	Fisons Plc	05807	
Focus Lawn Spot Weeder	Focus DIY Ltd	07468	
Focus Lawn Weedkiller	Focus DIY Ltd	06096	
Focus Nettle Gun	Focus DIY Ltd	06896	
Great Mills Lawn Feed&Weed	Great Mills (Retail) Ltd	08367	
Great Mills Lawn Spot Weeder	Great Mills (Retail) Ltd	05014	
Great Mills Lawn Weedkiller	Great Mills (Retail) Ltd	09296	
Homebase Lawn Weedkiller Liquid	Homebase Ltd	07539	
Homebase Lawn Weedkiller Ready To Use Sprayer	Sainsbury's Homebase House and Garden Centres	07536	
Homebase Liquid Lawn Feed and Weed	Homebase Ltd	07989	
Homebase Nettle Gun	Sainsbury's Homebase House and Garden Centres	07538	
Lawn Weedkiller	L C Solutions Ltd	09736	
Lawn Weedkiller Concentrate	L C Solutions Ltd	09740	
Levington Ready to Use Lawn Weedkiller	Levington Horticulture Ltd	07497	
Levington Water-on Lawn Weedkiller	Levington Horticulture Ltd	07495	
Murphy Clover-Kil	Fisons Plc	05271	
Tumbleweed Clover	Levington Horticulture Ltd	08325	
Tumbleweed Clover Ready to Use	Levington Horticulture Ltd	08546	
Tumbleweed Lawns	Levington Horticulture Ltd	08088	
Tumbleweed Lawns Ready to Use	Levington Horticulture Ltd	08089	
Wilko Lawn Spot Weeder	Wilkinson Home & Garden Stores	05130	
Wilko Lawn Weed and Feed	Wilkinson Group of Companies	07473	
Wilko Lawn Weedkiller	Wilkinson Group of Companies	03749	
Wilko Nettle Gun	Wilkinson Group of Companies	06898	
Woolworths Lawn Spot Weeder	Woolworths Plc	07105	

C   These products are "approved for agricultural use". For further details refer to page vii.
A   These products are approved for use in or near water. For further details refer to page vii.

Product Name	Marketing Company	Reg. No.	Expiry Date

991 2,4-D + Dichlorprop (Herbicide)—continued

Product Name	Marketing Company	Reg. No.	Expiry Date
Woolworths Lawn Weed and Feed	Woolworths Plc	07474	
Woolworths Lawn Weedkiller	Woolworths Plc	07064	

992 2,4-D + Ferrous sulphate + Mecoprop (Herbicide)

Product Name	Marketing Company	Reg. No.	Expiry Date
Asda Lawn Weed and Feed and Mosskiller	Asda Stores Ltd	03819	
Green Up Feed and Weed Plus Mosskiller	Vitax Ltd	06513	31/05/2001
Supergreen Feed, Weed & Mosskiller	Pan Britannica Industries Ltd	08118	31/01/2001
Weed 'N' Feed Extra	Vitax Ltd	06506	31/05/2001

993 2,4-D + Ferrous sulphate + Mecoprop-P (Herbicide)

Product Name	Marketing Company	Reg. No.	Expiry Date
ASB Greenworld Lawn Weed, Feed and Mosskiller Granular	ASB Greenworld Ltd	09354	
B & Q Lawn Feed, Weed and Mosskiller Granular	B & Q Plc	09355	
B & Q Summer Lawncare Feed Weed Mosskiller	B & Q Plc	10286	
Doff Granular Lawn Feed, Weed and Mosskiller	Doff Portland Ltd	08876	
Elliott's Lawn and Fine Turf Feed and Weed plus MossKiller	Thomas Elliott Ltd	09464	
Gem Lawn Weed and Feed + Mosskiller	Gem Gardening	07087	
Green Up Feed and Weed Plus Mosskiller	Vitax Ltd	09046	
Proctors Lawn Feed Weed and Mosskiller	H & T Proctor (Division of Willett & Son (Bristol) Ltd)	07808	
Supergreen Feed, Weed and Mosskiller 2	pbi Home & Garden Ltd	09224	
Weed 'N' Feed Extra	Vitax Ltd	09045	
Westland Triple Action Weed, Feed And Mosskill	Westland Horticulture	09640	

994 2,4-D + Mecoprop (Herbicide)

Product Name	Marketing Company	Reg. No.	Expiry Date
Homebase Lawn Feed and Weed Soluble	Sainsbury's Homebase	07690	30/09/2001
Lawn Feed and Weed Granules	pbi Home & Garden Ltd	05902	30/09/2001

C   These products are "approved for agricultural use". For further details refer to page vii.
A   These products are approved for use in or near water. For further details refer to page vii.

241

Product Name	Marketing Company	Reg. No.	Expiry Date

994 2,4-D + Mecoprop (Herbicide)—continued

Lawn Spot Weed Granules	Pan Britannica Industries Ltd	05903	
Supergreen and Weed	Pan Britannica Industries Agrochemicals Ltd	08911	30/09/2001
Supergreen Double (Feed and Weed)	Pan Britannica Industries Ltd	05949	28/02/2001
Supertox Spot	pbi Home & Garden Ltd	05951	30/11/2001
Toplawn Lawn Weedkiller	pbi Home & Garden Ltd	08694	31/01/2002
Toplawn Ready to Use Lawn Weedkiller	pbi Home & Garden Ltd	08695	31/12/2001
Verdone	Miracle Garden Care	08657	30/06/2001
Verdone 2	Imperial Chemical Industries Plc	03271	31/05/2001

995 2,4-D + Mecoprop-P (Herbicide)

ASB Greenworld Lawn Weed, And Feed Granular	ASB Greenworld Ltd	09356	
B & Q Lawn Weed and Feed Granular	B & Q Plc	09357	
Bio Toplawn Spot Weeder	pbi Home & Garden Ltd	09672	
Gem Lawn Weed & Feed	Gem Gardening	07086	
Green Up Spot Lawn Weedkiller	Vitax Ltd	06028	
Homebase Lawn Feed and Weed Soluble	Sainsbury's Homebase	09217	
Lawn Feed and Weed Granules	pbi Home & Garden Ltd	09225	
Supergreen and Weed	pbi Home & Garden Ltd	09218	
Supertox Spot	pbi Home & Garden Ltd	09306	
Toplawn Lawn Weedkiller	pbi Home & Garden Ltd	09589	
Toplawn Lawn Weedkiller	pbi Home & Garden Ltd	09408	31/05/2002
Toplawn Ready to Use Lawn Weedkiller	pbi Home & Garden Ltd	09348	31/07/2002
Verdone	The Scotts Company (UK) Limited	09642	
Verdone	Miracle Garden Care Ltd	09097	
Vitax 'Green Up' Granular Lawn Feed and Weed	Vitax Ltd	06158	
Wilko Lawn Feed & Weed Liquid	Wilkinson Home & Garden Stores	08789	

996 2,4-D + 2,3,6-TBA (Herbicide)

Touchweeder	Thomas Elliott Ltd	02864	

C   These products are "approved for agricultural use". For further details refer to page vii.
A   These products are approved for use in or near water. For further details refer to page vii.

Product Name	Marketing Company	Reg. No.	Expiry Date

997 Dicamba + 2,4-D (Herbicide)

B & Q Granular Weed and Feed	B & Q Plc	05294	
B & Q Lawnweed Spray	B & Q Plc	05804	
Betterware Spot Weedkiller Spray	Betterware UK Ltd	08377	
Bio Lawn Weedkiller	pbi Home & Garden Ltd	00268	
Bio Toplawn Feed and Weed	pbi Home & Garden Ltd	09958	
Elliott Easyweeder	Thomas Elliott Ltd	07693	31/07/2001
Elliott's Touchweeder	Thomas Elliott Ltd	09160	
Green Up Lawn Feed and Weed	Vitax Ltd	06419	
Green Up Weedfree Lawn Weedkiller	Vitax Ltd	05322	
Green Up Weedfree Spot Weedkiller for Lawns	Vitax Ltd	06321	
Lawn Builder Plus Weed Control	Miracle Garden Care Ltd	07936	
Lawn Weed Gun	Zeneca Garden Care	06838	31/05/2001
Lawn Weed Gun	ICI Garden & Professional Products	04407	31/05/2001
Toplawn Feed with Weedkiller	Pan Britannica Industries Ltd	08289	28/02/2003
Verdone Gun!	The Scotts Company (UK) Limited	09643	
Verdone Gun!	Miracle Garden Care Ltd	07671	

998 Dicamba + 2,4-D + Ferrous sulphate (Herbicide)

Bio Supergreen 123	pbi Home & Garden Ltd	07962	31/01/2001
Toplawn Feed with Weed & Mosskiller	pbi Home & Garden Ltd	08120	31/12/2001

999 Dicamba + 2,4-D + Ferrous sulphate heptahydrate (Herbicide)

Toplawn Feed, Weed & Mosskiller	pbi Home & Garden Ltd	09350	

1000 Dicamba + 2,4-D + Mecoprop (Herbicide)

B & Q Tree Stump and Brushwood Killer	B & Q Plc	07645	

1001 Dicamba + 2,4-D + Mecoprop-P (Herbicide)

New Formulation SBK Brushwood Killer	Vitax Ltd	08737	

C   These products are "approved for agricultural use". For further details refer to page vii.
A   These products are approved for use in or near water. For further details refer to page vii.

Product Name	Marketing Company	Reg. No.	Expiry Date

1002 Dicamba + Dichlorprop + Ferrous sulphate + MCPA (Herbicide)

B & Q Granular Weed and Feed and Mosskiller For Lawns	B & Q Plc	08381	
Evergreen Grasshopper	The Scotts Company (UK) Limited	09639	
'Grasshopper' Triple Action	Miracle Garden Care	08643	
'Greensward' 2	Miracle Garden Care	08380	30/11/2001
Miracle-Gro Lawn Food with Weed and Moss Control	Miracle Garden Care	08609	

1003 Dicamba + Dichlorprop + MCPA (Herbicide)

B & Q Granular Weed and Feed for Lawns	B & Q Plc	07850	
B & Q Liquid Weed and Feed	B & Q Plc	05293	
Boots Nettle and Bramble Weedkiller	The Boots Company Plc	03455	
Grasshopper Weed & Feed	The Scotts Company (UK) Limited	09785	
Grasshopper Weed & Feed	Miracle Garden Care Ltd	07849	30/06/2003
Greensward Triple Action	The Scotts Company (UK) Limited	09324	
Groundclear	pbi Home & Garden Ltd	05953	
Lawnsman Liquid Weed and Feed	Imperial Chemical Industries Plc	03610	31/05/2001
Miracle-Gro Weed & Feed	Miracle Garden Care	08550	31/12/2002
'Verdone Plus'	Miracle Garden Care	08689	
Woolworths Liquid Lawn Feed and Weed	Woolworths Plc	07060	

1004 Dicamba + Dichlorprop + Mecoprop (Herbicide)

New Supertox	pbi Home & Garden Ltd	06128	30/09/2001

1005 Dicamba + Dichlorprop + Mecoprop-P (Herbicide)

New Supertox	pbi Home & Garden Ltd	09209	

1006 Dicamba + MCPA (Herbicide)

Green Up Liquid Lawn Feed'n Weed	Vitax Ltd	08196	

1007 Dicamba + MCPA + Mecoprop (Herbicide)

Bio Spot	pbi Home & Garden Ltd	05071	31/12/2001
Bio Weed Pencil	pbi Home & Garden Ltd	04054	31/12/2001
Homebase Weed Pen	Sainsbury's Homebase	07648	30/06/2002

1008 Dicamba + MCPA + Mecoprop-P (Herbicide)

Bio Lawn Weed Pencil	pbi Home & Garden Ltd	09656	

C   These products are "approved for agricultural use". For further details refer to page vii.
A   These products are approved for use in or near water. For further details refer to page vii.

Product Name	Marketing Company	Reg. No.	Expiry Date

1008 Dicamba + MCPA + Mecoprop-P (Herbicide)—continued

Bio Spot	pbi Home & Garden Ltd	09210	
Bio Weed Pencil	pbi Home & Garden Ltd	09212	30/06/2002
Evergreen Feed and Weed Liquid	The Scotts Company (UK) Ltd	07766	
Homebase Lawn Feed and Weed Liquid	Homebase Ltd	07765	
Homebase Weed Pen	Sainsbury's Homebase	09211	

1009 Dichlobenil (Herbicide)

Bio Weed Ban	pbi Home & Garden Ltd	09705	
Casoron G4	Vitax Ltd	06866	
Casoron G4 Weed Block	Vitax Ltd	09371	
Path and Shrub Guard	Miracle Garden Care	08055	

1010 Dichlorophen (Fungicide) (Herbicide)

Bio Mosskiller Ready-to-Use	pbi Home & Garden Ltd	07734	

1011 Dichlorophen (Herbicide)

Moss Gun!	Miracle Garden Care	07787	28/02/2001
Tumbleweed Moss Ready to Use	Levington Horticulture Ltd	08081	

1012 Dichlorophen + 1-Naphthylacetic acid (Fungicide) (Herbicide)

Homebase Rooting Hormone Liquid	Sainsbury's Homebase	07698	

1013 Dichlorophen + 1-Naphthylacetic acid (Fungicide) (Miscellaneous)

Baby Bio Roota	pbi Home & Garden Ltd	09305	

1014 Dichlorophen + 1-Naphthylacetic acid (Herbicide)

Bio Roota	pbi Home & Garden Ltd	00271	30/11/2001

1015 Dichlorprop + 2,4-D (Herbicide)

AgrEvo Lawn Weed Killer	AgrEvo UK Ltd	08137	
AgrEvo Lawn Weedkiller Concentrate	AgrEvo UK Ltd	08912	
B & Q Lawn Feed and Weed	B & Q Plc	05487	
B & Q Lawn Spot Weeder	B & Q Plc	05486	
B & Q Lawn Weedkiller	B & Q Plc	05324	
B & Q Nettle Gun	B & Q Plc	06897	
Do It All Lawn Feed&Weed	Do-It-All Ltd	09311	
Do It All Wipeout! Lawn Weed Killer	Do-It-All Ltd	08197	

C   These products are "approved for agricultural use". For further details refer to page vii.
A   These products are approved for use in or near water. For further details refer to page vii.

Product Name	Marketing Company	Reg. No.	Expiry Date

1015 Dichlorprop + 2,4-D (Herbicide)—continued

Product Name	Marketing Company	Reg. No.	Expiry Date
Do-It-All Lawn Weedkiller	Do-It-All Ltd	08106	
Doff Lawn Feed and Weed	Doff Portland Ltd	05117	
Doff Lawn Spot Weeder	Doff Portland Ltd	03995	
Doff Nettle Gun	Doff Portland Ltd	07158	
Doff New Formula Lawn Weedkiller	Doff Portland Ltd	05666	
Fisons Ready-to-Use Lawn Weedkiller	Fisons Plc	05868	
Fisons Water-on Lawn Weedkiller	Fisons Plc	05807	
Focus Lawn Spot Weeder	Focus DIY Ltd	07468	
Focus Lawn Weedkiller	Focus DIY Ltd	06096	
Focus Nettle Gun	Focus DIY Ltd	06896	
Great Mills Lawn Feed&Weed	Great Mills (Retail) Ltd	08367	
Great Mills Lawn Spot Weeder	Great Mills (Retail) Ltd	05014	
Great Mills Lawn Weedkiller	Great Mills (Retail) Ltd	09296	
Homebase Lawn Weedkiller Liquid	Homebase Ltd	07539	
Homebase Lawn Weedkiller Ready To Use Sprayer	Sainsbury's Homebase House and Garden Centres	07536	
Homebase Liquid Lawn Feed and Weed	Homebase Ltd	07989	
Homebase Nettle Gun	Sainsbury's Homebase House and Garden Centres	07538	
Lawn Weedkiller	L C Solutions Ltd	09736	
Lawn Weedkiller Concentrate	L C Solutions Ltd	09740	
Levington Ready to Use Lawn Weedkiller	Levington Horticulture Ltd	07497	
Levington Water-on Lawn Weedkiller	Levington Horticulture Ltd	07495	
Murphy Clover-Kil	Fisons Plc	05271	
Tumbleweed Clover	Levington Horticulture Ltd	08325	
Tumbleweed Clover Ready to Use	Levington Horticulture Ltd	08546	
Tumbleweed Lawns	Levington Horticulture Ltd	08088	
Tumbleweed Lawns Ready to Use	Levington Horticulture Ltd	08089	
Wilko Lawn Spot Weeder	Wilkinson Home & Garden Stores	05130	
Wilko Lawn Weed and Feed	Wilkinson Group of Companies	07473	
Wilko Lawn Weedkiller	Wilkinson Group of Companies	03749	
Wilko Nettle Gun	Wilkinson Group of Companies	06898	
Woolworths Lawn Spot Weeder	Woolworths Plc	07105	

C   These products are "approved for agricultural use". For further details refer to page vii.
A   These products are approved for use in or near water. For further details refer to page vii.

Product Name	Marketing Company	Reg. No.	Expiry Date

1015 Dichlorprop + 2,4-D (Herbicide)—continued

| Woolworths Lawn Weed and Feed | Woolworths Plc | 07474 | |
| Woolworths Lawn Weedkiller | Woolworths Plc | 07064 | |

1016 Dichlorprop + Dicamba + Ferrous sulphate + MCPA (Herbicide)

B & Q Granular Weed and Feed and Mosskiller For Lawns	B & Q Plc	08381	
Evergreen Grasshopper	The Scotts Company (UK) Limited	09639	
'Grasshopper' Triple Action	Miracle Garden Care	08643	
'Greensward' 2	Miracle Garden Care	08380	30/11/2001
Miracle-Gro Lawn Food with Weed and Moss Control	Miracle Garden Care	08609	

1017 Dichlorprop + Dicamba + MCPA (Herbicide)

B & Q Granular Weed and Feed for Lawns	B & Q Plc	07850	
B & Q Liquid Weed and Feed	B & Q Plc	05293	
Boots Nettle and Bramble Weedkiller	The Boots Company Plc	03455	
Grasshopper Weed & Feed	The Scotts Company (UK) Limited	09785	
Grasshopper Weed & Feed	Miracle Garden Care Ltd	07849	30/06/2003
Greensward Triple Action	The Scotts Company (UK) Limited	09324	
Groundclear	pbi Home & Garden Ltd	05953	
Lawnsman Liquid Weed and Feed	Imperial Chemical Industries Plc	03610	31/05/2001
Miracle-Gro Weed & Feed	Miracle Garden Care	08550	31/12/2002
'Verdone Plus'	Miracle Garden Care	08689	
Woolworths Liquid Lawn Feed and Weed	Woolworths Plc	07060	

1018 Dichlorprop + Dicamba + Mecoprop (Herbicide)

| New Supertox | pbi Home & Garden Ltd | 06128 | 30/09/2001 |

1019 Dichlorprop + Dicamba + Mecoprop-P (Herbicide)

| New Supertox | pbi Home & Garden Ltd | 09209 | |

1020 Dichlorprop + Ferrous sulphate + MCPA (Herbicide)

| J Arthur Bower's Granular Feed, Weed and Mosskiller | William Sinclair Horticulture Ltd | 07042 | |
| J Arthur Bower's Lawn Feed, Weed and Mosskiller | William Sinclair Horticulture Ltd | 09459 | |

C These products are "approved for agricultural use". For further details refer to page vii.
A These products are approved for use in or near water. For further details refer to page vii.

247

Product Name	Marketing Company	Reg. No.	Expiry Date

1020 Dichlorprop + Ferrous sulphate + MCPA (Herbicide)—continued

J Arthur Bower's Lawn Weed and Mosskiller	William Sinclair Horticulture Ltd	09460	
Wilko Granular Feed, Weed and Mosskiller	Wilkinson Group of Companies	07471	
Wilko Granular Feed, Weed and Mosskiller for Lawns	Wilkinson Group of Companies	09731	
Wilko Lawn Feed, Weed and Mosskiller	Wilkinson Home & Garden Stores	04602	

1021 Dichlorprop + MCPA (Herbicide)

Doff Lawn Weed and Feed Soluble Powder	Doff Portland Ltd	05708	
J Arthur Bower's Granular Feed and Weed	William Sinclair Horticulture Ltd	07164	
J Arthur Bower's Lawn Feed and Weed	William Sinclair Horticulture Ltd	09458	
J Arthur Bower's Lawn Weedkiller	William Sinclair Horticulture Ltd	09461	
Wilko Granular Feed and Weed for Lawns	Wilkinson Group of Companies	09732	
Wilko Lawn Feed and Weed	Wilkinson Group of Companies	08816	
Wilko Soluble Lawn Food and Weedkiller	Wilkinson Home & Garden Stores	04391	

1022 Difenacoum + Cholecalciferol (Vertebrate Control)

Sorexa CD Mouse Killer	Sorex Ltd	08208	25/07/2003

1023 Dikegulac (Herbicide)

Cutlass	The Scotts Company (UK) Limited	09543	
Cutlass	Miracle Garden Care	07783	
Cutlass	Zeneca Garden Care	06839	31/05/2001
Cutlass	Imperial Chemical Industries Plc	00617	31/05/2001

1024 Dimethoate (Insecticide)

Doff Systemic Insecticide	Doff Portland Ltd	02658	31/08/2002

1025 Diphacinone (Vertebrate Control)

Tomcat Rat and Mouse Blox	Antec International Ltd	07794	31/05/2001
Tomcat Rat and Mouse Pellets	Antec International Ltd	07825	31/05/2001

1026 Diquat (Herbicide)

Weedol Gun!	The Scotts Company (UK) Limited	09279	
Weedol Gun!	Miracle Garden Care Ltd	09079	30/09/2001

C   These products are "approved for agricultural use". For further details refer to page vii.
A   These products are approved for use in or near water. For further details refer to page vii.

Product Name	Marketing Company	Reg. No.	Expiry Date

1027 Diquat + Amitrole + Paraquat + Simazine (Herbicide)

Pathclear	The Scotts Company (UK) Limited	09287	
Pathclear	Miracle Garden Care	07789	

1028 Diquat + Paraquat (Herbicide)

Weedol	The Scotts Company (UK) Limited	09280	
Weedol	Miracle Garden Care	07750	30/09/2001
Weedol	Zeneca Garden Care	06863	31/05/2001
Weedol Gun!	Miracle Garden Care	08254	31/07/2001

1029 Diquat + Paraquat + Simazine (Herbicide)

'Pathclear' S	The Scotts Company (UK) Limited	09285	
'Pathclear' S	Miracle Garden Care Ltd	08992	31/10/2001

1030 Disodium hydrogen phosphate (Miscellaneous)

Algae Control No 2 Anti-hair Algae	Interpet Ltd	09067	

1031 Diuron + Amitrole + 2,4-D (Herbicide)

B & Q Path & Patio Weedkiller	B & Q Plc	09545	
Doff Long-Lasting Path Weedkiller	Doff Portland Ltd	09289	
Trik	Nufarm Whyte Agriculture Ltd	09788	

1032 Diuron + Amitrole + 2,4-D + Simazine (Herbicide)

Hytrol	Agrichem (International) Ltd	04540	30/10/2001

1033 Diuron + Amitrole + Simazine (Herbicide)

Hytrol Total	Agrichem (International) Ltd	08296	

1034 Diuron + Glufosinate-ammonium (Herbicide)

AgrEvo Path Weed Killer	AgrEvo Environmental Health Ltd	08616	31/08/2002
Tumbleweed Paths and Patios	Levington Horticulture Ltd	08613	

1035 Diuron + Glyphosate (Herbicide)

Pathclear Gun!	The Scotts Company (UK) Limited	09492	
Weedatak Path and Drive	The Scotts Company (UK) Limited	09814	
Weedatak Path and Drive	Monsanto Plc	07659	30/11/2002

1036 Fatty acids (Herbicide)

B & Q Organic Weedkiller	B & Q Plc	10243	
Bio SpeedWeed	pbi Home & Garden Ltd	06134	
Organic Weed Control	pbi Home & Garden Ltd	10242	

C   These products are "approved for agricultural use". For further details refer to page vii.
A   These products are approved for use in or near water. For further details refer to page vii.

Product Name	Marketing Company	Reg. No.	Expiry Date

1037 Fatty acids (Insecticide)

Product Name	Marketing Company	Reg. No.	Expiry Date
AgrEvo Rose & Flower Greenfly Killer	AgrEvo UK Ltd	09508	
AgrEvo Rose & Flower Insect Killer	AgrEvo UK Ltd	09507	
Agrevo Rose PestKiller	AgrEvo UK Ltd	09509	
B & Q Insect Spray for Houseplants	B & Q Plc	09416	
B & Q Insect Spray for Roses and Flowers	B & Q Plc	09417	
B & Q Organic Insecticide	B & Q Plc	10245	
Bio Pest Pistol	pbi Home & Garden Ltd	07233	
Bug Gun! For Roses&Flowers	The Scotts Company (UK) Limited	09453	
Do It All De-Bug!2	Focus DIY Ltd	09506	
Doff Houseplant Pest Spray	Doff Portland Ltd	09419	
Doff Rose & Flower Pest Spray	Doff Portland Ltd	09418	
Fruit and Vegetable Insecticide	pbi Home & Garden Ltd	10193	
Garden Insecticide Concentrate	pbi Home & Garden Ltd	10189	
General Insect Killer	Growing Success Organics Ltd	08808	
Get Off Insect	Pet and Garden Manufacturing Plc	09162	
Great Mills Insect Spray for Roses	Doff Portland Ltd	09420	
Greenco GR1	Doff Portland Ltd	09455	
Greenco GR1	Doff Portland Ltd	07544	
Greenco GR3 Pest Jet	Doff Portland Ltd	09178	
Greenco GR3 Pest Jet	Greenco	07621	31/08/2001
Greenco GR3 Pest Jet	Greenco	07545	
Greenfly & Blackfly Killer	Growing Success Organics Ltd	08807	
Home Base Pest Gun 2	Homebase Ltd	09422	
Homebase Houseplant Insecticide Spray 2	Homebase Ltd	09421	
House Plant Insecticide	pbi Home & Garden Ltd	10188	
Houseplant Pest Killer	L C Solutions Ltd	09774	
Nature's Answer Organic Insecticide Ready to Use	The Scotts Company (UK) Ltd	07627	28/02/2002
Organic Pest Control	pbi Home & Garden Ltd	10239	
Phostrogen House Plant Insecticide	Phostrogen Ltd	06538	
Phostrogen Safer's Garden Insecticide Concentrate	The Scotts Company & Subsidiaries	09832	
Phostrogen Safer's Garden Insecticide Concentrate	Monsanto Plc	09332	

C   These products are "approved for agricultural use". For further details refer to page vii.
A   These products are approved for use in or near water. For further details refer to page vii.

Product Name	Marketing Company	Reg. No.	Expiry Date

1037 Fatty acids (Insecticide)—continued

Product Name	Marketing Company	Reg. No.	Expiry Date
Phostrogen Safer's Garden Insecticide Concentrate	Phostrogen Ltd	05499	30/11/2001
Phostrogen Safer's Ready-To-Use Fruit & Vegetable Insecticide	The Scotts Company & Subsidiaries	09829	
Phostrogen Safer's Ready-to-Use Fruit & Vegetable Insecticide	Monsanto Plc	09329	
Phostrogen Safer's Ready-To-Use Fruit & Vegetable Insecticide	Phostrogen Ltd	04329	30/11/2001
Phostrogen Safer's ready-To-Use House Plant Insecticide	The Scotts Company & Subsidiaries	09830	
Phostrogen Safer's Ready-To-Use House Plant Insecticide	Monsanto Plc	09333	
Phostrogen Safer's Ready-To-Use House Plant House Plant Insecticide	Phostrogen Ltd	04328	30/11/2001
Phostrogen Safer's RTU Rose & Flower Insecticide	The Scotts Company & Subsidiaries	09828	
Phostrogen Safer's RTU Rose & Flower Insecticide	Monsanto Plc, Solaris - Garden Division	09331	
Phostrogen Safer's RTU Rose & Flower Insecticide	Phostrogen Ltd	04341	30/11/2001
Rose & Flower Insect Killer	L C Solutions Ltd	09739	
Rose & Flower Insecticide	Growing Success Organics Ltd	08809	
Rose and Flower Insecticide	pbi Home & Garden Ltd	10192	
Safer's Insecticidal Soap Ready-to-Use	Safer Ltd	06573	
Savona Rose Spray	Safer Ltd	06297	
Wilko Flower and Rose Insecticide Spray	Wilkinson Group of Companies	09424	
Wilko Houseplant Insect Spray	Wilkinson Group of Companies	09921	
Wilko Insecticidal Houseplant Spray	Wilkinson Group of Companies	09423	28/02/2003
Woolworths Greenfly Killer for Flowers	Woolworths Plc	09427	
Woolworths Insecticide Spray for Flowers	Woolworths Plc	09426	
Woolworths Insecticide Spray for Houseplants	Woolworths Plc	09425	

1038 Fatty acids + Glufosinate-ammonium (Herbicide)

Product Name	Marketing Company	Reg. No.	Expiry Date
Bio Kills Weeds Dead Fast	pbi Home & Garden Ltd	09173	

C   These products are "approved for agricultural use". For further details refer to page vii.
A   These products are approved for use in or near water. For further details refer to page vii.

Product Name	Marketing Company	Reg. No.	Expiry Date

1038 Fatty acids + Glufosinate-ammonium (Herbicide)—continued

Bio Kills Weeds Dead Fast Concentrate	pbi Home & Garden Ltd	09752	
Bio Speedweed Ultra	pbi Home & Garden Ltd	09186	

1039 Fatty acids + Pyrethrins (Insecticide)

Phostrogen Safer's All Purpose Insecticde Ready to Use	Phostrogen Ltd	08093	
Safer's Indoor Trounce Insecticide Concentrate	Safer Ltd	08097	
Safer's Indoor Trounce Insecticide Ready to Use	Safer Ltd	08094	
Safer's Trounce Insecticide Concentrate	Safer Ltd	08098	
Safer's Trounce Insecticide Ready to Use	Safer Ltd	08095	

1040 Fatty acids + Sulphur (Fungicide) (Insecticide)

Nature's Answer Fungicide and Insect Killer	The Scotts Company (UK) Limited	07628	

1041 Fenitrothion (Insecticide)

Bio Fruit Spray	pbi Home & Garden Ltd	07994	20/04/2001

1042 Ferrous sulphate (Herbicide)

Asda Lawn Sand	Asda Stores Ltd	06062	
Asda Lawn Sand	Asda Stores Ltd	04520	
B & Q Granular Autumn Lawn Food with Mosskiller	B & Q Plc	07887	
B & Q Granular Lawn Feed and Mosskiller	B & Q Plc	07915	
B & Q Mosskiller for Lawns	B & Q Plc	05827	
B & Q Autumn Lawn Care	B & Q Plc	10251	
Bio Velvas	pbi Home & Garden Ltd	08545	
Boots Lawn Moss Killer and Fertiliser	The Boots Company Plc	02494	
Chempak Lawn Sand	Chempak Ltd	05723	
Cooke's Lawn Mosskiller	Nehra Cooke's Chemicals Ltd	08114	
Country Gardens Lawn Sand	Country Garden Centres Ltd	07969	
Doff Lawn Mosskiller and Fertilizer	Doff Portland Ltd	05689	
Evergreen Autumn	The Scotts Company (UK) Limited	08586	
Evergreen Lawn Sand	The Scotts Company (UK) Ltd	08629	

C   These products are "approved for agricultural use". For further details refer to page vii.
A   These products are approved for use in or near water. For further details refer to page vii.

Product Name	Marketing Company	Reg. No.	Expiry Date

1042 Ferrous sulphate (Herbicide)—continued

Product Name	Marketing Company	Reg. No.	Expiry Date
Evergreen Mosskil Extra	The Scotts Company (UK) Ltd	08669	
Evergreen Mosskill	The Scotts Company (UK) Ltd	08632	
Fisons Lawn Sand	Fisons Plc	00885	
Gem Lawn Sand	Gem Gardening	04555	
Green Up Mossfree	Vitax Ltd	05639	
Homebase Autumn Lawn Feed and Mosskiller	Homebase Ltd	08607	
Homebase Lawn Feed and Mosskiller	Homebase Ltd	09334	
Homebase Lawn Feed and Mosskiller	Homebase Ltd	07476	30/11/2001
Homebase Lawn Feed and Mosskiller	Homebase Ltd	06107	
Homebase Lawn Sand	Sainsbury's Homebase	07667	
J Arthur Bower's Lawn Sand	William Sinclair Horticulture Ltd	07028	
J Arthur Bowers Lawn Mosskiller	William Sinclair Horticulture Ltd	09457	
Lawn Builder Plus Moss Control	Miracle Garden Care Ltd	08612	
Lawn Sand	Wessex Horticultural Products Ltd	08936	
Levington Autumn Extra	The Scotts Company (UK) Ltd	07477	
Levington Lawn Sand	Levington Horticulture Ltd	07499	
Maxicrop Mosskiller and Lawn Tonic	Maxicrop (UK) Ltd	04661	
Mosskil Extra	Levington Horticulture Ltd	07478	
Mosskiller and Lawn Tonic	pbi Home & Garden Ltd	10191	
Mosskiller for Lawns	Miracle Garden Care	07869	
Murphy Lawn Feed and Mosskiller	Murphy Home and Garden Ltd	07513	
Phostrogen Soluble Mosskiller and Lawn Tonic	The Scotts Company & Subsidiaries	09833	
Phostrogen Soluble Mosskiller and Lawn Tonic	Monsanto Plc, Solaris - Garden Division	09335	
Phostrogen Soluble Mosskiller and Lawn Tonic	Phostrogen Ltd	07112	30/11/2001
Premier Autumn Lawn Feed with Mosskiller	Premier Way Ltd	07049	
Toplawn Feed with Mosskiller	pbi Home & Garden Ltd	08104	
Vitax Lawn Sand	Vitax Ltd	04352	
Wilko Lawn Sand	Wilkinson Home & Garden Stores	04084	

1043 Ferrous sulphate + Copper sulphate (Miscellaneous)

Product Name	Marketing Company	Reg. No.	Expiry Date
Snailaway	Interpet Ltd	02457	

C   These products are "approved for agricultural use". For further details refer to page vii.
A   These products are approved for use in or near water. For further details refer to page vii.

Product Name	Marketing Company	Reg. No.	Expiry Date

1044 Ferrous sulphate + Copper sulphate + Magnesium sulphate (Herbicide) (Miscellaneous)

Anti-Snail	Interpet Ltd	10117	

1045 Ferrous sulphate + Copper sulphate + Magnesium sulphate (Miscellaneous)

Snail Away	Interpet Ltd	08868	31/03/2003

1046 Ferrous sulphate + 2,4-D + Dicamba (Herbicide)

Bio Supergreen 123	pbi Home & Garden Ltd	07962	31/01/2001
Toplawn Feed with Weed & Mosskiller	pbi Home & Garden Ltd	08120	31/12/2001

1047 Ferrous sulphate + 2,4-D + Mecoprop (Herbicide)

Asda Lawn Weed and Feed and Mosskiller	Asda Stores Ltd	03819	
Green Up Feed and Weed Plus Mosskiller	Vitax Ltd	06513	31/05/2001
Supergreen Feed, Weed & Mosskiller	Pan Britannica Industries Ltd	08118	31/01/2001
Weed 'N' Feed Extra	Vitax Ltd	06506	31/05/2001

1048 Ferrous sulphate + 2,4-D + Mecoprop-P (Herbicide)

ASB Greenworld Lawn Weed, Feed and Mosskiller Granular	ASB Greenworld Ltd	09354	
B & Q Lawn Feed, Weed and Mosskiller Granular	B & Q Plc	09355	
B & Q Summer Lawncare Feed Weed Mosskiller	B & Q Plc	10286	
Doff Granular Lawn Feed, Weed and Mosskiller	Doff Portland Ltd	08876	
Elliott's Lawn and Fine Turf Feed and Weed plus Mosskiller	Thomas Elliott Ltd	09464	
Gem Lawn Weed and Feed + Mosskiller	Gem Gardening	07087	
Green Up Feed and Weed Plus Mosskiller	Vitax Ltd	09046	
Proctors Lawn Feed Weed and Mosskiller	H & T Proctor (Division of Willett & Son (Bristol) Ltd)	07808	
Supergreen Feed, Weed and Mosskiller 2	pbi Home & Garden Ltd	09224	
Weed 'N' Feed Extra	Vitax Ltd	09045	

C These products are "approved for agricultural use". For further details refer to page vii.

A These products are approved for use in or near water. For further details refer to page vii.

Product Name	Marketing Company	Reg. No.	Expiry Date

1048 Ferrous sulphate + 2,4-D + Mecoprop-P (Herbicide)—continued

| Westland Triple Action Weed, Feed And Mosskill | Westland Horticulture | 09640 | |

1049 Ferrous sulphate + Dicamba + Dichlorprop + MCPA (Herbicide)

B & Q Granular Weed and Feed and Mosskiller For Lawns	B & Q Plc	08381	
Evergreen Grasshopper	The Scotts Company (UK) Limited	09639	
'Grasshopper' Triple Action	Miracle Garden Care	08643	
'Greensward' 2	Miracle Garden Care	08380	30/11/2001
Miracle-Gro Lawn Food with Weed and Moss Control	Miracle Garden Care	08609	

1050 Ferrous sulphate + Dichlorprop + MCPA (Herbicide)

J Arthur Bower's Granular Feed, Weed and Mosskiller	William Sinclair Horticulture Ltd	07042	
J Arthur Bower's Lawn Feed, Weed and Mosskiller	William Sinclair Horticulture Ltd	09459	
J Arthur Bower's Lawn Weed and Mosskiller	William Sinclair Horticulture Ltd	09460	
Wilko Granular Feed, Weed and Mosskiller	Wilkinson Group of Companies	07471	
Wilko Granular Feed, Weed and Mosskiller for Lawns	Wilkinson Group of Companies	09731	
Wilko Lawn Feed, Weed and Mosskiller	Wilkinson Home & Garden Stores	04602	

1051 Ferrous sulphate + MCPA + Mecoprop (Herbicide)

| Fisons Evergreen Extra | Fisons Plc | 03890 | |

1052 Ferrous sulphate + MCPA + Mecoprop-P (Herbicide)

B & Q Granular Lawn Feed, Weed and Mosskiller	B & Q Plc	07762	
B & Q Spring Lawn Care Feed Weed Mosskiller	B & Q Plc	10248	
Do-It-All Complete Lawn Care	Do-It-All Ltd	08123	
Evergreen Complete	The Scotts Company (UK) Limited	10091	
Evergreen Easy	The Scotts Company (UK) Ltd	08103	
Evergreen Extra	The Scotts Company (UK) Limited	07549	
Evergreen Triple Action	The Scotts Company (UK) Ltd	09199	
Great Mills Lawn Feed, Weed and Mosskiller	Great Mills (Retail) Ltd	07552	

C   These products are "approved for agricultural use". For further details refer to page vii.
A   These products are approved for use in or near water. For further details refer to page vii.

Product Name	Marketing Company	Reg. No.	Expiry Date

1052 Ferrous sulphate + MCPA + Mecoprop-P (Herbicide)—continued

Homebase Lawn Feed, Weed and Mosskiller	Homebase Ltd	07551
Levington Gold	Levington Horticulture Ltd	07548
Murphy Ultra Lawn Feed, Weed and Mosskiller	Murphy Home and Garden Ltd	07553
Wickes Lawn Feed, Weed and Mosskiller	Wickes Building Supplies Ltd	08800
Wilko Granular Lawn Feed, Weed and Mosskiller	Wilkinson Group of Companies	07981
Woolworths Lawn Feed, Weed and Mosskiller	Woolworths Plc	07554

1053 Ferrous sulphate heptahydrate (Herbicide)

Autumn Toplawn Feed and Mosskiller	pbi Home & Garden Ltd	09161

1054 Ferrous sulphate heptahydrate + 2,4-D + Dicamba (Herbicide)

Toplawn Feed, Weed & Mosskiller	pbi Home & Garden Ltd	09350

1055 Fluroxypyr + Clopyralid + MCPA (Herbicide)

Weed-B-Gon	Solaris, Garden Division of Monsanto	08926
Weed-B-Gon	Dow AgroSciences Ltd	08925
Weed-B-Gon Ready To Use	Solaris, Garden Division of Monsanto	09451
Weed-B-Gon Ready To Use	Dow AgroSciences Ltd	09450

1056 Fluroxypyr + Mecoprop-P (Herbicide)

Verdone Extra	Miracle Garden Care Ltd	08999
Verdone Extra	Dow AgroSciences Ltd	08998

1057 Flutriafol + Bifenthrin (Fungicide) (Insecticide)

Roseclear Gun !	The Scotts Company (UK) Limited	09923

1058 Garlic oil + Orange peel oil + Orange pith oil (Vertebrate Control)

6X Cat Repellent	Organic Concentrates Ltd	07979
Growing Success Cat Repellent	Growing Success Organics Ltd	06262

1059 Glufosinate-ammonium (Herbicide)

AgrEvo Garden Weedkiller Concentrate	AgrEvo UK Ltd	08200

C   These products are "approved for agricultural use". For further details refer to page vii.
A   These products are approved for use in or near water. For further details refer to page vii.

Product Name	Marketing Company	Reg. No.	Expiry Date

1059 Glufosinate-ammonium (Herbicide)—continued

Product Name	Marketing Company	Reg. No.	Expiry Date
AgrEvo Garden Weedkiller Spray	AgrEvo UK Ltd	08176	
AgrEvo Patio Weedkiller	AgrEvo UK Ltd	08177	
Do It All Liquid Weedkiller Concentrate	Do-It-All Ltd	08201	
Do It All Nettle Gun	Do-It-All Ltd	08180	31/03/2002
Do It All Topple Weed!	Do-It-All Ltd	08130	31/03/2002
Do-It-All Liquid Weedkiller Concentrate	Focus Do It All Limited	09962	
Doff Knockdown Weedkiller Concentrate	Doff Portland Ltd	08199	
Doff Knockdown Weedkiller Spray	Doff Portland Ltd	07769	
Doff New Formula Nettle Gun	Doff Portland Ltd	08178	31/03/2002
Doff Path and Patio Weedkiller	Doff Portland Ltd	08181	
Doff Rose Weedkiller	Doff Portland Ltd	09342	
Fito Ready To Use Patio & Garden Weedkiller	L C Solutions Ltd	10133	
Garden Weedkiller	L C Solutions Ltd	09735	
Garden Weedkiller Concentrate	L C Solutions Ltd	09741	
Great Mills Fast Action Weedkiller Ready to Use	Great Mills (Retail) Ltd	07114	
Homebase Contact Action Weedkiller	Homebase Ltd	06735	
Homebase Contact Action Weedkiller Ready to Use Sprayer	Homebase Ltd	06736	
Homebase Nettle Spray	Homebase Ltd	08179	31/03/2002
Patio Weedkiller	L C Solutions Ltd	09734	
Phytek Ready To Use Patio and Garden Weedkiller	Aventis Environmental Science	10131	
SBK Path & Patio Weedkiller	Vitax Ltd	10134	
Tumbleweed General Purpose	Levington Horticulture Ltd	07992	
Tumbleweed General Purpose Ready To Use	Levington Horticulture Ltd	07993	
Westland Kill All General Purpose Weedkiller RTU	Westland Horticulture	10132	
Westland Kill All Path & Patio Weedkiller RTU	Westland Horticulture	10135	
Wilko Complete Weedkiller Concentrate	Wilkinson Group of Companies	08968	

C   These products are "approved for agricultural use". For further details refer to page vii.
A   These products are approved for use in or near water. For further details refer to page vii.

Product Name	Marketing Company	Reg. No.	Expiry Date

1059 Glufosinate-ammonium (Herbicide)—continued

Wilko Complete Weedkiller Spray	Wilkinson Home & Garden Stores	07771	
Wilko Path & Patio Weedkiller Spray	Wilkinson Hardware Stores Limited	09922	
Woolworths Complete Weedkiller Concentrate	Woolworths Plc	08001	
Woolworths Complete Weedkiller Spray	Woolworths Plc	07770	

1060 Glufosinate-ammonium + Diuron (Herbicide)

AgrEvo Path Weed Killer	AgrEvo Environmental Health Ltd	08616	31/08/2002
Tumbleweed Paths and Patios	Levington Horticulture Ltd	08613	

1061 Glufosinate-ammonium + Fatty acids (Herbicide)

Bio Kills Weeds Dead Fast	pbi Home & Garden Ltd	09173	
Bio Kills Weeds Dead Fast Concentrate	pbi Home & Garden Ltd	09752	
Bio Speedweed Ultra	pbi Home & Garden Ltd	09186	

1062 Glyphosate (Herbicide)

B & Q Complete Weedkiller	B&Q Plc	05290	
B & Q Complete Weedkiller Ready To Use	B&Q Plc	06722	
Biactive Roundup Brushkiller	Solaris, Garden Division of Monsanto	08220	31/07/2001
Biactive Roundup Brushkiller	Monsanto Plc	07589	31/07/2001
Biactive Roundup GC	Solaris, Garden Division of Monsanto	08219	31/07/2001
Biactive Roundup GC	Monsanto Plc	07590	31/07/2001
Bio Glyphosate	pbi Home & Garden Ltd	09242	
Bio Glyphosate Pen	pbi Home & Garden Ltd	09663	
Bio Glyphosate Ready-to-Use	pbi Home & Garden Ltd	09241	
Bio WeedEasy	Pan Britannica Industries Ltd	08567	25/07/2003
Boots Systemic Weed & Grass Killer Ready To Use	The Boots Company Plc	05028	
Glypho	Miracle Garden Care	07737	
Glypho Gun!	Miracle Garden Care	07738	
Great Mills Systemic Action Weedkiller Ready To Use	Great Mills (Retail) Ltd	07080	
Greenscape Ready To Use	Solaris, Garden Division of Monsanto	08218	
Greenscape Ready To Use Weed Killer	Monsanto Plc	04676	31/08/2001

C   These products are "approved for agricultural use". For further details refer to page vii.
A   These products are approved for use in or near water. For further details refer to page vii.

Product Name	Marketing Company	Reg. No.	Expiry Date

1062 Glyphosate (Herbicide)—continued

Product Name	Marketing Company	Reg. No.	Expiry Date
Greenscape Weedkiller	Solaris, Garden Division of Monsanto	08216	
Greenscape Weedkiller	Monsanto Plc	04321	
High Strength Tough Weed Gun!	Miracle Garden Care Ltd	08967	
Homebase Systemic Action Weedkiller	Homebase Ltd	06622	
Homebase Systemic Action Weedkiller Ready to Use	Homebase Ltd	06623	
Knock Out	Premier Way Ltd	07846	
Knock Out Weedkiller Ready To Use	Premier Way Ltd	07948	
MON 44068 Garden Weedkiller	Solaris	08222	
Mon 44068 Garden Weedkiller	Monsanto Plc	07367	31/12/2001
MON 77020 Garden Weedkiller	Solaris	08223	
MON 77020 Garden Weedkiller	Monsanto Plc	07368	30/04/2002
MON 77021 Garden Weedkiller	Solaris	08224	
MON 77021 Garden Weedkiller	Monsanto Plc	07369	30/04/2002
Murphy Tumbleweed Gel	Murphy Home & Garden Products	04009	
New Improved Leaf Action Roundup Brushkiller RTU	Solaris, Garden Division of Monsanto	08265	30/06/2002
New Improved Leaf Action Roundup Weedkiller RTU	Monsanto Roundup Lawn & Garden	08264	
Nomix Weedkiller	Nomix-Chipman Ltd	09591	
Roundup Brushkiller	Solaris, Garden Division of Monsanto	09166	30/06/2002
Roundup Brushkiller	Pbi Home and Garden	08225	
Roundup Brushkiller	Monsanto Plc	05755	
Roundup Estate	Monsanto Plc	09399	
Roundup GC	Monsanto Roundup Lawn & Garden	09167	
Roundup GC	Solaris, Garden Division of Monsanto	08217	
Roundup GC	Monsanto Plc	05538	
Roundup Ready-to-Use Faster Acting Formula	Monsanto Roundup Lawn & Garden	09277	

C   These products are "approved for agricultural use". For further details refer to page vii.
A   These products are approved for use in or near water. For further details refer to page vii.

Product Name	Marketing Company	Reg. No.	Expiry Date

1062 Glyphosate (Herbicide)—continued

Product Name	Marketing Company	Reg. No.	Expiry Date
Roundup Tough Weedkiller	Monsanto Roundup Lawn & Garden	09627	
Roundup Tough Weedkiller Ready-To-Use	Monsanto Roundup Lawn & Garden	09628	
Roundup Ultra 3000	Monsanto Roundup Lawn & Garden	08172	
Tough Weed Gun!	Miracle Garden Care	07748	
Tough Weed Killer	Miracle Garden Care	07749	
Tumbleweed Original	Levington Horticulture Ltd	07974	
Tumbleweed Original Extra Strong	The Scotts Company (UK) Ltd + Levington Horticulture Ltd	07975	
Tumbleweed Original Extra Strong Gel	The Scotts Company (UK) Ltd + Levington Horticulture Ltd	08090	
Tumbleweed Original Extra Strong Ready To Use	The Scotts Company (UK) Ltd + Levington Horticulture Ltd	08091	
Tumbleweed Original Ultra Concentrated	Levington Horticulture Ltd	07976	
Weed-Easy	Nufarm Whyte Agriculture Ltd	10237	
Weedclear	Miracle Garden Care Ltd	08672	30/09/2002
Wickes General Purpose Weedkiller Ready To Use	Wickes Building Supplies Ltd	08825	31/12/2004
Woolworths Weedkiller	Woolworths Plc	07108	

1063 Glyphosate + Diuron (Herbicide)

Product Name	Marketing Company	Reg. No.	Expiry Date
Pathclear Gun!	The Scotts Company (UK) Limited	09492	
Weedatak Path and Drive	The Scotts Company (UK) Limited	09814	
Weedatak Path and Drive	Monsanto Plc	07659	30/11/2002

1064 Heptenophos + Permethrin (Insecticide)

Product Name	Marketing Company	Reg. No.	Expiry Date
Murphy Systemic Action Insecticide	Levington Horticulture Ltd	07557	30/04/2001
Murphy Tumblebug	Levington Horticulture Ltd	07571	30/04/2001

1065 Imidacloprid (Insecticide)

Product Name	Marketing Company	Reg. No.	Expiry Date
Bayer UK 720	pbi Home & Garden Ltd	10234	
Bio Provado Complete Pest Killer	pbi Home & Garden Ltd	09691	25/07/2003
Bio Provado Vine Weevil Killer	pbi Home & Garden Ltd	09660	
Levington Plant Protection	The Scotts Company (UK) Ltd + Levington Horticulture Ltd	09377	

C   These products are "approved for agricultural use". For further details refer to page vii.
A   These products are approved for use in or near water. For further details refer to page vii.

Product Name	Marketing Company	Reg. No.	Expiry Date

1065 Imidacloprid (Insecticide)—continued

Plant Protection Compost	The Scotts Company (UK) Ltd + Levington Horticulture Ltd	08365	
Provado Pest Free	pbi Home &Garden Ltd	10249	

1066 Imidacloprid + Methiocarb (Insecticide) (Miscellaneous)

Baby Bio Provado Insect Control	pbi Home & Garden Ltd	10232	
Provado Ultimate Bug Killer	pbi Home & Garden Ltd	10233	

1067 4-Indol-3-ylbutyric acid (Herbicide)

Clearcut II	SupaPlants Ltd	08901	
Clearcut II	Levington Horticulture Ltd	07827	31/01/2001
Clonex	Growth Technology	09441	

1067a 4-Indol-3ylbutric acid + 1-Naphthylacetic acid + Thiram (Fungicide) (Herbicide)

Boots Hormone Rooting Powder	The Boots Company Plc	01067	

1068 Lindane (Insecticide)

Doff Ant Killer	Doff Portland Ltd	00739	
Doff Gamma BHC Dust	Doff Portland Ltd	04868	
Doff Weevil Killer	Doff Portland Ltd	08138	

1069 Magnesium sulphate + Copper sulphate + Ferrous sulphate (Herbicide) (Miscellaneous)

Anti-Snail	Interpet Ltd	10117	

1070 Magnesium sulphate + Copper sulphate + Ferrous sulphate (Miscellaneous)

Snail Away	Interpet Ltd	08868	31/03/2003

1071 Malathion (Insecticide)

Duramitex	Harkers Ltd	08664	
Murphy Liquid Malathion	The Scotts Company (UK) Ltd + Levington Horticulture Ltd	07881	
Murphy Malathion Dust	Levington Horticulture Ltd	07880	30/04/2001

1072 Maleic hydrazide (Herbicide)

Stop Gro G8	Botanical Developments	05923	
Stop Gro G8	Synchemicals Ltd	02029	

C   These products are "approved for agricultural use". For further details refer to page vii.
A   These products are approved for use in or near water. For further details refer to page vii.

Product Name	Marketing Company	Reg. No.	Expiry Date

1073 Mancozeb (Fungicide)

PBI Dithane 945	pbi Home & Garden Ltd	00718	

1074 MCPA + Clopyralid + Fluroxypyr (Herbicide)

Weed-B-Gon	Solaris, Garden Division of Monsanto	08926	
Weed-B-Gon	Dow AgroSciences Ltd	08925	
Weed-B-Gon Ready To Use	Solaris, Garden Division of Monsanto	09451	
Weed-B-Gon Ready To Use	Dow AgroSciences Ltd	09450	

1075 MCPA + Dicamba (Herbicide)

Green Up Liquid Lawn Feed'n Weed	Vitax Ltd	08196	

1076 MCPA + Dicamba + Dichlorprop (Herbicide)

B & Q Granular Weed and Feed for Lawns	B & Q Plc	07850	
B & Q Liquid Weed and Feed	B & Q Plc	05293	
Boots Nettle and Bramble Weedkiller	The Boots Company Plc	03455	
Grasshopper Weed&Feed	The Scotts Company (UK) Limited	09785	
Grasshopper Weed&Feed	Miracle Garden Care Ltd	07849	30/06/2003
Greensward Triple Action	The Scotts Company (UK) Limited	09324	
Groundclear	pbi Home & Garden Ltd	05953	
Lawnsman Liquid Weed and Feed	Imperial Chemical Industries Plc	03610	31/05/2001
Miracle-Gro Weed & Feed	Miracle Garden Care	08550	31/12/2002
'Verdone Plus'	Miracle Garden Care	08689	
Woolworths Liquid Lawn Feed and Weed	Woolworths Plc	07060	

1077 MCPA + Dicamba + Dichlorprop + Ferrous sulphate (Herbicide)

B & Q Granular Weed and Feed and Mosskiller For Lawns	B & Q Plc	08381	
Evergreen Grasshopper	The Scotts Company (UK) Limited	09639	
'Grasshopper' Triple Action	Miracle Garden Care	08643	
'Greensward' 2	Miracle Garden Care	08380	30/11/2001
Miracle-Gro Lawn Food with Weed and Moss Control	Miracle Garden Care	08609	

1078 MCPA + Dicamba + Mecoprop (Herbicide)

Bio Spot	pbi Home & Garden Ltd	05071	31/12/2001

C   These products are "approved for agricultural use". For further details refer to page vii.
A   These products are approved for use in or near water. For further details refer to page vii.

Product Name	Marketing Company	Reg. No.	Expiry Date

1078 MCPA + Dicamba + Mecoprop (Herbicide)—continued

Bio Weed Pencil	pbi Home & Garden Ltd	04054	31/12/2001
Homebase Weed Pen	Sainsbury's Homebase	07648	30/06/2002

1079 MCPA + Dicamba + Mecoprop-P (Herbicide)

Bio Lawn Weed Pencil	pbi Home & Garden Ltd	09656	
Bio Spot	pbi Home & Garden Ltd	09210	
Bio Weed Pencil	pbi Home & Garden Ltd	09212	30/06/2002
Evergreen Feed and Weed Liquid	The Scotts Company (UK) Ltd	07766	
Homebase Lawn Feed and Weed Liquid	Homebase Ltd	07765	
Homebase Weed Pen	Sainsbury's Homebase	09211	

1080 MCPA + Dichlorprop (Herbicide)

Doff Lawn Weed and Feed Soluble Powder	Doff Portland Ltd	05708	
J Arthur Bower's Granular Feed and Weed	William Sinclair Horticulture Ltd	07164	
J Arthur Bower's Lawn Feed and Weed	William Sinclair Horticulture Ltd	09458	
J Arthur Bower's Lawn Weedkiller	William Sinclair Horticulture Ltd	09461	
Wilko Granular Feed and Weed for Lawns	Wilkinson Group of Companies	09732	
Wilko Lawn Feed and Weed	Wilkinson Group of Companies	08816	
Wilko Soluble Lawn Food and Weedkiller	Wilkinson Home & Garden Stores	04391	

1081 MCPA + Dichlorprop + Ferrous sulphate (Herbicide)

J Arthur Bower's Granular Feed, Weed and Mosskiller	William Sinclair Horticulture Ltd	07042	
J Arthur Bower's Lawn Feed, Weed and Mosskiller	William Sinclair Horticulture Ltd	09459	
J Arthur Bower's Lawn Weed and Mosskiller	William Sinclair Horticulture Ltd	09460	
Wilko Granular Feed, Weed and Mosskiller	Wilkinson Group of Companies	07471	
Wilko Granular Feed, Weed and Mosskiller for Lawns	Wilkinson Group of Companies	09731	
Wilko Lawn Feed, Weed and Mosskiller	Wilkinson Home & Garden Stores	04602	

C   These products are "approved for agricultural use". For further details refer to page vii.
A   These products are approved for use in or near water. For further details refer to page vii.

Product Name	Marketing Company	Reg. No.	Expiry Date

1082 MCPA + Ferrous sulphate + Mecoprop (Herbicide)

Fisons Evergreen Extra	Fisons Plc	03890	

1083 MCPA + Ferrous sulphate + Mecoprop-P (Herbicide)

B & Q Granular Lawn Feed, Weed and Mosskiller	B & Q Plc	07762	
B & Q Spring Lawn Care Feed Weed Mosskiller	B & Q Plc	10248	
Do-It-All Complete Lawn Care	Do-It-All Ltd	08123	
Evergreen Complete	The Scotts Company (UK) Limited	10091	
Evergreen Easy	The Scotts Company (UK) Ltd	08103	
Evergreen Extra	The Scotts Company (UK) Limited	07549	
Evergreen Triple Action	The Scotts Company (UK) Ltd	09199	
Great Mills Lawn Feed, Weed and Mosskiller	Great Mills (Retail) Ltd	07552	
Homebase Lawn Feed, Weed and Mosskiller	Homebase Ltd	07551	
Levington Gold	Levington Horticulture Ltd	07548	
Murphy Ultra Lawn Feed, Weed and Mosskiller	Murphy Home and Garden Ltd	07553	
Wickes Lawn Feed, Weed and Mosskiller	Wickes Building Supplies Ltd	08800	
Wilko Granular Lawn Feed, Weed and Mosskiller	Wilkinson Group of Companies	07981	
Woolworths Lawn Feed, Weed and Mosskiller	Woolworths Plc	07554	

1084 MCPA + Mecoprop-P (Herbicide)

B & Q Granular Lawn Feed and Weed	B & Q Plc	07793	
B & Q Spring Lawn Care Feed Weed	B & Q Plc	10267	
Do-It-All Granular Lawn Feed and Weed	Do-It-All Ltd	07971	
Evergreen Feed and Weed	The Scotts Company (UK) Ltd	07595	
Homebase Lawn Feed and Weed	Homebase Ltd	07592	
Murphy Lawn Feed and Weed	Murphy Home and Garden Ltd	07596	

1085 Mecoprop + 2,4-D (Herbicide)

Homebase Lawn Feed and Weed Soluble	Sainsbury's Homebase	07690	30/09/2001
Lawn Feed and Weed Granules	pbi Home & Garden Ltd	05902	30/09/2001

C   These products are "approved for agricultural use". For further details refer to page vii.

A   These products are approved for use in or near water. For further details refer to page vii.

Product Name	Marketing Company	Reg. No.	Expiry Date

1085 Mecoprop + 2,4-D (Herbicide)—continued

Lawn Spot Weed Granules	Pan Britannica Industries Ltd	05903	
Supergreen and Weed	Pan Britannica Industries Agrochemicals Ltd	08911	30/09/2001
Supergreen Double (Feed and Weed)	Pan Britannica Industries Ltd	05949	28/02/2001
Supertox Spot	pbi Home & Garden Ltd	05951	30/11/2001
Toplawn Lawn Weedkiller	pbi Home & Garden Ltd	08694	31/01/2002
Toplawn Ready to Use Lawn Weedkiller	pbi Home & Garden Ltd	08695	31/12/2001
Verdone	Miracle Garden Care	08657	30/06/2001
Verdone 2	Imperial Chemical Industries Plc	03271	31/05/2001

1086 Mecoprop + 2,4-D + Dicamba (Herbicide)

B & Q Tree Stump and Brushwood Killer	B & Q Plc	07645	

1087 Mecoprop + 2,4-D + Ferrous sulphate (Herbicide)

Asda Lawn Weed and Feed and Mosskiller	Asda Stores Ltd	03819	
Green Up Feed and Weed Plus Mosskiller	Vitax Ltd	06513	31/05/2001
Supergreen Feed, Weed & Mosskiller	Pan Britannica Industries Ltd	08118	31/01/2001
Weed 'N' Feed Extra	Vitax Ltd	06506	31/05/2001

1088 Mecoprop + Dicamba + Dichlorprop (Herbicide)

New Supertox	pbi Home & Garden Ltd	06128	30/09/2001

1089 Mecoprop + Dicamba + MCPA (Herbicide)

Bio Spot	pbi Home & Garden Ltd	05071	31/12/2001
Bio Weed Pencil	pbi Home & Garden Ltd	04054	31/12/2001
Homebase Weed Pen	Sainsbury's Homebase	07648	30/06/2002

1090 Mecoprop + Ferrous sulphate + MCPA (Herbicide)

Fisons Evergreen Extra	Fisons Plc	03890	

1091 Mecoprop-P + 2,4-D (Herbicide)

ASB Greenworld Lawn Weed, And Feed Granular	ASB Greenworld Ltd	09356	
B & Q Lawn Weed and Feed Granular	B & Q Plc	09357	
Bio Toplawn Spot Weeder	pbi Home & Garden Ltd	09672	
Gem Lawn Weed & Feed	Gem Gardening	07086	

C   These products are "approved for agricultural use". For further details refer to page vii.
A   These products are approved for use in or near water. For further details refer to page vii.

Product Name	Marketing Company	Reg. No.	Expiry Date

1091 Mecoprop-P + 2,4-D (Herbicide)—continued

Green Up Spot Lawn Weedkiller	Vitax Ltd	06028	
Homebase Lawn Feed and Weed Soluble	Sainsbury's Homebase	09217	
Lawn Feed and Weed Granules	pbi Home & Garden Ltd	09225	
Supergreen and Weed	pbi Home & Garden Ltd	09218	
Supertox Spot	pbi Home & Garden Ltd	09306	
Toplawn Lawn Weedkiller	pbi Home & Garden Ltd	09589	
Toplawn Lawn Weedkiller	pbi Home & Garden Ltd	09408	31/05/2002
Toplawn Ready to Use Lawn Weedkiller	pbi Home & Garden Ltd	09348	31/07/2002
Verdone	The Scotts Company (UK) Limited	09642	
Verdone	Miracle Garden Care Ltd	09097	
Vitax 'Green Up' Granular Lawn Feed and Weed	Vitax Ltd	06158	
Wilko Lawn Feed & Weed Liquid	Wilkinson Home & Garden Stores	08789	

1092 Mecoprop-P + 2,4-D + Dicamba (Herbicide)

New Formulation SBK Brushwood Killer	Vitax Ltd	08737	

1093 Mecoprop-P + 2,4-D + Ferrous sulphate (Herbicide)

ASB Greenworld Lawn Weed, Feed and Mosskiller Granular	ASB Greenworld Ltd	09354	
B&Q Lawn Feed, Weed and Mosskiller Granular	B & Q Plc	09355	
B&Q Summer Lawncare Feed Weed Mosskiller	B & Q Plc	10286	
Doff Granular Lawn Feed, Weed and Mosskiller	Doff Portland Ltd	08876	
Elliott's Lawn and Fine Turf Feed and Weed plus Mosskiller	Thomas Elliott Ltd	09464	
Gem Lawn Weed and Feed + Mosskiller	Gem Gardening	07087	
Green Up Feed and Weed Plus Mosskiller	Vitax Ltd	09046	
Proctors Lawn Feed Weed and Mosskiller	H&T Proctor (Division of Willett & Son (Bristol) Ltd)	07808	

C   These products are "approved for agricultural use". For further details refer to page vii.
A   These products are approved for use in or near water. For further details refer to page vii.

Product Name	Marketing Company	Reg. No.	Expiry Date

1093 Mecoprop-P + 2,4-D + Ferrous sulphate (Herbicide)—continued

Supergreen Feed, Weed and Mosskiller 2	pbi Home & Garden Ltd	09224	
Weed 'N' Feed Extra	Vitax Ltd	09045	
Westland Triple Action Weed, Feed And Mosskill	Westland Horticulture	09640	

1094 Mecoprop-P + Dicamba + Dichlorprop (Herbicide)

New Supertox	pbi Home & Garden Ltd	09209	

1095 Mecoprop-P + Dicamba + MCPA (Herbicide)

Bio Lawn Weed Pencil	pbi Home & Garden Ltd	09656	
Bio Spot	pbi Home & Garden Ltd	09210	
Bio Weed Pencil	pbi Home & Garden Ltd	09212	30/06/2002
Evergreen Feed and Weed Liquid	The Scotts Company (UK) Ltd	07766	
Homebase Lawn Feed and Weed Liquid	Homebase Ltd	07765	
Homebase Weed Pen	Sainsbury's Homebase	09211	

1096 Mecoprop-P + Ferrous sulphate + MCPA (Herbicide)

B & Q Granular Lawn Feed, Weed and Mosskiller	B & Q Plc	07762	
B & Q Spring Lawn Care Feed Weed Mosskiller	B & Q Plc	10248	
Do-It-All Complete Lawn Care	Do-It-All Ltd	08123	
Evergreen Complete	The Scotts Company (UK) Limited	10091	
Evergreen Easy	The Scotts Company (UK) Ltd	08103	
Evergreen Extra	The Scotts Company (UK) Limited	07549	
Evergreen Triple Action	The Scotts Company (UK) Ltd	09199	
Great Mills Lawn Feed, Weed and Mosskiller	Great Mills (Retail) Ltd	07552	
Homebase Lawn Feed, Weed and Mosskiller	Homebase Ltd	07551	
Levington Gold	Levington Horticulture Ltd	07548	
Murphy Ultra Lawn Feed, Weed and Mosskiller	Murphy Home and Garden Ltd	07553	
Wickes Lawn Feed, Weed and Mosskiller	Wickes Building Supplies Ltd	08800	
Wilko Granular Lawn Feed, Weed and Mosskiller	Wilkinson Group of Companies	07981	
Woolworths Lawn Feed, Weed and Mosskiller	Woolworths Plc	07554	

C   These products are "approved for agricultural use". For further details refer to page vii.
A   These products are approved for use in or near water. For further details refer to page vii.

Product Name	Marketing Company	Reg. No.	Expiry Date

1097 Mecoprop-P + Fluroxypyr (Herbicide)

Verdone Extra	Miracle Garden Care Ltd	08999	
Verdone Extra	Dow AgroSciences Ltd	08998	

1098 Mecoprop-P + MCPA (Herbicide)

B & Q Granular Lawn Feed and Weed	B & Q Plc	07793	
B & Q Spring Lawn Care Feed Weed	B & Q Plc	10267	
Do-It-All Granular Lawn Feed and Weed	Do-It-All Ltd	07971	
Evergreen Feed and Weed	The Scotts Company (UK) Ltd	07595	
Homebase Lawn Feed and Weed	Homebase Ltd	07592	
Murphy Lawn Feed and Weed	Murphy Home and Garden Ltd	07596	

1099 Metaldehyde (Miscellaneous)

AgrEvo Slug Killer Pellets	AgrEvo UK Ltd	09839	
AgrEvo Slug Killer Pellets	AgrEvo UK Ltd	09551	31/12/2002
Aro Slug Killer Blue Mini Pellets	Makro Self Serv Wholesalers Ltd	06289	31/10/2001
B & Q Slug Killer Blue Mini Pellets	B & Q Plc	09678	
B & Q Slug Killer Blue Mini Pellets	B & Q Plc	05607	31/07/2002
B & Q Slug Killer Blue Mini Pellets	B & Q Plc	10275	
Bio Slug Mini Pellets	pbi Home & Garden Ltd	09062	
Bio Slug Mini Pellets	pbi Home & Garden Ltd	08595	28/02/2003
Cookes 3% Metaldehyde Slug Killer	Devcol Morgan Ltd	09618	31/10/2001
Devcol Morgan 3% Metaldehyde Slug Killer	Devcol Morgan Ltd	07113	31/10/2001
Do-It-All Slug Killer Pellets	Do-It-All Ltd	09679	
Do-It-All Slug Killer Pellets	Do-It-All Ltd	04895	31/07/2002
Doff Slugoids Slug Killer Blue Mini Pellets	Doff Portland Ltd	09665	
Doff Slugoids Slug Killer Blue Mini Pellets	Doff Portland Ltd	00744	30/06/2002
Focus Slug Killer Pellets	Focus DIY Ltd	09680	
Focus Slug Killer Pellets	Focus DIY Ltd	06027	31/07/2002
Great Mills Slug Killer Blue Mini Pellets	Great Mills (Retail) Ltd	09681	
Great Mills Slug Killer Blue Mini Pellets	Great Mills (Retail) Ltd	06088	31/07/2002

C These products are "approved for agricultural use". For further details refer to page vii.
A These products are approved for use in or near water. For further details refer to page vii.

Product Name	Marketing Company	Reg. No.	Expiry Date

1099 Metaldehyde (Miscellaneous)—continued

Homebase Slug Killer Blue Mini Pellets Centres	Sainsbury's Homebase House and Garden	09682	
Homebase Slug Killer Blue Mini Pellets Centres	Sainsbury's Homebase House and Garden	06410	31/07/2002
Murphy Dilute Slugit Liquid	Levington Horticulture Ltd	09239	
Murphy Slugit Liquid	The Scotts Company (UK) Ltd + Levington Horticulture Ltd	07560	
Murphy Slugit Liquid	Fisons Plc	03633	31/05/2001
Murphy Slugits	The Scotts Company (UK) Limited	09576	
Murphy Slugits	The Scotts Company (UK) Limited	07559	31/05/2002
Murphy Slugits	Murphy Home & Garden Products	03634	31/05/2001
PBI Slug Mini Pellets	Pan Britannica Industries Ltd	02611	31/05/2001
Portland Brand Slug Killer Blue Mini Pellets	Doff Portland Ltd	09677	
Portland Brand Slug Killer Blue Mini Pellets	Doff Portland Ltd	06411	31/07/2002
Slug Clear	The Scotts Company (UK) Ltd	10247	
Slug Clear Mini Pellets	The Scotts Company (UK) Limited	10246	
Slug Xtra	Miracle Garden Care	07858	31/05/2001
Slugit Xtra	The Scotts Company (UK) Limited	09465	
Slugit Xtra	Miracle Garden Care Ltd	09195	
Super Slug and Snail Killer	Chiltern Farm Chemicals Ltd	09575	
Wilko Slug Killer Blue Mini Pellets	Wilkinson Home & Garden Stores	09683	
Wilko Slug Killer Blue Mini Pellets	Wilkinson Home & Garden Stores	05608	31/07/2002
Woolworths Slug Killer Pellets	Woolworths Plc	09684	
Woolworths Slug Killer Pellets	Woolworths Plc	06924	31/07/2002
Woolworths Slug Pellets	Woolworths Plc	07038	

1100 Metaldehyde + Copper chloride (Miscellaneous)

Sera Snailpur	Sera GmbH	09587	
Sera Snailpur	Sera Werke Heimtierbedarf	09064	30/04/2001

1101 Methiocarb (Miscellaneous)

Bio Slug Guard	pbi Home & Garden Ltd	09593	
Slug Gard	pbi Home & Garden Ltd	01963	

1102 Methiocarb + Imidacloprid (Insecticide) (Miscellaneous)

Baby Bio Provado Insect Control	pbi Home & Garden Ltd	10232	
Provado Ultimate Bug Killer	pbi Home & Garden Ltd	10233	

C   These products are "approved for agricultural use". For further details refer to page vii.
A   These products are approved for use in or near water. For further details refer to page vii.

Product Name	Marketing Company	Reg. No.	Expiry Date

1103 Methyl nonyl ketone (Vertebrate Control)

Get Off My Garden	Pet and Garden Manufacturing Plc	06614	
Get Off Spray	Pet and Garden Manufacturing Plc	08919	
Wash and Get Off Spray	Pet and Garden Manufacturing Plc	08918	

1104 Methyl nonyl ketone + Citronella oil (Vertebrate Control)

Secto Keep Off	Sinclair Animal & Household Care Ltd	07890	

1105 Monolinuron (Herbicide)

AlgoFin	Tetra	08862	

1106 Myclobutanil (Fungicide)

Bio Fungus Fighter	pbi Home & Garden Ltd	09624	
Bio Systhane	pbi Home & Garden Ltd	08570	31/10/2003
Systhane	Pan Britannica Industries Ltd	04522	
Systhane Fungus Fighter	pbi Home & Garden Ltd	10279	

1107 Myclobutanil + Bifenthrin (Fungicide) (Insecticide)

Bug & Fungus Free	pbi Home & Garden Ltd	10205	
Multirose ready-to-use	pbi Home & Garden Ltd	10206	

1108 Naphthalene (Vertebrate Control)

Scent Off Buds	Synchemicals Ltd	02907	
Scent off Gel	Vitax Ltd	07366	
Scent Off Pellets	Vitax Ltd	01888	

1109 1-Naphthylacetic acid (Herbicide)

Homebase Rooting Hormone Powder 2	Sainsbury's Homebase	09912	
Strike 2	pbi Home & Garden Ltd	05952	

1110 1-Naphthylacetic acid + Captan (Fungicide) (Herbicide)

Doff Hormone Rooting Powder	Doff Portland Ltd	01065	
Homebase Hormone Rooting Powder	Homebase Ltd	09798	
Homebase Rooting Hormone Powder	Sainsbury's Homebase	07630	
Murphy Hormone Rooting Powder	The Scotts Company (UK) Limited + Levington Horticulture Ltd	07923	

C   These products are "approved for agricultural use". For further details refer to page vii.
A   These products are approved for use in or near water. For further details refer to page vii.

Product Name	Marketing Company	Reg. No.	Expiry Date

1110 1-Naphthylacetic acid + Captan (Fungicide) (Herbicide)—continued

Murphy Hormone Rooting Powder	Fisons Plc	03618	
New Strike	pbi Home & Garden Ltd	05956	31/07/2002
Rooting Powder	Vitax Ltd	06334	

1111 1-Naphthylacetic acid + Captan (Herbicide)

Bio Strike	pbi Home & Garden Ltd	09674	

1112 1-Naphthylacetic acid + Dichlorophen (Fungicide) (Herbicide)

Homebase Rooting Hormone Liquid	Sainsbury's Homebase	07698	

1113 1-Naphthylacetic acid + Dichlorophen (Fungicide) (Miscellaneous)

Baby Bio Roota	pbi Home & Garden Ltd	09305	

1114 1-Naphthylacetic acid + Dichlorophen (Herbicide)

Bio Roota	pbi Home & Garden Ltd	00271	30/11/2001

1114a 1-Naphthylacetic acid + 4-Indol-3ylbutric acid + Thiram (Fungicide) (Herbicide)

Boots Hormone Rooting Powder	The Boots Company Plc	01067	

1115 Orange peel oil + Garlic oil + Orange pith oil (Vertebrate Control)

6X Cat Repellent	Organic Concentrates Ltd	07979	
Growing Success Cat Repellent	Growing Success Organics Ltd	06262	

1116 Orange pith oil + Garlic oil + Orange peel oil (Vertebrate Control)

6X Cat Repellent	Organic Concentrates Ltd	07979	
Growing Success Cat Repellent	Growing Success Organics Ltd	06262	

1117 P-[(Diiodomethyl)Sulfonyl]Toluol (Herbicide)

Barrel Feature Clear	Interpet Ltd	09074	

1118 Paraquat + Amitrole + Diquat + Simazine (Herbicide)

Pathclear	The Scotts Company (UK) Limited	09287	
Pathclear	Miracle Garden Care	07789	

1119 Paraquat + Diquat (Herbicide)

Weedol	The Scotts Company (UK) Limited	09280	

C   These products are "approved for agricultural use". For further details refer to page vii.
A   These products are approved for use in or near water. For further details refer to page vii.

Product Name	Marketing Company	Reg. No.	Expiry Date

1119 Paraquat + Diquat (Herbicide)—continued

Weedol	Miracle Garden Care	07750	30/09/2001
Weedol	Zeneca Garden Care	06863	31/05/2001
Weedol Gun!	Miracle Garden Care	08254	31/07/2001

1120 Paraquat + Diquat + Simazine (Herbicide)

'Pathclear' S	The Scotts Company (UK) Limited	09285	
'Pathclear' S	Miracle Garden Care Ltd	08992	31/10/2001

1121 Penconazole (Fungicide)

Fungus Clear	The Scotts Company (UK) Ltd	10156	
Fungus Clear Gun!	The Scotts Company (UK) Ltd	10154	
Murphy Tumbleblite II	Levington Horticulture Ltd	07573	
Murphy Tumbleblite II Ready to Use	Levington Horticulture Ltd	07572	

1122 Pepper (Vertebrate Control)

PBI Pepper Dust	pbi Home & Garden Ltd	01569	
Pepper Dust	Vitax Ltd	09635	
Pepper Dust	Synchemicals Ltd	01570	
Secto Pepper Dust	Sinclair Animal & Household Care Ltd	07891	

1123 Permethrin (Herbicide)

Miracle-Gro Bug Spray	Miracle Garden Care Ltd	09176	

1124 Permethrin (Insecticide)

Bio Flydown	pbi Home & Garden Ltd	00267	
Bio Kill	Jesmond Ltd	07735	31/01/2001
Bio Sprayday	pbi Home & Garden Ltd	00272	
Fumite Whitefly Smoke Cone	The Scotts Company (UK) Limited	09535	
Fumite Whitefly Smoke Cone	Miracle Garden Care	07839	
Homebase All-In-One Insecticide	Sainsbury's Homebase	07946	
Levington Insect Spray for Houseplants	Levington Horticulture Ltd	07466	
Picket	Miracle Garden Care	07740	

1125 Permethrin + Bioallethrin (Insecticide)

Bio Spraydex Greenfly Killer	pbi Home & Garden Ltd	07404	31/01/2001
Floracid	Perycut Insectengun Ltd	06798	30/09/2002
Longer Lasting Bug Gun	Miracle Garden Care	08021	30/09/2002

C   These products are "approved for agricultural use". For further details refer to page vii.
A   These products are approved for use in or near water. For further details refer to page vii.

Product Name	Marketing Company	Reg. No.	Expiry Date

1126 Permethrin + Heptenophos (Insecticide)

Murphy Systemic Action Insecticide	Levington Horticulture Ltd	07557	30/04/2001
Murphy Tumblebug	Levington Horticulture Ltd	07571	30/04/2001

1127 Permethrin + Sulphur + Triforine (Fungicide) (Insecticide)

Bio Multirose	pbi Home & Garden Ltd	05716	
Homebase Rose Care	Sainsbury's Homebase	07754	

1128 Phenothrin + Tetramethrin (Insecticide)

Pesguard House and Plant Spray	Sumitomo Chemical (UK) Plc	07873	30/04/2001

1129 Pirimicarb (Insecticide)

Rapid Aerosol	Miracle Garden Care	07741	20/04/2001
Rapid Aerosol	Imperial Chemical Industries Plc	01689	20/04/2001
Rapid Greenfly Killer	The Scotts Company (UK) Limited	09500	30/04/2002
Rapid Greenfly Killer	Miracle Garden Care	07742	30/09/2001
Rapid Greenfly Killer	Zeneca Garden Care	06848	20/04/2001
Rapid Greenfly Killer	Imperial Chemical Industries Plc	01690	20/04/2001

1130 Pirimicarb + Bupirimate + Triforine (Fungicide) (Insecticide)

'Roseclear' 2	The Scotts Company (UK) Limited	09498	
'Roseclear' 2	Miracle Garden Care Ltd	08736	

1131 Pirimiphos-methyl (Insecticide)

B & Q Antkiller Dust	B & Q Plc	05830	20/04/2001
Fumite General Purpose Insecticide Smoke Cone	The Scotts Company (UK) Limited	09501	
Fumite General Purpose Insecticide Smoke Cone	Miracle Garden Care	07838	25/07/2003
Sybol	Miracle Garden Care	07745	20/04/2001
Sybol Dust	Miracle Garden Care	07746	30/09/2001
Sybol Dust	Zeneca Garden Care	06851	20/04/2001

1132 Pirimiphos-methyl + Resmethrin + Tetramethrin (Insecticide)

Miracle-Gro Bug Spray	Miracle Garden Care	08690	20/04/2001
Sybol Aerosol	Miracle Garden Care	07790	20/04/2001

1133 Polymeric quaternary ammonium chloride (Herbicide)

Feature Clear	Interpet Ltd	09075	

C These products are "approved for agricultural use". For further details refer to page vii.
A These products are approved for use in or near water. For further details refer to page vii.

Product Name	Marketing Company	Reg. No.	Expiry Date

1134 Propiconazole (Fungicide)

Murphy Tumbleblite	Levington Horticulture Ltd	09327	
Tumbleblite	Levington Horticulture Ltd	07567	
Tumbleblite	Fisons Plc	04691	

1135 Propiconazole + Cypermethrin (Fungicide) (Insecticide)

Murphy Roseguard	Levington Horticulture Ltd	08564	

1136 Pyrethrins (Insecticide)

AgrEvo Garden Insect Killer	AgrEvo UK Ltd	08096	
AgrEvo House Plant Insect Killer	AgrEvo UK Ltd	08099	
Aquablast Bug Spray	Agropharm Ltd	03461	
B & Q Complete Insecticide Spray	B & Q Plc	06964	
B & Q Fruit and Vegetable Insecticide Spray	B & Q Plc	05766	
B & Q House Plant Insecticide Spray	B & Q Plc	05828	
B & Q Houseplant Insecticide Spray	B & Q Plc	07032	
B & Q Insecticide Spray for Fruit and Vegetable	B & Q Plc	05829	
B & Q Insecticide Spray For Roses and Flowers	B & Q Plc	05875	
B & Q Rose and Flower Insecticide Spray	B & Q Plc	05769	
Bio Friendly Anti-Ant Duster	pbi Home & Garden Ltd	05100	
Bug Gun!	The Scotts Company (UK) Limited	09538	
Bug Gun!	The Scotts Company (UK) Limited	09536	
Bug Gun!	Miracle Garden Care	07782	
Bug Gun!	Miracle Garden Care	07781	
Bug Gun!	Zeneca Garden Care	06837	31/05/2001
Bug Gun!	Zeneca Garden Care	06836	31/05/2001
Bug Gun!	Imperial Chemical Industries Plc	05966	31/05/2001
Bug Gun!	Imperial Chemical Industries Plc	05965	31/05/2001
Bug Gun! Attack	The Scotts Company (UK) Limited	09590	
Co-op Garden Maker Rose & Flower Insecticide Insecticide Spray	Co-Operative Wholesale Society Ltd	05609	
Devcol All Purpose Natural Insecticide Spray	Devcol Ltd	05802	
Do It All De-Bug! Insect Killer	Do-It-All Ltd	08184	
Do It All Fruit and Vegetable Insecticide Spray	Do-It-All Ltd	05767	

C   These products are "approved for agricultural use". For further details refer to page vii.
A   These products are approved for use in or near water. For further details refer to page vii.

PSD | *AMATEUR PRODUCTS*

Product Name	Marketing Company	Reg. No.	Expiry Date

1136 Pyrethrins (Insecticide)—continued

Product Name	Marketing Company	Reg. No.	Expiry Date
Do It All Rose and Flower Insecticide Spray	Do-It-All Ltd	05765	
Doff 'All in One' Insecticide Spray	Doff Portland Ltd	06069	
Doff Fruit and Vegetable Insecticide Spray	Doff Portland Ltd	04040	
Doff Greenfly Killer	Doff Portland Ltd	07030	
Doff Houseplant Insecticide Spray	Doff Portland Ltd	06066	
Doff Rose and Flower Insecticide Spray	Doff Portland Ltd	04041	
Doff Rose Insecticide Spray	Doff Portland Ltd	08747	
Fellside Green Fruit and Vegetable Insect Spray	Doff Portland Ltd	05008	
Fellside Green Rose and Flower Insect Spray	Doff Portland Ltd	05007	
Focus All-In-One Insecticide Spray	Focus DIY Ltd	07510	
Focus Fruit and Vegetable Insecticide Spray	Focus DIY Ltd	06067	
Focus Rose and Flower Insecticide Spray	Focus DIY Ltd	06068	
Garden Insect Killer	L C Solutions Ltd	09738	
Great Mills Complete Insecticide Spray	Great Mills (Retail) Ltd	07031	
Great Mills Fruit and Vegetable Insecticide Spray	Great Mills (Retail) Ltd	05772	
Great Mills Rose and Flower Insecticide Spray	Great Mills (Retail) Ltd	05771	
Homebase Houseplant Insecticide	Sainsbury's Homebase	07692	
Homebase Pest Gun	Sainsbury's Homebase House and Garden Centres	06962	
Houseplant Insect Killer	L C Solutions Ltd	09737	
Keri Insect Spray	The Scotts Company (UK) Limited	09539	
Keri Insect Spray	Miracle Garden Care	07784	
Keri Insect Spray	Zeneca Garden Care	06885	31/05/2001
Keri Insect Spray	ICI Garden & Professional Products	06155	31/05/2001
Miracle-Gro Houseplant Bug Spray	The Scotts Company (UK) Limited	10271	
Murphy Bugmaster	The Scotts Company (UK) Limited	07575	
Nature's Answer to Insect Pests	The Scotts Company (UK) Limited	07504	

C   These products are "approved for agricultural use". For further details refer to page vii.
A   These products are approved for use in or near water. For further details refer to page vii.

Product Name	Marketing Company	Reg. No.	Expiry Date

1136 Pyrethrins (Insecticide)—continued

Product Name	Marketing Company	Reg. No.	Expiry Date
Nature's Answer to Insect Pests Concentrate	The Scotts Company (UK) Limited	08683	
Natures "Bug Gun!"	The Scotts Company (UK) Limited + Miracle Garden Care Ltd	09537	
Natures 'Bug Gun!'	Miracle Garden Care	07788	
Natures Bug Gun	Zeneca Garden Care	07077	31/05/2001
PBI Anti-Ant Duster	pbi Home & Garden Ltd	00098	
Premier Multi-Pest Insecticide Spray	Premier Group Ltd	07424	
Py Powder	Vitax Ltd	05542	
Py Spray Garden Insect Killer	Vitax Ltd	06085	
Py Spray Insect Killer	Vitax Ltd	05543	
Rentokil Garden Insect Killer	Rentokil Initial plc	07852	
Secto Nature Care Garden Insect Powder	Sinclair Animal & Household Care Ltd	07894	
Secto Nature Care Garden Insect Spray	Sinclair Animal & Household Care Ltd	07892	25/07/2003
Secto Nature Care Houseplant Insect Spray	Sinclair Animal & Household Care Ltd	07893	25/07/2003
Texas All-In-One Insecticide Spray	Texas Homecare Ltd	07426	
Trappit	Agrisense -BCS Limited	09452	31/07/2002
Vapona All-In-One	Sara Lee Household and Personal Care	06965	
Vapona House and Garden Plant Insect Killer	Sara Lee Household and Personal Care	05996	
Wickes General Purpose Insecticide Ready to Use	Wickes Building Supplies Ltd	08787	
Wilko Fruit and Vegetable Insecticide Spray	Wilkinson Group of Companies	09577	
Wilko Greenfly Killer	Wilkinson Home & Garden Stores	08727	
Wilko Houseplant Insecticide Spray	Wilkinson Home & Garden Stores	07425	
Wilko Multi-Purpose Insecticide Spray	Wilkinson Home & Garden Stores	06963	
Wilko Rose and Flower Insecticide Spray	Wilkinson Home & Garden Stores	05768	
Woolworths Complete Insecticide Spray	Woolworths Plc	06961	

1137 Pyrethrins + Fatty acids (Insecticide)

Product Name	Marketing Company	Reg. No.	Expiry Date
Phostrogen Safer's All Purpose Insecticide Ready to Use	Phostrogen Ltd	08093	

C   These products are "approved for agricultural use". For further details refer to page vii.
A   These products are approved for use in or near water. For further details refer to page vii.

Product Name	Marketing Company	Reg. No.	Expiry Date

1137 Pyrethrins + Fatty acids (Insecticide)—continued

Safer's Indoor Trounce Insecticide Concentrate	Safer Ltd	08097	
Safer's Indoor Trounce Insecticide Ready to Use	Safer Ltd	08094	
Safer's Trounce Insecticide Concentrate	Safer Ltd	08098	
Safer's Trounce Insecticide Ready to Use	Safer Ltd	08095	

1138 Pyrethrins + Resmethrin (Insecticide)

| House Plant Pest Killer | Vitax Ltd | 06432 | |

1139 Resmethrin + Pirimiphos-methyl + Tetramethrin (Insecticide)

| Miracle-Gro Bug Spray | Miracle Garden Care | 08690 | 20/04/2001 |
| Sybol Aerosol | Miracle Garden Care | 07790 | 20/04/2001 |

1140 Resmethrin + Pyrethrins (Insecticide)

| House Plant Pest Killer | Vitax Ltd | 06432 | |

1141 Rotenone (Insecticide)

Bio Friendly Pest Duster	pbi Home & Garden Ltd	06811	
Bio Liquid Derris Plus	pbi Home & Garden Ltd	07059	
BioSect	pbi Home & Garden Ltd	07807	
Derris Dust	Vitax Ltd	05452	
Doff Derris Dust	Doff Portland Ltd	00740	
Murphy Derris Dust	Fisons Plc	04005	
Nature's Answer Derris Dust	Levington Horticulture Ltd	09015	
Stirling Rescue Wasp Nest Killer	STV International Ltd	08205	
Wasp Exterminator	Battle Hayward & Bower Ltd	06333	

1142 Rotenone + Sulphur (Fungicide) (Insecticide)

| Bio Back to Nature Pest & Disease Duster | Pan Britannica Industries Ltd | 00265 | |

1143 Simazine (Herbicide)

Algae Control No 1 Anti-Algae	Interpet Ltd	09068	
Algae Destroyer For Freshwater Aquariums	Aquarium Pharmaceuticals E.C. Inc	08856	
Algae Destroyer For Ponds	Aquarium Pharmaceuticals E.C. Inc	08858	
Algae Destroyer Liquid	Aquarium Pharmaceuticals E.C. Inc	08857	

C   These products are "approved for agricultural use". For further details refer to page vii.
A   These products are approved for use in or near water. For further details refer to page vii.

Product Name	Marketing Company	Reg. No.	Expiry Date

1143 Simazine (Herbicide)—continued

Aquarium Algicide	Technical Aquatic Products	08841	
BWG Aquarium Anti Algae	Balgdon/KFS	08848	
BWG Pool Clinic Algicide	Blagdon Garden Products	08846	
Feature Algae Control	Interpet Ltd	09069	
Pond Doctor Algicide	Technical Aquatic Products	08810	
PPI Clarity Plus	Technical Aquatic Products	08837	
Tetra Algimin	Tetra	08861	

1144 Simazine + Amitrole (Herbicide)

Homebase Path & Drive Weed Killer	Sainsbury's Homebase	07755	
Path and Drive Weedkiller	pbi Home & Garden Ltd	05958	

1145 Simazine + Amitrole + 2,4-D + Diuron (Herbicide)

Hytrol	Agrichem (International) Ltd	04540	30/10/2001

1146 Simazine + Amitrole + Diquat + Paraquat (Herbicide)

Pathclear	The Scotts Company (UK) Limited	09287	
Pathclear	Miracle Garden Care	07789	

1147 Simazine + Amitrole + Diuron (Herbicide)

Hytrol Total	Agrichem (International) Ltd	08296	

1148 Simazine + Copper sulphate (Herbicide)

PPI Clarity Excel	Pet Products International Ltd	09095	
Tap Algasan	Technical Aquatic Products	08845	

1149 Simazine + Diquat + Paraquat (Herbicide)

'Pathclear' S	The Scotts Company (UK) Limited	09285	
'Pathclear' S	Miracle Garden Care Ltd	08992	31/10/2001

1150 Sodium chlorate (Herbicide)

Barrettine Sodium chlorate (Fire Suppressed) Weedkiller	Cooke's Chemicals (Sales) Ltd	06617	
Battle, Hayward and Bower Sodium chlorate Weedkiller with Fire Depressant	Battle Hayward & Bower Ltd	05876	
Blanchard's Sodium chlorate Weedkiller (Fire Suppressed)	Blanchard Martin and Simmonds Ltd	05649	
Cooke's Liquid Sodium chlorate Weedkiller	Cooke's Chemicals (Sales) Ltd	04280	

C   These products are "approved for agricultural use". For further details refer to page vii.
A   These products are approved for use in or near water. For further details refer to page vii.

Product Name	Marketing Company	Reg. No.	Expiry Date

1150 Sodium chlorate (Herbicide)—continued

Cooke's Sodium chlorate Weedkiller with Fire Depressant	Cooke's Chemicals (Sales) Ltd	04281	
Devcol Path Weedkiller	Devcol Ltd	06580	
Devcol-Sodium chlorate Weedkiller	Devcol Ltd	05656	
Doff Path Weedkiller	Doff Portland Ltd	07044	
Doff Sodium chlorate Weedkiller	Doff Portland Ltd	00500	
Focus Sodium chlorate Weedkiller	Focus DIY Ltd	06000	
Gem Sodium chlorate Weedkiller	Joseph Metcalf Ltd	04159	
Great Mills Sodium chlorate Weedkiller	Great Mills (Retail) Ltd	07078	
Homebase Sodium chlorate Weedkiller	Sainsbury's Homebase House and Garden Centres	06620	
Morgan's Sodium chlorate Weedkiller	David Morgan (Nottingham) Ltd	08928	
Premier Sodium chlorate	Premier Way Ltd	07469	
Strathclyde Sodium chlorate Weedkiller	Strathclyde Chemical Co Ltd	07421	
Wilko Sodium chlorate Weedkiller	Wilkinson Home & Garden Stores	06281	

1151 Sulphur (Fungicide)

Green Sulphur	Vitax Ltd	05782	
Safer's Natural Garden Fungicide	Safer Ltd	06298	
Sulphur Candle	Growing Success Organics Ltd	08688	
Sulphur Candles	Battle Hayward & Bower Ltd	02039	
Yellow Sulphur	Vitax Ltd	05783	

1152 Sulphur (Vertebrate Control)

Murphy Mole Smokes	Fisons Plc	03615	

1153 Sulphur + Fatty acids (Fungicide) (Insecticide)

Nature's Answer Fungicide and Insect Killer	The Scotts Company (UK) Limited	07628	

1154 Sulphur + Permethrin + Triforine (Fungicide) (Insecticide)

Bio Multirose	pbi Home & Garden Ltd	05716	
Homebase Rose Care	Sainsbury's Homebase	07754	

C   These products are "approved for agricultural use". For further details refer to page vii.
A   These products are approved for use in or near water. For further details refer to page vii.

Product Name	Marketing Company	Reg. No.	Expiry Date

1155 Sulphur + Rotenone (Fungicide) (Insecticide)

Bio Back to Nature Pest & Disease Duster	Pan Britannica Industries Ltd	00265	

1156 Tar acids (Fungicide) (Herbicide)

Armillatox	Armillatox Ltd	06234	

1157 Tar acids (Miscellaneous)

Jeyes Fluid	Jeyes Ltd	04606	

1158 Tar oils (Insecticide)

Murphy Mortegg	The Scotts Company (UK) Limited	07879	

1159 Tar oils + Citronella oil (Fungicide) (Miscellaneous)

Bio Scat-A-Cat	pbi Home & Garden Ltd	10037	

1160 2,3,6-TBA + 2,4-D (Herbicide)

Touchweeder	Thomas Elliott Ltd	02864	

1161 Terbutryn (Herbicide)

Algae-Kit	Intercel UK	08383	
Algizin A	Waterlife Research Industries Ltd	08850	
Blanc-Kit	Intercel UK	08391	

1162 Tetramethrin + Phenothrin (Insecticide)

Pesguard House and Plant Spray	Sumitomo Chemical (UK) Plc	07873	30/04/2001

1163 Tetramethrin + Pirimiphos-methyl + Resmethrin (Insecticide)

Miracle-Gro Bug Spray	Miracle Garden Care	08690	20/04/2001
Sybol Aerosol	Miracle Garden Care	07790	20/04/2001

1164 Thiophanate-methyl (Fungicide)

Liquid Club Root Control	pbi Home & Garden Ltd	05957	

1165 Thiram + 1-Naphthylacetic acid + 4-Indol-3-ylbutric acid (Fungicide) (Herbicide)

Boots Hormone Rooting Powder	The Boots Company Plc	01067	

1166 Triclopyr (Herbicide)

Murphy Nettlemaster	Levington Horticulture Ltd	07561	
New Garlon RTU	Dow AgroSciences Ltd	06758	

C   These products are "approved for agricultural use". For further details refer to page vii.
A   These products are approved for use in or near water. For further details refer to page vii.

280

Product Name	Marketing Company	Reg. No.	Expiry Date

1166 Triclopyr (Herbicide)—continued

| Tumbleweed Brushwood Ready to Use | Levington Horticulture Ltd | 08645 | |

1167 Triforine + Bupirimate (Fungicide)

Nimrod T	The Scotts Company (UK) Limited	09502	
Nimrod T	Miracle Garden Care	07876	
Nimrod T	Zeneca Garden Care	06843	31/01/2001
Nimrod T	Imperial Chemical Industries Plc	03982	31/05/2001

1168 Triforine + Bupirimate + Pirimicarb (Fungicide) (Insecticide)

| 'Roseclear' 2 | The Scotts Company (UK) Limited | 09498 | |
| 'Roseclear' 2 | Miracle Garden Care Ltd | 08736 | |

1169 Triforine + Permethrin + Sulphur (Fungicide) (Insecticide)

| Bio Multirose | pbi Home & Garden Ltd | 05716 | |
| Homebase Rose Care | Sainsbury's Homebase | 07754 | |

C   These products are "approved for agricultural use". For further details refer to page vii.
A   These products are approved for use in or near water. For further details refer to page vii.

3

PSD PRODUCT TRADE NAME INDEX

The number after the Trade Name gives the Active Ingredient Code Number under which the Active Ingredient in the product occurs in the Professional or Amateur Sections.

100-Plus *753*
3C Chlormequat 460 *58*
3C Chlormequat 600 *58*
50-50 Liquid Mosskiller *139*
6X Cat Repellent *1058*
6X Slug Killer *936*
AAprotect *906*
AAterra WP *542*
ACP 105 *815*
AGROXONE 50 *268*
ALS Premier Selective Plus *133*
ASB Greenworld Lawn Weed, And Feed
Granular *995*
ASB Greenworld Lawn Weed, Feed and Mosskiller
Granular *993*
Accolade *545*
Accrue *707*
Ace-Sodium Chlorate (Fire Suppressed)
Weedkiller *395*
Aceto Chlorpropham 50M *914*
Acquit *791*
Acrion *216*
Actellic 2% Dust *869*
Actellic D *869*
Actellic Smoke Generator No 10 *869*
Actellic Smoke Generator No 20 *869*
Actellifog *869*
Activus *100*
Actrilawn 10 *226*
Acumen *18*
Adagio *496*
Adagio C *496*
Adjust *58*
Admiral *672*
Admire *844*
Advance *36*
Advizor *54*
Advizor S *54*
Afalon *261*
Affinity *48*
Afrisect 10 *811*
Afugan *699*
Agate *681*
AgrEvo Garden Insect Killer *1136*
AgrEvo Garden Systemic Fungicide *960*
AgrEvo Garden Weedkiller Concentrate *1059*
AgrEvo Garden Weedkiller Spray *1059*
AgrEvo House Plant Insect Killer *1136*
AgrEvo Lawn Weed Killer *991*
AgrEvo Lawn Weedkiller Concentrate *991*
AgrEvo Path Weed Killer *1034*
AgrEvo Patio Weedkiller *1059*
AgrEvo Rose & Flower Greenfly Killer *1037*
AgrEvo Rose & Flower Insect Killer *1037*
AgrEvo Slug Killer Pellets *1099*
Agrevo Rose PestKiller *1037*

Agrichem 2,4-D *104*
Agrichem DB Plus *120*
Agrichem Flowable Thiram *742*
Agrichem MCPA 50 *268*
Agrichem Mancozeb 80 *623*
Agrichem Maneb 80 *637*
Agricola Lenacil FL *257*
Agricola Lens *260*
Agricorn 2,4-D *104*
Agricorn 500 *268*
Agricorn 500 II *268*
Agricorn D *104*
Agricorn D11 *104*
Agriguard 5C Chlormequat 460 *58*
Agriguard Alpha-Cyper *791*
Agriguard Chlormequat 700 *58*
Agriguard Chlormequat 720 *58*
Agriguard Chlorothalonil *488*
Agriguard Cypermethrin *811*
Agriguard Cyproconazole *511*
Agriguard Epoxiconazole *534*
Agriguard Ethofumesate Flo *176*
Agriguard Fluroxypyr *199*
Agriguard Glyphosate 180 *216*
Agriguard Glyphosate 360 *216*
Agriguard Isoproturon *239*
Agriguard Lambda-Cyhalo *847*
Agriguard Metamitron *320*
Agriguard Metribuzin *331*
Agriguard Metsulfuron *334*
Agriguard Napropamide *344*
Agriguard Paraquat *350*
Agriguard Pirimicarb *866*
Agriguard Proyzamide 50 WP *378*
Agritox 50 *268*
Agritox D *268*
Agroxone *268*
Agroxone 40 *268*
Agroxone 75 *268*
Agroxone Combi *112*
Agrys *554*
Aitkens Lawn Sand *187*
Aitkens Lawn Sand Plus *140*
Alert *791*
Alfacron 10 WP *794*
Alfacron 10WP *794*
Alfacron 50 WP *794*
Alfacron Plus *795*
Alfadex *872*
Algae Control No 1 Anti-Algae *1143*
Algae Control No 2 Anti-hair Algae *1030*
Algae Control No 3 Anti-Slime Algae *966*
Algae Destroyer For Freshwater Aquariums *1143*
Algae Destroyer For Ponds *1143*
Algae Destroyer Liquid *1143*
Algae-Kit *1161*

Algalit *974*
Algimycin PLL *971*
Algizin A *1161*
Algizin P *947*
AlgoFin *1105*
Alias *386*
Aliette *591*
Aliette 80 WG *591*
Aligran WDG *239*
Alistell *119*
Allicur *725*
Allied Colloids Chlormequat 460 *58*
Allied Colloids Chlormequat 460:320 *58*
Allied Colloids Chlormequat 730 *58*
Ally *334*
Ally Express *50*
Ally WSB *334*
Alpha Atrazine 50 SC *10*
Alpha Atrazine 50 WP *10*
Alpha Briotril *37*
Alpha Briotril Plus 19/19 *37*
Alpha Bromacil 80 WP *24*
Alpha Bromolin 225 EC *28*
Alpha Bromotril P *28*
Alpha Bromotril PT *42*
Alpha Captan 50 WP *464*
Alpha Captan 80 WDG *464*
Alpha Captan 83 WP *464*
Alpha Chlormequat 460 *58*
Alpha Chlorotoluron 500 *69*
Alpha Chlorpyrifos 48EC *806*
Alpha Isoproturon 500 *239*
Alpha Isoproturon 650 *239*
Alpha Linuron 50 SC *261*
Alpha Linuron 50 WP *261*
Alpha Pentagan *58*
Alpha Pentagan Extra *58*
Alpha Prometryne 50 WP *372*
Alpha Prometryne 80 WP *372*
Alpha Propachlor 50 SC *374*
Alpha Propachlor 65 WP *374*
Alpha Protugan Plus *250*
Alpha Simazine 50 SC *389*
Alpha Simazine 50 WP *389*
Alpha Simazine 80 WP *389*
Alpha Simazol *6*
Alpha Simazol T *6*
Alpha Terbalin 35 SC *415*
Alpha Terbutryne 50 SC *410*
Alpha Terbutryne 50 WP *410*
Alpha Trifluralin 48 EC *439*
Alphabird *893*
Alphachloralose Technical *893*
Alphathrin *791*
Alto 100 SL *511*
Alto 240 EC *511*

Alto Combi *470*
Alto Eco *515*
Alto Elite *492*
Alto Major *519*
Amazon *83*
Amcide *7*
Amega Pro TMF *216*
Amigo *453*
Amistar *452*
Amistar Gem *453*
Anchor *487*
Angle *101*
Anti-Snail *979*
Antracol *697*
Apache *216*
Apex 5E *858*
Aphox *866*
Aplan *511*
Aplan 240 EC *511*
Apollo 50 SC *809*
Applaud *801*
Apres *703*
Apron Combi FS *651*
Apron Elite *469*
Apron T69 WS *650*
Aquablast Bug Spray *1136*
Aqualife Algae and Snail Killer *975*
Aqualife Pond Algicide *974*
Aquarium Algicide *1143*
Aramo *403*
Ardent *166*
Arelon *239*
Arelon 2 *239*
Arelon 500 *239*
Argylene *398*
Aristo *924*
Armillatox *1156*
Aro Slug Killer Blue Mini Pellets *1099*
Array *708*
Arresin *340*
Arsenal *220*
Arsenal 50 *220*
Artillery *249*
Artillery SC *249*
Asda Lawn Sand *1042*
Asda Lawn Weed and Feed and Mosskiller *992*
Ashlade 460 CCC *58*
Ashlade 5C *58*
Ashlade 700 5C *58*
Ashlade 700 CCC *58*
Ashlade Brevis *58*
Ashlade CP *56*
Ashlade Carbendazim Flowable *466*
Ashlade Cyclops *491*
Ashlade Cypermethrin 10 EC *811*
Ashlade Flotin 2 *574*

Ashlade Linuron FL *261*
Ashlade Mancarb FL *476*
Ashlade Mancarb Plus *468*
Ashlade Mancozeb FL *623*
Ashlade Maneb Flowable *637*
Ashlade SMC Flowable *505*
Ashlade Solace *509*
Ashlade Sulphur FL *709*
Ashlade Summit *415*
Ashlade Trifluralin *439*
Ashlade Trimaran *439*
Assert *219*
Asset *13*
Astix *302*
Astix K *302*
Asulox *9*
Atlacide Soluble Powder Weedkiller *395*
Atladox HI *115*
Atlas 2,4-D *104*
Atlas 3C:645 Chlormequat *58*
Atlas 5C Chlormequat *58*
Atlas Atrazine *10*
Atlas Brown *77*
Atlas CIPC 40 *74*
Atlas Chlormequat 46 *58*
Atlas Chlormequat 460:46 *58*
Atlas Chlormequat 700 *58*
Atlas Cropguard *488*
Atlas Diuron *170*
Atlas Fieldguard *239*
Atlas Herbon Pabrac *74*
Atlas MCPA *268*
Atlas Minerva *32*
Atlas Orange *374*
Atlas Prometryn 50 WP *372*
Atlas Propachlor *374*
Atlas Protrum K *362*
Atlas Quintacel *58*
Atlas Red *75*
Atlas Simazine *389*
Atlas Solan 40 *360*
Atlas Somon *396*
Atlas Steward *849*
Atlas Tecgran 100 *400*
Atlas Tecnazene 6% Dust *400*
Atlas Tecnazene 6% dust *400*
Atlas Terbine *58*
Atlas Thor *176*
Atlas Tricol *58*
Atlas Trifluralin *439*
Atlazin D-Weed *937*
Atol *69*
Atrazine 90WG *10*
Atrazol *10*
Atum *239*
Atum WDG *239*

Auger *239*
Auger WDG *239*
Aura *559*
Aurora *46*
Autumn Kite *251*
Autumn Toplawn Feed and Mosskiller *1053*
Avadex 15 G *424*
Avadex BW *424*
Avadex BW 480 *424*
Avadex BW Granular *424*
Avadex Excel 15G *424*
Avenge 2 *156*
Aventis Betanal Progress OF *122*
Aventis Cerone *64*
Aventis Challenge *215*
Aventis Cheetah Super *183*
Aventis Compass *618*
Aventis Decis *813*
Aventis Eagle *1*
Aventis Flamenco *578*
Aventis Foil *580*
Aventis Galtak 50 SC *12*
Aventis Harvest *215*
Aventis Merlin *498*
Aventis Mocap 10G *829*
Aventis Rovral Flo *616*
Aventis Tattoo *636*
Aventis Temik 10G *790*
Aventis Tigress Ultra *155*
Axit GR *256*
Aztec *890*
Azural *216*
B & Q Ant Killer *946*
B & Q Antkiller Dust *1131*
B & Q Complete Insecticide Spray *1136*
B & Q Complete Weedkiller *1062*
B & Q Complete Weedkiller Ready To Use *1062*
B & Q Fruit and Vegetable Insecticide Spray *1136*
B & Q Garden Fungicide Spray *973*
B & Q Granular Autumn Lawn Food with
  Mosskiller *1042*
B & Q Granular Lawn Feed and Mosskiller *1042*
B & Q Granular Lawn Feed and Weed *1084*
B & Q Granular Lawn Feed, Weed and
  Mosskiller *1052*
B & Q Granular Weed and Feed *986*
B & Q Granular Weed and Feed and Mosskiller For
  Lawns *1002*
B & Q Granular Weed and Feed for Lawns *1003*
B & Q House Plant Insecticide Spray *1136*
B & Q Houseplant Insecticide Spray *1136*
B & Q Insecticide Spray For Roses and
  Flowers *1136*
B & Q Insecticide Spray for Fruit and
  Vegetable *1136*
B & Q Lawn Feed and Weed *991*

B & Q Lawn Spot Weeder *991*
B & Q Lawn Weedkiller *991*
B & Q Lawnweed Spray *986*
B & Q Liquid Weed and Feed *1003*
B & Q Mosskiller for Lawns *1042*
B & Q Nettle Gun *991*
B & Q Rose and Flower Insecticide Spray *1136*
B & Q Slug Killer Blue Mini Pellets *1099*
B & Q Systemic Fungicide Concentrate *960*
B & Q Tree Stump and Brushwood Killer *989*
B & Q Woodlice Killer *946*
B & Q Autumn Lawn Care *1042*
B & Q Insect Spray for Houseplants *1037*
B & Q Insect Spray for Roses and Flowers *1037*
B & Q Lawn Feed, Weed and Mosskiller
   Granular *993*
B & Q Lawn Weed and Feed Granular *995*
B & Q Organic Insecticide *1037*
B & Q Organic Weedkiller *1036*
B & Q Path & Patio Weedkiller *938*
B & Q Slug Killer Blue Mini Pellets *1099*
B & Q Spring Lawn Care Feed Weed *1084*
B & Q Spring Lawn Care Feed Weed
   Mosskiller *1052*
B & Q Summer Lawncare Feed Weed
   Mosskiller *993*
B-Nine *116*
BAS 03729H *302*
BAS 421F *559*
BAS 438 *493*
BAS 46402F *572*
BAS 48500F *565*
BAS 91580F *725*
BASF 3C Chlormequat 600 *58*
BASF 3C Chlormequat 720 *58*
BASF 3C Chlormequat 750 *58*
BASF Dimethoate 40 *823*
BASF MCPA Amine 50 *268*
BASF Turf Systemic Fungicide *466*
BB Chlorothalonil *488*
BH MCPA 75 *268*
BL500 *74*
BR Destral *4*
BUK 84000F *571*
BWG Anti Snails *976*
BWG Aquarium Anti Algae *1143*
BWG Pool Clinic Algicide *1143*
Baby Bio Provado Insect Control *1066*
Baby Bio Roota *1013*
Bacara *162*
Bactura WP *907*
Ballad *806*
Banco *22*
Banlene Super *133*
Banvel M *131*
Banvel T *137*

Banweed *344*
Barbarian *216*
Barclay Amtrak *521*
Barclay Avalon *538*
Barclay Banshee *68*
Barclay Banshee XL *59*
Barclay Barbarian *216*
Barclay Bolt *685*
Barclay Busker *714*
Barclay Canter *99*
Barclay Cleanup *216*
Barclay Cleave *1*
Barclay Clinch II *806*
Barclay Cluster *530*
Barclay Cobbler *576*
Barclay Coolmore *64*
Barclay Corrib 500 *488*
Barclay Dart *216*
Barclay Desiquat *167*
Barclay Dimethosect *823*
Barclay Dodex *533*
Barclay Eyetak *672*
Barclay Eyetak 40 *672*
Barclay Eyetak 450 *672*
Barclay Fentin Flow *574*
Barclay Fentin Flow 532 *574*
Barclay Flotilla *783*
Barclay Flumen *334*
Barclay Gallup *216*
Barclay Gallup 360 *216*
Barclay Gallup Amenity *216*
Barclay Gallup Biograde 360 *216*
Barclay Gallup Biograde Amenity *216*
Barclay Garryowen *216*
Barclay Goalpost *182*
Barclay Goldpost *182*
Barclay Guideline 500 *239*
Barclay Hat-Trick *132*
Barclay Haybob II *104*
Barclay Holdup *58*
Barclay Holdup 600 *58*
Barclay Holdup 640 *58*
Barclay Hurler *199*
Barclay Hurler 200 *199*
Barclay Karaoke *86*
Barclay Keeper *176*
Barclay Keeper 200 *176*
Barclay Keeper 500 FL *176*
Barclay Liffey *58*
Barclay Lucan *58*
Barclay Manzeb 455 *623*
Barclay Manzeb Flow *631*
Barclay Meadowman *268*
Barclay Meadowman II *268*
Barclay Mecrop *297*
Barclay Melody *302*

Barclay Metaza   *324*
Barclay Mutiny   *28*
Barclay Pirimisect   *866*
Barclay Piza 400 FL   *378*
Barclay Poacher   *926*
Barclay Punter XL   *362*
Barclay Rascal   *586*
Barclay Rebel II   *377*
Barclay Riverdance   *536*
Barclay Seismic   *320*
Barclay Seismic XL   *320*
Barclay Shandon   *511*
Barclay Shelter   *466*
Barclay Slogan   *164*
Barclay Take 5   *58*
Barclay Total   *350*
Barclay Tremor   *354*
Barclay Trustee   *216*
Barclay Winner   *192*
Barclay ZX   *452*
Barleyquat B   *58*
Baron   *580*
Barrel Feature Clear   *1117*
Barrettine Sodium Chlorate (Fire Suppressed)
   Weedkiller   *1150*
Barrier H   *81*
Basagran   *16*
Basagran SG   *16*
Basamid   *915*
Base 50W   *378*
Basilex   *750*
Basilex Soluble Sachets   *750*
Bastion T   *205*
Batallion   *409*
Baton 500 SC   *488*
Baton SC   *488*
Baton WG   *488*
Battle, Hayward and Bower Sodium Chlorate
   Weedkiller with Fire Depressant   *1150*
Bavistin   *466*
Bavistin DF   *466*
Bavistin FL   *466*
Bay UK 292   *541*
Bayer UK 093   *331*
Bayer UK 397   *844*
Bayer UK 456   *751*
Bayer UK 477   *707*
Bayer UK 479   *844*
Bayer UK 540   *181*
Bayer UK 552   *708*
Bayer UK 590   *2*
Bayer UK 593   *557*
Bayer UK 683   *313*
Bayer UK 720   *1065*
Bayer UK 808   *856*
Bayer UK 809   *856*

Bayer UK 892   *856*
Bayer UK 935   *856*
Bayer UK226   *714*
Bayer UK368   *844*
Bayer UK413   *482*
Bayer UK506   *571*
Bayfidan   *754*
Bayleton   *753*
Bayleton 5   *753*
Baytan   *598*
Baytan CF   *598*
Baytan Flowable   *598*
Baytan IM   *596*
Baytan Secur   *597*
Baythroid   *810*
Beam   *708*
Beetomax   *362*
Beetup   *362*
Belcocel   *58*
Bellmac Plus   *291*
Bellmac Straight   *293*
Belvedere   *569*
Benazalox   *14*
Benlate Fungicide   *456*
Berelex   *214*
Beret Gold   *577*
Besiege   *509*
Besiege WSB   *509*
Best   *815*
Betanal Compact   *123*
Betanal Congress   *122*
Betanal E   *362*
Betanal Flo   *362*
Betanal Montage   *182*
Betanal Progress   *122*
Betanal Progress OF   *122*
Betanal Quorum   *123*
Betanal Rostrum   *123*
Betanal Tandem   *182*
Betanal Trio OF   *181*
Betanal Trio WG   *181*
Betanal Ultima   *122*
Betapal Concentrate   *343*
Betosip   *362*
Betosip 114   *362*
Betosip Combi   *182*
Betta Algae Clear   *974*
Betta Snail Clear   *976*
Bettaquat B   *58*
Better DF   *52*
Better Flowable   *52*
Betterware Spot Weedkiller Spray   *986*
Biactive 270   *216*
Biactive Roundup Brushkiller   *1062*
Biactive Roundup GC   *1062*
Bingo   *566*

Bio Back to Nature Pest & Disease Duster   *1142*
Bio Catapult   *935*
Bio Flydown   *1124*
Bio Friendly Anti-Ant Duster   *1136*
Bio Friendly Pest Duster   *1141*
Bio Fruit Spray   *1041*
Bio Fungus Fighter   *1106*
Bio Glyphosate   *1062*
Bio Glyphosate Pen   *1062*
Bio Glyphosate Ready-to-Use   *1062*
Bio Kill   *1124*
Bio Kills Weeds Dead Fast   *1038*
Bio Kills Weeds Dead Fast Concentrate   *1038*
Bio Lawn Weed Pencil   *1008*
Bio Lawn Weedkiller   *986*
Bio Liquid Derris Plus   *1141*
Bio Mosskiller Ready-to-Use   *1010*
Bio Multirose   *1127*
Bio New Greenfly Killer Plus   *948*
Bio New Sprayday   *948*
Bio Pest Pistol   *1037*
Bio Provado Complete Pest Killer   *1065*
Bio Provado Vine Weevil Killer   *1065*
Bio Roota   *1014*
Bio Scat-A-Cat   *965*
Bio Slug Guard   *1101*
Bio Slug Mini Pellets   *1099*
Bio SpeedWeed   *1036*
Bio Speedweed Ultra   *1038*
Bio Spot   *1007*
Bio Sprayday   *1124*
Bio Sprayday Greenfly Killer Plus   *948*
Bio Spraydex Greenfly Killer   *951*
Bio Strike   *959*
Bio Supergreen 123   *987*
Bio Systhane   *1106*
Bio Toplawn Feed and Weed   *986*
Bio Toplawn Spot Weeder   *995*
Bio Velvas   *1042*
Bio Weed Ban   *1009*
Bio Weed Pencil   *1007*
Bio WeedEasy   *1062*
Bio-Active Algaway   *967*
BioSect   *1141*
Bioglyce   *216*
Bion   *449*
Biotal Bio-Clear   *966*
Birlane 24   *805*
Bison   *239*
Bison 83 WG   *239*
Bison 83WG   *239*
Blagdon Pondsafe Bio Clear   *966*
Blanc-Kit   *1161*
Blanchard's Sodium Chlorate Weedkiller (Fire Suppressed)   *1150*
Blazer   *354*

Blex   *869*
Blitz Bug Gun   *948*
Bolda FL   *477*
Bolero   *165*
Bombardier   *488*
Bombardier FL   *488*
Bonzi   *348*
Boots Hormone Rooting Powder   *1067a*
Boots Lawn Moss Killer and Fertiliser   *1042*
Boots Nettle and Bramble Weedkiller   *1003*
Boots Systemic Weed & Grass Killer Ready To Use   *1062*
Bordeaux Mixture   *969*
Borocil K   *26*
Bos MH 180   *266*
Boscor   *554*
Boxer   *121*
Brasoran   *11*
Bravo 500   *488*
Bravo 720   *488*
Bravocarb   *467*
Bray's Emulsion   *712*
Brestan 60 SP   *573*
Brits   *924*
Broadsword   *108*
Brom-O-Gas   *927*
Bromabloc   *896*
Bromomethane 100%   *927*
Bronze   *708*
Bug & Fungus Free   *950*
Bug Clear   *948*
Bug Clear Gun!   *948*
Bug Free Concentrate   *948*
Bug Gun!   *1136*
Bug Gun! Attack   *1136*
Bug Gun! For Gardens   *948*
Bug Gun! For Roses & Flowers   *1037*
Bug-Free Concentrate   *948*
Bug-Free Ready To Use   *948*
Buggy SG   *216*
Bullet   *100*
Bulwark Flo   *378*
Bumper 250 EC   *685*
Bumper P   *680*
Burex 430 SC   *52*
Butisan S   *324*
Bye Bye 20EC   *793*
Bygran F   *400*
Bygran S   *726*
CDA Rovral   *616*
CDA Vanquish   *216*
CYREN   *806*
Cadence   *125*
Calibre   *422*
Calidan   *474*
Calirus   *455*

Calixin 765
Camber 135
Camco Insect Powder 946
Campbell's CMPP 297
Campbell's Disulfoton FE 10 824
Campbell's MCPA 50 268
Camppex 110
Capricorn 471
Capture 33
Caramba 655
Carbate Flowable 466
Carbetamex 44
Cardel Egret 216
Cardel Glyphosate 216
Casoron G 138
Casoron G4 138
Casoron G4 Weed Block 1009
Casoron GSR 138
Castaway Plus 850
Cavalier 802
Celest 577
Centra 313
Cercobin Liquid 740
Cereline 459
Cereline Secur 458
Cerone 64
Challenge 215
Challenge 2 215
Challenge 60 215
Chandor 265
Charger 64
Check Turf II 316
Checkmate 388
Cheetah Super 183
Chempak Lawn Sand 1042
Chemtech Cypermethrin 10 EC 811
Chemtech Nuarimol 657
Chess 871
Chieftain 521
Childion 821
Chiltern Blues 924
Chiltern Hardy 924
Chiltern Hundreds 924
Chipko Diuron 80 170
Chipman Diuron 80 170
Chipman Diuron Flowable 170
Chipman Garlon 4 433
Chloropicrin Fumigant 913
Chlortoluron 500 69
Choir 806
Chryzoplus Grey 0.8% 224
Chryzopon Rose 0.1% 224
Chryzosan White 0.6% 224
Chryzotek Beige 224
Chryzotop Green 224
Ciba Chlormequat 460 58

Ciba Chlormequat 5C 460:320 58
Ciba Chlormequat 730 58
Circium II 268
Cirrus 362
Cirsium 268
Citadel 192
Citation 331
Citation 70 331
Clarlon 216
Clarosan 410
Clartex 924
Claymore 354
Clayton Benson 326
Clayton CCC 750 58
Clayton Chizm 466
Clayton Chloron 69
Clayton Cyprocon 511
Clayton Diquat 167
Clayton Fenican 625 164
Clayton Fluroxypyr 199
Clayton Gantry 538
Clayton Gazette 99
Clayton Glyphosate 216
Clayton IPU 239
Clayton Krypton 509
Clayton Lenacil 80W 257
Clayton Manquat 58
Clayton Metazachlor 324
Clayton Metsulfuron 334
Clayton Musket 924
Clayton Pendalin 354
Clayton Pendalin-IPU 37 249
Clayton Pendalin-IPU 47 249
Clayton Pirimicarb 50SG 866
Clayton Propel 378
Clayton Quatrow 167
Clayton Rhizeup 216
Clayton Siptu 50 FL 239
Clayton Siptu 500 239
Clayton Standup 58
Clayton Stobik 452
Clayton Swath 216
Clayton Tebucon 714
Clayton Tranche 194
Clayton Turret 488
Clean-Up-360 216
CleanCrop Aldicarb 10 G 790
CleanCrop Chlorothalonil 720 488
CleanCrop Cyprodinil 521
CleanCrop Cyprothal 492
CleanCrop Diquat 167
CleanCrop Fenpro 559
CleanCrop Fenpropimorph 559
CleanCrop Fonic 64
CleanCrop Fonic M 68
CleanCrop Kresoxazole 538

CleanCrop Kresoxazole Plus   537
CleanCrop MTZ 500   324
CleanCrop Metribuzin   331
CleanCrop Phenmedipham   362
CleanCrop PropaQ   377
CleanCrop Xanilite   509
Cleancrop Chlormequat 700   58
Cleancrop Lenflow   257
Cleancrop scope   791
Cleanrun 2   129
Clearcut II   1067
Clenecorn   297
Clenecorn Super   302
Clinic   216
Clinic Pro TMF   216
Cliophar   86
Clonex   1067
Clortosip   488
Clortosip 500   488
Clovotox   302
Club   856
Co-op Garden Maker Rose & Flower Insecticide
   Spray   1136
CoPilot   386
Cogito   694
Colstar   565
Combinex   671
Commando   191
Comodor 600   399
Compass   618
Compete Forte   193
Competitor   193
Compitox Extra P   302
Compitox Plus   302
Comrade   74
Condox   134
Contact 75   488
Contest   791
Contest Eco   791
Contrast   472
Controller Fungicide   685
Convoy   433
Cooke's Lawn Mosskiller   1042
Cooke's Liquid Sodium Chlorate Weedkiller   1150
Cooke's Professional Sodium Chlorate Weedkiller
   with Fire Depressant   395
Cooke's Sodium Chlorate Weedkiller with Fire
   Depressant   1150
Cooke's Weedclear   395
Cookes 3% Metaldehyde Slug Killer   1099
Cookes 6% Metaldehyde Slug Killer   924
Corbel   559
Corbel CL   493
Cordelia   239
Cordelia 2   239
Corniche   155

Corral   192
Cosmic FL   478
Cougar   164
Country Gardens Lawn Sand   1042
Croneton   828
Cropsafe 5C Chlormequat   58
Croptex Bronze   360
Croptex Chrome   75
Croptex Fungex   502
Croptex Pewter   74
Croptex Steel   397
Crossfire 480   806
Crusader S   31
Cultar   348
Cuprokylt   504
Cuprokylt FL   504
Cuprosana H   504
Curalan CL   501
Curb   894
Curb (Garden Pack)   935
Curzate M68   509
Curzate M68 WSB   509
Cutlass   1023
Cyanamid DD Soil Fumigant   916
Cyclade   61
Cymag   902
Cyperkill 10   811
Cyperkill 25   811
Cyperkill 5   811
Cypertox   811
Cyter   63
Cytro-Lane   855
DAPT Isoproturon 500 FL   239
DAPT Linuron 50 SC   261
DAPT Trifluralin 48 EC   439
DB Straight   117
DFF + IPU WDG   164
DOE   1762 518
DP 353   207
DP 911 WSB   339
DP 928   338
DPX-KE459 WSB   194
DUK 110   422
DUK 118   417
DUK 44   491
DUK 440   447
DUK 51   566
DUK 550   447
DUK 747   581
DUK 7876   565
DUK-880   260
Daconil Turf   488
Dacthal W75   78
Dagger   219
Danadim   823
Danadim Dimethoate 40   823

Dancer *362*
Dapt Atrazine 50 SC *10*
Darmycel Agarifume Smoke Generator *861*
Dash *215*
Dazide *116*
Dazzel *869*
Deadline Place Packs *896*
Debut *447*
Debut WSB *447*
Decade *569*
Decimate *79*
Decis *813*
Decis Micro *813*
Decisquick *814*
Decoy *856*
Decoy Plus *856*
Decoy Wetex *856*
Deep Root *943*
Deep Root Path & Patio Weedkiller *943*
Deeweed *937*
Deftor *330*
Degesch Plates *851*
Deleet *813*
Deloxil *37*
Delphi *496*
Delsene 50 Flo *466*
Delta Insect Powder *946*
Deny WP *623*
Deosan Chlorate Weedkiller *395*
Depitox *104*
Dequiman MZ *623*
Derosal Liquid *466*
Derosal WDG *466*
Derris Dust *1141*
Detia Gas Ex-P *792*
Detia Gas Ex-T *792*
Detia Gas-Ex-B *792*
Deuce *359*
Devcol All Purpose Natural Insecticide
    Spray *1136*
Devcol Liquid Derris *877*
Devcol Morgan 3% Metaldehyde Slug Killer *1099*
Devcol Morgan 6% Metaldehyde Slug Killer *924*
Devcol Path Weedkiller *1150*
Devcol-Sodium Chlorate Weedkiller *1150*
Devrinol *344*
Dextrone X *350*
Dexuron *175*
Di-Farmon R *135*
DiPel WP *907*
Diablo *578*
Dicofen *834*
Dicotox Extra *104*
Dicurane *69*
Dicurane Duo 446 SC *21*
Difenard *899*

Digermin *439*
Digit *729*
Dimilin 25 WP *822*
Dimilin Flo *822*
Dimilin WP *822*
Dioweed 50 *104*
Dipterex 80 *891*
Dispatch *806*
Disulfoton P10 *824*
Disyston FE-10 *824*
Disyston P-10 *824*
Dithane 945 *623*
Dithane Dry Flowable *623*
Dithane Dry Flowable Newtec *623*
Dithane NT Dry Flowable *623*
Dithane Superflo *623*
Dithianon Flowable *530*
Ditrac All-Weather Bait Bar *900*
Ditrac Rat and Mouse Bait *900*
Diuron 50 FL *170*
Diuron 80 WG *170*
Diuron 80 WP *170*
Diuron 80% WP *170*
Diuron 80WP *170*
Divora *518*
Dixie 6 *924*
Do It All Ant Killer *946*
Do It All De-Bug! Insect Killer *1136*
Do It All De-Bug!2 *1037*
Do It All Fruit and Vegetable Insecticide
    Spray *1136*
Do It All Lawn Feed & Weed *991*
Do It All Liquid Weedkiller Concentrate *1059*
Do It All Nettle Gun *1059*
Do It All Plant Disease Control *960*
Do It All Rose and Flower Insecticide Spray *1136*
Do It All Topple Weed! *1059*
Do It All Wipeout! Lawn Weed Killer *991*
Do-Away *216*
Do-It-All Complete Lawn Care *1052*
Do-It-All Granular Lawn Feed and Weed *1084*
Do-It-All Lawn Weedkiller *991*
Do-It-All Liquid Weedkiller Concentrate *1059*
Do-It-All Path and Patio Weedkiller *937*
Do-It-All Slug Killer Pellets *1099*
Do-It-All Woodlice Killer *946*
Dockmaster *135*
Doff 'All in One' Insecticide Spray *1136*
Doff Agricultural Slug Killer with Animal
    Repellent *924*
Doff Ant Control Powder *946*
Doff Ant Killer *1068*
Doff Derris Dust *1141*
Doff Fruit and Vegetable Insecticide Spray *1136*
Doff Gamma BHC Dust *1068*

Doff Granular Lawn Feed, Weed and
   Mosskiller *993*
Doff Greenfly Killer *1136*
Doff Hormone Rooting Powder *958*
Doff Horticultural Slug Killer Blue Mini Pellets *924*
Doff Houseplant Insecticide Spray *1136*
Doff Houseplant Pest Spray *1037*
Doff Knockdown Weedkiller Concentrate *1059*
Doff Knockdown Weedkiller Spray *1059*
Doff Lawn Feed and Weed *991*
Doff Lawn Mosskiller and Fertilizer *1042*
Doff Lawn Spot Weeder *991*
Doff Lawn Weed and Feed Soluble Powder *1021*
Doff Long-Lasting Path Weedkiller *938*
Doff Metaldehyde Slug Killer Mini Pellets *924*
Doff Nettle Gun *991*
Doff New Formula Lawn Weedkiller *991*
Doff New Formula Metaldehyde Slug Killer Mini
   Pellets *924*
Doff New Formula Nettle Gun *1059*
Doff Path Weedkiller *1150*
Doff Path and Patio Weedkiller *1059*
Doff Plant Disease Control *960*
Doff Rose & Flower Pest Spray *1037*
Doff Rose Insecticide Spray *1136*
Doff Rose Weedkiller *1059*
Doff Rose and Flower Insecticide Spray *1136*
Doff Slugoids Slug Killer Blue Mini Pellets *1099*
Doff Sodium Chlorate Weedkiller *395*
Doff Systemic Insecticide *1024*
Doff Total Path Weedkiller *937*
Doff Wasp Nest Killer *946*
Doff Weevil Killer *1068*
Doff Woodlice Killer *946*
Dorado *700*
Dorin *762*
Dormone *104*
Dosaflo *330*
Dovetail *848*
Dow Shield *86*
Doxstar *208*
Draco *708*
Draza *856*
Draza 2 *856*
Draza Plus *856*
Draza ST *926*
Draza Wetex *856*
Dreadnought Flo *496*
Druid *1*
Duckweed Control *966*
Duplosan *302*
Duplosan 500 *302*
Duplosan KV *302*
Duplosan KV 500 *302*
Duplosan New System CMPP *302*
Duramitex *1071*

Dursban 4 *806*
Dursban WG *806*
Dynamec *789*
EF 1077 *328*
EF 1166 *206*
EF-1288 *570*
EM 1617/01 *924*
ESP *924*
EXP 31165 *163*
EXP 31502A *163*
EXP03149D *64*
Eagle *1*
Early Impact *473*
Easel *104*
Eclipse *536*
Economix *216*
Effect *193*
Elliott Easyweeder *986*
Elliott's Lawn Sand *187*
Elliott's Lawn and Fine Turf Feed and Weed plus
   Moss Killer *993*
Elliott's Mosskiller *187*
Elliott's Touchweeder *986*
Elvaron M *751*
Elvaron Multi *751*
Elvaron WG *523*
Elvitox *856*
Embark *316*
Embark Lite *316*
Eminent *729*
Empal *268*
Encore *249*
Encore SC *249*
Endeavour *694*
Endspray II *330*
Enforcer *139*
Ensign *567*
Epic *534*
Epox *856*
Epsilon *170*
Erysto *703*
Escar-Go Z *924*
Escar-go 6 *924*
Escort *95*
Ethosan *176*
Ethrel C *64*
Evade *208*
Everest *385*
Evergreen Autumn *1042*
Evergreen Complete *1052*
Evergreen Easy *1052*
Evergreen Extra *1052*
Evergreen Feed and Weed *1084*
Evergreen Feed and Weed Liquid *1008*
Evergreen Grasshopper *1002*
Evergreen Lawn Sand *1042*

Evergreen Mosskil Extra 1042
Evergreen Mosskill 1042
Evergreen Triple Action 1052
Evict 883
Evidence 815
Exit 856
Exit Wetex 856
Extratect Flowable 611
F238 532
FCC Agricorn 50M 268
FCC Trigard 439
FP 107 924
FPE 120 183
FPE 55 183
Fairy Ring Destroyer 779
Falcon 377
Fargro Chlormequat 58
Farm Fly Spray 10 WP 794
Farmatin 560 574
Fastac 791
Fastac Dry 791
Favour 600 SC 652
Fazor 266
Feature Algae Control 1143
Feature Clear 1133
Fellside Green Fruit and Vegetable Insect
    Spray 1136
Fellside Green Rose and Flower Insect
    Spray 1136
Fernpath Banjo 811
Ferrax 540
Fertosan Slug and Snail Killer 936
Fertosan Slug and Snail Powder 936
Ficam Insect Powder 946
Ficam ULV 797
Field Marshal 133
Fielder 860
Fiesta T 57
Filex 682
Finale 215
Fisons Evergreen Extra 1051
Fisons Greenmaster Autumn 187
Fisons Greenmaster Mosskiller 187
Fisons Helarion 924
Fisons Lawn Sand 1042
Fisons Ready-to-Use Lawn Weedkiller 991
Fisons Water-on Lawn Weedkiller 991
Fito Ready To Use Patio & Garden
    Weedkiller 1059
Flagon 400 EC 28
Flamenco 578
Flamenco Plus 580
Flex 211
Flexidor 252
Flexidor 125 252
Flint 776

Floracid 951
Flute 488
Focal Liquid 466
Focal WDG 466
Focus All-In-One Insecticide Spray 1136
Focus Ant Killer 946
Focus Fruit and Vegetable Insecticide
    Spray 1136
Focus Lawn Spot Weeder 991
Focus Lawn Weedkiller 991
Focus Nettle Gun 991
Focus Rose and Flower Insecticide Spray 1136
Focus Slug Killer Pellets 1099
Focus Sodium Chlorate Weedkiller 1150
Foil 580
Folicur 714
Folio 497
Fongarid 599
For-ester 104
Force ST 883
Fortress 703
Fortrol 99
Foundation 135
Framolene 137
Freeway 170
Fruit and Vegetable Insecticide 1037
Fubol 58 WP 632
Fubol 75 WP 632
Fubol Gold 633
Fubol Gold WG 633
Fumetham 925
Fumispore 665
Fumite Dicloran Smoke 525
Fumite General Purpose Insecticide Smoke
    Cone 1131
Fumite Lindane 10 849
Fumite Lindane 40 849
Fumite Lindane Pellets 849
Fumite Permethrin Smoke 861
Fumite Pirimiphos methyl Smoke 869
Fumite Propoxur Smoke 698
Fumite Whitefly Smoke Cone 1124
Fumitoxin 792
Fumyl-O-Gas 927
Fungaflor 605
Fungaflor Smoke 605
Fungazil 100 SL 605
Fungus Clear 1121
Fungus Clear Gun! 1121
Fury 10 EW 892
Fusarex 10% Granules 726
Fusarex 10G 400
Fusarex 6% Dust 726
Fusilade 250 EW 192
Fusilade 5 192
Fusion 585

Fusonil Turf *488*
Fytospore 68 *509*
GEM 353 *207*
GEM 44 *491*
GEX 353 *207*
GEX 44 *491*
Gainer *714*
Gala *199*
Galben M *454*
Galis *2*
Gallery 125 *252*
Galtak 50 SC *12*
Gambit *551*
Garden Insect Killer *1136*
Garden Insecticide Concentrate *1037*
Garden Weedkiller *1059*
Garden Weedkiller Concentrate *1059*
Garlon 2 *433*
Garlon 4 *433*
Garnet *723*
Garvox 3G *797*
Gaucho *844*
Gaucho FS *844*
Gavel *164*
Gazelle *183*
Geest Fumite MK2 *861*
Geestline Fumite MK2 *861*
Gem 690 *334*
Gem Lawn Sand *1042*
Gem Lawn Weed & Feed *995*
Gem Lawn Weed and Feed + Mosskiller *993*
Gem Sodium Chlorate Weedkiller *395*
Gemini *572*
General Insect Killer *1037*
Genesis *928*
Genesis ST *928*
Genie *581*
Genie 25 *581*
Gesagard *372*
Gesaprim *10*
Gesatop *389*
Get Off Insect *1037*
Get Off My Garden *1103*
Get Off Spray *1103*
Gladiator *52*
Gladiator DF *52*
Gladio *557*
Glint *569*
Gly-480 *216*
Glyfonex 480 *216*
Glyfos *216*
Glyfos 480 *216*
Glyfos Proactive *216*
Glyfosate-360 *216*
Glyper *216*
Glypho *1062*

Glypho Gun! *1062*
Glyphogan *216*
Glyphosate 120B *216*
Glyphosate 360 *216*
Glyphosate 360A *216*
Glyphosate 360B *216*
Glyphosate Biactive *216*
Glytex *254*
Golden Malrin *857*
Goltix 90 *320*
Goltix Flowable *320*
Goltix WG *320*
Gorgon *216*
Gramoxone 100 *350*
Granit *460*
Granit M *461*
Graphic *547*
Grasp *423*
'Grasshopper' Triple Action *1002*
Grasshopper Weed & Feed *1003*
Grazon 90 *97*
Great Mills Ant Killer *946*
Great Mills Complete Insecticide Spray *1136*
Great Mills Fast Action Weedkiller Ready to Use *1059*
Great Mills Fruit and Vegetable Insecticide Spray *1136*
Great Mills Insect Spray for Roses *1037*
Great Mills Lawn Feed & Weed *991*
Great Mills Lawn Feed, Weed and Mosskiller *1052*
Great Mills Lawn Spot Weeder *991*
Great Mills Lawn Weedkiller *991*
Great Mills Rose and Flower Insecticide Spray *1136*
Great Mills Slug Killer Blue Mini Pellets *1099*
Great Mills Sodium Chlorate Weedkiller *1150*
Great Mills Systemic Action Weedkiller Ready to Use *1062*
Great Mills Woodlice Killer *946*
Green Algae Control *974*
Green Sulphur *1151*
Green Up Feed and Weed Plus Mosskiller *992*
Green Up Lawn Feed and Weed *986*
Green Up Liquid Lawn Feed'n Weed *1006*
Green Up Mossfree *1042*
Green Up Spot Lawn Weedkiller *995*
Green Up Weedfree Lawn Weedkiller *986*
Green Up Weedfree Spot Weedkiller for Lawns *986*
Greenco GR1 *1037*
Greenco GR3 Pest Jet *1037*
Greencrop Boomerang *167*
Greencrop Boulevard *82*
Greencrop Carna *58*
Greencrop Cong 750 *58*

Greencrop Coolfin   *58*
Greencrop Galore   *536*
Greencrop Gentian   *511*
Greencrop Glenroe   *866*
Greencrop Gweedore   *423*
Greencrop Gypsy   *216*
Greencrop Monsoon   *567*
Greencrop Mooncoin   *466*
Greencrop Orchid   *488*
Greencrop Pelethon   *823*
Greencrop Pontoon   *806*
Greencrop Reaper   *199*
Greencrop Reaper 2   *199*
Greencrop Rosette   *574*
Greencrop Saffron FL   *378*
Greencrop Tassle   *28*
Greencrop Twinstar   *680*
Greencrop Tycoon   *59*
Greencrop Valentia   *102*
Greenfly & Blackfly Killer   *1037*
Greenmaster Autumn   *187*
Greenmaster Extra   *292*
Greenmaster Mosskiller   *187*
Greenor   *94*
Greenscape Ready To Use Weed Killer   *1062*
Greenscape Ready to Use   *1062*
Greenscape Weedkiller   *1062*
Greenshield   *467*
'Greensward' 2   *1002*
Greensward Triple Action   *1003*
Gremlin   *53*
Grenadier   *164*
Grey Squirrel Liquid Concentrate   *904*
Griffin Ethofumesate 200   *176*
Griffin Phenmedipham 114   *362*
Gro-Tard II   *316*
GroWell 2,4-D Amine   *104*
Groundclear   *1003*
Growing Success Cat Repellent   *1058*
Growing Success Slug Killer   *936*
Guardsman B   *894*
Guardsman M   *894*
Guardsman STP   *894*
HY-CARB   *466*
HY-D   *104*
HY-MCPA   *268*
HY-TL   *739*
HY-VIC   *739*
Hallmark   *847*
Hallmark with Zeon Technology   *847*
Halo   *495*
Halophen RE 49   *524*
Halt   *714*
Hardy Z   *924*
Harlequin 500 SC   *250*
Harmony Express   *51*

Harmony M   *338*
Harvest   *215*
Hawk   *85*
Headland Addstem   *466*
Headland Addstem DF   *466*
Headland Archer   *18*
Headland Cedar   *120*
Headland Dual   *476*
Headland Inorganic Liquid Copper   *504*
Headland Judo   *378*
Headland Maple   *475*
Headland Neptune   *314*
Headland Polo   *112*
Headland Regain   *466*
Headland Relay P   *133*
Headland Relay Turf   *133*
Headland Spear   *268*
Headland Spirit   *637*
Headland Staff   *104*
Headland Sulphur   *709*
Headland Sword   *215*
Headland Venus   *709*
Headland Zebra Flo   *623*
Headland Zebra WP   *623*
Helimax   *924*
Helm 75 WG Newtec   *623*
Helm 75WG   *623*
Helmet   *521*
Helmsman   *45*
Helocarb Granule 500   *866*
Helosate   *216*
Herbasan   *362*
Herboxone   *104*
Herboxone 60   *104*
Hero   *847*
Herrisol New   *133*
Hi-Shot   *754*
Hickson Phenmedipham   *362*
Hickstor 10   *400*
Hickstor 3   *400*
Hickstor 5   *400*
Hickstor 6   *400*
Hickstor 6 Plus MBC   *43*
High Strength Tough Weed Gun!   *1062*
Hilite   *216*
Hispor 45 WP   *481*
Hoegrass   *154*
Holdfast D   *136*
Home Base Pest Gun 2   *1037*
Homebase All-In-One Insecticide   *1124*
Homebase Ant Killer   *946*
Homebase Autumn Lawn Feed and Moss
     Killer   *1042*
Homebase Contact Action Weedkiller   *1059*
Homebase Contact Action Weedkiller Ready to
     Use Sprayer   *1059*

Homebase Hormone Rooting Powder *958*
Homebase Houseplant Insecticide *1136*
Homebase Houseplant Insecticide Spray 2 *1037*
Homebase Lawn Feed and Moss killer *1042*
Homebase Lawn Feed and Mosskiller *1042*
Homebase Lawn Feed and Weed *1084*
Homebase Lawn Feed and Weed Liquid *1008*
Homebase Lawn Feed and Weed Soluble *994*
Homebase Lawn Feed, Weed and
    Mosskiller *1052*
Homebase Lawn Sand *1042*
Homebase Lawn Weedkiller Liquid *991*
Homebase Lawn Weedkiller Ready To Use
    Sprayer *991*
Homebase Liquid Lawn Feed and Weed *991*
Homebase Nettle Gun *991*
Homebase Nettle Spray *1059*
Homebase Path & Drive Weed Killer *942*
Homebase Pest Gun *1136*
Homebase Rooting Hormone Liquid *1012*
Homebase Rooting Hormone Powder *958*
Homebase Rooting Hormone Powder 2 *1109*
Homebase Rose Care *1127*
Homebase Slug Killer Blue Mini Pellets *1099*
Homebase Sodium Chlorate Weedkiller *1150*
Homebase Systemic Action Weedkiller *1062*
Homebase Systemic Action Weedkiller Ready to
    Use *1062*
Homebase Wasp Nest Killer *946*
Homebase Weed Pen *1007*
Homebase Woodlice Killer *946*
Homer *320*
Hortag Tecnacarb *43*
Hortag Tecnazene 10% Granules *726*
Hortag Tecnazene Double Dust *726*
Hortag Tecnazene Potato Dust *726*
Hortag Tecnazene Potato Granules *726*
Hortichem 2-Aminobutane *450*
Hostaquick *842*
Hotspur *93*
House Plant Insecticide *1037*
House Plant Pest Killer *1138*
Houseplant Insect Killer *1136*
Houseplant Pest Killer *1037*
Huron *856*
Hy-Flo *742*
Hyban *134*
Hyban P *135*
Hycamba Plus *133*
Hydon *27*
Hydroperfect Algalit HP-352 *974*
Hydroperfect Algalit HP-355 *974*
Hydroperfect Algalit HP-357 *974*
Hydroperfect Lemnalit HP-372 *974*
Hydroperfect Lemnalit HP-375 *974*
Hydroperfect Lemnalit HP-377 *974*

Hygrass *134*
Hygrass P *135*
Hykeep *731*
Hymec *297*
Hymec Trio *153*
Hymec Triple *153*
Hymush *731*
Hyprone *132*
Hyprone P *133*
Hyquat 70 *58*
Hystore 10 *400*
Hystore 10G *726*
Hysward *132*
Hysward P *133*
Hytane 500 SC *239*
Hytec *400*
Hytec 6 *400*
Hytec Super *402*
Hytrol *939*
Hytrol Total *941*
Hyvar X *24*
I T Clopyralid *86*
I.P.U Sun *239*
I.T. Cyanazine *99*
IPSO *248*
IPU 500 *239*
IPU 500 WDG *239*
ISK 375 *488*
ISO 500 *239*
IT Alpha-Cypermethrin 100 EC *791*
IT Alpha-cypermethrin *791*
IT Asulam *9*
IT Bentazone *16*
IT Bentazone 48 *16*
IT Glyphosate *216*
IT Iprodione *616*
Ice *60*
Impact *586*
Impact Excel *495*
Impuls *20*
Indar 5EC *544*
Indar 5EW *544*
Indar CG *547*
Ingot *163*
Insegar *836*
Insegar WG *836*
Inter-Metribuzin WG *331*
Inter-Pendimethalin *354*
Intercept 5GR *844*
Intercept 70 WG *844*
Interfarm Base *378*
Interfarm Quaver *378*
Intracrop Balance *58*
Intracrop MCCC *58*
Intrepid *129*
Invader *528*

Invader WDG   528
Ipifluor   439
Ipimethalin   354
Isoguard   239
Isoguard 83WG   239
Isoguard SVR   239
Isomec   302
Isoproturon 500   239
Isoproturon 553   239
Isotop SC   239
J Arthur Bower's Granular Feed and Weed   1021
J Arthur Bower's Granular Feed, Weed and
   Mosskiller   1020
J Arthur Bower's Lawn Feed and Weed   1021
J Arthur Bower's Lawn Feed, Weed and
   Mosskiller   1020
J Arthur Bower's Lawn Sand   1042
J Arthur Bower's Lawn Weed and
   Mosskiller.   1020
J Arthur Bower's Lawn Weedkiller   1021
J Arthur Bowers Lawn Mosskiller   1042
Jasper   762
Javelin   164
Javelin Gold   164
Jester   41
Jeyes Fluid   1157
Jockey   580
Jockey F   578
Jockey Flexi   578
Jockey Plus   198
Jolt   249
Jolt SC   249
Joust   695
Jubilee   334
Jubilee 20DF   334
Judge   928
Juggler   729
Jupital   488
Jupital DG   488
Justice   566
K & S Chlorofume   913
KE 459 DF   194
Karamate Dry Flo   623
Karamate Dry Flo Newtec   623
Karamate N   623
Karan   856
Karate Ready to Use Rat and Mouse Bait   897
Karathane Liquid   529
Karmex   170
Katamaran   325
Keetak   559
Kelthane   820
Kemifam E   362
Kemifam Flow   362
Kerb 50 W   378
Kerb 80 EDF   378

Kerb Flo   378
Kerb Granules   378
Kerb Pro Flo   378
Kerb Pro Granules   378
Keri Insect Spray   1136
Keytin   574
Killgerm Rat Rod   899
Killgerm Wax Bait   899
King British Formula KB7   976
Knock Out   1062
Knock Out Weedkiller Ready To Use   1062
Knot Out   252
Kombat WDG   475
Konker   484
Kor DF   623
Kor DF Newtec   623
Kor Flo   623
Kor NT Dry Flowable   623
Krenite   213
Kruga 5 EC   544
Kruga 5EC   544
Kubist   176
Kubist 2   176
Kumulus DF   709
Kumulus S   709
Landgold Aldicarb 10G   790
Landgold Amidosulfuron   1
Landgold CC 720   58
Landgold Carbosulfan 10 G   804
Landgold Clodinafop   82
Landgold Cycloxydim   102
Landgold Cyproconazole 100   511
Landgold DFF 625   164
Landgold Deltaland   813
Landgold Difenoconazole   526
Landgold Diquat   167
Landgold Epoxiconazole   534
Landgold Epoxiconazole FM   536
Landgold Ethofumesate 200   176
Landgold Fenpropidin 750   552
Landgold Fluazinam   576
Landgold Flusilazole MBC   472
Landgold Glyphosate 360   216
Landgold Iprodione 250   616
Landgold Isoproturon   239
Landgold Lambda-C   847
Landgold Mancozeb 80 W   623
Landgold Mecoprop-P   302
Landgold Mecoprop-P 600   302
Landgold Metamitron   320
Landgold Metazachlor 50   324
Landgold Metsulfuron   334
Landgold PQF 100   377
Landgold Pendimethalin 400   354
Landgold Pirimicarb 50   866
Landgold Propiconazole   685

Landgold Propyzamide 50 *378*
Landgold Rimsulfuron *387*
Landgold Strobilurin 250 *452*
Landgold Strobilurin KE *538*
Landgold Strobilurin KF *567*
Landgold TFS 50 *447*
Landgold Thifensulfuron *417*
Landgold Tralkoxydim *423*
Landgold Triadimefon 25 *753*
Landgold Triallate 480 *424*
Landgold Vinclozolin DG *783*
Landgold Vinclozolin SC *783*
Landmark *538*
Lanslide *259*
Larke *58*
Laser *102*
Lattice *549*
Lawn Builder Plus Moss Control *1042*
Lawn Builder Plus Weed Control *107*
Lawn Feed and Weed Granules *994*
Lawn Sand *1042*
Lawn Spot Weed Granules *994*
Lawn Weed Gun *986*
Lawn Weedkiller *991*
Lawn Weedkiller Concentrate *991*
Lawnsman Liquid Weed and Feed *1003*
Leap *663*
Legacy *576*
Legion S *477*
Legumex Extra *15*
Lektan *320*
Lemnalit *974*
Lentagran EC *382*
Lentagran WP *382*
Lentipur CL 500 *69*
Levi *167*
Levington Autumn Extra *1042*
Levington Gold *1052*
Levington Insect Spray for Houseplants *1124*
Levington Lawn Sand *1042*
Levington Octave *672*
Levington Plant Protection *1065*
Levington Professional Plus Intercept *844*
Levington Ready to Use Lawn Weedkiller *991*
Levington Water-on Lawn Weedkiller *991*
Lexone 70 DF *331*
Lexus 50 DF *194*
Lexus Class *47*
Lexus Class WSB *47*
Lexus Millenium *197*
Lexus Millenium WSB *197*
Lexus XPE *196*
Lexus XPE WSB *196*
Leyclene *34*
Lindane Flowable *849*
Linnet *265*

Linurex 50SC *261*
Linuron Flowable *261*
Liquid Club Root Control *1164*
Liquid Curb Crop Spray *894*
Liquid Derris *877*
Liquid Linuron *261*
Lo-Gran *425*
Loft *3*
Longer Lasting Bug Gun *951*
Longlife Clearun 2 *129*
Longlife Renovator 2 *128*
Lonpar *90*
Lontrel 100 *86*
Lontrel Plus *91*
Lorate 20DF *334*
Lorsban 480 *806*
Lorsban T *806*
Lorsban WG *806*
Lotus *80*
Lotus Algicide *947*
Lotus Fungicide *947*
Lucifer *83*
Lupus *856*
Luxan 2,4-D *104*
Luxan 9363 *924*
Luxan Chloridazon *52*
Luxan Chlorotoluron 500 Flowable *69*
Luxan Cypermethrin 10 *811*
Luxan Dichlobenil Granules *138*
Luxan Dichlorvos 600 *819*
Luxan Dichlorvos Aerosol 15 *819*
Luxan Gro-Stop 300 EC *74*
Luxan Gro-Stop Basis *74*
Luxan Gro-Stop Fog *74*
Luxan Gro-Stop HN *74*
Luxan Isoproturon 500 Flowable *239*
Luxan MCPA 500 *268*
Luxan Mancozeb Flowable *623*
Luxan Maneb 80 *637*
Luxan Metaldehyde *924*
Luxan Micro-Sulphur *709*
Luxan Phenmedipham *362*
Luxan Talunex *895*
Lynx *924*
Lyric *581*
MC Flowable *476*
MCC 25 EC *103*
MCPA 500 *268*
MM70 *320*
MON 10EC *729*
MON 44068 Garden Weedkiller *1062*
MON 44068 Pro *216*
MON 52276 *216*
MON 77020 Garden Weedkiller *1062*
MON 77021 Garden Weedkiller *1062*
MON 77158A *216*

MSS 2,4-D Amine *104*
MSS 2,4-D Ester *104*
MSS 2,4-DB + MCPA *120*
MSS 2,4-DP *141*
MSS 2,4-DP + MCPA *150*
MSS Aminotriazole Technical *3*
MSS Atrazine 50 FL *10*
MSS Atrazine 80 WP *10*
MSS Betaren Flow *362*
MSS CIPC 30M *74*
MSS CIPC 40 EC *74*
MSS CIPC 50 LF *74*
MSS CIPC 50M *74*
MSS CIPC 5G *74*
MSS Chlormequat 40 *58*
MSS Chlormequat 460 *58*
MSS Chlormequat 60 *58*
MSS Chlormequat 70 *58*
MSS Chlorothalonil *488*
MSS Chlortoluron 500 *69*
MSS Diuron 500 FL *170*
MSS Diuron 50FL *170*
MSS Diuron 80 WP *170*
MSS Flotin 480 *574*
MSS Glyfield *216*
MSS Iprofile *239*
MSS Linuron 500 *261*
MSS MCPA 50 *268*
MSS MCPB + MCPA *291*
MSS Mircam *135*
MSS Mircam Plus *133*
MSS Mircarb *466*
MSS Mircell *58*
MSS Mirprop *302*
MSS Mirquat *58*
MSS Optica *302*
MSS Primasan *176*
MSS Protrum G *362*
MSS Simazine 50 FL *389*
MSS Sulphur 80 *709*
MSS Sulphur 80 WP *709*
MSS Thor *176*
MSS Trifluralin 48 EC *439*
MTM CIPC 40 *74*
MTM Grassland Herbicide *133*
MTM Trifluralin *439*
Magnum *53*
Mainstay *488*
Malathion 60 *852*
Mallard *552*
Manconex *623*
Mancozeb 80 *623*
Mandate 75 WDG *623*
Mandate 80 WP *623*
Mandolin *362*
Mandops Barleyquat B *58*

Mandops Bettaquat B *58*
Mandops Chlormequat 460 *58*
Mandops Chlormequat 700 *58*
Maneb 80 *637*
Manex *637*
Manex II *623*
Manipulator *58*
Mantis *685*
Mantle *569*
Mantra *537*
Manzate 200 DF *623*
Manzate 200 PI *623*
Marathon *82*
Maraud *806*
Marks 2,4-D-A *104*
Marks 2,4-DB *117*
Marks 2,4-DB Extra *120*
Marks MCPA 50A *268*
Marks MCPA P30 *268*
Marks MCPA S25 *268*
Marks MCPA SP *268*
Marks MCPB *293*
Marks Polytox-K *141*
Marks Polytox-M *109*
Marnoch Chlorothalonil *488*
Marnoch Clodim *102*
Marnoch Glyphosate *216*
Marnoch Kempo *538*
Marnoch Lamoc *847*
Marnoch Mancym *509*
Marnoch Metazachlor *324*
Marnoch Penor D *669*
Marnoch Phonet *68*
Marnoch Vinol *783*
Marquise *320*
Marshal 10G *804*
Marshal Soil Insecticide suSCon CR
    granules *804*
Marshall 10G *804*
Masai *881*
Masai G *881*
Mascot Mosskiller *139*
Mascot Selective-P *114*
Mascot Super Selective-P *133*
Master Sward *15*
Masterspray *40*
Matador 200 SC *832*
Match *527*
Matrikerb *96*
Mavrik *880*
Mavrik Aquaflow *880*
Maxicrop Mosskiller and Conditioner *187*
Maxicrop Mosskiller and Lawn Tonic *1042*
Mazide 25 *266*
Mazide Selective *130*
Me^2 Aldee *790*

Me² Aducksbackside *387*
Me² Azoxystrobin *452*
Me² Cymoxeb *509*
Me² Exodus *928*
Me² KME *538*
Me² Lambda *847*
Me² Tebuconazole *714*
Mebrom 100 *927*
Mebrom 98 *927*
Meld *585*
Menace 80 EDF *378*
Menara *517*
Meothrin *837*
Merit *359*
Merlin *498*
Metam *510 925*
Metarex *924*
Metarex Green *924*
Metarex RG *924*
Metarex RGS *924*
Meteor *62*
Metham Sodium 400 *925*
Methyl Bromide *927*
Methyl Bromide 100% *927*
Methyl Bromide 98% *927*
Mextrol-Biox *37*
Micene 80 *623*
Micene DF *623*
Microsul Flowable Sulphur *709*
Microthiol Special *709*
Midstream *167*
Mifaslug *924*
Mildothane Liquid *740*
Mildothane Turf Liquid *740*
Minuet EW *892*
Miracle-Gro Bug Spray *1123*
Miracle-Gro Houseplant Bug Spray *1136*
Miracle-Gro Lawn Food with Weed and Moss
    Control *1002*
Miracle-Gro Weed & Feed *1003*
Mirage 40 EC *672*
Mirage Extra *546*
Mircam *135*
Mircam Plus *133*
Miros DF *488*
Mission *385*
Mister Green Algalit MG-352 *974*
Mister Green Algalit MG-355 *974*
Mister Green Algalit MG-357 *974*
Mister Green Lemnalit MG-372 *974*
Mister Green Lemnalit MG-375 *974*
Mister Green Lemnalit MG-377 *974*
Mistral *559*
Mitac 20 *793*
Mitac HF *793*
Mithras 80 EDF *378*

Mitron WG *320*
Mocap 10G *829*
Moddus *448*
Mogul *216*
Molluzin *976*
Molotov *924*
Mon 240 *216*
Mon 44068 Garden Weedkiller *1062*
Monceren DS *669*
Monceren FS *669*
Monceren IM *610*
Monceren IM Flowable *610*
Monicle *558*
Monty *216*
Moot *519*
Morestan *702*
Morgan's Sodium Chlorate Weedkiller *395*
Mortar Flo *623*
Moss Control Plus Lawn Fertilizer *187*
Moss Gun! *1011*
Mossicide *139*
Mosskil Extra *1042*
Mosskiller and Lawn Tonic *1042*
Mosskiller for Lawns *1042*
Multi-W FL *476*
Multirose ready-to-use *950*
Murphy Bugmaster *1136*
Murphy Clover-Kil *991*
Murphy Derris Dust *1141*
Murphy Dilute Slugit Liquid *1099*
Murphy Hormone Rooting Powder *958*
Murphy Lawn Feed and Mosskiller *1042*
Murphy Lawn Feed and Weed *1084*
Murphy Liquid Malathion *1071*
Murphy Malathion Dust *1071*
Murphy Mole Smokes *1152*
Murphy Mortegg *1158*
Murphy Nettlemaster *1166*
Murphy Path Weedkiller *937*
Murphy Roseguard *983*
Murphy Slugit Liquid *1099*
Murphy Slugits *1099*
Murphy Systemic Action Fungicide *960*
Murphy Systemic Action Insecticide *1064*
Murphy Traditional Copper Fungicide *972*
Murphy Tumbleblite *1134*
Murphy Tumbleblite II *1121*
Murphy Tumbleblite II Ready to Use *1121*
Murphy Tumblebug *1064*
Murphy Tumbleweed Gel *1062*
Murphy Ultra Lawn Feed, Weed and
    Mosskiller *1052*
Muster *216*
Muster LA *216*
Mycoguard *488*
Mycotal *909*

Mysen *239*
Mysen WDG *239*
NNI-850 5SC *838*
NOV 1390 *518*
NOV 219 *521*
NWA CTU 500 *69*
NX 3083 *174*
NX 8049 *216*
Nature's Answer Derris Dust *1141*
Nature's Answer Fungicide and Insect Killer *1040*
Nature's Answer Organic Insecticide Ready to
    Use *1037*
Nature's Answer to Insect Pests *1136*
Nature's Answer to Insect Pests
    Concentrate *1136*
Natures 'Bug Gun!' *1136*
Natures 'Bug Gun!' *1136*
Natures Bug Gun *1136*
Nebulin *726*
Nectec Paste *451*
Nemathorin 10G *839*
Neminfest *265*
Nemispor *623*
Nemolt *882*
Neon *707*
Neporex 2 SG *812*
New 5C Cycocel *58*
New Arena Plus *43*
New Arena TBZ 6 *402*
New Chlorophos *961*
New Estermone *107*
New Formulation SBK Brushwood Killer *990*
New Garlon RTU *1166*
New Hickstor 6 *400*
New Hickstor 6 Plus MBC *43*
New Hystore *726*
New Improved Leaf Action Roundup Brushkiller
    RTU *1062*
New Improved Leaf Action Roundup Weedkiller
    RTU *1062*
New Strike *958*
New Supertox *1004*
Nex *803*
Nico Soap *859*
Nicotine 40% Shreds *859*
Nimrod *462*
Nimrod T *463*
Nippon Ant Killer Liquid *953*
No Scald DPA 31 *918*
No-FID *859*
Nomad *59*
Nomix 2,4-D Herbicide *104*
Nomix G *216*
Nomix Nova *216*
Nomix Revenge *216*
Nomix Supernova *216*

Nomix Touchweed *215*
Nomix Weedkiller *1062*
Nomix-Chipman Mosskiller *139*
Nortron *176*
Nortron Flo *176*
Novagib *214*
Novak *480*
Novosol FC *907a*
Novosol Flowable Concentrate *907*
Nufarm MCPA DMA 480 *268*
Nufarm MCPA DMA 500 *268*
Nufarm Nu-Shot *108*
Nufarm Triflur *439*
Nufosate *216*
Nuvan 500 EC *819*
Octave *672*
Octolan *492*
Olympus *452*
Omen *729*
Oncol 10G *798*
Opie 80 WP *623*
Opogard *409*
Optica *302*
Optica 50 *302*
Optica Combi *292*
Optica DP *151*
Optica Forte *135*
Optica Plus *313*
Optica Trio *153*
Optimol *924*
Option *585*
Opus *534*
Opus Plus *539*
Opus Team *536*
Organic Pest Control *1037*
Organic Weed Control *1036*
Orient *249*
Orient SC *249*
Orka *570*
Ortiva *452*
Osprey 58 WP *632*
Ovasyn *793*
Oxytril CM *37*
P-Weed *373*
PA Chlormequat 400 *58*
PA Chlormequat 460 *58*
PA Dimethoate 40 *823*
PBI Anti-Ant Duster *1136*
PBI Cheshunt Compound *973*
PBI Dithane 945 *1073*
PBI Pepper Dust *1122*
PBI Racumin Mouse Bait *982*
PBI Slug Mini Pellets *1099*
PBI Supercarb *960*
PDQ *168*
PG Suspension *908*

PMG 216
PP 007 192
PP 450 586
PP Captan 80 WG 464
PP Captan 83 464
PPI Clarity Excel 981
PPI Clarity Plus 1143
Pacer 473
Palette 473
Palormone D 104
Panacide M 139
Panacide Technical Solution 139
Pancil T 658
Panoctine 601
Panoctine Plus 602
Panther 164
Pastor 94a
Pasturol D 133
Pasturol Plus 133
Pasturol-P 133
Patafol 634
Path and Drive Weedkiller 942
Path and Shrub Guard 1009
Pathclear 940
Pathclear Gun! 1035
Pathclear S 169
'Pathclear' S 1029
Patio Weedkiller 1059
Patriot EC 815
Patrol 552
pbi Slug Pellets 924
pbi Turf Fungicide 466
pbi Turf Systemic Fungicide 466
Pearl Micro 813
Penncozeb 623
Penncozeb WDG 623
Pepper Dust 1122
Percept 37
Permasect 10 EC 861
Permasect 25 EC 861
Permasect C 811
Permethrin 12 ED 861
Permit 861
Pesguard House and Plant Spray 1128
Pesta 924
Phantom 866
Phorate 10G 864
Phosfleur 1.5 73
Phosfleur 10% Liquid 73
Phostek 792
Phostoxin 895
Phostoxin I 792
Phostrogen House Plant Insecticide 1037
Phostrogen Safer's All Purpose Insecticide Ready
  to Use 1039

Phostrogen Safer's Garden Insecticide
  Concentrate 1037
Phostrogen Safer's RTU Rose & Flower
  Insecticide 1037
Phostrogen Safer's Ready-To-Use Fruit &
  Vegetable Insecticide 1037
Phostrogen Safer's Ready-To-Use House Plant
  Insecticide 1037
Phostrogen Safer's Ready-to-Use Fruit &
  Vegetable Insecticide 1037
Phostrogen Safer's ready-To-Use House Plant
  Insecticide 1037
Phostrogen Soluble Mosskiller and Lawn
  Tonic 1042
Phytek Ready To Use Patio and Garden
  Weedkiller 1059
Picket 1124
Pilot 385
Pilot D 386
Pinion 185
Pirimate 866
Pirimor 866
Plant Protection Compost 1065
Plantvax 20 664
Plantvax 75 664
Platform 46
Platform S 49
Plenum 871
Plinth 354
Plover 526
Podquat 58
Pointer 586
Poise 216
Polycote Select 650
Polycote Universal 645
Polymone X 109
Polyram DF 787
Polysect Insecticide 948
Polysect Insecticide Ready To Use 948
Pond Doctor Algicide 1143
Pond Pride Green Algae Control 974
Portland Brand Ant Killer 946
Portland Brand Slug Killer Blue Mini Pellets 1099
Portman Brasson 374
Portman Chlormequat 700 58
Portman Chlortoluron 69
Portman Glider 216
Portman Glyphosate 216
Portman Glyphosate 360 216
Portman Glyphosate 480 216
Portman Isotop 239
Portman Propachlor 50 FL 374
Portman Propachlor SC 374
Portman Supaquat 58
Portman Trifluralin 439
Portman Weedmaster 52

Posse 10 G   *804*
Prebane   *410*
Precis   *378*
Precis Flo   *378*
Prefix D   *138*
Prelude 20 LF   *672*
Prelude Universal LS   *486*
Premier Amenity Selective   *133*
Premier Ant Killer   *946*
Premier Autumn Lawn Feed with Mosskiller   *1042*
Premier Multi-Pest Insecticide Spray   *1136*
Premier Sodium Chlorate   *1150*
Premier Triple Selective   *132*
Premier Triple Selective P   *133*
Premiere Granules   *256*
Premis   *603*
Previcur N   *682*
Primus 25 SC   *121*
Priori   *452*
Proctors Lawn Feed Weed and Mosskiller   *993*
Profalon   *76*
Profile   *516*
Prophet   *556*
Proplant   *682*
Prospect   *417*
Protector   *476*
Protugan   *239*
Provado Pest Free   *1065*
Provado Ultimate Bug Killer   *1066*
Provax   *486*
Pulsar   *19*
Puma X   *185*
Punch C   *472*
Pursuit   *1*
Pursuit 50   *1*
Py Powder   *1136*
Py Spray Garden Insect Killer   *1136*
Py Spray Insect Killer   *1136*
Pynosect 30 Water Miscible   *873*
Pyramin DF   *52*
Pyramin FL   *52*
Pyrinex 48EC   *806*
Python   *331*
Quadrangle Chlormequat 700   *58*
Quadrangle Cyper 10   *811*
Quadrangle Hinge   *466*
Quadrangle MCPA 50   *268*
Quadrangle Quadban   *133*
Quantum   *429*
Quantum 75 DF   *429*
Quartz   *320*
Quaver Flo   *378*
Quell Flo   *623*
Questar   *52*
Quintil 500   *239*
Quintozene WP   *706*

Quintozene Wettable Powder   *706*
Quitt SL   *17*
RH Maneb 80   *637*
RP 4169   *22*
RP 500   *239*
RP 500 WDG   *239*
Racumin Master Mix   *898*
Radar   *685*
Radius   *514*
Radspor FL   *533*
Ragox   *104*
Rampart   *803*
Ramrod 20 Granular   *374*
Ramrod Flowable   *374*
Rapid Aerosol   *1129*
Rapid Greenfly Killer   *1129*
Rapier   *378*
Rat Eyre Rat Bait   *897*
Rat Eyre Rat and Mouse Bait   *897*
Raven   *315*
Ravine   *601*
Ravine Plus   *602*
Raxil   *714*
Raxil S   *724*
Raxil S CF   *724*
Raxil Secur   *615*
Recoil   *635*
Redeem Flo   *378*
Redipon   *141*
Redipon Extra   *150*
Redlegor   *120*
Reflex T   *212*
Regel   *924*
Reglone   *167*
Regulex   *214*
Regulox K   *266*
Reldan 22   *808*
Reldan 50   *808*
Reliance   *216*
Remtal SC   *394*
Renardine 72-2   *952*
Renardine 72/2   *952*
Renovator 2   *128*
Renown   *58*
Rentokil Alphachloralose   *893*
Rentokil Desyst   *844*
Rentokil Difenard   *899*
Rentokil Garden Insect Killer   *1136*
Rentokil Methyl Bromide   *927*
Rentokil Phostoxin   *895*
Reply   *99*
Repulse   *488*
Rescind   *170*
Rescur   *926*
Reserve   *85*
Restraint DF   *623*

Restraint DF Newtec *623*
Reward 5EC *544*
Rezist *894*
Rhizopon A Powder *223*
Rhizopon A Tablets *223*
Rhizopon AA Powder (0.5%) *224*
Rhizopon AA Powder (1 %) *224*
Rhizopon AA Powder (2%) *224*
Rhizopon AA Powder 2% *224*
Rhizopon AA Tablets *224*
Rhizopon B Powder (0.1%) *341*
Rhizopon B Powder (0.2 %) *341*
Rhizopon B Powder (0.2%) *341*
Rhizopon B Tablets *341*
Rhythm *509*
Ridomil MBC 60 WP 479
Ridomil MZ 75 WP *632*
Ridomil Plus *506*
Ridomil Plus 50 WP *506*
Rimidin *543*
Ringmaster *664*
Ripcord *811*
Ripost Pepite *510*
Ritefeed 2,4-D Amine *104*
Ritefeed Dichlorophen *139*
Rival *216*
Rivet *856*
Rizolex *750*
Rizolex 50 WP *750*
Rizolex 50 WP in Soluble Sachets *750*
Rizolex Flowable *750*
Robinson's Thiram 60 *742*
Robust *612*
Rogor L40 *823*
Rombus *696*
Ronilan DF *783*
Ronilan FL *783*
Ronstar 2G *345*
Ronstar Liquid *345*
Root-Out *943*
Rooting Powder *958*
Rose & Flower Insect Killer *1037*
Rose & Flower Insecticide *1037*
Rose and Flower Insecticide *1037*
'Roseclear' 2 *954*
Roseclear Gun ! *949*
Roundup *216*
Roundup 2000 *216*
Roundup A *216*
Roundup Accuply *216*
Roundup Amenity *216*
Roundup Biactive *216*
Roundup Biactive Dry *216*
Roundup Brushkiller *1062*
Roundup CF *216*
Roundup Estate *1062*

Roundup Four 80 *216*
Roundup GC *1062*
Roundup GT *216*
Roundup MX *216*
Roundup Pro *216*
Roundup Pro Biactive *216*
Roundup Rapide *216*
Roundup Ready-to-Use Faster Acting
    Formula *1062*
Roundup Tough Weedkiller *1062*
Roundup Tough Weedkiller Ready-To-Use *1062*
Roundup Ultra 3000 *1062*
Roundup Ultra ST *216*
Rover DF *488*
Rovral Flo *616*
Rovral Green *616*
Rovral Liquid FS *616*
Rovral WP *616*
Royal MH 180 *266*
Royal MH180 *266*
Rubenal Flow *362*
Rubigan *543*
Ruby *723*
SAN 619F 240 EC *511*
SAN 703 *492*
SAN 710 *516*
SAN 710 HF *516*
SAN 735 *519*
SBK Path & Patio Weedkiller *1059*
SHL Granular Feed & Weed *150*
SHL Granular Feed, Weed & Mosskiller *149*
SHL Lawn Sand *187*
SHL Lawn Sand Plus *140*
SHL Turf Feed & Weed *150*
SHL Turf Feed and Weed and Mosskiller *149*
SL 471 200 EC *700*
SL 520 *84*
SL 552A *555*
SL 567A *653*
SL 600 *380*
SL-556 500 EC *558*
SL571A *568*
Sabre *239*
Sabre 500 *239*
Safer's Indoor Trounce Insecticide
    Concentrate *1039*
Safer's Indoor Trounce Insecticide Ready to
    Use *1039*
Safer's Insecticidal Soap *831*
Safer's Insecticidal Soap Ready-to-Use *1037*
Safer's Natural Garden Fungicide *1151*
Safer's Trounce Insecticide Concentrate *1039*
Safer's Trounce Insecticide Ready to Use *1039*
Safran *575*
Sage *708*
Salute *176*

PSD
PRODUCT INDEX

Salvo 576
Sambarin 312.5 SC 499
Samurai 216
San 735 519
San 845H 125
Sanaphen M 268
Sanction 581
Sanction 25 581
Sanuron 170
Saprol 779
Saracen 528
Saracen WDG 528
Satellite 60
Satisfar 830
Satisfar Dust 830
Savona 831
Savona Rose Spray 1037
Scala 701
Scarab 5EW 544
Scent Off Buds 1108
Scent Off Pellets 1108
Scent off Gel 1108
Sceptre 386
Scoot 935
Scorpion 216
Scotts Octave 672
Sculptor 52
Scuttle 903
Scythe LC 350
Seconal 901
Secto Ant and Crawling Insect Powder 946
Secto Keep Off 964
Secto Nature Care Garden Insect Powder 1136
Secto Nature Care Garden Insect Spray 1136
Secto Nature Care Houseplant Insect Spray 1136
Secto Pepper Dust 1122
Secto Wasp Killer Powder 946
Sector 823
Seedox SC 797
Semeron 124
Senate 414
Sencorex WG 331
Sentinel 2 374
Septico Slug Killer 936
Sequel 838
Sera Algopur 974
Sera Snailpur 970
Seradix 1 224
Seradix 2 224
Seradix 3 224
Seritox 50 150
Seritox Turf 150
Setter 33 15
Sheen 556
Shield DPA 15% 918
Shire 164

Shirlan 576
Shirlan Programme 576
Shogun 377
Shortcut 930
sHYlin 739
Sibutol 457
Sibutol CF 457
Sibutol LS 457
Sibutol New Formula 457
Sibutol Secur 910
Sickle 35
Sierraron G 138
Sigma PCT 58
Silvacur 723
Silvacur 300 723
Silvapron D 104
Simba 20 DF 334
Sinbar 404
Sipcam Echo 75 488
Sipcam Flo 496
Sipcam Simazine Flowable 389
Sipcam UK Rover 488
Sipcam UK Rover 500 488
Sirocco 23
Sistan 925
Sistan 38 925
Sistan 51 925
Skater 320
Skirmish 255
Skua 521
Slug Clear 1099
Slug Clear Mini Pellets 1099
Slug Gard 1101
Slug Pellets 924
Slug Xtra 1099
Slugit Xtra 1099
Smart 360 216
Snail Away 980
Snail Control 976
Snail Smasher 976
Snailaway 978
Snooker 618
Sobrom BM 100 927
Sobrom BM 98 927
Sodium Chlorate 395
Solfa 709
Sorexa CD Mouse Killer 962
Source II 266
Sovereign 354
Spannit 806
Spannit Granules 806
Sparkle 45 WP 481
Spasor 216
Spasor Biactive 216
Spearhead 92
Spectron 53

Speedway *350*
Speedway 2 *168*
Speedway Liquid *350*
Sphere *520*
Sphinx *605*
Spin-Out *98*
Spinnaker *754*
Sponsor *555*
Sporgon 50 WP *672*
Sportak 45 EW *672*
Sportak 45 HF *672*
Sportak Alpha *480*
Sportak Alpha HF *480*
Sportak Delta 460 *516*
Sportak Delta 460 HF *516*
Sportak Focus EW *672*
Sportak Sierra EW *672*
Sportak Sierra HF *672*
Spotless *960*
Sprint *568*
Sprint HF *568*
Squire *1*
Stabilan 460 *58*
Stabilan 5C *58*
Stabilan 640 *58*
Stabilan 670 *58*
Stabilan 700 *58*
Stabilan 750 *58*
Stacato *216*
Stampede *216*
Stampout *847*
Standon Aldicarb 10G *790*
Standon Alpha-C10 *791*
Standon Azoxystrobin *452*
Standon Bentazone *16*
Standon Bentazone S *16*
Standon CIPC 300 HN *74*
Standon Chlorothalonil 500 *488*
Standon Chlorpyrifos 48 *806*
Standon Clodinafop 240 *82*
Standon Cyanazine 50 *99*
Standon Cycloxydim *102*
Standon Cymoxanil Extra *509*
Standon Cyproconazole *511*
Standon Cyprodinil *521*
Standon DFF-IPU *164*
Standon Dichlobenil 6G *138*
Standon Diflufenican IPU *164*
Standon Diquat *167*
Standon Epoxiconazole *534*
Standon Epoxifen *536*
Standon Ethofumesate 200 *176*
Standon Etridiazole 35 *542*
Standon Fenpropimorph 750 *559*
Standon Fluazinam 500 *576*
Standon Flupyrsulfuron MM *196*

Standon Fluroxypyr *199*
Standon Flusilazole Plus *472*
Standon Glyphosate 360 *216*
Standon IPU *239*
Standon Imazaquin 5C *62*
Standon Kresoxim FM *567*
Standon Kresoxim Super *537*
Standon Kresoxim-Epoxiconazole *538*
Standon Lambda-C *847*
Standon Mepiquat Plus *68*
Standon Metamitron *320*
Standon Metazachlor-Q *325*
Standon Metsulfuron *334*
Standon Pencycuron DP *669*
Standon Pirimicarb 50 *866*
Standon Pirimicarb H *866*
Standon Propaquizafop *377*
Standon Propiconazole *685*
Standon Propyzamide 400 SC *378*
Standon Propyzamide 50 *378*
Standon Rimsulfuron *387*
Standon Tebuconazole *714*
Standon Tralkoxydim *423*
Standon Tridemorph 750 *765*
Standon Triflusulfuron *447*
Standon Vinclozolin *783*
Stantion *64*
Star Cyproconazole *511*
Starane 2 *199*
Starane Super *207*
Starion *799*
Starter Flowable *52*
Status *37*
Stay Up *58*
Stay-Off *935*
Stefes 7G *320*
Stefes 880 *260*
Stefes Banlene Super *133*
Stefes Blytin *574*
Stefes C-Flo 2 *466*
Stefes CCC *58*
Stefes CCC 640 *58*
Stefes CCC 700 *58*
Stefes CCC 720 *58*
Stefes Carbendazim Flo *466*
Stefes Chloridazon *52*
Stefes Clover Ley *15*
Stefes Competitor *193*
Stefes Cypermethrin 3 *811*
Stefes Deny *623*
Stefes Derosal Liquid *466*
Stefes Derosal WDG *466*
Stefes Docklene Super *133*
Stefes Duplex *180*
Stefes Excel *260*
Stefes Forte *362*

PSD
PRODUCT INDEX

Stefes Forte 2 *362*
Stefes Fortune *546*
Stefes Fumat 2 *176*
Stefes Glyphosate *216*
Stefes IPU *239*
Stefes IPU 500 *239*
Stefes IPU 700 *239*
Stefes Inception *546*
Stefes K2 *58*
Stefes Kickdown *216*
Stefes Kickdown 2 *216*
Stefes Legumex Extra *15*
Stefes Leyclene *34*
Stefes Linuron *261*
Stefes Medimat *182*
Stefes Medimat 2 *182*
Stefes Medipham 2 *362*
Stefes Mepiquat *68*
Stefes Metamitron *320*
Stefes Phenoxylene 50 *268*
Stefes Poraz *672*
Stefes Pride *378*
Stefes Pride Flo *378*
Stefes Restraint *623*
Stefes ST/STE *180*
Stefes Spectron *53*
Stefes Swelland Super *133*
Stefes Tandem *182*
Stefes Toluron *69*
Stefes Union *251*
Stellox 380 EC *37*
Sterilite Hop Defoliant *8*
Sterilite Tar Oil Winter Wash 60% Stock
    Emulsion *713*
Sterilite Tar Oil Winter Wash 80% Miscible
    Quality *713*
Sting *216*
Sting CT *216*
Sting Eco *216*
Stirling Rescue Wasp Nest Killer *1141*
Stirrup *216*
Stoller Flowable Sulphur *709*
Stomp 400 SC *354*
Stop Gro G8 *1072*
Storite Clear Liquid *731*
Storite Excel *731*
Storite Flowable *731*
Strada *488*
Strada 500 *488*
Strate *59*
Strathclyde Sodium Chlorate Weedclear *395*
Strathclyde Sodium Chlorate Weedkiller *1150*
Stratos *102*
Stride *216*
Strike 2 *1109*
Stroby WG *619*

Stronghold *63*
Stryper *605*
Sulphur Candle *1151*
Sulphur Candles *1151*
Sulphur Flowable *709*
Sumi-Alpha *827*
Supaquat *58*
Super Mosstox *139*
Super Slug and Snail Killer *1099*
Super-Flor 6% Metaldehyde Slug Killer Mini
    Pellets *924*
Super-Tin 4L *574*
Super-Tin 80WP *574*
Super-flor 6% Metaldehyde Slug Killer Mini
    Pellets *924*
SuperMosstox *139*
Supercarb *466*
Supergreen Double (Feed and Weed) *994*
Supergreen Feed, Weed & Mosskiller *992*
Supergreen Feed, Weed and Mosskiller 2 *993*
Supergreen and Weed *994*
Supertin 4L *574*
Supertin 80 WP *574*
Supertox 30 *113*
Supertox Spot *994*
Surpass 5EC *544*
Suscon Green *806*
Suscon Green Soil Insecticide *806*
suSCon Indigo Soil Insecticide *806*
Swipe P *38*
Sybol *1131*
Sybol Aerosol *1132*
Sybol Dust *1131*
Sybol Extra *948*
Sydex *114*
Syford *104*
Synergol *225*
Sypex *59*
Sypex M *68*
Systemic Insecticide Pins *957*
Systhane *1106*
Systhane 20 EW *656*
Systhane 6 Flo *656*
Systhane 6W *656*
Systhane Fungus Fighter *1106*
Systhane ST *656*
Systol M *509*
Systol M WSB *509*
TC Elite *500*
TWK-Total Weedkiller *395*
Tachigaren 70 WP *604*
Tairel *454*
Takron *52*
Taktic Buildings Spray *793*
Talat *550*
Talisman *69*

Talon 806
Talstar 799
Talvin 707
Tap Algasan 981
Tariff 75 WG Newtec 623
Tariff 75WG 623
Tarot 387
Tattoo 636
Tattoo C 498
Taylor's Lawn Sand 187
Teal 39
Teal-M 39
Tecto Flowable Turf Fungicide 731
Tecto Superflowable Turf Fungicide 731
Tedion V-18 EC 884
Teldor 549
Telone 2000 817
Telone II 817
Temik 10G 790
Templar 42
Terbel Triple 108
Tern 552
Terpal 68
Terpal C 59
Terpitz 68
Terraclor 20D 706
Terraclor Flo 706
Terrazole 35 WP 542
Tetra 5634 729
Tetra Algimin 1143
Texas All-In-One Insecticide Spray 1136
The Ultimate Cat Repellent 963
Thinsec 802
Thiodan 20 EC 826
Thiovit 709
Thiraflo 742
Thripstick 813
Throttle 803
Thuricide HP 907
Tigress Ultra 155
Tilt 685
Tilt Turbo 475 EC 695
Timbrel 433
Tipoff 341
Tiptor 516
Titus 387
Tolkan 239
Tolkan Liquid 239
Tolkan Turbo 164
Tolkan WDG 239
Tolugan 700 69
Tolugan Extra 71
Tomahawk 199
Tomahawk 2000 199
Tomahawk Plus 35
Tomcat All-Weather Rodent Bar 900

Tomcat Rat and Mouse Bait 900
Tomcat Rat and Mouse Blox 1025
Tomcat Rat and Mouse Pellets 1025
Tomcat Super Blox All-Weather 900
Top Farm Carbendazim - 435 466
Top Farm Fluazinam 576
Top Farm Propyzamide 500 378
Top Farm Toluron 500 69
Topas 666
Topas C 50 WP 465
Topas D275 SC 531
Topik 82
Toplawn Feed with Mosskiller 1042
Toplawn Feed with Weed & Mosskiller 987
Toplawn Feed with Weedkiller 986
Toplawn Feed, Weed & Mosskiller 988
Toplawn Lawn Weedkiller 994
Toplawn Ready to Use Lawn Weedkiller 994
Toppel 10 811
Torch 707
Tordon 101 115
Tordon 22K 368
Torq 833
Torque 833
Total 174
Totem 72
Totril 226
Touchdown 216
Touchdown LA 216
Touche 174
Touchweeder 996
Tough Weed Gun! 1062
Tough Weed Killer 1062
Trappit 1136
Trecatol 454
Treflan 439
Tribel 480 433
Tribunil 326
Tribunil WG 326
Tribute 133
Tribute Plus 133
Tricur 482
Trident 494
Tridex 623
Triflur 439
Triflurex 48EC 439
Trifolex-Tra 291
Trifsan 439
Trigard 439
Triherbicide CIPC 74
Trik 5
Trilogy 439
Trimangol 80 637
Trimangol WDG 637
Trimanzone 623
Triminol 657

Trio *58*
Tripart 147 *476*
Tripart 5C *58*
Tripart Accendo *320*
Tripart Arena 10G *400*
Tripart Arena 3 *400*
Tripart Arena 5G *400*
Tripart Arena Plus *43*
Tripart Audax *806*
Tripart Beta *362*
Tripart Beta S *362*
Tripart Brevis *58*
Tripart Brevis 2 *58*
Tripart Chlormequat 460 *58*
Tripart Culmus *69*
Tripart Defensor FL *466*
Tripart Defensor Liq *466*
Tripart Defensor WDG *466*
Tripart Faber *488*
Tripart Gladiator 2 *52*
Tripart Imber *709*
Tripart MCPA 50 *268*
Tripart New Arena 6 *400*
Tripart Pugil 5 *239*
Tripart Sentinel *374*
Tripart Trifluralin 48 EC *439*
Tripart Victor *468*
Triplen *439*
Triptic 48 EC *433*
Tristar *439*
Tritoftorol *786*
Tritox *133*
Triumph *183*
Tropotox *293*
Tropotox Plus *291*
Trump *249*
Trump SC *249*
Trustan WDG *510*
Tumbleblite *1134*
Tumbleweed Brushwood Ready to Use *1166*
Tumbleweed Clover *991*
Tumbleweed Clover Ready to Use *991*
Tumbleweed General Purpose *1059*
Tumbleweed General Purpose Ready To
  Use *1059*
Tumbleweed Lawns *991*
Tumbleweed Lawns Ready to Use *991*
Tumbleweed Moss Ready to Use *1011*
Tumbleweed Original *1062*
Tumbleweed Original Extra Strong *1062*
Tumbleweed Original Extra Strong Gel *1062*
Tumbleweed Original Extra Strong Ready To
  Use *1062*
Tumbleweed Original Ultra Concentrated *1062*
Tumbleweed Paths and Patios *1034*
Turbair Flydown *872*

Turbair Grain Store Insecticide *835*
Turbair Kilsect Short Life Grade *872*
Turbair Permethrin *861*
Turbair Resmethrin Extra *874*
Turbair Rovral *616*
Turbair Super Flydown *872*
Turfclear *466*
Twin *182*
Twinspan *807*
Twist *776*
Twister Flow *911*
Typhoon 360 *216*
UK 200 *714*
UPL Camppex *111*
UPL Carbendazim 500 FL *466*
UPL Diuron 80 *170*
UPL Grassland Herbicide *133*
UPL Linuron 45% Flowable *261*
Ultrafaber *488*
Unicrop 6% Mini Slug Pellets *924*
Unicrop Atrazine 50 *10*
Unicrop Atrazine FL *10*
Unicrop Fenitrothion 50 *834*
Unicrop Flowable Atrazine *10*
Unicrop Flowable Diuron *170*
Unicrop Flowable Simazine *389*
Unicrop Mancozeb *623*
Unicrop Mancozeb 80 *623*
Unicrop Maneb 80 *637*
Unicrop Maneb FL *637*
Unicrop Simazine 50 *389*
Unicrop Simazine FL *389*
Unicrop Thianosan DG *742*
Unicrop Zineb *786*
Unison *548*
Unistar Ethephon 480 *64*
Unistar Glyfosate 360 *216*
Unistar Glyphosate 360 *216*
Unistar Pirimicarb 500 *866*
Unite *84*
Unix *521*
Upgrade *59*
Uplift *58*
Upright *62*
Valiant *320*
Vapam *925*
Vapona All-In-One *1136*
Vapona House and Garden Plant Insect
  Killer *1136*
Varmint *54*
Vassgro Cypermethrin Insecticide *811*
Venzar 80 WP *257*
Venzar Flowable *257*
Verdone *994*
Verdone 2 *994*
Verdone Extra *1056*

Verdone Gun! *986*
'Verdone Plus' *1003*
Vertalec *909*
Veto F *723*
Vindex *30*
Visclor 500 SC *488*
Visclor 75 DF *488*
Viscount *82*
Vista CT *494*
Vitaflo Extra *485*
Vitagrow Lawn Sand *187*
Vitagrow Mini-Granular Feed, Weed & Mosskiller *149*
Vitax 'Green Up' Granular Lawn Feed and Weed *995*
Vitax Lawn Sand *1042*
Vitax MCPA Ester *268*
Vitax Microgran 2 *187*
Vitax Turf Tonic *187*
Vitesse *474*
Vizor *257*
Volcan *320*
Volcan Combi *55*
Voodoo *500*
Vydate 10G *860*
Warefog 25 *74*
Wash and Get Off Spray *1103*
Wasp Exterminator *1141*
Waterscene Coldwater - Algae Control *974*
Weed 'N' Feed Extra *992*
Weed master SC *52*
Weed-B-Gon *968*
Weed-B-Gon Ready To Use *968*
Weed-Easy *1062*
Weedatak Path and Drive *1035*
Weedazol-TL *3*
Weedclear *1062*
Weedol *1028*
Weedol Gun! *1026*
Westland Kill All General Purpose Weedkiller RTU *1059*
Westland Kill All Path & Patio Weedkiller RTU *1059*
Westland Triple Action Weed, Feed And Mosskill *993*
Wetcol 3 Copper Fungicide *503*
Whip *185*
Whip X *185*
Whitefly Gun! *956*
Whyte Chlormequat 700 *58*
Whyte Trifluralin *439*
Wickes General Purpose Insecticide Ready to Use *1136*
Wickes General Purpose Weedkiller Ready To Use *1062*
Wickes Lawn Feed, Weed and Mosskiller *1052*

Widgeon *559*
Wildcat *183*
Wilko Ant Destroyer *946*
Wilko Complete Weedkiller Concentrate *1059*
Wilko Complete Weedkiller Spray *1059*
Wilko Flower and Rose Insecticide Spray *1037*
Wilko Fruit and Vegetable Insecticide Spray *1136*
Wilko Granular Feed and Weed for Lawns *1021*
Wilko Granular Feed, Weed and Mosskiller *1020*
Wilko Granular Feed, Weed and Mosskiller for Lawns *1020*
Wilko Granular Lawn Feed, Weed and Mosskiller *1052*
Wilko Greenfly Killer *1136*
Wilko Houseplant Insect Spray *1037*
Wilko Houseplant Insecticide Spray *1136*
Wilko Insecticidal Houseplant Spray *1037*
Wilko Lawn Feed & Weed Liquid *995*
Wilko Lawn Feed and Weed *1021*
Wilko Lawn Feed, Weed and Mosskiller *1020*
Wilko Lawn Sand *1042*
Wilko Lawn Spot Weeder *991*
Wilko Lawn Weed and Feed *991*
Wilko Lawn Weedkiller *991*
Wilko Multi-Purpose Insecticide Spray *1136*
Wilko Nettle Gun *991*
Wilko Path & Patio Weedkiller Spray *1059*
Wilko Path Weedkiller *937*
Wilko Plant Disease Control *960*
Wilko Rose and Flower Insecticide Spray *1136*
Wilko Slug Killer Blue Mini Pellets *1099*
Wilko Sodium Chlorate Weedkiller *1150*
Wilko Soluble Lawn Food and Weedkiller *1021*
Wilko Woodlice Killer *946*
Wizzard *192*
Woolworths Ant Killer *946*
Woolworths Complete Insecticide Spray *1136*
Woolworths Complete Weedkiller Concentrate *1059*
Woolworths Complete Weedkiller Spray *1059*
Woolworths Greenfly Killer for Flowers *1037*
Woolworths Insecticide Spray for Flowers *1037*
Woolworths Insecticide Spray for Houseplants *1037*
Woolworths Lawn Feed, Weed and Mosskiller *1052*
Woolworths Lawn Spot Weeder *991*
Woolworths Lawn Weed and Feed *991*
Woolworths Lawn Weedkiller *991*
Woolworths Liquid Lawn Feed and Weed *1003*
Woolworths Path Weedkiller *937*
Woolworths Slug Killer Pellets *1099*
Woolworths Slug Pellets *1099*
Woolworths Weedkiller *1062*
Woolworths Woodlice Killer *946*
X-Spor SC *637*

311

XL All Insecticide   *859*
XL All Nicotine 95%   *859*
Xanadu   *174*
Yaltox   *803*
Yellow Sulphur   *1151*
Yeoman   *924*
ZP Rodent Bait   *905*
ZP Rodent Pellets   *905*
Zapper   *218*
Zenon   *707*
Zolone Liquid   *865*
Zulu   *556*

4

PSD ACTIVE SUBSTANCE INDEX

The number after the Active Substance gives the Substance Code Number for the Professional or Amateur Sections.

Abamectin *789*
Acibenzolar-S-methyl *449*
Aldicarb *790*
Alphachloralose *893*
Alphacypermethrin *791*
Aluminium ammonium sulphate *894, 935*
Aluminium phosphide *792, 895*
Aluminium sulphate *936*
Amidosulfuron *1*
  + Metribuzin *2*
2-Aminobutane *450*
Amitraz *793*
Amitrole *3*
  + Atrazine *937*
  + Bromacil + Diuron *4*
  + 2,4-D + Diuron *5, 938*
  + 2,4-D + Diuron + Simazine *939*
  + Diquat + Paraquat + Simazine *940*
  + Diuron + Simazine *941*
  + Simazine *6, 942*
Ammonium sulphamate *7, 943*
Anthracene oils *8*
Asulam *9*
Atrazine *10*
  + Amitrole *944*
Azaconazole
  + Imazalil *451*
Azamethiphos *794*
  + (Z)-9-Tricosene *795*
Aziprotryne *11*
Azoxystrobin *452*
Azoxystrobin
  + Flutriafol *453*
Bacillus thuringiensis *907, 945*
Bacillus thuringiensis berliner var
  kur *907a*
Benalaxyl
  + Mancozeb *454*
Benazolin *12*
  + Bromoxynil + Ioxynil *13*
  + Clopyralid *14*
  + 2,4-DB + MCPA *15*
Bendiocarb *797, 946*
Benfuracarb *798*
Benodanil *455*
Benomyl *456*
Bentazone *16*
Bentazone
  + Dichlorprop-P *17*
  + MCPA + MCPB *18*
  + MCPB *19*
  + Pendimethalin *20*
Benzalkonium chloride
  + Copper sulphate *947*
Bifenox
  + Chlorotoluron *21*

  + Isoproturon *22*
  + MCPA + Mecoprop-P *23*
Bifenthrin *799, 948*
  + Flutriafol *949*
  + Myclobutanil *950*
Bioallethrin
  + Permethrin *951*
Bitertanol
  + Fuberidazole *457*
  + Fuberidazole + Imidacloprid *910*
  + Fuberidazole + Imidacloprid +
    Triadimenol *458, 800*
  + Fuberidazole + Triadimenol *459*
Bone oil *952*
Borax *953*
Bromacil *24*
  + Amitrole + Diuron *25*
  + Diuron *26*
  + Picloram *27*
Bromadiolone *896*
Bromoxynil *28*
  + Benazolin + Ioxynil *29*
  + Clopyralid *30*
  + Clopyralid + Fluroxypyr + Ioxynil *31*
  + Dichlorprop + Ioxynil + MCPA *32*
  + Diflufenican + Ioxynil *33*
  + Ethofumesate + Ioxynil *34*
  + Fluroxypyr *35*
  + Fluroxypyr + Ioxynil *36*
  + Ioxynil *37*
  + Ioxynil + Mecoprop-P *38*
  + Ioxynil + Triasulfuron *39*
  + Ioxynil + Trifluralin *40*
  + Prosulfuron *41*
  + Terbuthylazine *42*
Bromuconazole *460*
  + Fenpropimorph *461*
Bupirimate *462*
  + Pirimicarb + Triforine *954*
  + Triforine *463*
  + Triforine *955*
Buprofezin *801, 956*
Butoxycarboxim *957*
Captan *464*
  + 1-Naphthylacetic acid *958, 959*
  + Penconazole *465*
Carbaryl *802, 911*
Carbendazim *466, 912, 960*
  + Chlorothalonil *467*
  + Chlorothalonil + Maneb *468*
  + Cymoxanil + Oxadixyl + Thiram *469*
  + Cyproconazole *470*
  + Epoxiconazole *471*
  + Flusilazole *472*
  + Flutriafol *473*
  + Iprodione *474*

+ Mancozeb *475*
+ Maneb *476*
+ Maneb + Sulphur *477*
+ Maneb + Tridemorph *478*
+ Metalaxyl *479*
+ Prochloraz *480*
+ Propiconazole *481*
+ Tebuconazole *482*
+ Tecnazene *483*
+ Tecnazene *43*
+ Vinclozolin *484*
Carbetamide *44*
+ Diflufenican + Oxadiazon *45*
Carbofuran *803*
Carbosulfan *804*
Carboxin
+ Imazalil + Thiabendazole *485*
+ Prochloraz *486*
+ Thiram *487*
Carfentrazone-ethyl *46*
+ Flupyrsulfuron-methyl *47*
+ Isoproturon *48*
+ Mecoprop-P *49*
+ Metsulfuron-methyl *50*
+ Thifensulfuron-methyl *51*
Chlorfenvinphos *805*
Chloridazon *52*
+ Ethofumesate *53*
+ Lenacil *54*
+ Metamitron *55*
+ Propachlor *56*
+ Quinmerac *57*
Chlormequat *58*
+ 2-Chloroethylphosphonic acid *59*
+ 2-Chloroethylphosphonic acid +
  Imazaquin *60*
+ 2-Chloroethylphosphonic acid + Mepiquat *61*
+ Imazaquin *62*
+ Mepiquat *63*
2-Chloroethylphosphonic acid *64*
+ Chlormequat *65*
+ Chlormequat + Imazaquin *66*
+ Chlormequat + Mepiquat *67*
+ Mepiquat *68*
Chlorophacinone *897*
Chloropicrin *913*
Chlorothalonil *488*
+ Carbendazim *489*
+ Carbendazim + Maneb *490*
+ Cymoxanil *491*
+ Cyproconazole *492*
+ Fenpropimorph *493*
+ Fluquinconazole *494*
+ Flutriafol *495*
+ Mancozeb *496*
+ Metalaxyl *497*

+ Propamocarb hydrochloride *498*
+ Propiconazole *499*
+ Tetraconazole *500*
+ Vinclozolin *501*
Chlorotoluron *69*
+ Bifenox *70*
+ Isoproturon *71*
+ Pendimethalin *72*
Chlorphonium *73*
Chlorpropham *74, 914*
+ Fenuron *75*
+ Linuron *76*
+ Pentanochlor *77*
Chlorpyrifos *806, 961*
Disulfoton *807*
Chlorpyrifos-methyl *808*
Chlorthal-dimethyl *78*
+ Propachlor *79*
Cholecalciferol
+ Difenacoum *962*
Cinidon-ethyl *80*
Citronella oil *81, 963*
+ Methyl nonyl ketone *964*
+ Tar oils *965*
Citrus Extract *966, 967*
Clodinafop-propargyl *82*
+ Diflufenican *83*
+ Diflufenican + Isoproturon *84*
+ Trifluralin *85*
Clofentezine *809*
Clopyralid *86*
+ Benazolin *87*
+ Bromoxynil *88*
+ Bromoxynil + Fluroxypyr + Ioxynil *89*
+ 2,4-D + MCPA *90*
+ Dichlorprop + MCPA *91*
+ Diflufenican + MCPA *92*
+ Fluroxypyr + Ioxynil *93*
+ Fluroxypyr + MCPA *94, 968*
+ Fluroxypyr + Triclopyr *94a*
+ Ioxynil *95*
+ Propyzamide *96*
+ Triclopyr *97*
Copper *969*
Copper Ammonium Carbonate
502
Copper chloride
+ Metaldehyde *970*
Copper complex - Bordeaux *503*
Copper hydrate *971*
Copper hydroxide *98*
Copper oxychloride *504, 972*
+ Maneb + Sulphur *505*
+ Metalaxyl *506*
Copper sulphate *973, 974, 975, 976*
+ Benzalkonium chloride *977*

+ Ferrous sulphate *978*
+ Ferrous sulphate + Magnesium
    sulphate *979, 980*
+ Simazine *981*
Coumatetralyl *898, 982*
Cyanazine *99*
+ Pendimethalin *100*
+ Terbuthylazine *101*
Cycloxydim *102*
Cyfluthrin *810*
Cymoxanil
+ Carbendazim + Oxadixyl + Thiram *507*
+ Chlorothalonil *508*
+ Mancozeb *509*
+ Mancozeb + Oxadixyl *510*
Cypermethrin *103, 811*
+ Propiconazole *983*
Cyproconazole *511*
+ Carbendazim *512*
+ Chlorothalonil *513*
+ Cyprodinil *514*
+ Mancozeb *515*
+ Prochloraz *516*
+ Propiconazole *517*
+ Quinoxyfen *518*
+ Tridemorph *519*
+ Trifloxystrobin *520*
Cyprodinil *521*
+ Cyproconazole *522*
Cyromazine *812*
2,4-D *104*
+ Amitrole + Diuron *105, 984*
+ Amitrole + Diuron + Simazine *985*
+ Clopyralid + MCPA *106*
+ Dicamba *107, 986*
+ Dicamba + Ferrous sulphate *987*
+ Dicamba + Ferrous sulphate
    heptahydrate *988*
+ Dicamba + Mecoprop *989*
+ Dicamba + Mecoprop-P *990*
+ Dicamba + Triclopyr *108*
+ Dichlorprop *109, 991*
+ Dichlorprop + MCPA + Mecoprop *110*
+ Dichlorprop + MCPA + Mecoprop-P *111*
+ Ferrous sulphate + Mecoprop *992*
+ Ferrous sulphate + Mecoprop-P *993*
+ MCPA *112*
+ Mecoprop *113, 994*
+ Mecoprop-P *114, 995*
+ Picloram *115*
+ 2,3,6-TBA *996*
Daminozide *116*
Dazomet *915*
2,4-DB *117*
+ Benazolin + MCPA *118*
+ Linuron + MCPA *119*

+ MCPA *120*
DE-570 *121*
Deltamethrin *813*
+ Heptenophos *814*
+ Pirimicarb *815*
Desmedipham
+ Ethofumesate + Phenmedipham *122*
+ Phenmedipham *123*
Desmetryn *124*
Dicamba *125*
+ 2,4-D *126, 997*
+ 2,4-D + Ferrous sulphate *998*
+ 2,4-D + Ferrous sulphate heptahydrate *999*
+ 2,4-D + Mecoprop *1000*
+ 2,4-D + Mecoprop-P *1001*
+ 2,4-D + Triclopyr *127*
+ Dichlorprop + Ferrous sulphate + MCPA *128, 1002*
+ Dichlorprop + MCPA *129, 816, 1003*
+ Dichlorprop + Mecoprop *1004*
+ Dichlorprop + Mecoprop-P *1005*
+ Maleic hydrazide + MCPA *130*
+ MCPA *131, 1006*
+ MCPA + Mecoprop *132, 1007*
+ MCPA + Mecoprop-P *133, 1008*
+ Mecoprop *134*
+ Mecoprop-P *135*
+ Paclobutrazol *136*
+ Triasulfuron *137*
Dichlobenil *138, 1009*
Dichlofluanid *523*
Dichlorophen *139, 524, 1010, 1011*
+ Ferrous sulphate *140*
+ 1-Naphthylacetic acid *1012, 1013, 1014*
1,2-Dichloropropane
+ 1,3-Dichloropropene *916*
1,3-Dichloropropene *817*
+ 1,2-Dichloropropane *917*
Dichlorprop *141*
+ Bromoxynil + Ioxynil + MCPA *142*
+ Clopyralid + MCPA *143*
+ 2,4-D *144, 1015*
+ 2,4-D + MCPA + Mecoprop *145*
+ 2,4-D + MCPA + Mecoprop-P *146*
+ Dicamba + Ferrous sulphate + MCPA *147, 1016*
+ Dicamba + MCPA *148, 818, 1017*
+ Dicamba + Mecoprop *1018*
+ Dicamba + Mecoprop-P *1019*
+ Ferrous sulphate + MCPA *149, 1020*
+ MCPA *150, 1021*
Dichlorprop-P *151*
+ Bentazone *152*
+ MCPA + Mecoprop-P *153*
Dichlorvos *819*
Diclofop-methyl *154*

+ Fenoxaprop-P-ethyl   *155*
Dicloran   *525*
Dicofol   *820*
+ Tetradifon   *821*
Difenacoum   *899*
+ Cholecalciferol   *1022*
Difenoconazole   *526*
Difenzoquat   *156*
Difenzoquat   *527*
Diflubenzuron   *822*
Diflufenican
+ Bromoxynil + Ioxynil   *157*
+ Carbetamide + Oxadiazon   *158*
+ Clodinafop-propargyl   *159*
+ Clodinafop-propargyl + Isoproturon   *160*
+ Clopyralid + MCPA   *161*
+ Flurtamone   *162*
+ Flurtamone + Isoproturon   *163*
+ Isoproturon   *164*
+ Terbuthylazine   *165*
+ Trifluralin   *166*
Dikegulac   *1023*
Dimethoate   *823, 1024*
Dimethomorph
+ Mancozeb   *528*
Dinocap   *529*
Diphacinone   *900, 1025*
Diphenylamine   *918*
Diquat   *167, 1026*
+ Amitrole + Paraquat + Simazine   *1027*
+ Paraquat   *168, 1028*
+ Paraquat + Simazine   *169, 1029*
Disodium hydrogen phosphate   *1030*
Disulfoton   *824*
+ Chlorpyrifos   *825*
Dithianon   *530*
+ Penconazole   *531*
Diuron   *170*
+ Amitrole + Bromacil   *171*
+ Amitrole + 2,4-D   *172, 1031*
+ Amitrole + 2,4-D + Simazine   *1032*
+ Amitrole + Simazine   *1033*
+ Bromacil   *173*
+ Glufosinate-ammonium   *1034*
+ Glyphosate   *174, 1035*
+ Paraquat   *175*
Dodemorph   *532*
Dodine   *533*
Endosulfan   *826*
Epoxiconazole   *534*
+ Carbendazim   *535*
+ Fenpropimorph   *536*
+ Fenpropimorph + Kresoxim-methyl   *537*
+ Kresoxim-methyl   *538*
+ Tridemorph   *539*
Esfenvalerate   *827*

Ethiofencarb   *828*
Ethirimol
+ Flutriafol + Thiabendazole   *540*
+ Fuberidazole + Triadimenol   *541*
Ethofumesate   *176*
+ Bromoxynil + Ioxynil   *177*
+ Chloridazon   *178*
+ Desmedipham + Phenmedipham   *179*
+ Metamitron   *180*
+ Metamitron + Phenmedipham   *181*
+ Phenmedipham   *182*
Ethoprophos   *829*
Etridiazole   *542*
Etrimfos   *830*
Fatty acids   *831, 1036, 1037*
+ Glufosinate-ammonium   *1038*
+ Pyrethrins   *1039*
+ Sulphur   *1040*
Fenarimol   *543*
Fenazaquin   *832*
Fenbuconazole   *544*
+ Fenpropidin   *545*
+ Prochloraz   *546*
+ Propiconazole   *547*
+ Tridemorph   *548*
Fenbutatin oxide   *833*
Fenhexamid   *549*
+ Tolylfluanid   *550*
Fenitrothion   *834, 1041*
+ Permethrin + Resmethrin   *835*
Fenoxaprop-P-ethyl   *183*
+ Diclofop-methyl   *184*
+ Isoproturon   *185*
Fenoxycarb   *836*
Fenpiclonil   *551*
Fenpropathrin   *837*
Fenpropidin   *552*
+ Fenbuconazole   *553*
+ Fenpropimorph   *554*
+ Prochloraz   *555*
+ Propiconazole   *556*
+ Propiconazole + Tebuconazole   *557*
+ Tebuconazole   *558*
Fenpropimorph   *559*
+ Bromuconazole   *560*
+ Chlorothalonil   *561*
+ Epoxiconazole   *562*
+ Epoxiconazole + Kresoxim-methyl   *563*
+ Fenpropidin   *564*
+ Flusilazole   *565*
+ Flusilazole + Tridemorph   *566*
+ Kresoxim-methyl   *567*
+ Prochloraz   *568*
+ Propiconazole   *569*
+ Quinoxyfen   *570*
+ Tebuconazole   *571*

+ Tridemorph *572*
Fenpyroximate *838*
Fentin acetate
+ Maneb *573*
Fentin hydroxide *574*
+ Glufosinate-ammonium *575*
Fenuron
+ Chlorpropham *186*
Ferrous sulphate *187, 1042*
+ Copper sulphate *1043*
+ Copper sulphate + Magnesium
   sulphate *1044, 1045*
+ 2,4-D + Dicamba *1046*
+ 2,4-D + Mecoprop *1047*
+ 2,4-D + Mecoprop-P *1048*
+ Dicamba + Dichlorprop + MCPA *188, 1049*
+ Dichlorophen *189*
+ Dichlorprop + MCPA *190, 1050*
+ MCPA + Mecoprop *1051*
+ MCPA + Mecoprop-P *1052*
Ferrous sulphate heptahydrate *1053*
+ 2,4-D + Dicamba *1054*
Flamprop-M-isopropyl *191*
Fluazifop-P-butyl *192*
Fluazinam *576*
Fludioxonil *577*
Fluoroglycofen-ethyl
+ Isoproturon *193*
Flupyrsulfuron-methyl *194*
+ Carfentrazone-ethyl *195*
+ Metsulfuron-methyl *196*
+ Thifensulfuron-methyl *197*
Fluquinconazole *578*
+ Chlorothalonil *579*
+ Prochloraz *580, 198*
Fluroxypyr *199*
+ Bromoxynil *200*
+ Bromoxynil + Clopyralid + Ioxynil *201*
+ Bromoxynil + Ioxynil *202*
+ Clopyralid + Ioxynil *203*
+ Clopyralid + MCPA *204, 1055*
+ Clopyralid + Triclopyr *204a*
+ Mecoprop-P *205, 1056*
+ Metosulam *206*
+ Thifensulfuron-methyl + Tribenuron-
   methyl *207*
+ Triclopyr *208*
Flurtamone
+ Diflufenican *209*
+ Diflufenican + Isoproturon *210*
Flusilazole *581*
+ Carbendazim *582*
+ Fenpropimorph *583*
+ Fenpropimorph + Tridemorph *584*
+ Tridemorph *585*
Flutriafol *586*

+ Azoxystrobin *587*
+ Bifenthrin *1057*
+ Carbendazim *588*
+ Chlorothalonil *589*
+ Ethirimol + Thiabendazole *590*
Fomesafen *211*
+ Terbutryn *212*
Fosamine-ammonium *213*
Fosetyl-aluminium *591*
Fosthiazate *839*
Fuberidazole
+ Bitertanol *592*
+ Bitertanol + Imidacloprid *919*
+ Bitertanol + Imidacloprid + Triadimenol *593,*
   840
+ Bitertanol + Triadimenol *594*
+ Ethirimol + Triadimenol *595*
+ Imazalil + Triadimenol *596*
+ Imidacloprid + Triadimenol *597, 841*
+ Triadimenol *598*
Furalaxyl *599*
Garlic oil
+ Orange peel oil + Orange pith oil *1058*
Gibberellins *214*
Glufosinate-ammonium *215, 1059*
+ Diuron *1060*
+ Fatty acids *1061*
+ Fentin hydroxide *600*
Glyphosate *216, 1062*
+ Diuron *217, 1063*
+ Oxadiazon *218*
Guazatine *601*
+ Imazalil *602*
+ Triticonazole *603, 920*
Heptenophos *842*
+ Deltamethrin *843*
+ Permethrin *1064*
Hymexazol *604*
Imazalil *605*
+ Azaconazole *606*
+ Carboxin + Thiabendazole *607*
+ Fuberidazole + Triadimenol *608*
+ Guazatine *609*
+ Pencycuron *610*
+ Thiabendazole *611*
+ Triticonazole *612, 921*
Imazamethabenz-methyl *219*
Imazapyr *220*
Imazaquin
+ Chlormequat *221*
+ 2-Chloroethylphosphonic acid +
   Chlormequat *222*
Imidacloprid *844, 1065*
+ Bitertanol + Fuberidazole *922*
+ Bitertanol + Fuberidazole + Triadimenol *613,*
   845

+ Fuberidazole + Triadimenol *614, 846*
+ Methiocarb *1066*
+ Tebuconazole + Triazoxide *615*
Indol-3-ylacetic acid *223, 224, 1067*
4-Indol-3-ylbutyric acid
+ 1-Naphthylacetic acid *225*
+ 1-Naphthylacetic acid + Thiram *1067a*
Ioxynil *226*
+ Benazolin + Bromoxynil *227*
+ Bromoxynil *228*
+ Bromoxynil + Clopyralid + Fluroxypyr *229*
+ Bromoxynil + Dichlorprop + MCPA *230*
+ Bromoxynil + Diflufenican *231*
+ Bromoxynil + Ethofumesate *232*
+ Bromoxynil + Fluroxypyr *233*
+ Bromoxynil + Mecoprop-P *234*
+ Bromoxynil + Triasulfuron *235*
+ Bromoxynil + Trifluralin *236*
+ Clopyralid *237*
+ Clopyralid + Fluroxypyr *238*
Iprodione *616*
+ Carbendazim *617*
+ Thiophanate-methyl *618*
Isoproturon *239*
+ Bifenox *240*
+ Carfentrazone-ethyl *241*
+ Chlorotoluron *242*
+ Clodinafop-propargyl + Diflufenican *243*
+ Diflufenican *244*
+ Diflufenican + Flurtamone *245*
+ Fenoxaprop-P-ethyl *246*
+ Fluoroglycofen-ethyl *247*
+ Isoxaben *248*
+ Pendimethalin *249*
+ Simazine *250*
+ Trifluralin *251*
Isoxaben *252*
+ Isoproturon *253*
+ Methabenzthiazuron *254*
+ Terbuthylazine *255*
+ Trifluralin *256*
Kresoxim-methyl *619*
+ Epoxiconazole *620*
+ Epoxiconazole + Fenpropimorph *621*
+ Fenpropimorph *622*
Lambda-cyhalothrin *847*
+ Pirimicarb *848*
Lenacil *257*
+ Chloridazon *258*
+ Linuron *259*
+ Phenmedipham *260*
Lindane *849, 1068*
+ Thiophanate-methyl *923*
+ Thiophanate-methyl *850*
Linuron *261*
+ Chlorpropham *262*

+ 2,4-DB + MCPA *263*
+ Lenacil *264*
+ Trifluralin *265*
Magnesium Phosphide *851*
Magnesium sulphate
+ Copper sulphate + Ferrous sulphate *1069, 1070*
Malathion *852, 1071*
Maleic hydrazide *266, 1072*
+ Dicamba + MCPA *267*
Mancozeb *623, 1073*
+ Benalaxyl *624*
+ Carbendazim *625*
+ Chlorothalonil *626*
+ Cymoxanil *627*
+ Cymoxanil + Oxadixyl *628*
+ Cyproconazole *629*
+ Dimethomorph *630*
+ Maneb *631*
+ Metalaxyl *632*
+ Metalaxyl-M *633*
+ Ofurace *634*
+ Oxadixyl *635*
+ Propamocarb hydrochloride *636*
Maneb *637*
+ Carbendazim *638*
+ Carbendazim + Chlorothalonil *639*
+ Carbendazim + Sulphur *640*
+ Carbendazim + Tridemorph *641*
+ Copper oxychloride + Sulphur *642*
+ Fentin acetate *643*
+ Mancozeb *644*
MCPA *268*
+ Bentazone + MCPB *269*
+ Bifenox + Mecoprop-P *270*
+ Bromoxynil + Dichlorprop + Ioxynil *271*
+ Clopyralid + Dichlorprop *272*
+ Clopyralid + Diflufenican *273*
+ Clopyralid + Fluroxypyr *274, 1074*
+ 2,4-D *275*
+ 2,4-D + Clopyralid *276*
+ 2,4-D + Dichlorprop + Mecoprop *277*
+ 2,4-D + Dichlorprop + Mecoprop-P *278*
+ 2,4-DB *279*
+ 2,4-DB + Benazolin *280*
+ 2,4-DB + Linuron *281*
+ Dicamba *282, 1075*
+ Dicamba + Dichlorprop *283, 853, 1076*
+ Dicamba + Dichlorprop + Ferrous sulphate *284, 1077*
+ Dicamba + Maleic hydrazide *285*
+ Dicamba + Mecoprop *286, 1078*
+ Dicamba + Mecoprop-P *287, 1079*
+ Dichlorprop *288, 1080*
+ Dichlorprop + Ferrous sulphate *289, 1081*

+ Dichlorprop-P + Mecoprop-P *290*
+ Ferrous sulphate + Mecoprop *1082*
+ Ferrous sulphate + Mecoprop-P *1083*
+ MCPB *291*
+ Mecoprop-P *292, 1084*
MCPB *293*
+ Bentazone *294*
+ Bentazone + MCPA *295*
+ MCPA *296*
Mecoprop *297*
+ 2,4-D *298, 1085*
+ 2,4-D + Dicamba *1086*
+ 2,4-D + Dichlorprop + MCPA *299*
+ 2,4-D + Ferrous sulphate *1087*
+ Dicamba *300*
+ Dicamba + Dichlorprop *1088*
+ Dicamba + MCPA *301, 1089*
+ Ferrous sulphate + MCPA *1090*
Mecoprop-P *302*
+ Bifenox + MCPA *303*
+ Bromoxynil + Ioxynil *304*
+ Carfentrazone-ethyl *305*
+ 2,4-D *306, 1091*
+ 2,4-D + Dicamba *1092*
+ 2,4-D + Dichlorprop + MCPA *307*
+ 2,4-D + Ferrous sulphate *1093*
+ Dicamba *308*
+ Dicamba + Dichlorprop *1094*
+ Dicamba + MCPA *309, 1095*
+ Dichlorprop-P + MCPA *310*
+ Ferrous sulphate + MCPA *1096*
+ Fluroxypyr *311, 1097*
+ MCPA *312, 1098*
+ Metribuzin *313*
+ Metsulfuron-methyl *314*
+ Triasulfuron *315*
Mefluidide *316*
Mephosfolan *855*
Mepiquat
+ Chlormequat *317*
+ 2-Chloroethylphosphonic acid *318*
+ 2-Chloroethylphosphonic acid +
  Chlormequat *319*
Metalaxyl *645*
+ Carbendazim *646*
+ Chlorothalonil *647*
+ Copper oxychloride *648*
+ Mancozeb *649*
+ Thiabendazole *650*
+ Thiabendazole + Thiram *651*
+ Thiram *652*
Metalaxyl-M *653*
+ Mancozeb *654*
Metaldehyde *924, 1099*
+ Copper chloride *1100*
Metam-Sodium *925*

Metamitron *320*
+ Chloridazon *321*
+ Ethofumesate *322*
+ Ethofumesate + Phenmedipham *323*
Metazachlor *324*
+ Quinmerac *325*
Metconazole *655*
Methabenzthiazuron *326*
+ Isoxaben *327*
Methiocarb *856, 926, 1101*
+ Imidacloprid *1102*
Methomyl *857*
Methoprene *858*
Methyl bromide *927*
Methyl nonyl ketone *1103*
+ Citronella oil *1104*
Metosulam *328*
+ Fluroxypyr *329*
Metoxuron *330*
Metribuzin *331*
+ Amidosulfuron *332*
+ Mecoprop-P *333*
Metsulfuron-methyl *334*
+ Carfentrazone-ethyl *335*
+ Flupyrsulfuron-methyl *336*
+ Mecoprop-P *337*
+ Thifensulfuron-methyl *338*
+ Tribenuron-methyl *339*
Monolinuron *340, 1105*
Myclobutanil *656, 1106*
+ Bifenthrin *1107*
Naphthalene *1108*
1-Naphthylacetic acid *341, 1109*
+ Captan *1110, 1111*
+ Dichlorophen *1112, 1113, 1114*
+ 4-Indol-3-ylbutyric acid *342*
+ 4-Indol-3-ylbutyric acid + Thiram *1114a*
2-Naphthyloxyacetic acid *343*
Napropamide *344*
Nicotine *859*
Nuarimol *657*
Octhilinone *658*
Ofurace
+ Mancozeb *659*
Orange peel oil
+ Garlic oil + Orange pith oil *1115*
Orange pith oil
+ Garlic oil + Orange peel oil *1116*
Oxadiazon *345*
+ Carbetamide + Diflufenican *346*
+ Glyphosate *347*
Oxadixyl
+ Carbendazim + Cymoxanil + Thiram *660*
+ Cymoxanil + Mancozeb *661*
+ Mancozeb *662*
+ Thiabendazole + Thiram *663*

Oxamyl *860*
Oxycarboxin *664*
P-[(Diiodomethyl)Sulfonyl]Toluol *1117*
Paclobutrazol *348*
  + Dicamba *349*
Parahydroxyphenylsalicylamide *665*
Paraquat *350*
  + Amitrole + Diquat + Simazine *1118*
  + Diquat *351, 1119*
  + Diquat + Simazine *352, 1120*
  + Diuron *353*
Penconazole *666, 1121*
  + Captan *667*
  + Dithianon *668*
Pencycuron *669*
  + Imazalil *670*
Pendimethalin *354*
  + Bentazone *355*
  + Chlorotoluron *356*
  + Cyanazine *357*
  + Isoproturon *358*
  + Simazine *359*
Peniophora Gigantea *908*
Pentanochlor *360*
  + Chlorpropham *361*
Pepper *1122*
Permethrin *861, 1123, 1124*
  + Bioallethrin *1125*
  + Fenitrothion + Resmethrin *862*
  + Heptenophos *1126*
  + Sulphur + Triforine *1127*
  + Thiram *671, 863*
Phenmedipham *362*
  + Desmedipham *363*
  + Desmedipham + Ethofumesate *364*
  + Ethofumesate *365*
  + Ethofumesate + Metamitron *366*
  + Lenacil *367*
Phenothrin
  + Tetramethrin *1128*
Phorate *864*
Phosalone *865*
Picloram *368*
  + Bromacil *369*
  + 2,4-D *370*
Pirimicarb *866, 1129*
  + Bupirimate + Triforine *1130*
  + Deltamethrin *867*
  + Lambda-cyhalothrin *868*
Pirimiphos-methyl *869, 1131*
  + Resmethrin + Tetramethrin *1132*
Polymeric quaternary ammonium chloride *1133*
Prochloraz *672*
  + Carbendazim *673*
  + Carboxin *674*

  + Cyproconazole *675, 676*
  + Fenpropidin *677*
  + Fenpropimorph *678*
  + Fluquinconazole *371, 679*
  + Propiconazole *680*
  + Tebuconazole *681*
Prometryn *372*
  + Terbutryn *373*
Propachlor *374*
  + Chloridazon *375*
  + Chlorthal-dimethyl *376*
Propamocarb hydrochloride *682*
  + Chlorothalonil *683*
  + Mancozeb *684*
Propaquizafop *377*
Propiconazole *685, 1134*
  + Carbendazim *686*
  + Chlorothalonil *687*
  + Cypermethrin *1135*
  + Cyproconazole *688, 689*
  + Fenpropidin *690*
  + Fenpropidin + Tebuconazole *691*
  + Fenpropimorph *692*
  + Prochloraz *693*
  + Tebuconazole *694*
  + Tridemorph *695*
  + Trifloxystrobin *696*
Propineb *697*
Propoxur *698, 870*
Propyzamide *378*
Propyzamide
  + Clopyralid *379*
Prosulfuron *380*
Prosulfuron
  + Bromoxynil *381*
Pymetrozine *871*
Pyrazophos *699*
Pyrethrins *872, 1136*
  + Fatty acids *1137*
  + Resmethrin *873, 1138*
Pyridate *382*
Pyrifenox *700*
Pyrimethanil *701*
Quinmerac
  + Chloridazon *383*
  + Metazachlor *384*
Quinomethionate *702*
Quinoxyfen *703*
  + Cyproconazole *704*
  + Fenpropimorph *705*
Quintozene *706*
Quizalofop-ethyl *385*
Quizalofop-P-ethyl *386*
Resmethrin *874*
  + Fenitrothion + Permethrin *875*
  + Pirimiphos-methyl + Tetramethrin *1139*

+ Pyrethrins   *876, 1140*
Rimsulfuron   *387*
Rotenone   *877, 1141*
+ Sulphur   *1142*
Seconal   *901*
Sethoxydim   *388*
Simazine   *389, 1143*
+ Amitrole   *390, 1144*
+ Amitrole + 2,4-D + Diuron   *1145*
+ Amitrole + Diquat + Paraquat   *1146*
+ Amitrole + Diuron   *1147*
+ Copper sulphate   *1148*
+ Diquat + Paraquat   *391, 1149*
+ Isoproturon   *392*
+ Pendimethalin   *393*
+ Trietazine   *394*
Sodium chlorate   *395, 1150*
Sodium chloroacetate   *396*
Sodium cyanide   *902*
Sodium monochloroacetate   *397*
Sodium silver thiosulphate   *398*
Spiroxamine   *707*
+ Tebuconazole   *708*
Sulphonated cod liver oil   *903*
Sulphur   *878, 709, 1151, 1152*
+ Carbendazim + Maneb   *710*
+ Copper oxychloride + Maneb   *711*
+ Fatty acids   *1153*
+ Permethrin + Triforine   *1154*
+ Rotenone   *1155*
Tar acids   *712, 1156, 1157*
Tar oils   *713, 879, 1158*
Citronella oil   *1159*
Tau-fluvalinate   *880*
2,3,6-TBA + 2,4-D   *1160*
Tebuconazole   *714*
+ Carbendazim   *715*
+ Fenpropidin   *716*
+ Fenpropidin + Propiconazole   *717*
+ Fenpropimorph   *718*
+ Imidacloprid + Triazoxide   *719*
+ Prochloraz   *720*
+ Propiconazole   *721*
+ Spiroxamine   *722*
+ Triadimenol   *723*
+ Triazoxide   *724*
+ Tridemorph   *725*
Tebufenpyrad   *881*
Tebutam   *399*
Tecnazene   *400, 726*
+ Carbendazim   *401, 727*
+ Thiabendazole   *402, 728*
Teflubenzuron   *882*
Tefluthrin   *883*
Tepraloxydim   *403*
Terbacil   *404*

Terbuthylazine
+ Bromoxynil   *405*
+ Cyanazine   *406*
+ Diflufenican   *407*
+ Isoxaben   *408*
+ Terbutryn   *409*
Terbutryn   *410, 1161*
+ Fomesafen   *411*
+ Prometryn   *412*
+ Terbuthylazine   *413*
+ Trietazine   *414*
+ Trifluralin   *415*
Tetraconazole   *729*
+ Chlorothalonil   *730*
Tetradifon   *884*
+ Dicofol   *885*
Tetramethrin
+ Phenothrin   *1162*
+ Pirimiphos-methyl + Resmethrin   *1163*
Thiabendazole   *731*
+ Carboxin + Imazalil   *732*
+ Ethirimol + Flutriafol   *733*
+ Imazalil   *734*
+ Metalaxyl   *735*
+ Metalaxyl + Thiram   *736, 737*
+ Tecnazene   *416, 738*
+ Thiram   *739*
Thifensulfuron-methyl   *417*
+ Carfentrazone-ethyl   *418*
+ Flupyrsulfuron-methyl   *419*
+ Fluroxypyr + Tribenuron-methyl   *420*
+ Metsulfuron-methyl   *421*
+ Tribenuron-methyl   *422*
Thiodicarb   *928*
Thiophanate-methyl   *740, 1164*
+ Iprodione   *741*
+ Lindane   *886, 929*
Thiram   *742*
+ 1-Naphthylacetic acid + 4-Indol-3-ylbutric acid   *1165*
+ Carbendazim + Cymoxanil + Oxadixyl   *743*
+ Carboxin   *744*
+ Metalaxyl   *745*
+ Metalaxyl + Thiabendazole   *746*
+ Oxadixyl + Thiabendazole   *747*
+ Permethrin   *748, 887*
+ Thiabendazole   *749*
Tolclofos-methyl   *750*
Tolylfluanid   *751*
+ Fenhexamid   *752*
Tralkoxydim   *423*
Tri-allate   *424*
Triadimefon   *753, 754*
+ Bitertanol + Fuberidazole   *755*
+ Bitertanol + Fuberidazole + Imidacloprid   *756, 888*

+ Ethirimol + Fuberidazole   *757*
+ Fuberidazole   *758*
+ Fuberidazole + Imazalil   *759*
+ Fuberidazole + Imidacloprid   *760, 889*
+ Tebuconazole   *761*
+ Tridemorph   *762*
Triasulfuron   *425*
+ Bromoxynil + Ioxynil   *426*
+ Dicamba   *427*
+ Mecoprop-P   *428*
Triazamate   *890*
Triazoxide
+ Imidacloprid + Tebuconazole   *763*
+ Tebuconazole   *764*
Tribenuron-methyl   *429*
+ Fluroxypyr + Thifensulfuron-methyl   *430*
+ Metsulfuron-methyl   *431*
+ Thifensulfuron-methyl   *432*
Trichlorfon   *891*
Triclopyr   *433, 1166*
+ Clopyralid   *434*
+ Clopyralid + Fluroxypyr   *434a*
+ 2,4-D + Dicamba   *435*
+ Fluroxypyr   *436*
(Z)-9-Tricosene
+ Azamethiphos   *903a*
Tridemorph   *765*
+ Carbendazim + Maneb   *766*
+ Cyproconazole   *767*
+ Epoxiconazole   *768*
+ Fenbuconazole   *769*
+ Fenpropimorph   *770*
+ Fenpropimorph + Flusilazole   *771*
+ Flusilazole   *772*
+ Propiconazole   *773*
+ Tebuconazole   *774*

+ Triadimenol   *775*
Trietazine
+ Simazine   *437*
+ Terbutryn   *438*
Trifloxystrobin   *776*
+ Cyproconazole   *777*
+ Propiconazole   *778*
Trifluralin   *439*
+ Bromoxynil + Ioxynil   *440*
+ Clodinafop-propargyl   *441*
+ Diflufenican   *442*
+ Isoproturon   *443*
+ Isoxaben   *444*
+ Linuron   *445*
+ Terbutryn   *446*
Triflusulfuron-methyl   *447*
Triforine   *779*
+ Bupirimate   *780, 1167*
+ Bupirimate + Pirimicarb   *1168*
+ Permethrin + Sulphur   *1169*
Trinexapac-ethyl   *448, 930*
Triticonazole
+ Guazatine   *781, 931*
+ Imazalil   *782, 932*
Verticillium Lecanii   *909*
Vinclozolin   *783*
+ Carbendazim   *784*
+ Chlorothalonil   *785*
Warfarin   *904*
Zetacypermethrin   *892*
Zinc phosphide   *905*
Zineb   *786*
**Zineb-ethylene thiuram disulphide
    adduct**   *787*
Ziram   *906*

PART C

HSE Registered Products

HSE

1

ANTIFOULING PRODUCTS

Product Name	Marketing Company	Use	HSE No.

Antifouling Products

1 Copper

Avonclad	Avon Technical Products	Professional	6396
Copperbot	Copperbot 98 Ltd	Amateur Professional	6860
Copperbot 2000	Wessex Resins And Adhesives Ltd	Amateur Professional	6680
Copperguard	Synthetic Solutions Ltd	Amateur Professional	6670
Miricoat A.F. Coating	Miricoat Ltd	Professional	5587
VC 17M	International Coatings Ltd	Amateur Professional	7061
VC 17M-EP	International Coatings Ltd	Amateur Professional Professional (Aquaculture)	6102

2 Copper + Dichlorophenyl Dimethylurea

Coppercoat	Aquarius Marine Coatings Ltd	Amateur Professional	6428

3 Copper + Zinc Pyrithione + 2-Methylthio-4-Tertiary-Butylamino-6-Cyclopropylamino-S-Triazine

VC 17M-HS	International Coatings Ltd	Amateur Professional	5960

4 Copper Metal

D Amercoat 67E	Ameron BV	Amateur Professional	3201
D Amercoat 70E	Ameron BV	Amateur Professional Professional (Aquaculture)	3202
W Amercoat 70ESP	Ameron BV	Amateur Professional Professional (Aquaculture)	3203
D Copperbot	Copperbot '98 Ltd	Amateur Professional	5262
CU15	Hippo Marine Products Ltd	Amateur Professional	5872
VC17M	International Coatings Ltd	Amateur Professional Professional (Aquaculture)	4780

Product Name	Marketing Company	Use	HSE No.

5 Copper Metal + 2-Methylthio-4-Tertiary-Butylamino-6-Cyclopropylamino-S-Triazine

VC 17M Tropicana	International Coatings Ltd	Amateur Professional	4218

6 Copper Naphthenate + Cuprous Oxide + Dichlofluanid + Zinc Naphthenate

D Teamac Killa Copper Plus	Teal And Mackrill Ltd	Amateur Professional	4659

7 Copper Resinate + Cuprous Oxide

Double Shield Antifouling	Indestructible Paint Company Ltd	Amateur Professional	6040
W Double Shield Antifouling	Llewellyn Ryland Ltd	Amateur Professional	3471

8 Copper Resinate + Cuprous Oxide + Zineb

Sigma Pilot Ecol Antifouling	Sigma Coatings Ltd	Amateur Professional	4933

9 Copper Sulphate + 2,4,5,6-Tetrachloro Isophthalonitrile

W Flexgard VI Waterbase Preservative	Flexabar Aquatech Corporation	Professional Professional (Aquaculture)	3319

10 Cuprous Oxide

Algicide Antifouling	Blakes Marine Paints	Amateur Professional	3219
AMC Sport Antifouling	Aquarius Marine Coatings Ltd	Amateur Professional	6395
D Amercoat 275	Ameron BV	Amateur Professional	3204
D Amercoat 277	Ameron BV	Amateur Professional	3205
W Amercoat 279	Ameron BV	Amateur Professional	3206
Anti-Fouling Paint 161P (Red And Chocolate To TS10240)	Witham Oil & Paint (Lowestoft) Ltd	Amateur Professional	3503
Antifouling Paint 161P (Red And Chocolate To TS10240)	International Coatings Ltd	Professional	3401
Antifouling Sargasso	Jotun-Henry Clark Ltd	Amateur Professional	6073
W Antifouling Sargasso Non-Tin	Jotun-Henry Clark Ltd	Amateur Professional	3418
Antifouling Seaguardian	Jotun-Henry Clark Ltd	Amateur Professional	3856
Antifouling Seaguardian (Black, Blue And MD).	Jotun-Henry Clark Ltd	Professional	4273

Product Name	Marketing Company	Use	HSE No.

10 Cuprous Oxide—continued

Product Name	Marketing Company	Use	HSE No.
Antifouling Seaquantum FB	Jotun-Henry Clark Ltd	Professional	7047
Antifouling Seavictor 40	Jotun-Henry Clark Ltd	Amateur Professional	4957
Antifouling Super	Jotun-Henry Clark Ltd	Amateur Professional	5812
Antifouling Super Tropic	Jotun-Henry Clark Ltd	Amateur Professional	3413
Antifouling Supertropic	Jotun-Henry Clark Ltd	Amateur Professional	6470
Aqua-Net 089	Steen-Hansen Maling AS	Professional	6897
Aquacleen	Mariner	Amateur Professional	5667
Aquagard (Flexgard XI)	Marineware Ltd	Amateur Professional	6589
Aquasafe	Gjoco A/S	Professional	5983
Aquasafe W	Gjoco A/S	Professional Professional (Aquaculture)	6353
Armachlor AF275	Johnstone's Paints PLC	Amateur Professional	5929
Armacote AF259	Johnstone's Paints PLC	Amateur Professional	5928
Armarine AF259	Johnstone's Paints PLC	Amateur Professional	5926
Armarine AF275	Johnstone's Paints PLC	Amateur Professional	5927
Awlgrip Awlstar Gold Label Antifouling	NOF Europe NV	Amateur Professional	5065
Blueline Copper SBA100	International Paint Ltd	Amateur Professional	5140
Boatguard	International Coatings Ltd	Amateur Professional Professional (Aquaculture)	3399
Bottomkote	International Coatings Ltd	Amateur Professional Professional (Aquaculture)	5903
Bradite CA21	Bradite Ltd	Amateur Professional	6147
Broads Freshwater	Blakes Marine Paints	Amateur Professional	3220
C-Clean 100	Camrex Holdings BV	Amateur Professional	5946
C-Clean 200	Camrex Holdings BV	Amateur Professional	5943

Product Name	Marketing Company	Use	HSE No.

10 Cuprous Oxide—continued

Product Name	Marketing Company	Use	HSE No.
C-Worthy	Benfleet Marine Wholesale	Amateur Professional	5476
D Classica 3786/093 Red	Ernesto Stoppani SPA	Amateur Professional	4797
Cobra V	Valiant Marine	Amateur Professional	5194
Cooper's Copolymer Antifouling	Coopers Marine Paints	Amateur Professional	5609
Copperpaint	International Paint Ltd	Amateur Professional	4119
Cupron Plus T.F	New Guard Coatings Ltd	Amateur Professional	5661
Envoy TF100	W And J Leigh And Company	Amateur Professional	3951
Flagspeed Antifouling	C W Wastnage Ltd	Amateur Professional	5825
Flexgard VI-II Waterbase Preservative	Aquatess Ltd	Professional Professional (Aquaculture)	6543
Hard Racing	International Paint Ltd	Amateur Professional	3393
D Hempel's Antifouling 761GB	Hempel Paints Ltd	Amateur Professional	3339
D Hempel's Antifouling Nordic 7133	Hempel Paints Ltd	Amateur Professional	3325
D Hempel's Antifouling Paint 161P (Red And Chocolate To TS10240)	Hempel Paints Ltd	Amateur Professional	3355
Hempel's Copper Bottom Paint 7116	Hempel Paints Ltd	Amateur Professional	4274
Hempel's Net Antifouling 715GB	Hempel Paints Ltd	Professional Professional (Aquaculture)	6342
D Hempel's Tin Free Antifouling 744GB	Hempel Paints Ltd	Amateur Professional	3330
Hempel's Tin Free Antifouling 7660	Hempel Paints Ltd	Amateur Professional	3338
Inflatable Boat Antifouling	Polymarine Ltd	Amateur Professional	6647
Interclene Extra BAA100 Series	International Paint Ltd	Amateur Professional	3371
Interclene Premium BCA300 Series	International Coatings Ltd	Amateur Professional	3372
Interclene Super BCA400 Series (BCA400 Red)	International Coatings Ltd	Amateur Professional	4084
Interclene Underwater Premium BCA468 Red	International Coatings Ltd	Amateur Professional	5059

Product Name	Marketing Company	Use	HSE No.

10 Cuprous Oxide—continued

Product Name	Marketing Company	Use	HSE No.
International TBT Free Copolymer Antifouling BQA100 Series	International Coatings Ltd	Amateur Professional	3375
Interspeed System 2 BRO142/240 Series	International Coatings Ltd	Amateur Professional	5634
Marclear Full Strength EU45 Antifouling	Marclear Espana SI	Amateur Professional	5987
Marclear High Strength Antifouling	Marclear Espana SI	Amateur Professional	5264
Micron 400 Series	International Coatings Ltd	Amateur Professional	5728
Micron CSC 100 Series	International Coatings Ltd	Amateur Professional	5731
Netrex AF	Tulloch Enterprises	Professional Professional (Aquaculture)	5684
Noa-Noa Rame Black/Red	Ernesto Stoppani SPA	Amateur Professional	4795
Nordrift Antifouling	Norland Distributors	Amateur Professional	5993
Norimp 2000 Black	Jotun-Henry Clark Ltd	Professional Professional (Aquaculture)	3404
Penguin Racing	Marine And Industrial Sealants	Amateur Professional	5673
Pilot Antifouling	Blakes Marine Paints	Amateur Professional	3226
Professional	M B Marine Coatings Ltd	Amateur Professional	5981
Ravax AF	Camrex Chugoku Ltd	Amateur Professional	5319
Scotwest Antifouling	A And M Paints	Amateur Professional	6120
Seaprince	Jotun-Henry Clark Ltd	Amateur Professional	4957
Seatender 10	Camrex Chugoku Ltd	Amateur Professional	5321
Seatender 7	Camrex Holdings BV	Amateur Professional	5320
Shearwater Racing Antifouling	Blakes Marine Paints	Amateur Professional	3228
Shiprite Sailing	Bradite Ltd	Amateur Professional	6302
Shiprite Traditional	Bradite Ltd	Amateur Professional	6178
Speedclean Antifouling	Mariner Paints	Amateur Professional	5077

Product Name	Marketing Company	Use	HSE No.

10 Cuprous Oxide—continued

Product Name	Marketing Company	Use	HSE No.
Standard Antifouling	Skipper (UK) Ltd	Amateur Professional	6194
Super Tropical Antifouling	Blakes Marine Paints	Amateur Professional	3229
Super Tropical Extra Antifouling	Blakes Marine Paints	Amateur Professional	3230
W Superspeed	D R Margetson	Amateur Professional	5191
Superspeed	Mariner Ltd	Amateur Professional	6210
Teamac Tropical Copper Antifouling (C/260/65)	Teal And Mackrill Ltd	Amateur Professional	3496
Tiger Cruising	Blakes Marine Paints	Amateur Professional	5099
Transocean Cleanship Antifouling 2.90	Spencer Coatings Ltd	Professional	7135
D Tropical Super Service Antifouling Paint	Devoe Coatings BV	Amateur Professional	3316
TS 10240 Antifouling ADA160 Series	International Coatings Ltd	Amateur Professional	3386
Unitas Antifouling Paint Chocolate	Witham Oil & Paint (Lowestoft) Ltd	Amateur Professional	3499
Unitas Antifouling Paint Red	Witham Oil & Paint (Lowestoft) Ltd	Amateur Professional	3498
Viniline	Skipper (UK) Ltd	Amateur Professional	6193
Vinilstop 9926 Red.	Ernesto Stoppani SPA	Professional	4798
Vinyl Antifouling 2000	Akzo Coatings BV	Amateur Professional	5633

11 Cuprous Oxide + 2,3,5,6-Tetrachloro-4-(Methyl Sulphonyl)Pyridine

Product Name	Marketing Company	Use	HSE No.
W Grassline TF Anti-Fouling Type M396	W And J Leigh And Company	Amateur Professional	3462

12 Cuprous Oxide + 2,4,5,6-Tetrachloro Isophthalonitrile

Product Name	Marketing Company	Use	HSE No.
Flexgard VI	Flexabar Aquatech Corporation	Professional Professional (Aquaculture)	6035
TFA 10 LA	Camrex Chugoku Ltd	Amateur Professional	5361

13 Cuprous Oxide + 2,4,5,6-Tetrachloro Isophthalonitrile + 4,5-Dichloro-2-N-Octyl-4-Isothiazolin-3-One

Product Name	Marketing Company	Use	HSE No.
C-Clean 400	Camrex Holdings BV	Professional	5947
Sea Grandprix 500 TCI	Chugoku Paints BV	Professional	7107
Seatender 15	Camrex Chugoku Ltd	Professional	5348

Product Name	Marketing Company	Use	HSE No.

13 Cuprous Oxide + 2,4,5,6-Tetrachloro Isophthalonitrile + 4,5-Dichloro-2-N-Octyl-4-Isothiazolin-3-One—continued

D TFA 10 HG	Camrex Chugoku Ltd	Amateur Professional	5366
D TFA 10G	Camrex Chugoku Ltd	Amateur Professional	5347

14 Cuprous Oxide + 2-(Thiocyanomethylthio)Benzothiazole

A3 Antifouling	Nautix SA	Professional	4367
A3 Teflon Antifouling	Nautix SA	Professional	4368
W Titan Tin Free	Blakes Marine Paints	Amateur Professional	3556

15 Cuprous Oxide + 2-Methylthio-4-Tertiary-Butylamino-6-Cyclopropylamino-S-Triazine

Algicide	Blakes Marine Paints	Amateur Professional	5738
Aquaspeed	Blakes Marine Paints	Amateur Professional	4511
D Aquaspeed Antifouling	Blakes Marine Paints	Amateur Professional	5817
Broads Antifouling Red	Blakes Marine Paints	Amateur Professional	6878
Broads Black Antifouling	Blakes Marine Paints	Amateur Professional	5739
Broads Freshwater Red	Blakes Marine Paints	Amateur Professional	5736
Challenger Antifouling	Blakes Marine Paints	Amateur Professional	4099
W Even TF	Veneziani SPA	Amateur Professional	5166
W Even Tin Free	Mark Dowland Marine Ltd	Amateur Professional	5841
Hard Racing Antifouling	Blakes Marine Paints	Amateur Professional	5704
Hempel's Antifouling Classic 7611 Red (Tin Free) 5000	Hempel Paints Ltd	Amateur Professional	5064
D Hempel's Antifouling Classic 76540	Hempel Paints Ltd	Amateur Professional	5291
D Hempel's Antifouling Combic 71990	Hempel Paints Ltd	Amateur Professional	5600
D Hempel's Antifouling Combic 71992	Hempel Paints Ltd	Amateur Professional	5601
D Hempel's Antifouling Forte Tin Free 7625	Hempel Paints Ltd	Amateur Professional	3351
D Hempel's Antifouling Mille Dynamic	Hempel Paints Ltd	Amateur	4440

Product Name	Marketing Company	Use	HSE No.

15 Cuprous Oxide + 2-Methylthio-4-Tertiary-Butylamino-6-Cyclopropylamino-S-Triazine—continued

	Product Name	Marketing Company	Use	HSE No.
D	Hempel's Antifouling Nautic 71900	Hempel Paints Ltd	Amateur Professional	5290
W	Hempel's Antifouling Nautic 71902	Hempel Paints Ltd	Amateur Professional	5605
	Hempel's Antifouling Nautic 8190C	Hempel Paints Ltd	Amateur Professional	6043
	Hempel's Antifouling Olympic HI-7661	Hempel Paints Ltd	Amateur Professional	4898
D	Hempel's Antifouling Olympic Tin Free 7154	Hempel Paints Ltd	Amateur Professional	3718
D	Hempel's Antifouling Tin Free 745GB	Hempel Paints Ltd	Amateur Professional	3331
D	Hempel's Antifouling Tin Free 750GB	Hempel Paints Ltd	Amateur Professional	3332
	Hempel's Hard Racing 76480	Hempel Paints Ltd	Amateur Professional	5538
	Hempel's Mille Dynamic 71700	Hempel Paints Ltd	Amateur Professional	5574
	Hempels Antifouling Bravo Tin Free 7610	Hempel Paints Ltd	Amateur Professional	4482
	Interspeed Antifouling BWO900 Series	International Coatings Ltd	Amateur Professional	5636
	Interspeed System 2 BRA143 Brown	International Paint Ltd	Amateur Professional	4301
	Interviron BQA450 Series	International Paint Ltd	Amateur Professional	4657
	Interviron Super Tin-Free Polishing Antifouling BQO400 Series	International Coatings Ltd	Amateur Professional	5637
D	Long Life T.F.	Veneziani SPA	Amateur Professional	4936
	Longlife Antifouling	Skipper (UK) Ltd	Amateur Professional	6195
	Micron CSC 200 Series	International Coatings Ltd	Amateur Professional	5732
W	Patente Laxe	Teais SA	Amateur Professional	5497
	Pilot	Blakes Marine Paints	Amateur Professional	5959
D	Raffaello Plus Tin Free	Veneziani SPA	Amateur Professional	3642
	Seatech	Blakes Marine Paints	Amateur Professional	7117
	Shiprite Speed	Bradite Ltd	Amateur Professional	6603
	Tigerline	Blakes Marine Paints	Amateur Professional	6872

335

Product Name	Marketing Company	Use	HSE No.

15 Cuprous Oxide + 2-Methylthio-4-Tertiary-Butylamino-6-Cyclopropylamino-S-Triazine—continued

Titan FGA Antifouling	Blakes Marine Paints	Amateur Professional	5681
Titan Ultra	Blakes Marine Paints	Amateur Professional	7096
VC Offshore	International Coatings Ltd	Amateur Professional	4777
Waterline	Blakes Marine Paints	Amateur Professional	7099
XM Anti-Fouling C2000 Cruising Self Eroding	X M Yachting	Amateur	6176
XM Anti-Fouling P4000 Hard	X M Yachting	Amateur	6175
XM Antifouling HS3000 High Performance Self Eroding	X M Yachting	Amateur	6124

16 Cuprous Oxide + 2-Methylthio-4-Tertiary-Butylamino-6-Cyclopropylamino-S-Triazine + 4,5-Dichloro-2-N-Octyl-4-Isothiazolin-3-One

Grassline Type M396 Antifouling	W And J Leigh And Company	Professional	7075

17 Cuprous Oxide + 4,5-Dichloro-2-N-Octyl-4-Isothiazolin-3-One

ABC#3E Antifouling	Ameron BV	Professional	7051
Antifouling Seavictor 50	Jotun-Henry Clark Ltd	Professional	4958
Hempel's Antifouling Globic SP-Eco 81900	Hempel Paints Ltd	Professional	6531
Hempel's Antifouling Globic SP-Eco 81990	Hempel Paints Ltd	Professional	6532
D Hempel's Antifouling Nautic Tin Free 7190E	Hempel Paints Ltd	Amateur Professional	5145
Hempel's Antifouling Tin Free 743GB	Hempel Paints Ltd	Amateur Professional	3329
Hempel's Antifouling Tin Free 751GB	Hempel Paints Ltd	Amateur Professional	3333
Hempel's Antifouling Tin Free 7662	Hempel Paints Ltd	Amateur Professional	3336
Hempel's Tin Free Antifouling 7626	Hempel Paints Ltd	Amateur Professional	3337
Hempels Antifouling Globic SP-Eco 81920	Hempel Paints Ltd	Professional	6877
Interviron Super Tin-Free Polishing Antifouling BQO420 Series	International Coatings Ltd	Professional	5642
Micron CSC 300 Series	International Coatings Ltd	Professional	5724
Sigma Alphagen 20	Sigma Coatings BV	Professional	7089

Product Name	Marketing Company	Use	HSE No.

18 Cuprous Oxide + Copper Resinate

Double Shield Antifouling	Indestructible Paint Company Ltd	Amateur Professional	6040
W Double Shield Antifouling	Llewellyn Ryland Ltd	Amateur Professional	3471

19 Cuprous Oxide + Cuprous Sulphide

D Black Antifouling Paint 317 (To TS10239)	Hempel Paints Ltd	Amateur Professional	3370
D Hempel's Antifouling 762GB	Hempel Paints Ltd	Amateur Professional	3340

20 Cuprous Oxide + Cuprous Thiocyanate + 2,3,5,6-Tetrachloro-4-(Methyl Sulphonyl)Pyridine + 2-Methylthio-4-Tertiary-Butylamino-6-Cyclopropylamino-S-Triazine

W Envoy TF 400	W And J Leigh And Company	Amateur Professional	4432
W Envoy TF 500	W And J Leigh And Company	Amateur Professional	5599
W Meridian MP40 Antifouling	Deangate Marine Products	Amateur Professional	5027
W Micro-Tech	Marine (Chemi-Technics) Ltd	Amateur Professional	6316

21 Cuprous Oxide + Cuprous Thiocyanate + 2-Methylthio-4-Tertiary-Butylamino-6-Cyclopropylamino-S-Triazine + 4,5-Dichloro-2-N-Octyl-4-Isothiazolin-3-One

Envoy TF 400 Antifouling	W And J Leigh And Company	Professional	7072
Envoy TF 500 Antifouling	W And J Leigh And Company	Professional	7073

22 Cuprous Oxide + Cuprous Thiocyanate + Dichlorophenyl Dimethylurea + Zineb + 2,3,5,6-Tetrachloro-4-(Methyl Sulphonyl)Pyridine

W Even Extreme	New Guard Coatings Ltd	Amateur Professional	6625

23 Cuprous Oxide + Dichlofluanid

Aqua-Net	Steen-Hansen Maling AS	Professional (Aquaculture)	5845
Copper Net	Steen-Hansen Maling AS	Professional (Aquaculture)	6034
Halcyon 5000 (Base)	Waterline	Amateur Professional	5396

Product Name	Marketing Company	Use	HSE No.

23 Cuprous Oxide + Dichlofluanid—continued

Hempel's Antifouling Rennot 7150	Hempel Paints Ltd	Professional Professional (Aquaculture)	3364
D Hempel's Antifouling Rennot 7177	Hempel Paints Ltd	Professional Professional (Aquaculture)	3365
D Hempel's Antifouling Tin Free 742GB	Hempel Paints Ltd	Amateur Professional	3328
Interspeed Ultra	International Coatings Ltd	Amateur Professional	6660
Micron Extra	International Coatings Ltd	Amateur Professional	6663
Net-Guard	Steen-Hansen Maling AS	Professional (Aquaculture)	5657
New Improved Cruiser Premium	International Coatings Ltd	Amateur Professional	7095
Penguin Non-Stop	Marine And Industrial Sealants	Amateur Professional	5671
Seashield (Base)	Waterline	Amateur Professional	5438
Slipstream Antifouling	Witham Oil & Paint (Lowestoft) Ltd	Amateur Professional	3721

24 Cuprous Oxide + Dichlofluanid + 2-Methylthio-4-Tertiary-Butylamino-6-Cyclopropylamino-S-Triazine

Antifouling 1.2	Plastimo International	Amateur Professional	6159

25 Cuprous Oxide + Dichlofluanid + Dichlorophenyl Dimethylurea + 2-Methylthio-4-Tertiary-Butylamino-6-Cyclopropylamino-S-Triazine

Antifouling 1.3 Black And Red (400-03 / Ind.3)	Plastimo International	Amateur Professional	6350

26 Cuprous Oxide + Dichlofluanid + Zinc Naphthenate + Copper Naphthenate

D Teamac Killa Copper Plus	Teal And Mackrill Ltd	Amateur Professional	4659

27 Cuprous Oxide + Dichlorophenyl Dimethylurea

A4 Antifouling	Nautix SA	Amateur Professional	4369
Aquarius Extra Strong	International Paint Ltd	Amateur Professional	4280
Blakes Hard Racing	Blakes Marine Paints	Amateur Professional	5764

Product Name	Marketing Company	Use	HSE No.

27 Cuprous Oxide + Dichlorophenyl Dimethylurea—continued

Product Name	Marketing Company	Use	HSE No.
Blueline Tropical SBA300	International Paint Ltd	Amateur Professional	5139
Boatgard Antifouling	International Coatings Ltd	Amateur Professional	6854
Cruiser Premium	International Coatings Ltd	Amateur Professional	5127
Drake Antifouling	M B Marine Coatings Ltd	Amateur Professional	6671
W Even Tin Free Light Grey	Mark Dowland Marine Ltd	Amateur Professional	5839
Giraglia	M B Marine Coatings Ltd	Amateur Professional	5980
D Grafo Anti-Foul SW	Grafo Coatings Ltd	Amateur Professional	5380
D Hempel's Antifouling 8199D	Hempel Paints Ltd	Amateur Professional	6177
D Hempel's Antifouling 8199F	Hempel Paints Ltd	Amateur Professional	6179
D Hempel's Antifouling Classic 7654B	Hempel Paints Ltd	Amateur Professional	5285
Hempel's Antifouling Combic 7199B	Hempel Paints Ltd	Amateur Professional	5274
Hempel's Antifouling Nautic 7190B	Hempel Paints Ltd	Amateur Professional	5273
Hempel's Antifouling Nautic 8190H	Hempel Paints Ltd	Amateur Professional	6042
Hempel's Bravo 7610A	Hempel Paints Ltd	Amateur Professional	5603
Hempel's Hard Racing 7648A	Hempel Paints Ltd	Amateur Professional	5535
Hempel's Mille Dynamic 7170A	Hempel Paints Ltd	Amateur Professional	5536
Hempel's Seatech Antifouling 7820A	Hempel Paints Ltd	Amateur Professional	6098
International Tin Free SPC BNA100 Series	International Paint Ltd	Amateur Professional	5186
Intersmooth Hisol Tin Free BGA620 Series	International Paint Ltd	Amateur Professional	4787
Intersmooth Tin Free BGA530 Series	International Paint Ltd	Amateur Professional	4611
Interspeed Extra BWA500 Red	International Coatings Ltd	Amateur Professional	4303
Interspeed Extra Strong	International Coatings Ltd	Amateur Professional	4819
Interspeed Super BWA900 Red	International Coatings Ltd	Amateur Professional	4884

Product Name	Marketing Company	Use	HSE No.

27 Cuprous Oxide + Dichlorophenyl Dimethylurea—continued

Product Name	Marketing Company	Use	HSE No.
Interspeed Super BWA909 Black	International Coatings Ltd	Amateur Professional	5058
Interspeed System 2 BRA142 Brown	International Paint Ltd	Amateur Professional	4302
Interswift Tin Free BQA400 Series	International Coatings Ltd	Amateur Professional	4842
Interswift Tin-Free SPC BTA540 Series	International Paint Ltd	Amateur Professional	5078
Interviron Super BNA400 Series	International Coatings Ltd	Amateur Professional	5690
Interviron Super BQA400 Series	International Coatings Ltd	Amateur Professional	5409
W Long Life Tin Free	Mark Dowland Marine Ltd	Amateur Professional	5840
Micron 500 Series	International Coatings Ltd	Amateur Professional	5729
Micron CSC	International Paint Ltd	Amateur Professional	4775
Micron CSC Extra	International Coatings Ltd	Amateur Professional	6263
Micron Plus Antifouling	International Paint Ltd	Amateur Professional	5133
Noa-Noa Rame	Ernesto Stoppani SPA	Amateur Professional	4796
Prima	International Coatings Ltd	Amateur Professional	6683
Professional Soft Antifouling	International Coatings Ltd	Amateur Professional	7093
Seajet 033	Camrex Chugoku Ltd	Amateur Professional	5331
Seatech Antifouling	Blakes Marine Paints	Amateur Professional	6125
Slippy Bottom	Thincoat Technology International Ltd	Amateur Professional	6156
Soft Antifouling	International Coatings Ltd	Amateur Professional	6067
Teamac Antifouling D (C/258 Series)	Teal And Mackrill Ltd	Amateur Professional	6418
Teamac Killa Copper Premium Antifouling (C/262/24)	Teal And Mackrill Ltd	Amateur Professional	6410
Titan FGA	Hempel Paints Ltd	Amateur Professional	5767
Transocean Optima Antifouling 2.34	Spencer Coatings Ltd	Amateur Professional	6912
Waterways Antifouling	International Coatings Ltd	Amateur Professional	5565

28 Cuprous Oxide + Dichlorophenyl Dimethylurea + 2,3,5,6-Tetrachloro-4-(Methyl Sulphonyl)Pyridine

W Raffaello 3	New Guard Coatings Ltd	Amateur Professional	5826

29 Cuprous Oxide + Dichlorophenyl Dimethylurea + 2,4,5,6-Tetrachloro Isophthalonitrile

C-Clean 300	Camrex Holdings BV	Amateur Professional	5942
Seatender 12	Camrex Chugoku Ltd	Amateur Professional	5324
TFA 10	Camrex Chugoku Ltd	Amateur Professional	5346
D TFA 10 H	Camrex Chugoku Ltd	Amateur Professional	5358

30 Cuprous Oxide + Dichlorophenyl Dimethylurea + 2,4,5,6-Tetrachloro Isophthalonitrile + 2-Methylthio-4-Tertiary-Butylamino-6-Cyclopropylamino-S-Triazine

Sigmaplane Ecol HA 120 Antifouling	Sigma Coatings Ltd	Amateur Professional	5788

31 Cuprous Oxide + Dichlorophenyl Dimethylurea + 2-Methylthio-4-Tertiary-Butylamino-6-Cyclopropylamino-S-Triazine

A3 Antifouling 072015	Nautix SA	Amateur Professional	5224
A4 Antifouling 072017	Nautix SA	Amateur Professional	5226
Le Marin	Nautix SA	Amateur Professional	5480

32 Cuprous Oxide + Dichlorophenyl Dimethylurea + 4,5-Dichloro-2-N-Octyl-4-Isothiazolin-3-One

TFA-10 LA Q	Camrex Chugoku Ltd	Professional	6581

33 Cuprous Oxide + Dichlorophenyl Dimethylurea + Tributyltin Methacrylate + Tributyltin Oxide

A11 Antifouling 072014	Nautix SA	Professional	5376
Interswift BKO000/700 Series	International Coatings Ltd	Professional	5640

34 Cuprous Oxide + Tributyltin Acrylate

D Amercoat 697	Ameron BV	Professional	3513

Product Name	Marketing Company	Use	HSE No.

35 Cuprous Oxide + Tributyltin Methacrylate

ABC # 1 Antifouling	Ameron BV	Professional	3214
Devran MCP Antifouling Red	Devoe Coatings BV	Professional	3311
Devran MCP Antifouling Red Brown	Devoe Coatings BV	Professional	3310
D Takata III Antifouling	NOF Europe NV	Professional	4050
D Takata III Antifouling Hi-Solid	NOF Europe NV	Professional	4052
D Takata III Antifouling LS	NOF Europe NV	Professional	4051
D Takata III Antifouling Ls Hi-Solid	NOF Europe NV	Professional	4053
D Takata III Antifouling No 2001	NOF Europe NV	Professional	4054

36 Cuprous Oxide + Tributyltin Methacrylate + Tributyltin Oxide

AF Seaflo Mark 2-1	Camrex Chugoku Ltd	Professional	5316
D AF Seaflo Z-100 HS-1	Camrex Chugoku Ltd	Professional	5318
AF Seaflo Z-100 LE-HS-1	Camrex Chugoku Ltd	Professional	5313
W Antifouling Seaconomy	Jotun-Henry Clark Ltd	Professional	5791
Antifouling Seaconomy	Jotun-Henry Clark Ltd	Professional	6261
D Antifouling Seaconomy 200	Jotun-Henry Clark Ltd	Professional	4436
D Antifouling Seaconomy 300	Jotun-Henry Clark Ltd	Professional	4272
Antifouling Seamate FB30	Jotun-Henry Clark Ltd	Professional	6506
D Antifouling Seamate HB 33	Jotun-Henry Clark Ltd	Professional	4270
Antifouling Seamate HB 99	Jotun-Henry Clark Ltd	Professional	3411
Antifouling Seamate HB 99 Black	Jotun-Henry Clark Ltd	Professional	3410
Antifouling Seamate HB 99 Dark Red	Jotun-Henry Clark Ltd	Professional	3412
W Antifouling Seamate HB22	Jotun-Henry Clark Ltd	Professional	4271
Antifouling Seamate HB22	Jotun-Henry Clark Ltd	Professional	6301
Antifouling Seamate HB33	Jotun-Henry Clark Ltd	Professional	5811
Antifouling Seamate HB66	Jotun-Henry Clark Ltd	Professional	5698
Antifouling Seamate HB66	Jotun-Henry Clark Ltd	Professional	6475
Antifouling Seamate SB33	Jotun-Henry Clark Ltd	Professional	6517
C-Clean 6000	Camrex Holdings BV	Professional	5945
Grassline ABL Antifouling Type M349	W And J Leigh And Company	Professional	5374
D Hempel's Antifouling Combic 76990	Hempel Paints Ltd	Professional	3348
Hempel's Antifouling Nautic HI 76900	Hempel Paints Ltd	Professional	3360
D Hempel's Antifouling Nautic HI 76910	Hempel Paints Ltd	Professional	3361
Hempel's Antifouling Nautic SP-Ace 79031	Hempel Paints Ltd	Professional	6452
Hempel's Antifouling Nautic SP-Ace 79051	Hempel Paints Ltd	Professional	6453
Intersmooth SPC Antifouling BFO250 Series	International Coatings Ltd	Professional	5635
Intersmooth SPC BFA090/BFA190 Series	International Paint Ltd	Professional	3377
Seaflo 15	Camrex Chugoku Ltd	Professional	6284

Product Name	Marketing Company	Use	HSE No.

37 Cuprous Oxide + Tributyltin Methacrylate + Tributyltin Oxide + 4,5-Dichloro-2-N-Octyl-4-Isothiazolin-3-One

Intersmooth Hisol BFO270/950/970 Series	International Coatings Ltd	Professional	5641

38 Cuprous Oxide + Tributyltin Methacrylate + Tributyltin Oxide + Zineb

D	Hempel's Antifouling 7690D	Hempel Paints Ltd	Professional	6056
D	Hempel's Antifouling Economic SP-Sea 74030	Hempel Paints Ltd	Professional	5581
D	Hempel's Antifouling Nautic 7691B	Hempel Paints Ltd	Professional	5586
	Hempel's Antifouling Nautic SP-Ace 79030	Hempel Paints Ltd	Professional	6331
	Hempel's Antifouling Nautic SP-Ace 79050	Hempel Paints Ltd	Professional	6325
	Hempel's Antifouling Nautic SP-Ace 79070	Hempel Paints Ltd	Professional	6328
	Hempel's Economic SP-Sea 74010	Hempel Paints Ltd	Professional	6330
	Intersmooth 110 Standard	International Coatings Ltd	Professional	5907
	Intersmooth 120 Premium	International Coatings Ltd	Professional	5913
	Intersmooth 130 Ultra	International Coatings Ltd	Professional	5914
	Intersmooth 220 Premium	International Coatings Ltd	Professional	5911
	Intersmooth 230 Ultra	International Coatings Ltd	Professional	5912
	Intersmooth 320 Premium	International Coatings Ltd	Professional	5910
	Intersmooth 330 Ultra	International Coatings Ltd	Professional	5924
	Intersmooth Hisol 2000 BFA270 Series	International Coatings Ltd	Professional	3844
	Intersmooth Hisol 9000 BFA970 Series	International Coatings Ltd	Professional	5461
	Seaflo 10	Camrex Chugoku Ltd	Professional	6283
	Superyacht 900	International Coatings Ltd	Professional	6278
	Transocean Masterline Antifouling 2.84	Spencer Coatings Ltd	Professional	6934

39 Cuprous Oxide + Tributyltin Methacrylate + Zineb

Product Name	Marketing Company	Use	HSE No.
Antifouling Seamate HB22	Jotun-Henry Clark Ltd	Professional	5362
Antifouling Seamate HB33	Jotun-Henry Clark Ltd	Professional	5363
Nu Wave A/F	Wilckens Farben GmbH	Professional	6689
Rabamarine A/F 2500	Wilckens Farben GmbH	Professional	6690
Sigmaplane HA Antifouling	Sigma Coatings BV	Professional	4345
Sigmaplane HB Antifouling	Sigma Coatings Ltd	Professional	5721
Sigmaplane TA Antifouling	Sigma Coatings BV	Professional	4346

Product Name	Marketing Company	Use	HSE No.

40 Cuprous Oxide + Zinc Naphthenate

D Teamac Killa Copper	Teal And Mackrill Ltd	Amateur Professional	3494
D Teamac Super Tropical	Teal And Mackrill Ltd	Amateur Professional	3497

41 Cuprous Oxide + Zinc Pyrithione

Blakes Antifouling 87910	Blakes Marine Paints	Amateur Professional	6450
Intersmooth 360 Ecoloflex	International Coatings Ltd	Amateur Professional	6057
Intersmooth 365 Ecoloflex	International Coatings Ltd	Amateur Professional	6557
Intersmooth 460 Ecoloflex	International Coatings Ltd	Amateur Professional	6276
Intersmooth 465 Ecoloflex	International Coatings Ltd	Amateur Professional	6556
D Micron 600 Series	International Paint Ltd	Amateur Professional	5733
Micron Optima	International Coatings Ltd	Amateur Professional	5941
VC Offshore Extra 100 Series	International Coatings Ltd	Amateur Professional	5730

42 Cuprous Oxide + Zineb

Blueline SPC Tin Free SBA700 Series	International Coatings Ltd	Amateur Professional	5214
Equatorial	International Paint Ltd	Amateur Professional	4121
D Hempel's Antifouling Combic 7199F	Hempel Paints Ltd	Amateur Professional	5744
Interclene 245	International Coatings Ltd	Professional	7019
Interspeed 340	International Coatings Ltd	Amateur Professional	6089
Interspeed 340	International Coatings Ltd	Professional	7020
Interspeed System 2 BRA140/ BRA240 Series	International Coatings Ltd	Amateur Professional	3847
Interviron BQA200 Series	International Coatings Ltd	Amateur Professional	3846

43 Cuprous Oxide + Zineb + Copper Resinate

Sigma Pilot Ecol Antifouling	Sigma Coatings Ltd	Amateur Professional	4933

44 Cuprous Sulphide + Cuprous Oxide

D Black Antifouling Paint 317 (To TS10239)	Hempel Paints Ltd	Amateur Professional	3370

Product Name	Marketing Company	Use	HSE No.

44 Cuprous Sulphide + Cuprous Oxide—continued

| D Hempel's Antifouling 762GB | Hempel Paints Ltd | Amateur
Professional | 3340 |

45 Cuprous Thiocyanate

Antifouling Broken White DL-2253	International Paint Ltd	Amateur Professional	3400
D Hempel's Antifouling 763GB	Hempel Paints Ltd	Amateur Professional	3537
Interspeed 2001	International Coatings Ltd	Amateur Professional	5727
Unitas Antifouling Paint White	Witham Oil & Paint (Lowestoft) Ltd	Amateur Professional	3500

46 Cuprous Thiocyanate + 2,3,5,6-Tetrachloro-4-(Methyl Sulphonyl)Pyridine + 2-Methylthio-4-Tertiary-Butylamino-6-Cyclopropylamino-S-Triazine + Cuprous Oxide

W Envoy TF 400	W And J Leigh And Company	Amateur Professional	4432
W Envoy TF 500	W And J Leigh And Company	Amateur Professional	5599
W Meridian MP40 Antifouling	Deangate Marine Products	Amateur Professional	5027
W Micro-Tech	Marine (Chemi-Technics) Ltd	Amateur Professional	6316

47 Cuprous Thiocyanate + 2-Methylthio-4-Tertiary-Butylamino-6-Cyclopropylamino-S-Triazine

AL-27	International Paint Ltd	Amateur Professional	3613
Aquaspeed White	Blakes Marine Paints	Amateur Professional	4513
Boot Top Plus (Gull White)	International Paint Ltd	Amateur Professional	3389
Cruiser Superior 100 Series	International Coatings Ltd	Amateur Professional	5723
D Even TF Light Grey	Veneziani SPA	Amateur Professional	5167
Hard Racing Antifouling White	Blakes Marine Paints	Amateur Professional	5705
Hempel's Mille Dynamic 71600	Hempel Paints Ltd	Amateur Professional	5576
Hempel's Tin Free Hard Racing 76380	Hempel Paints Ltd	Amateur Professional	5540
D Hempel's Tin Free Hard Racing 7648	Hempel Paints Ltd	Amateur Professional	3367

Product Name	Marketing Company	Use	HSE No.

47 Cuprous Thiocyanate + 2-Methylthio-4-Tertiary-Butylamino-6-Cyclopropylamino-S-Triazine—continued

Interspeed 2000	International Coatings Ltd	Amateur Professional	4148
Marclear Powerboat Antifouling	Marclear Marine Products Ltd	Amateur Professional	6586
MPX	International Coatings Ltd	Amateur Professional	4818
Propeller T.F.	New Guard Coatings Ltd	Amateur Professional	5660
Quicksilver Antifouling	International Coatings Ltd	Amateur Professional	5938
W Raffaello Alloy	Mark Dowland Marine Ltd	Amateur Professional	5492
Tiger Cruising White	Blakes Marine Paints	Amateur Professional	5712
Tigerline Antifouling	Blakes Marine Paints	Amateur Professional	3842
Titan FGA Antifouling White	Blakes Marine Paints	Amateur Professional	5680
VC Offshore	International Coatings Ltd	Amateur Professional	4779
VC Prop-O-Drev	International Coatings Ltd	Amateur Professional	4217
Waterline White	Blakes Marine Paints	Amateur Professional	7097

48 Cuprous Thiocyanate + 2-Methylthio-4-Tertiary-Butylamino-6-Cyclopropylamino-S-Triazine + 4,5-Dichloro-2-N-Octyl-4-Isothiazolin-3-One + Cuprous Oxide

Envoy TF 400 Antifouling	W And J Leigh And Company	Professional	7072
Envoy TF 500 Antifouling	W And J Leigh And Company	Professional	7073

49 Cuprous Thiocyanate + 4,5-Dichloro-2-N-Octyl-4-Isothiazolin-3-One

Interspeed 2003	International Coatings Ltd	Professional	5726

50 Cuprous Thiocyanate + Dichlofluanid

Interspeed 2000 White	International Coatings Ltd	Amateur Professional	6631
Penguin Non-Stop White	Marine And Industrial Sealants	Amateur Professional	5672
Penguin Racing White	Marine And Industrial Sealants	Amateur Professional	5674

Product Name	Marketing Company	Use	HSE No.

50 Cuprous Thiocyanate + Dichlofluanid—continued

Selfpolishing Antifouling 2000	Akzo Nobel Coatings BV	Amateur Professional	6521
Trilux	International Coatings Ltd	Amateur Professional	6661

51 Cuprous Thiocyanate + Dichlofluanid + 2-Methylthio-4-Tertiary-Butylamino-6-Cyclopropylamino-S-Triazine

Antifouling 1.2 (W)	Plastimo International	Amateur Professional	6160
D Hempel's Antifouling Mille Dynamic 717 GB	Hempel Paints Ltd	Amateur Professional	4437

52 Cuprous Thiocyanate + Dichlofluanid + Dichlorophenyl Dimethylurea + 2-Methylthio-4-Tertiary-Butylamino-6-Cyclopropylamino-S-Triazine

Antifouling 1.3	Plastimo International	Amateur Professional	6161
Antifouling 1.3 White (327-04)	Plastimo International	Amateur Professional	6349

53 Cuprous Thiocyanate + Dichlofluanid + Zinc Pyrithione + 2-Methylthio-4-Tertiary-Butylamino-6-Cyclopropylamino-S-Triazine

Antifouling 1.3 (W)	Plastimo International	Amateur Professional	6162

54 Cuprous Thiocyanate + Dichlorophenyl Dimethylurea

Antialga	New Guard Coatings Ltd	Amateur Professional	5662
Aqua 12	International Coatings Ltd	Amateur Professional	4802
Aquarius AL	International Paint Ltd	Amateur Professional	4295
Hempel's Hard Racing 7638A	Hempel Paints Ltd	Amateur Professional	5539
Hempel's Mille ALU 71602	Hempel Paints Ltd	Amateur Professional	5578
Hempel's Mille Dynamic 7160A	Hempel Paints Ltd	Amateur Professional	5575
Seatech Antifouling White	Blakes Marine Paints	Amateur Professional	6140
D Tiger Cruising-A Antifouling	Blakes Marine Paints	Amateur Professional	5765

55 Cuprous Thiocyanate + Dichlorophenyl Dimethylurea + 2,3,5,6-Tetrachloro-4-(Methyl Sulphonyl)Pyridine

W Raffaello Racing	New Guard Coatings Ltd	Amateur Professional	6451

Product Name	Marketing Company	Use	HSE No.

56 Cuprous Thiocyanate + Dichlorophenyl Dimethylurea + 2-Methylthio-4-Tertiary-Butylamino-6-Cyclopropylamino-S-Triazine

A3 Antifouling 072016	Nautix SA	Amateur Professional	5225
A3 Teflon Antifouling 062015	Nautix SA	Amateur Professional	5481
A4 Antifouling 072018	Nautix SA	Amateur Professional	5227
A4 Teflon Antifouling 062017	Nautix SA	Amateur Professional	5482

57 Cuprous Thiocyanate + Dichlorophenyl Dimethylurea + 4,5-Dichloro-2-N-Octyl-4-Isothiazolin-3-One

Sprint	New Guard Coatings Ltd	Professional	6449

58 Cuprous Thiocyanate + Dichlorophenyl Dimethylurea + Zineb + 2,3,5,6-Tetrachloro-4-(Methyl Sulphonyl)Pyridine + Cuprous Oxide

W Even Extreme	New Guard Coatings Ltd	Amateur Professional	6625

59 Cuprous Thiocyanate + Tributyltin Methacrylate + Tributyltin Oxide

Antifouling HB 66 Ocean Green	Jotun-Henry Clark Ltd	Professional	3408
D Antifouling Seamate HB Green	Jotun-Henry Clark Ltd	Professional	3430
Antifouling Seamate HB22 Roundel Blue	Jotun-Henry Clark Ltd	Professional	4872
Antifouling Seamate HB66 Black	Jotun-Henry Clark Ltd	Professional	4871
Intersmooth SPC BFA040/BFA050 Series	International Coatings Ltd	Professional	3376

60 Cuprous Thiocyanate + Tributyltin Methacrylate + Tributyltin Oxide + Zineb

Cruiser Copolymer	International Paint Ltd	Professional	3514
Intersmooth Hisol BFA948 Orange	International Coatings Ltd	Professional	4281
Intersmooth Hisol SPC Antifouling BFA949 Red	International Coatings Ltd	Professional	4949
Micron 25 Plus	International Coatings Ltd	Professional	3402
Superyacht Antifouling	International Coatings Ltd	Professional	5462

61 Cuprous Thiocyanate + Zinc Pyrithione

D Hempel's Antifouling Combic 7199C	Hempel Paints Ltd	Amateur Professional	5742
Interspeed 2002	International Coatings Ltd	Amateur Professional	5725
D Titan FGA-C Antifouling	Blakes Marine Paints	Amateur Professional	5766

Product Name	Marketing Company	Use	HSE No.

62 Dichlofluanid

| Bayer AFC | Bayer PLC | Amateur
Professional | 5163 |

63 Dichlofluanid + 2-Methylthio-4-Tertiary-Butylamino-6-Cyclopropylamino-S-Triazine + Cuprous Oxide

| Antifouling 1.2 | Plastimo International | Amateur
Professional | 6159 |

64 Dichlofluanid + 2-Methylthio-4-Tertiary-Butylamino-6-Cyclopropylamino-S-Triazine + Cuprous Thiocyanate

| Antifouling 1.2 (W) | Plastimo International | Amateur
Professional | 6160 |
| D Hempel's Antifouling Mille Dynamic 717 GB | Hempel Paints Ltd | Amateur
Professional | 4437 |

65 Dichlofluanid + Cuprous Oxide

Aqua-Net	Steen-Hansen Maling AS	Professional Professional (Aquaculture)	5845
Copper Net	Steen-Hansen Maling AS	Professional (Aquaculture)	6034
Halcyon 5000 (BASE)	Waterline	Amateur Professional	5396
Hempel's Antifouling Rennot 7150	Hempel Paints Ltd	Professional Professional (Aquaculture)	3364
D Hempel's Antifouling Rennot 7177	Hempel Paints Ltd	Professional Professional (Aquaculture)	3365
D Hempel's Antifouling Tin Free 742GB	Hempel Paints Ltd	Amateur Professional	3328
Interspeed Ultra	International Coatings Ltd	Amateur Professional	6660
Micron Extra	International Coatings Ltd	Amateur Professional	6663
Net-Guard	Steen-Hansen Maling AS	Professional (Aquaculture)	5657
New Improved Cruiser Premium	International Coatings Ltd	Amateur Professional	7095
Penguin Non-Stop	Marine And Industrial Sealants	Amateur Professional	5671
Seashield (BASE)	Waterline	Amateur Professional	5438
Slipstream Antifouling	Witham Oil & Paint (Lowestoft) Ltd	Amateur Professional	3721

Product Name	Marketing Company	Use	HSE No.

66 Dichlofluanid + Cuprous Thiocyanate

Interspeed 2000 White	International Coatings Ltd	Amateur Professional	6631
Penguin Non-Stop White	Marine And Industrial Sealants	Amateur Professional	5672
Penguin Racing White	Marine And Industrial Sealants	Amateur Professional	5674
Selfpolishing Antifouling 2000	Akzo Nobel Coatings BV	Amateur Professional	6521
Trilux	International Coatings Ltd	Amateur Professional	6661

67 Dichlofluanid + Dichlorophenyl Dimethylurea + 2-Methylthio-4-Tertiary-Butylamino-6-Cyclopropylamino-S-Triazine + Cuprous Oxide

Antifouling 1.3 Black And Red (400-03 / Ind.3)	Plastimo International	Amateur Professional	6350

68 Dichlofluanid + Dichlorophenyl Dimethylurea + 2-Methylthio-4-Tertiary-Butylamino-6-Cyclopropylamino-S-Triazine + Cuprous Thiocyanate

Antifouling 1.3	Plastimo International	Amateur Professional	6161
Antifouling 1.3 White (327-04)	Plastimo International	Amateur Professional	6349

69 Dichlofluanid + Dichlorophenyl Dimethylurea + Zinc Pyrithione

Antifouling 1.4	Plastimo International	Professional	6163

70 Dichlofluanid + Zinc Naphthenate + Copper Naphthenate + Cuprous Oxide

D Teamac Killa Copper Plus	Teal And Mackrill Ltd	Amateur Professional	4659

71 Dichlofluanid + Zinc Pyrithione + 2-Methylthio-4-Tertiary-Butylamino-6-Cyclopropylamino-S-Triazine + Cuprous Thiocyanate

Antifouling 1.3 (W)	Plastimo International	Amateur Professional	6162

72 4,5-Dichloro-2-N-Octyl-4-Isothiazolin-3-One + 2-Methylthio-4-Tertiary-Butylamino-6-Cyclopropylamino-S-Triazine

Envoy TCF 600 Antifouling	W And J Leigh And Company	Professional	7074

73 4,5-Dichloro-2-N-Octyl-4-Isothiazolin-3-One + Cuprous Oxide

ABC#3E Antifouling	Ameron BV	Professional	7051
Antifouling Seavictor 50	Jotun-Henry Clark Ltd	Professional	4958

Product Name	Marketing Company	Use	HSE No.

73 4,5-Dichloro-2-N-Octyl-4-Isothiazolin-3-One + Cuprous Oxide—continued

Product Name	Marketing Company	Use	HSE No.
Hempel's Antifouling Globic SP-Eco 81900	Hempel Paints Ltd	Professional	6531
Hempel's Antifouling Globic SP-Eco 81990	Hempel Paints Ltd	Professional	6532
D Hempel's Antifouling Nautic Tin Free 7190E	Hempel Paints Ltd	Amateur Professional	5145
Hempel's Antifouling Tin Free 743GB	Hempel Paints Ltd	Amateur Professional	3329
Hempel's Antifouling Tin Free 751GB	Hempel Paints Ltd	Amateur Professional	3333
Hempel's Antifouling Tin Free 7662	Hempel Paints Ltd	Amateur Professional	3336
Hempel's Tin Free Antifouling 7626	Hempel Paints Ltd	Amateur Professional	3337
Hempels Antifouling Globic SP-Eco 81920	Hempel Paints Ltd	Professional	6877
Interviron Super Tin-Free Polishing Antifouling Bqo420 Series	International Coatings Ltd	Professional	5642
Micron CSC 300 Series	International Coatings Ltd	Professional	5724
Sigma Alphagen 20	Sigma Coatings BV	Professional	7089

74 4,5-Dichloro-2-N-Octyl-4-Isothiazolin-3-One + Cuprous Oxide + 2,4,5,6-Tetrachloro Isophthalonitrile

Product Name	Marketing Company	Use	HSE No.
C-Clean 400	Camrex Holdings BV	Professional	5947
Sea Grandprix 500 TCI	Chugoku Paints BV	Professional	7107
Seatender 15	Camrex Chugoku Ltd	Professional	5348
D TFA 10 HG	Camrex Chugoku Ltd	Amateur Professional	5366
D TFA 10G	Camrex Chugoku Ltd	Amateur Professional	5347

75 4,5-Dichloro-2-N-Octyl-4-Isothiazolin-3-One + Cuprous Oxide + 2-Methylthio-4-Tertiary-Butylamino-6-Cyclopropylamino-S-Triazine

Product Name	Marketing Company	Use	HSE No.
Grassline Type M396 Antifouling	W And J Leigh And Company	Professional	7075

76 4,5-Dichloro-2-N-Octyl-4-Isothiazolin-3-One + Cuprous Oxide + Cuprous Thiocyanate + 2-Methylthio-4-Tertiary-Butylamino-6-Cyclopropylamino-S-Triazine

Product Name	Marketing Company	Use	HSE No.
Envoy TF 400 Antifouling	W And J Leigh And Company	Professional	7072
Envoy TF 500 Antifouling	W And J Leigh And Company	Professional	7073

Product Name	Marketing Company	Use	HSE No.

77 4,5-Dichloro-2-N-Octyl-4-Isothiazolin-3-One + Cuprous Oxide + Dichlorophenyl Dimethylurea

TFA-10 LA Q	Camrex Chugoku Ltd	Professional	6581

78 4,5-Dichloro-2-N-Octyl-4-Isothiazolin-3-One + Cuprous Oxide + Tributyltin Methacrylate + Tributyltin Oxide

Intersmooth Hisol BFO270/950/970 Series	International Coatings Ltd	Professional	5641

79 4,5-Dichloro-2-N-Octyl-4-Isothiazolin-3-One + Cuprous Thiocyanate

Interspeed 2003	International Coatings Ltd	Professional	5726

80 4,5-Dichloro-2-N-Octyl-4-Isothiazolin-3-One + Cuprous Thiocyanate + Dichlorophenyl Dimethylurea

Sprint	New Guard Coatings Ltd	Professional	6449

81 4,5-Dichloro-2-N-Octyl-4-Isothiazolin-3-One + Tributyltin Methacrylate + Tributyltin Oxide

Antifouling Alusea Turbo	Jotun-Henry Clark Ltd	Professional	6464

82 4,5-Dichloro-2-N-Octyl-4-Isothiazolin-3-One + Zinc Pyrithione

D Hempel's Antifouling Nautic 7180E	Hempel Paints Ltd	Amateur Professional	5771
D Lynx-E	Blakes Marine Paints	Amateur Professional	5793

83 Dichlorophenyl Dimethylurea + 2,3,5,6-Tetrachloro-4-(Methyl Sulphonyl)Pyridine + Cuprous Oxide

W Raffaello 3	New Guard Coatings Ltd	Amateur Professional	5826

84 Dichlorophenyl Dimethylurea + 2,3,5,6-Tetrachloro-4-(Methyl Sulphonyl)Pyridine + Cuprous Thiocyanate

W Raffaello Racing	New Guard Coatings Ltd	Amateur Professional	6451

85 Dichlorophenyl Dimethylurea + 2,4,5,6-Tetrachloro Isophthalonitrile + 2-Methylthio-4-Tertiary-Butylamino-6-Cyclopropylamino-S-Triazine + Cuprous Oxide

Sigmaplane Ecol HA 120 Antifouling	Sigma Coatings Ltd	Amateur Professional	5788

86 Dichlorophenyl Dimethylurea + 2,4,5,6-Tetrachloro Isophthalonitrile + Cuprous Oxide

C-Clean 300	Camrex Holdings BV	Amateur Professional	5942

Product Name	Marketing Company	Use	HSE No.

86 Dichlorophenyl Dimethylurea + 2,4,5,6-Tetrachloro Isophthalonitrile + Cuprous Oxide—continued

Seatender 12	Camrex Chugoku Ltd	Amateur Professional	5324
TFA 10	Camrex Chugoku Ltd	Amateur Professional	5346
D TFA 10 H	Camrex Chugoku Ltd	Amateur Professional	5358

87 Dichlorophenyl Dimethylurea + 2-Methylthio-4-Tertiary-Butylamino-6-Cyclopropylamino-S-Triazine + Cuprous Oxide

A3 Antifouling 072015	Nautix SA	Amateur Professional	5224
A4 Antifouling 072017	Nautix SA	Amateur Professional	5226
Le Marin	Nautix SA	Amateur Professional	5480

88 Dichlorophenyl Dimethylurea + 2-Methylthio-4-Tertiary-Butylamino-6-Cyclopropylamino-S-Triazine + Cuprous Oxide + Dichlofluanid

Antifouling 1.3 Black And Red (400-03 / Ind.3)	Plastimo International	Amateur Professional	6350

89 Dichlorophenyl Dimethylurea + 2-Methylthio-4-Tertiary-Butylamino-6-Cyclopropylamino-S-Triazine + Cuprous Thiocyanate

A3 Antifouling 072016	Nautix SA	Amateur Professional	5225
A3 Teflon Antifouling 062015	Nautix SA	Amateur Professional	5481
A4 Antifouling 072018	Nautix SA	Amateur Professional	5227
A4 Teflon Antifouling 062017	Nautix SA	Amateur Professional	5482

90 Dichlorophenyl Dimethylurea + 2-Methylthio-4-Tertiary-Butylamino-6-Cyclopropylamino-S-Triazine + Cuprous Thiocyanate + Dichlofluanid

Antifouling 1.3	Plastimo International	Amateur Professional	6161
Antifouling 1.3 White (327-04)	Plastimo International	Amateur Professional	6349

91 Dichlorophenyl Dimethylurea + 4,5-Dichloro-2-N-Octyl-4-Isothiazolin-3-One + Cuprous Oxide

TFA-10 LA Q	Camrex Chugoku Ltd	Professional	6581

Product Name	Marketing Company	Use	HSE No.

92 Dichlorophenyl Dimethylurea + 4,5-Dichloro-2-N-Octyl-4-Isothiazolin-3-One + Cuprous Thiocyanate

Sprint	New Guard Coatings Ltd	Professional	6449

93 Dichlorophenyl Dimethylurea + Copper

Coppercoat	Aquarius Marine Coatings Ltd	Amateur Professional	6428

94 Dichlorophenyl Dimethylurea + Cuprous Oxide

A4 Antifouling	Nautix SA	Amateur Professional	4369
Aquarius Extra Strong	International Paint Ltd	Amateur Professional	4280
Blakes Hard Racing	Blakes Marine Paints	Amateur Professional	5764
Blueline Tropical SBA300	International Paint Ltd	Amateur Professional	5139
Boatgard Antifouling	International Coatings Ltd	Amateur Professional	6854
Cruiser Premium	International Coatings Ltd	Amateur Professional	5127
Drake Antifouling	M B Marine Coatings Ltd	Amateur Professional	6671
W Even Tin Free Light Grey	Mark Dowland Marine Ltd	Amateur Professional	5839
Giraglia	M B Marine Coatings Ltd	Amateur Professional	5980
D Grafo Anti-Foul SW	Grafo Coatings Ltd	Amateur Professional	5380
D Hempel's Antifouling 8199D	Hempel Paints Ltd	Amateur Professional	6177
D Hempel's Antifouling 8199F	Hempel Paints Ltd	Amateur Professional	6179
D Hempel's Antifouling Classic 7654B	Hempel Paints Ltd	Amateur Professional	5285
Hempel's Antifouling Combic 7199B	Hempel Paints Ltd	Amateur Professional	5274
Hempel's Antifouling Nautic 7190B	Hempel Paints Ltd	Amateur Professional	5273
Hempel's Antifouling Nautic 8190H	Hempel Paints Ltd	Amateur Professional	6042
Hempel's Bravo 7610A	Hempel Paints Ltd	Amateur Professional	5603
Hempel's Hard Racing 7648A	Hempel Paints Ltd	Amateur Professional	5535
Hempel's Mille Dynamic 7170A	Hempel Paints Ltd	Amateur Professional	5536

Product Name	Marketing Company	Use	HSE No.

94 Dichlorophenyl Dimethylurea + Cuprous Oxide—continued

Product Name	Marketing Company	Use	HSE No.
Hempel's Seatech Antifouling 7820A	Hempel Paints Ltd	Amateur Professional	6098
International Tin Free SPC BNA100 Series	International Paint Ltd	Amateur Professional	5186
Intersmooth Hisol Tin Free BGA620 Series	International Paint Ltd	Amateur Professional	4787
Intersmooth Tin Free BGA530 Series	International Paint Ltd	Amateur Professional	4611
Interspeed Extra BWA500 Red	International Coatings Ltd	Amateur Professional	4303
Interspeed Extra Strong	International Coatings Ltd	Amateur Professional	4819
Interspeed Super BWA900 Red	International Coatings Ltd	Amateur Professional	4884
Interspeed Super BWA909 Black	International Coatings Ltd	Amateur Professional	5058
Interspeed System 2 BRA142 Brown	International Paint Ltd	Amateur Professional	4302
Interswift Tin Free BQA400 Series	International Coatings Ltd	Amateur Professional	4842
Interswift Tin-Free SPC BTA540 Series	International Paint Ltd	Amateur Professional	5078
Interviron Super BNA400 Series	International Coatings Ltd	Amateur Professional	5690
Interviron Super BQA400 Series	International Coatings Ltd	Amateur Professional	5409
W Long Life Tin Free	Mark Dowland Marine Ltd	Amateur Professional	5840
Micron 500 Series	International Coatings Ltd	Amateur Professional	5729
Micron CSC	International Paint Ltd	Amateur Professional	4775
Micron CSC Extra	International Coatings Ltd	Amateur Professional	6263
Micron Plus Antifouling	International Paint Ltd	Amateur Professional	5133
Noa-Noa Rame	Ernesto Stoppani SPA	Amateur Professional	4796
Prima	International Coatings Ltd	Amateur Professional	6683
Professional Soft Antifouling	International Coatings Ltd	Amateur Professional	7093
Seajet 033	Camrex Chugoku Ltd	Amateur Professional	5331
Seatech Antifouling	Blakes Marine Paints	Amateur Professional	6125

Product Name	Marketing Company	Use	HSE No.

94 Dichlorophenyl Dimethylurea + Cuprous Oxide—continued

Slippy Bottom	Thincoat Technology International Ltd	Amateur Professional	6156
Soft Antifouling	International Coatings Ltd	Amateur Professional	6067
Teamac Antifouling "D" (C/258 Series)	Teal And Mackrill Ltd	Amateur Professional	6418
Teamac Killa Copper Premium Antifouling (C/262/24)	Teal And Mackrill Ltd	Amateur Professional	6410
Titan FGA	Hempel Paints Ltd	Amateur Professional	5767
Transocean Optima Antifouling 2.34	Spencer Coatings Ltd	Amateur Professional	6912
Waterways Antifouling	International Coatings Ltd	Amateur Professional	5565

95 Dichlorophenyl Dimethylurea + Cuprous Thiocyanate

Antialga	New Guard Coatings Ltd	Amateur Professional	5662
Aqua 12	International Coatings Ltd	Amateur Professional	4802
Aquarius AL	International Paint Ltd	Amateur Professional	4295
Hempel's Hard Racing 7638A	Hempel Paints Ltd	Amateur Professional	5539
Hempel's Mille ALU 71602	Hempel Paints Ltd	Amateur Professional	5578
Hempel's Mille Dynamic 7160A	Hempel Paints Ltd	Amateur Professional	5575
Seatech Antifouling White	Blakes Marine Paints	Amateur Professional	6140
D Tiger Cruising-A Antifouling	Blakes Marine Paints	Amateur Professional	5765

96 Dichlorophenyl Dimethylurea + Tributyltin Methacrylate + Tributyltin Oxide + Cuprous Oxide

A11 Antifouling 072014	Nautix SA	Professional	5376
Interswift BKO000/700 Series	International Coatings Ltd	Professional	5640

97 Dichlorophenyl Dimethylurea + Zinc Pyrithione

D Hempel's Antifouling Nautic 7180B	Hempel Paints Ltd	Amateur Professional	5770
D Lynx-A	Blakes Marine Paints	Amateur Professional	5794

Product Name	Marketing Company	Use	HSE No.

98 Dichlorophenyl Dimethylurea + Zinc Pyrithione + 2-(Thiocyanomethylthio)Benzothiazole + 2-Methylthio-4-Tertiary-Butylamino-6-Cyclopropylamino-S-Triazine

A2 Antifouling	Nautix SA	Professional	4832
A2 Teflon Antifouling	Nautix SA	Professional	4833

99 Dichlorophenyl Dimethylurea + Zinc Pyrithione + 2-Methylthio-4-Tertiary-Butylamino-6-Cyclopropylamino-S-Triazine

A6 Antifouling	Nautix SA	Professional	4944
A7 Teflon Antifouling	Nautix SA	Amateur Professional	4945
A7 Teflon Antifouling 072019	Nautix SA	Professional	5228
A8 Antifouling	Nautix SA	Professional	5982

100 Dichlorophenyl Dimethylurea + Zinc Pyrithione + Dichlofluanid

Antifouling 1.4	Plastimo International	Professional	6163

101 Dichlorophenyl Dimethylurea + Zineb + 2,3,5,6-Tetrachloro-4-(Methyl Sulphonyl)Pyridine + Cuprous Oxide + Cuprous Thiocyanate

W Even Extreme	New Guard Coatings Ltd	Amateur Professional	6625

102 2-Methylthio-4-Tertiary-Butylamino-6-Cyclopropylamino-S-Triazine + 2,3,5,6-Tetrachloro-4-(Methyl Sulphonyl)Pyridine

W Envoy TCF 600 Copper/Tin Free Antifouling	W And J Leigh And Company	Professional	6482

103 2-Methylthio-4-Tertiary-Butylamino-6-Cyclopropylamino-S-Triazine + 4,5-Dichloro-2-N-Octyl-4-Isothiazolin-3-One

Envoy TCF 600 Antifouling	W And J Leigh And Company	Professional	7074

104 2-Methylthio-4-Tertiary-Butylamino-6-Cyclopropylamino-S-Triazine + 4,5-Dichloro-2-N-Octyl-4-Isothiazolin-3-One + Cuprous Oxide

Grassline Type M396 Antifouling	W And J Leigh And Company	Professional	7075

105 2-Methylthio-4-Tertiary-Butylamino-6-Cyclopropylamino-S-Triazine + 4,5-Dichloro-2-N-Octyl-4-Isothiazolin-3-One + Cuprous Oxide + Cuprous Thiocyanate

Envoy TF 400 Antifouling	W And J Leigh And Company	Professional	7072
Envoy TF 500 Antifouling	W And J Leigh And Company	Professional	7073

HSE
ANTIFOULING PRODUCTS

Product Name	Marketing Company	Use	HSE No.

106 2-Methylthio-4-Tertiary-Butylamino-6-Cyclopropylamino-S-Triazine + Copper + Zinc Pyrithione

| VC 17M-HS | International Coatings Ltd | Amateur Professional | 5960 |

107 2-Methylthio-4-Tertiary-Butylamino-6-Cyclopropylamino-S-Triazine + Copper Metal

| VC 17M Tropicana | International Coatings Ltd | Amateur Professional | 4218 |

108 2-Methylthio-4-Tertiary-Butylamino-6-Cyclopropylamino-S-Triazine + Cuprous Oxide

Algicide	Blakes Marine Paints	Amateur Professional	5738
Aquaspeed	Blakes Marine Paints	Amateur Professional	4511
D Aquaspeed Antifouling	Blakes Marine Paints	Amateur Professional	5817
Broads Antifouling Red	Blakes Marine Paints	Amateur Professional	6878
Broads Black Antifouling	Blakes Marine Paints	Amateur Professional	5739
Broads Freshwater Red	Blakes Marine Paints	Amateur Professional	5736
Challenger Antifouling	Blakes Marine Paints	Amateur Professional	4099
W Even TF	Veneziani SPA	Amateur Professional	5166
W Even Tin Free	Mark Dowland Marine Ltd	Amateur Professional	5841
Hard Racing Antifouling	Blakes Marine Paints	Amateur Professional	5704
Hempel's Antifouling Classic 7611 Red (Tin Free) 5000	Hempel Paints Ltd	Amateur Professional	5064
D Hempel's Antifouling Classic 76540	Hempel Paints Ltd	Amateur Professional	5291
D Hempel's Antifouling Combic 71990	Hempel Paints Ltd	Amateur Professional	5600
D Hempel's Antifouling Combic 71992	Hempel Paints Ltd	Amateur Professional	5601
D Hempel's Antifouling Forte Tin Free 7625	Hempel Paints Ltd	Amateur Professional	3351
D Hempel's Antifouling Mille Dynamic	Hempel Paints Ltd	Amateur	4440
D Hempel's Antifouling Nautic 71900	Hempel Paints Ltd	Amateur Professional	5290
W Hempel's Antifouling Nautic 71902	Hempel Paints Ltd	Amateur Professional	5605

Product Name	Marketing Company	Use	HSE No.

108 2-Methylthio-4-Tertiary-Butylamino-6-Cyclopropylamino-S-Triazine + Cuprous Oxide—continued

Product Name	Marketing Company	Use	HSE No.
Hempel's Antifouling Nautic 8190C	Hempel Paints Ltd	Amateur Professional	6043
Hempel's Antifouling Olympic HI-7661	Hempel Paints Ltd	Amateur Professional	4898
D Hempel's Antifouling Olympic Tin Free 7154	Hempel Paints Ltd	Amateur Professional	3718
D Hempel's Antifouling Tin Free 745GB	Hempel Paints Ltd	Amateur Professional	3331
D Hempel's Antifouling Tin Free 750GB	Hempel Paints Ltd	Amateur Professional	3332
Hempel's Hard Racing 76480	Hempel Paints Ltd	Amateur Professional	5538
Hempel's Mille Dynamic 71700	Hempel Paints Ltd	Amateur Professional	5574
Hempels Antifouling Bravo Tin Free 7610	Hempel Paints Ltd	Amateur Professional	4482
Interspeed Antifouling BWO900 Series	International Coatings Ltd	Amateur Professional	5636
Interspeed System 2 BRA143 Brown	International Paint Ltd	Amateur Professional	4301
Interviron BQA450 Series	International Paint Ltd	Amateur Professional	4657
Interviron Super Tin-Free Polishing Antifouling BQO400 Series	International Coatings Ltd	Amateur Professional	5637
D Long Life T.F.	Veneziani SPA	Amateur Professional	4936
Longlife Antifouling	Skipper (UK) Ltd	Amateur Professional	6195
Micron CSC 200 Series	International Coatings Ltd	Amateur Professional	5732
W Patente Laxe	Teais SA	Amateur Professional	5497
Pilot	Blakes Marine Paints	Amateur Professional	5959
D Raffaello Plus Tin Free	Veneziani SPA	Amateur Professional	3642
Seatech	Blakes Marine Paints	Amateur Professional	7117
Shiprite Speed	Bradite Ltd	Amateur Professional	6603
Tigerline	Blakes Marine Paints	Amateur Professional	6872
Titan FGA Antifouling	Blakes Marine Paints	Amateur Professional	5681
Titan Ultra	Blakes Marine Paints	Amateur Professional	7096

Product Name	Marketing Company	Use	HSE No.

**108 2-Methylthio-4-Tertiary-Butylamino-6-Cyclopropylamino-S-Triazine +
Cuprous Oxide**—continued

VC Offshore	International Coatings Ltd	Amateur Professional	4777
Waterline	Blakes Marine Paints	Amateur Professional	7099
XM Anti-Fouling C2000 Cruising Self Eroding	X M Yachting	Amateur	6176
XM Anti-Fouling P4000 Hard	X M Yachting	Amateur	6175
XM Antifouling HS300 High Performance Self Eroding	X M Yachting	Amateur	6124

**109 2-Methylthio-4-Tertiary-Butylamino-6-Cyclopropylamino-S-Triazine +
Cuprous Oxide + Cuprous Thiocyanate + 2,3,5,6-Tetrachloro-4-(Methyl
Sulphonyl)Pyridine**

W Envoy TF 400	W And J Leigh And Company	Amateur Professional	4432
W Envoy TF 500	W And J Leigh And Company	Amateur Professional	5599
W Meridian MP40 Antifouling	Deangate Marine Products	Amateur Professional	5027
W Micro-Tech	Marine (Chemi-Technics) Ltd	Amateur Professional	6316

**110 2-Methylthio-4-Tertiary-Butylamino-6-Cyclopropylamino-S-Triazine +
Cuprous Oxide + Dichlofluanid**

Antifouling 1.2	Plastimo International	Amateur Professional	6159

**111 2-Methylthio-4-Tertiary-Butylamino-6-Cyclopropylamino-S-Triazine +
Cuprous Oxide + Dichlofluanid + Dichlorophenyl Dimethylurea**

Antifouling 1.3 Black And Red (400-03 / Ind.3)	Plastimo International	Amateur Professional	6350

**112 2-Methylthio-4-Tertiary-Butylamino-6-Cyclopropylamino-S-Triazine +
Cuprous Oxide + Dichlorophenyl Dimethylurea**

A3 Antifouling 072015	Nautix SA	Amateur Professional	5224
A4 Antifouling 072017	Nautix SA	Amateur Professional	5226
Le Marin	Nautix SA	Amateur Professional	5480

Product Name	Marketing Company	Use	HSE No.

113 2-Methylthio-4-Tertiary-Butylamino-6-Cyclopropylamino-S-Triazine + Cuprous Oxide + Dichlorophenyl Dimethylurea + 2,4,5,6-Tetrachloro Isophthalonitrile

| Sigmaplane Ecol HA 120 Antifouling | Sigma Coatings Ltd | Amateur Professional | 5788 |

114 2-Methylthio-4-Tertiary-Butylamino-6-Cyclopropylamino-S-Triazine + Cuprous Thiocyanate

Product Name	Marketing Company	Use	HSE No.
AL-27	International Paint Ltd	Amateur Professional	3613
Aquaspeed White	Blakes Marine Paints	Amateur Professional	4513
Boot Top Plus (Gull White)	International Paint Ltd	Amateur Professional	3389
Cruiser Superior 100 Series	International Coatings Ltd	Amateur Professional	5723
D Even TF Light Grey	Veneziani SPA	Amateur Professional	5167
Hard Racing Antifouling White	Blakes Marine Paints	Amateur Professional	5705
Hempel's Mille Dynamic 71600	Hempel Paints Ltd	Amateur Professional	5576
Hempel's Tin Free Hard Racing 76380	Hempel Paints Ltd	Amateur Professional	5540
D Hempel's Tin Free Hard Racing 7648	Hempel Paints Ltd	Amateur Professional	3367
Interspeed 2000	International Coatings Ltd	Amateur Professional	4148
Marclear Powerboat Antifouling	Marclear Marine Products Ltd	Amateur Professional	6586
MPX	International Coatings Ltd	Amateur Professional	4818
Propeller T.F.	New Guard Coatings Ltd	Amateur Professional	5660
Quicksilver Antifouling	International Coatings Ltd	Amateur Professional	5938
W Raffaello Alloy	Mark Dowland Marine Ltd	Amateur Professional	5492
Tiger Cruising White	Blakes Marine Paints	Amateur Professional	5712
Tigerline Antifouling	Blakes Marine Paints	Amateur Professional	3842
Titan FGA Antifouling White	Blakes Marine Paints	Amateur Professional	5680
VC Offshore	International Coatings Ltd	Amateur Professional	4779
VC Prop-O-Drev	International Coatings Ltd	Amateur Professional	4217

Product Name	Marketing Company	Use	HSE No.

114 2-Methylthio-4-Tertiary-Butylamino-6-Cyclopropylamino-S-Triazine +
Cuprous Thiocyanate—continued

Waterline White	Blakes Marine Paints	Amateur Professional	7097

115 2-Methylthio-4-Tertiary-Butylamino-6-Cyclopropylamino-S-Triazine +
Cuprous Thiocyanate + Dichlofluanid

Antifouling 1.2 (W)	Plastimo International	Amateur Professional	6160
D Hempel's Antifouling Mille Dynamic 717 GB	Hempel Paints Ltd	Amateur Professional	4437

116 2-Methylthio-4-Tertiary-Butylamino-6-Cyclopropylamino-S-Triazine +
Cuprous Thiocyanate + Dichlofluanid + Dichlorophenyl Dimethylurea

Antifouling 1.3	Plastimo International	Amateur Professional	6161
Antifouling 1.3 White (327-04)	Plastimo International	Amateur Professional	6349

117 2-Methylthio-4-Tertiary-Butylamino-6-Cyclopropylamino-S-Triazine +
Cuprous Thiocyanate + Dichlofluanid + Zinc Pyrithione

Antifouling 1.3 (W)	Plastimo International	Amateur Professional	6162

118 2-Methylthio-4-Tertiary-Butylamino-6-Cyclopropylamino-S-Triazine +
Cuprous Thiocyanate + Dichlorophenyl Dimethylurea

A3 Antifouling 072016	Nautix SA	Amateur Professional	5225
A3 Teflon Antifouling 062015	Nautix SA	Amateur Professional	5481
A4 Antifouling 072018	Nautix SA	Amateur Professional	5227
A4 Teflon Antifouling 062017	Nautix SA	Amateur Professional	5482

119 2-Methylthio-4-Tertiary-Butylamino-6-Cyclopropylamino-S-Triazine +
Dichlorophenyl Dimethylurea + Zinc Pyrithione

A6 Antifouling	Nautix SA	Professional	4944
A7 Teflon Antifouling	Nautix SA	Amateur Professional	4945
A7 Teflon Antifouling 072019	Nautix SA	Professional	5228
A8 Antifouling	Nautix SA	Professional	5982

120 2-Methylthio-4-Tertiary-Butylamino-6-Cyclopropylamino-S-Triazine +
Dichlorophenyl Dimethylurea + Zinc Pyrithione + 2-
(Thiocyanomethylthio)Benzothiazole

A2 Antifouling	Nautix SA	Professional	4832

Product Name	Marketing Company	Use	HSE No.

120 2-Methylthio-4-Tertiary-Butylamino-6-Cyclopropylamino-S-Triazine + Dichlorophenyl Dimethylurea + Zinc Pyrithione + 2-(Thiocyanomethylthio)Benzothiazole—continued

A2 Teflon Antifouling	Nautix SA	Professional	4833

121 2-Methylthio-4-Tertiary-Butylamino-6-Cyclopropylamino-S-Triazine + Tributyltin Methacrylate + Tributyltin Oxide

Antifouling Alusea Classic	Jotun-Henry Clark Ltd	Professional	6463

122 2-Methylthio-4-Tertiary-Butylamino-6-Cyclopropylamino-S-Triazine + Zinc Pyrithione

Hempel's Antifouling Nautic 71800	Hempel Paints Ltd	Amateur Professional	5769
Lynx Plus	Blakes Marine Paints	Amateur Professional	5792
White Tiger Cruising	Blakes Marine Paints	Amateur Professional	6873

123 No Recognised Active Ingredient

D Bioclean DX	Camrex Chugoku Ltd	Amateur Professional Professional (Aquaculture)	5317
Interclene AQ HZA700 Series (Base)	International Paint Ltd	Amateur Professional Professional (Aquaculture)	4765
Intersleek BXA 810/820 (Base)	International Coatings Ltd	Amateur Professional Professional (Aquaculture)	3403
Intersleek BXA580 Series (Base)	International Coatings Ltd	Amateur Professional	4786
Intersleek FCS HKA560 Series (Base)	International Paint Ltd	Amateur Professional Professional (Aquaculture)	4767
Intersleek FCS HKA580 Series (Base)	International Paint Ltd	Amateur Professional Professional (Aquaculture)	4766
Penguin Aqualine	Jotun-Henry Clark Ltd	Amateur Professional	6196

HSE
ANTIFOULING PRODUCTS

ANTIFOULING PRODUCTS

Product Name	Marketing Company	Use	HSE No.

124 2,4,5,6-Tetrachloro Isophthalonitrile

D Flexgard IV Waterbase Preservative	Flexabar Aquatech Corporation	Professional Professional (Aquaculture)	3321
D Flexgard V Waterbase Preservative	Flexabar Aquatech Corporation	Professional Professional (Aquaculture)	3320

125 2,4,5,6-Tetrachloro Isophthalonitrile + 2-Methylthio-4-Tertiary-Butylamino-6-Cyclopropylamino-S-Triazine + Cuprous Oxide + Dichlorophenyl Dimethylurea

Sigmaplane Ecol HA 120 Antifouling	Sigma Coatings Ltd	Amateur Professional	5788

126 2,4,5,6-Tetrachloro Isophthalonitrile + 4,5-Dichloro-2-N-Octyl-4-Isothiazolin-3-One + Cuprous Oxide

C-Clean 400	Camrex Holdings BV	Professional	5947
Sea Grandprix 500 TCI	Chugoku Paints BV	Professional	7107
Seatender 15	Camrex Chugoku Ltd	Professional	5348
D TFA 10 HG	Camrex Chugoku Ltd	Amateur Professional	5366
D TFA 10G	Camrex Chugoku Ltd	Amateur Professional	5347

127 2,4,5,6-Tetrachloro Isophthalonitrile + Copper Sulphate

W Flexgard VI Waterbase Preservative	Flexabar Aquatech Corporation	Professional Professional (Aquaculture)	3319

128 2,4,5,6-Tetrachloro Isophthalonitrile + Cuprous Oxide

Flexgard VI	Flexabar Aquatech Corporation	Professional Professional (Aquaculture)	6035
TFA 10 LA	Camrex Chugoku Ltd	Amateur Professional	5361

129 2,4,5,6-Tetrachloro Isophthalonitrile + Cuprous Oxide + Dichlorophenyl Dimethylurea

C-Clean 300	Camrex Holdings BV	Amateur Professional	5942
Seatender 12	Camrex Chugoku Ltd	Amateur Professional	5324
TFA 10	Camrex Chugoku Ltd	Amateur Professional	5346
D TFA 10 H	Camrex Chugoku Ltd	Amateur Professional	5358

Product Name	Marketing Company	Use	HSE No.

130 2,3,5,6-Tetrachloro-4-(Methyl Sulphonyl)Pyridine + 2-Methylthio-4-Tertiary-Butylamino-6-Cyclopropylamino-S-Triazine

W Envoy TCF 600 Copper/Tin Free Antifouling	W And J Leigh And Company	Professional	6482

131 2,3,5,6-Tetrachloro-4-(Methyl Sulphonyl)Pyridine + 2-Methylthio-4-Tertiary-Butylamino-6-Cyclopropylamino-S-Triazine + Cuprous Oxide + Cuprous Thiocyanate

W Envoy TF 400	W And J Leigh And Company	Amateur Professional	4432
W Envoy TF 500	W And J Leigh And Company	Amateur Professional	5599
W Meridian MP40 Antifouling	Deangate Marine Products	Amateur Professional	5027
W Micro-Tech	Marine (Chemi-Technics) Ltd	Amateur Professional	6316

132 2,3,5,6-Tetrachloro-4-(Methyl Sulphonyl)Pyridine + Cuprous Oxide

W Grassline TF Anti-Fouling Type M396	W And J Leigh And Company	Amateur Professional	3462

133 2,3,5,6-Tetrachloro-4-(Methyl Sulphonyl)Pyridine + Cuprous Oxide + Cuprous Thiocyanate + Dichlorophenyl Dimethylurea + Zineb

W Even Extreme	New Guard Coatings Ltd	Amateur Professional	6625

134 2,3,5,6-Tetrachloro-4-(Methyl Sulphonyl)Pyridine + Cuprous Oxide + Dichlorophenyl Dimethylurea

W Raffaello 3	New Guard Coatings Ltd	Amateur Professional	5826

135 2,3,5,6-Tetrachloro-4-(Methyl Sulphonyl)Pyridine + Cuprous Thiocyanate + Dichlorophenyl Dimethylurea

W Raffaello Racing	New Guard Coatings Ltd	Amateur Professional	6451

136 2-(Thiocyanomethylthio)Benzothiazole + 2-Methylthio-4-Tertiary-Butylamino-6-Cyclopropylamino-S-Triazine + Dichlorophenyl Dimethylurea + Zinc Pyrithione

A2 Antifouling	Nautix SA	Professional	4832
A2 Teflon Antifouling	Nautix SA	Professional	4833

137 2-(Thiocyanomethylthio)Benzothiazole + Cuprous Oxide

A3 Antifouling	Nautix SA	Professional	4367
A3 Teflon Antifouling	Nautix SA	Professional	4368

Product Name	Marketing Company	Use	HSE No.

137 2-(Thiocyanomethylthio)Benzothiazole + Cuprous Oxide—continued

W Titan Tin Free	Blakes Marine Paints	Amateur Professional	3556

138 Tributyltin Acrylate + Cuprous Oxide

D Amercoat 697	Ameron BV	Professional	3513

139 Tributyltin Methacrylate + Cuprous Oxide

ABC # 1 Antifouling	Ameron BV	Professional	3214
Devran MCP Antifouling Red	Devoe Coatings BV	Professional	3311
Devran MCP Antifouling Red Brown	Devoe Coatings BV	Professional	3310
D Takata LLL Antifouling	NOF Europe NV	Professional	4050
D Takata LLL Antifouling Hi-Solid	NOF Europe NV	Professional	4052
D Takata LLL Antifouling LS	NOF Europe NV	Professional	4051
D Takata LLL Antifouling LS Hi-Solid	NOF Europe NV	Professional	4053
D Takata LLL Antifouling No 2001	NOF Europe NV	Professional	4054

140 Tributyltin Methacrylate + Tributyltin Oxide

Hempel's Antifouling Nautic 76800	Hempel Paints Ltd	Professional	6025

141 Tributyltin Methacrylate + Tributyltin Oxide + 2-Methylthio-4-Tertiary-Butylamino-6-Cyclopropylamino-S-Triazine

Antifouling Alusea Classic	Jotun-Henry Clark Ltd	Professional	6463

142 Tributyltin Methacrylate + Tributyltin Oxide + 4,5-Dichloro-2-N-Octyl-4-Isothiazolin-3-One

Antifouling Alusea Turbo	Jotun-Henry Clark Ltd	Professional	6464

143 Tributyltin Methacrylate + Tributyltin Oxide + 4,5-Dichloro-2-N-Octyl-4-Isothiazolin-3-One + Cuprous Oxide

Intersmooth Hisol BFO270/950/970 Series	International Coatings Ltd	Professional	5641

144 Tributyltin Methacrylate + Tributyltin Oxide + Cuprous Oxide

AF Seaflo Mark 2-1	Camrex Chugoku Ltd	Professional	5316
D AF Seaflo Z-100 HS-1	Camrex Chugoku Ltd	Professional	5318
AF Seaflo Z-100 LE-HS-1	Camrex Chugoku Ltd	Professional	5313
W Antifouling Seaconomy	Jotun-Henry Clark Ltd	Professional	5791
Antifouling Seaconomy	Jotun-Henry Clark Ltd	Professional	6261
D Antifouling Seaconomy 200	Jotun-Henry Clark Ltd	Professional	4436
D Antifouling Seaconomy 300	Jotun-Henry Clark Ltd	Professional	4272
Antifouling Seamate FB30	Jotun-Henry Clark Ltd	Professional	6506
D Antifouling Seamate HB 33	Jotun-Henry Clark Ltd	Professional	4270
Antifouling Seamate HB 99	Jotun-Henry Clark Ltd	Professional	3411
Antifouling Seamate HB 99 Black	Jotun-Henry Clark Ltd	Professional	3410

Product Name	Marketing Company	Use	HSE No.

144 Tributyltin Methacrylate + Tributyltin Oxide + Cuprous Oxide—continued

Antifouling Seamate HB 99 Dark Red	Jotun-Henry Clark Ltd	Professional	3412
W Antifouling Seamate HB22	Jotun-Henry Clark Ltd	Professional	4271
Antifouling Seamate HB22	Jotun-Henry Clark Ltd	Professional	6301
Antifouling Seamate HB33	Jotun-Henry Clark Ltd	Professional	5811
Antifouling Seamate HB66	Jotun-Henry Clark Ltd	Professional	5698
Antifouling Seamate HB66	Jotun-Henry Clark Ltd	Professional	6475
Antifouling Seamate SB33	Jotun-Henry Clark Ltd	Professional	6517
C-Clean 6000	Camrex Holdings BV	Professional	5945
Grassline ABL Antifouling Type M349	W And J Leigh And Company	Professional	5374
D Hempel's Antifouling Combic 76990	Hempel Paints Ltd	Professional	3348
Hempel's Antifouling Nautic HI 76900	Hempel Paints Ltd	Professional	3360
D Hempel's Antifouling Nautic HI 76910	Hempel Paints Ltd	Professional	3361
Hempel's Antifouling Nautic SP-ACE 79031	Hempel Paints Ltd	Professional	6452
Hempel's Antifouling Nautic SP-ACE 79051	Hempel Paints Ltd	Professional	6453
Intersmooth SPC Antifouling BFO250 Series	International Coatings Ltd	Professional	5635
Intersmooth SPC BFA090/BFA190 Series	International Paint Ltd	Professional	3377
Seaflo 15	Camrex Chugoku Ltd	Professional	6284

145 Tributyltin Methacrylate + Tributyltin Oxide + Cuprous Oxide + Dichlorophenyl Dimethylurea

A11 Antifouling 072014	Nautix SA	Professional	5376
Interswift BKO000/700 Series	International Coatings Ltd	Professional	5640

146 Tributyltin Methacrylate + Tributyltin Oxide + Cuprous Thiocyanate

Antifouling HB 66 Ocean Green	Jotun-Henry Clark Ltd	Professional	3408
D Antifouling Seamate HB Green	Jotun-Henry Clark Ltd	Professional	3430
Antifouling Seamate HB22 Roundel Blue	Jotun-Henry Clark Ltd	Professional	4872
Antifouling Seamate HB66 Black	Jotun-Henry Clark Ltd	Professional	4871
Intersmooth SPC BFA040/BFA050 Series	International Coatings Ltd	Professional	3376

147 Tributyltin Methacrylate + Tributyltin Oxide + Zinc Pyrithione

Hempel's Antifouling Nautic 7680B	Hempel Paints Ltd	Professional	6172

148 Tributyltin Methacrylate + Tributyltin Oxide + Zineb + Cuprous Oxide

D Hempel's Antifouling 7690D	Hempel Paints Ltd	Professional	6056

HSE
ANTIFOULING PRODUCTS

Product Name	Marketing Company	Use	HSE No.

148 Tributyltin Methacrylate + Tributyltin Oxide + Zineb + Cuprous Oxide— continued

D Hempel's Antifouling Economic SP-Sea 74030	Hempel Paints Ltd	Professional	5581
D Hempel's Antifouling Nautic 7691B	Hempel Paints Ltd	Professional	5586
Hempel's Antifouling Nautic SP-ACE 79030	Hempel Paints Ltd	Professional	6331
Hempel's Antifouling Nautic SP-ACE 79050	Hempel Paints Ltd	Professional	6325
Hempel's Antifouling Nautic SP-ACE 79070	Hempel Paints Ltd	Professional	6328
Hempel's Economic SP-SEA 74010	Hempel Paints Ltd	Professional	6330
Intersmooth 110 Standard	International Coatings Ltd	Professional	5907
Intersmooth 120 Premium	International Coatings Ltd	Professional	5913
Intersmooth 130 Ultra	International Coatings Ltd	Professional	5914
Intersmooth 220 Premium	International Coatings Ltd	Professional	5911
Intersmooth 230 Ultra	International Coatings Ltd	Professional	5912
Intersmooth 320 Premium	International Coatings Ltd	Professional	5910
Intersmooth 330 Ultra	International Coatings Ltd	Professional	5924
Intersmooth Hisol 2000 BFA270 Series	International Coatings Ltd	Professional	3844
Intersmooth Hisol 9000 BFA970 Series	International Coatings Ltd	Professional	5461
Seaflo 10	Camrex Chugoku Ltd	Professional	6283
Superyacht 900	International Coatings Ltd	Professional	6278
Transocean Masterline Antifouling 2.84	Spencer Coatings Ltd	Professional	6934

149 Tributyltin Methacrylate + Tributyltin Oxide + Zineb + Cuprous Thiocyanate

Cruiser Copolymer	International Paint Ltd	Professional	3514
Intersmooth Hisol BFA948 Orange	International Coatings Ltd	Professional	4281
Intersmooth Hisol SPC Antifouling BFA949 Red	International Coatings Ltd	Professional	4949
Micron 25 Plus	International Coatings Ltd	Professional	3402
Superyacht Antifouling	International Coatings Ltd	Professional	5462

150 Tributyltin Methacrylate + Zineb + Cuprous Oxide

Antifouling Seamate HB22	Jotun-Henry Clark Ltd	Professional	5362
Antifouling Seamate HB33	Jotun-Henry Clark Ltd	Professional	5363
Nu Wave A/F	Wilckens Farben GmbH	Professional	6689
Rabamarine A/F 2500	Wilckens Farben GmbH	Professional	6690
Sigmaplane HA Antifouling	Sigma Coatings BV	Professional	4345
D Sigmaplane HB	Sigma Coatings BV	Professional	3487
Sigmaplane HB Antifouling	Sigma Coatings Ltd	Professional	5721
Sigmaplane TA Antifouling	Sigma Coatings BV	Professional	4346

151 Tributyltin Oxide + 2-Methylthio-4-Tertiary-Butylamino-6-Cyclopropylamino-S-Triazine + Tributyltin Methacrylate

Antifouling Alusea Classic	Jotun-Henry Clark Ltd	Professional	6463

152 Tributyltin Oxide + 4,5-Dichloro-2-N-Octyl-4-Isothiazolin-3-One + Cuprous Oxide + Tributyltin Methacrylate

Intersmooth Hisol BFO270/950/970 Series	International Coatings Ltd	Professional	5641

153 Tributyltin Oxide + 4,5-Dichloro-2-N-Octyl-4-Isothiazolin-3-One + Tributyltin Methacrylate

Antifouling Alusea Turbo	Jotun-Henry Clark Ltd	Professional	6464

154 Tributyltin Oxide + Cuprous Oxide + Dichlorophenyl Dimethylurea + Tributyltin Methacrylate

A11 Antifouling 072014	Nautix SA	Professional	5376
Interswift BKO000/700 Series	International Coatings Ltd	Professional	5640

155 Tributyltin Oxide + Cuprous Oxide + Tributyltin Methacrylate

	AF Seaflo Mark 2-1	Camrex Chugoku Ltd	Professional	5316
D	AF Seaflo Z-100 HS-1	Camrex Chugoku Ltd	Professional	5318
	AF Seaflo Z-100 LE-HS-1	Camrex Chugoku Ltd	Professional	5313
W	Antifouling Seaconomy	Jotun-Henry Clark Ltd	Professional	5791
	Antifouling Seaconomy	Jotun-Henry Clark Ltd	Professional	6261
D	Antifouling Seaconomy 200	Jotun-Henry Clark Ltd	Professional	4436
D	Antifouling Seaconomy 300	Jotun-Henry Clark Ltd	Professional	4272
	Antifouling Seamate FB30	Jotun-Henry Clark Ltd	Professional	6506
D	Antifouling Seamate HB 33	Jotun-Henry Clark Ltd	Professional	4270
	Antifouling Seamate HB 99	Jotun-Henry Clark Ltd	Professional	3411
	Antifouling Seamate HB 99 Black	Jotun-Henry Clark Ltd	Professional	3410
	Antifouling Seamate HB 99 Dark Red	Jotun-Henry Clark Ltd	Professional	3412
W	Antifouling Seamate HB22	Jotun-Henry Clark Ltd	Professional	4271
	Antifouling Seamate HB22	Jotun-Henry Clark Ltd	Professional	6301
	Antifouling Seamate HB33	Jotun-Henry Clark Ltd	Professional	5811
	Antifouling Seamate HB66	Jotun-Henry Clark Ltd	Professional	5698
	Antifouling Seamate HB66	Jotun-Henry Clark Ltd	Professional	6475
	Antifouling Seamate SB33	Jotun-Henry Clark Ltd	Professional	6517
	C-Clean 6000	Camrex Holdings BV	Professional	5945
	Grassline ABL Antifouling Type M349	W And J Leigh And Company	Professional	5374
D	Hempel's Antifouling Combic 76990	Hempel Paints Ltd	Professional	3348
	Hempel's Antifouling Nautic HI 76900	Hempel Paints Ltd	Professional	3360
D	Hempel's Antifouling Nautic HI 76910	Hempel Paints Ltd	Professional	3361

Product Name	Marketing Company	Use	HSE No.

155 Tributyltin Oxide + Cuprous Oxide + Tributyltin Methacrylate—continued

Hempel's Antifouling Nautic SP-ACE 79031	Hempel Paints Ltd	Professional	6452
Hempel's Antifouling Nautic SP-ACE 79051	Hempel Paints Ltd	Professional	6453
Intersmooth SPC Antifouling BFO250 Series	International Coatings Ltd	Professional	5635
Intersmooth SPC BFA090/BFA190 Series	International Paint Ltd	Professional	3377
Seaflo 15	Camrex Chugoku Ltd	Professional	6284

156 Tributyltin Oxide + Cuprous Thiocyanate + Tributyltin Methacrylate

Antifouling HB 66 Ocean Green	Jotun-Henry Clark Ltd	Professional	3408
D Antifouling Seamate HB Green	Jotun-Henry Clark Ltd	Professional	3430
Antifouling Seamate HB22 Roundel Blue	Jotun-Henry Clark Ltd	Professional	4872
Antifouling Seamate HB66 Black	Jotun-Henry Clark Ltd	Professional	4871
Intersmooth SPC BFA040/BFA050 Series	International Coatings Ltd	Professional	3376

157 Tributyltin Oxide + Tributyltin Methacrylate

Hempel's Antifouling Nautic 76800	Hempel Paints Ltd	Professional	6025

158 Tributyltin Oxide + Zinc Pyrithione + Tributyltin Methacrylate

Hempel's Antifouling Nautic 7680B	Hempel Paints Ltd	Professional	6172

159 Tributyltin Oxide + Zineb + Cuprous Oxide + Tributyltin Methacrylate

D Hempel's Antifouling 7690D	Hempel Paints Ltd	Professional	6056
D Hempel's Antifouling Economic SP-Sea 74030	Hempel Paints Ltd	Professional	5581
D Hempel's Antifouling Nautic 7691B	Hempel Paints Ltd	Professional	5586
Hempel's Antifouling Nautic SP-ACE 79030	Hempel Paints Ltd	Professional	6331
Hempel's Antifouling Nautic SP-ACE 79050	Hempel Paints Ltd	Professional	6325
Hempel's Antifouling Nautic SP-ACE 79070	Hempel Paints Ltd	Professional	6328
Hempel's Economic SP-SEA 74010	Hempel Paints Ltd	Professional	6330
Intersmooth 110 Standard	International Coatings Ltd	Professional	5907
Intersmooth 120 Premium	International Coatings Ltd	Professional	5913
Intersmooth 130 Ultra	International Coatings Ltd	Professional	5914
Intersmooth 220 Premium	International Coatings Ltd	Professional	5911
Intersmooth 230 Ultra	International Coatings Ltd	Professional	5912
Intersmooth 320 Premium	International Coatings Ltd	Professional	5910
Intersmooth 330 Ultra	International Coatings Ltd	Professional	5924
Intersmooth Hisol 2000 BFA270 Series	International Coatings Ltd	Professional	3844

Product Name	Marketing Company	Use	HSE No.

159 Tributyltin Oxide + Zineb + Cuprous Oxide + Tributyltin Methacrylate—
continued

Intersmooth Hisol 9000 BFA970 Series	International Coatings Ltd	Professional	5461
Seaflo 10	Camrex Chugoku Ltd	Professional	6283
Superyacht 900	International Coatings Ltd	Professional	6278
Transocean Masterline Antifouling 2.84	Spencer Coatings Ltd	Professional	6934

160 Tributyltin Oxide + Zineb + Cuprous Thiocyanate + Tributyltin Methacrylate

Cruiser Copolymer	International Paint Ltd	Professional	3514
Intersmooth Hisol BFA948 Orange	International Coatings Ltd	Professional	4281
Intersmooth Hisol SPC Antifouling BFA949 Red	International Coatings Ltd	Professional	4949
Micron 25 Plus	International Coatings Ltd	Professional	3402
Superyacht Antifouling	International Coatings Ltd	Professional	5462

161 Zinc Naphthenate + Copper Naphthenate + Cuprous Oxide + Dichlofluanid

D Teamac Killa Copper Plus	Teal And Mackrill Ltd	Amateur Professional	4659

162 Zinc Naphthenate + Cuprous Oxide

D Teamac Killa Copper	Teal And Mackrill Ltd	Amateur Professional	3494
D Teamac Super Tropical	Teal And Mackrill Ltd	Amateur Professional	3497

163 Zinc Pyrithione

No Foul-Zo	E Paint Company Inc	Amateur Professional	5789

164 Zinc Pyrithione + 2-(Thiocyanomethylthio)Benzothiazole + 2-Methylthio-4-Tertiary-Butylamino-6-Cyclopropylamino-S-Triazine + Dichlorophenyl Dimethylurea

A2 Antifouling	Nautix SA	Professional	4832
A2 Teflon Antifouling	Nautix SA	Professional	4833

165 Zinc Pyrithione + 2-Methylthio-4-Tertiary-Butylamino-6-Cyclopropylamino-S-Triazine

Hempel's Antifouling Nautic 71800	Hempel Paints Ltd	Amateur Professional	5769
Lynx Plus	Blakes Marine Paints	Amateur Professional	5792
White Tiger Cruising	Blakes Marine Paints	Amateur Professional	6873

Product Name	Marketing Company	Use	HSE No.

166 Zinc Pyrithione + 2-Methylthio-4-Tertiary-Butylamino-6-Cyclopropylamino-S-Triazine + Copper

VC 17M-HS	International Coatings Ltd	Amateur Professional	5960

167 Zinc Pyrithione + 2-Methylthio-4-Tertiary-Butylamino-6-Cyclopropylamino-S-Triazine + Cuprous Thiocyanate + Dichlofluanid

Antifouling 1.3 (W)	Plastimo International	Amateur Professional	6162

168 Zinc Pyrithione + 2-Methylthio-4-Tertiary-Butylamino-6-Cyclopropylamino-S-Triazine + Dichlorophenyl Dimethylurea

A6 Antifouling	Nautix SA	Professional	4944
A7 Teflon Antifouling	Nautix SA	Amateur Professional	4945
A7 Teflon Antifouling 072019	Nautix SA	Professional	5228
A8 Antifouling	Nautix SA	Professional	5982

169 Zinc Pyrithione + 4,5-Dichloro-2-N-Octyl-4-Isothiazolin-3-One

D Hempel's Antifouling Nautic 7180E	Hempel Paints Ltd	Amateur Professional	5771
D Lynx-E	Blakes Marine Paints	Amateur Professional	5793

170 Zinc Pyrithione + Cuprous Oxide

Blakes Antifouling 87910	Blakes Marine Paints	Amateur Professional	6450
Intersmooth 360 Ecoloflex	International Coatings Ltd	Amateur Professional	6057
Intersmooth 365 Ecoloflex	International Coatings Ltd	Amateur Professional	6557
Intersmooth 460 Ecoloflex	International Coatings Ltd	Amateur Professional	6276
Intersmooth 465 Ecoloflex	International Coatings Ltd	Amateur Professional	6556
D Micron 600 Series	International Paint Ltd	Amateur Professional	5733
Micron Optima	International Coatings Ltd	Amateur Professional	5941
VC Offshore Extra 100 Series	International Coatings Ltd	Amateur Professional	5730

171 Zinc Pyrithione + Cuprous Thiocyanate

D Hempel's Antifouling Combic 7199C	Hempel Paints Ltd	Amateur Professional	5742
Interspeed 2002	International Coatings Ltd	Amateur Professional	5725

Product Name	Marketing Company	Use	HSE No.

171 Zinc Pyrithione + Cuprous Thiocyanate—continued

D Titan FGA-C Antifouling	Blakes Marine Paints	Amateur Professional	5766

172 Zinc Pyrithione + Dichlofluanid + Dichlorophenyl Dimethylurea

Antifouling 1.4	Plastimo International	Professional	6163

173 Zinc Pyrithione + Dichlorophenyl Dimethylurea

D Hempel's Antifouling Nautic 7180B	Hempel Paints Ltd	Amateur Professional	5770
D Lynx-A	Blakes Marine Paints	Amateur Professional	5794

174 Zinc Pyrithione + Tributyltin Methacrylate + Tributyltin Oxide

Hempel's Antifouling Nautic 7680B	Hempel Paints Ltd	Professional	6172

175 Zineb + 2,3,5,6-Tetrachloro-4-(Methyl Sulphonyl)Pyridine + Cuprous Oxide + Cuprous Thiocyanate + Dichlorophenyl Dimethylurea

W Even Extreme	New Guard Coatings Ltd	Amateur Professional	6625

176 Zineb + Copper Resinate + Cuprous Oxide

Sigma Pilot Ecol Antifouling	Sigma Coatings Ltd	Amateur Professional	4933

177 Zineb + Cuprous Oxide

Blueline SPC Tin Free SBA700 Series	International Coatings Ltd	Amateur Professional	5214
Equatorial	International Paint Ltd	Amateur Professional	4121
D Hempel's Antifouling Combic 7199F	Hempel Paints Ltd	Amateur Professional	5744
Interclene 245	International Coatings Ltd	Professional	7019
Interspeed 340	International Coatings Ltd	Amateur Professional	6089
Interspeed 340	International Coatings Ltd	Professional	7020
Interspeed System 2 BRA140/ BRA240 Series	International Coatings Ltd	Amateur Professional	3847
Interviron BQA200 Series	International Coatings Ltd	Amateur Professional	3846

178 Zineb + Cuprous Oxide + Tributyltin Methacrylate

Antifouling Seamate HB22	Jotun-Henry Clark Ltd	Professional	5362
Antifouling Seamate HB33	Jotun-Henry Clark Ltd	Professional	5363
Nu Wave A/F	Wilckens Farben GmbH	Professional	6689
Rabamarine A/F 2500	Wilckens Farben GmbH	Professional	6690

HSE ANTIFOULING PRODUCTS

Product Name	Marketing Company	Use	HSE No.

178 Zineb + Cuprous Oxide + Tributyltin Methacrylate—continued

Sigmaplane HA Antifouling	Sigma Coatings BV	Professional	4345
D Sigmaplane HB	Sigma Coatings BV	Professional	3487
Sigmaplane HB Antifouling	Sigma Coatings Ltd	Professional	5721
Sigmaplane TA Antifouling	Sigma Coatings BV	Professional	4346

179 Zineb + Cuprous Oxide + Tributyltin Methacrylate + Tributyltin Oxide

D Hempel's Antifouling 7690D	Hempel Paints Ltd	Professional	6056
D Hempel's Antifouling Economic SP-SEA 74030	Hempel Paints Ltd	Professional	5581
D Hempel's Antifouling Nautic 7691B	Hempel Paints Ltd	Professional	5586
Hempel's Antifouling Nautic SP-ACE 79030	Hempel Paints Ltd	Professional	6331
Hempel's Antifouling Nautic SP-ACE 79050	Hempel Paints Ltd	Professional	6325
Hempel's Antifouling Nautic SP-ACE 79070	Hempel Paints Ltd	Professional	6328
Hempel's Economic SP-SEA 74010	Hempel Paints Ltd	Professional	6330
Intersmooth 110 Standard	International Coatings Ltd	Professional	5907
Intersmooth 120 Premium	International Coatings Ltd	Professional	5913
Intersmooth 130 Ultra	International Coatings Ltd	Professional	5914
Intersmooth 220 Premium	International Coatings Ltd	Professional	5911
Intersmooth 230 Ultra	International Coatings Ltd	Professional	5912
Intersmooth 320 Premium	International Coatings Ltd	Professional	5910
Intersmooth 330 Ultra	International Coatings Ltd	Professional	5924
Intersmooth Hisol 2000 BFA270 Series	International Coatings Ltd	Professional	3844
Intersmooth Hisol 9000 BFA970 Series	International Coatings Ltd	Professional	5461
Seaflo 10	Camrex Chugoku Ltd	Professional	6283
Superyacht 900	International Coatings Ltd	Professional	6278
Transocean Masterline Antifouling 2.84	Spencer Coatings Ltd	Professional	6934

180 Zineb + Cuprous Thiocyanate + Tributyltin Methacrylate + Tributyltin Oxide

Cruiser Copolymer	International Paint Ltd	Professional	3514
Intersmooth Hisol BFA948 Orange	International Coatings Ltd	Professional	4281
Intersmooth Hisol SPC Antifouling BFA949 Red	International Coatings Ltd	Professional	4949
Micron 25 Plus	International Coatings Ltd	Professional	3402
Superyacht Antifouling	International Coatings Ltd	Professional	5462

Product Name	Marketing Company	Use	HSE No.

Biocidal Paints

181 Carbendazim

Glidden Trade Fungicidal Vinyl Silk	Imperial Chemical Industries PLC	Amateur Professional	6659
Glidden Trade Fungicidal Acrylic Eggshell	Imperial Chemical Industries PLC	Amateur Professional	6662
Glidden Trade Fungicidal Vinyl Matt	Imperial Chemical Industries PLC	Amateur Professional	6658

182 Carbendazim + Dichlorophenyl Dimethylurea + 2-Octyl-2h-Isothiazolin-3-One

Anti-Mould Gloss	Plascon International	Amateur	6524
Aquaguard Anti-Mould Acrylic Matt Finish	Leyland Paint Company Ltd	Amateur Professional	6338
Biocheck A.C.C.	Mould Growth Consultants Ltd	Amateur Professional	6571
Biocheck Allweather Pliolite Resin Based Masonry Paint	Mould Growth Consultants Ltd	Amateur Professional	6907
Biocheck Water Based Gloss	Mould Growth Consultants Ltd	Amateur Professional	6573
Dulux Trade Mouldshield Fungicidal Quick Drying Eggshell	Imperial Chemical Industries PLC	Amateur Professional	6981
Dulux Trade Mouldshield Fungicidal Vinyl Matt	Imperial Chemical Industries PLC	Amateur Professional	6980
Manders Stop Mould	Manders Paints Ltd	Amateur Professional	6459
Sterashield	Johnstone's Paints PLC	Amateur Professional	6333
Wickes Mouldkill Emulsion	Wickes Building Supplies Ltd	Amateur Professional	6458

183 Carbendazim + Dichlorophenyl Dimethylurea + Dithio-2,2'-Bis(Benzmethylamide) + 2-Octyl-2h-Isothiazolin-3-One

Biocheck 01	Mould Growth Consultants Ltd	Professional	6536
Biocheck SP	Mould Growth Consultants Ltd	Amateur Professional	6600
Biocure Emulsion	Biotech Environmental (UK) Ltd	Amateur Professional	6850

184 Dichlofluanid

Dulux Trade Mouldshield Cellar Paint	ICI Paints	Amateur Professional	6481

Product Name	Marketing Company	Use	HSE No.

185 Dichlorophen

Roofcoat TP15	Mould Growth Consultants Ltd	Professional	6578

186 Dichlorophenyl Dimethylurea + 2-Octyl-2h-Isothiazolin-3-One + Carbendazim

Anti-Mould Gloss	Plascon International	Amateur	6524
Aquaguard Anti-Mould Acrylic Matt Finish	Leyland Paint Company Ltd	Amateur Professional	6338
Biocheck A.C.C.	Mould Growth Consultants Ltd	Amateur Professional	6571
Biocheck Allweather Pliolite Resin Based Masonry Paint	Mould Growth Consultants Ltd	Amateur Professional	6907
Biocheck Water Based Gloss	Mould Growth Consultants Ltd	Amateur Professional	6573
Dulux Trade Mouldshield Fungicidal Quick Drying Eggshell	Imperial Chemical Industries PLC	Amateur Professional	6981
Dulux Trade Mouldshield Fungicidal Vinyl Matt	Imperial Chemical Industries PLC	Amateur Professional	6980
Manders Stop Mould	Manders Paints Ltd	Amateur Professional	6459
Sterashield	Johnstone's Paints PLC	Amateur Professional	6333
Wickes Mouldkill Emulsion	Wickes Building Supplies Ltd	Amateur Professional	6458

187 Dichlorophenyl Dimethylurea + Dithio-2,2'-Bis(Benzmethylamide) + 2-Octyl-2h-Isothiazolin-3-One + Carbendazim

Biocheck 01	Mould Growth Consultants Ltd	Professional	6536
Biocheck SP	Mould Growth Consultants Ltd	Amateur Professional	6600
Biocure Emulsion	Biotech Environmental (UK) Ltd	Amateur Professional	6850

188 Dithio-2,2'-Bis(Benzmethylamide)

Biocheck 2507	Mould Growth Consultants Ltd	Professional	6610
Biocheck C	Mould Growth Consultants Ltd	Amateur Professional	6596
Biocheck Matt	Mould Growth Consultants Ltd	Amateur Professional	6599
Biocheck Silk	Mould Growth Consultants Ltd	Amateur Professional	6597
Bioshield	Mould Growth Consultants Ltd	Amateur Professional	6604

HSE
BIOCIDAL PAINTS

189 Dithio-2,2'-Bis(Benzmethylamide) + 2-Octyl-2h-Isothiazolin-3-One + Carbendazim + Dichlorophenyl Dimethylurea

Biocheck 01	Mould Growth Consultants Ltd	Professional	6536
Biocheck SP	Mould Growth Consultants Ltd	Amateur Professional	6600
Biocure Emulsion	Biotech Environmental (UK) Ltd	Amateur Professional	6850

190 3-Iodo-2-Propynyl-N-Butyl Carbamate

ACS Dry Rot Paint	Advanced Chemical Specialties Ltd	Amateur Professional	6310
W Anti-Mould Emulsion	Plascon International	Amateur	6473
Biotech Fungicidal Additive	Biotech Environmental (UK) Ltd	Amateur Professional	6650
Dulux Trade Mouldshield Fungicidal Eggshell	ICI Paints	Amateur Professional	6480
Dulux Trade Mouldshield Fungicidal Matt	ICI Paints	Amateur Professional	6494
Lectros Dry Rot Paint	Advanced Chemical Specialties Ltd	Amateur (Biocidal Paint) Professional (Biocidal Paint) Amateur (Wood Treatment) Professional (Wood Treatment)	6520
MGC Fungicidal Additive	Mould Growth Consultants Ltd	Amateur (Biocidal Paint) Professional (Biocidal Paint)	6605

191 3-Iodo-2-Propynyl-N-Butyl Carbamate + Propiconazole

Silexine Anticon	Remtox (Chemicals) Ltd	Amateur Professional	6476
Silexine Fungi-Chek Emulsion	Remtox (Chemicals) Ltd	Amateur Professional	6477
Silexine Fungi-Chek Oil Matt Paint	Remtox (Chemicals) Ltd	Amateur Professional	6478

192 2-Octyl-2h-Isothiazolin-3-One

ACS Antimould Emulsion	Advanced Chemical Specialties Ltd	Amateur Professional	6308

Product Name	Marketing Company	Use	HSE No.

192 2-Octyl-2h-Isothiazolin-3-One—continued

Product Name	Marketing Company	Use	HSE No.
ACS Antimould Microporous Replastering Paint	Advanced Chemical Specialties Ltd	Amateur Professional	6355
Azygo Antimould Coating (New Works)	Azygo International Direct Ltd	Amateur Professional	6483
Azygo Antimould Paint	Azygo International Direct Ltd	Amateur Professional	6406
Halocell 221	Mould Growth Consultants Ltd	Professional	6587
Lectros Antimould Breathable Paint	Lectros International Ltd	Amateur Professional	6496
Lectros Antimould Emulsion	Lectros International Ltd	Amateur Professional	6505

193 2-Octyl-2h-Isothiazolin-3-One + Carbendazim + Dichlorophenyl Dimethylurea

Product Name	Marketing Company	Use	HSE No.
Anti-Mould Gloss	Plascon International	Amateur	6524
Aquaguard Anti-Mould Acrylic Matt Finish	Leyland Paint Company Ltd	Amateur Professional	6338
Biocheck A.C.C.	Mould Growth Consultants Ltd	Amateur Professional	6571
Biocheck Allweather Pliolite Resin Based Masonry Paint	Mould Growth Consultants Ltd	Amateur Professional	6907
Biocheck Water Based Gloss	Mould Growth Consultants Ltd	Amateur Professional	6573
Dulux Trade Mouldshield Fungicidal Quick Drying Eggshell	Imperial Chemical Industries PLC	Amateur Professional	6981
Dulux Trade Mouldshield Fungicidal Vinyl Matt	Imperial Chemical Industries PLC	Amateur Professional	6980
Manders Stop Mould	Manders Paints Ltd	Amateur Professional	6459
Sterashield	Johnstone's Paints PLC	Amateur Professional	6333
Wickes Mouldkill Emulsion	Wickes Building Supplies Ltd	Amateur Professional	6458

194 2-Octyl-2h-Isothiazolin-3-One + Carbendazim + Dichlorophenyl Dimethylurea + Dithio-2,2'-Bis(Benzmethylamide)

Product Name	Marketing Company	Use	HSE No.
Biocheck 01	Mould Growth Consultants Ltd	Professional	6536
Biocheck SP	Mould Growth Consultants Ltd	Amateur Professional	6600
Biocure Emulsion	Biotech Environmental (UK) Ltd	Amateur Professional	6850

Product Name	Marketing Company	Use	HSE No.

195 Permethrin + Pyrethrins + 2,3,5,6-Tetrachloro-4-(Methyl Sulphonyl)Pyridine

W Stericide AM	Resin Surfaces Ltd	Amateur Professional	6513

196 Propiconazole + 3-Iodo-2-Propynyl-N-Butyl Carbamate

Silexine Anticon	Remtox (Chemicals) Ltd	Amateur Professional	6476
Silexine Fungi-Chek Emulsion	Remtox (Chemicals) Ltd	Amateur Professional	6477
Silexine Fungi-Chek Oil Matt Paint	Remtox (Chemicals) Ltd	Amateur Professional	6478

197 Pyrethrins + 2,3,5,6-Tetrachloro-4-(Methyl Sulphonyl)Pyridine + Permethrin

W Stericide Am	Resin Surfaces Ltd	Amateur Professional	6513

198 2,3,5,6-Tetrachloro-4-(Methyl Sulphonyl)Pyridine

W Haloseal	Mould Growth Consultants Ltd	Amateur Professional	6572
W Stericide	Resin Surfaces Ltd	Amateur Professional	6454

199 2,3,5,6-Tetrachloro-4-(Methyl Sulphonyl)Pyridine + Permethrin + Pyrethrins

W Stericide Am	Resin Surfaces Ltd	Amateur Professional	6513

3

INSECT REPELLENTS

Product Name	Marketing Company	Use	HSE No.

Insect repellents

200 Citronella Oil

Aztec B.B.Q. Patio Candles	Chartan Aldred Ltd	Amateur	5613
B.B.Q Fly Repellent Terracotta Pot Candle	Eclipse Candles Ltd	Amateur	5627
B.B.Q. Fly Repellant Candle	Eclipse Candles Ltd	Amateur	5625
B.B.Q. Patio Candles	Sherwood Promark Ltd	Amateur	5556
Betterware Adjustable Insect Repellent	Betterware UK Ltd	Amateur	6883
Bouchard Citronella Insect Repellant	IBA UK Ltd	Amateur	6938
Buzz Off Candle	Carberry Candles Ltd	Amateur	6908
Candle Pot - Scottish Midge	Thistle Products Limited	Amateur	6923
Citronella Bug Spray	STV International Ltd	Amateur	6369
Citronella Candle (Code D10)	Colony Gift Corporation Ltd	Amateur	6005
Citronella Floater Candle	Colony Gift Corporation Ltd	Amateur	6004
Citronella Insect Repellent Oil	Home And Leisure UK Ltd	Amateur	5423
Citronella Scented Candle	Colony Gift Corporation Ltd	Amateur	6006
Floral Bouquet (Oil Of Citronella)	Parlour Products PLC	Amateur	5886
Killgerm Repel	Killgerm Chemicals Ltd	Amateur Professional	5790
Killgerm Repel RTU	Killgerm Chemicals Ltd	Amateur Professional	5971
Vapona Citronella Candle	Ashe Ltd	Amateur	6861

201 Deltamethrin

K-Othrine Moustiquaire S.C.	Aventis Environmental Science	Amateur Professional	6129

202 1,4-Dichlorobenzene

Bouchard Anti Moth Proofer Pouches	IBA UK Ltd	Amateur	6403
D Moth Repellent	The Boots Company PLC	Amateur	5079

203 1,4-Dichlorobenzene + Naphthalene

Mothaks	Sara Lee Household And Body Care UK Ltd	Amateur	5124
Vapona Mothaks	Ashe Consumer Products Ltd	Amateur	5682

204 Naphthalene

Dragon Brand Moth Balls	R A Davies And Partners Ltd	Amateur	5385
Jertox Moth Balls	Thornton And Ross Ltd	Amateur	5057
Phernal Brand Moth Balls	Harrow Drug Company	Amateur	5740

Product Name	Marketing Company	Use	HSE No.

205 Naphthalene + 1,4-Dichlorobenzene

Mothaks	Sara Lee Household And Body Care UK Ltd	Amateur	5124
Vapona Mothaks	Ashe Consumer Products Ltd	Amateur	5682

206 Permethrin

Peripel	Aventis Environmental Science	Professional	5265
Raid Mothproofer	Johnson Wax Ltd	Amateur	5646

207 Pyrethrins

Repel Fabric Spray	Ibis Products Ltd	Amateur	6212
X-Gnat Advanced Insect Repellent - Fabric Spray	X-Gnat Laboratories Ltd	Amateur	6155

4
INSECTICIDES

Product Name	Marketing Company	Use	HSE No.

Insecticides

208 Allethrin

W Spira 'No-Bite' Outdoor Mosquito Coils	Masters London Ltd	Amateur	4421
Travel Masters Anti-Mosquito Coils	Cork International	Amateur	6940

209 D-Allethrin

D Baygon	Scholl Consumer Products Ltd	Amateur	5752
Floret Fast Knock Down	Reckitt And Colman Products Ltd	Amateur	4396
Gelert Mosquito Repellent	Bryncir Products Ltd	Amateur	5465
Jeyes Expel Plug-In Flying Insect Killer	Jeyes Group Ltd	Amateur	5653
Moskil Mosquito Repellent	I & M Steiner Ltd	Amateur	4342
Pynamin Forte 40mg Mat	Sumitomo Chemical (UK) PLC	Amateur	5852
Pynamin Forte Mat 120	Sumitomo Chemical (UK) PLC	Amateur	4417
Shelltox Mat 2	Temana International Ltd	Amateur	4742
Vapona Plug-In Flying Insect Killer	Ashe Consumer Products Ltd	Amateur	5708

210 D-Allethrin + Cypermethrin

Shelltox Crawling Insect Killer Liquid Spray	Temana International Ltd	Amateur Professional	3565
Vapona Ant And Crawling Insect Spray	Ashe Consumer Products Ltd	Amateur	3593

211 D-Allethrin + Cypermethrin + Permethrin + Tetramethrin

Elf Insecticide	Elf Oil UK Ltd	Amateur	6669

212 D-Allethrin + Cypermethrin + Tetramethrin

"Bop" Flying Insect Killer	Mcbride International	Amateur Professional	4141
"Bop" Flying Insect Killer (MC)	Mcbride International	Amateur Professional	4140
"Bop" Flying And Crawling Insect Killer (MC)	Mcbride International	Amateur Professional	4135
"Bop" Flying And Crawling Insect Killer (Water Based)	Mcbride International	Amateur Professional	4137
"Bop" Flying And Crawling Insect Killer	Mcbride International	Amateur Professional	4136

Product Name	Marketing Company	Use	HSE No.

213 D-Allethrin + Dichlorvos + Permethrin + Tetramethrin

Bop	Mcbride International	Amateur	4773

214 D-Allethrin + Permethrin

New Vapona Fly And Wasp Killer	Ashe Consumer Products Ltd	Amateur	4201
Raid Outdoor Insectguard	Johnson Wax Ltd	Amateur	5459

215 D-Allethrin + Permethrin + Tetramethrin

Hail Plus Crawling Insect Killer	Mcbride International	Amateur	5567
New Super Raid Insecticide	Johnson Wax Ltd	Amateur Professional	4583

216 D-Allethrin + Tetramethrin

Boots Dry Fly And Wasp Killer	The Boots Company PLC	Amateur	4037
D Boots Dry Fly And Wasp Killer 3	The Boots Company PLC	Amateur	4241
Fly, Wasp And Mosquito Killer	Rentokil Initial UK Ltd	Amateur	4882
New Vapona Fly Killer Dry Formulation	Ashe Consumer Products Ltd	Amateur	4202
Raid Fly And Wasp Killer	Johnson Wax Ltd	Amateur	3000
W Sainsbury's Fly, Wasp And Mosquito Killer	J Sainsbury PLC	Amateur	5164

217 D-Allethrin + D-Phenothrin

Boots New Improved Fly And Wasp Killer	The Boots Company PLC	Amateur	6432
Boots Pump Fly And Wasp Killer	The Boots Company PLC	Amateur	7030
Fly And Ant Killer	Rentokil Initial UK Ltd	Amateur	5561
Jeyes Crawling Insect Spray	Jeyes Group Ltd	Amateur	5270
Kleeneze New Improved Fly Killer	Kleeneze Ltd	Amateur	6431
Kleenoff Crawling Insect Spray	Jeyes Group Ltd	Amateur	5269
Kwik Insecticide	Hand Associates Ltd	Amateur	4290
Nippon Ready For Use Fly Killer Spray	Vitax Ltd	Amateur	4631
Pedigree Care Bedding Spray	Thomas's Europe	Amateur	5427
Pesguard Ps 102	Sumitomo Chemical (UK) PLC	Amateur	5312
Pesguard Ps 102a	Sumitomo Chemical (UK) PLC	Amateur	4173
Pesguard Ps 102b	Sumitomo Chemical (UK) PLC	Amateur	4174
Pesguard Ps 102c	Sumitomo Chemical (UK) PLC	Amateur	4175
Pesguard Ps 102d	Sumitomo Chemical (UK) PLC	Amateur	4176
W Secto Ant And Crawling Insect Spray.	Sinclair Animal And Household Care Ltd	Amateur	5714

HSE
INSECTICIDES

Product Name	Marketing Company	Use	HSE No.

217 D-Allethrin + D-Phenothrin—continued

W Target Flying Insect Killer	Reckitt And Colman Products Ltd	Amateur	4178
Whiskas Care Bedding Spray	Thomas's Europe	Amateur	4837
Whiskas Care Bedding Pest Control Spray	Thomas's Europe	Amateur	4596

218 Alphacypermethrin

Alpha 15sc	Hand Associates Ltd	Professional	7031
Cytrol Alpha 20	Pelgar International Ltd	Professional	7112
Cytrol Alpha 25 EC	Pelgar International Ltd	Professional	6677
Cytrol Alpha 5 SC	Pelgar International Ltd	Professional	6423
Fendona 1.5 SC	Sorex Ltd	Professional	4092
Fendona 6SC	Sorex Ltd	Professional	4455
Fendona ASC	Rentokil Initial UK Ltd	Professional	4946
Fendona Lacquer	Sorex Ltd	Amateur Professional	4278
Fendona WP	Sorex Ltd	Professional	4299
Littac	Sorex Ltd	Professional Professional (Animal Husbandry)	5176

219 Alphacypermethrin + Flufenoxuron

Tenopa	Sorex Ltd	Professional	6206

220 Alphacypermethrin + Tetramethrin

Cytrol Alpha Super 50/50 SE	Pelgar International Ltd	Professional	6678

221 Azamethiphos

W AL63 Crawling Insect Killer	Reckitt And Colman Products Ltd	Professional	5450
Raid Flyguard	Johnson Wax Ltd	Amateur	5713
Vapona Fly Killer Window Sticker	Ashe Ltd	Amateur	7106

222 Bacillus Thuringiensis Var Israelensis

Bactimos Flowable Concentrate	Valent Biosciences	Professional	4792
Bactimos Wettable Powder	Valent Biosciences	Professional	4793
W Skeetal Flowable Concentrate	Industrial Pesticides (N.W)	Professional	5706
Teknar HP-D	Killgerm Chemicals Ltd	Professional	5355
Teknar HP-FC	Agrisense Bcs Ltd	Professional	7104
Vectobac 12as	Univar PLC	Professional	6205

223 Bendiocarb

Ant Bait	Rentokil Initial UK Ltd	Amateur	6896
Ant Killer Powder	Lc Solutions Ltd	Amateur	6846

Product Name	Marketing Company	Use	HSE No.

223 Bendiocarb—continued

Ant Stop Powder	The Scotts Company (UK) Limited	Amateur	7088
Aventis Ant Killer Powder	Aventis Environmental Science	Amateur	6071
Aventis Wasp Nest Destroyer	Aventis Environmental Science	Amateur	6070
Aventis Woodlice Killer	Aventis Environmental Science	Amateur	6069
B & Q Ant Killer	B & Q PLC	Amateur	7010
B & Q Woodlice Killer	B & Q PLC	Amateur	7011
Bendiocarb Dusting Powder	Rentokil Initial UK Ltd	Professional	5125
Bendiocarb Wettable Powder	Rentokil Initial UK Ltd	Professional	5416
Bio Ant Kill Plus Powder	PBI Home & Garden Ltd	Amateur	6512
Bob Martin Home Flea Powder	The Bob Martin Company	Amateur	6656
Camco Ant Powder	Aventis Cropscience UK Ltd	Amateur	5218
Camco Insect Powder	Aventis Cropscience UK Ltd	Amateur	6995
Combat Ant & Crawling Insect Powder	Sinclair Animal And Household Care Ltd	Amateur	6925
Combat Ant Bait	Sinclair Animal And Household Care Ltd	Amateur	6947
Combat Woodlice Killer	Sinclair Animal And Household Care Ltd	Amateur	6946
Do It All Ant Killer	Focus Do It All	Amateur	7015
Do It All Woodlice Killer	Focus Do It All	Amateur	7007
Doff Ant Control Powder	Doff Portland Ltd	Amateur	6868
Doff Wasp Nest Killer	Doff Portland Ltd	Amateur	7005
Doff Woodlice Killer	Doff Portland Ltd	Amateur	7013
W Ficam 20w	AgrEvo UK Ltd	Professional	3682
Ficam W	Aventis Environmental Science	Professional	5390
Focus Ant Killer	Focus Do It All	Amateur	7014
Great Mills Ant Killer	Great Mills (Retail) Ltd	Amateur	7009
Great Mills Woodlice Killer	Great Mills (Retail) Ltd	Amateur	7008
Homebase Ant Killer	Homebase Ltd	Amateur	7000
Homebase Wasp Nest Killer	Homebase Ltd	Amateur	6999
Homebase Woodlice Killer	Homebase Ltd	Amateur	6998
Kil-Ant Ant Killer Powder	The Scotts Company (UK) Limited	Amateur	6522
Killgerm Insect Powder	Killgerm Chemicals Ltd	Amateur	6899
Murphy Kil-Ant Powder	The Scotts Company (UK) Limited	Amateur	5237
Portland Brand Ant Killer	Doff Portland Ltd	Amateur	7004
Premier Ant Killer	Premier Way Limited	Amateur	7003
Rentokil Ant And Insect Killer	Rentokil Initial UK Ltd	Amateur	5810
Rentokil Ant And Insect Powder Professional	Rentokil Initial UK Ltd	Professional	5386
Rentokil Wasp Killer	Rentokil Initial UK Ltd	Amateur	5813

HSE
INSECTICIDES

Product Name	Marketing Company	Use	HSE No.

223 Bendiocarb—continued

Secto Ant Bait	Sinclair Animal And Household Care Ltd	Amateur	5088
Wasp Nest Destroyer	LC Solutions Ltd	Amateur	6844
Wasp Nest Killer Professional	Rentokil Initial UK Ltd	Professional	5651
Westland Kill-All Ant Killer Powder	Westland Horticulture	Amateur	6851
Wilko Ant Destroyer	Wilkinson Group Of Companies	Amateur	7002
Wilko Woodlice Killer	Wilkinson Group Of Companies	Amateur	7001
Woodlice Killer	LC Solutions Ltd	Amateur	6845
Woolworths Ant Killer	Woolworths PLC	Amateur	6996
Woolworths Woodlice Killer	Woolworths PLC	Amateur	7012

224 Bendiocarb + Pyrethrins

Ficam Plus	Aventis Environmental Science	Professional	4830

225 Bendiocarb + Tetramethrin

W Camco Insect Spray	AgrEvo UK Ltd	Amateur	4222
Devcol Household Insect Spray	Devcol Ltd	Amateur	5512

226 Benzalkonium Chloride + Naphthalene + Permethrin + Pyrethrins

D Roxem D	Horton Hygiene Company	Professional	5107

227 Benzyl Benzoate

W Acarosan Foam	Crawford Pharmaceuticals	Amateur	5750
W Acarosan Moist Powder	Crawford Pharmaceuticals	Amateur	5751

228 Bioallethrin

Betterware Flying Insect Killer	Betterware UK Ltd	Amateur	6137
Boots Electric Mosquito Killer	The Boots Company PLC	Amateur	3871
Boots UK Flying Insect Killer	The Boots Company PLC	Amateur	4145
Buzz Off 1	Volex Accessories Ltd	Amateur	5823
D Buzz-Off	Ross Consumer Electronics	Amateur	3854
Far And Away	Dencon Accessories Ltd	Amateur	5753
Globol Pyrethrum Electrical Evaporator	Globol Chemicals (UK) Ltd	Amateur	4439
Haden Mosquito And Flying Insect Killer	D H Haden PLC	Amateur	5716
Lyvia Mosquito Killer	Lyvia Electrical Ltd	Amateur	4212
Mosqui - Go Electric	Design Go Travel Emporium	Amateur	4083
Mosquito Killer Travel Pack	Culmstock Ltd	Amateur	4926
Mosquito Repellent	Cometform Ltd	Amateur	5746
Nippon Flying Insect Killer Tablets	Vitax Ltd	Amateur	5199
Pif Paf Mosquito Mats	Reckitt And Colman Products Ltd	Amateur	5156

Product Name	Marketing Company	Use	HSE No.

228 Bioallethrin—continued

Rentokil Flying Insect Killer	Rentokil Initial UK Ltd	Amateur	5976
Shelltox Mat 1	Temana International Ltd	Amateur	4740
W Spira 'No-Bite' Mosquito Killer	Masters London Ltd	Amateur	3727
W Stradz Mosquito Killer	Stradz Ltd	Amateur	4144
Superdrug Mosquito Killer	Superdrug Stores PLC	Amateur	4078
Travel Masters No Bite Mosquito Killer Replacement Tablets	Cork International	Amateur	6930
Vapona Plug-In	Ashe Consumer Products Ltd	Amateur	5773
Wahl Envoyage Mosquito Killer	Wahl Europe Ltd	Amateur	5108
Woolworth Mosquito Killer	F W Woolworth Ltd	Amateur	4210

229 S-Bioallethrin

Esbiol 200	Aventis Environmental Science	Professional	6932

230 S-Bioallethrin + Bioallethrin

Actomite	Ceuta Healthcare Ltd	Amateur	4182

231 Bioallethrin + Bioresmethrin

Days Fly Spray	Day And Sons (Crewe) Ltd	Professional Professional (Animal Husbandry)	6457
Flyclear BHB	Battle, Hayward And Bower Ltd	Professional	6435
Pybuthrin 33 Bb	Aventis Environmental Science	Professional	5162
Trilanco Fly Spray	Trilanco	Professional Professional (Animal Husbandry)	6456

232 S-Bioallethrin + Bioresmethrin

Procontrol Fik H20	Aventis Environmental Science	Amateur Professional	6579
Procontrol Fik Super	Aventis Environmental Science	Amateur Professional	6580
Wasp Killer	LC Solutions Ltd	Amateur	6853

233 S-Bioallethrin + Deltamethrin

Crackdown Rapide	Aventis Environmental Science	Professional	5996

234 S-Bioallethrin + Dichlorvos + Permethrin

Farco Rapid Kill	F A Richard & Company Ltd	Amateur	6152

Product Name	Marketing Company	Use	HSE No.

234 S-Bioallethrin + Dichlorvos + Permethrin—continued

W Pif Paf Insecticide	Reckitt And Colman Products Ltd	Amateur Professional	5150

235 Bioallethrin + Methoprene + Permethrin

Arrest	Arnolds Veterinary Products Ltd	Amateur	6389
Bob Martin Home Flea Fogger Plus	The Bob Martin Company	Amateur	6630
Bob Martin Home Flea Spray Plus	The Bob Martin Company	Amateur	6223
R.I.P. Fleas	The Bob Martin Company	Amateur	6207

236 Bioallethrin + Permethrin

Agrevo Procontrol Flying Insect Killer	Aventis Environmental Science	Amateur Professional	6117
Ant Gun! 2	Miracle Garden Care Ltd	Amateur Professional	5629
Antkiller Spray 2	B & Q PLC	Amateur Professional	5185
Aventis Procontrol Crawling Insect Killer	Aventis Environmental Science	Amateur Professional	6116
Baby Bio Houseplant Spray	PBI Home & Garden Ltd	Amateur Professional	5828
Bio Spraydex Ant And Insect Killer	PBI Home & Garden Ltd	Amateur Professional	5829
Crawling Insect And Ant Killer	Rentokil Initial UK Ltd	Amateur	4650
Creepy Crawly Gun!	Miracle Garden Care Ltd	Amateur Professional	5183
Defest II	Sherley's Ltd	Amateur	6002
Farco Flying Insect Killer	F A Richard & Company Ltd	Amateur	6151
Fleegard	Bayer PLC	Amateur	5564
Hi-Craft Insecticidal Household Spray	Harry Irving & Co Ltd	Amateur	6612
Homebase Ant And Crawling Insect Killer	Sainsbury's Homebase House And Garden Centres	Amateur	5846
Insectrol	Rentokil Initial UK Ltd	Amateur	4568
Insectrol Professional	Rentokil Initial UK Ltd	Professional	4624
Nippon Fly Killer Pads	Vitax Ltd	Amateur	5426
Pc Insect Killer	Perycut Chemie Ag	Amateur Professional	4415
W Perycut Cockroach Mat	Perycut Chemie Ag	Amateur Professional	5467
Pif Paf Crawling Insect Killer	Reckitt And Colman Products Ltd	Amateur	5153
Pif Paf Flying Insect Killer	Reckitt And Colman Products Ltd	Amateur Professional	5151

Product Name	Marketing Company	Use	HSE No.

236 Bioallethrin + Permethrin—continued

Product Name	Marketing Company	Use	HSE No.
Pif Paf Flying Insect Killer Aerosol	Reckitt And Colman Products Ltd	Amateur	5152
Pybuthrin Fly Killer	AgrEvo UK Ltd	Amateur Professional	5141
Pybuthrin Fly Spray.	AgrEvo UK Ltd	Amateur	5116
W Sainsbury's Crawling Insect And Ant Killer	J Sainsbury PLC	Amateur	5231
D Spraydex Ant & Insect Killer	Pan Britannica Industries Ltd	Amateur	5645
D Spraydex Houseplant Spray	Spraydex Ltd	Amateur	3540
D Spraydex Insect Killer	Spraydex Ltd	Amateur	3834
Vapona Ant And Crawling Insect Killer	Ashe Consumer Products Ltd	Amateur Professional	5206
Vapona Fly And Wasp Killer	Ashe Consumer Products Ltd	Amateur	5205

237 S-Bioallethrin + Permethrin

Product Name	Marketing Company	Use	HSE No.
Agrevo Procontrol Yellow Flying Insect Killer Aerosol	AgrEvo UK Ltd	Professional	6232
Aqua Reslin Premium	AgrEvo UK Ltd	Professional	5172
Aqua Reslin Super	AgrEvo UK Ltd	Professional	5175
Atta-X Flying Insect Killer	Sanderson Curtis Ltd	Amateur	6348
Bug Wars	Design Go Travel Emporium	Amateur	6361
D Coopex S25% EC	AgrEvo UK Ltd	Professional	5146
D Imperator Fog S 18/6	Zeneca Public Health	Professional	4960
D Imperator Fog Super S 30/15	Zeneca Public Health	Professional	4955
D Imperator ULV S 6/3	Zeneca Public Health	Professional	4954
D Imperator ULV Super S 10/4	Zeneca Public Health	Professional	4953
Pif Paf Yellow Flying Insect Killer	Reckitt And Colman Products Ltd	Amateur	5937
Purge	Aztec Aerosols Limited	Amateur	6634
Resigen	Aventis Environmental Science	Professional	5143
Reslin Premium	Aventis Environmental Science	Professional	5174
D Reslin Super	AgrEvo UK Ltd	Professional	5173

238 Bioallethrin + Permethrin + Pyrethrins

Product Name	Marketing Company	Use	HSE No.
Tyrax	DBC	Amateur	6829

239 Bioallethrin + S-Bioallethrin

Product Name	Marketing Company	Use	HSE No.
Actomite	Ceuta Healthcare Ltd	Amateur	4182

240 S-Bioallethrin + Tetramethrin

Product Name	Marketing Company	Use	HSE No.
Vapona Wasp Killer Aerosol	Ashe Consumer Products Ltd	Amateur	6063

Product Name	Marketing Company	Use	HSE No.

241 S-Bioallethrin + Tetramethrin + Permethrin

Aventis Procontrol Crawling Insect Killer Aerosol	Aventis Environmental Science	Professional	6231
Pif Paf Crawling Insect Killer Aerosol	Reckitt And Colman Products Ltd	Amateur	5595

242 Bioresmethrin

Biosol RTU	Microsol	Professional	5306
Blade	Chemsearch	Professional	4839
Safe Kill RTU	Friendly Systems	Professional	5563

243 Bioresmethrin + Bioallethrin

Days Fly Spray	Day And Sons (Crewe) Ltd	Professional Professional (Animal Husbandry)	6457
Flyclear BHB	Battle, Hayward And Bower Ltd	Professional	6435
Pybuthrin 33 BB	Aventis Environmental Science	Professional	5162
Trilanco Fly Spray	Trilanco	Professional Professional (Animal Husbandry)	6456

244 Bioresmethrin + S-Bioallethrin

Procontrol Fik H20	Aventis Environmental Science	Amateur Professional	6579
Procontrol Fik Super	Aventis Environmental Science	Amateur Professional	6580
Wasp Killer	LC Solutions Ltd	Amateur	6853

245 Boric Acid

Baracaf Cockroach Control Sticker (Domestic)	Laboratoire Baracaf Sarl	Amateur Professional	4964
Boric Acid Concentrate	Rentokil Initial UK Ltd	Professional	4420
Boric Acid Dust	Blyth Valley Pest Control Prod	Professional	7069
Boric Acid Powder	Rentokil Initial UK Ltd	Professional	4373
Cockroach Control	Cockroach Control Services Ltd	Professional	5979
Dr Moss's Liquid Bait System	Blyth Valley Pest Control Prod	Professional	7059
Drax - Pf Gel	Killgerm Chemicals Ltd	Professional	6936
Drax Dual Gel	Killgerm Chemicals Ltd	Professional	6937
Drax Gel	Killgerm Chemicals Ltd	Professional	6935

Product Name	Marketing Company	Use	HSE No.

245 Boric Acid—continued

Enviromite	Environetics International Ltd	Amateur Professional	6203
Flea Ban	Animal Care Ltd	Amateur Professional (Animal Husbandry)	6201
Instasective	Instafoam & Fibre Ltd	Professional	5109
Killgerm Boric Acid Powder	Killgerm Chemicals Ltd	Professional	4527
Paragon Formula 10 Cockroach Bait	Paragon	Amateur Professional	6438
Pharaohkill	Sabre Pest Control Services	Professional	6565
Roachbuster	Roachbuster UK Ltd	Amateur Professional	5215
Sherley's Flea Busters	Sherley's Ltd	Amateur	6377
VF For Fleas	Veterinarian Formulae Ltd	Amateur Professional (Animal Husbandry)	5973

246 Carbon Dioxide

Rentokil Carbon Dioxide	Rentokil Initial UK Ltd	Professional	5964

247 Ceto-Stearyl Diethoxylate + Oleyl Monoethoxylate

Larvex-100	Accotec Ltd	Professional	5393
Larvex-15	Accotec Ltd	Professional	5091

248 Chlorpyrifos

Ant Stop	The Scotts Company (UK) Limited	Amateur	5905
W Bob Martin Microshield Household Flea Killing Spray	The Bob Martin Company	Amateur	6015
Chlorpyrifos	Rentokil Initial UK Ltd	Professional	6401
Contra Insect: The Vermin Bait	Frunol Delicia GmbH	Amateur	6299
Contra-Ants Bait Box	Frunol Delicia GmbH	Amateur	6298
Contra-Insect 480 Tec	Frunol Delicia GmbH	Professional	6224
W Duratrol	3m Health Care Ltd	Amateur	5968
Dursban 4tc	Dow Agrosciences Ltd	Professional	6288
D Dursban Lo	Dowelanco Ltd	Professional	4771
Empire 20	Dow Agrosciences Ltd	Professional	4844
Etisso Ant-Ex Bait Box	Frunol Delicia GmbH	Amateur	6300
Gett	Dow Agrosciences Ltd	Amateur	5863
Killgerm Terminate	Killgerm Chemicals Ltd	Professional	3553
Raid Ant Bait	Johnson Wax Ltd	Amateur	5585
Rentokil Chlorpyrifos Gel	Rentokil Initial UK Ltd	Professional	5862
W Swat Gel	Dow Agrosciences Ltd	Professional	6332
Vapona Ant Bait	Ashe Ltd	Amateur	6931

Product Name	Marketing Company	Use	HSE No.
248 Chlorpyrifos—continued			
Vapona Micro-Tech	Ashe Ltd	Amateur	6591
249 Chlorpyrifos + Cypermethrin			
Contra-Insect 200/20 EC	Frunol Delicia GmbH	Professional	6225
Duo 1	Hand Associates Ltd	Professional	6548
250 Chlorpyrifos + Cypermethrin + Pyrethrins			
New Tetracide	Killgerm Chemicals Ltd	Professional	3724
251 Chlorpyrifos + Pyrethrins			
Contra-Insect Aerosol	Frunol Delicia GmbH	Amateur	6372
Contra-Insect Universal	Frunol Delicia GmbH	Amateur	6405
252 Chlorpyrifos + Tetramethrin			
Raid Wasp Nest Destroyer	Johnson Wax Ltd	Amateur	5597
253 Chlorpyrifos-Methyl			
W Smite	AgrEvo UK Ltd	Professional	5142
254 Chlorpyrifos-Methyl + Permethrin + Pyrethrins			
D Multispray	AgrEvo UK Ltd	Professional	5165
255 Cyfluthrin + Pyriproxyfen			
Fleegard Plus	Bayer PLC	Amateur	6582
256 Cyfluthrin + Transfluthrin			
Baygon Insect Spray	Bayer PLC	Amateur	6546
257 Cypermethrin			
B And Q New Formula Ant Killer Spray	B & Q PLC	Amateur	5606
Combat Cockroach And Woodlice Killer	Sinclair Animal And Household Care Ltd	Amateur	6924
Cymperator	Killgerm Chemicals Ltd	Professional	3970
Cyperkill 10	Mitchell Cotts Chemicals	Professional	4025
Cyperkill 10 WP	Mitchell Cotts Chemicals	Professional	3649
Cypermethrin 10% EC	Killgerm Chemicals Ltd	Professional	3136
Cypermethrin 10% WP	Killgerm Chemicals Ltd	Professional	3137
Cypermethrin Lacquer	Killgerm Chemicals Ltd	Professional	3164
D Cypermethrin PH-10EC	Zeneca Public Health	Professional	4961
Cytrol Forte WP	Pelgar International Ltd	Professional	6424
Demon 40 WP	Zeneca Public Health	Professional	5457
Doff Ant Killer Spray	Doff Portland Ltd	Amateur	5588
Great Mills Ant Killer Spray	Great Mills (Retail) Ltd	Amateur	5572

Product Name	Marketing Company	Use	HSE No.
257 Cypermethrin—continued			
Growing Success Ant & Crawling Insect Killer	Growing Success Organics Ltd	Amateur Professional	7098
Homebase Antkiller Spray	Sainsbury's Homebase House And Garden Centres	Amateur	5573
Major Kil	Anglo Net	Amateur Professional	5874
Maximus RTU	Certified Laboratories	Amateur Professional (Animal Husbandry)	5837
Morgan Ant & Crawling Insect Killer	David Morgan (Nottingham) Ltd	Amateur Professional	6628
Murphy Kil-Ant Ready To Use	Levington Horticulture Ltd	Amateur	5168
New Siege II	Chemsearch	Amateur Professional	5638
Patriot CIK	Agropharm Ltd	Amateur Professional	7068
Pure Zap	Pure-Solve Hygiene	Amateur Professional (Animal Husbandry)	6061
Pyrasol C RTU	Microsol	Amateur Professional	5336
Pyrasol CP	Microsol	Professional	5478
Ready Kill	Friendly Systems	Amateur Professional	5566
Rid-Ant	Nehra Cooke's Chemicals Ltd	Amateur Professional	5748
Secto Ant And Crawling Insect Spray	Sinclair Animal And Household Care Ltd	Amateur	5994
Siege II	Chemsearch	Amateur Professional	5251
D Texas Ant Gun	Texas Homecare Ltd	Amateur	5679
Vapona Ant Chalk	Ashe Consumer Products Ltd	Amateur Professional	6104
Vapona Antpen	Ashe Consumer Products Ltd	Amateur Professional	3592
Vapona Flypen	Ashe Consumer Products Ltd	Amateur	3606
Vulcan C	Pelgar International Ltd	Professional	7121
Wilko Ant Killer Spray	Wilkinson Home And Garden Stores	Amateur	5583
258 Cypermethrin + Chlorpyrifos			
Contra-Insect 200/20 EC	Frunol Delicia GmbH	Professional	6225
Duo 1	Hand Associates Ltd	Professional	6548

Product Name	Marketing Company	Use	HSE No.

259 Cypermethrin + Methoprene

Killgerm Precor ULV III	Killgerm Chemicals Ltd	Professional	3523

260 Cypermethrin + Permethrin + Tetramethrin + D-Allethrin

Elf Insecticide	Elf Oil UK Ltd	Amateur	6669

261 Cypermethrin + Pyrethrins

Bolt Crawling Insect Killer	Johnson Wax Ltd	Amateur	3609
Raid Cockroach Killer Formula 2	Johnson Wax Ltd	Amateur	4239

262 Cypermethrin + Pyrethrins + Chlorpyrifos

New Tetracide	Killgerm Chemicals Ltd	Professional	3724

263 Cypermethrin + Tetramethrin

Cyperkill 10t	Mitchell Cotts Chemicals	Professional	6914
Cyperkill Plus Wp	Mitchell Cotts Chemicals	Professional	4855
Cytrol Universal	Pelgar International Ltd	Professional	6675
Cytrol XL	Pelgar International Ltd	Professional	6679
New Vapona Ant Killer	Ashe Consumer Products Ltd	Amateur	4298
Raid Ant And Crawling Insect Killer	Johnson Wax Ltd	Amateur	3001
Raid Residual Crawling Insect Killer	Johnson Wax Ltd	Amateur Professional	4584
S C Johnson Wax Raid Ant And Cockroach Killer	Johnson Wax Ltd	Amateur	5216
Shelltox Cockroach And Crawling Insect Killer 3	Temana International Ltd	Amateur Professional	4031
Shelltox Cockroach And Crawling Insect Killer 4	Temana International Ltd	Amateur	4739
Super Shelltox Crawling Insect Killer	Kortman Intradal BV	Amateur	5454
Vapona Ant And Crawling Insect Killer Aerosol	Ashe Consumer Products Ltd	Amateur	5282

264 Cypermethrin + Tetramethrin + D-Allethrin

"Bop" Flying Insect Killer	Mcbride International	Amateur Professional	4141
"Bop" Flying Insect Killer (MC)	Mcbride International	Amateur Professional	4140
"Bop" Flying And Crawling Insect Killer (MC)	Mcbride International	Amateur Professional	4135
"Bop" Flying And Crawling Insect Killer (Water Based)	Mcbride International	Amateur Professional	4137
"Bop" Flying And Crawling Insect Killer	Mcbride International	Amateur Professional	4136

Product Name	Marketing Company	Use	HSE No.

265 Cypermethrin + d-Allethrin

Product Name	Marketing Company	Use	HSE No.
Shelltox Crawling Insect Killer Liquid Spray	Temana International Ltd	Amateur Professional	3565
Vapona Ant And Crawling Insect Spray	Ashe Consumer Products Ltd	Amateur	3593

266 Cyromazine + Permethrin

Product Name	Marketing Company	Use	HSE No.
Flego Trigger Pack	Novartis Animal Health UK Ltd	Amateur	6051
Pet Rescue Household Flea Spray	Pet Rescue	Amateur Professional	7063
Sherley's Flego	Sherley's Ltd	Amateur Professional	5803
Sherley's Flego Trigger Pack	Sherley's Ltd	Amateur	6065
Staykil Household Flea Spray	Novartis Animal Health UK Ltd	Amateur Professional	5784

267 Deltamethrin

Product Name	Marketing Company	Use	HSE No.
Agrevo Ant Killer	Aventis Cropscience UK Ltd	Amateur Professional	5877
Ant Killer Spray	LC Solutions Ltd	Amateur Professional	6843
Ant-Off!	Pet And Garden Manufacturing PLC	Amateur Professional	5918
Aqua K-Othrine	Protim Solignum Ltd	Professional	6027
B & Q Ant And Crawling Insect Spray.	Miracle Garden Care Ltd	Amateur	5966
B & Q New Ant Killer Spray	B & Q PLC	Amateur Professional	5895
Bio Ant Kill Plus Spray	PBI Home & Garden Ltd	Amateur Professional	6516
Combat Ant And Crawling Insect Spray	Sinclair Animal And Household Care Ltd	Amateur	6837
Crackdown	Aventis Environmental Science	Professional	5097
Crackdown D	Aventis Environmental Science	Professional	6054
Deth-Zap Ant Killer	Gerhardt Pharmaceuticals Ltd	Amateur Professional	6869
Dethlac Insect Lacquer	Gerhardt Pharmaceuticals Ltd	Amateur	6891
Do It All Blitz! Ant Killer	Do It All Ltd	Amateur Professional	6072
Doff Ant And Crawling Insect Killer	Doff Portland Ltd	Amateur	6866
Doff New Ant Killer Spray	Doff Portland Ltd	Amateur Professional	5894

Product Name	Marketing Company	Use	HSE No.

267 Deltamethrin—continued

Flea-Off!	Pet And Garden Manufacturing PLC	Amateur Professional	5917
Focus New Ant Killer Spray	Focus DIY Ltd	Amateur Professional	5896
Great Mills New Ant Killer Spray	Great Mills (Retail) Ltd	Amateur Professional	5897
Homebase New Ant Killer Spray	Sainsbury's Homebase House And Garden Centres	Professional	5898
Johnson's Home Flea Guard Trigger Spray	Johnson's Veterinary Products Ltd	Amateur	6106
K-Othrine Crawling Insect Powder	Aventis Environmental Science	Amateur	6055
Nippon Ready-To-Use Ant And Crawling Insect Killer Spray	Vitax Ltd	Amateur Professional	6905
Pro Control CIK Super	Aventis Environmental Science	Amateur	6819
Procontrol CIK Super (P)	Aventis Environmental Science	Professional	6820
S C Johnson Wax Raid Ant Killer Powder	Johnson Wax Ltd	Amateur	6267
Selkil Crawling Insect Killer	Selden Research Ltd	Professional	6008
Vapona Ant And Woodlice Killer Powder	Ashe Ltd	Amateur	6238
Westland Ant Killer RTU	Westland Horticulture	Amateur Professional	6852
Wilko New Ant Killer Spray	Wilkinson Home And Garden Stores	Amateur Professional	5899
Woolworths New Ant Killer Spray	Woolworths PLC	Amateur Professional	5900

268 Deltamethrin + Pyrethrins

Agrevo Pro Control CIK Superfast	Aventis Environmental Science	Amateur	6831
Doom Ant & Cockroach Killer	Napa Products Ltd	Amateur	6875
Doom Flea Killer	Napa Products Ltd	Amateur	6881
Pro Control CIK Superfast (P)	Aventis Environmental Science	Professional	6836

269 Deltamethrin + S-Bioallethrin

| Crackdown Rapide | Aventis Environmental Science | Professional | 5996 |

270 Diazinon

| W B And Q Ant Killer Lacquer | B & Q PLC | Amateur | 4750 |

Product Name	Marketing Company	Use	HSE No.

270 Diazinon—continued

W Dethlac Insecticidal Lacquer	Gerhardt Pharmaceuticals Ltd	Amateur	4423
W Doff Antlak	Doff Portland Ltd	Amateur	3591
W Knox Out 2 FM	Rentokil Initial UK Ltd	Professional	4464
W Secto Ant And Crawling Insect Lacquer	Sinclair Animal And Household Care Ltd	Amateur	4400
W Wilko Ant Killer Lacquer	Wilkinson Home And Garden Stores	Amateur	4965

271 Diazinon + Dichlorvos

D Vijurrax Spray	Vijusa UK Ltd	Amateur	5717

272 1,4-Dichlorobenzene + Dichlorvos

Combat Moth Killer	Sinclair Animal And Household Care Ltd	Amateur	7086
Secto Moth Killer	Sinclair Animal And Household Care Ltd	Amateur	4430

273 Dichlorophen + Dichlorvos + Tetramethrin

D Wintox	Mould Growth Consultants Ltd	Professional	4236

274 Dichlorvos

Betterware Flying Insect Killer (Slow Release)	Betterware UK Ltd	Amateur	5795
Betterware Slow Release Fly Killer	Betterware UK Ltd	Amateur	6959
Betterware Small Space Insect Killer	Betterware UK Ltd	Amateur	5990
W Boots Moth Killer	The Boots Company PLC	Amateur	5543
W Boots Slow Release Fly And Wasp Killer	The Boots Company PLC	Amateur	5545
Boots Slow Release Fly Killer	The Boots Company PLC	Amateur	4922
Boots Small Space Moth Killer	The Boots Company PLC	Amateur	5272
Bouchard Slow Release Fly Killer	IBA UK Ltd	Amateur	6958
Combat Fly Killer	Sinclair Animal And Household Care Ltd	Amateur	7085
Cromessol Fly-Away	Cromessol Company Ltd	Amateur	4338
Culmstock Mothproofer	Culmstock Ltd	Amateur	4027
Culmstock Slow Release Fly Killer	Culmstock Ltd	Amateur	3961
DDVP (Toxicant) Strip	International Pheromone Systems Ltd	Professional	5517
Fly Killer	Rentokil Initial UK Ltd	Amateur	5676
Flying Insect Killer	Culmstock Ltd	Amateur	4442
Freshways Slow Release Insect Killer	Freshways Of York	Amateur	4618
Funnel Trap Insecticidal Strip	Agrisense BCS Ltd	Professional	4292

Product Name	Marketing Company	Use	HSE No.
274 Dichlorvos—continued			
Globol Small Space Fly And Moth Strip	Globol Chemicals (UK) Ltd	Amateur	4692
Industrial Hygiene Controllable Cassette Insect Killer	Industrial Hygiene Company	Amateur	4617
W Jeyes Expel Moth Killer	Jeyes Group Ltd	Amateur	5504
W Jeyes Expel Slow Release Fly And Wasp Killer	Jeyes Group Ltd	Amateur	5505
W Jeyes Kontrol Moth Killer	J Sainsbury PLC	Amateur	5546
W Jeyes Kontrol Slow Release Fly And Wasp Killer	J Sainsbury PLC	Amateur	5544
Jeyes Slow Release Fly Killer Controllable Cassette	Jeyes Group Ltd	Amateur	5259
Jeyes Small Space Fly And Moth Strip	Jeyes Group Ltd	Amateur	5289
Kleenoff Slow Release Fly Killer Controllable Cassette	Jeyes Group Ltd	Amateur	5261
Kleenoff Small SPAce Fly And Moth Strip	Jeyes Group Ltd	Amateur	5288
Kontrol Kitchen Size Fly Killer	Sinclair Animal And Household Care Ltd	Amateur	5598
W Kontrol Slow Release Fly Killer	J Sainsbury PLC	Amateur	5804
Lloyds Supersave Slow Release Fly And Wasp Killer.	Lloyds	Amateur	4920
Lloyds Supersave Small Space Fly And Moth Strip	Lloyds	Amateur	4921
Moth And Fly Killer	Rentokil Initial UK Ltd	Amateur	5677
Murphys Long Lasting Fly And Wasp Killer	Levington Horticulture Ltd	Amateur	6268
Secto Fly Killer Living Room Size	Sinclair Animal And Household Care Ltd	Amateur	4397
Secto Mini-SPAce Insect Killers	Sinclair Animal And Household Care Ltd	Amateur	4398
Secto Slow Release Fly Killer Kitchen Size	Sinclair Animal And Household Care Ltd	Amateur	4399
Slow Release Fly Killer Controllable Cassette	Globol Chemicals (UK) Ltd	Amateur	4339
Superdrug Slow Release Fly Killer	Superdrug Stores PLC	Amateur	3962
Superdrug Small Space Fly/Moth Strip	Superdrug Stores PLC	Amateur	4068
Teepol Products Vapona Fly Killer	Teepol Products Ltd	Amateur Professional	3673
Terminator Slow Release Fly Killer	Sanoda Ltd	Amateur	6168
Terminator Small Space Fly And Moth Killer	Sanoda Ltd	Amateur	6167
Vapona Fly Killer	Ashe Consumer Products Ltd	Amateur	4717
Vapona Moth Killer	Ashe Ltd	Amateur	4724

Product Name	Marketing Company	Use	HSE No.

274 Dichlorvos—continued

Vapona Professional Cockroach Killer	Ashe Consumer Products Ltd	Amateur	5685
Vapona Small Space Fly Killer	Ashe Consumer Products Ltd	Amateur	4723

275 Dichlorvos + 1,4-Dichlorobenzene

Combat Moth Killer	Sinclair Animal And Household Care Ltd	Amateur	7086
Secto Moth Killer	Sinclair Animal And Household Care Ltd	Amateur	4430

276 Dichlorvos + Diazinon

D Vijurrax Spray	Vijusa UK Ltd	Amateur	5717

277 Dichlorvos + Iodofenphos

W Defest Flea Free	Sherley's Ltd	Amateur	5715
W Nuvan Staykil	Novartis Animal Health UK Ltd	Amateur Professional	4017

278 Dichlorvos + Permethrin + S-Bioallethrin

Farco Rapid Kill	F A Richard & Company Ltd	Amateur	6152
W Pif Paf Insecticide	Reckitt And Colman Products Ltd	Amateur Professional	5150

279 Dichlorvos + Permethrin + Tetramethrin + D-Allethrin

Bop	Mcbride International	Amateur	4773

280 Dichlorvos + Tetramethrin

Shelltox Extra Flykiller	Temana International Ltd	Amateur Professional	3696
Shelltox Flykiller	Temana International Ltd	Amateur	3162
Vapona Fly And Wasp Killer Spray	Ashe Ltd	Amateur	4907

281 Dichlorvos + Tetramethrin + Dichlorophen

D Wintox	Mould Growth Consultants Ltd	Professional	4236

282 Disodium Tetraborate

Betterware Ant Gel	Betterware UK Ltd	Amateur	6928
Nippon Ant Killer Liquid	Vitax Ltd	Amateur	6915

283 Fenitrothion

Demise	Sorex Ltd	Professional	5084

Product Name	Marketing Company	Use	HSE No.
283 Fenitrothion—continued			
Fenitrothion 47EC	Hand Associates Ltd	Professional	6687
		Professional (Animal Husbandry)	
Fenitrothion Dusting Powder	Rentokil Initial UK Ltd	Professional	4372
W Fenitrothion Emulsion Concentrate	Rentokil Initial UK Ltd	Professional	4783
Fenitrothion Wettable Powder	Rentokil Initial UK Ltd	Professional	4827
Killgerm Fenitrothion 40 WP	Killgerm Chemicals Ltd	Professional	4858
		Professional (Animal Husbandry)	
Killgerm Fenitrothion 50 EC	Killgerm Chemicals Ltd	Professional	4722
Micromite	Micro-Biologicals Ltd	Professional	4480
Sectacide 50 EC	Anglian Farm Supplies Ltd	Professional	4939
Sumithion 20 MC	Pelgar International Ltd	Professional	6209
Sumithion 20% MC	Sumitomo Chemical (UK) PLC	Professional	4905
Sumithion 40 Wp	Vetrepharm Ltd	Professional	6334
284 Fenitrothion + Permethrin + Resmethrin			
W Turbair Beetle Killer	Pan Britannica Industries Ltd	Professional	4648
285 Fenitrothion + Permethrin + Tetramethrin			
Connect Insect Killer	Sun Oil Ltd	Amateur	6109
Motox	HVM Products Ltd	Amateur	5622
286 Fenitrothion + Tetramethrin			
D Big D Cockroach And Crawling Insect Killer	Domestic Fillers Ltd	Amateur	4098
Big D Cockroach And Crawling Insect Killer	Dylon International Ltd	Amateur	5834
W Big D Rocket Fly Killer	Domestic Fillers Ltd	Amateur	3552
Doom Ant And Crawling Insect Killer Aerosol	Napa Products Ltd	Amateur	4753
Flying Insect Killer	Keen (World Marketing) Ltd	Amateur	4313
Flying Insect Killer Faster Knock Down	Keen (World Marketing) Ltd	Amateur	4312
Killgerm Fenitrothion-Pyrethroid Concentrate	Killgerm Chemicals Ltd	Professional	4623
Motox Crawling Insect	HVM Products Ltd	Amateur	5940
Top Kill Insect Killer	Houdret And Co Ltd	Amateur	6188
Tox Exterminating Fly And Wasp Killer	Keen (World Marketing) Ltd	Amateur	4314

Product Name	Marketing Company	Use	HSE No.

287 Fipronil

Goliath Bait Station	Aventis Environmental Science	Professional	6515
Goliath Gel	Aventis Environmental Science	Professional	6514
Nexa Cockroach Bait Station	The Scotts Company (UK) Limited	Amateur	6588

288 Flufenoxuron

| Motto | Sorex Ltd | Professional | 5970 |

289 Flufenoxuron + Alphacypermethrin

| Tenopa | Sorex Ltd | Professional | 6206 |

290 Hydramethylnon

Faslane	Killgerm Chemicals Ltd	Professional	6380
Faslane	Killgerm Chemicals Ltd	Professional Professional (Animal Husbandry)	6471
Maxforce Bait Station	Aventis Environmental Science	Professional	5371
Maxforce Gel	Aventis Environmental Science	Professional	5365
Maxforce Pharaoh's Ant Killer	Aventis Environmental Science	Professional	5082
Maxforce Ultra	Aventis Environmental Science	Professional	6013

291 Hydroprene

| D Protrol | Novartis Animal Health UK Ltd | Professional | 5333 |

292 S-Hydroprene

| Protrol 9% | Novartis Animal Health UK Ltd | Professional | 5909 |

293 Iodofenphos

W Cockroach Bait	Rentokil Initial UK Ltd	Professional	4385
W Iodofenphos Granular Bait	Rentokil Initial UK Ltd	Professional	4387
W Nuvanol N 500 SC	Novartis Animal Health UK Ltd	Professional	4951
W Rentokil Iodofenphos Gel	Rentokil Initial UK Ltd	Professional	4359
W Waspex	Rentokil Initial UK Ltd	Professional	4386

Product Name	Marketing Company	Use	HSE No.
294 Iodofenphos + Dichlorvos			
W Defest Flea Free	Sherley's Ltd	Amateur	5715
W Nuvan Staykil	Novartis Animal Health UK Ltd	Amateur Professional	4017
295 Lambda-Cyhalothrin			
Demand CS	Zeneca Public Health	Professional	6287
Icon 2.5 EC	Zeneca Ltd	Professional	5614
296 Lindane			
W Doom Ant And Insect Powder	Napa Products Ltd	Amateur	4570
Fumite Lindane Generator Size 10	Hortichem Limited	Amateur Professional	4713
Fumite Lindane Generator Size 40	Hortichem Limited	Amateur Professional	4714
W Fumite Lindane Pellet No 3	Hortichem Limited	Professional	4733
Fumite Lindane Pellet No 4	Hortichem Limited	Professional	4734
297 Lindane + Tetramethrin			
W Doom Flea Killer	Napa Products Ltd	Amateur	4572
Doom Moth Proofer Aerosol	Napa Products Ltd	Amateur	4571
298 Methoprene			
Biopren BM Ready To Use Pharaoh Ant Killer Bait	Five Star Pest Control	Amateur Professional	6826
Pharorid	Novartis Animal Health UK Ltd	Professional	5332
299 S-Methoprene			
D Dianex	Novartis Animal Health UK Ltd	Professional	5709
Pharorid S	Novartis Animal Health UK Ltd	Professional	5479
300 Methoprene + Cypermethrin			
Killgerm Precor ULV III	Killgerm Chemicals Ltd	Professional	3523
301 Methoprene + Permethrin			
Norshield 150	Norbrook Laboratories (GB) Ltd	Amateur	6892
Pedigree And Whiskas Exelpet Household Flea Spray	Thomas's Europe	Amateur	6346
Pedigree Care Household Spray	Thomas's Europe	Amateur	6552
Precor ULV II	Killgerm Chemicals Ltd	Professional	3191
Whiskas Care Household Spray	Thomas's Europe	Amateur	6554

Product Name	Marketing Company	Use	HSE No.

302 S-Methoprene + Permethrin

Acclaim 2000	Sanofi Animal Health Ltd	Amateur	6448
Hartz RID Flea Control Home Spray	Novartis Animal Health UK Ltd	Amateur	5835
Room Fogger	Cork International	Amateur	6598
Total Home Guard	Cork International	Amateur	6594
Vet Kem Acclaim 2000	Sanofi Animal Health Ltd	Amateur	6127
Vet-Kem Pump Spray	Sanofi Animal Health Ltd	Amateur	5475
Zodiac + Flea Spray	Petlove	Amateur	5821
D Zodiac + Pump Spray	Grosvenor Pet Health Ltd	Amateur	5472
Zodiac Fogger	Interpet Ltd	Amateur	6128
Zodiac Maxi Household Flea Spray	Petlove	Amateur	6519

303 Methoprene + Permethrin + Bioallethrin

Arrest	Arnolds Veterinary Products Ltd	Amateur	6389
Bob Martin Home Flea Fogger Plus	The Bob Martin Company	Amateur	6630
Bob Martin Home Flea Spray Plus	The Bob Martin Company	Amateur	6223
R.I.P. Fleas	The Bob Martin Company	Amateur	6207

304 Methoprene + Permethrin + Pyrethrins

Biopren BM Residual Flea Killer Aerosol	Five Star Pest Control	Amateur Professional	6672
Friends Extra Long Lasting Household Flea Spray	Sinclair Animal And Household Care Ltd	Amateur	6685
Johnson's Household Extra Guard Flea And Insect Spray	Johnson's Veterinary Products Ltd	Amateur	6343

305 Methoprene + Permethrin + Tetramethrin

Johnson's Cage And Hutch Insect Spray	Johnson's Veterinary Products Ltd	Amateur	6864
Johnson's House Flea Spray Extra Guard	Johnson's Veterinary Products Ltd	Amateur	6649

306 Methoprene + Pyrethrins

Armitage Pet Care, Pet Bedding & Household Flea Spray	Armitage Brothers PLC	Amateur	6657
Biopren BM 0.68	Five Star Pest Control	Amateur Professional	6618
Canovel Pet Bedding And Household Spray	Smithkline Beecham Animal Health	Amateur	5330
Killgerm Precor RTU	Killgerm Chemicals Ltd	Professional	3165
Killgerm Precor ULV I	Killgerm Chemicals Ltd	Professional	3098
Pedigree Care And Whiskas Care Household Anti-Flea Spray	Thomas's Europe	Amateur	6347
Pedigree Care Household Anti Flea Spray	Thomas's Europe	Amateur	6553

HSE
INSECTICIDES

Product Name	Marketing Company	Use	HSE No.

306 Methoprene + Pyrethrins—continued

Protect B Flea Killer Aerosol	Five Star Pest Control	Amateur Professional	6445
Whiskas Care Household Anti Flea Spray	Thomas's Europe	Amateur	6551

307 S-Methoprene + Pyrethrins

Acclaim	Sanofi Animal Health Ltd	Amateur	5473
Canovel Pet Bedding And Household Spray Plus	Pfizer Animal Health	Amateur	5838
Raid Flea Killer Plus	Johnson Wax Ltd	Amateur	5967
Vet-Kem Acclaim	Sanofi Animal Health Ltd	Amateur	5474
D Zodiac + Household Flea Spray	Grosvenor Pet Health Ltd	Amateur	5471
Zodiac + Household Flea Spray	Petlove	Amateur	5822

308 S-Methoprene + Tetramethrin

Zodiac Household Flea Spray	Petlove	Amateur	6502

309 Naphthalene + Permethrin + Pyrethrins + Benzalkonium Chloride

D Roxem D	Horton Hygiene Company	Professional	5107

310 Oleyl Monoethoxylate + Ceto-Stearyl Diethoxylate

Larvex-100	Accotec Ltd	Professional	5393
Larvex-15	Accotec Ltd	Professional	5091

311 Permethrin

Amogas Ant Killer	Thornton And Ross Ltd	Amateur	5126
Ant & Crawling Insect Spray	Rentokil Initial UK Ltd	Amateur	6245
Ant Off Powder	Pet And Garden Manufacturing PLC	Amateur	6010
Armitages Good Boy Household Flea Trigger Spray	Armitage Brothers PLC	Amateur	5995
Battles Louse Powder	Battle, Hayward And Bower Ltd	Amateur	6029
Betterware Cyclone And Vac Fresh X2	IBA UK Ltd	Amateur	6973
Bio New Anti-Ant Duster	PBI Home & Garden Ltd	Amateur Professional	6041
Bio Wasp Nest Destroyer	PBI Home & Garden Ltd	Amateur	6021
Biokill	Bio-Industries Ltd	Professional	6870
Bob Martin Flea Bomb	The Bob Martin Company	Amateur	5891
Bob Martin Flea Kill	The Bob Martin Company	Amateur	5569
Bob Martin Home Flea Spray	The Bob Martin Company	Amateur	3914
Boots Powder Ant Killer	The Boots Company PLC	Amateur	5525
Bouchard Anti-Dust Mite Vacuum Freshener	IBA UK Ltd	Amateur	6972
Bug Free	Elite Force UK Ltd	Amateur	6148

Product Name	Marketing Company	Use	HSE No.
311 Permethrin—continued			
Bug Proof	Nomad Medical Ltd	Amateur	6525
Chirton Ant And Insect Killer	Ronson International Ltd	Amateur	3835
Constrain	Historyonics	Amateur	6149
Cooke's Liquid Insecticide	Nehra Cooke's Chemicals Ltd	Amateur	6110
D Coopex 25% EC	AgrEvo UK Ltd	Professional	5147
Coopex Insect Powder	Aventis Environmental Science	Professional	5052
Coopex Maxi Smoke Generator	Aventis Environmental Science	Professional	5131
Coopex Mini Smoke Generator	Aventis Environmental Science	Professional	5130
Coopex Smoke Generator	Aventis Environmental Science	Professional	5132
Coopex WP	Aventis Environmental Science	Professional	5096
Delta Insect Powder	Bennetts Company Ltd	Amateur	4664
Doff New Ant Killer	Doff Portland Ltd	Amateur	6601
Dyna-Mite	Stock Nutrition	Amateur Professional Professional (Animal Husbandry)	6411
Farco Rapid Kill Insecticide Powder	F A Richard & Company Ltd	Amateur	6919
Flea Off Powder	Pet And Garden Manufacturing PLC	Amateur	6009
Flea Powder	Rentokil Initial UK Ltd	Amateur	5974
D Flee Flea	Flee Flea International	Amateur	5644
Fresh-A-Pet Insecticidal Rug And Carpet Freshener	Bilaurand Laboratories Ltd	Amateur	4109
Friends Household Flea Powder	Sinclair Animal And Household Care Ltd	Amateur	6667
Gold Label Kennel And Stable Powder	Stockcare Ltd	Amateur	5699
Good Boy Insecticidal Carpet And Upholstery Powder	Armitage Brothers PLC	Amateur	5984
Hand Anti-Ant Powder	Hand Associates Ltd	Amateur Professional	6848
Insectaban	Rentokil Initial UK Ltd	Professional	6216
Jeyes Expel Ant Killer Powder	Jeyes Group Ltd	Amateur	5524
Jeyes Kontrol Ant Killer Powder	Jeyes Group Ltd	Amateur	5520
Johnson's Carpet Flea Guard	Johnson's Veterinary Products Ltd	Amateur	4943
Johnson's Household Flea Powder	Johnson's Veterinary Products Ltd	Amateur	5998
Kudos	AgrEvo UK Ltd	Professional	5263
Lanosol P	Microsol	Professional	5477

Product Name	Marketing Company	Use	HSE No.
311 Permethrin—continued			
Lanosol RTU	Microsol	Professional	5307
Lincoln Lice Control Powder	Battle, Hayward And Bower Ltd	Amateur	5956
Micrapor Environmental Flea Spray	Petlife	Amateur	5444
Mite-Kill	Net-Tex Agricultural Ltd	Amateur	6090
Mitetak	Stimvak Ltd	Professional	6929
Morgan's Ant Powder	Growing Success Organics Ltd	Amateur	6922
Mosiguard Shield	Medical Advisory Services For Travellers Abroad	Amateur	4515
Mostyn Permethrin 25 WP	Hockley International Ltd	Professional	7023
Multi-Shot Permethrin Spray	Greenhill Chemical Products Ltd	Professional	7119
Muscatrol	Rentokil Initial UK Ltd	Professional	4579
Nippon Ant Killer Powder	Vitax Ltd	Amateur	5499
Nippon Ready For Use Ant And Crawling Insect Killer	Vitax Ltd	Amateur	4328
Nippon Woodlice Killer Powder	Vitax Ltd	Amateur	5919
W No Fleas On Me	LA Compagne Sans Frontiere Ltd	Amateur	6180
Nomad Residex P	Nomad Pharmacy Ltd	Amateur	5177
Outright Household Flea Spray	Outright Pet Care Products	Amateur	5623
Patriot RTU	Agropharm Ltd	Amateur	7077
Pedigree Care Anti-Flea Carpet Powder	Thomas's Europe	Amateur	6154
Permasect 0.5 Dust	Mitchell Cotts Chemicals	Professional	3134
Permasect 10 WP	Mitchell Cotts Chemicals	Professional	4854
Permasect Powder	Mitchell Cotts Chemicals	Amateur	5498
Permethrin	Lifesystems Ltd	Amateur	5612
Permethrin Dusting Powder	Rentokil Initial UK Ltd	Professional	4688
Permethrin PH - 25WP	Zeneca Public Health	Professional	4956
Permethrin Wettable Powder	Rentokil Initial UK Ltd	Professional	4581
Permost 0.5% Dust Powder	Hockley International Ltd	Amateur Professional	6184
Permost 25 WP	Hockley International Ltd	Professional	6081
Pestrin	Hughes And Hughes Ltd	Professional	4499
Pif Paf Crawling Insect Powder	Reckitt And Colman Products Ltd	Amateur	5155
Powder Ant Killer	The Boots Company PLC	Amateur	7055
Pro Mite Solvent Based Fabric Protector	Active Chemical Products Ltd	Professional	6918
Pro Mite Water Based Carpet Protector	Active Chemical Products Ltd	Professional	6921
Pynosect 6	Mitchell Cotts Chemicals	Amateur Professional	4528
Pynosect Pco	Mitchell Cotts Chemicals	Professional	4566
Pynosect Pcp	Mitchell Cotts Chemicals	Professional	5805

Product Name	Marketing Company	Use	HSE No.

311 Permethrin—continued

Product Name	Marketing Company	Use	HSE No.
Pynosect Powder	Killgerm Chemicals Ltd	Professional	5849
Raid Ant Killer Powder	Johnson Wax Ltd	Amateur Professional	5796
Residex P	Agropharm Ltd	Professional	5170
Residual Powerkill	Reabrook Limited	Professional	5180
Safeclean Promite Solvent Protector	Lilly Industries (UK) Ltd	Professional	7022
Secto Extra Strength Insect Killer	Sinclair Animal And Household Care Ltd	Amateur	4329
Secto Flea Free Insecticidal Rug And Carpet Freshener	Sinclair Animal And Household Care Ltd	Amateur	4151
Sherley's Rug-De-Bug	Sherley's Ltd	Amateur	5128
Single-Shot Permethrin Spray	Greenhill	Professional	7118
Stapro Insecticide	Stapro Ltd	Professional	4526
Tent Treatment	Ibis Products Ltd	Amateur	6050
Vapona Ant And Crawling Insect Killer Powder	Ashe Consumer Products Ltd	Amateur	5207
Vapona Carpet And Household Flea Powder	Ashe Consumer Products Ltd	Amateur	5650
Vetpet Crusade	Alstoe Ltd	Amateur	6113
Vitax Kontrol Ant Killer Powder	Vitax Ltd	Amateur	5593
Whiskas Care Anti-Flea Carpet Powder	Thomas's Europe	Amateur	6153
D Z-Stop Anti-Wasp Strip	Thames Laboratories Ltd	Amateur	3532

312 Permethrin + Bioallethrin

Product Name	Marketing Company	Use	HSE No.
Ant Gun! 2	Miracle Garden Care Ltd	Amateur Professional	5629
Antkiller Spray 2	B & Q PLC	Amateur Professional	5185
Aventis Procontrol Crawling Insect Killer	Aventis Environmental Science	Amateur Professional	6116
Aventis Procontrol Flying Insect Killer	Aventis Environmental Science	Amateur Professional	6117
Aventis Procontrol Flying Insect Killer Aerosol	Aventis Environmental Science	Professional	6230
Baby Bio Houseplant Spray	PBI Home & Garden Ltd	Amateur Professional	5828
Bio Spraydex Ant And Insect Killer	PBI Home & Garden Ltd	Amateur Professional	5829
Crawling Insect And Ant Killer	Rentokil Initial UK Ltd	Amateur	4650
Creepy Crawly Gun!	Miracle Garden Care Ltd	Amateur Professional	5183
Defest II	Sherley's Ltd	Amateur	6002
Farco Flying Insect Killer	F A Richard & Company Ltd	Amateur	6151
Fleegard	Bayer PLC	Amateur	5564

Product Name	Marketing Company	Use	HSE No.

312 Permethrin + Bioallethrin—continued

Product Name	Marketing Company	Use	HSE No.
Hi-Craft Insecticidal Household Spray	Harry Irving & Co Ltd	Amateur	6612
Homebase Ant And Crawling Insect Killer	Sainsbury's Homebase House And Garden Centres	Amateur	5846
Insectrol	Rentokil Initial UK Ltd	Amateur	4568
Insectrol Professional	Rentokil Initial UK Ltd	Professional	4624
Nippon Fly Killer Pads	Vitax Ltd	Amateur	5426
PC Insect Killer	Perycut Chemie AG	Amateur Professional	4415
W Perycut Cockroach Mat	Perycut Chemie AG	Amateur Professional	5467
Pif Paf Crawling Insect Killer	Reckitt And Colman Products Ltd	Amateur	5153
Pif Paf Flying Insect Killer	Reckitt And Colman Products Ltd	Amateur Professional	5151
Pif Paf Flying Insect Killer Aerosol	Reckitt And Colman Products Ltd	Amateur	5152
Pybuthrin Fly Killer	AgrEvo UK Ltd	Amateur Professional	5141
Pybuthrin Fly Spray.	AgrEvo UK Ltd	Amateur	5116
W Sainsbury's Crawling Insect And Ant Killer	J Sainsbury PLC	Amateur	5231
D Spraydex Ant & Insect Killer	Pan Britannica Industries Ltd	Amateur	5645
D Spraydex Houseplant Spray	Spraydex Ltd	Amateur	3540
D Spraydex Insect Killer	Spraydex Ltd	Amateur	3834
Vapona Ant And Crawling Insect Killer	Ashe Consumer Products Ltd	Amateur Professional	5206
Vapona Fly And Wasp Killer	Ashe Consumer Products Ltd	Amateur	5205

313 Permethrin + Bioallethrin + Methoprene

Product Name	Marketing Company	Use	HSE No.
Arrest	Arnolds Veterinary Products Ltd	Amateur	6389
Bob Martin Home Flea Fogger Plus	The Bob Martin Company	Amateur	6630
Bob Martin Home Flea Spray Plus	The Bob Martin Company	Amateur	6223
R.I.P. Fleas	The Bob Martin Company	Amateur	6207

314 Permethrin + Cyromazine

Product Name	Marketing Company	Use	HSE No.
Flego Trigger Pack	Novartis Animal Health UK Ltd	Amateur	6051
Pet Rescue Household Flea Spray	Pet Rescue	Amateur Professional	7063
Sherley's Flego	Sherley's Ltd	Amateur Professional	5803
Sherley's Flego Trigger Pack	Sherley's Ltd	Amateur	6065

Product Name	Marketing Company	Use	HSE No.

314 Permethrin + Cyromazine —continued

| Staykil Household Flea Spray | Novartis Animal Health UK Ltd | Amateur Professional | 5784 |

315 Permethrin + Methoprene

Norshield 150	Norbrook Laboratories (GB) Ltd	Amateur	6892
Pedigree And Whiskas Exelpet Household Flea Spray	Thomas's Europe	Amateur	6346
Pedigree Care Household Spray	Thomas's Europe	Amateur	6552
Precor ULV II	Killgerm Chemicals Ltd	Professional	3191
Whiskas Care Household Spray	Thomas's Europe	Amateur	6554

316 Permethrin + Propiconazole

| Aqueous Universal | Protim Solignum Ltd | Amateur Professional | 5776 |

317 Permethrin + Pyrethrins

Detia Crawling Insect Spray	K D Pest Control Products	Amateur	3710
Flamil Finale	Flore-Chemie GmbH	Professional	5083
Home Insect Fogger	International Surplus (UK) Ltd	Amateur	6364
Insecticidal Aerosol	Aerosols International Ltd	Amateur	5118
Johnson's Household Flea Spray	Johnson's Veterinary Products Ltd	Amateur	5703
Johnson's Pet Housing Spray	Johnson's Veterinary Products Ltd	Amateur	5843
SC Johnson Wax Raid Ant & Cockroach Killer With Natural Pyrethrum	Johnson Wax Ltd	Amateur	6088
Secto Household Flea Spray	Sinclair Animal And Household Care Ltd	Amateur	4131
X-Gnat M/C (Mosquito Control)	X-Gnat Laboratories Ltd	Professional	6323

318 Permethrin + Pyrethrins + Benzalkonium Chloride + Naphthalene

| D Roxem D | Horton Hygiene Company | Professional | 5107 |

319 Permethrin + Pyrethrins + Bioallethrin

| Tyrax | DBC | Amateur | 6829 |

320 Permethrin + Pyrethrins + Chlorpyrifos-Methyl

| D Multispray | AgrEvo UK Ltd | Professional | 5165 |

321 Permethrin + Pyrethrins + Methoprene

| Biopren BM Residual Flea Killer Aerosol | Five Star Pest Control | Amateur Professional | 6672 |

HSE
INSECTICIDES

413

Product Name	Marketing Company	Use	HSE No.

321 Permethrin + Pyrethrins + Methoprene—continued

Friends Extra Long Lasting Household Flea Spray	Sinclair Animal And Household Care Ltd	Amateur	6685
Johnson's Household Extra Guard Flea And Insect Spray	Johnson's Veterinary Products Ltd	Amateur	6343

322 Permethrin + Pyriproxyfen

Friskies Pro Control Household Flea Spray	Alfamed S.A.	Amateur	7006
Indorex Press And Go Mist	Virbac UK Ltd	Amateur	6956
Indorex Spray	Virbac S A	Amateur	6626
Vetzyme Household Patrol	Seven Seas Ltd	Amateur	6191

323 Permethrin + Resmethrin + Fenitrothion

W Turbair Beetle Killer	Pan Britannica Industries Ltd	Professional	4648

324 Permethrin + S-Bioallethrin

Agrevo Procontrol Yellow Flying Insect Killer Aerosol	AgrEvo UK Ltd	Professional	6232
Aqua Reslin Premium	AgrEvo UK Ltd	Professional	5172
Aqua Reslin Super	AgrEvo UK Ltd	Professional	5175
Atta-X Flying Insect Killer	Sanderson Curtis Ltd	Amateur	6348
Bug Wars	Design Go Travel Emporium	Amateur	6361
D Coopex S25% EC	AgrEvo UK Ltd	Professional	5146
D Imperator Fog S 18/6	Zeneca Public Health	Professional	4960
D Imperator Fog Super S 30/15	Zeneca Public Health	Professional	4955
D Imperator ULV S 6/3	Zeneca Public Health	Professional	4954
D Imperator ULV Super S 10/4	Zeneca Public Health	Professional	4953
Pif Paf Yellow Flying Insect Killer	Reckitt And Colman Products Ltd	Amateur	5937
Purge	Aztec Aerosols Limited	Amateur	6634
Resigen	Aventis Environmental Science	Professional	5143
Reslin Premium	Aventis Environmental Science	Professional	5174
D Reslin Super	AgrEvo UK Ltd	Professional	5173

325 Permethrin + S-Bioallethrin + Dichlorvos

Farco Rapid Kill	F A Richard & Company Ltd	Amateur	6152
W Pif Paf Insecticide	Reckitt And Colman Products Ltd	Amateur Professional	5150

326 Permethrin + S-Bioallethrin + Tetramethrin

Aventis Procontrol Crawling Insect Killer Aerosol	Aventis Environmental Science	Professional	6231
Pif Paf Crawling Insect Killer Aerosol	Reckitt And Colman Products Ltd	Amateur	5595

Product Name	Marketing Company	Use	HSE No.
327 Permethrin + S-Methoprene			
Acclaim 2000	Sanofi Animal Health Ltd	Amateur	6448
Hartz Rid Flea Control Home Spray	Novartis Animal Health UK Ltd	Amateur	5835
Room Fogger	Cork International	Amateur	6598
Total Home Guard	Cork International	Amateur	6594
Vet Kem Acclaim 2000	Sanofi Animal Health Ltd	Amateur	6127
Vet-Kem Pump Spray	Sanofi Animal Health Ltd	Amateur	5475
Zodiac + Flea Spray	Petlove	Amateur	5821
D Zodiac + Pump Spray	Grosvenor Pet Health Ltd	Amateur	5472
Zodiac Fogger	Interpet Ltd	Amateur	6128
Zodiac Maxi Household Flea Spray	Petlove	Amateur	6519
328 Permethrin + Tetramethrin			
Aventis Procontrol Wasp Nest Destroyer	Aventis Environmental Science	Professional	6138
Atta-X Cockroach Killer	Sanderson Curtis Ltd	Amateur	6362
W Bolt Ant And Crawling Insect Killer	Johnson Wax Ltd	Amateur	3619
Bop Crawling Insect Killer	Mcbride International	Amateur Professional	4138
Bop Crawling Insect Killer (MC)	Mcbride International	Amateur Professional	4139
Canned Death	Pearson & Wilkinson	Professional	6112
Combat Fly And Wasp Killer	Sinclair Animal And Household Care Ltd	Amateur	7065
Combat Fly Killer	Sinclair Animal And Household Care Ltd	Amateur	5808
Corry's Fly And Wasp Killer	Vitax Ltd	Amateur	5322
D-Stroy	Future Developments (Manufacturing) Ltd	Professional	5411
Delta Cockroach And Ant Spray	Bennetts Company Ltd	Amateur	3950
Doom Tropical Strength Fly And Wasp Killer Aerosol	Napa Products Ltd	Amateur	4752
Dragon Flykiller	Zeneca Public Health	Amateur	4952
Fly And Wasp Killer	Rentokil Initial UK Ltd	Amateur Professional	4643
Fly Free Zone	Fly Away Ltd	Amateur Professional Professional (Animal Husbandry)	6394
Fly Spray	Merton Cleaning Supplies	Professional	5112
Fly-Kill	Stagchem Ltd	Professional	6202
Hail Flying Insect Killer	Mcbride International	Amateur	5568
Insecticide Aerosol	Aerosols International Ltd	Amateur	6567
Insectrol, Fly And Wasp Killer	Rentokil Initial UK Ltd	Amateur	5827
Johnson's House Flea Spray	Johnson's Veterinary Products Ltd	Amateur	6623

Product Name	Marketing Company	Use	HSE No.

328 Permethrin + Tetramethrin—continued

Product Name	Marketing Company	Use	HSE No.
Johnson's New Pet Housing Spray	Johnson's Veterinary Products Ltd	Amateur	6624
K02 Selkil	Selden Research Ltd	Professional	4828
Mostyn 8/64 TFC	Hockley International Ltd	Professional	7105
Nippon Ant And Crawling Insect Killer	Vitax Ltd	Amateur	4414
Nippon Fly Killer Spray	Vitax Ltd	Amateur	4254
Nippon Killaquer Crawling Insect Killer	Arrow Chemicals Ltd	Professional	3694
Nippon Wasp Nest Destroyer Foam	Vitax Ltd	Amateur	7100
Nisa Fly And Wasp Killer	Nisa	Amateur	4962
Permost Uni	Hockley International Ltd	Professional Professional (Animal Husbandry)	6399
Purge Insect Destroyer	Aztec Chemicals Ltd	Professional	5048
Pynosect 10	Mitchell Cotts Chemicals	Professional	3171
Pynosect Extra Fog	Mitchell Cotts Chemicals	Professional	3172
W Raid Ant And Roach Killer	Johnson Wax Ltd	Amateur	5182
Sactif Flying Insect Killer	Diverseylever Ltd	Professional	4924
W Sainsbury's Fly And Wasp Killer	J Sainsbury PLC	Amateur	5335
Sanmex Ant & Crawling Insect Killer	The British Products Sanmex Company Ltd	Amateur	6655
Sanmex Fly And Wasp Killer	The British Products Sanmex Company Ltd	Amateur	4006
Sanmex Supakil Insecticide	The British Products Sanmex Company Ltd	Amateur	4009
Scorpio Fly Spray For Flying Insects	Jangro Ltd	Professional	5049
Secto Fly Killer	Sinclair Animal And Household Care Ltd	Amateur	3862
Shelltox Insect Killer	Temana International Ltd	Amateur Professional	3697
Shelltox Insect Killer 2	Temana International Ltd	Amateur Professional	4061
Stop Insect Killer	Temana International Ltd	Amateur Professional	3163
Superdrug Fly Killer	Superdrug Stores PLC	Amateur	4185
Zappit	BLP Ltd	Amateur	6181

329 Permethrin + Tetramethrin + Fenitrothion

Product Name	Marketing Company	Use	HSE No.
Connect Insect Killer	Sun Oil Ltd	Amateur	6109
Motox	HVM Products Ltd	Amateur	5622

Product Name	Marketing Company	Use	HSE No.

330 Permethrin + Tetramethrin + Methoprene

Johnson's Cage And Hutch Insect Spray	Johnson's Veterinary Products Ltd	Amateur	6864
Johnson's House Flea Spray Extra Guard	Johnson's Veterinary Products Ltd	Amateur	6649

331 Permethrin + Tetramethrin + d-Allethrin

Hail Plus Crawling Insect Killer	Mcbride International	Amateur	5567
New Super Raid Insecticide	Johnson Wax Ltd	Amateur Professional	4583

332 Permethrin + Tetramethrin + d-Allethrin + Cypermethrin

Elf Insecticide	Elf Oil UK Ltd	Amateur	6669

333 Permethrin + Tetramethrin + d-Allethrin + Dichlorvos

Bop	Mcbride International	Amateur	4773

334 Permethrin + d-Allethrin

New Vapona Fly And Wasp Killer	Ashe Consumer Products Ltd	Amateur	4201
Raid Outdoor Insectguard	Johnson Wax Ltd	Amateur	5459

335 d-Phenothrin

Aircraft Aerosol Insect Control	Sumitomo Chemical (UK) PLC	Professional	5408
Aircraft Disinsectant	Arrow Chemicals Ltd	Professional	6254
Aircraft Disinsectant Multi	Greenhill Chemical Products Ltd	Professional	7016
Aircraft Insecticide	British Airways	Professional	7021
Boots Anti-Dust Mite Spray	The Boots Company PLC	Amateur	7091
Disinsecting Spray Multi	WK Thomas And Company Ltd	Professional	7024
Disinsecting Spray Single	WK Thomas And Company Ltd	Professional	6977
Extermisect	John Horsfall	Professional	6976
Extermisect Multi	John Horsfall	Professional	7027
One-Shot Aircraft Aerosol Insect Control	Sumitomo Chemical (UK) PLC	Professional	5407
Orvec Aircraft Insecticide	Orvec International	Professional	7028
Phillips Pestkill Karpet Killer	Seven Seas Ltd	Amateur	5604
Pygo	Rentokil Initial UK Ltd	Professional	7078
Roebuck Eyot Aircraft Disinsection Spray	Roebuck Eyot Ltd	Professional	6404
Secto Household Dust Mite Control.	Sinclair Animal And Household Care Ltd	Amateur	5870
Sergeant's Dust Mite Patrol	Allerayde Ltd	Amateur	6012
Sergeants Car Patrol	Seven Seas Ltd	Amateur	4503

HSE
INSECTICIDES

Product Name	Marketing Company	Use	HSE No.
335 d-Phenothrin—continued			
Sergeants Carpet Patrol	Seven Seas Ltd	Amateur	4763
W Sergeants Dust Mite Patrol	Seven Seas Ltd	Amateur	4862
Sergeants Rug Patrol	Seven Seas Ltd	Amateur	4504
Sumithrin 10 Sec	Sumitomo Chemical (UK) PLC	Professional	3762
Sumithrin 10 Sec Carpet Treatment	Sumitomo Chemical (UK) PLC	Amateur	5398
Vapona Anti Dust Mite	Ashe Ltd	Amateur	7048
Vetzyme Rug Patrol	Seven Seas Ltd	Amateur	6990
Willo-Zawb	Willows Francis Ltd	Amateur	3076
Willodorm	Willows Francis Ltd	Amateur	4782
336 d-Phenothrin + Pyrethrins			
Pyrbek	Hand Associates Ltd	Professional	6550
337 d-Phenothrin + Pyriproxyfen			
Nylar 650 ULV	Pelgar International Ltd	Professional	6593
338 d-Phenothrin + Tetramethrin			
"Flak"	Mcbride International	Amateur Professional	4143
Ant Destroyer Foam	Rentokil Initial UK Ltd	Amateur	6412
Betterware Foaming Ant & Wasp Nest Destroyer	Betterware UK Ltd	Amateur	6615
D Big D Ant And Crawling Insect Killer	Domestic Fillers Ltd	Amateur	4097
Big D Ant And Crawling Insect Killer	Dylon International Ltd	Amateur	5818
Bio Foaming Wasp Nest Destroyer	PBI Home & Garden Ltd	Amateur	6665
Boots Ant And Crawling Insect Killer	The Boots Company PLC	Amateur	5554
Boots Dry Fly And Wasp Killer	The Boots Company PLC	Amateur	7029
Boots Flea Spray	The Boots Company PLC	Amateur	5522
Bop Arabic	Mcbride International	Amateur Professional	4142
Bouchard Fly And Wasp Killer	IBA UK Ltd	Amateur	7053
Brimpex ULV 1500	Brimpex Metal Treatments	Professional Professional (Animal Husbandry)	4942
Bug Off	Dimex Ltd	Amateur	6356
Combat Wasp Killer Foam	Sinclair Animal And Household Care Ltd	Amateur	6974
Deosan Fly Spray	Diverseylever Ltd	Professional	5246
W Envar 6/14 EC	Agrimar (UK) PLC	Professional	4354
W Envar 6/14 Oil	Agrimar (UK) PLC	Professional	4355
Fly And Bug Spray	Jangro Ltd	Amateur	6933

Product Name	Marketing Company	Use	HSE No.

338 d-Phenothrin + Tetramethrin—continued

Product Name	Marketing Company	Use	HSE No.
Flying And Crawling Insect Killer	Betterware UK	Amateur	6607
Friends Household Flea Spray	Sinclair Animal And Household Care Ltd	Amateur	6664
GLP Flying And Crawling Insect Killer	Greenhill Chemical Products Ltd	Amateur	6303
Insect Killer	Rentokil Initial UK Ltd	Amateur	6244
Jeyes Expel Ant And Crawling Insect Killer	Jeyes Group Ltd	Amateur	5552
Jeyes Expel Flea Killer	Jeyes Group Ltd	Amateur	5523
Jeyes Kontrol Ant And Crawling Insect Killer	Jeyes Group Ltd	Amateur	5553
Jeyes Kontrol Flea Killer	Jeyes Group Ltd	Amateur	5521
Killgerm ULV 1500	Killgerm Chemicals Ltd	Professional	6961
Killgerm ULV 500	Killgerm Chemicals Ltd	Professional Professional (Animal Husbandry)	4647
W Killoff Aerosol Insect Killer	Ferosan Products	Amateur	3974
Kybosh 2	PBI Home & Garden Ltd	Amateur	6622
Murphy Foaming Wasp Nest Destroyer	Levington Horticulture Ltd	Amateur	6609
New Secto Household Flea Spray	Sinclair Animal And Household Care Ltd	Amateur	6504
Pesguard NS 6/14 Oil	Sumitomo Chemical (UK) PLC	Professional	4361
Pesguard OBA F 7305 C	Sumitomo Chemical (UK) PLC	Amateur	4064
Pesguard OBA F 7305 D	Sumitomo Chemical (UK) PLC	Amateur	4065
Pesguard OBA F 7305 E	Sumitomo Chemical (UK) PLC	Amateur	4066
Pesguard OBA F 7305 F	Sumitomo Chemical (UK) PLC	Amateur	4067
Pesguard OBA F 7305 G	Sumitomo Chemical (UK) PLC	Amateur	4063
Pesguard OBA F7305 B	Sumitomo Chemical (UK) PLC	Amateur	4419
Pesguard WBA F-2656	Sumitomo Chemical (UK) PLC	Amateur	4263
Pesguard WBA F-2692	Sumitomo Chemical (UK) PLC	Amateur	4262
Pestkill Household Spray	Phillips Yeast Products Ltd	Amateur	4938
Pet Bedding And Household Flea Spray	Armitage Brothers PLC	Amateur	4090
PY Kill 25 Plus	Killgerm Chemicals Ltd	Professional	4900
Raid Fly Wasp And Plant Insect Killer	Johnson Wax Ltd	Amateur	5432

Product Name	Marketing Company	Use	HSE No.

338 d-Phenothrin + Tetramethrin—continued

Product Name	Marketing Company	Use	HSE No.
Rentokil Wasp Nest Destroyer	Rentokil Initial UK Ltd	Amateur	6214
S C Johnson Professional Raid Flying Insect Killer	S C Johnson Professional Ltd	Amateur	6407
S C Johnson Wax Raid Fly And Wasp Killer	Johnson Wax Ltd	Amateur	5181
W Sergeant's Dust Mite Patrol Injector And Spray	Seven Seas Ltd	Amateur Professional	5513
W Sergeant's Dust Mite Patrol Spray	Seven Seas Ltd	Amateur	5041
Sergeant's Household Spray	Seven Seas Ltd	Amateur	6011
W Sergeant's Insect Patrol	Seven Seas Ltd	Amateur	5397
Shelltox Flying Insect Killer 2	Temana International Ltd	Amateur Professional	4032
Shelltox Flying Insect Killer 3	Temana International Ltd	Amateur	4741
Sorex Fly Spray RTU	Sorex Ltd	Professional	5718
Sorex Super Fly Spray	Sorex Ltd	Amateur	6297
W Super Fly Spray	Sorex Ltd	Amateur	4468
Super Raid II	Johnson Wax Ltd	Amateur	4015
Super Shelltox Flying Insect Killer	Kortman Intradal BV	Amateur	5440
Terminator	Aerosol Logistics	Amateur	6416
Terminator Fly And Wasp Killer	Sanoda Ltd	Amateur	6150
Terminator Wasp & Insect Nest Destroyer	Globol Chemicals (UK) Ltd	Amateur	6100
Terminator Wasp And Insect Nest Destroyer	Sanoda Ltd	Amateur	6144
Vapona House And Plant Fly Spray	Ashe Consumer Products Ltd	Amateur	5236
Vapona Wasp And Fly Killer	Ashe Consumer Products Ltd	Amateur	5754
Vapona Wasp And Fly Killer Spray	Ashe Consumer Products Ltd	Amateur	5280
Vetzyme Pestkill Household Spray	Seven Seas Ltd	Amateur	7034
Vitapet Flearid Household Spray	Seven Seas Ltd	Amateur	5434
Wasp Destroyer Foam	Rentokil Initial UK Ltd	Amateur	6455
Zappit Fly And Wasp Killer	BLP Ltd	Amateur	6336

339 d-Phenothrin + d-Allethrin

Product Name	Marketing Company	Use	HSE No.
Boots New Improved Fly And Wasp Killer	The Boots Company PLC	Amateur	6432
Boots Pump Fly And Wasp Killer	The Boots Company PLC	Amateur	7030
Fly And Ant Killer	Rentokil Initial UK Ltd	Amateur	5561
Jeyes Crawling Insect Spray	Jeyes Group Ltd	Amateur	5270
Kleeneze New Improved Fly Killer	Kleeneze Ltd	Amateur	6431
Kleenoff Crawling Insect Spray	Jeyes Group Ltd	Amateur	5269
Kwik Insecticide	Hand Associates Ltd	Amateur	4290
Nippon Ready For Use Fly Killer Spray	Vitax Ltd	Amateur	4631
Pedigree Care Bedding Spray	Thomas's Europe	Amateur	5427

Product Name	Marketing Company	Use	HSE No.

339 d-Phenothrin + d-Allethrin—continued

Pesguard PS 102	Sumitomo Chemical (UK) PLC	Amateur	5312
Pesguard PS 102A	Sumitomo Chemical (UK) PLC	Amateur	4173
Pesguard PS 102B	Sumitomo Chemical (UK) PLC	Amateur	4174
Pesguard PS 102C	Sumitomo Chemical (UK) PLC	Amateur	4175
Pesguard PS 102D	Sumitomo Chemical (UK) PLC	Amateur	4176
W Secto Ant And Crawling Insect Spray.	Sinclair Animal And Household Care Ltd	Amateur	5714
W Target Flying Insect Killer	Reckitt And Colman Products Ltd	Amateur	4178
Whiskas Care Bedding Spray	Thomas's Europe	Amateur	4837
Whiskas Care Bedding Pest Control Spray	Thomas's Europe	Amateur	4596

340 d-Phenothrin + d-Tetramethrin

Pesguard WBA F2714	Sumitomo Chemical (UK) PLC	Amateur	4418
Vapona Max Concentrated Fly And Wasp Killer	Ashe Consumer Products Ltd	Amateur	6062

341 Pirimiphos-Methyl

Actellic 25 EC	Zeneca Public Health	Professional Professional (Animal Husbandry) Professional (Food Storage Practice)	4880
Actellic Dust	Zeneca Public Health	Professional	4881
W Antkiller Dust 2	Miracle Garden Care Ltd	Amateur	6384

342 Pirimiphos-Methyl + Resmethrin + Tetramethrin

W Waspend	Miracle Garden Care Ltd	Amateur Professional	4857

343 Potassium Salts Of Fatty Acids

Devcol Liquid Ant Killer	Nehra Cooke's Chemicals Ltd	Amateur	5518
Devcol Wasp Killer	Nehra Cooke's Chemicals Ltd	Amateur	5519
Greenco Ant Killer	Doff Portland Ltd	Amateur	4260
Natural Ant Gun	Doff Portland Ltd	Amateur	4916

Product Name	Marketing Company	Use	HSE No.
343 Potassium Salts Of Fatty Acids—continued			
Natural Wasp Gun	Doff Portland Ltd	Amateur	5327
Natural Wasp Killer	Doff Portland Ltd	Amateur	5326
344 Prallethrin			
Boots Mosquito Killer	The Boots Company PLC	Amateur	6590
Etoc 10 Mg Mat	Sumitomo Chemical (UK) PLC	Amateur	5925
Etoc Liquid Vaporiser	Sumitomo Chemical (UK) PLC	Amateur	5618
Mosqui-Go Liquid	Design Go Travel Emporium	Amateur	6654
Mosqui-Go Tablets	Go Travel Products	Amateur	6668
Tourist Flying Insect Killer	Pik-A-Pak	Amateur	6447
Vape Mat	Travel Companions	Amateur	6666
Vapona Mosquito Travel Plug	Ashe Ltd	Amateur	7056
345 Propiconazole + Permethrin			
Aqueous Universal	Protim Solignum Ltd	Amateur Professional	5776
346 Propoxur			
W Killgerm Propoxur 20 EC	Killgerm Chemicals Ltd	Professional	4546
347 Pyrethrins			
Aquapy	Aventis Environmental Science	Professional	5799
Aquapy Micro	Aventis Environmental Science	Professional	6296
Aventis Procontrol Fly Spray	Aventis Environmental Science	Professional	5154
Bob Martin Natural Household Flea Spray	The Bob Martin Company	Amateur	6495
Bolt Dry Flying Insect Killer	Johnson Wax Ltd	Amateur	3581
Boots Pump Action Fly And Wasp Killer	The Boots Company PLC	Amateur	5292
W Boots Pump Action Fly And Wasp Killer	The Boots Company PLC	Amateur	5547
Cookes Ant Killer	Nehra Cooke's Chemicals Ltd	Amateur	5972
Dairy Fly Spray	Battle, Hayward And Bower Ltd	Professional Professional (Animal Husbandry)	5579
Days Farm Fly Spray	Day And Sons (Crewe) Ltd	Professional	4749
Detia Pyrethrum Spray	K D Pest Control Products	Amateur	3695
D Devcol Universal Pest Powder	Devcol Ltd	Amateur	4864

Product Name	Marketing Company	Use	HSE No.
347 Pyrethrins—continued			
Downland Dairy Fly Spray	Downland Marketing Ltd	Professional / Professional (Animal Husbandry)	6032
Drione	AgrEvo UK Ltd	Professional	5254
Emprasan Pendle Mist Fly Spray	Emprasan (Chemicals) Limited	Professional	6239
Flamil Finale Super	Flore-Chemie GmbH	Amateur / Professional	5615
Fortefog	Agropharm Ltd	Amateur / Professional	4820
Fortefog	Agropharm Ltd	Amateur / Professional	6486
Globol Shake And Spray Insect Killer	Globol Chemicals (UK) Ltd	Amateur	4883
Growing Success Ant Killer	Growing Success Organics Ltd	Amateur	5831
Growing Success Indoor And Outdoor Ant Killer	Growing Success Organics Ltd	Amateur	6442
W Jeyes Expel Pump Action Fly And Wasp Killer	Jeyes Group Ltd	Amateur	5506
Jeyes Flying Insect Spray	Jeyes Group Ltd	Amateur	5248
D Karate Dairy Fly Spray	Lever Industrial Ltd	Professional / Professional (Animal Husbandry)	5611
Killgerm Pyrethrum Spray	Killgerm Chemicals Ltd	Professional	4636
Killgerm ULV 400	Killgerm Chemicals Ltd	Professional / Professional (Animal Husbandry)	4838
Kleeneze Fly And Flying Insect Pump Spray	Kleeneze Ltd	Amateur	6122
Kleenoff Flying Insect Spray	Jeyes Group Ltd	Amateur	5247
Konk 1 (B) Flying Insect Killer	Pharvet Ltd	Professional / Professional (Animal Husbandry)	5486
Konk I	Airguard Control Inc	Professional / Professional (Animal Husbandry)	5085
W Patriot Flying And Crawling Insect Killer	Agropharm Ltd	Amateur / Professional	4959
Patriot Flying And Crawling Insect Killer Aerosol	Agropharm Ltd	Amateur / Professional	6166
Prevent	Agropharm Ltd	Amateur	4851

Product Name	Marketing Company	Use	HSE No.

347 Pyrethrins—continued

Product Name	Marketing Company	Use	HSE No.
Pybuthrin 2/16	Aventis Environmental Science	Professional	5115
Pybuthrin 33	Aventis Environmental Science	Professional	5106
Raid Dry Natural Flying Insect Killer	Johnson Wax Ltd	Amateur	4235
Residex	Agropharm Ltd	Amateur	4811
Samsung Natural Insect Spray	Samsung Deutchland GmbH	Amateur Professional	6121
SC Johnson Flying Insect Killer	Johnson Wax Ltd	Amateur	5094
Superdrug Fly And Wasp Killer Pump Spray	Superdrug Stores PLC	Amateur	5171
Swak Natural	Technical Concepts International Ltd	Professional	4982
Trappit Ant Powder	Agrisense Bcs Ltd	Amateur	6613
Trilanco Farm Fly Spray	Trilanco	Professional	4558
Trilanco Original Fly Spray	Trilanco	Professional	7102
Trilanco Super Fly Spray	Trilanco	Professional	6824
Vapona Green Arrow Fly And Wasp Killer	Ashe Consumer Products Ltd	Amateur	4460
Vapona House And Plant Fly Aerosol	Ashe Consumer Products Ltd	Amateur	5235

348 Pyrethrins + Bendiocarb

Product Name	Marketing Company	Use	HSE No.
Ficam Plus	Aventis Environmental Science	Professional	4830

349 Pyrethrins + Benzalkonium Chloride + Naphthalene + Permethrin

Product Name	Marketing Company	Use	HSE No.
D Roxem D	Horton Hygiene Company	Professional	5107

350 Pyrethrins + Bioallethrin + Permethrin

Product Name	Marketing Company	Use	HSE No.
Tyrax	DBC	Amateur	6829

351 Pyrethrins + Chlorpyrifos

Product Name	Marketing Company	Use	HSE No.
Contra-Insect Aerosol	Frunol Delicia GmbH	Amateur	6372
Contra-Insect Universal	Frunol Delicia GmbH	Amateur	6405

352 Pyrethrins + Chlorpyrifos + Cypermethrin

Product Name	Marketing Company	Use	HSE No.
New Tetracide	Killgerm Chemicals Ltd	Professional	3724

353 Pyrethrins + Chlorpyrifos-Methyl + Permethrin

Product Name	Marketing Company	Use	HSE No.
D Multispray	AgrEvo UK Ltd	Professional	5165

354 Pyrethrins + Cypermethrin

Product Name	Marketing Company	Use	HSE No.
Bolt Crawling Insect Killer	Johnson Wax Ltd	Amateur	3609
Raid Cockroach Killer Formula 2	Johnson Wax Ltd	Amateur	4239

Product Name	Marketing Company	Use	HSE No.
355 Pyrethrins + Deltamethrin			
Agrevo Pro Control CIK Superfast	Aventis Environmental Science	Amateur	6831
Doom Ant & Cockroach Killer	Napa Products Ltd	Amateur	6875
Doom Flea Killer	Napa Products Ltd	Amateur	6881
Pro Control CIK Superfast (P)	Aventis Environmental Science	Professional	6836
356 Pyrethrins + Methoprene			
Armitage Pet Care, Pet Bedding & Household Flea Spray	Armitage Brothers PLC	Amateur	6657
Biopren BM 0.68	Five Star Pest Control	Amateur Professional	6618
Canovel Pet Bedding And Household Spray	Smithkline Beecham Animal Health	Amateur	5330
Killgerm Precor RTU	Killgerm Chemicals Ltd	Professional	3165
Killgerm Precor ULV I	Killgerm Chemicals Ltd	Professional	3098
Pedigree Care And Whiskas Care Household Anti-Flea Spray	Thomas's Europe	Amateur	6347
Pedigree Care Household Anti Flea Spray	Thomas's Europe	Amateur	6553
Protect B Flea Killer Aerosol	Five Star Pest Control	Amateur Professional	6445
Whiskas Care Household Anti Flea Spray	Thomas's Europe	Amateur	6551
357 Pyrethrins + Methoprene + Permethrin			
Biopren BM Residual Flea Killer Aerosol	Five Star Pest Control	Amateur Professional	6672
Friends Extra Long Lasting Household Flea Spray	Sinclair Animal And Household Care Ltd	Amateur	6685
Johnson's Household Extra Guard Flea And Insect Spray	Johnson's Veterinary Products Ltd	Amateur	6343
358 Pyrethrins + Permethrin			
Detia Crawling Insect Spray	K D Pest Control Products	Amateur	3710
Flamil Finale	Flore-Chemie GmbH	Professional	5083
Home Insect Fogger	International Surplus (UK) Ltd	Amateur	6364
Insecticidal Aerosol	Aerosols International Ltd	Amateur	5118
Johnson's Household Flea Spray	Johnson's Veterinary Products Ltd	Amateur	5703
Johnson's Pet Housing Spray	Johnson's Veterinary Products Ltd	Amateur	5843
SC Johnson Wax Raid Ant & Cockroach Killer With Natural Pyrethrum	Johnson Wax Ltd	Amateur	6088

Product Name	Marketing Company	Use	HSE No.

358 Pyrethrins + Permethrin—continued

Secto Household Flea Spray	Sinclair Animal And Household Care Ltd	Amateur	4131
X-Gnat M/C (Mosquito Control)	X-Gnat Laboratories Ltd	Professional	6323

359 Pyrethrins + Resmethrin

Rentokil Houseplant Insect Killer	Rentokil Initial UK Ltd	Amateur	4433

360 Pyrethrins + S-Methoprene

Acclaim	Sanofi Animal Health Ltd	Amateur	5473
Canovel Pet Bedding And Household Spray Plus	Pfizer Animal Health	Amateur	5838
Raid Flea Killer Plus	Johnson Wax Ltd	Amateur	5967
Vet-Kem Acclaim	Sanofi Animal Health Ltd	Amateur	5474
D Zodiac + Household Flea Spray	Grosvenor Pet Health Ltd	Amateur	5471
Zodiac + Household Flea Spray	Petlove	Amateur	5822

361 Pyrethrins + d-Phenothrin

Pyrbek	Hand Associates Ltd	Professional	6550

362 Pyriproxyfen

Nylar 10 EC	Pelgar International Ltd	Professional	6219
Nylar 100	Pelgar International Ltd	Professional	6234
Nylar 4EW	Pelgar International Ltd	Professional	6373
Sumilarv 10 EC	Sumitomo Chemical (UK) PLC	Professional	5991

363 Pyriproxyfen + Cyfluthrin

Fleegard Plus	Bayer PLC	Amateur	6582

364 Pyriproxyfen + Permethrin

Friskies Pro Control Household Flea Spray	Alfamed S.A.	Amateur	7006
Indorex Press And Go Mist	Virbac UK Ltd	Amateur	6956
Indorex Spray	Virbac S A	Amateur	6626
Vetzyme Household Patrol	Seven Seas Ltd	Amateur	6191

365 Pyriproxyfen + d-Phenothrin

Nylar 650 ULV	Pelgar International Ltd	Professional	6593

366 Resmethrin + Fenitrothion + Permethrin

W Turbair Beetle Killer	Pan Britannica Industries Ltd	Professional	4648

367 Resmethrin + Pyrethrins

Rentokil Houseplant Insect Killer	Rentokil Initial UK Ltd	Amateur	4433

Product Name	Marketing Company	Use	HSE No.
368 Resmethrin + Tetramethrin			
W 151 Dry Fly And Wasp Killer	SB Ltd	Amateur	5169
Apex Fly-Killer	Apex Industrial Chemicals Ltd	Amateur Professional	4300
Chem-Kil	Inter-Chem	Amateur Professional	4266
W Chirton Dry Fly And Wasp Killer	Ronson International Ltd	Amateur	5287
D Cromessol Insect Killer	Cromessol Company Ltd	Amateur	5067
Cromessol Multi-Purpose Insect Killer	Wallace Cameron And Co Ltd	Amateur Professional	6045
Fly And Maggot Killer	Net-Tex Agricultural Ltd	Amateur	6097
D Fly Killer	Kalon Agricultural Division	Amateur	4817
D Hospital Flying Insect Spray	Kalon Chemicals	Amateur	4652
D Janisolve Fly Killer	Johmar Enterprises Company	Amateur	4228
K.O. Fly Spray	Genuine Solutions	Amateur Professional	5879
Maggot & Fly Attack	Battle, Hayward And Bower Ltd	Amateur Professional	6997
Momar Fly Killer	Momar	Amateur	6264
D Penetone Fly Killer	Penetone Ltd	Amateur	4072
Pynosect 2	Mitchell Cotts Chemicals	Amateur Professional	4654
Pynosect 4	Mitchell Cotts Chemicals	Amateur Professional	4655
Quantum Hygiene Fly Killer	Quantum Hygiene	Amateur Professional	5066
Rapid Exit Fly Spray	LMA Services	Amateur Professional	5787
Sorex Wasp Nest Destroyer	Sorex Ltd	Professional	6294
Swat	Solvitol Ltd	Amateur Professional	3564
Swat A	Solvitol Ltd	Professional	5051
Swot	Southern Chemical Services	Amateur Professional	6007
D Trilanco Fly Killer	Trilanco	Amateur	4816
Wasp Nest Destroyer	Sorex Ltd	Professional	4410
369 Resmethrin + Tetramethrin + Pirimiphos-Methyl			
W Waspend	Miracle Garden Care Ltd	Amateur Professional	4857
370 Silicon Dioxide(Amorphous Non Coated Precipitated)			
Rid Insect Powder	Rentokil Initial UK Ltd	Professional	5772
371 Sodium Tetraborate			
Betterware Ant Killer	Betterware UK Ltd	Amateur	6352

HSE
INSECTICIDES

Product Name	Marketing Company	Use	HSE No.

371 Sodium Tetraborate—continued

| Bouchard Ant Killer | IBA UK Ltd | Amateur | 6939 |

372 Tetramethrin

Big D Fly And Wasp Killer	Dylon International Ltd	Amateur	5422
Bolt Flying Insect Killer	Johnson Wax Ltd	Amateur	3791
Boots Fly And Wasp Killer	The Boots Company PLC	Amateur	5632
Di-Fly	Lever Industrial Ltd	Amateur	4509
Dry Fly Killer	Domestic Fillers Ltd	Amateur	4237
Fly And Wasp Killer	Lloyds	Amateur	5639
Jeyes Expel Fly And Wasp Killer	Jeyes Group Ltd	Amateur	5516
Jeyes Kontrol Fly And Wasp Killer	Jeyes Group Ltd	Amateur	5359
Killgerm PY-Kill W	Killgerm Chemicals Ltd	Professional / Professional (Animal Husbandry)	4632
Kleeneze Fly Killer	Kleeneze Ltd	Amateur	4225
Murphy Fly And Wasp Killer	Levington Horticulture Ltd	Amateur	6252
D Odex Fly Spray	Odex Ltd	Amateur	4814
Premiere Fly And Wasp Killer	Premiere Products	Amateur / Professional	4516
Provence Fly And Wasp Killer	Haventrail Ltd	Amateur	4382
Pyrakill Flying Insect Spray	Forward Chemicals Ltd	Amateur	5357
Raid Flying Insect Killer	Johnson Wax Ltd	Amateur / Professional	4585
Red Can Fly Killer	Yellow Can Company Ltd	Amateur	4508
SC Johnson Raid Fly And Wasp Killer	Johnson Wax Ltd	Amateur	5113
St Michael Flyspray	Marks And Spencer PLC	Amateur	4177
Super Raid Insect Killer Formula 2	Johnson Wax Ltd	Amateur	4240

373 Tetramethrin + Alphacypermethrin

| Cytrol Alpha Super 50/50 SE | Pelgar International Ltd | Professional | 6678 |

374 Tetramethrin + Bendiocarb

| W Camco Insect Spray | AgrEvo UK Ltd | Amateur | 4222 |
| Devcol Household Insect Spray | Devcol Ltd | Amateur | 5512 |

375 Tetramethrin + Chlorpyrifos

| Raid Wasp Nest Destroyer | Johnson Wax Ltd | Amateur | 5597 |

376 Tetramethrin + Cypermethrin

Cyperkill 10T	Mitchell Cotts Chemicals	Professional	6914
Cyperkill Plus WP	Mitchell Cotts Chemicals	Professional	4855
Cytrol Universal	Pelgar International Ltd	Professional	6675
Cytrol XL	Pelgar International Ltd	Professional	6679

Product Name	Marketing Company	Use	HSE No.
376 Tetramethrin + Cypermethrin—continued			
New Vapona Ant Killer	Ashe Consumer Products Ltd	Amateur	4298
Raid Ant And Crawling Insect Killer	Johnson Wax Ltd	Amateur	3001
Raid Residual Crawling Insect Killer	Johnson Wax Ltd	Amateur Professional	4584
S C Johnson Wax Raid Ant And Cockroach Killer	Johnson Wax Ltd	Amateur	5216
Shelltox Cockroach And Crawling Insect Killer 3	Temana International Ltd	Amateur Professional	4031
Shelltox Cockroach And Crawling Insect Killer 4	Temana International Ltd	Amateur	4739
Super Shelltox Crawling Insect Killer	Kortman Intradal BV	Amateur	5454
Vapona Ant And Crawling Insect Killer Aerosol	Ashe Consumer Products Ltd	Amateur	5282
377 Tetramethrin + Dichlorophen + Dichlorvos			
D Wintox	Mould Growth Consultants Ltd	Professional	4236
378 Tetramethrin + Dichlorvos			
Shelltox Extra Flykiller	Temana International Ltd	Amateur Professional	3696
Shelltox Flykiller	Temana International Ltd	Amateur	3162
Vapona Fly And Wasp Killer Spray	Ashe Ltd	Amateur	4907
379 Tetramethrin + Fenitrothion			
D Big D Cockroach And Crawling Insect Killer	Domestic Fillers Ltd	Amateur	4098
Big D Cockroach And Crawling Insect Killer	Dylon International Ltd	Amateur	5834
W Big D Rocket Fly Killer	Domestic Fillers Ltd	Amateur	3552
Doom Ant And Crawling Insect Killer Aerosol	Napa Products Ltd	Amateur	4753
Flying Insect Killer	Keen (World Marketing) Ltd	Amateur	4313
Flying Insect Killer Faster Knock Down	Keen (World Marketing) Ltd	Amateur	4312
Killgerm Fenitrothion-Pyrethroid Concentrate	Killgerm Chemicals Ltd	Professional	4623
Motox Crawling Insect	HVM Products Ltd	Amateur	5940
Top Kill Insect Killer	Houdret And Co Ltd	Amateur	6188
Tox Exterminating Fly And Wasp Killer	Keen (World Marketing) Ltd	Amateur	4314
380 Tetramethrin + Fenitrothion + Permethrin			
Connect Insect Killer	Sun Oil Ltd	Amateur	6109

HSE
INSECTICIDES

Product Name	Marketing Company	Use	HSE No.

380 Tetramethrin + Fenitrothion + Permethrin—continued

Motox	HVM Products Ltd	Amateur	5622

381 Tetramethrin + Lindane

W Doom Flea Killer	Napa Products Ltd	Amateur	4572
Doom Moth Proofer Aerosol	Napa Products Ltd	Amateur	4571

382 Tetramethrin + Methoprene + Permethrin

Johnson's Cage And Hutch Insect Spray	Johnson's Veterinary Products Ltd	Amateur	6864
Johnson's House Flea Spray Extra Guard	Johnson's Veterinary Products Ltd	Amateur	6649

383 Tetramethrin + Permethrin

Aventis Procontrol Wasp Nest Destroyer	Aventis Environmental Science	Professional	6138
Atta-X Cockroach Killer	Sanderson Curtis Ltd	Amateur	6362
W Bolt Ant And Crawling Insect Killer	Johnson Wax Ltd	Amateur	3619
Bop Crawling Insect Killer	Mcbride International	Amateur Professional	4138
Bop Crawling Insect Killer (MC)	Mcbride International	Amateur Professional	4139
Canned Death	Pearson & Wilkinson	Professional	6112
Combat Fly And Wasp Killer	Sinclair Animal And Household Care Ltd	Amateur	7065
Combat Fly Killer	Sinclair Animal And Household Care Ltd	Amateur	5808
Corry's Fly And Wasp Killer	Vitax Ltd	Amateur	5322
D-Stroy	Future Developments (Manufacturing) Ltd	Professional	5411
Delta Cockroach And Ant Spray	Bennetts Company Ltd	Amateur	3950
Doom Tropical Strength Fly And Wasp Killer Aerosol	Napa Products Ltd	Amateur	4752
Dragon Flykiller	Zeneca Public Health	Amateur	4952
Fly And Wasp Killer	Rentokil Initial UK Ltd	Amateur Professional	4643
Fly Free Zone	Fly Away Ltd	Amateur Professional Professional (Animal Husbandry)	6394
Fly Spray	Merton Cleaning Supplies	Professional	5112
Fly-Kill	Stagchem Ltd	Professional	6202
Hail Flying Insect Killer	Mcbride International	Amateur	5568
Insecticide Aerosol	Aerosols International Ltd	Amateur	6567
Insectrol, Fly And Wasp Killer	Rentokil Initial UK Ltd	Amateur	5827

Product Name	Marketing Company	Use	HSE No.

383 Tetramethrin + Permethrin—continued

Product Name	Marketing Company	Use	HSE No.
Johnson's House Flea Spray	Johnson's Veterinary Products Ltd	Amateur	6623
Johnson's New Pet Housing Spray	Johnson's Veterinary Products Ltd	Amateur	6624
K02 Selkil	Selden Research Ltd	Professional	4828
Mostyn 8/64 TFC	Hockley International Ltd	Professional	7105
Nippon Ant And Crawling Insect Killer	Vitax Ltd	Amateur	4414
Nippon Fly Killer Spray	Vitax Ltd	Amateur	4254
Nippon Killaquer Crawling Insect Killer	Arrow Chemicals Ltd	Professional	3694
Nippon Wasp Nest Destroyer Foam	Vitax Ltd	Amateur	7100
Nisa Fly And Wasp Killer	Nisa	Amateur	4962
Permost Uni	Hockley International Ltd	Professional Professional (Animal Husbandry)	6399
Purge Insect Destroyer	Aztec Chemicals Ltd	Professional	5048
Pynosect 10	Mitchell Cotts Chemicals	Professional	3171
Pynosect Extra Fog	Mitchell Cotts Chemicals	Professional	3172
W Raid Ant And Roach Killer	Johnson Wax Ltd	Amateur	5182
Sactif Flying Insect Killer	Diverseylever Ltd	Professional	4924
W Sainsbury's Fly And Wasp Killer	J Sainsbury PLC	Amateur	5335
Sanmex Ant & Crawling Insect Killer	The British Products Sanmex Company Ltd	Amateur	6655
Sanmex Fly And Wasp Killer	The British Products Sanmex Company Ltd	Amateur	4006
Sanmex Supakil Insecticide	The British Products Sanmex Company Ltd	Amateur	4009
Scorpio Fly Spray For Flying Insects	Jangro Ltd	Professional	5049
Secto Fly Killer	Sinclair Animal And Household Care Ltd	Amateur	3862
Shelltox Insect Killer	Temana International Ltd	Amateur Professional	3697
Shelltox Insect Killer 2	Temana International Ltd	Amateur Professional	4061
Stop Insect Killer	Temana International Ltd	Amateur Professional	3163
Superdrug Fly Killer	Superdrug Stores PLC	Amateur	4185
Zappit	BLP Ltd	Amateur	6181

384 Tetramethrin + Permethrin + S-Bioallethrin

Product Name	Marketing Company	Use	HSE No.
Aventis Procontrol Crawling Insect Killer Aerosol	Aventis Environmental Science	Professional	6231
Pif Paf Crawling Insect Killer Aerosol	Reckitt And Colman Products Ltd	Amateur	5595

Product Name	Marketing Company	Use	HSE No.

385 Tetramethrin + Pirimiphos-Methyl + Resmethrin

W Waspend	Miracle Garden Care Ltd	Amateur Professional	4857

386 Tetramethrin + Resmethrin

W 151 Dry Fly And Wasp Killer	SB Ltd	Amateur	5169
Apex Fly-Killer	Apex Industrial Chemicals Ltd	Amateur Professional	4300
Chem-Kil	Inter-Chem	Amateur Professional	4266
W Chirton Dry Fly And Wasp Killer	Ronson International Ltd	Amateur	5287
D Cromessol Insect Killer	Cromessol Company Ltd	Amateur	5067
Cromessol Multi-Purpose Insect Killer	Wallace Cameron And Co Ltd	Amateur Professional	6045
Fly And Maggot Killer	Net-Tex Agricultural Ltd	Amateur	6097
D Fly Killer	Kalon Agricultural Division	Amateur	4817
D Hospital Flying Insect Spray	Kalon Chemicals	Amateur	4652
D Janisolve Fly Killer	Johmar Enterprises Company	Amateur	4228
K.O. Fly Spray	Genuine Solutions	Amateur Professional	5879
Maggot & Fly Attack	Battle, Hayward And Bower Ltd	Amateur Professional	6997
Momar Fly Killer	Momar	Amateur	6264
D Penetone Fly Killer	Penetone Ltd	Amateur	4072
Pynosect 2	Mitchell Cotts Chemicals	Amateur Professional	4654
Pynosect 4	Mitchell Cotts Chemicals	Amateur Professional	4655
Quantum Hygiene Fly Killer	Quantum Hygiene	Amateur Professional	5066
Rapid Exit Fly Spray	LMA Services	Amateur Professional	5787
Sorex Wasp Nest Destroyer	Sorex Ltd	Professional	6294
Swat	Solvitol Ltd	Amateur Professional	3564
Swat A	Solvitol Ltd	Professional	5051
Swot	Southern Chemical Services	Amateur Professional	6007
D Trilanco Fly Killer	Trilanco	Amateur	4816
Wasp Nest Destroyer	Sorex Ltd	Professional	4410

387 Tetramethrin + S-Bioallethrin

Vapona Wasp Killer Aerosol	Ashe Consumer Products Ltd	Amateur	6063

432

Product Name	Marketing Company	Use	HSE No.

388 Tetramethrin + S-Methoprene

Zodiac Household Flea Spray	Petlove	Amateur	6502

389 Tetramethrin + d-Allethrin

Boots Dry Fly And Wasp Killer	The Boots Company PLC	Amateur	4037
D Boots Dry Fly And Wasp Killer 3	The Boots Company PLC	Amateur	4241
Fly, Wasp And Mosquito Killer	Rentokil Initial UK Ltd	Amateur	4882
New Vapona Fly Killer Dry Formulation	Ashe Consumer Products Ltd	Amateur	4202
Raid Fly And Wasp Killer	Johnson Wax Ltd	Amateur	3000
W Sainsbury's Fly, Wasp And Mosquito Killer	J Sainsbury PLC	Amateur	5164

390 Tetramethrin + d-Allethrin + Cypermethrin

Bop Flying And Crawling Insect Killer	Mcbride International	Amateur Professional	4136
Bop Flying And Crawling Insect Killer (MC)	Mcbride International	Amateur Professional	4135
Bop Flying And Crawling Insect Killer (Water Based)	Mcbride International	Amateur Professional	4137
Bop Flying Insect Killer	Mcbride International	Amateur Professional	4141
Bop Flying Insect Killer (MC)	Mcbride International	Amateur Professional	4140

391 Tetramethrin + d-Allethrin + Cypermethrin + Permethrin

Elf Insecticide	Elf Oil UK Ltd	Amateur	6669

392 Tetramethrin + d-Allethrin + Dichlorvos + Permethrin

Bop	Mcbride International	Amateur	4773

393 Tetramethrin + d-Allethrin + Permethrin

Hail Plus Crawling Insect Killer	Mcbride International	Amateur	5567
New Super Raid Insecticide	Johnson Wax Ltd	Amateur Professional	4583

394 Tetramethrin + d-Phenothrin

Flak	Mcbride International	Amateur Professional	4143
Ant Destroyer Foam	Rentokil Initial UK Ltd	Amateur	6412
Betterware Foaming Ant & Wasp Nest Destroyer	Betterware UK Ltd	Amateur	6615
D Big D Ant And Crawling Insect Killer	Domestic Fillers Ltd	Amateur	4097
Big D Ant And Crawling Insect Killer	Dylon International Ltd	Amateur	5818
Bio Foaming Wasp Nest Destroyer	PBI Home & Garden Ltd	Amateur	6665

Product Name	Marketing Company	Use	HSE No.

394 Tetramethrin + d-Phenothrin—continued

Product Name	Marketing Company	Use	HSE No.
Boots Ant And Crawling Insect Killer	The Boots Company PLC	Amateur	5554
Boots Dry Fly And Wasp Killer	The Boots Company PLC	Amateur	7029
Boots Flea Spray	The Boots Company PLC	Amateur	5522
Bop Arabic	Mcbride International	Amateur Professional	4142
Bouchard Fly And Wasp Killer	IBA UK Ltd	Amateur	7053
Brimpex ULV 1500	Brimpex Metal Treatments	Professional Professional (Animal Husbandry)	4942
Bug Off	Dimex Ltd	Amateur	6356
Combat Wasp Killer Foam	Sinclair Animal And Household Care Ltd	Amateur	6974
Deosan Fly Spray	Diverseylever Ltd	Professional	5246
W Envar 6/14 EC	Agrimar (UK) PLC	Professional	4354
W Envar 6/14 Oil	Agrimar (UK) PLC	Professional	4355
Fly And Bug Spray	Jangro Ltd	Amateur	6933
Flying And Crawling Insect Killer	Betterware UK Ltd	Amateur	6607
Friends Household Flea Spray	Sinclair Animal And Household Care Ltd	Amateur	6664
GLP Flying And Crawling Insect Killer	Greenhill Chemical Products Ltd	Amateur	6303
Insect Killer	Rentokil Initial UK Ltd	Amateur	6244
Jeyes Expel Ant And Crawling Insect Killer	Jeyes Group Ltd	Amateur	5552
Jeyes Expel Flea Killer	Jeyes Group Ltd	Amateur	5523
Jeyes Kontrol Ant And Crawling Insect Killer	Jeyes Group Ltd	Amateur	5553
Jeyes Kontrol Flea Killer	Jeyes Group Ltd	Amateur	5521
Killgerm ULV 1500	Killgerm Chemicals Ltd	Professional	6961
Killgerm ULV 500	Killgerm Chemicals Ltd	Professional Professional (Animal Husbandry)	4647
W Killoff Aerosol Insect Killer	Ferosan Products	Amateur	3974
Kybosh 2	PBI Home & Garden Ltd	Amateur	6622
Murphy Foaming Wasp Nest Destroyer	Levington Horticulture Ltd	Amateur	6609
New Secto Household Flea Spray	Sinclair Animal And Household Care Ltd	Amateur	6504
Pesguard NS 6/14 Oil	Sumitomo Chemical (UK) PLC	Professional	4361
Pesguard OBA F 7305 C	Sumitomo Chemical (UK) PLC	Amateur	4064
Pesguard OBA F 7305 D	Sumitomo Chemical (UK) PLC	Amateur	4065

Product Name	Marketing Company	Use	HSE No.
394 Tetramethrin + d-Phenothrin—continued			
Pesguard OBA F 7305 E	Sumitomo Chemical (UK) PLC	Amateur	4066
Pesguard OBA F 7305 F	Sumitomo Chemical (UK) PLC	Amateur	4067
Pesguard OBA F 7305 G	Sumitomo Chemical (UK) PLC	Amateur	4063
Pesguard OBA F7305 B	Sumitomo Chemical (UK) PLC	Amateur	4419
Pesguard WBA F-2656	Sumitomo Chemical (UK) PLC	Amateur	4263
Pesguard WBA F-2692	Sumitomo Chemical (UK) PLC	Amateur	4262
Pestkill Household Spray	Phillips Yeast Products Ltd	Amateur	4938
Pet Bedding And Household Flea Spray	Armitage Brothers PLC	Amateur	4090
PY Kill 25 Plus	Killgerm Chemicals Ltd	Professional	4900
Raid Fly Wasp And Plant Insect Killer	Johnson Wax Ltd	Amateur	5432
Rentokil Wasp Nest Destroyer	Rentokil Initial UK Ltd	Amateur	6214
S C Johnson Professional Raid Flying Insect Killer	S C Johnson Professional Ltd	Amateur	6407
S C Johnson Wax Raid Fly And Wasp Killer	Johnson Wax Ltd	Amateur	5181
W Sergeant's Dust Mite Patrol Injector And Spray	Seven Seas Ltd	Amateur Professional	5513
W Sergeant's Dust Mite Patrol Spray	Seven Seas Ltd	Amateur	5041
Sergeant's Household Spray	Seven Seas Ltd	Amateur	6011
W Sergeant's Insect Patrol	Seven Seas Ltd	Amateur	5397
Shelltox Flying Insect Killer 2	Temana International Ltd	Amateur Professional	4032
Shelltox Flying Insect Killer 3	Temana International Ltd	Amateur	4741
Sorex Fly Spray RTU	Sorex Ltd	Professional	5718
Sorex Super Fly Spray	Sorex Ltd	Amateur	6297
W Super Fly Spray	Sorex Ltd	Amateur	4468
Super Raid II	Johnson Wax Ltd	Amateur	4015
Super Shelltox Flying Insect Killer	Kortman Intradal BV	Amateur	5440
Terminator	Aerosol Logistics	Amateur	6416
Terminator Fly And Wasp Killer	Sanoda Ltd	Amateur	6150
Terminator Wasp & Insect Nest Destroyer	Globol Chemicals (UK) Ltd	Amateur	6100
Terminator Wasp And Insect Nest Destroyer	Sanoda Ltd	Amateur	6144
Vapona House And Plant Fly Spray	Ashe Consumer Products Ltd	Amateur	5236
Vapona Wasp And Fly Killer	Ashe Consumer Products Ltd	Amateur	5754

Product Name	Marketing Company	Use	HSE No.

394 Tetramethrin + d-Phenothrin—continued

Vapona Wasp And Fly Killer Spray	Ashe Consumer Products Ltd	Amateur	5280
Vetzyme Pestkill Household Spray	Seven Seas Ltd	Amateur	7034
Vitapet Flearid Household Spray	Seven Seas Ltd	Amateur	5434
Wasp Destroyer Foam	Rentokil Initial UK Ltd	Amateur	6455
Zappit Fly And Wasp Killer	BLP Ltd	Amateur	6336

395 d-Tetramethrin + d-Phenothrin

| Pesguard WBA F2714 | Sumitomo Chemical (UK) PLC | Amateur | 4418 |
| Vapona Max Concentrated Fly And Wasp Killer | Ashe Consumer Products Ltd | Amateur | 6062 |

396 Transfluthrin

| Baygon Moth Paper | Bayer PLC | Amateur | 6228 |

397 Transfluthrin + Cyfluthrin

| Baygon Insect Spray | Bayer PLC | Amateur | 6546 |

398 Trichlorphon

W Ant Killer	Rentokil Initial UK Ltd	Amateur	5906
W Boots Ant Trap	The Boots Company PLC	Amateur	5542
W Detia Ant Bait	K D Pest Control Products	Amateur	3803
W Jeyes Expel Ant Trap	Jeyes Group Ltd	Amateur	5503
W Murphy Kil-Ant Trap	Levington Horticulture Ltd	Amateur	6258
W Vapona Ant Bait	Ashe Consumer Products Ltd	Amateur	6092
Vapona Ant Trap	Ashe Consumer Products Ltd	Amateur	3804

5
RODENTICIDE

Rodenticide

399 Alphachloralose

Alphachloralose Concentrate	Rentokil Initial UK Ltd	Professional	6693
Alphakil	Rentokil Initial UK Ltd	Amateur	6694
Alphamouse	Killgerm Chemicals Ltd	Professional	6692
Townex Mouse Poison	Town And Country Pest Services Ltd	Professional	6695

400 Brodifacoum

Brodifacoum Paste	Rentokil Initial UK Ltd	Professional	6696
Erasor	Sorex Ltd	Professional	6697
Klerat	Sorex Ltd	Professional	6701
Klerat Wax Blocks	Sorex Ltd	Professional	6703
Mouser	Sorex Ltd	Amateur	6704
Solo Blox	Bell Laboratories Inc	Professional	7038
Solo Mouse Blox	Bell Laboratories Inc	Professional	7040
Solo Rodenticide	Bell Laboratories Inc	Professional	7041
Solo Super Blox	Bell Laboratories Inc	Professional	7039
Sorex Brodifacoum Bait Blocks	Sorex Ltd	Professional	6705
Sorex Brodifacoum Rat And Mouse Bait	Sorex Ltd	Professional	6706
Sorex Brodifacoum Sewer Bait	Sorex Ltd	Professional	6707
Sorex Brodifacoum Sewer Bait Sachets	Sorex Ltd	Professional	6708
Sorexa Checkatube	Sorex Ltd	Professional	6702
Talon Rat And Mouse Bait (Cut Wheat)	Killgerm Chemicals Ltd	Professional	6709
Talon Rat And Mouse Bait (Whole Wheat)	Killgerm Chemicals Ltd	Professional	6710

401 Bromadiolone

Bromard	Rentokil Initial UK Ltd	Professional	6713
Bromatrol	Rentokil Initial UK Ltd	Professional	6714
Bromatrol Contact Dust	Rentokil Initial UK Ltd	Professional	6715
Bromatrol Mouse Blocks	Rentokil Initial UK Ltd	Professional	6716
Bromatrol Rat Blocks	Rentokil Initial UK Ltd	Professional	6698
Bromatrol Rat Blocks	Rentokil Initial UK Ltd	Professional	6717
Contrac Blox	Bell Laboratories Inc	Professional	6718
Contrac Rodenticide	Bell Laboratories Inc	Professional	6719
Contrac Super Blox	Bell Laboratories Inc	Professional	6720
Deadline	Rentokil Initial UK Ltd	Professional	6721
Deadline Contact Dust	Rentokil Initial UK Ltd	Professional	6722
Deadline Liquid Concentrate	Rentokil Initial UK Ltd	Professional	6723
Deadline Place Packs	Rentokil Initial UK Ltd	Professional	6724
Endorats Premium Rat Killer	Irish Drugs Ltd	Professional	6725

Product Name	Marketing Company	Use	HSE No.

401 Bromadiolone—continued

Liquid Bromatrol	Rentokil Initial UK Ltd	Professional	6726
Mouse Killer Bait	Rentokil Initial UK Ltd	Amateur	6699
Murphy Mouse Killer	Rentokil Initial UK Ltd	Amateur	6727
Rentokil Rat Killer	Rentokil Initial UK Ltd	Amateur	6711
Rentokil Rat Killer	Rentokil Initial UK Ltd	Professional	6712
Rodine	Rentokil Initial UK Ltd	Amateur	6728
Slaymor	Novartis Animal Health UK Ltd	Professional	6729
Slaymor Bait Bags	Novartis Animal Health UK Ltd	Professional	6730
Slaymor Bait Bags	Novartis Animal Health UK Ltd	Amateur	6731
Slaymor Mouse Bait	Novartis Animal Health UK Ltd	Amateur	6700
Slaymor Rat Blocks	Novartis Animal Health UK Ltd	Professional	7115
Tomcat 2 Blox	Antec International Ltd	Professional	6732
Tomcat 2 Mouse Bait Station	Antec International Ltd	Amateur	6733
Tomcat 2 Rat And Mouse Blox	Antec International Ltd	Amateur	6734
Tomcat 2 Rat And Mouse Killer	Antec International Ltd	Amateur	6735
Tomcat 2 Rodenticide	Antec International Ltd	Professional	6736
Tomcat 2 Super Blox	Antec International Ltd	Professional	6737
Total Mouse Killer System	Rentokil Initial UK Ltd	Amateur	6738

402 Calciferol

Deerat Concentrate	Rentokil Initial UK Ltd	Professional	6739

403 Calciferol + Difenacoum

Sorexa CD Concentrate	Sorex Ltd	Professional	6740

404 Chlorophacinone

Drat	Aventis Environmental Science	Professional	6741
Drat Bait	Aventis Environmental Science	Professional	6742
Drat Rat Bait	Battle, Hayward And Bower Ltd	Professional	6743
Endorats	American Products	Professional	6744
Farmers Ridento RTU Rat Bait	Wcf Ltd	Professional	6798
Karate Ready To Use Rat Bait	Diverseylever Ltd	Professional	6745
Karate Ready To Use Rodenticide Sachets	Diverseylever Ltd	Professional	6746
Rat Eyre Rat Bait	V.E.S. Pest Control Supplies	Professional	6747
Ridento Ready To Use Rat Bait	Ace Chemicals Ltd	Professional	6748

HSE
RODENTICIDE

Product Name	Marketing Company	Use	HSE No.
404 Chlorophacinone—continued			
Rout	Samuel Mccausland Ltd	Professional	6749
Ruby Rat	J V Heatherington Ltd	Professional	6750
405 Cholecalciferol			
Sorex Fatal	Sorex Ltd	Professional	7087
406 Cholecalciferol + Difenacoum			
Sorex CD Mouse Bait	Sorex Ltd	Professional	6751
Sorexa CD Mouse Killer	Sorex Ltd	Amateur	6752
407 Coumatetralyl			
Bio Racumin Rat Bait	PBI Home & Garden Ltd	Amateur	6753
Racumin Contact Powder	Bayer PLC	Professional	6755
Racumin Rat Bait	Bayer PLC	Professional	6754
Townex Sachets	Town And Country Pest Services Ltd	Professional	6756
408 Difenacoum			
Bio Mouse Killer	PBI Home & Garden Ltd	Amateur	6757
Bio Mouse Killer Bait Station	PBI Home & Garden Ltd	Amateur	7058
Bio Rat Killer	PBI Home & Garden Ltd	Amateur	6758
Bio Rat Killer Bait Bags	PBI Home & Garden Ltd	Amateur	6799
Deosan Rataway	Diverseylever Ltd	Professional	6759
Deosan Rataway Bait Bags	Diverseylever Ltd	Professional	6760
Deosan Rataway W	Diverseylever Ltd	Professional	6761
Difenard	Rentokil Initial UK Ltd	Professional	6762
Endem	Downland Marketing Ltd	Amateur	6763
Endem Bait Sachets	Downland Marketing Ltd	Amateur	6764
Fentrol	Rentokil Initial UK Ltd	Professional	6765
Fentrol Gel	Rentokil Initial UK Ltd	Professional	6766
Neo Sorexa Block In A Box	Sorex Ltd	Amateur	7057
Neokil	Sorex Ltd	Professional	6769
Neokil	Sorex Ltd	Amateur	6770
Neokil Bait Bags	Sorex Ltd	Professional	6771
Neokil Bait Bags	Sorex Ltd	Amateur	6772
Neosorexa	Sorex Ltd	Professional	6773
Neosorexa	Sorex Ltd	Amateur	6774
Neosorexa Bait Blocks	Sorex Ltd	Professional	6775
Neosorexa Bait Blocks	Sorex Ltd	Amateur	6817
Neosorexa Concentrate	Sorex Ltd	Professional	6776
Neosorexa Liquid Concentrate	Sorex Ltd	Professional	6777
Neosorexa Mixed Grain Bait	Sorex Ltd	Professional	6778
Neosorexa Mixed Grain Bait	Sorex Ltd	Amateur	6779
Neosorexa Pellet Packs	Sorex Ltd	Amateur	6804
Neosorexa Pellets	Sorex Ltd	Professional	6780
Neosorexa Pellets	Sorex Ltd	Amateur	6781

Product Name	Marketing Company	Use	HSE No.

408 Difenacoum—continued

Neosorexa Ratpacks	Sorex Ltd	Professional	6782
Neosorexa Ratpacks	Sorex Ltd	Amateur	6783
Ratak	Sorex Ltd	Amateur	6784
Ratak	Sorex Ltd	Professional	6785
Ratak Cut Wheat	Sorex Ltd	Professional	6786
Ratak Wax Blocks	Sorex Ltd	Professional	6787
Sakarat D (Whole Wheat)	Killgerm Chemicals Ltd	Professional	7109
Sorexa D	Sorex Ltd	Professional	6788
Sorexa D Mouse Killer	Sorex Ltd	Amateur	6789
Sorexa Gel	Sorex Ltd	Professional	6790

409 Difenacoum + Calciferol

Sorexa CD Concentrate	Sorex Ltd	Professional	6740

410 Difenacoum + Cholecalciferol

Sorex CD Mouse Bait	Sorex Ltd	Professional	6751
Sorexa CD Mouse Killer	Sorex Ltd	Amateur	6752

411 Diphacinone

Ditrac Blox	Bell Laboratories Inc	Professional	6791
Ditrac Super Blox	Bell Laboratories Inc	Professional	6792
Tomcat Blox	Antec International Ltd	Professional	6793

412 Flocoumafen

Storm	Sorex Ltd	Professional	6794
Storm Mouse Bait Blocks	Sorex Ltd	Professional	6795
Storm Sewer Bait	Sorex Ltd	Professional	6796
Storm Sewer Bait Sachets	Sorex Ltd	Professional	6797

413 Warfarin

Killgerm Sewercide Cut Wheat Rat Bait	Killgerm Chemicals Ltd	Professional	6805
Killgerm Sewercide Whole Wheat Rat Bait	Killgerm Chemicals Ltd	Professional	6806
Sakarat Ready To Use (Cut Wheat Base)	Killgerm Chemicals Ltd	Professional	6807
Sakarat Ready To Use (Whole Wheat)	Killgerm Chemicals Ltd	Professional	6808
Sakarat X Ready To Use Warfarin Rat Bait	Killgerm Chemicals Ltd	Professional	6809
Sewarin Extra	Killgerm Chemicals Ltd	Professional	6810
Sewarin P	Killgerm Chemicals Ltd	Professional	6811
Sorex Warfarin 250 PPM Rat Bait	Sorex Ltd	Professional	6812
Sorex Warfarin 500 PPM Rat Bait	Sorex Ltd	Professional	6813
Sorex Warfarin Sewer Bait	Sorex Ltd	Professional	6814

Product Name	Marketing Company	Use	HSE No.
413 Warfarin—continued			
Warfarin Concentrate	Battle, Hayward And Bower Ltd	Professional	6815
Warfarin Ready Mixed Bait	Battle, Hayward And Bower Ltd	Professional	6816

6

SURFACE BIOCIDES

Product Name	Marketing Company	Use	HSE No.

Surface Biocides

414 Alkylaryltrimethyl Ammonium Chloride

Abicide 82	Langlow Products Ltd	Amateur Professional	4470
Albany Fungicidal Wash	C Brewer And Sons Ltd	Amateur Professional	4471
Anti-Mould Solution	Macpherson Paints Ltd	Amateur (Surface Biocide) Amateur (Wood Preservative) Professional (Surface Biocide) Professional (Wood Preservative)	4873
Beeline Fungicidal Wash	Ward Bekker Ltd	Amateur Professional	4224
W Fungishield Sterilising Solution Concentrate GS36	Glixtone Ltd	Amateur Professional	4283
W Fungishield Sterilising Solution GS37	Glixtone Ltd	Amateur Professional	4284
ICI Decorator 1st Fungicidal Wash	ICI Paints	Amateur Professional	6479
Kibes Sterilising Solution Concentrate GS 36	Kibes (UK) Insulation Ltd	Amateur Professional	5493
W Mould Cleaner	J H Woodman Ltd	Amateur	4291

415 Alkylaryltrimethyl Ammonium Chloride + Boric Acid

Kleeneze Anti-Mould Spray	Kleeneze Ltd	Amateur Professional	4599

416 Alkylaryltrimethyl Ammonium Chloride + Disodium Octaborate

Boracol 10RH Surface Biocide	Advanced Chemical Specialties Ltd	Professional	4911
Remtox Borocol 10 RH Masonry Biocide	Remtox (Chemicals) Ltd	Professional	4100

Product Name	Marketing Company	Use	HSE No.

417 Azaconazole + Permethrin

Product Name	Marketing Company	Use	HSE No.
Fongix SE Total Treatment For Wood	Liberon Waxes Ltd	Amateur (Surface Biocide) Amateur (Wood Preservative) Professional (Surface Biocide) Professional (Wood Preservative)	5610
Woodworm Killer And Rot Treatment	Liberon Waxes Ltd	Amateur (Surface Biocide) Amateur (Wood Preservative) Professional (Surface Biocide) Professional (Wood Preservative)	4826

418 Benzalkonium Bromide

Product Name	Marketing Company	Use	HSE No.
X Moss	Thames Valley Specialist Products Ltd	Amateur Professional	6839

419 Benzalkonium Chloride

Product Name	Marketing Company	Use	HSE No.
A-Zygo 3 Sterilising Solution	Premier Condensation And Mould Control Ltd	Professional	5368
Albio B084	Millcliff Ltd	Professional	6059
Algae Remover	Akzo Nobel Woodcare	Amateur Professional	5337
Algaecide	Premkem Ltd	Professional	5842
B And Q Value Fungicide	B & Q PLC	Amateur	6834
W Bedclear	Fargro Ltd	Professional	5179
Belzona 9421 Fungicidal Wash	Belzona Polymerics Limited	Professional	6911
Bio-Kil Dentolite Solution	Remtox-Silexine Ltd	Amateur Professional	4593
Bio-Natura's Green Algae Remover	Bio-Natura Ltd	Amateur Professional	6503
Bio-Natura's Green Algae Remover Ready To Use	Bio-Natura Ltd	Amateur Professional	6688

Product Name	Marketing Company	Use	HSE No.

419 Benzalkonium Chloride—continued

Product Name	Marketing Company	Use	HSE No.
BN Algae Remover	Nehra Cooke's Chemicals Ltd	Amateur Professional	5381
D BN Mosskiller	Morgan Nehra Holdings Ltd	Amateur Professional	5080
W Conc Quat	Semitec Ltd	Professional	4904
Cuprinol Exterior Fungicide	Cuprinol Ltd	Amateur Professional	5526
Cuprinol Garden Decking Cleaner	Cuprinol Ltd	Amateur Professional	6857
Cuprinol Garden Furniture Algae Killer	Cuprinol Ltd	Amateur Professional	5922
Cuprinol Garden Timber Algae Killer	Cuprinol Ltd	Amateur Professional	5921
Cuprinol Greenhouse Algae Killer	Cuprinol Ltd	Amateur Professional	5939
Cuprinol Path And Patio Cleaner	Cuprinol Ltd	Amateur Professional	5527
Cuprotect Exterior Fungicide	Cuprinol Ltd	Amateur Professional	4165
Cuprotect Patio Cleaner	Cuprinol Ltd	Amateur Professional	4164
Doff Path & Patio Cleaner	Doff Portland Ltd	Amateur	6305
Doff Path And Patio Cleaner Concentrate	Doff Portland Ltd	Amateur	6248
FEB Fungicide	FEB MBT	Amateur	6833
FMB 451-5 Quat (50%)	Midland Europe	Professional (Surface Biocide) Professional (Wood Preservative) Industrial (Wood Preservative)	5734
Fungi-Shield Sterilising Solution Concentrate GS36	Glixtone Ltd	Amateur Professional	6292
Fungi-Shield Sterilising Solution GS37	Glixtone Ltd	Amateur Professional	6291
Granocryl Fungicidal Wash	Kalon Group PLC	Amateur Professional	5797
Green N Clean Algae Killer Concentrate	Phostrogen Ltd	Amateur Professional	6835
Green N Clean Algae Killer Ready To Use	Phostrogen Ltd	Amateur Professional	6830
Homebase Mould Cleaner	Homebase Ltd	Amateur Professional	5707

Product Name	Marketing Company	Use	HSE No.

419 Benzalkonium Chloride—continued

Product Name	Marketing Company	Use	HSE No.
Homebase Weathercoat Fungicidal Wash	Sainsbury's Homebase House And Garden Centres	Amateur Professional	4860
Howes Olympic Algaecide	Killgerm Chemicals Ltd	Professional	4845
W Interior Mould Remover Wipes	Cadismark Household Products Ltd	Amateur	5695
Jewson Fungicidal Wash	Jewson Ltd	Amateur Professional	6886
Jeyes Algae And Mould Killer	Jeyes Group Ltd	Amateur	7066
Leyland Sterilisation Wash	Leyland Paint Company Ltd	Professional	4751
D LPL Biocidal Wash	Liquid Plastics Ltd	Professional	4479
LPL Biowash	Liquid Plastics Ltd	Professional	5853
Mangers Fungicidal Spray	Mangers	Amateur Professional	7035
Mangers Sterilisation Wash	Mangers	Amateur Professional	7036
W Mould Killer	Cadismark Household Products Ltd	Amateur	5047
Palace Fungicidal Wash	Palace Chemicals Ltd	Amateur Professional	6253
Paramos	Chemsearch	Professional	4524
Path And Patio Cleaner	Otter Nurseries Ltd	Amateur	6060
W Remtox Remwash Extra	Remtox (Chemicals) Ltd	Professional	5178
Rhodaquat RP50	Rhodia HPCII UK Ltd	Professional (Surface Biocide) Professional (Wood Preservative)	5762
Rid Moss	Friendly Systems	Professional	6222
Sigma Fungicidal Solution	Sigma Coatings Ltd	Amateur Professional	6564
Snowcem Algicide	Snowcem	Amateur Professional	4806
Sovereign Dentolite Solution Concentrate	Sovereign Chemical Industries Ltd	Amateur Professional	6354
Texas Fungicidal Wash	Texas Homecare Ltd	Amateur Professional	5745
Thompson's Mould And Moss Killer	Ronseal Ltd	Amateur	6208
Thompson's Patio And Paving Cleaner	Ronseal Ltd	Amateur	5985
Travis Perkins Fungicidal Wash	Travis Perkins Trading Company Ltd	Amateur Professional	6619
Triflex Sterilising Solution Concentrate	Triflex (UK) Ltd	Amateur Professional	6295
Wickes Fungicidal Wash	Wickes Building Supplies Ltd	Amateur Professional	6039

HSE
SURFACE BIOCIDES

Product Name	Marketing Company	Use	HSE No.

420 Benzalkonium Chloride + 2-Phenylphenol

Product Name	Marketing Company	Use	HSE No.
Anti Fungus Wash	Craig & Rose PLC	Professional	6293
Blackfriar Anti-Mould Solution	E Parsons And Sons Ltd	Amateur	6676
Bradite Fungicidal Wash	Bradite Ltd	Amateur	6078
C/S Clean R833	Construction Specialties (UK) Ltd	Amateur Professional	6865
D Chelwash	Chelec Ltd	Amateur	3096
Crown Fungicidal Wall Solution	Crown Chemicals Ltd	Professional	5868
Deadly Nightshade	Andura Textured Masonry Coatings Ltd	Professional	4895
Fungicidal Algaecidal Bacteriacidal Wash	Philip Johnstone Group Ltd	Amateur Professional	4216
D Fungicidal Wash	C W Wastnage Ltd	Amateur	3072
Fungicidal Wash	Everbuild Building Products Ltd	Amateur Professional	6000
D Fungicidal Wash Solution	Johnstone's Paints PLC	Amateur Professional	3776
Fungiguard	Spaldings Ltd	Professional	5582
Houseplan Fungi Wash	Ruberoid Building Products Ltd	Amateur	6374
Johnstone's Fungicidal Wash	Johnstone's Paints PLC	Amateur Professional	5678
Larsen Concentrated Algicide 2	Larsen Manufacturing Ltd	Amateur Professional	6606
Micro FWS	Lectros International Ltd	Professional	6135
Micro Masonry Biocide	Grip-Fast Systems Ltd	Professional	6444
Microguard FWS	Permagard Products Ltd	Professional	5485
Microstel	Bedec Products Ltd	Professional	4198
Mouldcheck Sterilizer 2	Biotech Environmental (UK) Ltd	Amateur Professional	6385
Nitromors Mould Remover	Henkel Home Improvements And Adhesive Products	Amateur	5507
P.J. Fungicidal Solution	Philip Johnstone Group Ltd	Amateur Professional	6318
Permacide Masonry Fungicide	Triton Perma Industries Ltd	Professional	5570
Permarock Fungicidal Wash	Permarock Products Ltd	Professional	5193
Pob Mould Treatment	Darcy Industries Ltd	Amateur	4076
Ruberoid Fungicidal Wash Solution	Ruberoid Building Products Ltd	Amateur	6375
Solvex Rapid Mould Cleaner	Tetrosyl (Building Products) Ltd	Amateur Professional	4206
Spencer Stormcote Fungicidal Treatment	Spencer Coatings Ltd	Professional	4155
Z.144 Fungicidal Wash	Crosbie Coatings Ltd	Amateur Professional	3003

Product Name	Marketing Company	Use	HSE No.

421 Benzalkonium Chloride + Boric Acid

Barrettine Premier Multicide Fluid	Barrettine Products Ltd	Amateur (Surface Biocide) Amateur (Wood Preservative)	6312
Microguard Mouldicidal Wood Preserver	Permagard Products Ltd	Amateur (Surface Biocide) Amateur (Wood Preservative) Professional (Surface Biocide) Professional (Wood Preservative) Industrial (Wood Preservative)	5100

422 Benzalkonium Chloride + Carbendazim + Dialkyldimethyl Ammonium Chloride

Rentokil Mould Cure Spray	Rentokil Initial UK Ltd	Amateur	4227

423 Benzalkonium Chloride + Disodium Octaborate

Boracol B8.5 RH Mouldicide/Wood Preservative	Advanced Chemical Specialties Ltd	Amateur (Surface Biocide) Amateur (Wood Preservative) Professional (Surface Biocide) Professional (Wood Preservative)	4809
Boron Biocide 10	Philip Johnstone Group Ltd	Professional (Surface Biocide) Professional (Wood Preservative)	6485

Product Name	Marketing Company	Use	HSE No.

423 Benzalkonium Chloride + Disodium Octaborate—continued

Product Name	Marketing Company	Use	HSE No.
Brunosol Boron 10	Philip Johnstone Group Ltd	Professional (Surface Biocide) Professional (Wood Preservative)	6488
Cuprinol Fungicidal Spray	Cuprinol Ltd	Amateur	4221
Cuprinol Mould Killer	Cuprinol Ltd	Amateur Professional	4163
W Deepflow 11 Inorganic Boron Masonry Biocide.	Safeguard Chemicals Ltd	Professional	5528
Lectros Boracol 10 RH	Lectros International Ltd	Professional	6499
Permabor Boracol 10RH	Permagard Products Ltd	Professional	6358
Polycell 3 In 1 Mould Killer	Polycell Products Ltd	Amateur Professional	6157
Probor 10	Safeguard Chemicals Ltd	Professional	6595
Probor 50	Safeguard Chemicals Ltd	Professional (Surface Biocide) Professional (Wood Preservative)	5596
Safeguard Boracol 10 RH	Safeguard Chemicals Ltd	Professional	6397
Tribor 10	Triton Chemical Manufacturing Company Ltd	Professional	6286
Weathershield Mulit-Surface Fungicidal Wash	Imperial Chemical Industries PLC	Amateur Professional	6823

424 Benzalkonium Chloride + Naphthalene + Permethrin + Pyrethrins

Product Name	Marketing Company	Use	HSE No.
D Roxem D	Horton Hygiene Company	Professional	5107

425 Boric Acid

Product Name	Marketing Company	Use	HSE No.
Rentokil Dry Rot Fluid (E) For Bonded Warehouses	Rentokil Initial UK Ltd	Professional (Surface Biocide) Professional (Wood Preservative)	3986

426 Boric Acid + Alkylaryltrimethyl Ammonium Chloride

Product Name	Marketing Company	Use	HSE No.
Kleeneze Anti-Mould Spray	Kleeneze Ltd	Amateur Professional	4599

Product Name	Marketing Company	Use	HSE No.

427 Boric Acid + Benzalkonium Chloride

Barrettine Premier Multicide Fluid	Barrettine Products Ltd	Amateur (Surface Biocide) Amateur (Wood Preservative)	6312
Microguard Mouldicidal Wood Preserver	Permagard Products Ltd	Amateur (Surface Biocide) Amateur (Wood Preservative) Professional (Surface Biocide) Professional (Wood Preservative) Industrial (Wood Preservative)	5100

428 Boric Acid + Cypermethrin + Propiconazole

Rentokil Total Wood Treatment	Rentokil Initial UK Ltd	Amateur Professional	6053

429 Boric Acid + Disodium Tetraborate

Control Fluid SB	Rentokil Initial UK Ltd	Professional (Surface Biocide) Professional (Wood Preservative)	6018

430 Carbendazim + Dialkyldimethyl Ammonium Chloride + Benzalkonium Chloride

Rentokil Mould Cure Spray	Rentokil Initial UK Ltd	Amateur	4227

431 Cypermethrin + Propiconazole + Boric Acid

Rentokil Total Wood Treatment	Rentokil Initial UK Ltd	Amateur Professional	6053

432 Dialkyldimethyl Ammonium Chloride

W Biophen Barrier	Mycobio Consultants	Amateur Professional	6048

Product Name	Marketing Company	Use	HSE No.

432 Dialkyldimethyl Ammonium Chloride—continued

Product Name	Marketing Company	Use	HSE No.
Halophane Bonding Solution	Mould Growth Consultants Ltd	Professional	3929
Halophen BM1165L	Mould Growth Consultants Ltd	Professional	3930
HG Green Slime Remover	HG International B.V.	Amateur	6086
Lichenite	Mould Growth Consultants Ltd	Professional	3936
Lichenite B	Mould Growth Consultants Ltd	Professional	6134
Mosscheck	Biotech Environmental (UK) Ltd	Amateur Professional	4870
Mouldcheck Barrier 2	Biotech Environmental (UK) Ltd	Amateur Professional	4856
Mouldkill	Biotech Environmental (UK) Ltd	Amateur Professional	5192
Remwash Masonry Steriliser	Remtox (Chemicals) Ltd	Amateur Professional	6363
RLT Clearmould Spray	Mould Growth Consultants Ltd	Amateur Professional	4824
RLT Halophen	Mould Growth Consultants Ltd	Amateur Professional	4081
RLT Halophen DS	Mould Growth Consultants Ltd	Amateur Professional	4483
Sovereign Masonry Sterilising Wash	Sovereign Chemical Industries Ltd	Amateur Professional	6360
X Thallo	Thames Valley Specialist Products Ltd	Amateur Professional	6841

433 Dialkyldimethyl Ammonium Chloride + Benzalkonium Chloride + Carbendazim

Product Name	Marketing Company	Use	HSE No.
Rentokil Mould Cure Spray	Rentokil Initial UK Ltd	Amateur	4227

434 Dichlorophen

Product Name	Marketing Company	Use	HSE No.
W Bioclean Sterilizer	Mycobio Consultants	Amateur Professional	6047
D Denso Mouldshield Biocidal Cleanser	Winn And Coales (Denso) Ltd	Amateur Professional	4731
D Denso Mouldshield Surface Biocide	Winn And Coales (Denso) Ltd	Professional	4732
En-Tout-Cas Moss Killer	En-Tout-Cas Sports Equipment Ltd	Amateur Professional	5855
Halodec	Mould Growth Consultants Ltd	Professional	4454
W Halophane No.1 Aerosol	Mould Growth Consultants Ltd	Professional	3815

Product Name	Marketing Company	Use	HSE No.

434 Dichlorophen—continued

W Halophane No.3 Aerosol	Mould Growth Consultants Ltd	Professional	3816
Mouldcheck Cleaner	Biotech Environmental (UK) Ltd	Amateur Professional	6927
Mouldcheck Spray	Biotech Environmental (UK) Ltd	Amateur	4093
Mouldcheck Sterilizer	Biotech Environmental (UK) Ltd	Amateur Professional	4505
W Nipacide DP30	Nipa Laboratories Ltd	Professional	4925
Panacide M21	Coalite Chemicals Division	Professional	4923
Panaclean 736	Coalite Chemicals Division	Professional	5075
SP.153 Sterilising Detergent Wash	Mason Coatings PLC	Professional	3920
SP.154 Fungicidal And Bactericidal Treatment	Mason Coatings PLC	Professional	3919

435 Disodium Octaborate

ACS Borotreat 10P	Advanced Chemical Specialties Ltd	Professional (Surface Biocide) Professional (Wood Preservative)	6335
Azygo Boron Preservative PDR	Azygo International Direct Ltd	Professional (Surface Biocide) Professional (Wood Preservative)	6497
Eco - DB8	Restoration (UK) Ltd	Professional Industrial	6944

436 Disodium Octaborate + Alkylaryltrimethyl Ammonium Chloride

Boracol 10RH Surface Biocide	Advanced Chemical Specialties Ltd	Professional	4911
Remtox Borocol 10 RH Masonry Biocide	Remtox (Chemicals) Ltd	Professional	4100

Product Name	Marketing Company	Use	HSE No.

437 Disodium Octaborate + Benzalkonium Chloride

Product Name	Marketing Company	Use	HSE No.
Boracol B8.5 RH Mouldicide/Wood Preservative	Advanced Chemical Specialties Ltd	Amateur (Surface Biocide) Amateur (Wood Preservative) Professional (Surface Biocide) Professional (Wood Preservative)	4809
Boron Biocide 10	Philip Johnstone Group Ltd	Professional (Surface Biocide) Professional (Wood Preservative)	6485
Brunosol Boron 10	Philip Johnstone Group Ltd	Professional (Surface Biocide) Professional (Wood Preservative)	6488
Cuprinol Fungicidal Spray	Cuprinol Ltd	Amateur	4221
Cuprinol Mould Killer	Cuprinol Ltd	Amateur Professional	4163
W Deepflow 11 Inorganic Boron Masonry Biocide.	Safeguard Chemicals Ltd	Professional	5528
Lectros Boracol 10 RH	Lectros International Ltd	Professional	6499
Permabor Boracol 10RH	Permagard Products Ltd	Professional	6358
Polycell 3 In 1 Mould Killer	Polycell Products Ltd	Amateur Professional	6157
Probor 10	Safeguard Chemicals Ltd	Professional	6595
Probor 50	Safeguard Chemicals Ltd	Professional Professional (Surface Biocide) Professional (Wood Preservative)	5596
Safeguard Boracol 10 RH	Safeguard Chemicals Ltd	Professional	6397
Tribor 10	Triton Chemical Manufacturing Company Ltd	Professional	6286
Weathershield Mulit-Surface Fungicidal Wash	Imperial Chemical Industries PLC	Amateur Professional	6823

Product Name	Marketing Company	Use	HSE No.

438 Disodium Tetraborate + Boric Acid

| Control Fluid SB | Rentokil Initial UK Ltd | Professional (Surface Biocide) Professional (Wood Preservative) | 6018 |

439 Dodecylamine Lactate + Dodecylamine Salicylate

W Anti Fungus Wash	Craig & Rose PLC	Professional	4375
W Blackfriar Anti Mould Solution 195/9	E Parsons And Sons Ltd	Amateur	4231
W Durham BI 730	Durham Chemicals	Professional (Surface Biocide) Professional (Wood Preservative) Industrial (Wood Preservative)	4744
W Environmental Woodrot Treatment	SCI (Building Products)	Professional (Surface Biocide) Professional (Wood Preservative) Industrial (Wood Preservative)	5530
W FEB Fungicide	FEB Ltd	Amateur Professional	3009
W Fungicidal Wash	Polybond Ltd	Amateur (Surface Biocide) Amateur (Wood Preservative) Professional (Surface Biocide) Professional (Wood Preservative)	3601

HSE SURFACE BIOCIDES

Product Name	Marketing Company	Use	HSE No.

439 Dodecylamine Lactate + Dodecylamine Salicylate—continued

Product Name	Marketing Company	Use	HSE No.
D Fungicidal Wash	Smyth - Morris Chemicals Ltd	Professional (Surface Biocide) Professional (Wood Preservative)	3004
W Green Range Murosol 20	Cementone Beaver Ltd	Professional	4456
W Hyperion Mould Inhibiting Solution	Hird Hastie Paints Ltd	Professional	4412
W Kingfisher Fungicidal Wall Solution	Kingfisher Chemicals Ltd	Professional	4014
W Larsen Concentrated Algicide	Larsen Manufacturing Ltd	Professional	4968
D M-Tec Biocide	M-Protective Coatings Ltd	Professional	4381
W Macpherson Anti-Mould Solution	Macpherson Paints Ltd	Professional (Surface Biocide) Professional (Wood Preservative)	4791
W Metalife Fungicidal Wash	Metalife International Ltd	Amateur (Surface Biocide) Amateur (Wood Preservative) Professional (Surface Biocide) Professional (Wood Preservative)	5406
W Mouldrid	S And K Maintenance Products	Amateur (Surface Biocide) Amateur (Wood Preservative) Professional (Surface Biocide) Professional (Wood Preservative)	3671
W MRS Clear Mould Fungicidal Wash	MR (Polymer Cement Products) Ltd	Professional	3620

Product Name	Marketing Company	Use	HSE No.

439 Dodecylamine Lactate + Dodecylamine Salicylate—continued

W Nubex WDR	Nubex Ltd	Professional (Surface Biocide) Professional (Wood Preservative)	3810
W Palace Mould Remover	Palace Chemicals Ltd	Amateur Professional	5204
W Permadex Masonry Fungicide Concentrate	Permagard Products Ltd	Professional	4007
W Permoglaze Micatex Fungicidal Treatment	Permoglaze Paints Ltd	Professional (Surface Biocide) Professional (Wood Preservative)	5249
W Polycell Mould Cleaner	Polycell Products Ltd	Amateur Professional	3612
W Protim Wall Solution II	Protim Solignum Ltd	Amateur Professional	5006
W Protim Wall Solution II Concentrate	Protim Solignum Ltd	Professional	5007
W Ruberoid Fungicidal Wash	Ruberoid Building Products Ltd	Amateur	5949
D Safeguard Deepwood Surface Biocide Concentrate	Safeguard Chemicals Ltd	Professional	4616
D Safeguard Mould And Moss Killer	Safeguard Chemicals Ltd	Amateur Professional	4607
W Solignum Dry Rot Killer	Protim Solignum Ltd	Amateur Professional	5025
W Solignum Fungicide	Protim Solignum Ltd	Amateur Professional	5026
W Thompson's Interior And Exterior Fungicidal Spray	Ronseal Ltd	Amateur	5626
W Thompson's Interior Mould Killer	Ronseal Ltd	Amateur	5624
W Torkill Fungicidal Solution 'W'	Tor Coatings Ltd	Professional	4246
W Unisil S Silicone Waterproofing Solution	Witham Oil & Paint (Lowestoft) Ltd	Professional	4758
W Unitas Fungicidal Wash-Exterior (Solvent Based)	Witham Oil & Paint (Lowestoft) Ltd	Professional	4759
W Unitas Fungicidal Wash-Interior (Water Based)	Witham Oil & Paint (Lowestoft) Ltd	Professional	4760
W Vallance Fungicidal Wash	Siroflex Ltd	Amateur Professional	5375

Product Name	Marketing Company	Use	HSE No.

439 Dodecylamine Lactate + Dodecylamine Salicylate—continued

W Weathershield Fungicidal Wash	ICI Paints	Amateur (Surface Biocide) Amateur (Wood Preservative) Professional (Surface Biocide) Professional (Wood Preservative)	5134

440 Dodecylamine Salicylate + Dodecylamine Lactate

W Anti Fungus Wash	Craig & Rose PLC	Professional	4375
W Blackfriar Anti Mould Solution 195/9	E Parsons And Sons Ltd	Amateur	4231
W Durham BI 730	Durham Chemicals	Professional (Surface Biocide) Professional (Wood Preservative) Industrial (Wood Preservative)	4744
W Environmental Woodrot Treatment	SCI (Building Products)	Professional (Surface Biocide) Professional (Wood Preservative) Industrial (Wood Preservative)	5530
W FEB Fungicide	FEB Ltd	Amateur Professional	3009

Product Name	Marketing Company	Use	HSE No.

440 Dodecylamine Salicylate + Dodecylamine Lactate—continued

Product Name	Marketing Company	Use	HSE No.
W Fungicidal Wash	Polybond Ltd	Amateur (Surface Biocide) Amateur (Wood Preservative) Professional (Surface Biocide) Professional (Wood Preservative)	3601
D Fungicidal Wash	Smyth - Morris Chemicals Ltd	Professional (Surface Biocide) Professional (Wood Preservative)	3004
W Green Range Murosol 20	Cementone Beaver Ltd	Professional	4456
W Hyperion Mould Inhibiting Solution	Hird Hastie Paints Ltd	Professional	4412
W Kingfisher Fungicidal Wall Solution	Kingfisher Chemicals Ltd	Professional	4014
W Larsen Concentrated Algicide	Larsen Manufacturing Ltd	Professional	4968
D M-Tec Biocide	M-Protective Coatings Ltd	Professional	4381
W Macpherson Anti-Mould Solution	Macpherson Paints Ltd	Professional (Surface Biocide) Professional (Wood Preservative)	4791
W Metalife Fungicidal Wash	Metalife International Ltd	Amateur (Surface Biocide) Amateur (Wood Preservative) Professional (Surface Biocide) Professional (Wood Preservative)	5406

HSE
SURFACE BIOCIDES

Product Name	Marketing Company	Use	HSE No.

440 Dodecylamine Salicylate + Dodecylamine Lactate—continued

Product Name	Marketing Company	Use	HSE No.
W Mouldrid	S And K Maintenance Products	Amateur (Surface Biocide) Amateur (Wood Preservative) Professional (Surface Biocide) Professional (Wood Preservative)	3671
W MRS Clear Mould Fungicidal Wash	MR (Polymer Cement Products) Ltd	Professional	3620
W Nubex WDR	Nubex Ltd	Professional (Surface Biocide) Professional (Wood Preservative)	3810
W Palace Mould Remover	Palace Chemicals Ltd	Amateur Professional	5204
W Permadex Masonry Fungicide Concentrate	Permagard Products Ltd	Professional	4007
W Permoglaze Micatex Fungicidal Treatment	Permoglaze Paints Ltd	Professional (Surface Biocide) Professional (Wood Preservative)	5249
W Polycell Mould Cleaner	Polycell Products Ltd	Amateur Professional	3612
W Protim Wall Solution II	Protim Solignum Ltd	Amateur Professional	5006
W Protim Wall Solution II Concentrate	Protim Solignum Ltd	Professional	5007
W Ruberoid Fungicidal Wash	Ruberoid Building Products Ltd	Amateur	5949
D Safeguard Deepwood Surface Biocide Concentrate	Safeguard Chemicals Ltd	Professional	4616
D Safeguard Mould And Moss Killer	Safeguard Chemicals Ltd	Amateur Professional	4607
W Solignum Dry Rot Killer	Protim Solignum Ltd	Amateur Professional	5025
W Solignum Fungicide	Protim Solignum Ltd	Amateur Professional	5026
W Thompson's Interior And Exterior Fungicidal Spray	Ronseal Ltd	Amateur	5626

Product Name	Marketing Company	Use	HSE No.

440 Dodecylamine Salicylate + Dodecylamine Lactate—continued

W Thompson's Interior Mould Killer	Ronseal Ltd	Amateur	5624
W Torkill Fungicidal Solution 'W'	Tor Coatings Ltd	Professional	4246
W Unisil S Silicone Waterproofing Solution	Witham Oil & Paint (Lowestoft) Ltd	Professional	4758
W Unitas Fungicidal Wash-Exterior (Solvent Based)	Witham Oil & Paint (Lowestoft) Ltd	Professional	4759
W Unitas Fungicidal Wash-Interior (Water Based)	Witham Oil & Paint (Lowestoft) Ltd	Professional	4760
W Vallance Fungicidal Wash	Siroflex Ltd	Amateur Professional	5375
W Weathershield Fungicidal Wash	ICI Paints	Amateur (Surface Biocide) Amateur (Wood Preservative) Professional (Surface Biocide) Professional (Wood Preservative)	5134

441 3-Iodo-2-Propynyl-N-Butyl Carbamate

Antel FWS	Antel UK Ltd	Professional	6602
BPC-FS Fungicidal Solution Concentrate	Building Preservation Centre	Professional (Surface Biocide) Professional (Wood Preservative)	6484
W CME 30/F Concentrate	Terminix Peter Cox Ltd	Professional (Surface Biocide) Professional (Wood Preservative)	5948
EQ109 Surface Biocide	Enviroquest UK Ltd	Professional	6383
Excel Masonry Biocide	Restoration (UK) Ltd	Professional	6942
D Fungicidal Wall Solution	Sovereign Chemical Industries Ltd	Professional	5387
KF-18 Masonry Biocide (Microemulsion)	Kingfisher Chemicals Ltd	Professional	6529
Lectro FWS	Lectros International Ltd	Professional	6388
Masonry Biocide 3	Terminix Peter Cox Ltd	Professional	6402
Masonry Sterilant	Deal Direct Ltd	Professional	6247
Micro FWS	Construction Chemicals	Professional	6643

Product Name	Marketing Company	Use	HSE No.
441 3-Iodo-2-Propynyl-N-Butyl Carbamate—continued			
Microtech Biocide 25	Philip Johnstone Group Ltd	Professional	5997
Microtreat Biocide 25X	Philip Johnstone Group Ltd	Professional	6916
Omega Biocide	Restoration (UK) Ltd	Professional (Surface Biocide) Professional (Wood Preservative)	6381
Palace FWS (Microemulsion)	Palace Chemicals Ltd	Professional	6637
Protim Wall Solution 250	Protim Solignum Ltd	Professional	6986
W Remtox Dry Rot F. W. S.	Remtox (Chemicals) Ltd	Professional	5188
Remtox Dry Rot Paint	Remtox (Chemicals) Ltd	Amateur (Surface Biocide) Amateur (Wood Preservative) Professional (Surface Biocide) Professional (Wood Preservative)	5187
Remtox Fungicidal Wall Solution RS	Remtox (Chemicals) Ltd	Professional	5484
Remtox Micro Fungicidal Wall Solution RS	Remtox Ltd	Professional	7032
W Remtox Microactive FWS W6	Remtox (Chemicals) Ltd	Professional	5589
Restor-MFC	Restoration (UK) Ltd	Professional (Surface Biocide) Professional (Wood Preservative)	6246
Sovaq Micro FWS	Sovereign Chemical Industries Ltd	Professional	5800
Sprytech 3	Spry Chemicals Ltd	Professional	6491
Stanhope Biocide 25X	Philip Johnstone Group Ltd	Professional	6309
Trisol 23	Triton Chemical Manufacturing Company Ltd	Professional	6274
UKC-2 Dry Rot Killer	UK Chemicals Ltd	Professional	6378

Product Name	Marketing Company	Use	HSE No.

441 3-Iodo-2-Propynyl-N-Butyl Carbamate—continued

UKC-2 Dry Rot Killer Concentrate	UK Chemicals Ltd	Professional (Surface Biocide) Professional (Wood Preservative)	6107
Ultra Tech 2000M	Crown Chemicals Ltd	Professional	6164
Wall Fungicide	Timberwise Preservation Scot	Professional	6642

442 Naphthalene + Permethrin + Pyrethrins + Benzalkonium Chloride

D Roxem D	Horton Hygiene Company	Professional	5107

443 2-Octyl-2h-Isothiazolin-3-One

Halostain	Mould Growth Consultants Ltd	Professional	6559

444 Permethrin + Azaconazole

Fongix Se Total Treatment For Wood	Liberon Waxes Ltd	Amateur (Surface Biocide) Amateur (Wood Preservative) Professional (Surface Biocide) Professional (Wood Preservative)	5610
Woodworm Killer And Rot Treatment	Liberon Waxes Ltd	Amateur (Surface Biocide) Amateur (Wood Preservative) Professional (Surface Biocide) Professional (Wood Preservative)	4826

445 Permethrin + Pyrethrins + Benzalkonium Chloride + Naphthalene

D Roxem D	Horton Hygiene Company	Professional	5107

HSE
SURFACE BIOCIDES

Product Name	Marketing Company	Use	HSE No.

446 Permethrin + Zinc Octoate

W Premium Wood Treatment	Rentokil Initial UK Ltd	Amateur (Surface Biocide)	5350
Amateur (Wood Preservative)	Amateur	Professional (Surface Biocide) Professional (Wood Preservative)	

447 2-Phenylphenol + Benzalkonium Chloride

Anti Fungus Wash	Craig & Rose PLC	Professional	6293
Blackfriar Anti-Mould Solution	E Parsons And Sons Ltd	Amateur	6676
Bradite Fungicidal Wash	Bradite Ltd	Amateur	6078
C/S Clean R833	Construction Specialties (UK) Ltd	Amateur Professional	6865
D Chelwash	Chelec Ltd	Amateur	3096
Crown Fungicidal Wall Solution	Crown Chemicals Ltd	Professional	5868
Deadly Nightshade	Andura Textured Masonry Coatings Ltd	Professional	4895
Fungicidal Algaecidal Bacteriacidal Wash	Philip Johnstone Group Ltd	Amateur Professional	4216
D Fungicidal Wash	C W Wastnage Ltd	Amateur	3072
Fungicidal Wash	Everbuild Building Products Ltd	Amateur Professional	6000
D Fungicidal Wash Solution	Johnstone's Paints PLC	Amateur Professional	3776
Fungiguard	Spaldings Ltd	Professional	5582
Houseplan Fungi Wash	Ruberoid Building Products Ltd	Amateur	6374
Johnstone's Fungicidal Wash	Johnstone's Paints PLC	Amateur Professional	5678
Larsen Concentrated Algicide 2	Larsen Manufacturing Ltd	Amateur Professional	6606
Micro FWS	Lectros International Ltd	Professional	6135
Micro Masonry Biocide	Grip-Fast Systems Ltd	Professional	6444
Microguard FWS	Permagard Products Ltd	Professional	5485
Microstel	Bedec Products Ltd	Professional	4198
Mouldcheck Sterilizer 2	Biotech Environmental (UK) Ltd	Amateur Professional	6385
Nitromors Mould Remover	Henkel Home Improvements And Adhesive Products	Amateur	5507
P.J. Fungicidal Solution	Philip Johnstone Group Ltd	Amateur Professional	6318
Permacide Masonry Fungicide	Triton Perma Industries Ltd	Professional	5570
Permarock Fungicidal Wash	Permarock Products Ltd	Professional	5193
Pob Mould Treatment	Darcy Industries Ltd	Amateur	4076

Product Name	Marketing Company	Use	HSE No.

447 2-Phenylphenol + Benzalkonium Chloride—continued

Ruberoid Fungicidal Wash Solution	Ruberoid Building Products Ltd	Amateur	6375
Solvex Rapid Mould Cleaner	Tetrosyl (Building Products) Ltd	Amateur Professional	4206
Spencer Stormcote Fungicidal Treatment	Spencer Coatings Ltd	Professional	4155
Z.144 Fungicidal Wash	Crosbie Coatings Ltd	Amateur Professional	3003

448 Propiconazole

Trisol 22	Triton Chemical Manufacturing Company Ltd	Professional	6111
Wocosen 100 SL	Janssen Pharmaceutica NV	Professional (Surface Biocide) Professional (Wood Preservative) Industrial (Wood Preservative)	5743

449 Propiconazole + Boric Acid + Cypermethrin

Rentokil Total Wood Treatment	Rentokil Initial UK Ltd	Amateur Professional	6053

450 Pyrethrins + Benzalkonium Chloride + Naphthalene + Permethrin

D Roxem D	Horton Hygiene Company	Professional	5107

451 Sodium 2-Phenylphenoxide

Green Range Fungicidal Concentrate	Philip Johnstone Group Ltd	Professional	3689
Mar-Cide	Marcher Chemicals Ltd	Professional (Surface Biocide) Professional (Wood Preservative) Industrial (Wood Preservative)	5631
PC-D	Terminix Peter Cox Ltd	Professional	5719
PC-D Concentrate	Terminix Peter Cox Ltd	Professional	5720
PLA Products Dry Rot Killer	Sealocrete PLA Ltd	Amateur Professional	5339

HSE
SURFACE BIOCIDES

465

Product Name	Marketing Company	Use	HSE No.

451 Sodium 2-Phenylphenoxide—continued

D Safeguard Fungicidal Wall Solution	Safeguard Chemicals Ltd	Professional	3903

452 Sodium Dichlorophen

Bactdet D	Mould Growth Consultants Ltd	Professional	3809
Fungo	Dax Products Ltd	Amateur	4768
RLT Bactdet	Mould Growth Consultants Ltd	Amateur Professional	3928

453 Sodium Hypochlorite

Anti-Moss	Keychem Ltd	Professional	6064
Brolac Fungicidal Solution	Akzo Nobel Decorative Coatings Ltd	Amateur	4413
Crown Trade Stronghold Fungicidal Solution	Akzo Nobel Decorative Coatings Ltd	Professional	5383
D Palace Fungicidal Wash	Palace Chemicals Ltd	Amateur	4091
Sandtex Fungicide	Akzo Nobel Decorative Coatings Ltd	Amateur Professional	5258
Tetra Construction Fungicidal Wash	Tetrosyl (Building Products) Ltd	Amateur	6250
Tetra Fungicidal Wash	Tetrosyl (Building Products) Ltd	Amateur	6240
Thompson's 60 Second Cleaner	Ronseal Ltd	Amateur Professional	6985

454 Sodium Pentachlorophenoxide

W Protim Plug Compound	Protim Solignum Ltd	Professional (Surface Biocide) Professional (Wood Preservative)	5012

455 Thiophanate-Methyl

Deemold	Mr Leiper	Amateur Professional	3852

Product Name	Marketing Company	Use	HSE No.

456 Zinc Octoate

W Rentokil Dry Rot And Wet Rot Treatment	Rentokil Initial UK Ltd	Amateur (Surface Biocide) Amateur (Wood Preservative) Professional (Surface Biocide) Professional (Wood Preservative)	4378

457 Zinc Octoate + Permethrin

W Premium Wood Treatment	Rentokil Initial UK Ltd	Amateur (Surface Biocide) Amateur (Wood Preservative) Professional (Surface Biocide) Professional (Wood Preservative) Wood Preservative	5350

7

WOOD PRESERVATIVES

Product Name	Marketing Company	Use	HSE No.

Wood Preservatives

458 Acypetacs Copper

Product Name	Marketing Company	Use	HSE No.
Cuprinol Low Odour Wood Preserver Green	Cuprinol Ltd	Amateur Professional Industrial	5448
D Cuprinol Wood Preserver Green S	Cuprinol Ltd	Amateur Professional	4697

459 Acypetacs Copper + Acypetacs Zinc + Permethrin

Product Name	Marketing Company	Use	HSE No.
Cuprinol Low Odour End Cut	Cuprinol Ltd	Amateur Professional Industrial	6114
Cuprisol P	Cuprinol Ltd	Industrial	3634
Protim 800P	Protim Solignum Ltd	Industrial	5702

460 Acypetacs Zinc

Product Name	Marketing Company	Use	HSE No.
Cuprinol Garden Shed And Fence Preserver	Cuprinol Ltd	Amateur Professional	5873
Cuprinol Low Odour Wet And Dry Rot Killer For Timber	Cuprinol Ltd	Amateur Professional Industrial	5446
D Cuprinol Wet And Dry Rot Killer For Timber(S)	Cuprinol Ltd	Amateur	3589
Cuprinol Wood Preserver	Cuprinol Ltd	Amateur Professional	4698
D Cuprinol Wood Preserver Clear S	Cuprinol Ltd	Amateur Professional Industrial	4708
Cuprisol F	Cuprinol Ltd	Industrial	3734
Cuprisol XQD	Cuprinol Ltd	Industrial	3733
Exterior Wood Preserver S	Cuprinol Ltd	Amateur Professional	4716
Green End Coat	Richard Burbidge Ltd	Amateur Professional	6960
Protim FDR 800	Protim Solignum Ltd	Industrial	4963
Protim JP 800	Protim Solignum Ltd	Industrial	4987
Protim Paste 800 F	Protim Solignum Ltd	Professional Industrial	4989
Weathershield Timber Preservative	Imperial Chemical Industries PLC	Amateur Professional	7018
Wickes Exterior Wood Preserver	Wickes Building Supplies Ltd	Amateur Professional	5697
Wickes Wood Preserver Clear	Wickes Building Supplies Ltd	Amateur Professional	5694

Product Name	Marketing Company	Use	HSE No.

461 Acypetacs Zinc (Equivalent To 3% Zinc Metal) + Dichlofluanid

Cuprinol Garden Furniture Preserver	Cuprinol Ltd	Amateur	7123

462 Acypetacs Zinc + Dichlofluanid

W Cuprinol Decorative Preserver	Cuprinol Ltd	Amateur Professional Industrial	5343
Cuprinol Decorative Preserver	Cuprinol Ltd	Amateur Professional Industrial	6277
W Cuprinol Decorative Preserver Red Cedar	Cuprinol Ltd	Amateur Professional Industrial	5342
Cuprinol Decorative Preserver Red Cedar	Cuprinol Ltd	Amateur Professional Industrial	6280
Cuprinol Garden Decking Protector	Cuprinol Ltd	Amateur Professional	6862
Cuprinol Preservative Base	Cuprinol Ltd	Amateur Professional	4929
Weathershield Exterior Preservative Primer	Imperial Chemical Industries PLC	Amateur Professional	6983
Weathershield Preservative Basecoat	ICI Paints	Amateur Professional	6982
Wickes Decorative Wood Preserver	Wickes Building Supplies Ltd	Amateur Professional Industrial	6419
Wickes Decorative Wood Preserver Rich Cedar	Wickes Building Supplies Ltd	Amateur Professional Industrial	6420

463 Acypetacs Zinc + Permethrin

D Cuprinol 5 Star Complete Wood Treatment S	Cuprinol Ltd	Amateur Professional	4710
Cuprinol Combination Grade S	Cuprinol Ltd	Amateur Professional Industrial	3588
Cuprinol Low Odour 5 Star Complete Wood Treatment	Cuprinol Ltd	Amateur Professional	5445
Cuprisol FN	Cuprinol Ltd	Industrial	3632
Cuprisol WR	Palace Chemicals Ltd	Industrial	3633
Cuprisol XQD Special	Cuprinol Ltd	Industrial	3732
Cut End	Protim Solignum Ltd	Amateur Professional Industrial	5490
Protim 800	Protim Solignum Ltd	Industrial	4991

Product Name	Marketing Company	Use	HSE No.

463 Acypetacs Zinc + Permethrin—continued

Product Name	Marketing Company	Use	HSE No.
Protim 800 C	Protim Solignum Ltd	Industrial	4990
Protim 800 CWR	Protim Solignum Ltd	Industrial	4993
Protim 800 WR	Protim Solignum Ltd	Industrial	4992
Protim Brown CDB	Protim Solignum Ltd	Amateur Professional Industrial	6472
Protim Paste 800	Protim Solignum Ltd	Professional Industrial	4986
Universal	Protim Solignum Ltd	Amateur Professional Industrial	5491
Wickes All Purpose Wood Treatment	Wickes Building Supplies Ltd	Amateur Professional	5696

464 Acypetacs Zinc + Permethrin + Acypetacs Copper

Product Name	Marketing Company	Use	HSE No.
Cuprinol Low Odour End Cut	Cuprinol Ltd	Amateur Professional Industrial	6114
Cuprisol P	Cuprinol Ltd	Industrial	3634
Protim 800P	Protim Solignum Ltd	Industrial	5702

465 Alkylaryltrimethyl Ammonium Chloride

Product Name	Marketing Company	Use	HSE No.
Anti-Mould Solution	Macpherson Paints Ltd	Amateur (Surface Biocide) Amateur (Wood Preservative) Professional (Surface Biocide) Professional (Wood Preservative)	4873
Barrettine Timberguard	Barrettine Products Ltd	Amateur	3089
Timberguard 1 Plus 7 Concentrate	Barrettine Products Ltd	Professional Industrial	5211

466 Alkyltrimethyl Ammonium Chloride + Sodium Tetraborate

Product Name	Marketing Company	Use	HSE No.
Sinesto B	Finnmex Co	Industrial	5136

467 Ammonium Bifluoride + Sodium Dichromate + Sodium Fluoride

Product Name	Marketing Company	Use	HSE No.
W Rentex	Rentokil Initial UK Ltd	Industrial	4756

468 Arsenic Pentoxide + Chromium Trioxide + Copper Oxide

Product Name	Marketing Company	Use	HSE No.
Celcure AO	Protim Solignum Ltd	Industrial	4220

Product Name	Marketing Company	Use	HSE No.

468 Arsenic Pentoxide + Chromium Trioxide + Copper Oxide—continued

Celcure C6099	Protim Solignum Ltd	Industrial	6825
Celcure CCA Type C	Protim Solignum Ltd	Industrial	5104
Celcure CCA Type C-40%	Protim Solignum Ltd	Industrial	6028
Celcure CCA Type C-50%	Protim Solignum Ltd	Industrial	6030
Celcure CCA Type C-60%	Protim Solignum Ltd	Industrial	5934
Celcure Type C6099	Protim Solignum Ltd	Industrial	6920
Injecta Osmose K33 C50	Injecta Osmose Ltd	Industrial	5889
Injecta Osmose K33 C58.7	Injecta Osmose Ltd	Industrial	6498
Injecta Osmose K33 C72	Injecta Osmose Ltd	Industrial	5890
Laporte CCA Awpa Type C	Laporte Wood Preservation	Industrial	3179
Laporte CCA Oxide Type 1	Laporte Wood Preservation	Industrial	3177
Laporte CCA Oxide Type 2	Laporte Wood Preservation	Industrial	3178
Osmose K33 C50	Protim Solignum Ltd	Industrial	6874
Osmose K33 C58.7	Protim Solignum Ltd	Industrial	6889
Osmose K33 C72	Protim Solignum Ltd	Industrial	6888
Protim CCA Oxide 50	Protim Solignum Ltd	Industrial	5537
Protim CCA Oxide 58	Protim Solignum Ltd	Industrial	5686
Protim CCA Oxide 72	Protim Solignum Ltd	Industrial	5594
Tanalith 3302	Hickson Timber Products Ltd	Industrial	5098
Tanalith 3313	Hickson Timber Products Ltd	Industrial	4422
D Tanalith Oxide C3309	Hickson Timber Products Ltd	Industrial	4668
Tanalith Oxide C3310	Hickson Timber Products Ltd	Industrial	4669
D Tanalith Oxide C3314	Hickson Timber Products Ltd	Industrial	4774
Tecca LQ1	Tecca Ltd	Industrial	5999

469 Arsenic Pentoxide + Copper Sulphate + Sodium Dichromate

Celcure A Concentrate	Protim Solignum Ltd	Industrial	5458
Celcure A Fluid 10	Protim Solignum Ltd	Industrial	3764
Celcure A Fluid 6	Protim Solignum Ltd	Industrial	3155
Celcure A Paste	Protim Solignum Ltd	Industrial	4523
Kemwood CCA Type BS	Laporte Kemwood AB	Industrial	5534
Laporte CCA Type 1	Laporte Wood Preservation	Industrial	3175
Laporte CCA Type 2	Laporte Wood Preservation	Industrial	3176
Laporte Permawood CCA	Laporte Wood Preservation	Industrial	3529
W Protim CCA Salts Type 2	Protim Solignum Ltd	Industrial	4972
Tanalith 3357	Hickson Timber Products Ltd	Industrial	4431
D Tanalith CL (3354)	Hickson Timber Products Ltd	Industrial	4196
Tanalith CP 3353	Hickson Timber Products Ltd	Industrial	4667
Tecca P2	Tecca Ltd	Industrial	3958

470 Azaconazole

Rodewod 50 SL	Janssen Pharmaceutica NV	Amateur Professional Industrial	3839
Safetray SL	Progress Products	Professional	5464

Product Name	Marketing Company	Use	HSE No.
471 Azaconazole + Dichlofluanid			
D Xyladecor Matt U 404	NWE Distributors Ltd	Amateur Professional	4445
D Xylamon Primer Dipping Stain U 415	Venilia Ltd	Professional	4444
472 Azaconazole + Permethrin			
Fongix SE Total Treatment For Wood	Liberon Waxes Ltd	Amateur (Surface Biocide) Amateur (Wood Preservative) Professional (Surface Biocide) Professional (Wood Preservative)	5610
Woodworm Killer And Rot Treatment	Liberon Waxes Ltd	Amateur (Surface Biocide) Amateur (Wood Preservative) Professional (Surface Biocide) Professional (Wood Preservative)	4826
D Xylamon Brown U 101 C	Venilia Ltd	Amateur Professional	4443
D Xylamon Curative U 152 G/H	NWE Distributors Ltd	Professional	4446
473 Benzalkonium Bromide			
X Moss	Thames Valley Specialist Products Ltd	Amateur Professional	6839
474 Benzalkonium Chloride			
Algae Remover	Akzo Nobel Woodcare	Amateur Professional	5337
Belzona 9421 Fungicidal Wash	Belzona Polymerics Limited	Professional	6911

Product Name	Marketing Company	Use	HSE No.
474 Benzalkonium Chloride—continued			
FMB 451-5 Quat (50%)	Midland Europe	Professional (Surface Biocide) Professional (Wood Preservative) Industrial (Wood Preservative) Industrial (Surface Biocide)	5734
Glen Wood Care Wood Preservative	Glen Wood Care	Amateur Professional Industrial	4096
Langlow Timbershield	Langlow Products Ltd	Amateur Professional	4394
Rhodaquat RP50	Rhodia HPCII UK Ltd	Professional (Surface Biocide) Professional (Wood Preservative)	5762
Timberdip	Langlow Products Ltd	Professional Industrial	4393
475 Benzalkonium Chloride + 2-Phenylphenol			
D Chel Dry Rot Killer For Masonry And Brickwork	Chelec Ltd	Professional	3067
476 Benzalkonium Chloride + Boric Acid			
Barrettine Premier Multicide Fluid	Barrettine Products Ltd	Amateur (Surface Biocide) Amateur (Wood Preservative)	6312

Product Name	Marketing Company	Use	HSE No.

476 Benzalkonium Chloride + Boric Acid—continued

| Microguard Mouldicidal Wood Preserver | Permagard Products Ltd | Amateur (Surface Biocide) Amateur (Wood Preservative) Professional (Surface Biocide) Professional (Wood Preservative) Industrial (Wood Preservative) | 5100 |

477 Benzalkonium Chloride + Boric Acid + Dialkyldimethyl Ammonium Chloride + Methylene Bis(Thiocyanate)

| Celbrite MT | Protim Solignum Ltd | Industrial | 4550 |

478 Benzalkonium Chloride + Copper Oxide

| Laporte ACQ 1900 | Laporte Wood Preservation | Industrial | 6487 |

479 Benzalkonium Chloride + Dialkyldimethyl Ammonium Chloride

ABL Aqueous Wood Preserver Concentrate 1:9	Advanced Bitumens Ltd	Professional Industrial	4199
Aqueous Wood Preserver	Advanced Bitumens Ltd	Amateur Professional Industrial	4727
Celbrite M	Protim Solignum Ltd	Industrial	4535
Celbronze B	Protim Solignum Ltd	Professional Industrial	4549
D Colourfast Protector	Rentokil Initial UK Ltd	Amateur Professional	4609

480 Benzalkonium Chloride + Disodium Octaborate

| Boracol 20 RH | Advanced Chemical Specialties Ltd | Professional | 4019 |

Product Name	Marketing Company	Use	HSE No.

480 Benzalkonium Chloride + Disodium Octaborate—continued

Product Name	Marketing Company	Use	HSE No.
Boracol B8.5 RH Mouldicide/Wood Preservative	Advanced Chemical Specialties Ltd	Amateur (Surface Biocide) Amateur (Wood Preservative) Professional (Surface Biocide) Professional (Wood Preservative)	4809
Boron Biocide 10	Philip Johnstone Group Ltd	Professional (Surface Biocide) Professional (Wood Preservative)	6485
Brunosol Boron 10	Philip Johnstone Group Ltd	Professional (Surface Biocide) Professional (Wood Preservative)	6488
Deepwood 20 Inorganic Boron Wood Preservative	Safeguard Chemicals Ltd	Professional	5514
Preservative Gel	Protim Solignum Ltd	Professional	6528
Probor 10	Safeguard Chemicals Ltd	Professional	6595
Probor 50	Safeguard Chemicals Ltd	Professional (Surface Biocide) Professional (Wood Preservative)	5596
Weathershield Mulit-Surface Fungicidal Wash	Imperial Chemical Industries PLC	Amateur Professional	6823

481 Benzalkonium Chloride + Disodium Octaborate + 3-Iodo-2-Propynyl-N-Butyl Carbamate

Antiblu Select	Hickson Timber Products Ltd	Industrial	6087

482 Benzalkonium Chloride + Lindane

W Biokil	Jaymar Chemicals	Professional	4247

483 Benzalkonium Chloride + Permethrin

W Biokil Emulsion	Jaymar Chemicals	Professional	4562

Product Name	Marketing Company	Use	HSE No.

484 Benzalkonium Chloride + Permethrin + Tebuconazole

Vacsol Aqua 6105	Hickson Timber Products Ltd	Industrial	5892

485 Benzalkonium Chloride + Tebuconazole

Vacsol Aqua 6101	Hickson Timber Products Ltd	Industrial	5557

486 Boric Acid

Celbor M	Protim Solignum Ltd	Industrial	4912
Control Fluid FB	Rentokil Initial UK Ltd	Professional	5323
Dricon	Hickson Timber Products Ltd	Industrial	5688
D Dricon	Lambson Ltd	Industrial	5315
PJG Boron Rods	Philip Johnstone Group Ltd	Professional	6616
Pyro-Red	Hoover Treated Wood Products Inc	Industrial	5867
Rentokil Dry Rot Fluid (E) For Bonded Warehouses	Rentokil Initial UK Ltd	Professional (Surface Biocide) Professional (Wood Preservative)	3986
Rentokil Wood Preserver	Rentokil Initial UK Ltd	Amateur Professional	6052

487 Boric Acid + 2-(Thiocyanomethylthio)Benzothiazole

Protim Fentex Europa 1 RFU	Protim Solignum Ltd	Industrial	5119
Protim Fentex Europa I	Protim Solignum Ltd	Industrial	4984

488 Boric Acid + 2-Phenylphenol

Protim Fentex P	Protim Solignum Ltd	Industrial	6095

489 Boric Acid + 3-Iodo-2-Propynyl-N-Butyl Carbamate

W Protim Fentex I	Protim Solignum Ltd	Industrial	6096

490 Boric Acid + Benzalkonium Chloride

Barrettine Premier Multicide Fluid	Barrettine Products Ltd	Amateur (Surface Biocide) Amateur (Wood Preservative)	6312

Product Name	Marketing Company	Use	HSE No.

490 Boric Acid + Benzalkonium Chloride—continued

| Microguard Mouldicidal Wood Preserver | Permagard Products Ltd | Amateur (Surface Biocide) Amateur (Wood Preservative) Professional (Surface Biocide) Professional (Wood Preservative) Industrial (Wood Preservative) | 5100 |

491 Boric Acid + Chromium Acetate + Copper Sulphate + Sodium Dichromate

| Celgard CF | Protim Solignum Ltd | Industrial | 4608 |

492 Boric Acid + Chromium Trioxide + Copper Oxide

| Tanalith (3419) CBC | Hickson Timber Products Ltd | Industrial | 4022 |

493 Boric Acid + Copper Carbonate Hydroxide

| W Tanalith 3487 | Hickson Timber Products Ltd | Industrial | 5305 |

494 Boric Acid + Copper Carbonate Hydroxide + Propiconazole

Ensele 3450	Hickson Timber Products Ltd	Amateur Professional	6992
Noah Gold	Protim Solignum Ltd	Industrial	6965
Thompson's End Grain Preserver	Ronseal Ltd	Amateur Professional	6991
Wolmanit CX-P	Dr Wolman GmbH	Industrial	6566

495 Boric Acid + Copper Carbonate Hydroxide + Propiconazole + Tebuconazole

| Tanalith E (3492) | Hickson Timber Products Ltd | Industrial | 6269 |
| Tanalith E (3494) | Hickson Timber Products Ltd | Industrial | 6434 |

496 Boric Acid + Copper Carbonate Hydroxide + Tebuconazole

| Ensele 3428 | Hickson Timber Products Ltd | Professional Industrial | 5878 |
| Tanalith E (3485) | Hickson Timber Products Ltd | Industrial | 5562 |

497 Boric Acid + Copper Oxide

| D Tanalith 3422 | Hickson Timber Products Ltd | Industrial | 4403 |

HSE
WOOD PRESERVATIVES

Product Name	Marketing Company	Use	HSE No.

498 Boric Acid + Copper Sulphate

Product Name	Marketing Company	Use	HSE No.
Ensele 3426	Hickson Timber Products Ltd	Professional Industrial	4364
D Ensele 3427	Hickson Timber Products Ltd	Amateur	5053
Ensele 3429	Hickson Timber Products Ltd	Amateur Professional	6392
Laporte End Grain Treatment	Laporte Wood Preservation	Amateur Professional Industrial	6885

499 Boric Acid + Copper Sulphate + Potassium Dichromate

Tecca CCB 1	Tecca Ltd	Industrial	5584

500 Boric Acid + Copper Sulphate + Sodium Dichromate

Celcure CB90	Protim Solignum Ltd	Industrial	5117
Laporte CCB	Laporte Wood Preservation	Industrial	5875
Tanalith CBC Paste 3402	Hickson Timber Products Ltd	Industrial	3833

501 Boric Acid + Cypermethrin + Propiconazole

Rentokil Total Wood Treatment	Rentokil Initial UK Ltd	Amateur Professional	6053

502 Boric Acid + Dialkyldimethyl Ammonium Chloride

Celbor P	Protim Solignum Ltd	Industrial	5923
Celbor P 25% Solution	Rentokil Initial UK Ltd	Industrial	6033
Celbor P25	Protim Solignum Ltd	Industrial	6563
Celbor P5	Protim Solignum Ltd	Industrial	6558
Protim E470	Protim Solignum Ltd	Industrial	6386

503 Boric Acid + Dialkyldimethyl Ammonium Chloride + Methylene Bis(Thiocyanate) + Benzalkonium Chloride

Celbrite MT	Protim Solignum Ltd	Industrial	4550

504 Boric Acid + Disodium Octaborate

ACS Woodkeeper Paste Preservative	Advanced Chemical Specialties Ltd	Professional	6221

505 Boric Acid + Disodium Tetraborate

Control Fluid SB	Rentokil Initial UK Ltd	Professional (Surface Biocide) Professional (Wood Preservative)	6018
PC-K	Terminix Peter Cox Ltd	Professional	4409
Pyrolith 3505 Ready To Use	Hickson Timber Products Ltd	Industrial	3636

Product Name	Marketing Company	Use	HSE No.

505 Boric Acid + Disodium Tetraborate—continued

| Rentokil Control Paste SB | Rentokil Initial UK Ltd | Professional | 6501 |

506 Boric Acid + Methylene Bis(Thiocyanate)

| Hickson Antiblu 3737 | Hickson Timber Products Ltd | Industrial | 4745 |
| Hickson Antiblu 3738 | Hickson Timber Products Ltd | Industrial | 4746 |

507 Boric Acid + Methylene Bis(Thiocyanate) + 2-(Thiocyanomethylthio)Benzothiazole

| Fentex Elite | Protim Solignum Ltd | Industrial | 5532 |

508 Boric Acid + Permethrin

Activ-8-II	Restoration (UK) Ltd	Professional Industrial	6170
Micro 8 FI	Lectros International Ltd	Professional Industrial	6131
Micro Dual Purpose 8	Grip-Fast Systems Ltd	Professional Industrial	6408
D Microguard FI Concentrate	Permagard Products Ltd	Professional Industrial	5101
Microguard FI Concentrate	Permagard Products Ltd	Professional Industrial	5854

509 Boric Acid + Permethrin + Zinc Octoate

D Celpruf BZP	Rentokil Initial UK Ltd	Industrial	4190
D Celpruf BZP WR	Rentokil Initial UK Ltd	Industrial	4191
W Cut'N'treat	Rentokil Initial UK Ltd	Amateur Professional	5802
W Premium Grade Wood Treatment	Rentokil Initial UK Ltd	Amateur Professional	4193

510 Boric Acid + Propiconazole

| Celbor PR | Protim Solignum Ltd | Industrial | 6192 |

511 Boric Acid + Sodium 2,4,6-Trichlorophenoxide

| Protim Fentex TWR Plus | Protim Solignum Ltd | Industrial | 6547 |

512 Boric Acid + Sodium Tetraborate

| Celbor | Protim Solignum Ltd | Industrial | 4614 |
| Diffusit | Dr Wolman GmbH | Professional | 5992 |

513 Boric Acid + Zinc Octoate

D Celpruf BZ	Rentokil Initial UK Ltd	Industrial	4188
D Celpruf BZ WR	Rentokil Initial UK Ltd	Industrial	4189
W Rentokil Dry Rot Paste (D)	Rentokil Initial UK Ltd	Professional	3987

HSE WOOD PRESERVATIVES

Product Name	Marketing Company	Use	HSE No.

514 Boric Oxide

ACS Boron Rods	Advanced Chemical Specialties Ltd	Amateur Professional	6115

515 Carbendazim

Injecta Osmose ABS 33 A	Injecta Osmose Ltd	Industrial	6533
Osmose ABS33A	Protim Solignum Ltd	Industrial	6887

516 Carbendazim + Disodium Octaborate

Weathershield Aquatech Preservative Basecoat	ICI Paints	Amateur Professional	5757

517 5-Chloro-2-Methyl-4-Isothiazolin-3-One + 2-Methyl-4-Isothiazolin-3-One

Celkil 90	Protim Solignum Ltd	Industrial	5428
Kathon 886F	Rohm & Haas (UK) Ltd	Industrial	5431
Laporte Mould-Ex	Laporte Wood Preservation	Industrial	5430
Osmose ABS33	Protim Solignum Ltd	Industrial	6909
Tanamix 3743	Hickson Timber Products Ltd	Industrial	5429

518 5-Chloro-2-Methyl-4-Isothiazolin-3-One + Copper Sulphate + 2-Methyl-4-Isothiazolin-3-One

Injecta Osmose ABS33	Injecta Osmose Ltd	Industrial	5978

519 Chromium Acetate + Copper Sulphate + Sodium Dichromate

Celcure B	Protim Solignum Ltd	Industrial	4541
Celcure O	Protim Solignum Ltd	Industrial	4539

520 Chromium Acetate + Copper Sulphate + Sodium Dichromate + Boric Acid

Celgard CF	Protim Solignum Ltd	Industrial	4608

521 Chromium Trioxide + Copper Oxide

Osmose K55	Protim Solignum Ltd	Industrial	6867

522 Chromium Trioxide + Copper Oxide + Arsenic Pentoxide

Celcure AO	Protim Solignum Ltd	Industrial	4220
Celcure C6099	Protim Solignum Ltd	Industrial	6825
Celcure CCA Type C	Protim Solignum Ltd	Industrial	5104
Celcure CCA Type C-40%	Protim Solignum Ltd	Industrial	6028
Celcure CCA Type C-50%	Protim Solignum Ltd	Industrial	6030
Celcure CCA Type C-60%	Protim Solignum Ltd	Industrial	5934
Celcure Type C6099	Protim Solignum Ltd	Industrial	6920
Injecta Osmose K33 C50	Injecta Osmose Ltd	Industrial	5889
Injecta Osmose K33 C58.7	Injecta Osmose Ltd	Industrial	6498
Injecta Osmose K33 C72	Injecta Osmose Ltd	Industrial	5890
Laporte CCA AWPA Type C	Laporte Wood Preservation	Industrial	3179
Laporte CCA Oxide Type 1	Laporte Wood Preservation	Industrial	3177

Product Name	Marketing Company	Use	HSE No.

522 Chromium Trioxide + Copper Oxide + Arsenic Pentoxide—continued

Laporte CCA Oxide Type 2	Laporte Wood Preservation	Industrial	3178
Osmose K33 C50	Protim Solignum Ltd	Industrial	6874
Osmose K33 C58.7	Protim Solignum Ltd	Industrial	6889
Osmose K33 C72	Protim Solignum Ltd	Industrial	6888
Protim CCA Oxide 50	Protim Solignum Ltd	Industrial	5537
Protim CCA Oxide 58	Protim Solignum Ltd	Industrial	5686
Protim CCA Oxide 72	Protim Solignum Ltd	Industrial	5594
Tanalith 3302	Hickson Timber Products Ltd	Industrial	5098
Tanalith 3313	Hickson Timber Products Ltd	Industrial	4422
D Tanalith Oxide C3309	Hickson Timber Products Ltd	Industrial	4668
Tanalith Oxide C3310	Hickson Timber Products Ltd	Industrial	4669
D Tanalith Oxide C3314	Hickson Timber Products Ltd	Industrial	4774
Tecca LQ1	Tecca Ltd	Industrial	5999

523 Chromium Trioxide + Copper Oxide + Boric Acid

Tanalith (3419) CBC	Hickson Timber Products Ltd	Industrial	4022

524 Coal Tar Creosote

ABL Brown Creosote	Advanced Bitumens Ltd	Professional Industrial	3859
Bartoline Dark And Light Creosote	Bartoline Ltd	Amateur Professional	4459
Bitmac Creosote	Bitmac Ltd	Professional Industrial	5882
BS 144 Creosote	Oakmere Technical Services Ltd	Industrial	3712
Carbo Creosote	Talke Chemical Company Ltd	Amateur Professional Industrial	4362
D Coal Tar Creosote	Creohaul And Company	Amateur	3898
Coal Tar Creosote	Great Marsh Ltd	Professional Industrial	4574
W Coal Tar Creosote	Hardmans Of Hull Ltd	Amateur Professional	4917
Coal Tar Creosote	James Of Bedlington Ltd	Amateur Professional Industrial	6031
W Coal Tar Creosote	William Mathwin And Son (Newcastle) Ltd	Amateur Professional Industrial	4602
Creosote	Bagnalls Haulage Ltd	Amateur Professional	5975
Creosote	Bartoline Ltd	Amateur Professional Industrial	7026

Product Name	Marketing Company	Use	HSE No.
524 Coal Tar Creosote—continued			
W Creosote	Blanchard Martin And Simmonds Ltd	Amateur	3884
D Creosote	C W Wastnage Ltd	Amateur Professional	5267
D Creosote	G And B Fuels Ltd	Amateur Professional Industrial	4469
Creosote	Great Mills (Retail) Ltd	Amateur Professional	5666
Creosote	Liver Grease Oil And Chemical Co Ltd	Amateur Professional Industrial	4467
Creosote	PMC Holdings Ltd	Amateur Professional Industrial	5495
Creosote	R K And J Jones Ltd	Amateur Professional Industrial	4384
Creosote	T K Bird Ltd	Amateur Professional Industrial	4903
Creosote	Witham Oil & Paint (Lowestoft) Ltd	Amateur Professional	6901
Creosote Blend	Laybond Products Ltd	Amateur Professional Industrial	4948
Creosote Blend Mk1 (Medium Dark)	Great Marsh Ltd	Amateur Professional	4588
Creosote Blend Mk3 (Light Golden)	Great Marsh Ltd	Amateur Professional	4590
Creosote Blended Wood Preservative	Rye Oil Ltd	Amateur Professional	3730
Creosote BS 144	Bitmac Ltd	Amateur Professional Industrial	5885
Creosote BS 144 (Type I And Type II)	Coalite Chemicals Division	Industrial	3811
Creosote BS 144 Type III	Coalite Chemicals Division	Amateur Professional Industrial	3812
Creosote BS 144(3)	Great Marsh Ltd	Professional Industrial	5279
Greenhills Creosote	Greenhills (Wessex) Ltd	Amateur Professional Industrial	4205
W Langlow Creosote	Langlow Products Ltd	Amateur Professional	4639

Product Name	Marketing Company	Use	HSE No.

524 Coal Tar Creosote—continued

Product Name	Marketing Company	Use	HSE No.
Larsen Creosote	Larsen Manufacturing Ltd	Amateur Professional Industrial	4930
D Light Creosote Blend	Bitmac Ltd	Amateur Professional	5806
D Medium Brown Creosote	C W Wastnage Ltd	Amateur Professional	4691
Middletons Creosote	E W Middleton And Sons	Amateur Professional	4441
Middletons Enviro. Shed & Fence Wood Preservative	E W Middleton And Sons	Amateur Professional	4289
Ovoline 275 Golden Creosote	Bretts Oils Ltd	Amateur Professional Industrial	3651
Solignum Dark Brown	Protim Solignum Ltd	Amateur Professional Industrial	5023
Wickes Creosote	Wickes Building Supplies Ltd	Amateur Professional	6101
Wilko Creosote	T K Bird Ltd	Amateur Professional Industrial	5043
Wood Preservative Type 3	Great Marsh Ltd	Professional Industrial	5252

525 Copper Carbonate Hydroxide + Boric Acid

Product Name	Marketing Company	Use	HSE No.
W Tanalith 3487	Hickson Timber Products Ltd	Industrial	5305

526 Copper Carbonate Hydroxide + Propiconazole + Boric Acid

Product Name	Marketing Company	Use	HSE No.
Ensele 3450	Hickson Timber Products Ltd	Amateur Professional	6992
Noah Gold	Protim Solignum Ltd	Industrial	6965
Thompson's End Grain Preserver	Ronseal Ltd	Amateur Professional	6991
Wolmanit CX-P	Dr Wolman GmbH	Industrial	6566

527 Copper Carbonate Hydroxide + Propiconazole + Tebuconazole + Boric Acid

Product Name	Marketing Company	Use	HSE No.
Tanalith E (3492)	Hickson Timber Products Ltd	Industrial	6269
Tanalith E (3494)	Hickson Timber Products Ltd	Industrial	6434

528 Copper Carbonate Hydroxide + Tebuconazole + Boric Acid

Product Name	Marketing Company	Use	HSE No.
Ensele 3428	Hickson Timber Products Ltd	Professional Industrial	5878
Tanalith E (3485)	Hickson Timber Products Ltd	Industrial	5562

Product Name	Marketing Company	Use	HSE No.

529 Copper Naphthenate

Product Name	Marketing Company	Use	HSE No.
W Bio-Kil Cunap Pole Wrap	Bio-Kil Chemicals Ltd	Professional	5341
W Blackfriar Wood Preserver (WP Green)	E Parsons And Sons Ltd	Amateur Professional	6427
W Blackfriars Green Wood Preserver	E Parsons And Sons Ltd	Amateur Professional Industrial	4209
W Carbo Wood Preservative Green	Talke Chemical Company Ltd	Amateur Professional Industrial	3820
W Cromar Wood Preserver Green	Cromar Building Products Ltd	Amateur Professional	6257
W Flag Brand Wood Preservative Green	C W Wastnage Ltd	Amateur Professional Industrial	3073
W Glen Wood Care Green	Glen Wood Care	Amateur Professional Industrial	4095
W Green Plus Wood Preserver	Philip Johnstone Group Ltd	Amateur Professional	5040
W Green Wood Preservative	Advanced Bitumens Ltd	Amateur Professional	4728
W Green Wood Preservative	Strathbond Ltd	Professional	4660
W Langlow Wood Preservative Green	Langlow Products Ltd	Amateur Professional Industrial	4395
W Larsen Green Wood Preservative	Larsen Manufacturing Ltd	Amateur Professional Industrial	3708
W Laybond Green Wood Preserver	Laybond Products Ltd	Amateur Professional	5372
W Leyland Timbrene Green Environmental Formula	Leyland Paint Company Ltd	Amateur Professional	4799
W Premier Barrettine Green Wood Preserver	Barrettine Products Ltd	Amateur Professional Industrial	4159
W Preservative For Wood Green	Rentokil Initial UK Ltd	Amateur Professional	4613
W Protim Green WR	Protim Solignum Ltd	Amateur Professional Industrial	4976
W Ronseal's All-Purpose Wood Preserver Green	Ronseal Ltd	Amateur Professional	6187
W Supergrade Wood Preserver Green	Rentokil Initial UK Ltd	Professional Industrial	3717
W Teamac Woodtec Green	Teal And Mackrill Ltd	Amateur Professional	4429

Product Name	Marketing Company	Use	HSE No.

529 Copper Naphthenate—continued

W Thompson's All Purpose Wood Preserver Green	Ronseal Ltd	Amateur Professional	5989
W Timber Preservative Green	Antel UK Ltd	Professional	4847
W Timbrene Green Wood Preserver	Kalon Group PLC	Amateur Professional	4967
W Wood Preserver Green	Philip Johnstone Group Ltd	Amateur Professional	4927

530 Copper Naphthenate + 3-Iodo-2-Propynyl-N-Butyl Carbamate

| W Osmose Endcoat Green | Injecta Osmose Ltd | Amateur Professional Industrial | 6562 |
| W Sovereign Green Timber Preservative/Dipcoat | Sovereign Chemical Industries Ltd | Amateur Professional Industrial | 6341 |

531 Copper Naphthenate + Disodium Octaborate

| W Bio-Kil SR Pole Wrap | Bio-Kil Chemicals Ltd | Professional | 5675 |

532 Copper Naphthenate + Permethrin

W Green	Protim Solignum Ltd	Amateur Professional Industrial	5483
W Protim Green E	Protim Solignum Ltd	Amateur Professional Industrial	4975
W Solignum Green	Protim Solignum Ltd	Amateur Professional Industrial	5020

533 Copper Naphthenate + Tri(Hexylene Glycol)Biborate

W BCR Green Wood Preserver	Building Chemical Research Ltd	Amateur Professional Industrial	6285
W Dark Green Wood Preservative	Palace Chemicals Ltd	Amateur Professional Industrial	5243
W Kingfisher Wood Preservative	Kingfisher Chemicals Ltd	Professional	4011
W Langlow Green Wood Preserver	Langlow Products Ltd	Amateur Professional Industrial	6279
W Langlow Pale Green Wood Preserver	Langlow Products Ltd	Amateur Professional Industrial	6068

HSE WOOD PRESERVATIVES

Product Name	Marketing Company	Use	HSE No.

533 Copper Naphthenate + Tri(Hexylene Glycol)Biborate—continued

W Pale Green Wood Preservative	Palace Chemicals Ltd	Amateur Professional Industrial	5241
W Sovereign Timber Preservative Pale Green	Sovereign Chemical Industries Ltd	Professional	3806

534 Copper Naphthenate + Zinc Octoate

W Weathershield Exterior Timber Preservative	ICI Paints	Amateur Professional	4915

535 Copper Oxide + Arsenic Pentoxide + Chromium Trioxide

Celcure AO	Protim Solignum Ltd	Industrial	4220
Celcure C6099	Protim Solignum Ltd	Industrial	6825
Celcure CCA Type C	Protim Solignum Ltd	Industrial	5104
Celcure CCA Type C-40%	Protim Solignum Ltd	Industrial	6028
Celcure CCA Type C-50%	Protim Solignum Ltd	Industrial	6030
Celcure CCA Type C-60%	Protim Solignum Ltd	Industrial	5934
Celcure Type C6099	Protim Solignum Ltd	Industrial	6920
Injecta Osmose K33 C50	Injecta Osmose Ltd	Industrial	5889
Injecta Osmose K33 C58.7	Injecta Osmose Ltd	Industrial	6498
Injecta Osmose K33 C72	Injecta Osmose Ltd	Industrial	5890
Laporte CCA AWPA Type C	Laporte Wood Preservation	Industrial	3179
Laporte CCA Oxide Type 1	Laporte Wood Preservation	Industrial	3177
Laporte CCA Oxide Type 2	Laporte Wood Preservation	Industrial	3178
Osmose K33 C50	Protim Solignum Ltd	Industrial	6874
Osmose K33 C58.7	Protim Solignum Ltd	Industrial	6889
Osmose K33 C72	Protim Solignum Ltd	Industrial	6888
Protim CCA Oxide 50	Protim Solignum Ltd	Industrial	5537
Protim CCA Oxide 58	Protim Solignum Ltd	Industrial	5686
Protim CCA Oxide 72	Protim Solignum Ltd	Industrial	5594
Tanalith 3302	Hickson Timber Products Ltd	Industrial	5098
Tanalith 3313	Hickson Timber Products Ltd	Industrial	4422
D Tanalith Oxide C3309	Hickson Timber Products Ltd	Industrial	4668
Tanalith Oxide C3310	Hickson Timber Products Ltd	Industrial	4669
D Tanalith Oxide C3314	Hickson Timber Products Ltd	Industrial	4774
Tecca LQ1	Tecca Ltd	Industrial	5999

536 Copper Oxide + Benzalkonium Chloride

Laporte ACQ 1900	Laporte Wood Preservation	Industrial	6487

537 Copper Oxide + Boric Acid

D Tanalith 3422	Hickson Timber Products Ltd	Industrial	4403

538 Copper Oxide + Boric Acid + Chromium Trioxide

Tanalith (3419) CBC	Hickson Timber Products Ltd	Industrial	4022

Product Name	Marketing Company	Use	HSE No.

539 Copper Oxide + Chromium Trioxide

Osmose K55	Protim Solignum Ltd	Industrial	6867

540 Copper Oxide + Dialkyldimethyl Ammonium Chloride

Laporte ACQ 2100	Laporte Wood Preservation	Industrial	6490

541 Copper Sulphate + 2-Methyl-4-Isothiazolin-3-One + 5-Chloro-2-Methyl-4-Isothiazolin-3-One

Injecta Osmose ABS33	Injecta Osmose Ltd	Industrial	5978

542 Copper Sulphate + Boric Acid

Ensele 3426	Hickson Timber Products Ltd	Professional Industrial	4364
D Ensele 3427	Hickson Timber Products Ltd	Amateur	5053
Ensele 3429	Hickson Timber Products Ltd	Amateur Professional	6392
Laporte End Grain Treatment	Laporte Wood Preservation	Amateur Professional Industrial	6885

543 Copper Sulphate + Potassium Dichromate + Boric Acid

Tecca CCB 1	Tecca Ltd	Industrial	5584

544 Copper Sulphate + Sodium Dichromate

Laporte Cut End Preservative	Laporte Wood Preservation	Industrial	3528

545 Copper Sulphate + Sodium Dichromate + Arsenic Pentoxide

Celcure A Concentrate	Protim Solignum Ltd	Industrial	5458
Celcure A Fluid 10	Protim Solignum Ltd	Industrial	3764
Celcure A Fluid 6	Protim Solignum Ltd	Industrial	3155
Celcure A Paste	Protim Solignum Ltd	Industrial	4523
Kemwood CCA Type BS	Laporte Kemwood AB	Industrial	5534
Laporte CCA Type 1	Laporte Wood Preservation	Industrial	3175
Laporte CCA Type 2	Laporte Wood Preservation	Industrial	3176
Laporte Permawood CCA	Laporte Wood Preservation	Industrial	3529
W Protim CCA Salts Type 2	Protim Solignum Ltd	Industrial	4972
Tanalith 3357	Hickson Timber Products Ltd	Industrial	4431
D Tanalith CL (3354)	Hickson Timber Products Ltd	Industrial	4196
Tanalith CP 3353	Hickson Timber Products Ltd	Industrial	4667
Tecca P2	Tecca Ltd	Industrial	3958

546 Copper Sulphate + Sodium Dichromate + Boric Acid

Celcure CB90	Protim Solignum Ltd	Industrial	5117
Laporte CCB	Laporte Wood Preservation	Industrial	5875
Tanalith CBC Paste 3402	Hickson Timber Products Ltd	Industrial	3833

Product Name	Marketing Company	Use	HSE No.

547 Copper Sulphate + Sodium Dichromate + Boric Acid + Chromium Acetate

Celgard CF	Protim Solignum Ltd	Industrial	4608

548 Copper Sulphate + Sodium Dichromate + Chromium Acetate

Celcure B	Protim Solignum Ltd	Industrial	4541
Celcure O	Protim Solignum Ltd	Industrial	4539

549 Creosote

B And Q Creosote	B & Q PLC	Amateur Professional	5093
Barrettine Creosote	Barrettine Products Ltd	Amateur Professional Industrial	3146
Chelec Creosote	Chelec Ltd	Amateur Professional	5412
Cindu Creosote	Cindu Chemicals BV	Amateur Professional Industrial	6001
D Creosote	Hilbre Building Chemicals Ltd	Amateur Professional	5301
Creosote	Langlow Products Ltd	Amateur Professional	4034
Creosote	Palace Chemicals Ltd	Amateur Professional Industrial	3841
W Creosote	Suffolk & Essex Supplies Ltd	Amateur Professional	3141
Creosote Blend	Langlow Products Ltd	Amateur Professional	4392
Creosote Coke Oven Oil	Great Marsh Ltd	Professional	4586
D Creosote Emulsion	Oakmere Technical Services Ltd	Industrial	5345
Creosote MK2	Great Marsh Ltd	Professional Industrial	4589
DIY Time Creosote	Nurdin And Peacock	Amateur Professional	5314
Golden Creosote	A-Chem Ltd	Amateur Professional	4334
D Golden Creosote	C W Wastnage Ltd	Amateur Professional	4690
Homebase Creosote	Sainsbury's Homebase House And Garden Centres	Amateur Professional	5054
Jewson Creosote	Jewson Ltd	Amateur Professional	5072
Kalon Creosote	Kalon Group PLC	Amateur Professional	5496

Product Name	Marketing Company	Use	HSE No.

549 Creosote—continued

Keyline Creosote	Keyline	Amateur Professional	7054
Laybond Creosote Blend	Laybond Products Ltd	Amateur Professional Industrial	6211
Leyland Creosote	Leyland Paint Company Ltd	Amateur Professional	5074
Maxim Creosote	S J Dixon	Amateur Professional	5848
Nitromors Creosote	Kalon Group PLC	Amateur Professional	5129
Nut Brown Creosote	A-Chem Ltd	Amateur Professional	4335
Oakmere Creosote Type 2	Oakmere Technical Services Ltd	Amateur Professional Industrial	4390
D Signpost Creosote	Macpherson Paints Ltd	Amateur Professional	4507
Solignum Fencing Fluid	Protim Solignum Ltd	Industrial	5016
W Southdown Creosote	C Brewer And Sons Ltd	Amateur Professional	4310
Texas Creosote	Texas Homecare Ltd	Amateur Professional	5092
Travis Perkins Creosote	Travis Perkins Trading Company Ltd	Amateur Professional	4941
W Valspar Creosote	Akzo Coatings PLC	Amateur Professional	5469
Woodman Creosote And Light Brown Creosote	J H Woodman Ltd	Amateur Professional Industrial	5120

550 Cyfluthrin

Bayer WPC-5	Bayer PLC	Professional	6534

551 Cypermethrin

Cementone Woodworm Killer	Philip Johnstone Group Ltd	Amateur	3850
Crown Micro Insecticide	Crown Chemicals Ltd	Professional	6413
Crown Woodworm Concentrate	Crown Chemicals Ltd	Professional	5470
Devatern 0.5 L	Sorex Ltd	Amateur	3874
Devatern 1.0 L	Sorex Ltd	Professional Industrial	3876
Devatern EC	Sorex Ltd	Professional Industrial	3875
Green Range Woodworm Killer	Philip Johnstone Group Ltd	Professional	3711
Green Range Woodworm Killer AQ	Philip Johnstone Group Ltd	Professional	3688
Hickson Antiborer 3767	Hickson Timber Products Ltd	Industrial	3583

HSE
WOOD PRESERVATIVES

491

Product Name	Marketing Company	Use	HSE No.

551 Cypermethrin—continued

Product Name	Marketing Company	Use	HSE No.
Killgerm Woodworm Killer	Killgerm Chemicals Ltd	Professional	3126
Lignosol P	Microsol	Professional	5832
Microtech Woodworm Killer AQ	Philip Johnstone Group Ltd	Professional	5452
Nubex Emulsion Concentrate C (Low Odour)	Philip Johnstone Group Ltd	Professional	5268
Palace Microfine Insecticide Concentrate	Palace Chemicals Ltd	Amateur Professional Industrial	5550
D PC-H/3	Terminix Peter Cox Ltd	Professional	4408
W PCX-12/P	Terminix Peter Cox Ltd	Professional	4407
W PCX-12/P Concentrate	Terminix Peter Cox Ltd	Professional	4406
Water Based Woodworm Killer	Philip Johnstone Group Ltd	Amateur Professional	4874
Waterbased Woodworm Killer 5X	Philip Johnstone Group Ltd	Amateur Professional	5861
Woodworm Killer	Oakmere Technical Services Ltd	Amateur Professional Industrial	4371
Woodworm Killer	Palace Chemicals Ltd	Amateur Professional	3769

552 Cypermethrin + 2-Phenylphenol

Product Name	Marketing Company	Use	HSE No.
Laybond Timber Protector	Laybond Products Ltd	Amateur Professional	5369

553 Cypermethrin + 3-Iodo-2-Propynyl-N-Butyl Carbamate

Product Name	Marketing Company	Use	HSE No.
Microtech Dual Purpose AQ	Philip Johnstone Group Ltd	Professional	5453
W Water Based Wood Preserver (Concentrate)	Philip Johnstone Group Ltd	Amateur Professional	5630
Water-Based Wood Preserver	Philip Johnstone Group Ltd	Amateur Professional	4902

554 Cypermethrin + Dichlofluanid

Product Name	Marketing Company	Use	HSE No.
Universal Wood Preservative	Oakmere Technical Services Ltd	Amateur Professional Industrial	4170

555 Cypermethrin + Propiconazole

Product Name	Marketing Company	Use	HSE No.
Celpruf B	Protim Solignum Ltd	Industrial	5935
Cuprinol Quick Drying Clear Preserver	Cuprinol Ltd	Amateur Professional Industrial	6014
W Cuprinol Wood Defender	Cuprinol Ltd	Amateur Professional Industrial	5798

Product Name	Marketing Company	Use	HSE No.

555 Cypermethrin + Propiconazole—continued

Wocosen 100 SL-C	Janssen Pharmaceutica NV	Professional Industrial	5807

556 Cypermethrin + Propiconazole + Boric Acid

Rentokil Total Wood Treatment	Rentokil Initial UK Ltd	Amateur Professional	6053

557 Cypermethrin + Tri(Hexylene Glycol)Biborate

BCR Universal Wood Preserver	Building Chemical Research Ltd	Amateur Professional	6290
Cementone Multiplus	Philip Johnstone Group Ltd	Amateur Professional	3849
Crown Fungicide Insecticide	Crown Chemicals Ltd	Professional	6003
Crown Fungicide Insecticide Concentrate	Crown Chemicals Ltd	Professional	5463
Crown Micro Dual Purpose	Crown Chemicals Ltd	Professional	6414
Devatern Wood Preserver	Sorex Ltd	Professional Industrial	4073
Ecology Fungicide Insecticide Concentrate	Palace Chemicals Ltd	Professional	3792
Fungicide Insecticide	Palace Chemicals Ltd	Amateur Professional	3795
Green Range Dual Purpose AQ	Philip Johnstone Group Ltd	Professional	3725
Green Range Wykamol Plus	Philip Johnstone Group Ltd	Professional	3731
W Langlow Clear Wood Preserver	Langlow Products Ltd	Amateur Professional Industrial	3802
Nubex Emulsion Concentrate CB (Low Odour)	Philip Johnstone Group Ltd	Professional	5328
Nubex Woodworm All Purpose CB	Philip Johnstone Group Ltd	Professional	3189
Palace Microfine Fungicide Insecticide (Concentrate)	Palace Chemicals Ltd	Professional Industrial	5421
D PC-H/4	Terminix Peter Cox Ltd	Amateur Professional	3740
W PCX-122	Terminix Peter Cox Ltd	Professional	3999
W PCX-122 Concentrate	Terminix Peter Cox Ltd	Professional	4000

558 Cypermethrin + Zinc Versatate

W Dipsar G R	Philip Johnstone Group Ltd	Industrial	4449
W Everclear	Injecta Osmose Ltd	Industrial	6058

559 Deltamethrin

K-Otec AL	Aventis Environmental Science	Amateur Professional	6570
K-Otek EC	Aventis Environmental Science	Professional	6568

HSE
WOOD PRESERVATIVES

Product Name	Marketing Company	Use	HSE No.

559 Deltamethrin—continued

| K-Otek SL | Aventis Environmental Science | Professional | 6569 |

560 Dialkyldimethyl Ammonium Chloride

| X Thallo | Thames Valley Specialist Products Ltd | Amateur Professional | 6841 |

561 Dialkyldimethyl Ammonium Chloride + 3-Iodo-2-Propynyl-N-Butyl Carbamate

| W Hickson NP-1 | Hickson Timber Products Ltd | Industrial | 5401 |

562 Dialkyldimethyl Ammonium Chloride + Benzalkonium Chloride

ABL Aqueous Wood Preserver Concentrate 1:9	Advanced Bitumens Ltd	Professional Industrial	4199
Aqueous Wood Preserver	Advanced Bitumens Ltd	Amateur Professional Industrial	4727
Celbrite M	Protim Solignum Ltd	Industrial	4535
Celbronze B	Protim Solignum Ltd	Professional Industrial	4549
D Colourfast Protector	Rentokil Initial UK Ltd	Amateur Professional	4609

563 Dialkyldimethyl Ammonium Chloride + Boric Acid

Celbor P	Protim Solignum Ltd	Industrial	5923
Celbor P 25% Solution	Rentokil Initial UK Ltd	Industrial	6033
Celbor P25	Protim Solignum Ltd	Industrial	6563
Celbor P5	Protim Solignum Ltd	Industrial	6558
Protim E470	Protim Solignum Ltd	Industrial	6386

564 Dialkyldimethyl Ammonium Chloride + Copper Oxide

| Laporte ACQ 2100 | Laporte Wood Preservation | Industrial | 6490 |

565 Dialkyldimethyl Ammonium Chloride + Disodium Octaborate

| Ecogel | Restoration (UK) Ltd | Professional | 6847 |
| Probor 20 Brushable Gel | Safeguard Chemicals Ltd | Professional | 6422 |

566 Dialkyldimethyl Ammonium Chloride + Methylene Bis(Thiocyanate) + Benzalkonium Chloride + Boric Acid

| Celbrite MT | Protim Solignum Ltd | Industrial | 4550 |

567 Dialkyldimethyl Ammonium Chloride + Permethrin + Propiconazole

| Traditional Preservative Oil | Oakmasters Of Sussex | Amateur Professional Industrial | 6023 |

Product Name	Marketing Company	Use	HSE No.

568 Dichlofluanid

Product Name	Marketing Company	Use	HSE No.
ABL Wood Preservative (D)	Advanced Bitumens Ltd	Amateur Professional Industrial	3744
Cedarwood Protector	Rentokil Initial UK Ltd	Amateur Professional Industrial	4587
Cedarwood Special	Cuprinol Ltd	Amateur Professional Industrial	4729
Cromar Wood Preserver	Cromar Building Products Ltd	Amateur Professional Industrial	6243
Cuprinol Hardwood Basecoat	Cuprinol Ltd	Amateur Professional Industrial	5015
Cuprinol Hardwood Basecoat Meranti	Cuprinol Ltd	Industrial	4707
Cuprinol Preservative-Wood Hardener	Cuprinol Ltd	Amateur Professional	3590
D Dualprime F	Cuprinol Ltd	Industrial	4706
Flexarb Timber Coating	Macpherson Paints Ltd	Amateur Professional	4861
Forsham's Re-Treat	Forsham Cottage Arks	Amateur	6281
Hardwood Protector	Rentokil Initial UK Ltd	Amateur Professional Industrial	4580
Impra-Color	Impra Systems Ltd	Amateur Professional	5403
Intertox	International Coatings Ltd	Amateur	5217
Langlow Wood Preserver Formulation A	Langlow Products Ltd	Amateur Professional Industrial	3753
Lister Teak Dressing	Lister Lutyens Company Ltd	Amateur Professional	4628
New Formula Cedarwood	Cuprinol Ltd	Amateur Professional Industrial	4699
Regency Garden Buildings Wood Preserver	Regency Garden Buildings	Amateur Professional Industrial	6530
RWN Building Products Wood Preserver	RWN Building Products	Amateur Professional	6523
Sigmalife Impregnant	Philip Johnstone Group Ltd	Professional	6204
Timber Preservative	Philip Johnstone Group Ltd	Amateur Professional	6317
Timberlife Extra	Palace Chemicals Ltd	Amateur Professional	5352

Product Name	Marketing Company	Use	HSE No.

568 Dichlofluanid—continued

Product Name	Marketing Company	Use	HSE No.
Ultrabond Wood Preservative	Philip Johnstone Group Ltd	Amateur Professional	4343
W Wood Preservative	Langlow Products Ltd	Amateur Professional Industrial	4600
Wood Preservative	Oakmere Technical Services Ltd	Amateur Professional Industrial	4370

569 Dichlofluanid + Acypetacs Zinc

Product Name	Marketing Company	Use	HSE No.
W Cuprinol Decorative Preserver	Cuprinol Ltd	Amateur Professional Industrial	5343
Cuprinol Decorative Preserver	Cuprinol Ltd	Amateur Professional Industrial	6277
W Cuprinol Decorative Preserver Red Cedar	Cuprinol Ltd	Amateur Professional Industrial	5342
Cuprinol Decorative Preserver Red Cedar	Cuprinol Ltd	Amateur Professional Industrial	6280
Cuprinol Garden Decking Protector	Cuprinol Ltd	Amateur Professional	6862
Cuprinol Preservative Base	Cuprinol Ltd	Amateur Professional	4929
Weathershield Exterior Preservative Primer	Imperial Chemical Industries PLC	Amateur Professional	6983
Weathershield Preservative Basecoat	ICI Paints	Amateur Professional	6982
Wickes Decorative Wood Preserver	Wickes Building Supplies Ltd	Amateur Professional Industrial	6419
Wickes Decorative Wood Preserver Rich Cedar	Wickes Building Supplies Ltd	Amateur Professional Industrial	6420

570 Dichlofluanid + Acypetacs Zinc (Equivalent To 3% Zinc Metal)

Product Name	Marketing Company	Use	HSE No.
Cuprinol Garden Furniture Preserver	Cuprinol Ltd	Amateur	7123

571 Dichlofluanid + Azaconazole

Product Name	Marketing Company	Use	HSE No.
D Xyladecor Matt U 404	NWE Distributors Ltd	Amateur Professional	4445
D Xylamon Primer Dipping Stain U 415	Venilia Ltd	Professional	4444

Product Name	Marketing Company	Use	HSE No.

572 Dichlofluanid + Cypermethrin

Universal Wood Preservative	Oakmere Technical Services Ltd	Amateur Professional Industrial	4170

573 Dichlofluanid + Permethrin + Tebuconazole

Impra-Holzschutzgrund (Primer)	Impra Systems Ltd	Professional Industrial	5735

574 Dichlofluanid + Permethrin + Tri(Hexylene Glycol)Biborate

Blackfriar Wood Preserver WP Gold Star Clear	E Parsons And Sons Ltd	Amateur Professional	6417
Blackfriars Gold Star Clear	E Parsons And Sons Ltd	Amateur Professional Industrial	4149
Johnstone's Woodworks Interior/ Exterior All Purpose Preserver	Kalon Decorative Products	Amateur Professional	7071
Leyland Timbrene Supreme Environmental Formula	Leyland Paint Company Ltd	Amateur Professional	4800
D Nitromors Timbrene Supreme Environmental Formula	Henkel Home Improvements And Adhesive Products	Amateur Professional Industrial	4107
Premier Barrettine New Universal Fluid D	Barrettine Products Ltd	Amateur Professional Industrial	4004
Timber Master 6 In 1 Wood Treatment	Sealocrete PLA Ltd	Amateur Professional	7080
Timbrene Supreme	Kalon Group PLC	Amateur Professional	4995

575 Dichlofluanid + Permethrin + Zinc Octoate

W OS Color Wood Stain And Preservative	Ostermann And Scheiwe GmbH	Amateur	5531
W OS Color WR	Ostermann And Scheiwe GmbH	Amateur	5364

576 Dichlofluanid + Tri(Hexylene Glycol)Biborate

ABL Universal Woodworm Killer Db	Advanced Bitumens Ltd	Amateur Professional Industrial	3858
Blackfriar Wood Preserver	E Parsons And Sons Ltd	Amateur Professional	6443
Blackfriars New Wood Preserver	E Parsons And Sons Ltd	Amateur Professional Industrial	4150
Double Action Timber Preservative For Doors	International Paint Ltd	Amateur	3707

HSE
WOOD PRESERVATIVES

Product Name	Marketing Company	Use	HSE No.

576 Dichlofluanid + Tri(Hexylene Glycol)Biborate—continued

Product Name	Marketing Company	Use	HSE No.
Double Action Wood Preservative	International Paint Ltd	Amateur	3706
Homebase Weathercoat Deep Penetrating Preservative Primer	Homebase Ltd	Amateur Professional	7084
Johnstone's Woodworks Exterior Universal Primer	Kalon Decorative Products	Amateur Professional	7070
D Nitromors Timbrene Clear Environmental Formula	Henkel Home Improvements And Adhesive Products	Amateur Professional Industrial	4106
Premier Barrettine New Wood Preserver	Barrettine Products Ltd	Amateur Professional Industrial	4038
Premier Pro-Tec (S)	Premier Coatings	Amateur Professional	6978
W Ranch Preservative	International Paint Ltd	Amateur	4085
Ranch Preservative	Plascon International	Amateur	6085
Sadolin New Base	Sadolin UK Ltd	Amateur Professional Industrial	4341
Sealocrete Timber Master Wood Preserver	Sealocrete PLA Ltd	Amateur Professional	7052
Ultrabond Woodworm Killer	Philip Johnstone Group Ltd	Amateur Professional	4344

577 Dichlofluanid + Tri(Hexylene Glycol)Biborate + Zinc Naphthenate

Product Name	Marketing Company	Use	HSE No.
W Weathershield Exterior Preservative Basecoat	ICI Paints	Amateur Professional	5245
W Weathershield Preservative Primer	ICI Paints	Amateur Professional	5244

578 Dichlofluanid + Tri(Hexylene Glycol)Biborate + Zinc Octoate

Product Name	Marketing Company	Use	HSE No.
W Timber Preservative Clear TFP7	Johnstone's Paints PLC	Amateur Professional	4016

579 Dichlofluanid + Tributyltin Oxide

Product Name	Marketing Company	Use	HSE No.
D Celpruf CP Special	Rentokil Initial UK Ltd	Industrial	4582
Hickson Woodex	Hickson Timber Products Ltd	Industrial	5405
Protim Cedar	Protim Solignum Ltd	Industrial	5042
Protim Solignum Softwood Basestain CS	Protim Solignum Ltd	Industrial	5647
D Wood Preservative AA 155/00	Becker Acroma Kemira Ltd	Industrial	5308
D Wood Preservative AA 155/03	Becker Acroma Kemira Ltd	Industrial	5310
D Wood Preservative AA155	Becker Acroma Kemira Ltd	Industrial	5311

580 Dichlofluanid + Zinc Naphthenate

Product Name	Marketing Company	Use	HSE No.
W Bradite Timber Preservative	Bradite Ltd	Amateur	6091

Product Name	Marketing Company	Use	HSE No.

580 Dichlofluanid + Zinc Naphthenate—continued

Product Name	Marketing Company	Use	HSE No.
W Masterstroke Wood Preserver	Akzo Nobel Decorative Coatings Ltd	Amateur	5689
W Sikkens Wood Preserver	Akzo Nobel Woodcare	Amateur Professional	6365

581 Dichlofluanid + Zinc Octoate

Product Name	Marketing Company	Use	HSE No.
W Exterior Preservative Primer	Windeck Paints Ltd	Amateur	3857
W Timbercare Microporous Exterior Preservative	Manders Paints Ltd	Amateur Professional	5157
D Wilko Exterior System Preservative Primer	Wilkinson Home And Garden Stores	Amateur	3686

582 Disodium Octaborate

Product Name	Marketing Company	Use	HSE No.
ACS Borate Rod	Advanced Chemical Specialties Ltd	Amateur Professional	6545
ACS Boron Wood Preservative Paste	Advanced Chemical Specialties Ltd	Professional	6093
ACS Borotreat 10P	Advanced Chemical Specialties Ltd	Professional (Surface Biocide) Professional (Wood Preservative)	6335
Advance Guard	Protim Solignum Ltd	Industrial	6674
Azygo Boron Preservative Pdr	Azygo International Direct Ltd	Professional (Surface Biocide) Professional (Wood Preservative)	6497
Bio-Kil Boron Paste	Bio-Kil Chemicals Ltd	Professional Industrial	3866
W Biokil Timbor Rods	Remtox-Silexine Ltd	Amateur Professional	4502
Boracol 20	Advanced Chemical Specialties Ltd	Professional	4018
Boracol 20	Advanced Chemical Specialties Ltd	Professional	6339
Boracol B40	Advanced Chemical Specialties Ltd	Professional	4788
Borax Wood Preservative Clear Odourless	Auro Organic Paints Supplies Ltd	Professional Industrial	6271
Boron 510 Powder Preservative	Stanhope Chemical Products Ltd	Professional Industrial	7083
Boron Gel 40	Philip Johnstone Group Ltd	Professional	6460

Product Name	Marketing Company	Use	HSE No.

582 Disodium Octaborate—continued

Product Name	Marketing Company	Use	HSE No.
W Channelwood Boracol B40 Wood Preservative Gel	Advanced Chemical Specialties Ltd	Professional	6094
Channelwood Boracol B40/1 Wood Preservative Gel	Advanced Chemical Specialties Ltd	Professional	6133
Dry Pin	Window Care Systems Ltd	Professional Industrial	5144
Eco - DB8	Restoration (UK) Ltd	Professional Industrial	6944
Ecobar II	Advanced Chemical Specialties Ltd	Professional	6838
Ensele 3430	Hickson Timber Products Ltd	Amateur Professional	4671
Lectrobor Gel	Lectros International Ltd	Professional	6441
Lectros Boracol 20	Lectros International Ltd	Professional	6493
Lectros Boracol B40 Gel	Lectros International Ltd	Professional	6492
Lectros Lectrobor II	Lectros International Ltd	Professional	6898
Pandrol Timbershield Rods	Pandrol International Ltd	Professional	3901
Permabor Boracol 20	Permagard Products Ltd	Professional	6357
Permabor Boracol 40 Paste	Permagard Products Ltd	Professional	6359
Permadip 9 Concentrate	Permagard Products Ltd	Professional Industrial	4160
Probor Powder	Safeguard Chemicals Ltd	Professional Industrial	6967
Protek 9 Star Wood Protection	Protek Products (Sun Europa) Ltd	Amateur Professional Industrial	3104
Protek Double 9 Star Wood Protection	Protek Products (Sun Europa) Ltd	Amateur Professional Industrial	3101
Protek Shedstar	Protek Products (Sun Europa) Ltd	Amateur Professional Industrial	3103
Protek Wood Protection Fencegrade 9:1	Ashby Timber Treatments	Amateur	3019
Protek Wood Protection Shed Grade	Ashby Timber Treatments	Amateur	3020
Protek Woodstar	Protek Products (Sun Europa) Ltd	Amateur Professional Industrial	3102
Protim B10	Protim Solignum Ltd	Amateur Professional	4971
Protim E480	Protim Solignum Ltd	Industrial	6425
Protim Solignum D.O.T.	Protim Solignum Ltd	Industrial	6429
Pyro-Black	Timber Treatments	Industrial	5866
Remtox Borocol 20 Wood Preservative	Remtox (Chemicals) Ltd	Professional	4094

Product Name	Marketing Company	Use	HSE No.

582 Disodium Octaborate—continued

Product Name	Marketing Company	Use	HSE No.
Remtox-Silexine B40 Paste Wood Preservative	Remtox (Chemicals) Ltd	Professional	4161
Remtox-Silexine Boron Rods	Remtox-Silexine Ltd	Professional	5560
Ronseal Wet Rot Prevention Tablets	Ronseal Ltd	Amateur Professional	6627
Ronseal Wood Preservative Tablets	Ronseal Ltd	Amateur Professional	5271
Sadolin WB 100	Akzo Nobel Woodcare	Industrial	7079
Safeguard Antiflame 4050 WD Wood Preservative	Safeguard Chemicals Ltd	Professional	5656
Safeguard Boracol 20	Safeguard Chemicals Ltd	Professional	6393
Safeguard Boracol 40 Paste	Safeguard Chemicals Ltd	Professional	6400
Sovereign Timbor Rod	Sovereign Chemical Industries Ltd	Amateur Professional	4627
W Super Andy Man Wood Protection	Amb Products Ltd	Amateur Professional	3535
W Tim-Bor	Borax Europe Ltd	Professional Industrial	5687
Timbertex Pi	Marcher Chemicals Ltd	Professional Industrial	3611
Timbor Paste	Protim Solignum Ltd	Industrial	5402
Tribor 20	Triton Chemical Manufacturing Company Ltd	Professional	6242
W Tribor 20	Triton Chemicals International Ltd	Professional	6183
Tribor Gel	Triton Chemical Manufacturing Company Ltd	Professional	6099
Woodtreat Boron 40	Philip Johnstone Group Ltd	Professional	6461

583 Disodium Octaborate + 3-Iodo-2-Propynyl-N-Butyl Carbamate + Benzalkonium Chloride

Product Name	Marketing Company	Use	HSE No.
Antiblu Select	Hickson Timber Products Ltd	Industrial	6087

584 Disodium Octaborate + Benzalkonium Chloride

Product Name	Marketing Company	Use	HSE No.
Boracol 20 RH	Advanced Chemical Specialties Ltd	Professional	4019

Product Name	Marketing Company	Use	HSE No.

584 Disodium Octaborate + Benzalkonium Chloride—continued

Product Name	Marketing Company	Use	HSE No.
Boracol B8.5 RH Mouldicide/Wood Preservative	Advanced Chemical Specialties Ltd	Amateur (Surface Biocide) Amateur (Wood Preservative) Professional (Surface Biocide) Professional (Wood Preservative)	4809
Boron Biocide 10	Philip Johnstone Group Ltd	Professional (Surface Biocide) Professional (Wood Preservative)	6485
Brunosol Boron 10	Philip Johnstone Group Ltd	Professional (Surface Biocide) Professional (Wood Preservative)	6488
Deepwood 20 Inorganic Boron Wood Preservative	Safeguard Chemicals Ltd	Professional	5514
Preservative Gel	Protim Solignum Ltd	Professional	6528
Probor 10	Safeguard Chemicals Ltd	Professional	6595
Probor 50	Safeguard Chemicals Ltd	Professional (Surface Biocide) Professional (Wood Preservative)	5596
Weathershield Mulit-Surface Fungicidal Wash	Imperial Chemical Industries PLC	Amateur Professional	6823

585 Disodium Octaborate + Boric Acid

Product Name	Marketing Company	Use	HSE No.
ACS Woodkeeper Paste Preservative	Advanced Chemical Specialties Ltd	Professional	6221

586 Disodium Octaborate + Carbendazim

Product Name	Marketing Company	Use	HSE No.
Weathershield Aquatech Preservative Basecoat	ICI Paints	Amateur Professional	5757

Product Name	Marketing Company	Use	HSE No.

587 Disodium Octaborate + Copper Naphthenate

W Bio-Kil SR Pole Wrap	Bio-Kil Chemicals Ltd	Professional	5675

588 Disodium Octaborate + Dialkyldimethyl Ammonium Chloride

Ecogel	Restoration (UK) Ltd	Professional	6847
Probor 20 Brushable Gel	Safeguard Chemicals Ltd	Professional	6422

589 Disodium Octaborate + Methylene Bis(Thiocyanate) + 2-(Thiocyanomethylthio)Benzothiazole

Timbertex PD	Marcher Chemicals Ltd	Industrial	5876

590 Disodium Octaborate + Permethrin + Propiconazole + Tebuconazole

Protim E410	Protim Solignum Ltd	Industrial	6218
Protim E410(I)	Protim Solignum Ltd	Industrial	6145
Protim E415	Protim Solignum Ltd	Industrial	5908
Protim E415 (I)	Protim Solignum Ltd	Industrial	6017

591 Disodium Octaborate + Propiconazole + Tebuconazole

Protim B610	Protim Solignum Ltd	Industrial	6366

592 Disodium Octaborate + Sodium Pentachlorophenoxide

W Gallwey ABS	Protim Solignum Ltd	Industrial	5014
W Protim Fentex Green Concentrate	Protim Solignum Ltd	Industrial	5122
W Protim Fentex Green RFU	Protim Solignum Ltd	Industrial	5123
W Protim Fentex M	Protim Solignum Ltd	Industrial	4973
D Protim Kleen II	Protim Solignum Ltd	Industrial	5203

593 Disodium Tetraborate + Boric Acid

Control Fluid SB	Rentokil Initial UK Ltd	Professional (Surface Biocide) Professional (Wood Preservative)	6018
PC-K	Terminix Peter Cox Ltd	Professional	4409
Pyrolith 3505 Ready To Use	Hickson Timber Products Ltd	Industrial	3636
Rentokil Control Paste SB	Rentokil Initial UK Ltd	Professional	6501

Product Name	Marketing Company	Use	HSE No.
594 Dodecylamine Lactate + Dodecylamine Salicylate			
W Durham Bl 730	Durham Chemicals	Professional (Surface Biocide) Professional (Wood Preservative) Industrial (Wood Preservative)	4744
W Environmental Woodrot Treatment	SCI (Building Products)	Professional (Surface Biocide) Professional (Wood Preservative) Industrial (Wood Preservative)	5530
W Fungicidal Wash	Polybond Ltd	Amateur (Surface Biocide) Amateur (Wood Preservative) Professional (Surface Biocide) Professional (Wood Preservative)	3601
D Fungicidal Wash	Smyth - Morris Chemicals Ltd	Professional (Surface Biocide) Professional (Wood Preservative)	3004
W MacPherson Anti-Mould Solution	MacPherson Paints Ltd	Professional (Surface Biocide) Professional (Wood Preservative)	4791

Product Name	Marketing Company	Use	HSE No.

594 Dodecylamine Lactate + Dodecylamine Salicylate—continued

Product Name	Marketing Company	Use	HSE No.
W Metalife Fungicidal Wash	Metalife International Ltd	Amateur (Surface Biocide) Amateur (Wood Preservative) Professional (Surface Biocide) Professional (Wood Preservative)	5406
W Mouldrid	S And K Maintenance Products	Amateur (Surface Biocide) Amateur (Wood Preservative) Professional (Surface Biocide) Professional (Wood Preservative)	3671
W Nubex WDR	Nubex Ltd	Professional (Surface Biocide) Professional (Wood Preservative)	3810
W Permoglaze Micatex Fungicidal Treatment	Permoglaze Paints Ltd	Professional (Surface Biocide) Professional (Wood Preservative)	5249

HSE
WOOD PRESERVATIVES

Product Name	Marketing Company	Use	HSE No.

594 Dodecylamine Lactate + Dodecylamine Salicylate—continued

W Weathershield Fungicidal Wash	ICI Paints	Amateur (Surface Biocide) Amateur (Wood Preservative) Professional (Surface Biocide) Professional (Wood Preservative)	5134

595 Dodecylamine Lactate + Dodecylamine Salicylate + Permethrin

W Kingfisher Timber Paste	Kingfisher Chemicals Ltd	Professional	4327
W Kingfisher Timberpaste	Kingfisher Chemicals Ltd	Professional	5931
W Safeguard Deepwood Paste	Safeguard Chemicals Ltd	Professional	5758

596 Dodecylamine Salicylate + Dodecylamine Lactate

W Durham Bi 730	Durham Chemicals	Professional (Surface Biocide) Professional (Wood Preservative) Industrial (Wood Preservative)	4744
W Environmental Woodrot Treatment	SCI (Building Products)	Professional (Surface Biocide) Professional (Wood Preservative) Industrial (Wood Preservative)	5530

Product Name	Marketing Company	Use	HSE No.

596 Dodecylamine Salicylate + Dodecylamine Lactate—continued

Product Name	Marketing Company	Use	HSE No.
W Fungicidal Wash	Polybond Ltd	Amateur (Surface Biocide) Amateur (Wood Preservative) Professional (Surface Biocide) Professional (Wood Preservative)	3601
D Fungicidal Wash	Smyth - Morris Chemicals Ltd	Professional (Surface Biocide) Professional (Wood Preservative)	3004
W MacPherson Anti-Mould Solution	MacPherson Paints Ltd	Professional (Surface Biocide) Professional (Wood Preservative)	4791
W Metalife Fungicidal Wash	Metalife International Ltd	Amateur (Surface Biocide) Amateur (Wood Preservative) Professional (Surface Biocide) Professional (Wood Preservative)	5406

Product Name	Marketing Company	Use	HSE No.

596 Dodecylamine Salicylate + Dodecylamine Lactate—continued

Product Name	Marketing Company	Use	HSE No.
W Mouldrid	S And K Maintenance Products	Amateur (Surface Biocide) Amateur (Wood Preservative) Professional (Surface Biocide) Professional (Wood Preservative)	3671
W Nubex WDR	Nubex Ltd	Professional (Surface Biocide) Professional (Wood Preservative)	3810
W Permoglaze Micatex Fungicidal Treatment	Permoglaze Paints Ltd	Professional (Surface Biocide) Professional (Wood Preservative)	5249
W Weathershield Fungicidal Wash	ICI Paints	Amateur (Surface Biocide) Amateur (Wood Preservative) Professional (Surface Biocide) Professional (Wood Preservative)	5134

597 Dodecylamine Salicylate + Permethrin + Dodecylamine Lactate

Product Name	Marketing Company	Use	HSE No.
W Kingfisher Timber Paste	Kingfisher Chemicals Ltd	Professional	4327
W Kingfisher Timberpaste	Kingfisher Chemicals Ltd	Professional	5931
W Safeguard Deepwood Paste	Safeguard Chemicals Ltd	Professional	5758

598 Flufenoxuron

Product Name	Marketing Company	Use	HSE No.
M8 Super Concentrate	Remtox (Chemicals) Ltd	Professional	6507
Sovaq FLX I	Sovereign Chemical Industries Ltd	Professional	6510

Product Name	Marketing Company	Use	HSE No.

599 Flufenoxuron + Propiconazole + 3-Iodo-2-Propynyl-N-Butyl Carbamate

Product Name	Marketing Company	Use	HSE No.
Remtox Dual Purpose F/1 K7	Remtox Ltd	Professional	6871
Remtox M9 Dual Purpose Super Concentrate	Remtox (Chemicals) Ltd	Professional	6508
Sovaq FLX F/I	Sovereign Chemical Industries Ltd	Professional	6509

600 3-Iodo-2-Propynyl-N-Butyl Carbamate

Product Name	Marketing Company	Use	HSE No.
W Bio-Kil Board Preservative	Bio-Kil Chemicals Ltd	Professional Industrial	4899
BPC-FS Fungicidal Solution Concentrate	Building Preservation Centre	Professional (Surface Biocide) Professional (Wood Preservative)	6484
W CME 30/F Concentrate	Terminix Peter Cox Ltd	Professional (Surface Biocide) Professional (Wood Preservative)	5948
W Conductive Pilt 80 RFU	PPG Industries (UK) Ltd	Industrial	4500
W Crysolite Wood Preservative	Crysolite Protective Coatings	Amateur Professional	5847
W Deepkill F	Sovereign Chemical Industries Ltd	Professional Industrial	5250
Omega Biocide	Restoration (UK) Ltd	Professional (Surface Biocide) Professional (Wood Preservative)	6381
Osmose Endcoat Brown	Injecta Osmose Ltd	Amateur Professional Industrial	6561
Osmose Endcoat Clear	Injecta Osmose Ltd	Amateur Professional Industrial	6560
W Pilt 80 RFU	PPG Industries (UK) Ltd	Industrial	4877
W Pilt NF4 R.F.U.	PPG Industries (UK) Ltd	Industrial	5541
Polyphase Solvent Based Ready For Use	Troy UK	Amateur Professional Industrial	3943
Polyphase Water Based Concentrate	Troy UK	Professional	3944
Polyphase Water Based Ready For Use	Troy UK	Amateur Professional	3945

Product Name	Marketing Company	Use	HSE No.
600 3-Iodo-2-Propynyl-N-Butyl Carbamate—continued			
Protim FDR 250	Protim Solignum Ltd	Industrial	5095
Protim Wall Solution 250	Protim Solignum Ltd	Professional	6986
W Remecology Spirit Based Fungicide R5	Remtox (Chemicals) Ltd	Professional Industrial	4803
W Remtox AQ Fungicide R7	Remtox (Chemicals) Ltd	Professional Industrial	5410
Remtox Dry Rot Paint	Remtox (Chemicals) Ltd	Amateur (Surface Biocide) Amateur (Wood Preservative) Professional (Surface Biocide) Professional (Wood Preservative)	5187
W Remtox Fungicide Microemulsion M7	Remtox (Chemicals) Ltd	Professional Industrial	5419
W Remtox Fungicide Paste K6	Remtox (Chemicals) Ltd	Professional Industrial	5400
W Remtox Microactive Fungicide W7	Remtox (Chemicals) Ltd	Professional	5590
Restor-MFC	Restoration (UK) Ltd	Professional (Surface Biocide) Professional (Wood Preservative)	6246
Sadolin Shed And Fence Preserver	Akzo Nobel Woodcare	Amateur	6821
Sovac F	Sovereign Chemical Industries Ltd	Amateur Professional Industrial	5233
D Sovac FWR	Sovereign Chemical Industries Ltd	Professional Industrial	5344
D Sovaq Micro F	Sovereign Chemical Industries Ltd	Professional Industrial	5668
W Sovaq Micro F	Sovereign Chemical Industries Ltd	Professional Industrial	5824
D Sovereign AQF	Sovereign Chemical Industries Ltd	Professional Industrial	5229
Sovereign Coloured Wood Preservatives	Sovereign Chemical Industries Ltd	Professional Industrial	6337
Timber Water Repellent	Terminix Peter Cox Ltd	Professional	6621

Product Name	Marketing Company	Use	HSE No.

600 3-Iodo-2-Propynyl-N-Butyl Carbamate—continued

| UKC-2 Dry Rot Killer Concentrate | UK Chemicals Ltd | Professional (Surface Biocide) Professional (Wood Preservative) | 6107 |

601 3-Iodo-2-Propynyl-N-Butyl Carbamate + Benzalkonium Chloride + Disodium Octaborate

| Antiblu Select | Hickson Timber Products Ltd | Industrial | 6087 |

602 3-Iodo-2-Propynyl-N-Butyl Carbamate + Boric Acid

| W Protim Fentex I | Protim Solignum Ltd | Industrial | 6096 |

603 3-Iodo-2-Propynyl-N-Butyl Carbamate + Copper Naphthenate

| W Osmose Endcoat Green | Injecta Osmose Ltd | Amateur Professional Industrial | 6562 |
| W Sovereign Green Timber Preservative/Dipcoat | Sovereign Chemical Industries Ltd | Amateur Professional Industrial | 6341 |

604 3-Iodo-2-Propynyl-N-Butyl Carbamate + Cypermethrin

Microtech Dual Purpose AQ	Philip Johnstone Group Ltd	Professional	5453
W Water Based Wood Preserver (Concentrate)	Philip Johnstone Group Ltd	Amateur Professional	5630
Water-Based Wood Preserver	Philip Johnstone Group Ltd	Amateur Professional	4902

605 3-Iodo-2-Propynyl-N-Butyl Carbamate + Dialkyldimethyl Ammonium Chloride

| W Hickson NP-1 | Hickson Timber Products Ltd | Industrial | 5401 |

606 3-Iodo-2-Propynyl-N-Butyl Carbamate + Flufenoxuron + Propiconazole

Remtox Dual Purpose F/1 K7	Remtox Ltd	Professional	6871
Remtox M9 Dual Purpose Super Concentrate	Remtox (Chemicals) Ltd	Professional	6508
Sovaq FLX F/I	Sovereign Chemical Industries Ltd	Professional	6509

607 3-Iodo-2-Propynyl-N-Butyl Carbamate + Permethrin

| Brunol Dual 8 | Philip Johnstone Group Ltd | Professional | 5857 |
| Brunol PP | Philip Johnstone Group Ltd | Professional Industrial | 5414 |

HSE
WOOD PRESERVATIVES

Product Name	Marketing Company	Use	HSE No.

607 3-Iodo-2-Propynyl-N-Butyl Carbamate + Permethrin—continued

Product Name	Marketing Company	Use	HSE No.
Brunol SPI	Philip Johnstone Group Ltd	Professional Industrial	5415
W Deepkill	Sovereign Chemical Industries Ltd	Professional Industrial	5238
Deepkill	Sovereign Chemical Industries Ltd	Professional Industrial	6084
Microtreat Dual	Philip Johnstone Group Ltd	Professional	5858
Profilan - Primer	Impra Systems Ltd	Amateur Professional Industrial	6511
Protim 250	Protim Solignum Ltd	Industrial	5070
Protim 250 WR	Protim Solignum Ltd	Industrial	5071
Protim Universal AQ 250	Protim Solignum Ltd	Professional	5844
Remtox AQ Fungicide/Insecticide R9	Remtox (Chemicals) Ltd	Professional Industrial	5370
W Remtox Dual Purpose Paste K9	Remtox (Chemicals) Ltd	Professional Industrial	5399
Remtox Dual Purpose Paste K9	Remtox (Chemicals) Ltd	Professional Industrial	6182
Remtox Fungicide/Insecticide Microemulsion M9	Remtox (Chemicals) Ltd	Professional Industrial	5774
W Remtox Spirit Based F/I K7	Remtox (Chemicals) Ltd	Professional Industrial	5391
Remtox Spirit Based F/I K7	Remtox (Chemicals) Ltd	Professional Industrial	6173
Sadolin Wood Preserver	Akzo Nobel Woodcare	Amateur Professional	5649
W Sovac F/I	Sovereign Chemical Industries Ltd	Professional Industrial	5232
Sovac F/I	Sovereign Chemical Industries Ltd	Professional Industrial	6262
D Sovac F/I WR	Sovereign Chemical Industries Ltd	Professional Industrial	5300
D Sovereign AQ F/I	Sovereign Chemical Industries Ltd	Professional Industrial	5221
Sovereign Sovaq Micro F/I	Sovereign Chemical Industries Ltd	Professional Industrial	5756
Timber-Treat Ecology	Palace Chemicals Ltd	Professional	7110
Tripaste	Triton Chemical Manufacturing Company Ltd	Professional	6251
Tripaste PP	Triton Chemical Manufacturing Company Ltd	Amateur Professional	4595
Woodtreat 25	Philip Johnstone Group Ltd	Professional	5413

Product Name	Marketing Company	Use	HSE No.

608 3-Iodo-2-Propynyl-N-Butyl Carbamate + Permethrin + Propiconazole

EQ159 RTU Microemulsion Timber Fluid	Enviroquest UK Ltd	Professional	6832
Microtreat Dual 12.5X	Philip Johnstone Group Ltd	Professional	6917
Timber-Gel Ready To Use Microemulsion Timber Fluid	Terminix Peter Cox Ltd	Professional	6840
Universal P.I.	Protim Solignum Ltd	Amateur Professional Industrial	7045

609 3-Iodo-2-Propynyl-N-Butyl Carbamate + Permethrin + Propiconazole + Tebuconazole

Vacsol 2718 2:1 Conc	Hickson Timber Products Ltd	Industrial	5884
Vacsol 2719 RTU	Hickson Timber Products Ltd	Industrial	5883
Vacsol Aqua 6107	Hickson Timber Products Ltd	Industrial	6082
Vacsol Azure 2722	Hickson Timber Products Ltd	Industrial	6462

610 3-Iodo-2-Propynyl-N-Butyl Carbamate + Propiconazole

EQ158 Timber Fungicide	Enviroquest UK Ltd	Professional	6859
Sadolin Quick Drying Wood Preserver	Akzo Nobel Woodcare	Amateur	6617

611 3-Iodo-2-Propynyl-N-Butyl Carbamate + Propiconazole + Tebuconazole

Protim 418J	Protim Solignum Ltd	Industrial	6046
Protim FDR418J	Protim Solignum Ltd	Industrial	6467
Vacsol Azure 2720	Hickson Timber Products Ltd	Industrial	6344
Vacsol Azure 2721	Hickson Timber Products Ltd	Industrial	5820

612 Lindane

Fumite Lindane Generator Size 10	Hortichem Limited	Amateur Professional	4713
Fumite Lindane Generator Size 40	Hortichem Limited	Amateur Professional	4714

613 Lindane + Benzalkonium Chloride

W Biokil	Jaymar Chemicals	Professional	4247

614 Lindane + Pentachlorophenol

D Supergrade Wood Preserver	Rentokil Initial UK Ltd	Professional	4641
D Supergrade Wood Preserver Black	Rentokil Initial UK Ltd	Professional Industrial	3713
D Supergrade Wood Preserver Brown	Rentokil Initial UK Ltd	Professional Industrial	3715
D Supergrade Wood Preserver Clear	Rentokil Initial UK Ltd	Professional Industrial	3716
D Woodtreat	Stanhope Chemical Products Ltd	Professional	4475

Product Name	Marketing Company	Use	HSE No.

615 Lindane + Pentachlorophenol + Tributyltin Oxide

D Celpruf PK	Rentokil Initial UK Ltd	Industrial	4540
D Celpruf PK WR	Rentokil Initial UK Ltd	Industrial	4538
D Lar-Vac 100	Larsen Manufacturing Ltd	Industrial	3723
W Protim 80	Protim Solignum Ltd	Industrial	5029
W Protim 80 C	Protim Solignum Ltd	Industrial	5036
W Protim 80 CWR	Protim Solignum Ltd	Industrial	5031
W Protim 80 WR	Protim Solignum Ltd	Industrial	5030
W Protim Brown	Protim Solignum Ltd	Industrial	5000

616 Lindane + Pentachlorophenyl Laurate

W Brunol ATP	Stanhope Chemical Products Ltd	Professional	4498
W Mystox BTL	Stanhope Chemical Products Ltd	Professional	4425
W Mystox BTV	Stanhope Chemical Products Ltd	Professional	4424

617 Lindane + Tributyltin Naphthenate

W Protim 230L	Protim Solignum Ltd	Industrial	6964

618 Lindane + Tributyltin Oxide

W Cedasol 2320	Hickson Timber Products Ltd	Industrial	4365
D Dipsar	Cementone Beaver Ltd	Industrial	3604
D Hickson Timbercare WRQD	Hickson Timber Products Ltd	Industrial	3594
D Lar-Vac 400	Larsen Manufacturing Ltd	Industrial	5782
W Protim 210	Protim Solignum Ltd	Industrial	5033
W Protim 210 C	Protim Solignum Ltd	Industrial	5038
W Protim 210 CWR	Protim Solignum Ltd	Industrial	5035
W Protim 210 WR	Protim Solignum Ltd	Industrial	5039
W Protim 215	Protim Solignum Ltd	Industrial	6016
W Protim 23 WR	Protim Solignum Ltd	Industrial	5037
W Protim 90	Protim Solignum Ltd	Industrial	5276
W Protim Paste	Protim Solignum Ltd	Professional	4979
W Vacsol MWR Concentrate 2203	Hickson Timber Products Ltd	Industrial	3170
Vacsol MWR Ready To Use 2204	Hickson Timber Products Ltd	Industrial	3167
W Vacsol WR Ready To Use 2116	Hickson Timber Products Ltd	Industrial	3168

619 2-Methyl-4-Isothiazolin-3-One + 5-Chloro-2-Methyl-4-Isothiazolin-3-One

Celkil 90	Protim Solignum Ltd	Industrial	5428
Kathon 886F	Rohm & Haas (UK) Ltd	Industrial	5431
Laporte Mould-EX	Laporte Wood Preservation	Industrial	5430
Osmose ABS33	Protim Solignum Ltd	Industrial	6909
Tanamix 3743	Hickson Timber Products Ltd	Industrial	5429

Product Name	Marketing Company	Use	HSE No.

620 2-Methyl-4-Isothiazolin-3-One + 5-Chloro-2-Methyl-4-Isothiazolin-3-One + Copper Sulphate

Injecta Osmose ABS33	Injecta Osmose Ltd	Industrial	5978

621 Methylene Bis(Thiocyanate)

Timbertone T10 Preservative	Timbertone	Industrial	4118

622 Methylene Bis(Thiocyanate) + 2-(Thiocyanomethylthio)Benzothiazole

Busan 1009	Buckman Laboratories SA	Industrial	5256
Celbrite TC	Protim Solignum Ltd	Industrial	3921
Hickson Antiblu 3739	Hickson Timber Products Ltd	Industrial	4841
Injecta Osmose S-11	Injecta Osmose Ltd	Industrial	5957
Laporte Antisapstain	Laporte Wood Preservation	Industrial	5977
Mect	Buckman Laboratories SA	Industrial	5255
Protim Stainguard	Protim Solignum Ltd	Industrial	5009

623 Methylene Bis(Thiocyanate) + 2-(Thiocyanomethylthio)Benzothiazole + Boric Acid

Fentex Elite	Protim Solignum Ltd	Industrial	5532

624 Methylene Bis(Thiocyanate) + 2-(Thiocyanomethylthio)Benzothiazole + Disodium Octaborate

Timbertex PD	Marcher Chemicals Ltd	Industrial	5876

625 Methylene Bis(Thiocyanate) + Benzalkonium Chloride + Boric Acid + Dialkyldimethyl Ammonium Chloride

Celbrite MT	Protim Solignum Ltd	Industrial	4550

626 Methylene Bis(Thiocyanate) + Boric Acid

Hickson Antiblu 3737	Hickson Timber Products Ltd	Industrial	4745
Hickson Antiblu 3738	Hickson Timber Products Ltd	Industrial	4746

627 Oxine-Copper

W Mitrol PQ 8	Laporte Kemwood AB	Industrial	5533

628 Pentachlorophenol

D Protim GC	Protim Solignum Ltd	Industrial	5202

629 Pentachlorophenol + Lindane

D Supergrade Wood Preserver	Rentokil Initial UK Ltd	Professional	4641
D Supergrade Wood Preserver Black	Rentokil Initial UK Ltd	Professional Industrial	3713
D Supergrade Wood Preserver Brown	Rentokil Initial UK Ltd	Professional Industrial	3715
D Supergrade Wood Preserver Clear	Rentokil Initial UK Ltd	Professional Industrial	3716

Product Name	Marketing Company	Use	HSE No.

629 Pentachlorophenol + Lindane—continued

D Woodtreat	Stanhope Chemical Products Ltd	Professional	4475

630 Pentachlorophenol + Tributyltin Oxide

D Celpruf JP	Rentokil Initial UK Ltd	Industrial	4548
D Celpruf JP WR	Rentokil Initial UK Ltd	Industrial	4543
W Protim FDR-H	Protim Solignum Ltd	Industrial	4985
W Protim JP	Protim Solignum Ltd	Industrial	4978
D Protim R Clear	Protim Solignum Ltd	Industrial	5013

631 Pentachlorophenol + Tributyltin Oxide + Lindane

D Celpruf PK	Rentokil Initial UK Ltd	Industrial	4540
D Celpruf PK WR	Rentokil Initial UK Ltd	Industrial	4538
D Lar-Vac 100	Larsen Manufacturing Ltd	Industrial	3723
W Protim 80	Protim Solignum Ltd	Industrial	5029
W Protim 80 C	Protim Solignum Ltd	Industrial	5036
W Protim 80 CWR	Protim Solignum Ltd	Industrial	5031
W Protim 80 WR	Protim Solignum Ltd	Industrial	5030
W Protim Brown	Protim Solignum Ltd	Industrial	5000

632 Pentachlorophenyl Laurate + Lindane

W Brunol ATP	Stanhope Chemical Products Ltd	Professional	4498
W Mystox BTL	Stanhope Chemical Products Ltd	Professional	4425
W Mystox BTV	Stanhope Chemical Products Ltd	Professional	4424

633 Permethrin

Activ-8-I	Restoration (UK) Ltd	Professional Industrial	6169
Antel Woodworm Killer (P) Concentrate (Water Dilutable)	Antel UK Ltd	Professional	4044
D Barrettine New Woodworm Fluid	Barrettine Products Ltd	Amateur Professional Industrial	3622
BPC-8	Safeguard Chemicals Ltd	Professional Industrial	6282
Brunol Insecticidal 8	Philip Johnstone Group Ltd	Professional	5856
Brunol PC	Philip Johnstone Group Ltd	Professional	3863
Brunol PY	Philip Johnstone Group Ltd	Professional	4476
CME 20/P Concentrate	Terminix Peter Cox Ltd	Professional	5785
Colron Woodworm Killer	Ronseal Ltd	Amateur	6324
Crown Micro Woodworm	Crown Chemicals Ltd	Professional	6894
D Cuprinol Insecticidal Emulsion Concentrate	Cuprinol Ltd	Professional	4709

Product Name	Marketing Company	Use	HSE No.

633 Permethrin—continued

Cuprinol Low Odour Woodworm Killer	Cuprinol Ltd	Amateur Professional Industrial	5449
Deal Direct Woodworm Treatment Concentrate	Deal Direct Ltd	Professional Industrial	6249
Deepkill I	Sovereign Chemical Industries Ltd	Professional Industrial	5239
D Deepwood "Clear" Insecticide Emulsion Concentrate	Safeguard Chemicals Ltd	Professional Industrial	4950
Deepwood 8 Micro Emulsifiable Insecticide Concentrate	Safeguard Chemicals Ltd	Professional Industrial	5664
Deepwood Standard Emulsion Concentrate	Safeguard Chemicals Ltd	Professional	5665
Environmental Timber Treatment	SCI (Building Products)	Professional	5451
Environmental Timber Treatment Paste	SCI (Building Products)	Professional	5529
Excel 8 Insecticide	Restoration (UK) Ltd	Professional	6941
Hickson Antiborer 3768	Hickson Timber Products Ltd	Industrial	3149
Impra-Sanol	Impra Systems Ltd	Amateur Professional	4325
Inaccessible Woodworm Killer	Euro-Japan International Ltd	Amateur Professional	7033
Insecticide 1	Terminix Peter Cox Ltd	Professional	6390
W Kenwood Micro Emulsion Woodworm Treatment (Concentrate)	Kenwood Damp Proofing PLC	Professional Industrial	6220
Kenwood Micro Emulsion Woodworm Treatment (Concentrate)	Kenwood Damp Proofing PLC	Professional	6313
Kenwood Woodworm Treatment Concentrate	Kenwood Damp Proofing PLC	Professional Industrial	6103
KF8 Insecticide Microemulsion Concentrate	Kingfisher Chemicals Ltd	Professional	6827
Kingfisher KF-8 Micro Emulsion Insecticide Concentrate	Kingfisher Chemicals Ltd	Professional Industrial	5786
Lectro 8WW	Lectros International Ltd	Professional	6398
Lectros 8P	Lectros International Ltd	Professional	6818
Leyland Timbrene Woodworm Killer	Leyland Paint Company Ltd	Amateur Professional	4825
Low Odour Woodworm Killer	Cuprinol Ltd	Amateur Professional	5435
Micro 8 WW	Lectros International Ltd	Professional Industrial	6132
Micro 8I	Construction Chemicals	Professional	6645
Micro Woodworm 8	Grip-Fast Systems Ltd	Professional Industrial	6415

517

Product Name	Marketing Company	Use	HSE No.

633 Permethrin—continued

Product Name	Marketing Company	Use	HSE No.
Micro-Emulsion Woodworm Treatment Concentrate	Kenwood Damp Proofing PLC	Professional	6387
D Microguard Permethrin Concentrate	Permagard Products Ltd	Professional Industrial	5102
Microguard Permethrin Concentrate	Permagard Products Ltd	Professional Industrial	5741
Microguard Woodworm Fluid	Permagard Products Ltd	Amateur Professional Industrial	5103
Microtech Woodworm Killer 25X	Philip Johnstone Group Ltd	Professional	5888
Microtreat Insecticide	Philip Johnstone Group Ltd	Professional	5859
Microtreat Insecticide 25X	Philip Johnstone Group Ltd	Professional	6913
Omega	Restoration (UK) Ltd	Professional Industrial	6379
Palace Microfine 8 -Insecticide Concentrate	Palace Chemicals Ltd	Professional Industrial	5809
Palace Woodworm Killer 8 (Microemulsion)	Palace Chemicals Ltd	Professional	6629
Permethrin WW Conc	Permagard Products Ltd	Professional	4020
Premier Barrettine New Woodworm Killer	Barrettine Products Ltd	Amateur Professional Industrial	4005
Protect And Shine Spray	Rentokil Initial UK Ltd	Amateur	6123
Protim 600WR	Protim Solignum Ltd	Industrial	5833
Protim AQ	Protim Solignum Ltd	Industrial	5201
Protim Aquachem-Insecticidal Emulsion P	Protim Solignum Ltd	Amateur Professional	5161
Protim Insecticidal AQ8	Protim Solignum Ltd	Professional	6526
Protim Insecticidal Emulsion 8	Protim Solignum Ltd	Professional Industrial	5701
Protim Insecticidal Emulsion P	Protim Solignum Ltd	Professional Industrial	5008
Protim Paste P	Protim Solignum Ltd	Amateur Professional	4980
Protim Woodworm Killer P	Protim Solignum Ltd	Amateur Professional Industrial	5011
Remecology Insecticide R8	Remtox (Chemicals) Ltd	Professional	4233
W Remecology Spirit Based Insecticide R6	Remtox (Chemicals) Ltd	Professional	4836
D Remtox Insecticide Microemulsion M8	Remtox (Chemicals) Ltd	Professional Industrial	5710
Remtox Insecticide Microemulsion M8	Remtox (Chemicals) Ltd	Professional Industrial	5775
W Remtox Insecticide Paste R4	Remtox (Chemicals) Ltd	Professional Industrial	5389
W Remtox Microactive Insecticide W8	Remtox (Chemicals) Ltd	Professional	5555

Product Name	Marketing Company	Use	HSE No.

633 Permethrin—continued

Rentokil Classic Wax Polish	Rentokil Initial UK Ltd	Amateur	3954
Rentokil Permethrin 2.5 EC	Rentokil Initial UK Ltd	Professional	6979
Rentokil Woodworm Killer	Rentokil Initial UK Ltd	Amateur Professional	4208
Rentokil Woodworm Treatment	Rentokil Initial UK Ltd	Amateur Professional	3911
Restor-8-Insecticide Concentrate	Restoration (UK) Ltd	Professional Industrial	5962
Safeguard Deepwood I Insecticide Emulsion Concentrate	Safeguard Chemicals Ltd	Professional	5304
Safeguard Deepwood IV Woodworm Killer	Safeguard Chemicals Ltd	Professional	3946
Screwfix Woodworm Killer	Screwfix Direct Ltd	Amateur Professional Industrial	7064
Solignum Insecticidal AQ8	Protim Solignum Ltd	Professional	6527
Solignum Woodworm Killer	Protim Solignum Ltd	Amateur Professional Industrial	5488
Solignum Woodworm Killer Concentrate	Protim Solignum Ltd	Professional Industrial	5466
D Sov AQ Micro I	Sovereign Chemical Industries Ltd	Professional Industrial	5220
D Sovac I	Sovereign Chemical Industries Ltd	Professional Industrial	5230
Sovaq Micro I	Sovereign Chemical Industries Ltd	Professional	6272
D Sovereign Aqueous Insecticide 2	Sovereign Chemical Industries Ltd	Professional	4620
W Sovereign Sovaq Micro I	Sovereign Chemical Industries Ltd	Professional Industrial	5755
Sprytech 1	Spry Chemicals Ltd	Professional Industrial	6489
Timbrene Woodworm Killer	Kalon Group PLC	Amateur Professional	4996
Trimethrin 20S	Triton Chemical Manufacturing Company Ltd	Amateur Professional	4621
D Trimethrin 2AQ	Triton Chemicals International Ltd	Professional	4625
D Trimethrin 3AQ	Triton Chemicals International Ltd	Professional	4626
D Trimethrin 3OS	Triton Chemicals International Ltd	Amateur Professional	4622
Tritec	Triton Chemical Manufacturing Company Ltd	Professional	5424

Product Name	Marketing Company	Use	HSE No.
633 Permethrin—continued			
Tritec 120	Triton Chemicals International Ltd	Professional	6189
Tritec 121	Triton Chemical Manufacturing Company Ltd	Professional	6541
Triton 120	Triton Chemical Manufacturing Company Ltd	Professional	6539
Triton Woodworm Killer	Triton Chemical Manufacturing Company Ltd	Amateur	6226
UKC-8 Woodworm Killer Concentrate	UK Chemicals Ltd	Professional	6108
UKC-8 Woodworm Killer Microemulsion Concentrate	UK Chemicals Ltd	Professional	6376
Ultra Tech 2000 I	Crown Chemicals Ltd	Professional	6143
Water Based Woodworm Fluid	Barrettine Products Ltd	Amateur Professional	6311
Wickes Woodworm Killer	Wickes Building Supplies Ltd	Amateur Professional	5692
Woodworm 8 Micro Emulsifiable Concentrate	Crown Chemicals Ltd	Professional Industrial	5819
D Woodworm Fluid (B) EC	Rentokil Initial UK Ltd	Professional	4213
Woodworm Fluid (B) For Bonded Warehouses	Rentokil Initial UK Ltd	Professional	4725
Woodworm Fluid B	Rentokil Initial UK Ltd	Professional	4743
Woodworm Fluid FP	Rentokil Initial UK Ltd	Professional	4577
Woodworm Fluid Z Emulsion Concentrate	Rentokil Initial UK Ltd	Professional	4264
Woodworm Killer	Langlow Products Ltd	Amateur Professional Industrial	3691
Woodworm Killer	Protim Solignum Ltd	Amateur Professional Industrial	5487
Woodworm Killer (Grade P)	Barrettine Products Ltd	Amateur Professional Industrial	5213
Woodworm Killer P8	Antel UK Ltd	Professional	6583
Woodworm Paste B	Rentokil Initial UK Ltd	Professional	4578
Woodworm Treatment Spray	Rentokil Initial UK Ltd	Amateur	5619
634 Permethrin + 2-Phenylphenol			
Brunol OPA	Philip Johnstone Group Ltd	Professional	4473
Cromar Dry Rot And Woodworm Killer	Cromar Building Products Ltd	Amateur Professional	6421

Product Name	Marketing Company	Use	HSE No.

635 Permethrin + 3-Iodo-2-Propynyl-N-Butyl Carbamate

Product Name	Marketing Company	Use	HSE No.
Brunol Dual 8	Philip Johnstone Group Ltd	Professional	5857
Brunol PP	Philip Johnstone Group Ltd	Professional Industrial	5414
Brunol SPI	Philip Johnstone Group Ltd	Professional Industrial	5415
W Deepkill	Sovereign Chemical Industries Ltd	Professional Industrial	5238
Deepkill	Sovereign Chemical Industries Ltd	Professional Industrial	6084
Microtreat Dual	Philip Johnstone Group Ltd	Professional	5858
Profilan - Primer	Impra Systems Ltd	Amateur Professional Industrial	6511
Protim 250	Protim Solignum Ltd	Industrial	5070
Protim 250 WR	Protim Solignum Ltd	Industrial	5071
Protim Universal AQ 250	Protim Solignum Ltd	Professional	5844
Remtox AQ Fungicide/Insecticide R9	Remtox (Chemicals) Ltd	Professional Industrial	5370
W Remtox Dual Purpose Paste K9	Remtox (Chemicals) Ltd	Professional Industrial	5399
Remtox Dual Purpose Paste K9	Remtox (Chemicals) Ltd	Professional Industrial	6182
Remtox Fungicide/Insecticide Microemulsion M9	Remtox (Chemicals) Ltd	Professional Industrial	5774
W Remtox Spirit Based F/I K7	Remtox (Chemicals) Ltd	Professional Industrial	5391
Remtox Spirit Based F/I K7	Remtox (Chemicals) Ltd	Professional Industrial	6173
Sadolin Wood Preserver	Akzo Nobel Woodcare	Amateur Professional	5649
W Sovac F/I	Sovereign Chemical Industries Ltd	Professional Industrial	5232
Sovac F/I	Sovereign Chemical Industries Ltd	Professional Industrial	6262
D Sovac F/I WR	Sovereign Chemical Industries Ltd	Professional Industrial	5300
D Sovereign AQ F/I	Sovereign Chemical Industries Ltd	Professional Industrial	5221
Sovereign Sovaq Micro F/I	Sovereign Chemical Industries Ltd	Professional Industrial	5756
Timber-Treat Ecology	Palace Chemicals Ltd	Professional	7110
Tripaste	Triton Chemical Manufacturing Company Ltd	Professional	6251
Tripaste PP	Triton Chemical Manufacturing Company Ltd	Amateur Professional	4595

HSE
WOOD PRESERVATIVES

Product Name	Marketing Company	Use	HSE No.

635 Permethrin + 3-Iodo-2-Propynyl-N-Butyl Carbamate—continued

Woodtreat 25	Philip Johnstone Group Ltd	Professional	5413

636 Permethrin + Acypetacs Copper + Acypetacs Zinc

Cuprinol Low Odour End Cut	Cuprinol Ltd	Amateur Professional Industrial	6114
Cuprisol P	Cuprinol Ltd	Industrial	3634
Protim 800P	Protim Solignum Ltd	Industrial	5702

637 Permethrin + Acypetacs Zinc

D Cuprinol 5 Star Complete Wood Treatment S	Cuprinol Ltd	Amateur Professional	4710
Cuprinol Combination Grade S	Cuprinol Ltd	Amateur Professional Industrial	3588
Cuprinol Low Odour 5 Star Complete Wood Treatment	Cuprinol Ltd	Amateur Professional	5445
Cuprisol FN	Cuprinol Ltd	Industrial	3632
Cuprisol WR	Palace Chemicals Ltd	Industrial	3633
Cuprisol XQD Special	Cuprinol Ltd	Industrial	3732
Cut End	Protim Solignum Ltd	Amateur Professional Industrial	5490
Protim 800	Protim Solignum Ltd	Industrial	4991
Protim 800 C	Protim Solignum Ltd	Industrial	4990
Protim 800 CWR	Protim Solignum Ltd	Industrial	4993
Protim 800 WR	Protim Solignum Ltd	Industrial	4992
Protim Brown CDB	Protim Solignum Ltd	Amateur Professional Industrial	6472
Protim Paste 800	Protim Solignum Ltd	Professional Industrial	4986
Universal	Protim Solignum Ltd	Amateur Professional Industrial	5491
Wickes All Purpose Wood Treatment	Wickes Building Supplies Ltd	Amateur Professional	5696

Product Name	Marketing Company	Use	HSE No.
638 Permethrin + Azaconazole			
Fongix SE Total Treatment For Wood	Liberon Waxes Ltd	Amateur (Surface Biocide) Amateur (Wood Preservative) Professional (Surface Biocide) Professional (Wood Preservative)	5610
Woodworm Killer And Rot Treatment	Liberon Waxes Ltd	Amateur (Surface Biocide) Amateur (Wood Preservative) Professional (Surface Biocide) Professional (Wood Preservative)	4826
D Xylamon Brown U 101 C	Venilia Ltd	Amateur Professional	4443
D Xylamon Curative U 152 G/H	NWE Distributors Ltd	Professional	4446
639 Permethrin + Benzalkonium Chloride			
W Biokil Emulsion	Jaymar Chemicals	Professional	4562
640 Permethrin + Boric Acid			
Activ-8-II	Restoration (UK) Ltd	Professional Industrial	6170
Micro 8 FI	Lectros International Ltd	Professional Industrial	6131
Micro Dual Purpose 8	Grip-Fast Systems Ltd	Professional Industrial	6408
D Microguard FI Concentrate	Permagard Products Ltd	Professional Industrial	5101
Microguard FI Concentrate	Permagard Products Ltd	Professional Industrial	5854

HSE
WOOD PRESERVATIVES

Product Name	Marketing Company	Use	HSE No.

641 Permethrin + Copper Naphthenate

W Green	Protim Solignum Ltd	Amateur Professional Industrial	5483
W Protim Green E	Protim Solignum Ltd	Amateur Professional Industrial	4975
W Solignum Green	Protim Solignum Ltd	Amateur Professional Industrial	5020

642 Permethrin + Dodecylamine Lactate + Dodecylamine Salicylate

W Kingfisher Timber Paste	Kingfisher Chemicals Ltd	Professional	4327
W Kingfisher Timberpaste	Kingfisher Chemicals Ltd	Professional	5931
W Safeguard Deepwood Paste	Safeguard Chemicals Ltd	Professional	5758

643 Permethrin + Propiconazole

Aqueous Universal	Protim Solignum Ltd	Amateur Professional	5776
Celpruf WB11	Protim Solignum Ltd	Industrial	6682
Dual Purpose Fi8 (Microemulsion)	Antel UK Ltd	Professional	6585
EQ116 Dual Purpose Fl	Enviroquest UK Ltd	Professional	6555
Lectro Micro 8 Fl	Lectros International Ltd	Professional	6584
Micro 8 Fl	Construction Chemicals	Professional	6644
Palace Dual Purpose Fl 8 (Microemulsion)	Palace Chemicals Ltd	Professional	6646
Protim 340 WR	Protim Solignum Ltd	Industrial	5509
Protim 340E	Protim Solignum Ltd	Industrial	5511
Tritec 120 Plus	Triton Chemical Manufacturing Company Ltd	Professional	6289
Tritec 120 Plus	Triton Chemical Manufacturing Company Ltd	Professional	6537
Tritec 121 Plus	Triton Chemical Manufacturing Company Ltd	Professional	6538
Ultra Tech 2000 D	Crown Chemicals Ltd	Professional	6304
Vacsele Aqua 6106	Hickson Timber Products Ltd	Amateur Professional Industrial	5893
W Wocosen 12 OL	Janssen Pharmaceutica NV	Amateur Professional Industrial	5404
Wocosen S-P	Janssen Pharmaceutica NV	Amateur Professional Industrial	5871

Product Name	Marketing Company	Use	HSE No.

643 Permethrin + Propiconazole—continued

Wolsit KD 20	Dr Wolman GmbH	Industrial	6307
Wolsit KD 20C	Dr Wolman GmbH	Industrial	6966

644 Permethrin + Propiconazole + 3-Iodo-2-Propynyl-N-Butyl Carbamate

EQ159 RTU Microemulsion Timber Fluid	Enviroquest UK Ltd	Professional	6832
Microtreat Dual 12.5X	Philip Johnstone Group Ltd	Professional	6917
Timber-Gel Ready To Use Microemulsion Timber Fluid	Terminix Peter Cox Ltd	Professional	6840
Universal P.I.	Protim Solignum Ltd	Amateur Professional Industrial	7045

645 Permethrin + Propiconazole + Dialkyldimethyl Ammonium Chloride

Traditional Preservative Oil	Oakmasters Of Sussex	Amateur Professional Industrial	6023

646 Permethrin + Propiconazole + Tebuconazole

Protim 415	Protim Solignum Ltd	Industrial	5815
Protim 415 T Special Solvent	Protim Solignum Ltd	Industrial	6077
Protim 415T	Protim Solignum Ltd	Industrial	5880
Protim 415TWR	Protim Solignum Ltd	Industrial	5916
Protim 418	Protim Solignum Ltd	Industrial	6466
Protim 418V	Protim Solignum Ltd	Industrial	6611
Protim 418WR	Protim Solignum Ltd	Industrial	6842
Vacsol 2627 Azure	Hickson Timber Products Ltd	Industrial	5558
Vacsol Aqua 6108	Hickson Timber Products Ltd	Industrial	6083
Vacsol Aqua 6108 RTU	Hickson Timber Products Ltd	Industrial	6994
Vacsol Aqua 6109	Hickson Timber Products Ltd	Industrial	6430

647 Permethrin + Propiconazole + Tebuconazole + 3-Iodo-2-Propynyl-N-Butyl Carbamate

Vacsol 2718 2:1 Conc	Hickson Timber Products Ltd	Industrial	5884
Vacsol 2719 RTU	Hickson Timber Products Ltd	Industrial	5883
Vacsol Aqua 6107	Hickson Timber Products Ltd	Industrial	6082
Vacsol Azure 2722	Hickson Timber Products Ltd	Industrial	6462

648 Permethrin + Propiconazole + Tebuconazole + Disodium Octaborate

Protim E410	Protim Solignum Ltd	Industrial	6218
Protim E410(I)	Protim Solignum Ltd	Industrial	6145
Protim E415	Protim Solignum Ltd	Industrial	5908
Protim E415 (I)	Protim Solignum Ltd	Industrial	6017

Product Name	Marketing Company	Use	HSE No.

649 Permethrin + Tebuconazole

Brunol ATP New	Philip Johnstone Group Ltd	Professional Industrial	5620
Brunol STP	Philip Johnstone Group Ltd	Professional Industrial	5621

650 Permethrin + Tebuconazole + Benzalkonium Chloride

Vacsol Aqua 6105	Hickson Timber Products Ltd	Industrial	5892

651 Permethrin + Tebuconazole + Dichlofluanid

Impra-Holzschutzgrund (Primer)	Impra Systems Ltd	Professional Industrial	5735

652 Permethrin + Tri(Hexylene Glycol)Biborate

Antel Dual Purpose (BP) Concentrate (Water Dilutable)	Antel UK Ltd	Professional	4046
D Barrettine New Universal Fluid	Barrettine Products Ltd	Amateur Professional Industrial	3623
BPC8-DP	Building Preservation Centre	Professional	6306
Brunol PBO	Philip Johnstone Group Ltd	Professional	4474
Brunol Special P	Philip Johnstone Group Ltd	Professional	3864
D Chel-Wood Preserver/Woodworm Dry Rot Killer BP	Chelec Ltd	Amateur Professional	4897
Crown Micro Dual Purpose F/I	Crown Chemicals Ltd	Professional	6895
Crown Timber Paste	Crown Chemicals Ltd	Professional	5869
Deepwood FI Dual Purpose Emulsion Concentrate	Safeguard Chemicals Ltd	Professional Industrial	5087
Dual 8	Deal Direct Ltd	Professional	6255
W Ecology Fungicide Insecticide Aqueous (Concentrate) Wood Preservative Dual Purpose	Kingfisher Chemicals Ltd	Professional	4039
Ecology Fungicide Insecticide Aqueous (Concentrate) Wood Preservative Dual Purpose	Kingfisher Chemicals Ltd	Professional	5930
Everbuild Lumberjack Universal Rot And Woodworm Killer	Palace Chemicals Ltd	Amateur Professional Industrial	6906
D Insecticide Fungicide Wood Preservative	Kingfisher Chemicals Ltd	Professional	4048
Insecticide Fungicide Wood Preservative	Kingfisher Chemicals Ltd	Professional	5933
J B Ward & Son Dual Purpose Timber Treatment Concentrate	J B Ward And Son	Professional Industrial	7050
KF-8 Dual Purpose Micro Emulsion Concentrate	Kingfisher Chemicals Ltd	Professional Industrial	6075
Kingfisher Timberpaste	Kingfisher Chemicals Ltd	Professional	6686

Product Name	Marketing Company	Use	HSE No.

652 Permethrin + Tri(Hexylene Glycol)Biborate—continued

Product Name	Marketing Company	Use	HSE No.
Larsen Wood Preservative Clear 2 And Brown 2	Larsen Manufacturing Ltd	Amateur Professional Industrial	4287
Larsen Woodworm Killer 2	Larsen Manufacturing Ltd	Amateur Professional Industrial	4285
Lectro Paste PB	Lectros International Ltd	Professional	6130
Mar-Kil S	Marcher Chemicals Ltd	Professional Industrial	5234
Mar-Kil W	Marcher Chemicals Ltd	Professional Industrial	5219
Micron 8 Dual Purpose Emulsion Concentrate	Safeguard Chemicals Ltd	Professional Industrial	5851
Omega Plus	Restoration (UK) Ltd	Professional	6382
Perma AQ Dual Purpose	Triton Perma Industries Ltd	Professional	5617
Permapaste PB	Triton Perma Industries Ltd	Professional	5616
Permethrin F And I Concentrate	Permagard Products Ltd	Professional	4008
Protim CB	Protim Solignum Ltd	Professional Industrial	5864
Protim CDB	Protim Solignum Ltd	Amateur Professional Industrial	4999
Protim WR 260	Protim Solignum Ltd	Amateur Professional Industrial	4997
D Remecology Spirit Based K7	Remtox (Chemicals) Ltd	Professional	4248
Restor-8-Dual Purpose Concentrate	Restoration (UK) Ltd	Professional Industrial	5963
Restor-Paste	Restoration (UK) Ltd	Professional	6079
D Safeguard Deepwood II Dual Purpose Emulsion Concentrate	Safeguard Chemicals Ltd	Professional	5303
Safeguard Deepwood II Dual Purpose Emulsion Concentrate	Safeguard Chemicals Ltd	Professional	5850
Safeguard Deepwood III Timber Treatment	Safeguard Chemicals Ltd	Amateur	3937
D Safeguard Deepwood Paste	Safeguard Chemicals Ltd	Professional	4110
Sandhill Aqueous Fungicide - Insecticide	Sandhill Building Products	Professional	5950
Sandhill Fungicide - Insecticide	Sandhill Building Products	Amateur	5952
Solignum Cut End	Protim Solignum Ltd	Amateur Professional Industrial	7046
Solignum Remedial Concentrate	Protim Solignum Ltd	Professional	5021
Solignum Remedial PB	Protim Solignum Ltd	Amateur Professional Industrial	5607

Product Name	Marketing Company	Use	HSE No.

652 Permethrin + Tri(Hexylene Glycol)Biborate—continued

Product Name	Marketing Company	Use	HSE No.
Solignum Universal	Protim Solignum Ltd	Amateur Professional Industrial	7042
Sovereign Aqueous Fungicide/ Insecticide 2	Sovereign Chemical Industries Ltd	Professional Industrial	4619
Sovereign Insecticide/Fungicide 2	Sovereign Chemical Industries Ltd	Professional	4649
Sprytech 2	Spry Chemicals Ltd	Professional Industrial	6500
Timber Paste	Grip-Fast Systems Ltd	Professional	6409
D Trimethrin AQ Plus	Triton Chemicals International Ltd	Professional	3522
Trimethrin OS Plus	Triton Chemical Manufacturing Company Ltd	Amateur Professional	3521
Tripaste PB	Triton Chemicals International Ltd	Professional	6080
Tritec Plus	Triton Chemical Manufacturing Company Ltd	Professional	5425
D Universal Fluid (Grade PB)	Barrettine Products Ltd	Amateur Professional Industrial	5212
Universal Wood Preserver	Langlow Products Ltd	Amateur Professional Industrial	3692
Woodworm Killer	Sealocrete PLA Ltd	Amateur Professional	5028

653 Permethrin + Tri(Hexylene Glycol)Biborate + Dichlofluanid

Product Name	Marketing Company	Use	HSE No.
Blackfriar Wood Preserver WP Gold Star Clear	E Parsons And Sons Ltd	Amateur Professional	6417
Blackfriars Gold Star Clear	E Parsons And Sons Ltd	Amateur Professional Industrial	4149
Johnstone's Woodworks Interior/ Exterior All Purpose Preserver	Kalon Decorative Products	Amateur Professional	7071
Leyland Timbrene Supreme Environmental Formula	Leyland Paint Company Ltd	Amateur Professional	4800
D Nitromors Timbrene Supreme Environmental Formula	Henkel Home Improvements And Adhesive Products	Amateur Professional Industrial	4107
Premier Barrettine New Universal Fluid D	Barrettine Products Ltd	Amateur Professional Industrial	4004
Timber Master 6 In 1 Wood Treatment	Sealocrete PLA Ltd	Amateur Professional	7080

Product Name	Marketing Company	Use	HSE No.

653 Permethrin + Tri(Hexylene Glycol)Biborate + Dichlofluanid—continued

Product Name	Marketing Company	Use	HSE No.
Timbrene Supreme	Kalon Group PLC	Amateur Professional	4995

654 Permethrin + Tributyltin Naphthenate

Product Name	Marketing Company	Use	HSE No.
Celpruf TNM	Protim Solignum Ltd	Industrial	4704
Celpruf TNM WR	Protim Solignum Ltd	Industrial	4703
Protim 230	Protim Solignum Ltd	Industrial	5044
Protim 230 WR	Protim Solignum Ltd	Industrial	5045
Vacsol 2622/2623 WR	Hickson Timber Products Ltd	Industrial	4645
Vacsol 2625/2626 WR 2:1 Concentrate	Hickson Timber Products Ltd	Industrial	5329
Vacsol 2713/2714	Hickson Timber Products Ltd	Industrial	4808

655 Permethrin + Zinc Octoate

Product Name	Marketing Company	Use	HSE No.
W Green Range Timber Treatment Paste	Philip Johnstone Group Ltd	Professional	4651
W Larvac 300	Larsen Manufacturing Ltd	Industrial	4461
W Palace Timbertreat Ecology	Palace Chemicals Ltd	Professional	4931
W Premium Wood Treatment	Rentokil Initial UK Ltd	Amateur (Surface Biocide) Amateur (Wood Preservative) Professional (Surface Biocide) Professional (Wood Preservative)	5350
D Rentokil Dual Purpose Fluid	Rentokil Initial UK Ltd	Professional	4074
W Rentokil Wood Preservative	Rentokil Initial UK Ltd	Amateur Professional	4195
W Ronseal's Universal Wood Treatment	Ronseal Ltd	Amateur Professional	6186
W Solignum Colourless	Protim Solignum Ltd	Amateur Professional Industrial	5549
W Solignum Wood Preservative Paste	Protim Solignum Ltd	Professional	5001
W Thompson's Universal Wood Treatment	Ronseal Ltd	Amateur Professional	5986
W Wood Preservative Clear	Rentokil Initial UK Ltd	Amateur Professional	4077
W Woodtreat BP	Philip Johnstone Group Ltd	Professional	3085

HSE
WOOD PRESERVATIVES

Product Name	Marketing Company	Use	HSE No.

656 Permethrin + Zinc Octoate + Boric Acid

D Celpruf BZP	Rentokil Initial UK Ltd	Industrial	4190
D Celpruf BZP WR	Rentokil Initial UK Ltd	Industrial	4191
W Cut'N'Treat	Rentokil Initial UK Ltd	Amateur Professional	5802
W Premium Grade Wood Treatment	Rentokil Initial UK Ltd	Amateur Professional	4193

657 Permethrin + Zinc Octoate + Dichlofluanid

W OS Color Wood Stain And Preservative	Ostermann And Scheiwe GmbH	Amateur	5531
W OS Color WR	Ostermann And Scheiwe GmbH	Amateur	5364

658 Permethrin + Zinc Versatate

W Cedasol Ready To Use (2306)	Hickson Timber Products Ltd	Industrial	4117
W Celpruf Zop	Rentokil Initial UK Ltd	Industrial	4681
W Celpruf Zop WR	Rentokil Initial UK Ltd	Industrial	4682
W Imersol 2410	Hickson Timber Products Ltd	Industrial	4674
W Protim 220	Protim Solignum Ltd	Industrial	5198
W Protim 220 CWR	Protim Solignum Ltd	Amateur Professional Industrial	5196
W Protim 220 WR	Protim Solignum Ltd	Industrial	5195
W Protim 220C	Protim Solignum Ltd	Industrial	5780
W Protim Paste 220	Protim Solignum Ltd	Professional Industrial	4983
W Solignum Universal	Protim Solignum Ltd	Amateur Professional Industrial	5019
W Vacsele 2611	Hickson Timber Products Ltd	Professional	4675
W Vacsele P2312	Hickson Timber Products Ltd	Professional Industrial	4363
W Vacsol (2:1 Conc) 2711/2712	Hickson Timber Products Ltd	Industrial	4685
W Vacsol 2709/2710	Hickson Timber Products Ltd	Industrial	4678
W Vacsol WR (2:1 Conc) 2614/2615	Hickson Timber Products Ltd	Industrial	4679
W Vacsol WR 2612/2613	Hickson Timber Products Ltd	Industrial	4676

659 2-Phenylphenol

D Basiment 560	Venilia Ltd	Professional Industrial	4411
Laybond Wood Preserver	Laybond Products Ltd	Amateur Professional	5382
Preventol OF	Bayer PLC	Professional Industrial	5200
Protim Joinery Lining 280	Protim Solignum Ltd	Professional	5135

Product Name	Marketing Company	Use	HSE No.

660 2-Phenylphenol + Benzalkonium Chloride

D Chel Dry Rot Killer For Masonry And Brickwork	Chelec Ltd	Professional	3067

661 2-Phenylphenol + Boric Acid

Protim Fentex P	Protim Solignum Ltd	Industrial	6095

662 2-Phenylphenol + Cypermethrin

Laybond Timber Protector	Laybond Products Ltd	Amateur Professional	5369

663 2-Phenylphenol + Permethrin

Brunol OPA	Philip Johnstone Group Ltd	Professional	4473
Cromar Dry Rot And Woodworm Killer	Cromar Building Products Ltd	Amateur Professional	6421

664 Pirimiphos-Methyl

Actellic 25 EC	Zeneca Public Health	Professional (Animal Husbandry) Professional (Food Storage Practice)	4880

665 Potassium Dichromate + Boric Acid + Copper Sulphate

Tecca CCB 1	Tecca Ltd	Industrial	5584

666 Propiconazole

CME 25/F Concentrate	Terminix Peter Cox Ltd	Professional	6136
Fungicide 2	Terminix Peter Cox Ltd	Professional	6391
Impralan Grund G 200	Impra Systems Ltd	Professional Industrial	6620
Impralan-Grund G 100	Impra Systems Ltd	Professional Industrial	6549
Microtech Fungicide 10X	Philip Johnstone Group Ltd	Professional	5887
OS Color Wood Stain And Protector	Ostermann And Scheiwe GmbH	Amateur Professional Industrial	7113
Safetray P	Progress Products	Professional	6275
Teak Oil Extra	Swan Hattersley Ltd	Amateur Professional	7092
Timber Fungicide	Timberwise Preservation Scot	Professional	6641
Timber Treatment Plus	Alexander Rose Ltd	Amateur Professional	6648

Product Name	Marketing Company	Use	HSE No.

666 Propiconazole—continued

Trisol 120	Triton Chemical Manufacturing Company Ltd	Professional	6190
Trisol 120	Triton Chemical Manufacturing Company Ltd	Professional	6540
Triton Timber Rot Killer	Triton Chemical Manufacturing Company Ltd	Amateur	6227
Ultra Tech 2000 F	Crown Chemicals Ltd	Professional	6142
Wocosen 100 SL	Janssen Pharmaceutica NV	Professional (Surface Biocide) Professional (Wood Preservative) Industrial (Wood Preservative)	5743
Wocosen S	Janssen Pharmaceutica NV	Amateur Professional Industrial	5394

667 Propiconazole + 3-Iodo-2-Propynyl-N-Butyl Carbamate

| EQ158 Timber Fungicide | Enviroquest UK Ltd | Professional | 6859 |
| Sadolin Quick Drying Wood Preserver | Akzo Nobel Woodcare | Amateur | 6617 |

668 Propiconazole + 3-Iodo-2-Propynyl-N-Butyl Carbamate + Flufenoxuron

Remtox Dual Purpose F/1 K7	Remtox Ltd	Professional	6871
Remtox M9 Dual Purpose Super Concentrate	Remtox (Chemicals) Ltd	Professional	6508
Sovaq FLX F/I	Sovereign Chemical Industries Ltd	Professional	6509

669 Propiconazole + 3-Iodo-2-Propynyl-N-Butyl Carbamate + Permethrin

EQ159 RTU Microemulsion Timber Fluid	Enviroquest UK Ltd	Professional	6832
Microtreat Dual 12.5X	Philip Johnstone Group Ltd	Professional	6917
Timber-Gel Ready To Use Microemulsion Timber Fluid	Terminix Peter Cox Ltd	Professional	6840
Universal P.I.	Protim Solignum Ltd	Amateur Professional Industrial	7045

Product Name	Marketing Company	Use	HSE No.

670 Propiconazole + Boric Acid

Celbor PR	Protim Solignum Ltd	Industrial	6192

671 Propiconazole + Boric Acid + Copper Carbonate Hydroxide

Ensele 3450	Hickson Timber Products Ltd	Amateur Professional	6992
Noah Gold	Protim Solignum Ltd	Industrial	6965
Thompson's End Grain Preserver	Ronseal Ltd	Amateur Professional	6991
Wolmanit CX-P	Dr Wolman GmbH	Industrial	6566

672 Propiconazole + Boric Acid + Cypermethrin

Rentokil Total Wood Treatment	Rentokil Initial UK Ltd	Amateur Professional	6053

673 Propiconazole + Cypermethrin

Celpruf B	Protim Solignum Ltd	Industrial	5935
Cuprinol Quick Drying Clear Preserver	Cuprinol Ltd	Amateur Professional Industrial	6014
W Cuprinol Wood Defender	Cuprinol Ltd	Amateur Professional Industrial	5798
Wocosen 100 SL-C	Janssen Pharmaceutica NV	Professional Industrial	5807

674 Propiconazole + Dialkyldimethyl Ammonium Chloride + Permethrin

Traditional Preservative Oil	Oakmasters Of Sussex	Amateur Professional Industrial	6023

675 Propiconazole + Permethrin

Aqueous Universal	Protim Solignum Ltd	Amateur Professional	5776
Celpruf WB11	Protim Solignum Ltd	Industrial	6682
Dual Purpose FI8 (Microemulsion)	Antel UK Ltd	Professional	6585
EQ116 Dual Purpose FI	Enviroquest UK Ltd	Professional	6555
Lectro Micro 8 FI	Lectros International Ltd	Professional	6584
Micro 8 FI	Construction Chemicals	Professional	6644
Palace Dual Purpose FI 8 (Microemulsion)	Palace Chemicals Ltd	Professional	6646
Protim 340 WR	Protim Solignum Ltd	Industrial	5509
Protim 340E	Protim Solignum Ltd	Industrial	5511
Tritec 120 Plus	Triton Chemical Manufacturing Company Ltd	Professional	6289

Product Name	Marketing Company	Use	HSE No.

675 Propiconazole + Permethrin—continued

Tritec 120 Plus	Triton Chemical Manufacturing Company Ltd	Professional	6537
Tritec 121 Plus	Triton Chemical Manufacturing Company Ltd	Professional	6538
Ultra Tech 2000 D	Crown Chemicals Ltd	Professional	6304
Vacsele Aqua 6106	Hickson Timber Products Ltd	Amateur Professional Industrial	5893
W Wocosen 12 OL	Janssen Pharmaceutica NV	Amateur Professional Industrial	5404
Wocosen S-P	Janssen Pharmaceutica NV	Amateur Professional Industrial	5871
Wolsit KD 20	Dr Wolman GmbH	Industrial	6307
Wolsit KD 20C	Dr Wolman GmbH	Industrial	6966

676 Propiconazole + Tebuconazole

Protim FDR418	Protim Solignum Ltd	Industrial	6465
Protim FDR418V	Protim Solignum Ltd	Industrial	6652

677 Propiconazole + Tebuconazole + 3-Iodo-2-Propynyl-N-Butyl Carbamate

Protim 418J	Protim Solignum Ltd	Industrial	6046
Protim FDR 418J	Protim Solignum Ltd	Industrial	6467
Vacsol Azure 2720	Hickson Timber Products Ltd	Industrial	6344
Vacsol Azure 2721	Hickson Timber Products Ltd	Industrial	5820

678 Propiconazole + Tebuconazole + 3-Iodo-2-Propynyl-N-Butyl Carbamate + Permethrin

Vacsol 2718 2:1 Conc	Hickson Timber Products Ltd	Industrial	5884
Vacsol 2719 RTU	Hickson Timber Products Ltd	Industrial	5883
Vacsol Aqua 6107	Hickson Timber Products Ltd	Industrial	6082
Vacsol Azure 2722	Hickson Timber Products Ltd	Industrial	6462

679 Propiconazole + Tebuconazole + Boric Acid + Copper Carbonate Hydroxide

Tanalith E (3492)	Hickson Timber Products Ltd	Industrial	6269
Tanalith E (3494)	Hickson Timber Products Ltd	Industrial	6434

680 Propiconazole + Tebuconazole + Disodium Octaborate

Protim B610	Protim Solignum Ltd	Industrial	6366

681 Propiconazole + Tebuconazole + Disodium Octaborate + Permethrin

Protim E410	Protim Solignum Ltd	Industrial	6218

Product Name	Marketing Company	Use	HSE No.

681 Propiconazole + Tebuconazole + Disodium Octaborate + Permethrin—
continued

Protim E410 (i)	Protim Solignum Ltd	Industrial	6145
Protim E415	Protim Solignum Ltd	Industrial	5908
Protim E415 (i)	Protim Solignum Ltd	Industrial	6017

682 Propiconazole + Tebuconazole + Permethrin

Protim 415	Protim Solignum Ltd	Industrial	5815
Protim 415 T Special Solvent	Protim Solignum Ltd	Industrial	6077
Protim 415T	Protim Solignum Ltd	Industrial	5880
Protim 415TWR	Protim Solignum Ltd	Industrial	5916
Protim 418	Protim Solignum Ltd	Industrial	6466
Protim 418V	Protim Solignum Ltd	Industrial	6611
Protim 418WR	Protim Solignum Ltd	Industrial	6842
Vacsol 2627 Azure	Hickson Timber Products Ltd	Industrial	5558
Vacsol Aqua 6108	Hickson Timber Products Ltd	Industrial	6083
Vacsol Aqua 6108 RTU	Hickson Timber Products Ltd	Industrial	6994
Vacsol Aqua 6109	Hickson Timber Products Ltd	Industrial	6430

683 Sodium 2,4,6-Trichlorophenoxide + Boric Acid

| Protim Fentex TWR Plus | Protim Solignum Ltd | Industrial | 6547 |

684 Sodium 2-Phenylphenoxide

Aquaguard Wood Preserver	Langlow Products Ltd	Industrial	5936
D Fence Protector	Barrettine Products Ltd	Amateur Professional Industrial	5208
Mar-Cide	Marcher Chemicals Ltd	Professional (Surface Biocide) Professional (Wood Preservative) Industrial (Wood Preservative)	5631
Timbertek	Catomance PLC	Amateur	5920
Timbertek Conc	Advanced Bitumens Ltd	Industrial	5904
Timbertex PI/2	Marcher Chemicals Ltd	Professional Industrial	5628
W.S.P. Wood Treatment	Barrettine Products Ltd	Amateur Professional Industrial	5209

685 Sodium Dichromate + Arsenic Pentoxide + Copper Sulphate

| Celcure A Concentrate | Protim Solignum Ltd | Industrial | 5458 |
| Celcure A Fluid 10 | Protim Solignum Ltd | Industrial | 3764 |

Product Name	Marketing Company	Use	HSE No.

685 Sodium Dichromate + Arsenic Pentoxide + Copper Sulphate—continued

Celcure A Fluid 6	Protim Solignum Ltd	Industrial	3155
Celcure A Paste	Protim Solignum Ltd	Industrial	4523
Kemwood CCA Type BS	Laporte Kemwood AB	Industrial	5534
Laporte CCA Type 1	Laporte Wood Preservation	Industrial	3175
Laporte CCA Type 2	Laporte Wood Preservation	Industrial	3176
Laporte Permawood CCA	Laporte Wood Preservation	Industrial	3529
W Protim CCA Salts Type 2	Protim Solignum Ltd	Industrial	4972
Tanalith 3357	Hickson Timber Products Ltd	Industrial	4431
D Tanalith CL (3354)	Hickson Timber Products Ltd	Industrial	4196
Tanalith CP 3353	Hickson Timber Products Ltd	Industrial	4667
Tecca P2	Tecca Ltd	Industrial	3958

686 Sodium Dichromate + Boric Acid + Chromium Acetate + Copper Sulphate

Celgard CF	Protim Solignum Ltd	Industrial	4608

687 Sodium Dichromate + Boric Acid + Copper Sulphate

Celcure CB90	Protim Solignum Ltd	Industrial	5117
Laporte CCB	Laporte Wood Preservation	Industrial	5875
Tanalith CBC Paste 3402	Hickson Timber Products Ltd	Industrial	3833

688 Sodium Dichromate + Chromium Acetate + Copper Sulphate

Celcure B	Protim Solignum Ltd	Industrial	4541
Celcure O	Protim Solignum Ltd	Industrial	4539

689 Sodium Dichromate + Copper Sulphate

Laporte Cut End Preservative	Laporte Wood Preservation	Industrial	3528

690 Sodium Dichromate + Sodium Fluoride + Ammonium Bifluoride

W Rentex	Rentokil Initial UK Ltd	Industrial	4756

691 Sodium Fluoride + Ammonium Bifluoride + Sodium Dichromate

W Rentex	Rentokil Initial UK Ltd	Industrial	4756

692 Sodium Pentachlorophenoxide

W Protim Fentex WR	Protim Solignum Ltd	Industrial	4974
W Protim Panelguard	Protim Solignum Ltd	Industrial	5010
W Protim Plug Compound	Protim Solignum Ltd	Professional (Surface Biocide) Professional (Wood Preservative)	5012

693 Sodium Pentachlorophenoxide + Disodium Octaborate

W Gallwey ABS	Protim Solignum Ltd	Industrial	5014

Product Name	Marketing Company	Use	HSE No.

693 Sodium Pentachlorophenoxide + Disodium Octaborate—continued

W Protim Fentex Green Concentrate	Protim Solignum Ltd	Industrial	5122
W Protim Fentex Green RFU	Protim Solignum Ltd	Industrial	5123
W Protim Fentex M	Protim Solignum Ltd	Industrial	4973
D Protim Kleen II	Protim Solignum Ltd	Industrial	5203

694 Sodium Tetraborate + Alkyltrimethyl Ammonium Chloride

Sinesto B	Finnmex Co	Industrial	5136

695 Sodium Tetraborate + Boric Acid

Celbor	Protim Solignum Ltd	Industrial	4614
Diffusit	Dr Wolman GmbH	Professional	5992

696 Tebuconazole

Bayer WPC 2-1.5	Bayer PLC	Professional Industrial	5501
Bayer WPC 2-25	Bayer PLC	Professional Industrial	5500
Bayer WPC 2-25-SB	Bayer PLC	Professional Industrial	5502

697 Tebuconazole + 3-Iodo-2-Propynyl-N-Butyl Carbamate + Permethrin + Propiconazole

Vacsol 2718 2:1 Conc	Hickson Timber Products Ltd	Industrial	5884
Vacsol 2719 RTU	Hickson Timber Products Ltd	Industrial	5883
Vacsol Aqua 6107	Hickson Timber Products Ltd	Industrial	6082
Vacsol Azure 2722	Hickson Timber Products Ltd	Industrial	6462

698 Tebuconazole + 3-Iodo-2-Propynyl-N-Butyl Carbamate + Propiconazole

Protim 418J	Protim Solignum Ltd	Industrial	6046
Protim FDR418J	Protim Solignum Ltd	Industrial	6467
Vacsol Azure 2720	Hickson Timber Products Ltd	Industrial	6344
Vacsol Azure 2721	Hickson Timber Products Ltd	Industrial	5820

699 Tebuconazole + Benzalkonium Chloride

Vacsol Aqua 6101	Hickson Timber Products Ltd	Industrial	5557

700 Tebuconazole + Benzalkonium Chloride + Permethrin

Vacsol Aqua 6105	Hickson Timber Products Ltd	Industrial	5892

701 Tebuconazole + Boric Acid + Copper Carbonate Hydroxide

Ensele 3428	Hickson Timber Products Ltd	Professional Industrial	5878
Tanalith E (3485)	Hickson Timber Products Ltd	Industrial	5562

HSE
WOOD PRESERVATIVES

Product Name	Marketing Company	Use	HSE No.

702 Tebuconazole + Boric Acid + Copper Carbonate Hydroxide + Propiconazole

Tanalith E (3492)	Hickson Timber Products Ltd	Industrial	6269
Tanalith E (3494)	Hickson Timber Products Ltd	Industrial	6434

703 Tebuconazole + Dichlofluanid + Permethrin

Impra-Holzschutzgrund (Primer)	Impra Systems Ltd	Professional Industrial	5735

704 Tebuconazole + Disodium Octaborate + Permethrin + Propiconazole

Protim E410	Protim Solignum Ltd	Industrial	6218
Protim E410 (i)	Protim Solignum Ltd	Industrial	6145
Protim E415	Protim Solignum Ltd	Industrial	5908
Protim E415 (i)	Protim Solignum Ltd	Industrial	6017

705 Tebuconazole + Disodium Octaborate + Propiconazole

Protim B610	Protim Solignum Ltd	Industrial	6366

706 Tebuconazole + Permethrin

Brunol ATP New	Philip Johnstone Group Ltd	Professional Industrial	5620
Brunol STP	Philip Johnstone Group Ltd	Professional Industrial	5621

707 Tebuconazole + Permethrin + Propiconazole

Protim 415	Protim Solignum Ltd	Industrial	5815
Protim 415 T Special Solvent	Protim Solignum Ltd	Industrial	6077
Protim 415T	Protim Solignum Ltd	Industrial	5880
Protim 415TWR	Protim Solignum Ltd	Industrial	5916
Protim 418	Protim Solignum Ltd	Industrial	6466
Protim 418V	Protim Solignum Ltd	Industrial	6611
Protim 418WR	Protim Solignum Ltd	Industrial	6842
Vacsol 2627 Azure	Hickson Timber Products Ltd	Industrial	5558
Vacsol Aqua 6108	Hickson Timber Products Ltd	Industrial	6083
Vacsol Aqua 6108 RTU	Hickson Timber Products Ltd	Industrial	6994
Vacsol Aqua 6109	Hickson Timber Products Ltd	Industrial	6430

708 Tebuconazole + Propiconazole

Protim FDR418	Protim Solignum Ltd	Industrial	6465
Protim FDR418V	Protim Solignum Ltd	Industrial	6652

709 2-(Thiocyanomethylthio)Benzothiazole

Bayer WPC-3	Bayer PLC	Industrial	4969
Bayer WPC-4	Bayer PLC	Industrial	4970
Busan 1111	Buckman Laboratories SA	Industrial	4764

Product Name	Marketing Company	Use	HSE No.

709 2-(Thiocyanomethylthio)Benzothiazole—continued

Crysolite Glaramara Concentrated Timber Treatment	Crysolite Protective Coatings	Industrial	5443
Fentex NP-UF	Protim Solignum Ltd	Industrial	5069
Paneltone	Protim Solignum Ltd	Industrial	4253

710 2-(Thiocyanomethylthio)Benzothiazole + Boric Acid

| Protim Fentex Europa 1 RFU | Protim Solignum Ltd | Industrial | 5119 |
| Protim Fentex Europa I | Protim Solignum Ltd | Industrial | 4984 |

711 2-(Thiocyanomethylthio)Benzothiazole + Boric Acid + Methylene Bis(Thiocyanate)

| Fentex Elite | Protim Solignum Ltd | Industrial | 5532 |

712 2-(Thiocyanomethylthio)Benzothiazole + Disodium Octaborate + Methylene Bis(Thiocyanate)

| Timbertex PD | Marcher Chemicals Ltd | Industrial | 5876 |

713 2-(Thiocyanomethylthio)Benzothiazole + Methylene Bis(Thiocyanate)

Busan 1009	Buckman Laboratories SA	Industrial	5256
Celbrite TC	Protim Solignum Ltd	Industrial	3921
Hickson Antiblu 3739	Hickson Timber Products Ltd	Industrial	4841
Injecta Osmose S-11	Injecta Osmose Ltd	Industrial	5957
Laporte Antisapstain	Laporte Wood Preservation	Industrial	5977
Mect	Buckman Laboratories SA	Industrial	5255
Protim Stainguard	Protim Solignum Ltd	Industrial	5009

714 Tri(Hexylene Glycol)Biborate

ABL Wood Preservative (B)	Advanced Bitumens Ltd	Amateur Professional Industrial	3743
D Barrettine New Preserver	Barrettine Products Ltd	Amateur Professional Industrial	3624
BCR Wood Preserver	Building Chemical Research Ltd	Amateur Professional Industrial	6315
Brown Wood Preservative	Palace Chemicals Ltd	Amateur Professional Industrial	5242
Clear Wood Preservative	Palace Chemicals Ltd	Amateur Professional Industrial	5240
Crown Timber Preservative	Crown Chemicals Ltd	Amateur Professional Industrial	6882

Product Name	Marketing Company	Use	HSE No.

714 Tri(Hexylene Glycol)Biborate—continued

Product Name	Marketing Company	Use	HSE No.
Everbuild Lumberjack Wood Preservative	Everbuild Building Products Ltd	Amateur Professional Industrial	6900
Kingfisher Wood Preservative	Kingfisher Chemicals Ltd	Professional	4012
Langlow Wood Preserver	Langlow Products Ltd	Amateur Professional Industrial	3690
Langlow Wood Preserver Formulation B	Langlow Products Ltd	Amateur Professional Industrial	3754
Larsen Joinery Grade 2	Larsen Manufacturing Ltd	Professional Industrial	4286
Osmose Endcoat	Protim Solignum Ltd	Amateur Professional	7049
Palace Base Coat Wood Preservative	Palace Chemicals Ltd	Amateur Professional Industrial	5351
Palace Microfine Fungicide Concentrate	Palace Chemicals Ltd	Professional Industrial	5551
D PC-F/B	Terminix Peter Cox Ltd	Amateur Professional	3739
D PC-XJ/2	Terminix Peter Cox Ltd	Amateur Professional	3737
Protim 260 F	Protim Solignum Ltd	Amateur Professional Industrial	4998
Richard Burbidge Green End Coat	Richard Burbidge Ltd	Amateur Professional	7037
Safeguard Deepwood Fungicide	Safeguard Chemicals Ltd	Amateur Professional	4183
Sandhill Wood Preservative Clear	Sandhill Building Products	Amateur Professional	5951
D Sovereign AQ/FT	Sovereign Chemical Industries Ltd	Professional Industrial	5433
Sovereign Coloured Wood Preservative	Sovereign Chemical Industries Ltd	Professional Industrial	6165
Sovereign Dipcoat	Sovereign Chemical Industries Ltd	Professional Industrial	5783
Sovereign Timber Preservative	Sovereign Chemical Industries Ltd	Professional	3807
Timber Preservative With Repellent	Antel UK Ltd	Professional	4848
Timeless Timber All Purpose End Coat And Preserver	BSW Timber PLC	Amateur Professional	7043

Product Name	Marketing Company	Use	HSE No.

715 Tri(Hexylene Glycol)Biborate + Copper Naphthenate

Product Name	Marketing Company	Use	HSE No.
W BCR Green Wood Preserver	Building Chemical Research Ltd	Amateur Professional Industrial	6285
W Dark Green Wood Preservative	Palace Chemicals Ltd	Amateur Professional Industrial	5243
W Kingfisher Wood Preservative	Kingfisher Chemicals Ltd	Professional	4011
W Langlow Green Wood Preserver	Langlow Products Ltd	Amateur Professional Industrial	6279
W Langlow Pale Green Wood Preserver	Langlow Products Ltd	Amateur Professional Industrial	6068
W Pale Green Wood Preservative	Palace Chemicals Ltd	Amateur Professional Industrial	5241
W Sovereign Timber Preservative Pale Green	Sovereign Chemical Industries Ltd	Professional	3806

716 Tri(Hexylene Glycol)Biborate + Cypermethrin

Product Name	Marketing Company	Use	HSE No.
BCR Universal Wood Preserver	Building Chemical Research Ltd	Amateur Professional	6290
Cementone Multiplus	Philip Johnstone Group Ltd	Amateur Professional	3849
Crown Fungicide Insecticide	Crown Chemicals Ltd	Professional	6003
Crown Fungicide Insecticide Concentrate	Crown Chemicals Ltd	Professional	5463
Crown Micro Dual Purpose	Crown Chemicals Ltd	Professional	6414
Devatern Wood Preserver	Sorex Ltd	Professional Industrial	4073
Ecology Fungicide Insecticide Concentrate	Palace Chemicals Ltd	Professional	3792
Fungicide Insecticide	Palace Chemicals Ltd	Amateur Professional	3795
Green Range Dual Purpose AQ	Philip Johnstone Group Ltd	Professional	3725
Green Range Wykamol Plus	Philip Johnstone Group Ltd	Professional	3731
W Langlow Clear Wood Preserver	Langlow Products Ltd	Amateur Professional Industrial	3802
Nubex Emulsion Concentrate CB (Low Odour)	Philip Johnstone Group Ltd	Professional	5328
Nubex Woodworm All Purpose CB	Philip Johnstone Group Ltd	Professional	3189
Palace Microfine Fungicide Insecticide (Concentrate)	Palace Chemicals Ltd	Professional Industrial	5421
D PC-H/4	Terminix Peter Cox Ltd	Amateur Professional	3740
W PCX-122	Terminix Peter Cox Ltd	Professional	3999

Product Name	Marketing Company	Use	HSE No.

716 Tri(Hexylene Glycol)Biborate + Cypermethrin—continued

| W PCX-122 Concentrate | Terminix Peter Cox Ltd | Professional | 4000 |

717 Tri(Hexylene Glycol)Biborate + Dichlofluanid

ABL Universal Woodworm Killer DB	Advanced Bitumens Ltd	Amateur Professional Industrial	3858
Blackfriar Wood Preserver	E Parsons And Sons Ltd	Amateur Professional	6443
Blackfriars New Wood Preserver	E Parsons And Sons Ltd	Amateur Professional Industrial	4150
Double Action Timber Preservative For Doors	International Paint Ltd	Amateur	3707
Double Action Wood Preservative	International Paint Ltd	Amateur	3706
Homebase Weathercoat Deep Penetrating Preservative Primer	Homebase Ltd	Amateur Professional	7084
Johnstone's Woodworks Exterior Universal Primer	Kalon Decorative Products	Amateur Professional	7070
D Nitromors Timbrene Clear Environmental Formula	Henkel Home Improvements And Adhesive Products	Amateur Professional Industrial	4106
Premier Barrettine New Wood Preserver	Barrettine Products Ltd	Amateur Professional Industrial	4038
Premier Pro-Tec (S)	Premier Coatings	Amateur Professional	6978
W Ranch Preservative	International Paint Ltd	Amateur	4085
Ranch Preservative	Plascon International	Amateur	6085
Sadolin New Base	Sadolin UK Ltd	Amateur Professional Industrial	4341
Sealocrete Timber Master Wood Preserver	Sealocrete PLA Ltd	Amateur Professional	7052
Ultrabond Woodworm Killer	Philip Johnstone Group Ltd	Amateur Professional	4344

718 Tri(Hexylene Glycol)Biborate + Dichlofluanid + Permethrin

Blackfriar Wood Preserver Wp Gold Star Clear	E Parsons And Sons Ltd	Amateur Professional	6417
Blackfriars Gold Star Clear	E Parsons And Sons Ltd	Amateur Professional Industrial	4149
Johnstone's Woodworks Interior/ Exterior All Purpose Preserver	Kalon Decorative Products	Amateur Professional	7071
Leyland Timbrene Supreme Environmental Formula	Leyland Paint Company Ltd	Amateur Professional	4800

Product Name	Marketing Company	Use	HSE No.

718 Tri(Hexylene Glycol)Biborate + Dichlofluanid + Permethrin—continued

D Nitromors Timbrene Supreme Environmental Formula	Henkel Home Improvements And Adhesive Products	Amateur Professional Industrial	4107
Premier Barrettine New Universal Fluid D	Barrettine Products Ltd	Amateur Professional Industrial	4004
Timber Master 6 In 1 Wood Treatment	Sealocrete PLA Ltd	Amateur Professional	7080
Timbrene Supreme	Kalon Group PLC	Amateur Professional	4995

719 Tri(Hexylene Glycol)Biborate + Permethrin

Antel Dual Purpose (BP) Concentrate (Water Dilutable)	Antel UK Ltd	Professional	4046
D Barrettine New Universal Fluid	Barrettine Products Ltd	Amateur Professional Industrial	3623
BPC8-DP	Building Preservation Centre	Professional	6306
Brunol PBO	Philip Johnstone Group Ltd	Professional	4474
Brunol Special P	Philip Johnstone Group Ltd	Professional	3864
D Chel-Wood Preserver/Woodworm Dry Rot Killer BP	Chelec Ltd	Amateur Professional	4897
Crown Micro Dual Purpose F/I	Crown Chemicals Ltd	Professional	6895
Crown Timber Paste	Crown Chemicals Ltd	Professional	5869
Deepwood FI Dual Purpose Emulsion Concentrate	Safeguard Chemicals Ltd	Professional Industrial	5087
Dual 8	Deal Direct Ltd	Professional	6255
W Ecology Fungicide Insecticide Aqueous (Concentrate) Wood Preservative Dual Purpose	Kingfisher Chemicals Ltd	Professional	4039
Ecology Fungicide Insecticide Aqueous (Concentrate) Wood Preservative Dual Purpose	Kingfisher Chemicals Ltd	Professional	5930
Everbuild Lumberjack Universal Rot And Woodworm Killer	Palace Chemicals Ltd	Amateur Professional Industrial	6906
D Insecticide Fungicide Wood Preservative	Kingfisher Chemicals Ltd	Professional	4048
Insecticide Fungicide Wood Preservative	Kingfisher Chemicals Ltd	Professional	5933
J B Ward & Son Dual Purpose Timber Treatment Concentrate	J B Ward And Son	Professional Industrial	7050
KF-8 Dual Purpose Micro Emulsion Concentrate	Kingfisher Chemicals Ltd	Professional Industrial	6075
Kingfisher Timberpaste	Kingfisher Chemicals Ltd	Professional	6686

Product Name	Marketing Company	Use	HSE No.

719 Tri(Hexylene Glycol)Biborate + Permethrin—continued

Product Name	Marketing Company	Use	HSE No.
Larsen Wood Preservative Clear 2 And Brown 2	Larsen Manufacturing Ltd	Amateur Professional Industrial	4287
Larsen Woodworm Killer 2	Larsen Manufacturing Ltd	Amateur Professional Industrial	4285
Lectro Paste PB	Lectros International Ltd	Professional	6130
Mar-Kil S	Marcher Chemicals Ltd	Professional Industrial	5234
Mar-Kil W	Marcher Chemicals Ltd	Professional Industrial	5219
Micron 8 Dual Purpose Emulsion Concentrate	Safeguard Chemicals Ltd	Professional Industrial	5851
Omega Plus	Restoration (UK) Ltd	Professional	6382
Perma AQ Dual Purpose	Triton Perma Industries Ltd	Professional	5617
Permapaste PB	Triton Perma Industries Ltd	Professional	5616
Permethrin F And I Concentrate	Permagard Products Ltd	Professional	4008
Protim CB	Protim Solignum Ltd	Professional Industrial	5864
Protim CDB	Protim Solignum Ltd	Amateur Professional Industrial	4999
Protim WR 260	Protim Solignum Ltd	Amateur Professional Industrial	4997
D Remecology Spirit Based K7	Remtox (Chemicals) Ltd	Professional	4248
Restor-8-Dual Purpose Concentrate	Restoration (UK) Ltd	Professional Industrial	5963
Restor-Paste	Restoration (UK) Ltd	Professional	6079
D Safeguard Deepwood II Dual Purpose Emulsion Concentrate	Safeguard Chemicals Ltd	Professional	5303
Safeguard Deepwood II Dual Purpose Emulsion Concentrate	Safeguard Chemicals Ltd	Professional	5850
Safeguard Deepwood III Timber Treatment	Safeguard Chemicals Ltd	Amateur	3937
D Safeguard Deepwood Paste	Safeguard Chemicals Ltd	Professional	4110
Sandhill Aqueous Fungicide - Insecticide	Sandhill Building Products	Professional	5950
Sandhill Fungicide - Insecticide	Sandhill Building Products	Amateur	5952
Solignum Cut End	Protim Solignum Ltd	Amateur Professional Industrial	7046
Solignum Remedial Concentrate	Protim Solignum Ltd	Professional	5021
Solignum Remedial PB	Protim Solignum Ltd	Amateur Professional Industrial	5607

Product Name	Marketing Company	Use	HSE No.

719 Tri(Hexylene Glycol)Biborate + Permethrin—continued

Product Name	Marketing Company	Use	HSE No.
Solignum Universal	Protim Solignum Ltd	Amateur Professional Industrial	7042
Sovereign Aqueous Fungicide/ Insecticide 2	Sovereign Chemical Industries Ltd	Professional Industrial	4619
Sovereign Insecticide/Fungicide 2	Sovereign Chemical Industries Ltd	Professional	4649
Sprytech 2	Spry Chemicals Ltd	Professional Industrial	6500
Timber Paste	Grip-Fast Systems Ltd	Professional	6409
D Trimethrin AQ Plus	Triton Chemicals International Ltd	Professional	3522
Trimethrin OS Plus	Triton Chemical Manufacturing Company Ltd	Amateur Professional	3521
Tripaste PB	Triton Chemicals International Ltd	Professional	6080
Tritec Plus	Triton Chemical Manufacturing Company Ltd	Professional	5425
D Universal Fluid (Grade PB)	Barrettine Products Ltd	Amateur Professional Industrial	5212
Universal Wood Preserver	Langlow Products Ltd	Amateur Professional Industrial	3692
Woodworm Killer	Sealocrete PLA Ltd	Amateur Professional	5028

720 Tri(Hexylene Glycol)Biborate + Zinc Naphthenate + Dichlofluanid

Product Name	Marketing Company	Use	HSE No.
W Weathershield Exterior Preservative Basecoat	ICI Paints	Amateur Professional	5245
W Weathershield Preservative Primer	ICI Paints	Amateur Professional	5244

721 Tri(Hexylene Glycol)Biborate + Zinc Octoate + Dichlofluanid

Product Name	Marketing Company	Use	HSE No.
W Timber Preservative Clear TFP7	Johnstone's Paints PLC	Amateur Professional	4016

722 Tributyltin Naphthenate

Product Name	Marketing Company	Use	HSE No.
Celpruf Primer TN	Protim Solignum Ltd	Industrial	5378
Celpruf TN	Protim Solignum Ltd	Industrial	4701
Celpruf TN WR	Protim Solignum Ltd	Industrial	4702
Protim FDR 230	Protim Solignum Ltd	Industrial	5046
Protim JP 230	Protim Solignum Ltd	Industrial	5779
Vacsol 2652/2653 JWR	Hickson Timber Products Ltd	Industrial	4646

HSE
WOOD PRESERVATIVES

Product Name	Marketing Company	Use	HSE No.

722 Tributyltin Naphthenate—continued
Vacsol 2746/2747 J	Hickson Timber Products Ltd	Industrial	4807

723 Tributyltin Naphthenate + Lindane
W Protim 230L	Protim Solignum Ltd	Industrial	6964

724 Tributyltin Naphthenate + Permethrin
Celpruf TNM	Protim Solignum Ltd	Industrial	4704
Celpruf TNM WR	Protim Solignum Ltd	Industrial	4703
Protim 230	Protim Solignum Ltd	Industrial	5044
Protim 230 WR	Protim Solignum Ltd	Industrial	5045
Vacsol 2622/2623 WR	Hickson Timber Products Ltd	Industrial	4645
Vacsol 2625/2626 WR 2:1 Concentrate	Hickson Timber Products Ltd	Industrial	5329
Vacsol 2713/2714	Hickson Timber Products Ltd	Industrial	4808

725 Tributyltin Oxide
D Celpruf Primer	Rentokil Initial UK Ltd	Industrial	3796
Protim 215 PP	Protim Solignum Ltd	Industrial	5034
Protim FDR 2125C	Protim Solignum Ltd	Industrial	6146
Protim FDR 215C	Protim Solignum Ltd	Industrial	6126
Protim FDR 210	Protim Solignum Ltd	Industrial	5278
Protim FDR 215	Protim Solignum Ltd	Industrial	6322
Protim JP 210	Protim Solignum Ltd	Industrial	5032
Vacsol 2234 J Conc	Hickson Timber Products Ltd	Industrial	4665
Vacsol 2243	Hickson Timber Products Ltd	Industrial	6215
Vacsol 2244	Hickson Timber Products Ltd	Industrial	6351
Vacsol J RTU	Hickson Timber Products Ltd	Industrial	3152
Vacsol JWR Concentrate	Hickson Timber Products Ltd	Industrial	3153
Vacsol JWR RTU	Hickson Timber Products Ltd	Industrial	3154
D Wood Preservative AA 155/01	Becker Acroma Kemira Ltd	Industrial	5309

726 Tributyltin Oxide + Dichlofluanid
D Celpruf CP Special	Rentokil Initial UK Ltd	Industrial	4582
Hickson Woodex	Hickson Timber Products Ltd	Industrial	5405
Protim Cedar	Protim Solignum Ltd	Industrial	5042
Protim Solignum Softwood Basestain CS	Protim Solignum Ltd	Industrial	5647
D Wood Preservative AA 155/00	Becker Acroma Kemira Ltd	Industrial	5308
D Wood Preservative AA 155/03	Becker Acroma Kemira Ltd	Industrial	5310
D Wood Preservative AA155	Becker Acroma Kemira Ltd	Industrial	5311

727 Tributyltin Oxide + Lindane
W Cedasol 2320	Hickson Timber Products Ltd	Industrial	4365
D Dipsar	Cementone Beaver Ltd	Industrial	3604
D Hickson Timbercare WRQD	Hickson Timber Products Ltd	Industrial	3594
D Lar-Vac 400	Larsen Manufacturing Ltd	Industrial	5782

Product Name	Marketing Company	Use	HSE No.

727 Tributyltin Oxide + Lindane—continued

W Protim 210	Protim Solignum Ltd	Industrial	5033
W Protim 210 C	Protim Solignum Ltd	Industrial	5038
W Protim 210 CWR	Protim Solignum Ltd	Industrial	5035
W Protim 210 WR	Protim Solignum Ltd	Industrial	5039
W Protim 215	Protim Solignum Ltd	Industrial	6016
W Protim 23 WR	Protim Solignum Ltd	Industrial	5037
W Protim 90	Protim Solignum Ltd	Industrial	5276
W Protim Paste	Protim Solignum Ltd	Professional	4979
W Vacsol MWR Concentrate 2203	Hickson Timber Products Ltd	Industrial	3170
Vacsol MWR Ready To Use 2204	Hickson Timber Products Ltd	Industrial	3167
W Vacsol WR Ready To Use 2116	Hickson Timber Products Ltd	Industrial	3168

728 Tributyltin Oxide + Lindane + Pentachlorophenol

D Celpruf PK	Rentokil Initial UK Ltd	Industrial	4540
D Celpruf PK WR	Rentokil Initial UK Ltd	Industrial	4538
D Lar-Vac 100	Larsen Manufacturing Ltd	Industrial	3723
W Protim 80	Protim Solignum Ltd	Industrial	5029
W Protim 80 C	Protim Solignum Ltd	Industrial	5036
W Protim 80 CWR	Protim Solignum Ltd	Industrial	5031
W Protim 80 WR	Protim Solignum Ltd	Industrial	5030
W Protim Brown	Protim Solignum Ltd	Industrial	5000

729 Tributyltin Oxide + Pentachlorophenol

D Celpruf JP	Rentokil Initial UK Ltd	Industrial	4548
D Celpruf JP WR	Rentokil Initial UK Ltd	Industrial	4543
W Protim FDR-H	Protim Solignum Ltd	Industrial	4985
W Protim JP	Protim Solignum Ltd	Industrial	4978
D Protim R Clear	Protim Solignum Ltd	Industrial	5013

730 Zinc Naphthenate

W Carbo Wood Preservative	Talke Chemical Company Ltd	Amateur Professional Industrial	3821
W Spencer Wood Preservative XP	Spencer Coatings Ltd	Amateur Professional	3947
W Teamac Wood Preservative	Teal And Mackrill Ltd	Amateur Professional	4428
W Wood Preservative	C W Wastnage Ltd	Amateur Professional Industrial	5284

731 Zinc Naphthenate + Dichlofluanid

W Bradite Timber Preservative	Bradite Ltd	Amateur	6091
W Masterstroke Wood Preserver	Akzo Nobel Decorative Coatings Ltd	Amateur	5689

HSE
WOOD PRESERVATIVES

Product Name	Marketing Company	Use	HSE No.

731 Zinc Naphthenate + Dichlofluanid—continued

Product Name	Marketing Company	Use	HSE No.
W Sikkens Wood Preserver	Akzo Nobel Woodcare	Amateur Professional	6365

732 Zinc Naphthenate + Dichlofluanid + Tri(Hexylene Glycol)Biborate

Product Name	Marketing Company	Use	HSE No.
W Weathershield Exterior Preservative Basecoat	ICI Paints	Amateur Professional	5245
W Weathershield Preservative Primer	ICI Paints	Amateur Professional	5244

733 Zinc Octoate

Product Name	Marketing Company	Use	HSE No.
W B And Q Clear Wood Preserver	B & Q PLC	Amateur	3013
W B And Q Exterior Wood Preservative	B & Q PLC	Amateur	4705
W B And Q Formula Wood Preserver	B & Q PLC	Amateur	3014
W Crown Timber Preservative	Crown Chemicals Ltd	Professional	5442
W Homebase Weathercoat Preservative Primer	Sainsbury's Homebase House And Garden Centres	Amateur Professional	5902
W Johnstone's Preservative Clear	Kalon Group PLC	Amateur Professional	6319
W Leyland Timbrene Environmental Formula	Leyland Paint Company Ltd	Amateur Professional	4801
W Leyland Universal Preservative Base	Kalon Group PLC	Amateur Professional	3760
D Nitromors Timbrene Environmental Formula Wood Preservative	Henkel Home Improvements And Adhesive Products	Amateur Professional Industrial	4108
W PLA Wood Preserver	Sealocrete PLA Ltd	Amateur Professional Industrial	5340
W Premier Pro-Tec(S) Environmental Formula Wood Preservative	Premier Decorative Products	Amateur Professional Industrial	4689
W Rentokil Dry Rot And Wet Rot Treatment	Rentokil Initial UK Ltd	Amateur (Surface Biocide) Amateur (Wood Preservative) Professional (Surface Biocide) Professional (Wood Preservative)	4378

Product Name	Marketing Company	Use	HSE No.

733 Zinc Octoate—continued

W Ronseal Low Odour Wood Preserver	Ronseal Ltd	Amateur	4919
W Ronseal's All-Purpose Wood Preserver	Ronseal Ltd	Amateur Professional	6185
W Square Deal Deep Protection Wood Preserver	Texas Homecare Ltd	Amateur Professional	4640
W Thompson's All Purpose Wood Preserver	Ronseal Ltd	Amateur Professional	5988
W Timber Preservative	Antel UK Ltd	Professional	4846
W Timbrene Wood Preserver	Kalon Group PLC	Amateur Professional	4966
W Trade Ronseal Low Odour Wood Preserver	Ronseal Ltd	Professional Industrial	4932

734 Zinc Octoate + Boric Acid

D Celpruf BZ	Rentokil Initial UK Ltd	Industrial	4188
D Celpruf BZ WR	Rentokil Initial UK Ltd	Industrial	4189
W Rentokil Dry Rot Paste (D)	Rentokil Initial UK Ltd	Professional	3987

735 Zinc Octoate + Boric Acid + Permethrin

D Celpruf BZP	Rentokil Initial UK Ltd	Industrial	4190
D Celpruf BZP WR	Rentokil Initial UK Ltd	Industrial	4191
W Cut'N'treat	Rentokil Initial UK Ltd	Amateur Professional	5802
W Premium Grade Wood Treatment	Rentokil Initial UK Ltd	Amateur Professional	4193

736 Zinc Octoate + Copper Naphthenate

W Weathershield Exterior Timber Preservative	ICI Paints	Amateur Professional	4915

737 Zinc Octoate + Dichlofluanid

W Exterior Preservative Primer	Windeck Paints Ltd	Amateur	3857
W Timbercare Microporous Exterior Preservative	Manders Paints Ltd	Amateur Professional	5157
D Wilko Exterior System Preservative Primer	Wilkinson Home And Garden Stores	Amateur	3686

738 Zinc Octoate + Dichlofluanid + Permethrin

W OS Color Wood Stain And Preservative	Ostermann And Scheiwe GmbH	Amateur	5531
W OS Color WR	Ostermann And Scheiwe GmbH	Amateur	5364

Product Name	Marketing Company	Use	HSE No.

739 Zinc Octoate + Dichlofluanid + Tri(Hexylene Glycol)Biborate

W Timber Preservative Clear TFP7	Johnstone's Paints PLC	Amateur Professional	4016

740 Zinc Octoate + Permethrin

W Green Range Timber Treatment Paste	Philip Johnstone Group Ltd	Professional	4651
W Larvac 300	Larsen Manufacturing Ltd	Industrial	4461
W Palace Timbertreat Ecology	Palace Chemicals Ltd	Professional	4931
W Premium Wood Treatment	Rentokil Initial UK Ltd	Amateur (Surface Biocide) Amateur (Wood Preservative) Professional (Surface Biocide) Professional (Wood Preservative)	5350
D Rentokil Dual Purpose Fluid	Rentokil Initial UK Ltd	Professional	4074
W Rentokil Wood Preservative	Rentokil Initial UK Ltd	Amateur Professional	4195
W Ronseal's Universal Wood Treatment	Ronseal Ltd	Amateur Professional	6186
W Solignum Colourless	Protim Solignum Ltd	Amateur Professional Industrial	5549
W Solignum Wood Preservative Paste	Protim Solignum Ltd	Professional	5001
W Thompson's Universal Wood Treatment	Ronseal Ltd	Amateur Professional	5986
W Wood Preservative Clear	Rentokil Initial UK Ltd	Amateur Professional	4077
W Woodtreat BP	Philip Johnstone Group Ltd	Professional	3085

741 Zinc Versatate

W Celpruf Primer ZV	Rentokil Initial UK Ltd	Professional Industrial	5379
W Celpruf ZO	Rentokil Initial UK Ltd	Industrial	4683
W Celpruf ZO WR	Rentokil Initial UK Ltd	Industrial	4684
W Protim FDR 220	Protim Solignum Ltd	Industrial	5777
W Protim JP 220	Protim Solignum Ltd	Industrial	5197
W Protim Paste 220 F	Protim Solignum Ltd	Professional Industrial	4988
W Vacsol J (2:1 Conc) 2744/2745	Hickson Timber Products Ltd	Industrial	4687
W Vacsol J 2742/2743	Hickson Timber Products Ltd	Industrial	4686

Product Name	Marketing Company	Use	HSE No.
741 Zinc Versatate —continued			
W Vacsol JWR (2:1 Conc) 2642/2643	Hickson Timber Products Ltd	Industrial	4677
W Vacsol JWR 2640/2641	Hickson Timber Products Ltd	Industrial	4680
D Vacsol P Ready To Use (2335)	Hickson Timber Products Ltd	Industrial	4116
742 Zinc Versatate + Cypermethrin			
W Dipsar G R	Philip Johnstone Group Ltd	Industrial	4449
W Everclear	Injecta Osmose Ltd	Industrial	6058
743 Zinc Versatate + Permethrin			
W Cedasol Ready To Use (2306)	Hickson Timber Products Ltd	Industrial	4117
W Celpruf ZOP	Rentokil Initial UK Ltd	Industrial	4681
W Celpruf ZOP WR	Rentokil Initial UK Ltd	Industrial	4682
W Imersol 2410	Hickson Timber Products Ltd	Industrial	4674
W Protim 220	Protim Solignum Ltd	Industrial	5198
W Protim 220 CWR	Protim Solignum Ltd	Amateur Professional Industrial	5196
W Protim 220 WR	Protim Solignum Ltd	Industrial	5195
W Protim 220C	Protim Solignum Ltd	Industrial	5780
W Protim Paste 220	Protim Solignum Ltd	Professional Industrial	4983
W Solignum Universal	Protim Solignum Ltd	Amateur Professional Industrial	5019
W Vacsele 2611	Hickson Timber Products Ltd	Professional	4675
W Vacsele P2312	Hickson Timber Products Ltd	Professional Industrial	4363
W Vacsol (2:1 Conc) 2711/2712	Hickson Timber Products Ltd	Industrial	4685
W Vacsol 2709/2710	Hickson Timber Products Ltd	Industrial	4678
W Vacsol WR (2:1 Conc) 2614/2615	Hickson Timber Products Ltd	Industrial	4679
W Vacsol WR 2612/2613	Hickson Timber Products Ltd	Industrial	4676

8

WOOD TREATMENT

Product Name	Marketing Company	Use	HSE No.

Wood Treatment

744 Boric Acid + Sodium 2,4,6-Trichlorophenoxide

Protim Fentex Phoenix	Protim Solignum Ltd	Industrial	7108
Protim Fentex T	Protim Solignum Ltd	Industrial	6367
Protim Fentex TG	Protim Solignum Ltd	Industrial	6440
Protim Fentex TP	Protim Solignum Ltd	Industrial	6370
Protim Fentex TPS	Protim Solignum Ltd	Industrial	6371
Protim Fentex TW	Protim Solignum Ltd	Industrial	6368
Protim Fentex TW	Protim Solignum Ltd	Industrial	6876

745 Carbendazim + Dichlorophenyl Dimethylurea + 2-Octyl-2h-Isothiazol-3-One

Cuprinol Garden Shades	Cuprinol Ltd	Amateur Professional	6326

746 Carbendazim + Dichlorophenyl Dimethylurea + 2-Octyl-2h-Isothiazolin-3-One

Cuprinol Ducksback	Cuprinol Ltd	Amateur Professional	6327
Cuprinol Garden Decking Seal	Cuprinol Ltd	Amateur Professional	6858

747 Dichlorophenyl Dimethylurea + 2-Octyl-2h-Isothiazol-3-One + Carbendazim

Cuprinol Garden Shades	Cuprinol Ltd	Amateur Professional	6326

748 Dichlorophenyl Dimethylurea + 2-Octyl-2h-Isothiazolin-3-One + Carbendazim

Cuprinol Ducksback	Cuprinol Ltd	Amateur Professional	6327
Cuprinol Garden Decking Seal	Cuprinol Ltd	Amateur Professional	6858

749 3-Iodo-2-Propynyl-N-Butyl Carbamate

ACS Dry Rot Paint	Advanced Chemical Specialties Ltd	Amateur Professional	6310
Lectros Dry Rot Paint	Advanced Chemical Specialties Ltd	Professional (Biocidal Paint) Amateur (Biocidal Paint) Amateur (Wood Treatment) Professional (Wood Treatment)	6520

Product Name	Marketing Company	Use	HSE No.

750 3-Iodo-2-Propynyl-N-Butyl Carbamate + Permethrin

Woodstain Aquarethane	Syntilor	Amateur	7101

751 2-Octyl-2h-Isothiazol-3-One + Carbendazim + Dichlorophenyl Dimethylurea

Cuprinol Garden Shades	Cuprinol Ltd	Amateur Professional	6326

752 2-Octyl-2h-Isothiazolin-3-One + Carbendazim + Dichlorophenyl Dimethylurea

Cuprinol Ducksback	Cuprinol Ltd	Amateur Professional	6327
Cuprinol Garden Decking Seal	Cuprinol Ltd	Amateur Professional	6858

753 Permethrin + 3-Iodo-2-Propynyl-N-Butyl Carbamate

Woodstain Aquarethane	Syntilor	Amateur	7101

754 Sodium 2,4,6-Trichlorophenoxide

Protek 9 Star L40	Protek Products (Sun Europa) Ltd	Industrial	7090

755 Sodium 2,4,6-Trichlorophenoxide + Boric Acid

Protim Fentex Phoenix	Protim Solignum Ltd	Industrial	7108
Protim Fentex T	Protim Solignum Ltd	Industrial	6367
Protim Fentex TG	Protim Solignum Ltd	Industrial	6440
Protim Fentex TP	Protim Solignum Ltd	Industrial	6370
Protim Fentex TPS	Protim Solignum Ltd	Industrial	6371
Protim Fentex TW	Protim Solignum Ltd	Industrial	6368
Protim Fentex TW	Protim Solignum Ltd	Industrial	6876

HSE
WOOD TREATMENT

9

HSE PRODUCT NAME INDEX

A-Zygo 3 Sterilising Solution 419
A11 Antifouling 072014 33
A11 Antifouling 072014 96
A11 Antifouling 072014 145
A11 Antifouling 072014 154
A2 Antifouling 98
A2 Antifouling 120
A2 Antifouling 136
A2 Antifouling 164
A2 Teflon Antifouling 98
A2 Teflon Antifouling 120
A2 Teflon Antifouling 136
A2 Teflon Antifouling 164
A3 Antifouling 14
A3 Antifouling 137
A3 Antifouling 072015 31
A3 Antifouling 072015 87
A3 Antifouling 072015 112
A3 Antifouling 072016 56
A3 Antifouling 072016 89
A3 Antifouling 072016 118
A3 Teflon Antifouling 14
A3 Teflon Antifouling 137
A3 Teflon Antifouling 062015 56
A3 Teflon Antifouling 062015 89
A3 Teflon Antifouling 062015 118
A4 Antifouling 27
A4 Antifouling 94
A4 Antifouling 072017 31
A4 Antifouling 072017 87
A4 Antifouling 072017 112
A4 Antifouling 072018 56
A4 Antifouling 072018 89
A4 Antifouling 072018 118
A4 Teflon Antifouling 062017 56
A4 Teflon Antifouling 062017 89
A4 Teflon Antifouling 062017 118
A6 Antifouling 99
A6 Antifouling 119
A6 Antifouling 168
A7 Teflon Antifouling 99
A7 Teflon Antifouling 119
A7 Teflon Antifouling 168
A7 Teflon Antifouling 072019 99
A7 Teflon Antifouling 072019 119
A7 Teflon Antifouling 072019 168
A8 Antifouling 99
A8 Antifouling 119
A8 Antifouling 168
Abc # 1 Antifouling 35
Abc # 1 Antifouling 139
Abc#3e Antifouling 17
Abc#3e Antifouling 73
Abicide 82 414
Abl Aqueous Wood Preserver Concentrate
 1:9 479

Abl Aqueous Wood Preserver Concentrate
 1:9 562
Abl Brown Creosote 524
Abl Universal Woodworm Killer Db 576
Abl Universal Woodworm Killer Db 717
Abl Wood Preservative (B) 714
Abl Wood Preservative (D) 568
Acarosan Foam 227
Acarosan Moist Powder 227
Acclaim 307
Acclaim 360
Acclaim 2000 302
Acclaim 2000 327
Acs Antimould Emulsion 192
Acs Antimould Microporous Replastering
 Paint 192
Acs Borate Rod 582
Acs Boron Rods 514
Acs Boron Wood Preservative Paste 582
Acs Borotreat 10p 435
Acs Borotreat 10p 582
Acs Dry Rot Paint 190
Acs Dry Rot Paint 749
Acs Woodkeeper Paste Preservative 504
Acs Woodkeeper Paste Preservative 585
Actellic 25 Ec 341
Actellic 25 Ec 664
Actellic Dust 341
Activ-8-I 633
Activ-8-Ii 508
Activ-8-Ii 640
Actomite 230
Actomite 239
Advance Guard 582
Af Seaflo Mark 2-1 36
Af Seaflo Mark 2-1 144
Af Seaflo Mark 2-1 155
Af Seaflo Z-100 Hs-1 36
Af Seaflo Z-100 Hs-1 144
Af Seaflo Z-100 Hs-1 155
Af Seaflo Z-100 Le-Hs-1 36
Af Seaflo Z-100 Le-Hs-1 144
Af Seaflo Z-100 Le-Hs-1 155
Agrevo Ant Killer 267
Agrevo Pro Control Cik Superfast 268
Agrevo Pro Control Cik Superfast 355
Agrevo Procontrol Fly Spray 347
Agrevo Procontrol Flying Insect Killer 236
Agrevo Procontrol Flying Insect Killer 312
Agrevo Procontrol Wasp Nest Destroyer 328
Agrevo Procontrol Wasp Nest Destroyer 383
Agrevo Procontrol Yellow Flying Insect Killer
 Aerosol 237
Agrevo Procontrol Yellow Flying Insect Killer
 Aerosol 324
Aircraft Aerosol Insect Control 335

Aircraft Disinsectant Multi *335*
Aircraft Insecticide *335*
Al-27 *47*
Al-27 *114*
Al63 Crawling Insect Killer *221*
Albany Fungicidal Wash *414*
Albio B084 *419*
Algae Remover *419*
Algae Remover *474*
Algaecide *419*
Algicide *15*
Algicide *108*
Algicide Antifouling *10*
Alpha 15sc *218*
Alphachloralose Concentrate *399*
Alphakil *399*
Alphamouse *399*
Amc Sport Antifouling *10*
Amercoat 275 *10*
Amercoat 277 *10*
Amercoat 279 *10*
Amercoat 67e *4*
Amercoat 697 *34*
Amercoat 697 *138*
Amercoat 70e *4*
Amercoat 70esp *4*
Amogas Ant Killer *311*
Ant & Crawling Insect Spray *311*
Ant Bait *223*
Ant Destroyer Foam *338*
Ant Destroyer Foam *394*
Ant Gun! 2 *236*
Ant Gun! 2 *312*
Ant Killer *398*
Ant Killer Powder *223*
Ant Killer Spray *267*
Ant Off Powder *311*
Ant Stop *248*
Ant Stop Powder *223*
Ant-Off! *267*
Antel Dual Purpose (Bp) Concentrate (Water Dilutable) *652*
Antel Dual Purpose (Bp) Concentrate (Water Dilutable) *719*
Antel Fws *441*
Antel Woodworm Killer (P) Concentrate (Water Dilutable) *633*
Anti Fungus Wash *420*
Anti Fungus Wash *439*
Anti Fungus Wash *440*
Anti Fungus Wash *447*
Anti-Fouling Paint 161p (Red And Chocolate To TS 10240) *10*
Anti-Moss *453*
Anti-Mould Emulsion *190*
Anti-Mould Gloss *182*

Anti-Mould Gloss *186*
Anti-Mould Gloss *193*
Anti-Mould Solution *414*
Anti-Mould Solution *465*
Antialga *54*
Antialga *95*
Antiblu Select *481*
Antiblu Select *583*
Antiblu Select *601*
Antifouling 1.2 *24*
Antifouling 1.2 *63*
Antifouling 1.2 *110*
Antifouling 1.2 (W) *51*
Antifouling 1.2 (W) *64*
Antifouling 1.2 (W) *115*
Antifouling 1.3 *52*
Antifouling 1.3 *68*
Antifouling 1.3 *90*
Antifouling 1.3 *116*
Antifouling 1.3 (W) *53*
Antifouling 1.3 (W) *71*
Antifouling 1.3 (W) *117*
Antifouling 1.3 (W) *167*
Antifouling 1.3 Black And Red (400-03 / Ind.3) *25*
Antifouling 1.3 Black And Red (400-03 / Ind.3) *67*
Antifouling 1.3 Black And Red (400-03 / Ind.3) *88*
Antifouling 1.3 Black And Red (400-03 / Ind.3) *111*
Antifouling 1.3 White (327-04) *52*
Antifouling 1.3 White (327-04) *68*
Antifouling 1.3 White (327-04) *90*
Antifouling 1.3 White (327-04) *116*
Antifouling 1.4 *69*
Antifouling 1.4 *100*
Antifouling 1.4 *172*
Antifouling Alusea Classic *121*
Antifouling Alusea Classic *141*
Antifouling Alusea Classic *151*
Antifouling Alusea Turbo *81*
Antifouling Alusea Turbo *142*
Antifouling Alusea Turbo *153*
Antifouling Broken White Dl-2253 *45*
Antifouling Hb 66 Ocean Green *59*
Antifouling Hb 66 Ocean Green *146*
Antifouling Hb 66 Ocean Green *156*
Antifouling Paint 161p (Red And Chocolate To TS 10240) *10*
Antifouling Sargasso *10*
Antifouling Sargasso Non-Tin *10*
Antifouling Seaconomy *36*
Antifouling Seaconomy *144*
Antifouling Seaconomy *155*
Antifouling Seaconomy 200 *36*
Antifouling Seaconomy 200 *144*
Antifouling Seaconomy 200 *155*
Antifouling Seaconomy 300 *36*

Antifouling Seaconomy 300 *144*
Antifouling Seaconomy 300 *155*
Antifouling Seaguardian *10*
Antifouling Seaguardian (Black, Blue And
  Md). *10*
Antifouling Seamate Fb30 *36*
Antifouling Seamate Fb30 *144*
Antifouling Seamate Fb30 *155*
Antifouling Seamate Hb 33 *36*
Antifouling Seamate Hb 33 *144*
Antifouling Seamate Hb 33 *155*
Antifouling Seamate Hb 99 *36*
Antifouling Seamate Hb 99 *144*
Antifouling Seamate Hb 99 *155*
Antifouling Seamate Hb 99 Black *36*
Antifouling Seamate Hb 99 Black *144*
Antifouling Seamate Hb 99 Black *155*
Antifouling Seamate Hb 99 Dark Red *36*
Antifouling Seamate Hb 99 Dark Red *144*
Antifouling Seamate Hb 99 Dark Red *155*
Antifouling Seamate Hb Green *59*
Antifouling Seamate Hb Green *146*
Antifouling Seamate Hb Green *156*
Antifouling Seamate Hb22 *36*
Antifouling Seamate Hb22 *39*
Antifouling Seamate Hb22 *144*
Antifouling Seamate Hb22 *150*
Antifouling Seamate Hb22 *155*
Antifouling Seamate Hb22 *178*
Antifouling Seamate Hb22 Roundel Blue *59*
Antifouling Seamate Hb22 Roundel Blue *146*
Antifouling Seamate Hb22 Roundel Blue *156*
Antifouling Seamate Hb33 *36*
Antifouling Seamate Hb33 *39*
Antifouling Seamate Hb33 *144*
Antifouling Seamate Hb33 *150*
Antifouling Seamate Hb33 *155*
Antifouling Seamate Hb33 *178*
Antifouling Seamate Hb66 *36*
Antifouling Seamate Hb66 *144*
Antifouling Seamate Hb66 *155*
Antifouling Seamate Hb66 Black *59*
Antifouling Seamate Hb66 Black *146*
Antifouling Seamate Hb66 Black *156*
Antifouling Seamate Sb33 *36*
Antifouling Seamate Sb33 *144*
Antifouling Seamate Sb33 *155*
Antifouling Seaquantum Fb *10*
Antifouling Seavictor 50 *17*
Antifouling Seavictor 50 *73*
Antifouling Super *10*
Antifouling Super Tropic *10*
Antifouling Supertropic *10*
Antkiller Dust 2 *341*
Antkiller Spray 2 *236*
Antkiller Spray 2 *312*

Apex Fly-Killer *368*
Apex Fly-Killer *386*
Aqua 12 *54*
Aqua 12 *95*
Aqua K-Othrine *267*
Aqua Reslin Premium *237*
Aqua Reslin Premium *324*
Aqua Reslin Super *237*
Aqua Reslin Super *324*
Aqua-Net *23*
Aqua-Net *65*
Aqua-Net 089 *10*
Aquacleen *10*
Aquagard (Flexgard Xi) *10*
Aquaguard Anti-Mould Acrylic Matt Finish *182*
Aquaguard Anti-Mould Acrylic Matt Finish *186*
Aquaguard Anti-Mould Acrylic Matt Finish *193*
Aquaguard Wood Preserver *684*
Aquapy *347*
Aquapy Micro *347*
Aquarius Al *54*
Aquarius Al *95*
Aquarius Extra Strong *27*
Aquarius Extra Strong *94*
Aquasafe *10*
Aquasafe W *10*
Aquaspeed *15*
Aquaspeed *108*
Aquaspeed Antifouling *15*
Aquaspeed Antifouling *108*
Aquaspeed White *47*
Aquaspeed White *114*
Aqueous Universal *316*
Aqueous Universal *345*
Aqueous Universal *643*
Aqueous Universal *675*
Aqueous Wood Preserver *479*
Aqueous Wood Preserver *562*
Armachlor Af275 *10*
Armacote Af259 *10*
Armarine Af259 *10*
Armarine Af275 *10*
Armitage Pet Care, Pet Bedding & Household Flea
  Spray *306*
Armitage Pet Care, Pet Bedding & Household Flea
  Spray *356*
Armitages Good Boy Household Flea Trigger
  Spray *311*
Arrest *235*
Arrest *303*
Arrest *313*
Atta-X Cockroach Killer *328*
Atta-X Cockroach Killer *383*
Atta-X Flying Insect Killer *237*
Atta-X Flying Insect Killer *324*
Aventis Ant Killer Powder *223*

Aventis Procontrol Crawling Insect Killer   236
Aventis Procontrol Crawling Insect Killer   312
Aventis Procontrol Crawling Insect Killer
   Aerosol   241
Aventis Procontrol Crawling Insect Killer
   Aerosol   326
Aventis Procontrol Crawling Insect Killer
   Aerosol   384
Aventis Wasp Nest Destroyer   223
Aventis Woodlice Killer   223
Avonclad   1
Awlgrip Awlstar Gold Label Antifouling   10
Aztec B.B.Q. Patio Candles   200
Azygo Antimould Coating (New Works)   192
Azygo Antimould Paint   192
Azygo Boron Preservative Pdr   435
Azygo Boron Preservative Pdr   582
B & Q Ant And Crawling Insect Spray.   267
B & Q Ant Killer   223
B & Q New Ant Killer Spray   267
B & Q Woodlice Killer   223
B And Q Ant Killer Lacquer   270
B And Q Clear Wood Preserver   733
B And Q Creosote   549
B And Q Exterior Wood Preservative   733
B And Q Formula Wood Preserver   733
B And Q New Formula Ant Killer Spray   257
B And Q Value Fungicide   419
B.B.Q Fly Repellent Terracotta Pot Candle   200
B.B.Q. Fly Repellant Candle   200
B.B.Q. Patio Candles   200
Baby Bio Houseplant Spray   236
Baby Bio Houseplant Spray   312
Bactdet D   452
Bactimos Flowable Concentrate   222
Bactimos Wettable Powder   222
Baracaf Cockroach Control Sticker
   (Domestic)   245
Barrettine Creosote   549
Barrettine New Preserver   714
Barrettine New Universal Fluid   652
Barrettine New Universal Fluid   719
Barrettine New Woodworm Fluid   633
Barrettine Premier Multicide Fluid   421
Barrettine Premier Multicide Fluid   427
Barrettine Premier Multicide Fluid   476
Barrettine Premier Multicide Fluid   490
Barrettine Timberguard   465
Bartoline Dark And Light Creosote   524
Basiment 560   659
Battles Louse Powder   311
Bayer Afc   62
Bayer Wpc 2-1.5   696
Bayer Wpc 2-25   696
Bayer Wpc 2-25-Sb   696
Bayer Wpc-3   709

Bayer Wpc-4   709
Bayer Wpc-5   550
Baygon   209
Baygon Insect Spray   256
Baygon Insect Spray   397
Baygon Moth Paper   396
Bcr Green Wood Preserver   533
Bcr Green Wood Preserver   715
Bcr Universal Wood Preserver   557
Bcr Universal Wood Preserver   716
Bcr Wood Preserver   714
Bedclear   419
Beeline Fungicidal Wash   414
Belzona 9421 Fungicidal Wash   419
Belzona 9421 Fungicidal Wash   474
Bendiocarb Dusting Powder   223
Bendiocarb Wettable Powder   223
Betterware Adjustable Insect Repellent   200
Betterware Ant Gel   282
Betterware Ant Killer   371
Betterware Cyclone And Vac Fresh X2   311
Betterware Flying Insect Killer   228
Betterware Flying Insect Killer (Slow
   Release)   274
Betterware Foaming Ant & Wasp Nest
   Destroyer   338
Betterware Foaming Ant & Wasp Nest
   Destroyer   394
Betterware Slow Release Fly Killer   274
Betterware Small Space Insect Killer   274
Big D Ant And Crawling Insect Killer   338
Big D Ant And Crawling Insect Killer   394
Big D Cockroach And Crawling Insect Killer   286
Big D Cockroach And Crawling Insect Killer   379
Big D Fly And Wasp Killer   372
Big D Rocket Fly Killer   286
Big D Rocket Fly Killer   379
Bio Ant Kill Plus Powder   223
Bio Ant Kill Plus Spray   267
Bio Foaming Wasp Nest Destroyer   338
Bio Foaming Wasp Nest Destroyer   394
Bio Mouse Killer   408
Bio Mouse Killer Bait Station   408
Bio New Anti-Ant Duster   311
Bio Racumin Rat Bait   407
Bio Rat Killer   408
Bio Rat Killer Bait Bags   408
Bio Spraydex Ant And Insect Killer   236
Bio Spraydex Ant And Insect Killer   312
Bio Wasp Nest Destroyer   311
Bio-Kil Board Preservative   600
Bio-Kil Boron Paste   582
Bio-Kil Cunap Pole Wrap   529
Bio-Kil Dentolite Solution   419
Bio-Kil Sr Pole Wrap   531
Bio-Kil Sr Pole Wrap   587

Bio-Natura's Green Algae Remover *419*
Biocheck 01 *183*
Biocheck 01 *187*
Biocheck 01 *189*
Biocheck 01 *194*
Biocheck 2507 *188*
Biocheck A.C.C. *182*
Biocheck A.C.C. *186*
Biocheck A.C.C. *193*
Biocheck Allweather Pliolite Resin Based Masonry
    Paint *182*
Biocheck Allweather Pliolite Resin Based Masonry
    Paint *186*
Biocheck Allweather Pliolite Resin Based Masonry
    Paint *193*
Biocheck C *188*
Biocheck Matt *188*
Biocheck Silk *188*
Biocheck Sp *183*
Biocheck Sp *187*
Biocheck Sp *189*
Biocheck Sp *194*
Biocheck Water Based Gloss *182*
Biocheck Water Based Gloss *186*
Biocheck Water Based Gloss *193*
Bioclean Dx *123*
Bioclean Sterilizer *434*
Biocure Emulsion *183*
Biocure Emulsion *187*
Biocure Emulsion *189*
Biocure Emulsion *194*
Biokil *482*
Biokil *613*
Biokil Emulsion *483*
Biokil Emulsion *639*
Biokil Timbor Rods *582*
Biokill *311*
Bionatura's Green Algae Remover *419*
Biophen Barrier *432*
Biopren Bm 0.68 *306*
Biopren Bm 0.68 *356*
Biopren Bm Ready To Use Pharaoh Ant Killer
    Bait *298*
Biopren Bm Residual Flea Killer Aerosol *304*
Biopren Bm Residual Flea Killer Aerosol *321*
Biopren Bm Residual Flea Killer Aerosol *357*
Bioshield *188*
Biosol Rtu *242*
Biotech Fungicidal Additive *190*
Bitmac Creosote *524*
Black Antifouling Paint 317 (To Ts10239) *19*
Black Antifouling Paint 317 (To Ts10239) *44*
Blackfriar Anti Mould Solution 195/9 *439*
Blackfriar Anti Mould Solution 195/9 *440*
Blackfriar Anti-Mould Solution *420*
Blackfriar Anti-Mould Solution *447*

Blackfriar Wood Preserver *576*
Blackfriar Wood Preserver *717*
Blackfriar Wood Preserver (Wp Green) *529*
Blackfriar Wood Preserver Wp Gold Star
    Clear *574*
Blackfriar Wood Preserver Wp Gold Star
    Clear *653*
Blackfriar Wood Preserver Wp Gold Star
    Clear *718*
Blackfriars Gold Star Clear *574*
Blackfriars Gold Star Clear *653*
Blackfriars Gold Star Clear *718*
Blackfriars Green Wood Preserver *529*
Blackfriars New Wood Preserver *576*
Blackfriars New Wood Preserver *717*
Blade *242*
Blakes Antifouling 87910 *41*
Blakes Antifouling 87910 *170*
Blakes Hard Racing *27*
Blakes Hard Racing *94*
Blueline Copper Sba100 *10*
Blueline Spc Tin Free Sba700 Series *42*
Blueline Spc Tin Free Sba700 Series *177*
Blueline Tropical Sba300 *27*
Blueline Tropical Sba300 *94*
Bn Algae Remover *419*
Bn Mosskiller *419*
Boatgard Antifouling *27*
Boatgard Antifouling *94*
Boatguard *10*
Bob Martin Flea Bomb *311*
Bob Martin Flea Kill *311*
Bob Martin Home Flea Fogger Plus *235*
Bob Martin Home Flea Fogger Plus *303*
Bob Martin Home Flea Fogger Plus *313*
Bob Martin Home Flea Powder *223*
Bob Martin Home Flea Spray *311*
Bob Martin Home Flea Spray Plus *235*
Bob Martin Home Flea Spray Plus *303*
Bob Martin Home Flea Spray Plus *313*
Bob Martin Microshield Household Flea Killing
    Spray *248*
Bob Martin Natural Household Flea Spray *347*
Bolt Ant And Crawling Insect Killer *328*
Bolt Ant And Crawling Insect Killer *383*
Bolt Crawling Insect Killer *261*
Bolt Crawling Insect Killer *354*
Bolt Dry Flying Insect Killer *347*
Bolt Flying Insect Killer *372*
Boot Top Plus (Gull White) *47*
Boot Top Plus (Gull White) *114*
Boots Ant And Crawling Insect Killer *338*
Boots Ant And Crawling Insect Killer *394*
Boots Ant Trap *398*
Boots Anti-Dust Mite Spray *335*
Boots Dry Fly And Wasp Killer *216*

Boots Dry Fly And Wasp Killer   *338*
Boots Dry Fly And Wasp Killer   *389*
Boots Dry Fly And Wasp Killer   *394*
Boots Dry Fly And Wasp Killer 3   *216*
Boots Dry Fly And Wasp Killer 3   *389*
Boots Electric Mosquito Killer   *228*
Boots Flea Spray   *338*
Boots Flea Spray   *394*
Boots Fly And Wasp Killer   *372*
Boots Mosquito Killer   *344*
Boots Moth Killer   *274*
Boots New Improved Fly And Wasp Killer   *217*
Boots New Improved Fly And Wasp Killer   *339*
Boots Powder Ant Killer   *311*
Boots Pump Action Fly And Wasp Killer   *347*
Boots Pump Fly And Wasp Killer   *217*
Boots Pump Fly And Wasp Killer   *339*
Boots Slow Release Fly And Wasp Killer   *274*
Boots Slow Release Fly Killer   *274*
Boots Small Space Moth Killer   *274*
Boots Uk Flying Insect Killer   *228*
Bop   *213*
Bop   *279*
Bop   *333*
Bop   *392*
Bop Arabic   *338*
Bop Arabic   *394*
Bop Crawling Insect Killer   *328*
Bop Crawling Insect Killer   *383*
Bop Crawling Insect Killer (Mc)   *328*
Bop Crawling Insect Killer (Mc)   *383*
Bop Flying And Crawling Insect Killer   *212*
Bop Flying And Crawling Insect Killer   *261*
Bop Flying And Crawling Insect Killer   *390*
Bop Flying And Crawling Insect Killer (Mc)   *212*
Bop Flying And Crawling Insect Killer (Mc)   *264*
Bop Flying And Crawling Insect Killer (Mc)   *390*
Bop Flying And Crawling Insect Killer (Water
   Based)   *212*
Bop Flying And Crawling Insect Killer (Water
   Based)   *264*
Bop Flying And Crawling Insect Killer (Water
   Based)   *390*
Bop Flying Insect Killer   *212*
Bop Flying Insect Killer   *264*
Bop Flying Insect Killer   *390*
Bop Flying Insect Killer (Mc)   *212*
Bop Flying Insect Killer (Mc)   *264*
Bop Flying Insect Killer (Mc)   *390*
Boracol 10rh Surface Biocide   *416*
Boracol 10rh Surface Biocide   *436*
Boracol 20   *582*
Boracol 20 Rh   *480*
Boracol 20 Rh   *584*
Boracol B40   *582*

Boracol B8.5 Rh Mouldicide/Wood
   Preservative   *423*
Boracol B8.5 Rh Mouldicide/Wood
   Preservative   *437*
Boracol B8.5 Rh Mouldicide/Wood
   Preservative   *480*
Boracol B8.5 Rh Mouldicide/Wood
   Preservative   *584*
Borax Wood Preservative Clear Odourless   *582*
Boric Acid Concentrate   *245*
Boric Acid Dust   *245*
Boric Acid Powder   *245*
Boron 510 Powder Preservative   *582*
Boron Biocide 10   *423*
Boron Biocide 10   *437*
Boron Biocide 10   *480*
Boron Biocide 10   *584*
Boron Gel 40   *582*
Bottomkote   *10*
Bouchard Ant Killer   *371*
Bouchard Anti Moth Proofer Pouches   *202*
Bouchard Anti-Dust Mite Vacuum Freshener   *311*
Bouchard Citronella Insect Repellant   *200*
Bouchard Fly And Wasp Killer   *338*
Bouchard Fly And Wasp Killer   *394*
Bouchard Slow Release Fly Killer   *274*
Bpc-8   *633*
Bpc-Fs Fungicidal Solution Concentrate   *441*
Bpc-Fs Fungicidal Solution Concentrate   *600*
Bpc8-Dp   *652*
Bpc8-Dp   *719*
Bradite Ca21   *10*
Bradite Fungicidal Wash   *420*
Bradite Fungicidal Wash   *447*
Bradite Timber Preservative   *580*
Bradite Timber Preservative   *731*
Brimpex Ulv 1500   *338*
Brimpex Ulv 1500   *394*
Broads Antifouling Red   *15*
Broads Antifouling Red   *108*
Broads Black Antifouling   *15*
Broads Black Antifouling   *108*
Broads Freshwater   *10*
Broads Freshwater Red   *15*
Broads Freshwater Red   *108*
Brodifacoum Paste   *400*
Brolac Fungicidal Solution   *453*
Bromard   *401*
Bromatrol   *401*
Bromatrol Contact Dust   *401*
Bromatrol Mouse Blocks   *401*
Bromatrol Rat Blocks   *401*
Brown Wood Preservative   *714*
Brunol Atp   *616*
Brunol Atp   *632*
Brunol Atp New   *649*

HSE
PRODUCT INDEX

Brunol Atp New 706
Brunol Dual 8 607
Brunol Dual 8 635
Brunol Insecticidal 8 633
Brunol Opa 634
Brunol Opa 663
Brunol Pbo 652
Brunol Pbo 719
Brunol Pc 633
Brunol Pp 607
Brunol Pp 635
Brunol Py 633
Brunol Special P 652
Brunol Special P 719
Brunol Spi 607
Brunol Spi 635
Brunol Stp 649
Brunol Stp 706
Brunosol Boron 10 423
Brunosol Boron 10 437
Brunosol Boron 10 480
Brunosol Boron 10 584
Bs 144 Creosote 524
Bug Free 311
Bug Off 338
Bug Off 394
Bug Proof 311
Bug Wars 237
Bug Wars 324
Busan 1009 622
Busan 1009 713
Busan 1111 709
Buzz Off 1 228
Buzz Off Candle 200
Buzz-Off 228
C-Clean 100 10
C-Clean 200 10
C-Clean 300 29
C-Clean 300 86
C-Clean 300 129
C-Clean 400 13
C-Clean 400 74
C-Clean 400 126
C-Clean 6000 36
C-Clean 6000 144
C-Clean 6000 155
C-Worthy 10
C/S Clean R833 420
C/S Clean R833 447
Camco Ant Powder 223
Camco Insect Powder 223
Camco Insect Spray 225
Camco Insect Spray 374
Candle Pot - Scottish Midge 200
Canned Death 328
Canned Death 383

Canovel Pet Bedding And Household Spray 306
Canovel Pet Bedding And Household Spray 356
Canovel Pet Bedding And Household Spray
    Plus 307
Canovel Pet Bedding And Household Spray
    Plus 360
Carbo Creosote 524
Carbo Wood Preservative 730
Carbo Wood Preservative Green 529
Cedarwood Protector 568
Cedarwood Special 568
Cedasol 2320 618
Cedasol 2320 727
Cedasol Ready To Use (2306) 658
Cedasol Ready To Use (2306) 743
Celbor 512
Celbor 695
Celbor M 486
Celbor P 502
Celbor P 563
Celbor P 25% Solution 502
Celbor P 25% Solution 563
Celbor P25 502
Celbor P25 563
Celbor P5 502
Celbor P5 563
Celbor Pr 510
Celbor Pr 670
Celbrite M 479
Celbrite M 562
Celbrite Mt 477
Celbrite Mt 503
Celbrite Mt 566
Celbrite Mt 625
Celbrite Tc 622
Celbrite Tc 713
Celbronze B 479
Celbronze B 562
Celcure A Concentrate 469
Celcure A Concentrate 545
Celcure A Concentrate 685
Celcure A Fluid 10 469
Celcure A Fluid 10 545
Celcure A Fluid 10 685
Celcure A Fluid 6 469
Celcure A Fluid 6 545
Celcure A Fluid 6 685
Celcure A Paste 469
Celcure A Paste 545
Celcure A Paste 685
Celcure Ao 468
Celcure Ao 522
Celcure Ao 535
Celcure B 519
Celcure B 548
Celcure B 688

Celcure C6099 *468*
Celcure C6099 *522*
Celcure C6099 *535*
Celcure Cb90 *500*
Celcure Cb90 *546*
Celcure Cb90 *687*
Celcure Cca Type C *468*
Celcure Cca Type C *522*
Celcure Cca Type C *535*
Celcure Cca Type C-40% *468*
Celcure Cca Type C-40% *522*
Celcure Cca Type C-40% *535*
Celcure Cca Type C-50% *468*
Celcure Cca Type C-50% *522*
Celcure Cca Type C-50% *535*
Celcure Cca Type C-60% *468*
Celcure Cca Type C-60% *522*
Celcure Cca Type C-60% *535*
Celcure O *519*
Celcure O *548*
Celcure O *688*
Celcure Type C6099 *468*
Celcure Type C6099 *522*
Celcure Type C6099 *535*
Celgard Cf *491*
Celgard Cf *520*
Celgard Cf *547*
Celgard Cf *686*
Celkil 90 *517*
Celkil 90 *619*
Celpruf B *555*
Celpruf B *673*
Celpruf Bz *513*
Celpruf Bz *734*
Celpruf Bz Wr *513*
Celpruf Bz Wr *734*
Celpruf Bzp *509*
Celpruf Bzp *656*
Celpruf Bzp *735*
Celpruf Bzp Wr *509*
Celpruf Bzp Wr *656*
Celpruf Bzp Wr *735*
Celpruf Cp Special *579*
Celpruf Cp Special *726*
Celpruf Jp *630*
Celpruf Jp *729*
Celpruf Jp Wr *630*
Celpruf Jp Wr *729*
Celpruf Pk *615*
Celpruf Pk *631*
Celpruf Pk *728*
Celpruf Pk Wr *615*
Celpruf Pk Wr *631*
Celpruf Pk Wr *728*
Celpruf Primer *725*
Celpruf Primer Tn *722*

Celpruf Primer Zv *741*
Celpruf Tn *722*
Celpruf Tn Wr *722*
Celpruf Tnm *654*
Celpruf Tnm *724*
Celpruf Tnm Wr *654*
Celpruf Tnm Wr *724*
Celpruf Wb11 *643*
Celpruf Wb11 *675*
Celpruf Zo *741*
Celpruf Zo Wr *741*
Celpruf Zop *658*
Celpruf Zop *743*
Celpruf Zop Wr *658*
Celpruf Zop Wr *743*
Cementone Multiplus *557*
Cementone Multiplus *716*
Cementone Woodworm Killer *551*
Challenger Antifouling *15*
Challenger Antifouling *108*
Channelwood Boracol B40 Wood Preservative
    Gel *582*
Channelwood Boracol B40/1 Wood Preservative
    Gel *582*
Chel Dry Rot Killer For Masonry And
    Brickwork *475*
Chel Dry Rot Killer For Masonry And
    Brickwork *660*
Chel-Wood Preserver/Woodworm Dry Rot Killer
    Bp *652*
Chel-Wood Preserver/Woodworm Dry Rot Killer
    Bp *719*
Chelec Creosote *549*
Chelwash *420*
Chelwash *447*
Chem-Kil *368*
Chem-Kil *386*
Chirton Ant And Insect Killer *311*
Chirton Dry Fly And Wasp Killer *368*
Chirton Dry Fly And Wasp Killer *386*
Chlorpyrifos *248*
Cindu Creosote *549*
Citronella Bug Spray *200*
Citronella Candle (Code D10) *200*
Citronella Floater Candle *200*
Citronella Insect Repellent Oil *200*
Citronella Scented Candle *200*
Classica 3786/093 Red *10*
Clear Wood Preservative *714*
Cme 20/P Concentrate *633*
Cme 25/F Concentrate *666*
Cme 30/F Concentrate *441*
Cme 30/F Concentrate *600*
Coal Tar Creosote *524*
Cobra V *10*
Cockroach Bait *293*

Cockroach Control   245
Colourfast Protector   479
Colourfast Protector   562
Colron Woodworm Killer   633
Combat Ant & Crawling Insect Powder   223
Combat Ant And Crawling Insect Spray   267
Combat Ant Bait   223
Combat Cockroach And Woodlice Killer   257
Combat Fly And Wasp Killer   328
Combat Fly And Wasp Killer   383
Combat Fly Killer   274
Combat Fly Killer   328
Combat Fly Killer   383
Combat Moth Killer   272
Combat Moth Killer   275
Combat Wasp Killer Foam   338
Combat Wasp Killer Foam   394
Combat Woodlice Killer   223
Conc Quat   419
Conductive Pilt 80 Rfu   600
Connect Insect Killer   285
Connect Insect Killer   329
Connect Insect Killer   380
Constrain   311
Contra Insect: The Vermin Bait   248
Contra-Ants Bait Box   248
Contra-Insect 200/20 Ec   249
Contra-Insect 200/20 Ec   258
Contra-Insect 480 Tec   248
Contra-Insect Aerosol   251
Contra-Insect Aerosol   351
Contra-Insect Universal   251
Contra-Insect Universal   351
Contrac Blox   401
Contrac Rodenticide   401
Contrac Super Blox   401
Control Fluid Fb   486
Control Fluid Sb   429
Control Fluid Sb   438
Control Fluid Sb   505
Control Fluid Sb   593
Cooke's Liquid Insecticide   311
Cookes Ant Killer   347
Cooper's Copolymer Antifouling   10
Coopex 25% Ec   311
Coopex Insect Powder   311
Coopex Maxi Smoke Generator   311
Coopex Mini Smoke Generator   311
Coopex S25% Ec   237
Coopex S25% Ec   324
Coopex Smoke Generator   311
Coopex Wp   311
Copper Net   23
Copper Net   65
Copperbot   1
Copperbot   4

Copperbot 2000   1
Coppercoat   2
Coppercoat   93
Copperguard   1
Copperpaint   10
Corry's Fly And Wasp Killer   328
Corry's Fly And Wasp Killer   383
Crackdown   267
Crackdown D   267
Crackdown Rapide   233
Crackdown Rapide   269
Crawling Insect And Ant Killer   236
Crawling Insect And Ant Killer   312
Creepy Crawly Gun!   236
Creepy Crawly Gun!   312
Creosote   524
Creosote   549
Creosote Blend   524
Creosote Blend   549
Creosote Blend Mk1 (Medium Dark)   524
Creosote Blend Mk3 (Light Golden)   524
Creosote Blended Wood Preservative   524
Creosote Bs 144   524
Creosote Bs 144 (Type I And Type Ii)   524
Creosote Bs 144 Type Iii   524
Creosote Bs 144(3)   524
Creosote Coke Oven Oil   549
Creosote Emulsion   549
Creosote Mk2   549
Cromar Dry Rot And Woodworm Killer   634
Cromar Dry Rot And Woodworm Killer   663
Cromar Wood Preserver   568
Cromar Wood Preserver Green   529
Cromessol Fly-Away   274
Cromessol Insect Killer   368
Cromessol Insect Killer   386
Cromessol Multi-Purpose Insect Killer   368
Cromessol Multi-Purpose Insect Killer   386
Crown Fungicidal Wall Solution   420
Crown Fungicidal Wall Solution   447
Crown Fungicide Insecticide   557
Crown Fungicide Insecticide   716
Crown Fungicide Insecticide Concentrate   557
Crown Fungicide Insecticide Concentrate   716
Crown Micro Dual Purpose   557
Crown Micro Dual Purpose   716
Crown Micro Dual Purpose F/I   652
Crown Micro Dual Purpose F/I   719
Crown Micro Insecticide   551
Crown Micro Woodworm   633
Crown Timber Paste   652
Crown Timber Paste   719
Crown Timber Preservative   714
Crown Timber Preservative   733
Crown Trade Stronghold Fungicidal Solution   453
Crown Woodworm Concentrate   551

Cruiser Copolymer   60
Cruiser Copolymer   149
Cruiser Copolymer   160
Cruiser Copolymer   180
Cruiser Premium   27
Cruiser Premium   94
Cruiser Superior 100 Series   47
Cruiser Superior 100 Series   114
Crysolite Glaramara Concentrated Timber
    Treatment   709
Crysolite Wood Preservative   600
Cu15   4
Culmstock Mothproofer   274
Culmstock Slow Release Fly Killer   274
Cuprinol 5 Star Complete Wood Treatment S   463
Cuprinol 5 Star Complete Wood Treatment S   637
Cuprinol Combination Grade S   463
Cuprinol Combination Grade S   637
Cuprinol Decorative Preserver   462
Cuprinol Decorative Preserver   569
Cuprinol Decorative Preserver Red Cedar   462
Cuprinol Decorative Preserver Red Cedar   569
Cuprinol Ducksback   746
Cuprinol Ducksback   748
Cuprinol Ducksback   752
Cuprinol Exterior Fungicide   419
Cuprinol Fungicidal Spray   423
Cuprinol Fungicidal Spray   437
Cuprinol Garden Decking Cleaner   419
Cuprinol Garden Decking Protector   462
Cuprinol Garden Decking Protector   569
Cuprinol Garden Decking Seal   746
Cuprinol Garden Decking Seal   748
Cuprinol Garden Decking Seal   752
Cuprinol Garden Furniture Algae Killer   419
Cuprinol Garden Furniture Preserver   461
Cuprinol Garden Furniture Preserver   570
Cuprinol Garden Shades   745
Cuprinol Garden Shades   747
Cuprinol Garden Shades   751
Cuprinol Garden Shed And Fence Preserver   460
Cuprinol Garden Timber Algae Killer   419
Cuprinol Greenhouse Algae Killer   419
Cuprinol Hardwood Basecoat   568
Cuprinol Hardwood Basecoat Meranti   568
Cuprinol Insecticidal Emulsion Concentrate   633
Cuprinol Low Odour 5 Star Complete Wood
    Treatment   463
Cuprinol Low Odour 5 Star Complete Wood
    Treatment   637
Cuprinol Low Odour End Cut   459
Cuprinol Low Odour End Cut   464
Cuprinol Low Odour End Cut   636
Cuprinol Low Odour Wet And Dry Rot Killer For
    Timber   460
Cuprinol Low Odour Wood Preserver Green   458

Cuprinol Low Odour Woodworm Killer   633
Cuprinol Mould Killer   423
Cuprinol Mould Killer   437
Cuprinol Path And Patio Cleaner   419
Cuprinol Preservative Base   462
Cuprinol Preservative Base   569
Cuprinol Preservative-Wood Hardener   568
Cuprinol Quick Drying Clear Preserver   555
Cuprinol Quick Drying Clear Preserver   673
Cuprinol Wet And Dry Rot Killer For
    Timber(S)   460
Cuprinol Wood Defender   555
Cuprinol Wood Defender   673
Cuprinol Wood Preserver   460
Cuprinol Wood Preserver Clear S   460
Cuprinol Wood Preserver Green S   458
Cuprisol F   460
Cuprisol Fn   463
Cuprisol Fn   637
Cuprisol P   459
Cuprisol P   464
Cuprisol P   636
Cuprisol Wr   463
Cuprisol Wr   637
Cuprisol Xqd   460
Cuprisol Xqd Special   463
Cuprisol Xqd Special   637
Cupron Plus T.F   10
Cuprotect Exterior Fungicide   419
Cuprotect Patio Cleaner   419
Cut End   463
Cut End   637
Cut'N'treat   509
Cut'N'treat   656
Cut'N'treat   735
Cymperator   257
Cyperkill 10   257
Cyperkill 10 Wp   257
Cyperkill 10t   263
Cyperkill 10t   376
Cyperkill Plus Wp   263
Cyperkill Plus Wp   376
Cypermethrin 10% Ec   257
Cypermethrin 10% Wp   257
Cypermethrin Lacquer   257
Cypermethrin Ph-10ec   257
Cytrol Alpha 20   218
Cytrol Alpha 25 Ec   218
Cytrol Alpha 5 Sc   218
Cytrol Alpha Super 50/50 Se   220
Cytrol Alpha Super 50/50 Se   373
Cytrol Forte Wp   257
Cytrol Universal   263
Cytrol Universal   376
Cytrol Xl   263
Cytrol Xl   376

HSE
PRODUCT INDEX

D-Stroy *328*
D-Stroy *383*
Dairy Fly Spray *347*
Dark Green Wood Preservative *533*
Dark Green Wood Preservative *715*
Days Farm Fly Spray *347*
Days Fly Spray *231*
Days Fly Spray *243*
Ddvp (Toxicant) Strip *274*
Deadline *401*
Deadline Contact Dust *401*
Deadline Liquid Concentrate *401*
Deadline Place Packs *401*
Deadly Nightshade *420*
Deadly Nightshade *447*
Deal Direct Woodworm Treatment
 Concentrate *633*
Deemold *455*
Deepflow 11 Inorganic Boron Masonry
 Biocide. *423*
Deepflow 11 Inorganic Boron Masonry
 Biocide. *437*
Deepkill *607*
Deepkill *635*
Deepkill F *600*
Deepkill I *633*
Deepwood "Clear" Insecticide Emulsion
 Concentrate" *633*
Deepwood 20 Inorganic Boron Wood
 Preservative *480*
Deepwood 20 Inorganic Boron Wood
 Preservative *584*
Deepwood 8 Micro Emulsifiable Insecticide
 Concentrate *633*
Deepwood Fi Dual Purpose Emulsion
 Concentrate *652*
Deepwood Fi Dual Purpose Emulsion
 Concentrate *719*
Deepwood Standard Emulsion Concentrate *633*
Deerat Concentrate *402*
Defest Flea Free *277*
Defest Flea Free *294*
Defest Ii *236*
Defest Ii *312*
Delta Cockroach And Ant Spray *328*
Delta Cockroach And Ant Spray *383*
Delta Insect Powder *311*
Demand Cs *295*
Demise *283*
Demon 40 Wp *257*
Denso Mouldshield Biocidal Cleanser *434*
Denso Mouldshield Surface Biocide *434*
Deosan Fly Spray *338*
Deosan Fly Spray *394*
Deosan Rataway *408*
Deosan Rataway Bait Bags *408*

Deosan Rataway W *408*
Deth-Zap Ant Killer *267*
Dethlac Insect Lacquer *267*
Dethlac Insecticidal Lacquer *270*
Detia Ant Bait *398*
Detia Crawling Insect Spray *317*
Detia Crawling Insect Spray *358*
Detia Pyrethrum Spray *347*
Devatern 0.5l *551*
Devatern 1.0 L *551*
Devatern Ec *551*
Devatern Wood Preserver *557*
Devatern Wood Preserver *716*
Devcol Household Insect Spray *225*
Devcol Household Insect Spray *374*
Devcol Liquid Ant Killer *343*
Devcol Universal Pest Powder *347*
Devcol Wasp Killer *343*
Devran Mcp Antifouling Red *35*
Devran Mcp Antifouling Red *139*
Devran Mcp Antifouling Red Brown *35*
Devran Mcp Antifouling Red Brown *139*
Di-Fly *372*
Dianex *299*
Difenard *408*
Diffusit *512*
Diffusit *695*
Dipsar *618*
Dipsar *727*
Dipsar G R *558*
Dipsar G R *742*
Disinsecting Spray Multi *335*
Ditrac Blox *411*
Ditrac Super Blox *411*
Diy Time Creosote *549*
Do It All Ant Killer *223*
Do It All Blitz! Ant Killer *267*
Do It All Woodlice Killer *223*
Doff Ant And Crawling Insect Killer *267*
Doff Ant Control Powder *223*
Doff Ant Killer Spray *257*
Doff Antlak *270*
Doff New Ant Killer *311*
Doff New Ant Killer Spray *267*
Doff Path & Patio Cleaner *419*
Doff Path And Patio Cleaner Concentrate *419*
Doff Wasp Nest Killer *223*
Doff Woodlice Killer *223*
Doom Ant & Cockroach Killer *268*
Doom Ant & Cockroach Killer *355*
Doom Ant And Crawling Insect Killer Aerosol *286*
Doom Ant And Crawling Insect Killer Aerosol *379*
Doom Ant And Insect Powder *296*
Doom Flea Killer *268*
Doom Flea Killer *297*
Doom Flea Killer *355*

Doom Flea Killer *381*
Doom Moth Proofer Aerosol *297*
Doom Moth Proofer Aerosol *381*
Doom Tropical Strength Fly And Wasp Killer
  Aerosol *328*
Doom Tropical Strength Fly And Wasp Killer
  Aerosol *383*
Double Action Timber Preservative For
  Doors *576*
Double Action Timber Preservative For
  Doors *717*
Double Action Wood Preservative *576*
Double Action Wood Preservative *717*
Double Shield Antifouling *7*
Double Shield Antifouling *18*
Downland Dairy Fly Spray *347*
Dr Moss's Liquid Bait System *245*
Dragon Brand Moth Balls *204*
Dragon Flykiller *328*
Dragon Flykiller *383*
Drake Antifouling *27*
Drake Antifouling *94*
Drat *404*
Drat Bait *404*
Drat Rat Bait *404*
Drax - Pf Gel *245*
Drax Dual Gel *245*
Drax Gel *245*
Dricon *486*
Drione *347*
151 Dry Fly And Wasp Killer *368*
151 Dry Fly And Wasp Killer *386*
Dry Fly Killer *372*
Dry Pin *582*
Dual 8 *652*
Dual 8 *719*
Dual Purpose Fi8 (Microemulsion) *643*
Dual Purpose Fi8 (Microemulsion) *675*
Dualprime F *568*
Dulux Trade Mouldshield Cellar Paint *184*
Dulux Trade Mouldshield Fungicidal
  Eggshell *190*
Dulux Trade Mouldshield Fungicidal Matt *190*
Dulux Trade Mouldshield Fungicidal Quick Drying
  Eggshell *182*
Dulux Trade Mouldshield Fungicidal Quick Drying
  Eggshell *186*
Dulux Trade Mouldshield Fungicidal Quick Drying
  Eggshell *193*
Dulux Trade Mouldshield Fungicidal Vinyl
  Matt *182*
Dulux Trade Mouldshield Fungicidal Vinyl
  Matt *186*
Dulux Trade Mouldshield Fungicidal Vinyl
  Matt *193*
Duo 1 *249*

Duo 1 *258*
Duratrol *248*
Durham Bi 730 *439*
Durham Bi 730 *440*
Durham Bi 730 *594*
Durham Bi 730 *596*
Dursban 4tc *248*
Dursban Lo *248*
Dyna-Mite *311*
Eco - Db8 *435*
Eco - Db8 *582*
Ecobar Ii *582*
Ecogel *565*
Ecogel *588*
Ecology Fungicide Insecticide Aqueous
  (Concentrate) Wood Preservative Dual
  Purpose *652*
Ecology Fungicide Insecticide Aqueous
  (Concentrate) Wood Preservative Dual
  Purpose *719*
Ecology Fungicide Insecticide Concentrate *557*
Ecology Fungicide Insecticide Concentrate *716*
Elf Insecticide *211*
Elf Insecticide *260*
Elf Insecticide *332*
Elf Insecticide *391*
Empire 20 *248*
Emprasan Pendle Mist Fly Spray *347*
En-Tout-Cas Moss Killer *434*
Endem *408*
Endem Bait Sachets *408*
Endorats *404*
Endorats Premium Rat Killer *401*
Ensele 3426 *498*
Ensele 3426 *542*
Ensele 3427 *498*
Ensele 3427 *542*
Ensele 3428 *496*
Ensele 3428 *528*
Ensele 3428 *701*
Ensele 3429 *498*
Ensele 3429 *542*
Ensele 3430 *582*
Ensele 3450 *494*
Ensele 3450 *526*
Ensele 3450 *671*
Envar 6/14 Ec *338*
Envar 6/14 Ec *394*
Envar 6/14 Oil *338*
Envar 6/14 Oil *394*
Enviromite *245*
Environmental Timber Treatment *633*
Environmental Timber Treatment Paste *633*
Environmental Woodrot Treatment *439*
Environmental Woodrot Treatment *440*
Environmental Woodrot Treatment *594*

Environmental Woodrot Treatment   596
Envoy Tcf 600 Antifouling   72
Envoy Tcf 600 Antifouling   103
Envoy Tcf 600 Copper/Tin Free Antifouling   102
Envoy Tcf 600 Copper/Tin Free Antifouling   130
Envoy Tf 400   20
Envoy Tf 400   46
Envoy Tf 400   109
Envoy Tf 400   131
Envoy Tf 400 Antifouling   21
Envoy Tf 400 Antifouling   48
Envoy Tf 400 Antifouling   76
Envoy Tf 400 Antifouling   105
Envoy Tf 500   20
Envoy Tf 500   46
Envoy Tf 500   109
Envoy Tf 500   131
Envoy Tf 500 Antifouling   21
Envoy Tf 500 Antifouling   48
Envoy Tf 500 Antifouling   76
Envoy Tf 500 Antifouling   105
Envoy Tf100   10
Eq109 Surface Biocide   441
Eq116 Dual Purpose Fi   643
Eq116 Dual Purpose Fi   675
Eq158 Timber Fungicide   610
Eq158 Timber Fungicide   667
Eq159 Rtu Microemulsion Timber Fluid   608
Eq159 Rtu Microemulsion Timber Fluid   644
Eq159 Rtu Microemulsion Timber Fluid   669
Equatorial   42
Equatorial   177
Erasor   400
Esbiol 200   229
Etisso Ant-Ex Bait Box   248
Etoc 10 Mg Mat   344
Etoc Liquid Vaporiser   344
Even Extreme   22
Even Extreme   58
Even Extreme   101
Even Extreme   133
Even Extreme   175
Even Tf   15
Even Tf   108
Even Tf Light Grey   47
Even Tf Light Grey   114
Even Tin Free   15
Even Tin Free   108
Even Tin Free Light Grey   27
Even Tin Free Light Grey   94
Everbuild Lumberjack Universal Rot And
    Woodworm Killer   652
Everbuild Lumberjack Universal Rot And
    Woodworm Killer   719
Everbuild Lumberjack Wood Preservative   714
Everclear   558

Everclear   742
Excel 8 Insecticide   633
Excel Masonry Biocide   441
Exterior Preservative Primer   581
Exterior Preservative Primer   737
Exterior Wood Preserver S   460
Extermisect Multi   335
Far And Away   228
Farco Flying Insect Killer   236
Farco Flying Insect Killer   312
Farco Rapid Kill   234
Farco Rapid Kill   278
Farco Rapid Kill   325
Farco Rapid Kill Insecticide Powder   311
Farmers Ridento Rtu Rat Bait   404
Faslane   290
Feb Fungicide   419
Feb Fungicide   439
Feb Fungicide   440
Fence Protector   684
Fendona 1.5 Sc   218
Fendona 6sc   218
Fendona Asc   218
Fendona Lacquer   218
Fendona Wp   218
Fenitrothion 47ec   283
Fenitrothion Dusting Powder   283
Fenitrothion Emulsion Concentrate   283
Fenitrothion Wettable Powder   283
Fentex Elite   507
Fentex Elite   623
Fentex Elite   711
Fentex Np-Uf   709
Fentrol   408
Fentrol Gel   408
Ficam 20w   223
Ficam Plus   224
Ficam Plus   348
Ficam W   223
Flag Brand Wood Preservative Green   529
Flagspeed Antifouling   10
Flak   338
Flak   394
Flamil Finale   317
Flamil Finale   358
Flamil Finale Super   347
Flea Ban   245
Flea Off Powder   311
Flea Powder   311
Flea-Off!   267
Flee Flea   311
Fleegard   236
Fleegard   312
Fleegard Plus   255
Fleegard Plus   363
Flego Trigger Pack   266

Flego Trigger Pack   *314*
Flexarb Timber Coating   *568*
Flexgard Iv Waterbase Preservative   *124*
Flexgard V Waterbase Preservative   *124*
Flexgard Vi   *12*
Flexgard Vi   *128*
Flexgard Vi Waterbase Preservative   *9*
Flexgard Vi Waterbase Preservative   *127*
Flexgard Vi-Ii Waterbase Preservative   *10*
Floral Bouquet (Oil Of Citronella)   *200*
Floret Fast Knock Down   *209*
Fly And Ant Killer   *217*
Fly And Ant Killer   *339*
Fly And Maggot Killer   *368*
Fly And Maggot Killer   *386*
Fly And Wasp Killer   *328*
Fly And Wasp Killer   *372*
Fly And Wasp Killer   *383*
Fly Free Zone   *328*
Fly Free Zone   *383*
Fly Killer   *274*
Fly Killer   *368*
Fly Killer   *386*
Fly Spray   *328*
Fly Spray   *383*
Fly, Wasp And Mosquito Killer   *216*
Fly, Wasp And Mosquito Killer   *389*
Fly-Kill   *328*
Fly-Kill   *383*
Flyclear Bhb   *231*
Flyclear Bhb   *243*
Flying Insect Killer   *274*
Flying Insect Killer   *286*
Flying Insect Killer   *379*
Flying Insect Killer Faster Knock Down   *286*
Flying Insect Killer Faster Knock Down   *379*
Fmb 451-5 Quat (50%)   *419*
Fmb 451-5 Quat (50%)   *474*
Focus Ant Killer   *223*
Focus New Ant Killer Spray   *267*
Fongix Se Total Treatment For Wood   *417*
Fongix Se Total Treatment For Wood   *444*
Fongix Se Total Treatment For Wood   *472*
Fongix Se Total Treatment For Wood   *638*
Forsham's Re-Treat   *568*
Fortefog   *347*
Fresh-A-Pet Insecticidal Rug And Carpet
    Freshener   *311*
Freshways Slow Release Insect Killer   *274*
Friends Extra Long Lasting Household Flea
    Spray   *304*
Friends Extra Long Lasting Household Flea
    Spray   *321*
Friends Extra Long Lasting Household Flea
    Spray   *357*
Friends Household Flea Powder   *311*

Friends Household Flea Spray   *338*
Friends Household Flea Spray   *394*
Friskies Pro Control Household Flea Spray   *322*
Friskies Pro Control Household Flea Spray   *364*
Fumite Lindane Generator Size 10   *296*
Fumite Lindane Generator Size 10   *612*
Fumite Lindane Generator Size 40   *296*
Fumite Lindane Generator Size 40   *612*
Fumite Lindane Pellet No 3   *296*
Fumite Lindane Pellet No 4   *296*
Fungi-Shield Sterilising Solution Concentrate
    Gs36   *419*
Fungi-Shield Sterilising Solution Gs37   *419*
Fungicidal Algaecidal Bacteriacidal Wash   *420*
Fungicidal Algaecidal Bacteriacidal Wash   *447*
Fungicidal Wall Solution   *441*
Fungicidal Wash   *420*
Fungicidal Wash   *439*
Fungicidal Wash   *440*
Fungicidal Wash   *447*
Fungicidal Wash   *594*
Fungicidal Wash   *596*
Fungicidal Wash Solution   *420*
Fungicidal Wash Solution   *447*
Fungicide 2   *666*
Fungicide Insecticide   *557*
Fungicide Insecticide   *716*
Fungiguard   *420*
Fungiguard   *447*
Fungishield Sterilising Solution Concentrate
    Gs36   *414*
Fungishield Sterilising Solution Gs37   *414*
Fungo   *452*
Funnel Trap Insecticidal Strip   *274*
Gallwey Abs   *592*
Gallwey Abs   *693*
Gelert Mosquito Repellent   *209*
Gett   *248*
Giraglia   *27*
Giraglia   *94*
Glen Wood Care Green   *529*
Glen Wood Care Wood Preservative   *474*
Glidden Trade Fungicidal Vinyl Silk   *181*
Glidden Trade Fungicidal Acrylic Eggshell   *181*
Glidden Trade Fungicidal Vinyl Matt   *181*
Globol Pyrethrum Electrical Evaporator   *228*
Globol Shake And Spray Insect Killer   *347*
Globol Small Space Fly And Moth Strip   *274*
Gold Label Kennel And Stable Powder   *311*
Golden Creosote   *549*
Goliath Bait Station   *287*
Goliath Gel   *287*
Good Boy Insecticidal Carpet And Upholstery
    Powder   *311*
Grafo Anti-Foul Sw   *27*
Grafo Anti-Foul Sw   *94*

Granocryl Fungicidal Wash *419*
Grassline Abl Antifouling Type M349 *36*
Grassline Abl Antifouling Type M349 *144*
Grassline Abl Antifouling Type M349 *155*
Grassline Tf Anti-Fouling Type M396 *11*
Grassline Tf Anti-Fouling Type M396 *132*
Grassline Type M396 Antifouling *16*
Grassline Type M396 Antifouling *75*
Grassline Type M396 Antifouling *104*
Great Mills Ant Killer *223*
Great Mills Ant Killer Spray *257*
Great Mills New Ant Killer Spray *267*
Great Mills Woodlice Killer *223*
Green *532*
Green *641*
Green End Coat *460*
Green N Clean Algae Killer Concentrate *419*
Green N Clean Algae Killer Ready To Use *419*
Green Plus Wood Preserver *529*
Green Range Dual Purpose Aq *557*
Green Range Dual Purpose Aq *716*
Green Range Fungicidal Concentrate *451*
Green Range Murosol 20 *439*
Green Range Murosol 20 *440*
Green Range Timber Treatment Paste *655*
Green Range Timber Treatment Paste *740*
Green Range Woodworm Killer *551*
Green Range Woodworm Killer Aq *551*
Green Range Wykamol Plus *557*
Green Range Wykamol Plus *716*
Green Wood Preservative *529*
Greenco Ant Killer *343*
Greenhills Creosote *524*
Growing Success Ant & Crawling Insect
   Killer *257*
Growing Success Ant Killer *347*
Growing Success Indoor And Outdoor Ant
   Killer *347*
Haden Mosquito And Flying Insect Killer *228*
Hail Flying Insect Killer *328*
Hail Flying Insect Killer *383*
Hail Plus Crawling Insect Killer *215*
Hail Plus Crawling Insect Killer *331*
Hail Plus Crawling Insect Killer *393*
Halcyon 5000 (Base) *23*
Halcyon 5000 (Base) *65*
Halocell 221 *192*
Halodec *434*
Halophane Bonding Solution *432*
Halophane No.1 Aerosol *434*
Halophane No.3 Aerosol *434*
Halophen Bm1165l *432*
Haloseal *198*
Halostain *443*
Hand Anti-Ant Powder *311*
Hard Racing *10*

Hard Racing Antifouling *15*
Hard Racing Antifouling *108*
Hard Racing Antifouling White *47*
Hard Racing Antifouling White *114*
Hardwood Protector *568*
Hartz Rid Flea Control Home Spray *302*
Hartz Rid Flea Control Home Spray *327*
Hempel's Antifouling 761gb *10*
Hempel's Antifouling 762gb *19*
Hempel's Antifouling 762gb *44*
Hempel's Antifouling 763gb *45*
Hempel's Antifouling 7690d *38*
Hempel's Antifouling 7690d *148*
Hempel's Antifouling 7690d *159*
Hempel's Antifouling 7690d *179*
Hempel's Antifouling 8199d *27*
Hempel's Antifouling 8199d *94*
Hempel's Antifouling 8199f *27*
Hempel's Antifouling 8199f *94*
Hempel's Antifouling Classic 7611 Red (Tin Free)
   5000 *15*
Hempel's Antifouling Classic 7611 Red (Tin Free)
   5000 *108*
Hempel's Antifouling Classic 76540 *15*
Hempel's Antifouling Classic 76540 *108*
Hempel's Antifouling Classic 7654b *27*
Hempel's Antifouling Classic 7654b *94*
Hempel's Antifouling Combic 71990 *15*
Hempel's Antifouling Combic 71990 *108*
Hempel's Antifouling Combic 71992 *15*
Hempel's Antifouling Combic 71992 *108*
Hempel's Antifouling Combic 7199b *27*
Hempel's Antifouling Combic 7199b *94*
Hempel's Antifouling Combic 7199c *61*
Hempel's Antifouling Combic 7199c *171*
Hempel's Antifouling Combic 7199f *42*
Hempel's Antifouling Combic 7199f *177*
Hempel's Antifouling Combic 76990 *36*
Hempel's Antifouling Combic 76990 *144*
Hempel's Antifouling Combic 76990 *155*
Hempel's Antifouling Economic Sp-Sea 74030 *38*
Hempel's Antifouling Economic Sp-Sea
   74030 *148*
Hempel's Antifouling Economic Sp-Sea
   74030 *159*
Hempel's Antifouling Economic Sp-Sea
   74030 *179*
Hempel's Antifouling Forte Tin Free 7625 *15*
Hempel's Antifouling Forte Tin Free 7625 *108*
Hempel's Antifouling Globic Sp-Eco 81900 *17*
Hempel's Antifouling Globic Sp-Eco 81900 *73*
Hempel's Antifouling Globic Sp-Eco 81990 *17*
Hempel's Antifouling Globic Sp-Eco 81990 *73*
Hempel's Antifouling Mille Dynamic *15*
Hempel's Antifouling Mille Dynamic *108*
Hempel's Antifouling Mille Dynamic 717 Gb *51*

Hempel's Antifouling Mille Dynamic 717 Gb   64
Hempel's Antifouling Mille Dynamic 717 Gb   115
Hempel's Antifouling Nautic 71800   122
Hempel's Antifouling Nautic 71800   165
Hempel's Antifouling Nautic 7180b   97
Hempel's Antifouling Nautic 7180b   173
Hempel's Antifouling Nautic 7180e   82
Hempel's Antifouling Nautic 7180e   169
Hempel's Antifouling Nautic 71900   15
Hempel's Antifouling Nautic 71900   108
Hempel's Antifouling Nautic 71902   15
Hempel's Antifouling Nautic 71902   108
Hempel's Antifouling Nautic 7190b   27
Hempel's Antifouling Nautic 7190b   94
Hempel's Antifouling Nautic 76800   140
Hempel's Antifouling Nautic 76800   157
Hempel's Antifouling Nautic 7680b   147
Hempel's Antifouling Nautic 7680b   158
Hempel's Antifouling Nautic 7680b   174
Hempel's Antifouling Nautic 7691b   38
Hempel's Antifouling Nautic 7691b   148
Hempel's Antifouling Nautic 7691b   159
Hempel's Antifouling Nautic 7691b   179
Hempel's Antifouling Nautic 8190c   15
Hempel's Antifouling Nautic 8190c   108
Hempel's Antifouling Nautic 8190h   27
Hempel's Antifouling Nautic 8190h   94
Hempel's Antifouling Nautic Hi 76900   36
Hempel's Antifouling Nautic Hi 76900   144
Hempel's Antifouling Nautic Hi 76900   155
Hempel's Antifouling Nautic Hi 76910   36
Hempel's Antifouling Nautic Hi 76910   144
Hempel's Antifouling Nautic Hi 76910   155
Hempel's Antifouling Nautic Sp-Ace 79030   38
Hempel's Antifouling Nautic Sp-Ace 79030   148
Hempel's Antifouling Nautic Sp-Ace 79030   159
Hempel's Antifouling Nautic Sp-Ace 79030   179
Hempel's Antifouling Nautic Sp-Ace 79031   36
Hempel's Antifouling Nautic Sp-Ace 79031   144
Hempel's Antifouling Nautic Sp-Ace 79031   155
Hempel's Antifouling Nautic Sp-Ace 79050   38
Hempel's Antifouling Nautic Sp-Ace 79050   148
Hempel's Antifouling Nautic Sp-Ace 79050   159
Hempel's Antifouling Nautic Sp-Ace 79050   179
Hempel's Antifouling Nautic Sp-Ace 79051   36
Hempel's Antifouling Nautic Sp-Ace 79051   144
Hempel's Antifouling Nautic Sp-Ace 79051   155
Hempel's Antifouling Nautic Sp-Ace 79070   38
Hempel's Antifouling Nautic Sp-Ace 79070   148
Hempel's Antifouling Nautic Sp-Ace 79070   159
Hempel's Antifouling Nautic Sp-Ace 79070   179
Hempel's Antifouling Nautic Tin Free 7190e   17
Hempel's Antifouling Nautic Tin Free 7190e   73
Hempel's Antifouling Nordic 7133   10
Hempel's Antifouling Olympic Hi-7661   15
Hempel's Antifouling Olympic Hi-7661   108

Hempel's Antifouling Olympic Tin Free 7154   15
Hempel's Antifouling Olympic Tin Free 7154   108
Hempel's Antifouling Paint 161p (Red And
    Chocolate To Ts10240)   10
Hempel's Antifouling Rennot 7150   23
Hempel's Antifouling Rennot 7150   65
Hempel's Antifouling Rennot 7177   23
Hempel's Antifouling Rennot 7177   65
Hempel's Antifouling Tin Free 742gb   23
Hempel's Antifouling Tin Free 742gb   65
Hempel's Antifouling Tin Free 743gb   17
Hempel's Antifouling Tin Free 743gb   73
Hempel's Antifouling Tin Free 745gb   15
Hempel's Antifouling Tin Free 745gb   108
Hempel's Antifouling Tin Free 750gb   15
Hempel's Antifouling Tin Free 750gb   108
Hempel's Antifouling Tin Free 751gb   17
Hempel's Antifouling Tin Free 751gb   73
Hempel's Antifouling Tin Free 7662   17
Hempel's Antifouling Tin Free 7662   73
Hempel's Bravo 7610a   27
Hempel's Bravo 7610a   94
Hempel's Copper Bottom Paint 7116   10
Hempel's Economic Sp-Sea 74010   38
Hempel's Economic Sp-Sea 74010   148
Hempel's Economic Sp-Sea 74010   159
Hempel's Economic Sp-Sea 74010   179
Hempel's Hard Racing 7638a   54
Hempel's Hard Racing 7638a   95
Hempel's Hard Racing 76480   15
Hempel's Hard Racing 76480   108
Hempel's Hard Racing 7648a   27
Hempel's Hard Racing 7648a   94
Hempel's Mille Alu 71602   54
Hempel's Mille Alu 71602   95
Hempel's Mille Dynamic 71600   47
Hempel's Mille Dynamic 71600   114
Hempel's Mille Dynamic 7160a   54
Hempel's Mille Dynamic 7160a   95
Hempel's Mille Dynamic 71700   15
Hempel's Mille Dynamic 71700   108
Hempel's Mille Dynamic 7170a   27
Hempel's Mille Dynamic 7170a   94
Hempel's Net Antifouling 715gb   10
Hempel's Seatech Antifouling 7820a   27
Hempel's Seatech Antifouling 7820a   94
Hempel's Tin Free Antifouling 744gb   10
Hempel's Tin Free Antifouling 7626   17
Hempel's Tin Free Antifouling 7626   73
Hempel's Tin Free Antifouling 7660   10
Hempel's Tin Free Hard Racing 76380   47
Hempel's Tin Free Hard Racing 76380   114
Hempel's Tin Free Hard Racing 7648   47
Hempel's Tin Free Hard Racing 7648   114
Hempels Antifouling Bravo Tin Free 7610   15
Hempels Antifouling Bravo Tin Free 7610   108

HSE
PRODUCT INDEX

Hempels Antifouling Globic Sp-Eco 81920   *17*
Hempels Antifouling Globic Sp-Eco 81920   *73*
Hg Green Slime Remover   *432*
Hi-Craft Insecticidal Household Spray   *236*
Hi-Craft Insecticidal Household Spray   *312*
Hickson Antiblu 3737   *506*
Hickson Antiblu 3737   *626*
Hickson Antiblu 3738   *506*
Hickson Antiblu 3738   *626*
Hickson Antiblu 3739   *622*
Hickson Antiblu 3739   *713*
Hickson Antiborer 3767   *551*
Hickson Antiborer 3768   *633*
Hickson Np-1   *561*
Hickson Np-1   *605*
Hickson Timbercare Wrqd   *618*
Hickson Timbercare Wrqd   *727*
Hickson Woodex   *579*
Hickson Woodex   *726*
Home Insect Fogger   *317*
Home Insect Fogger   *358*
Homebase Ant And Crawling Insect Killer   *236*
Homebase Ant And Crawling Insect Killer   *312*
Homebase Ant Killer   *223*
Homebase Antkiller Spray   *257*
Homebase Creosote   *549*
Homebase Mould Cleaner   *419*
Homebase New Ant Killer Spray   *267*
Homebase Wasp Nest Killer   *223*
Homebase Weathercoat Deep Penetrating
   Preservative Primer   *576*
Homebase Weathercoat Deep Penetrating
   Preservative Primer   *717*
Homebase Weathercoat Fungicidal Wash   *419*
Homebase Weathercoat Preservative
   Primer   *733*
Homebase Woodlice Killer   *223*
Hospital Flying Insect Spray   *368*
Hospital Flying Insect Spray   *386*
Houseplan Fungi Wash   *420*
Houseplan Fungi Wash   *447*
Howes Olympic Algaecide   *419*
Hyperion Mould Inhibiting Solution   *439*
Hyperion Mould Inhibiting Solution   *440*
Ici Decorator 1st Fungicidal Wash   *414*
Icon 2.5 Ec   *295*
Imersol 2410   *658*
Imersol 2410   *743*
Imperator Fog S 18/6   *237*
Imperator Fog S 18/6   *324*
Imperator Fog Super S 30/15   *237*
Imperator Fog Super S 30/15   *324*
Imperator Ulv S 6/3   *237*
Imperator Ulv S 6/3   *324*
Imperator Ulv Super S 10/4   *237*
Imperator Ulv Super S 10/4   *324*

Impra-Color   *568*
Impra-Holzschutzgrund (Primer)   *573*
Impra-Holzschutzgrund (Primer)   *651*
Impra-Holzschutzgrund (Primer)   *703*
Impra-Sanol   *633*
Impralan Grund G 200   *666*
Impralan-Grund G 100   *666*
Inaccessible Woodworm Killer   *633*
Indorex Press And Go Mist   *322*
Indorex Press And Go Mist   *364*
Indorex Spray   *322*
Indorex Spray   *364*
Industrial Hygiene Controllable Cassette Insect
   Killer   *274*
Inflatable Boat Antifouling   *10*
Injecta Osmose Abs 33 A   *515*
Injecta Osmose Abs33   *518*
Injecta Osmose Abs33   *541*
Injecta Osmose Abs33   *620*
Injecta Osmose K33 C50   *468*
Injecta Osmose K33 C50   *522*
Injecta Osmose K33 C50   *535*
Injecta Osmose K33 C58.7   *468*
Injecta Osmose K33 C58.7   *522*
Injecta Osmose K33 C58.7   *535*
Injecta Osmose K33 C72   *468*
Injecta Osmose K33 C72   *522*
Injecta Osmose K33 C72   *535*
Injecta Osmose S-11   *622*
Injecta Osmose S-11   *713*
Insect Killer   *338*
Insect Killer   *394*
Insectaban   *311*
Insecticidal Aerosol   *317*
Insecticidal Aerosol   *358*
Insecticide 1   *633*
Insecticide Aerosol   *328*
Insecticide Aerosol   *383*
Insecticide Fungicide Wood Preservative   *652*
Insecticide Fungicide Wood Preservative   *719*
Insectrol   *236*
Insectrol   *312*
Insectrol Professional   *236*
Insectrol Professional   *312*
Insectrol, Fly And Wasp Killer   *328*
Insectrol, Fly And Wasp Killer   *383*
Instasective   *245*
Interclene 245   *42*
Interclene 245   *177*
Interclene Aq Hza700 Series (Base)   *123*
Interclene Extra Baa100 Series   *10*
Interclene Premium Bca300 Series   *10*
Interclene Super Bca400 Series (Bca400
   Red)   *10*
Interclene Underwater Premium Bca468 Red   *10*
Interior Mould Remover Wipes   *419*

International Tbt Free Copolymer Antifouling
  Bqa100 Series   *10*
International Tin Free Spc Bna100 Series   *27*
International Tin Free Spc Bna100 Series   *94*
Intersleek Bxa 810/820 (Base)   *123*
Intersleek Bxa580 Series (Base)   *123*
Intersleek Fcs Hka560 Series (Base)   *123*
Intersleek Fcs Hka580 Series (Base)   *123*
Intersmooth 110 Standard   *38*
Intersmooth 110 Standard   *148*
Intersmooth 110 Standard   *159*
Intersmooth 110 Standard   *179*
Intersmooth 120 Premium   *38*
Intersmooth 120 Premium   *148*
Intersmooth 120 Premium   *159*
Intersmooth 120 Premium   *179*
Intersmooth 130 Ultra   *38*
Intersmooth 130 Ultra   *148*
Intersmooth 130 Ultra   *159*
Intersmooth 130 Ultra   *179*
Intersmooth 220 Premium   *38*
Intersmooth 220 Premium   *148*
Intersmooth 220 Premium   *159*
Intersmooth 220 Premium   *179*
Intersmooth 230 Ultra   *38*
Intersmooth 230 Ultra   *148*
Intersmooth 230 Ultra   *159*
Intersmooth 230 Ultra   *179*
Intersmooth 320 Premium   *38*
Intersmooth 320 Premium   *148*
Intersmooth 320 Premium   *159*
Intersmooth 320 Premium   *179*
Intersmooth 330 Ultra   *38*
Intersmooth 330 Ultra   *148*
Intersmooth 330 Ultra   *159*
Intersmooth 330 Ultra   *179*
Intersmooth 360 Ecoloflex   *41*
Intersmooth 360 Ecoloflex   *170*
Intersmooth 365 Ecoloflex   *41*
Intersmooth 365 Ecoloflex   *170*
Intersmooth 460 Ecoloflex   *41*
Intersmooth 460 Ecoloflex   *170*
Intersmooth 465 Ecoloflex   *41*
Intersmooth 465 Ecoloflex   *170*
Intersmooth Hisol 2000 Bfa270 Series   *38*
Intersmooth Hisol 2000 Bfa270 Series   *148*
Intersmooth Hisol 2000 Bfa270 Series   *159*
Intersmooth Hisol 2000 Bfa270 Series   *179*
Intersmooth Hisol 9000 Bfa970 Series   *38*
Intersmooth Hisol 9000 Bfa970 Series   *148*
Intersmooth Hisol 9000 Bfa970 Series   *159*
Intersmooth Hisol 9000 Bfa970 Series   *179*
Intersmooth Hisol Bfa948 Orange   *60*
Intersmooth Hisol Bfa948 Orange   *149*
Intersmooth Hisol Bfa948 Orange   *160*
Intersmooth Hisol Bfa948 Orange   *180*

Intersmooth Hisol Bfo270/950/970 Series   *37*
Intersmooth Hisol Bfo270/950/970 Series   *78*
Intersmooth Hisol Bfo270/950/970 Series   *143*
Intersmooth Hisol Bfo270/950/970 Series   *152*
Intersmooth Hisol Spc Antifouling Bfa949 Red   *60*
Intersmooth Hisol Spc Antifouling Bfa949
  Red   *149*
Intersmooth Hisol Spc Antifouling Bfa949
  Red   *160*
Intersmooth Hisol Spc Antifouling Bfa949
  Red   *180*
Intersmooth Hisol Tin Free Bga620 Series   *27*
Intersmooth Hisol Tin Free Bga620 Series   *94*
Intersmooth Spc Antifouling Bfo250 Series   *36*
Intersmooth Spc Antifouling Bfo250 Series   *144*
Intersmooth Spc Antifouling Bfo250 Series   *155*
Intersmooth Spc Bfa040/Bfa050 Series   *59*
Intersmooth Spc Bfa040/Bfa050 Series   *146*
Intersmooth Spc Bfa040/Bfa050 Series   *156*
Intersmooth Spc Bfa090/Bfa190 Series   *36*
Intersmooth Spc Bfa090/Bfa190 Series   *144*
Intersmooth Spc Bfa090/Bfa190 Series   *155*
Intersmooth Tin Free Bga530 Series   *27*
Intersmooth Tin Free Bga530 Series   *94*
Interspeed 2000   *47*
Interspeed 2000   *114*
Interspeed 2000 White   *50*
Interspeed 2000 White   *66*
Interspeed 2001   *45*
Interspeed 2002   *61*
Interspeed 2002   *171*
Interspeed 2003   *49*
Interspeed 2003   *79*
Interspeed 340   *42*
Interspeed 340   *177*
Interspeed Antifouling Bwo900 Series   *15*
Interspeed Antifouling Bwo900 Series   *108*
Interspeed Extra Bwa500 Red   *27*
Interspeed Extra Bwa500 Red   *94*
Interspeed Extra Strong   *27*
Interspeed Extra Strong   *94*
Interspeed Super Bwa900 Red   *27*
Interspeed Super Bwa900 Red   *94*
Interspeed Super Bwa909 Black   *27*
Interspeed Super Bwa909 Black   *94*
Interspeed System 2 Bra140/Bra240 Series   *42*
Interspeed System 2 Bra140/Bra240 Series   *177*
Interspeed System 2 Bra142 Brown   *27*
Interspeed System 2 Bra142 Brown   *94*
Interspeed System 2 Bra143 Brown   *15*
Interspeed System 2 Bra143 Brown   *108*
Interspeed System 2 Bro142/240 Series   *10*
Interspeed Ultra   *23*
Interspeed Ultra   *65*
Interswift Bko000/700 Series   *33*
Interswift Bko000/700 Series   *96*

Interswift Bko000/700 Series   *145*
Interswift Bko000/700 Series   *154*
Interswift Tin Free Bqa400 Series   *27*
Interswift Tin Free Bqa400 Series   *94*
Interswift Tin-Free Spc Bta540 Series   *27*
Interswift Tin-Free Spc Bta540 Series   *94*
Intertox   *568*
Interviron Bqa200 Series   *42*
Interviron Bqa200 Series   *177*
Interviron Bqa450 Series   *15*
Interviron Bqa450 Series   *108*
Interviron Super Bna400 Series   *27*
Interviron Super Bna400 Series   *94*
Interviron Super Bqa400 Series   *27*
Interviron Super Bqa400 Series   *94*
Interviron Super Tin-Free Polishing Antifouling
  Bqo400 Series   *15*
Interviron Super Tin-Free Polishing Antifouling
  Bqo400 Series   *108*
Interviron Super Tin-Free Polishing Antifouling
  Bqo420 Series   *17*
Interviron Super Tin-Free Polishing Antifouling
  Bqo420 Series   *73*
Iodofenphos Granular Bait   *293*
J B Ward & Son Dual Purpose Timber Treatment
  Concentrate   *652*
J B Ward & Son Dual Purpose Timber Treatment
  Concentrate   *719*
Janisolve Fly Killer   *368*
Janisolve Fly Killer   *386*
Jertox Moth Balls   *204*
Jewson Creosote   *549*
Jewson Fungicidal Wash   *419*
Jeyes Algae And Mould Killer   *419*
Jeyes Crawling Insect Spray   *217*
Jeyes Crawling Insect Spray   *339*
Jeyes Expel Ant And Crawling Insect Killer   *338*
Jeyes Expel Ant And Crawling Insect Killer   *394*
Jeyes Expel Ant Killer Powder   *311*
Jeyes Expel Ant Trap   *398*
Jeyes Expel Flea Killer   *338*
Jeyes Expel Flea Killer   *394*
Jeyes Expel Fly And Wasp Killer   *372*
Jeyes Expel Moth Killer   *274*
Jeyes Expel Plug-In Flying Insect Killer   *209*
Jeyes Expel Pump Action Fly And Wasp
  Killer   *347*
Jeyes Expel Slow Release Fly And Wasp
  Killer   *274*
Jeyes Flying Insect Spray   *347*
Jeyes Kontrol Ant And Crawling Insect Killer   *338*
Jeyes Kontrol Ant And Crawling Insect Killer   *394*
Jeyes Kontrol Ant Killer Powder   *311*
Jeyes Kontrol Flea Killer   *338*
Jeyes Kontrol Flea Killer   *394*
Jeyes Kontrol Fly And Wasp Killer   *372*

Jeyes Kontrol Moth Killer   *274*
Jeyes Kontrol Slow Release Fly And Wasp
  Killer   *274*
Jeyes Slow Release Fly Killer Controllable
  Cassette   *274*
Jeyes Small Space Fly And Moth Strip   *274*
Johnson's Cage And Hutch Insect Spray   *305*
Johnson's Cage And Hutch Insect Spray   *330*
Johnson's Cage And Hutch Insect Spray   *382*
Johnson's Carpet Flea Guard   *311*
Johnson's Home Flea Guard Trigger Spray   *267*
Johnson's House Flea Spray   *328*
Johnson's House Flea Spray   *383*
Johnson's House Flea Spray Extra Guard   *305*
Johnson's House Flea Spray Extra Guard   *330*
Johnson's House Flea Spray Extra Guard   *382*
Johnson's Household Extra Guard Flea And Insect
  Spray   *304*
Johnson's Household Extra Guard Flea And Insect
  Spray   *321*
Johnson's Household Extra Guard Flea And Insect
  Spray   *357*
Johnson's Household Flea Powder   *311*
Johnson's Household Flea Spray   *317*
Johnson's Household Flea Spray   *358*
Johnson's New Pet Housing Spray   *328*
Johnson's New Pet Housing Spray   *383*
Johnson's Pet Housing Spray   *317*
Johnson's Pet Housing Spray   *358*
Johnstone's Fungicidal Wash   *420*
Johnstone's Fungicidal Wash   *447*
Johnstone's Preservative Clear   *733*
Johnstone's Woodworks Exterior Universal
  Primer   *576*
Johnstone's Woodworks Exterior Universal
  Primer   *717*
Johnstone's Woodworks Interior/Exterior All
  Purpose Preserver   *574*
Johnstone's Woodworks Interior/Exterior All
  Purpose Preserver   *653*
Johnstone's Woodworks Interior/Exterior All
  Purpose Preserver   *718*
K-Otec Al   *559*
K-Otek Ec   *559*
K-Otek Sl   *559*
K-Othrine Crawling Insect Powder   *267*
K-Othrine Moustiquaire S.C.   *201*
K.O. Fly Spray   *368*
K.O. Fly Spray   *386*
K02 Selkil   *328*
K02 Selkil   *383*
Kalon Creosote   *549*
Karate Dairy Fly Spray   *347*
Karate Ready To Use Rat Bait   *404*
Karate Ready To Use Rodenticide Sachets   *404*
Kathon 886f   *517*

Kathon 886f   *619*
Kemwood Cca Type Bs   *469*
Kemwood Cca Type Bs   *545*
Kemwood Cca Type Bs   *685*
Kenwood Micro Emulsion Woodworm Treatment
   (Concentrate)   *633*
Kenwood Woodworm Treatment
   Concentrate   *633*
Keyline Creosote   *549*
Kf-18 Masonry Biocide (Microemulsion)   *441*
Kf-8 Dual Purpose Micro Emulsion
   Concentrate   *652*
Kf-8 Dual Purpose Micro Emulsion
   Concentrate   *719*
Kf8 Insecticide Microemulsion Concentrate   *633*
Kibes Sterilising Solution Concentrate Gs 36   *414*
Kil-Ant Ant Killer Powder   *223*
Killgerm Boric Acid Powder   *245*
Killgerm Fenitrothion 40 Wp   *283*
Killgerm Fenitrothion 50 Ec   *283*
Killgerm Fenitrothion-Pyrethroid
   Concentrate   *286*
Killgerm Fenitrothion-Pyrethroid
   Concentrate   *379*
Killgerm Insect Powder   *223*
Killgerm Precor Rtu   *306*
Killgerm Precor Rtu   *356*
Killgerm Precor Ulv I   *306*
Killgerm Precor Ulv I   *356*
Killgerm Precor Ulv Iii   *259*
Killgerm Precor Ulv Iii   *300*
Killgerm Propoxur 20 Ec   *346*
Killgerm Py-Kill W   *372*
Killgerm Pyrethrum Spray   *347*
Killgerm Repel   *200*
Killgerm Repel Rtu   *200*
Killgerm Sewercide Cut Wheat Rat Bait   *413*
Killgerm Sewercide Whole Wheat Rat Bait   *413*
Killgerm Terminate   *248*
Killgerm Ulv 1500   *338*
Killgerm Ulv 1500   *394*
Killgerm Ulv 400   *347*
Killgerm Ulv 500   *338*
Killgerm Ulv 500   *394*
Killgerm Woodworm Killer   *551*
Killoff Aerosol Insect Killer   *338*
Killoff Aerosol Insect Killer   *394*
Kingfisher Fungicidal Wall Solution   *439*
Kingfisher Fungicidal Wall Solution   *440*
Kingfisher Kf-8 Micro Emulsion Insecticide
   Concentrate   *633*
Kingfisher Timber Paste   *595*
Kingfisher Timber Paste   *597*
Kingfisher Timber Paste   *642*
Kingfisher Timberpaste   *595*
Kingfisher Timberpaste   *597*

Kingfisher Timberpaste   *642*
Kingfisher Timberpaste   *652*
Kingfisher Timberpaste   *719*
Kingfisher Wood Preservative   *533*
Kingfisher Wood Preservative   *714*
Kingfisher Wood Preservative   *715*
Kleeneze Anti-Mould Spray   *415*
Kleeneze Anti-Mould Spray   *426*
Kleeneze Fly And Flying Insect Pump Spray   *347*
Kleeneze Fly Killer   *372*
Kleeneze New Improved Fly Killer   *217*
Kleeneze New Improved Fly Killer   *339*
Kleenoff Crawling Insect Spray   *217*
Kleenoff Crawling Insect Spray   *339*
Kleenoff Flying Insect Spray   *347*
Kleenoff Slow Release Fly Killer Controllable
   Cassette   *274*
Kleenoff Small Space Fly And Moth Strip   *274*
Klerat   *400*
Klerat Wax Blocks   *400*
Knox Out 2 Fm   *270*
Konk 1 (B) Flying Insect Killer   *347*
Konk I   *347*
Kontrol Kitchen Size Fly Killer   *274*
Kontrol Slow Release Fly Killer   *274*
Kudos   *311*
Kwik Insecticide   *217*
Kwik Insecticide   *339*
Kybosh 2   *338*
Kybosh 2   *394*
Langlow Clear Wood Preserver   *557*
Langlow Clear Wood Preserver   *716*
Langlow Creosote   *524*
Langlow Green Wood Preserver   *533*
Langlow Green Wood Preserver   *715*
Langlow Pale Green Wood Preserver   *533*
Langlow Pale Green Wood Preserver   *715*
Langlow Timbershield   *474*
Langlow Wood Preservative Green   *529*
Langlow Wood Preserver   *714*
Langlow Wood Preserver Formulation A   *568*
Langlow Wood Preserver Formulation B   *714*
Lanosol P   *311*
Lanosol Rtu   *311*
Laporte Acq 1900   *478*
Laporte Acq 1900   *536*
Laporte Acq 2100   *540*
Laporte Acq 2100   *564*
Laporte Antisapstain   *622*
Laporte Antisapstain   *713*
Laporte Cca Awpa Type C   *468*
Laporte Cca Awpa Type C   *522*
Laporte Cca Awpa Type C   *535*
Laporte Cca Oxide Type 1   *468*
Laporte Cca Oxide Type 1   *522*
Laporte Cca Oxide Type 1   *535*

HSE
PRODUCT INDEX

Laporte Cca Oxide Type 2 *468*
Laporte Cca Oxide Type 2 *522*
Laporte Cca Oxide Type 2 *535*
Laporte Cca Type 1 *469*
Laporte Cca Type 1 *545*
Laporte Cca Type 1 *685*
Laporte Cca Type 2 *469*
Laporte Cca Type 2 *545*
Laporte Cca Type 2 *685*
Laporte Ccb *500*
Laporte Ccb *546*
Laporte Ccb *687*
Laporte Cut End Preservative *544*
Laporte Cut End Preservative *689*
Laporte End Grain Treatment *498*
Laporte End Grain Treatment *542*
Laporte Mould-Ex *517*
Laporte Mould-Ex *619*
Laporte Permawood Cca *469*
Laporte Permawood Cca *545*
Laporte Permawood Cca *685*
Lar-Vac 100 *615*
Lar-Vac 100 *631*
Lar-Vac 100 *728*
Lar-Vac 400 *618*
Lar-Vac 400 *727*
Larsen Concentrated Algicide *439*
Larsen Concentrated Algicide *440*
Larsen Concentrated Algicide 2 *420*
Larsen Concentrated Algicide 2 *447*
Larsen Creosote *524*
Larsen Green Wood Preservative *529*
Larsen Joinery Grade 2 *714*
Larsen Wood Preservative Clear 2 And Brown 2 *652*
Larsen Wood Preservative Clear 2 And Brown 2 *719*
Larsen Woodworm Killer 2 *652*
Larsen Woodworm Killer 2 *719*
Larvac 300 *655*
Larvac 300 *740*
Larvex-100 *247*
Larvex-100 *310*
Larvex-15 *247*
Larvex-15 *310*
Laybond Creosote Blend *549*
Laybond Green Wood Preserver *529*
Laybond Timber Protector *552*
Laybond Timber Protector *662*
Laybond Wood Preserver *659*
Le Marin *31*
Le Marin *87*
Le Marin *112*
Lectro 8ww *633*
Lectro Fws *441*
Lectro Micro 8 Fi *643*

Lectro Micro 8 Fi *675*
Lectro Paste Pb *652*
Lectro Paste Pb *719*
Lectrobor Gel *582*
Lectros 8p *633*
Lectros Antimould Breathable Paint *192*
Lectros Antimould Emulsion *192*
Lectros Boracol 10 Rh *423*
Lectros Boracol 10 Rh *437*
Lectros Boracol 20 *582*
Lectros Boracol B40 Gel *582*
Lectros Dry Rot Paint *190*
Lectros Dry Rot Paint *749*
Lectros Lectrobor Ii *582*
Leyland Creosote *549*
Leyland Sterilisation Wash *419*
Leyland Timbrene Environmental Formula *733*
Leyland Timbrene Green Environmental Formula *529*
Leyland Timbrene Supreme Environmental Formula *574*
Leyland Timbrene Supreme Environmental Formula *653*
Leyland Timbrene Supreme Environmental Formula *718*
Leyland Timbrene Woodworm Killer *633*
Leyland Universal Preservative Base *733*
Lichenite *432*
Lichenite B *432*
Light Creosote Blend *524*
Lignosol P *551*
Lincoln Lice Control Powder *311*
Liquid Bromatrol *401*
Lister Teak Dressing *568*
Littac *218*
Lloyds Supersave Slow Release Fly And Wasp Killer. *274*
Lloyds Supersave Small Space Fly And Moth Strip *274*
Long Life T.F. *15*
Long Life T.F. *108*
Long Life Tin Free *27*
Long Life Tin Free *94*
Longlife Antifouling *15*
Longlife Antifouling *108*
Low Odour Woodworm Killer *633*
Lpl Biocidal Wash *419*
Lpl Biowash *419*
Lynx Plus *122*
Lynx Plus *165*
Lynx-A *97*
Lynx-A *173*
Lynx-E *82*
Lynx-E *169*
Lyvia Mosquito Killer *228*
M-Tec Biocide *439*

M-Tec Biocide   *440*
M8 Super Concentrate   *598*
Macpherson Anti-Mould Solution   *439*
Macpherson Anti-Mould Solution   *440*
Macpherson Anti-Mould Solution   *594*
Macpherson Anti-Mould Solution   *596*
Maggot & Fly Attack   *368*
Maggot & Fly Attack   *386*
Major Kil   *257*
Manders Stop Mould   *182*
Manders Stop Mould   *186*
Manders Stop Mould   *193*
Mangers Fungicidal Spray   *419*
Mangers Sterilisation Wash   *419*
Mar-Cide   *451*
Mar-Cide   *684*
Mar-Kil S   *652*
Mar-Kil S   *719*
Mar-Kil W   *652*
Mar-Kil W   *719*
Marclear Full Strength Eu45 Antifouling   *10*
Marclear High Strength Antifouling   *10*
Marclear Powerboat Antifouling   *47*
Marclear Powerboat Antifouling   *114*
Masonry Biocide 3   *441*
Masonry Sterilant   *441*
Masterstroke Wood Preserver   *580*
Masterstroke Wood Preserver   *731*
Maxforce Bait Station   *290*
Maxforce Gel   *290*
Maxforce Pharaoh's Ant Killer   *290*
Maxforce Ultra   *290*
Maxim Creosote   *549*
Maximus Rtu   *257*
Mect   *622*
Mect   *713*
Medium Brown Creosote   *524*
Meridian Mp40 Antifouling   *20*
Meridian Mp40 Antifouling   *46*
Meridian Mp40 Antifouling   *109*
Meridian Mp40 Antifouling   *131*
Metalife Fungicidal Wash   *439*
Metalife Fungicidal Wash   *440*
Metalife Fungicidal Wash   *594*
Metalife Fungicidal Wash   *596*
Mgc Fungicidal Additive   *190*
Micrapor Environmental Flea Spray   *311*
Micro 8 Fi   *508*
Micro 8 Fi   *640*
Micro 8 Fi   *643*
Micro 8 Fi   *675*
Micro 8 Ww   *633*
Micro 8i   *633*
Micro Dual Purpose 8   *508*
Micro Dual Purpose 8   *640*
Micro Fws   *420*

Micro Fws   *441*
Micro Fws   *447*
Micro Masonry Biocide   *420*
Micro Masonry Biocide   *447*
Micro Woodworm 8   *633*
Micro-Emulsion Woodworm Treatment
   Concentrate   *633*
Micro-Tech   *20*
Micro-Tech   *46*
Micro-Tech   *109*
Micro-Tech   *131*
Microguard Fi Concentrate   *508*
Microguard Fi Concentrate   *640*
Microguard Fws   *420*
Microguard Fws   *447*
Microguard Mouldicidal Wood Preserver   *421*
Microguard Mouldicidal Wood Preserver   *427*
Microguard Mouldicidal Wood Preserver   *476*
Microguard Mouldicidal Wood Preserver   *490*
Microguard Permethrin Concentrate   *633*
Microguard Woodworm Fluid   *633*
Micromite   *283*
Micron 25 Plus   *60*
Micron 25 Plus   *149*
Micron 25 Plus   *160*
Micron 25 Plus   *180*
Micron 400 Series   *10*
Micron 500 Series   *27*
Micron 500 Series   *94*
Micron 600 Series   *41*
Micron 600 Series   *170*
Micron 8 Dual Purpose Emulsion
   Concentrate   *652*
Micron 8 Dual Purpose Emulsion
   Concentrate   *719*
Micron Csc   *27*
Micron Csc   *94*
Micron Csc 100 Series   *10*
Micron Csc 200 Series   *15*
Micron Csc 200 Series   *108*
Micron Csc 300 Series   *17*
Micron Csc 300 Series   *73*
Micron Csc Extra   *27*
Micron Csc Extra   *94*
Micron Extra   *23*
Micron Extra   *65*
Micron Optima   *41*
Micron Optima   *170*
Micron Plus Antifouling   *27*
Micron Plus Antifouling   *94*
Microstel   *420*
Microstel   *447*
Microtech Biocide 25   *441*
Microtech Dual Purpose Aq   *553*
Microtech Dual Purpose Aq   *604*
Microtech Fungicide 10x   *666*

PRODUCT INDEX

Microtech Woodworm Killer 25x   *633*
Microtech Woodworm Killer Aq   *551*
Microtreat Biocide 25x   *441*
Microtreat Dual   *607*
Microtreat Dual   *635*
Microtreat Dual 12.5x   *608*
Microtreat Dual 12.5x   *644*
Microtreat Dual 12.5x   *669*
Microtreat Insecticide   *633*
Microtreat Insecticide 25x   *633*
Middletons Creosote   *524*
Middletons Enviro. Shed & Fence Wood
   Preservative   *524*
Miricoat A.F. Coating   *1*
Mite-Kill   *311*
Mitetak   *311*
Mitrol Pq 8   *627*
Momar Fly Killer   *368*
Momar Fly Killer   *386*
Morgan Ant & Crawling Insect Killer   *257*
Morgan's Ant Powder   *311*
Mosiguard Shield   *311*
Moskil Mosquito Repellent   *209*
Mosqui - Go Electric   *228*
Mosqui-Go Liquid   *344*
Mosquito Killer Travel Pack   *228*
Mosquito Repellent   *228*
Mosscheck   *432*
Mostyn 8/64 Tfc   *328*
Mostyn 8/64 Tfc   *383*
Mostyn Permethrin 25 Wp   *311*
Moth And Fly Killer   *274*
Moth Repellent   *202*
Mothaks   *203*
Mothaks   *205*
Motox   *285*
Motox   *329*
Motox   *380*
Motox Crawling Insect   *286*
Motox Crawling Insect   *379*
Motto   *288*
Mould Cleaner   *414*
Mould Killer   *419*
Mouldcheck Barrier 2   *432*
Mouldcheck Cleaner   *434*
Mouldcheck Spray   *434*
Mouldcheck Sterilizer   *434*
Mouldcheck Sterilizer 2   *420*
Mouldcheck Sterilizer 2   *447*
Mouldkill   *432*
Mouldrid   *439*
Mouldrid   *440*
Mouldrid   *594*
Mouldrid   *596*
Mouse Killer Bait   *401*
Mouser   *400*

Mpx   *47*
Mpx   *114*
Mrs Clear Mould Fungicidal Wash   *439*
Mrs Clear Mould Fungicidal Wash   *440*
Multi-Shot Permethrin Spray   *311*
Multispray   *254*
Multispray   *320*
Multispray   *353*
Murphy Fly And Wasp Killer   *372*
Murphy Foaming Wasp Nest Destroyer   *338*
Murphy Foaming Wasp Nest Destroyer   *394*
Murphy Kil-Ant Powder   *223*
Murphy Kil-Ant Ready To Use   *257*
Murphy Kil-Ant Trap   *398*
Murphy Mouse Killer   *401*
Murphys Long Lasting Fly And Wasp Killer   *274*
Muscatrol   *311*
Mystox Btl   *616*
Mystox Btl   *632*
Mystox Btv   *616*
Mystox Btv   *632*
Natural Ant Gun   *343*
Natural Wasp Gun   *343*
Natural Wasp Killer   *343*
Neo Sorexa Block In A Box   *408*
Neokil   *408*
Neokil Bait Bags   *408*
Neosorexa   *408*
Neosorexa Bait Blocks   *408*
Neosorexa Concentrate   *408*
Neosorexa Liquid Concentrate   *408*
Neosorexa Mixed Grain Bait   *408*
Neosorexa Pellet Packs   *408*
Neosorexa Pellets   *408*
Neosorexa Ratpacks   *408*
Net-Guard   *23*
Net-Guard   *65*
Netrex Af   *10*
New Formula Cedarwood   *568*
New Improved Cruiser Premium   *23*
New Improved Cruiser Premium   *65*
New Secto Household Flea Spray   *338*
New Secto Household Flea Spray   *394*
New Siege Ii   *257*
New Super Raid Insecticide   *215*
New Super Raid Insecticide   *331*
New Super Raid Insecticide   *393*
New Tetracide   *250*
New Tetracide   *262*
New Tetracide   *352*
New Vapona Ant Killer   *263*
New Vapona Ant Killer   *376*
New Vapona Fly And Wasp Killer   *214*
New Vapona Fly And Wasp Killer   *334*
New Vapona Fly Killer Dry Formulation   *216*
New Vapona Fly Killer Dry Formulation   *389*

Nexa Cockroach Bait Station *287*
Nipacide Dp30 *434*
Nippon Ant And Crawling Insect Killer *328*
Nippon Ant And Crawling Insect Killer *383*
Nippon Ant Killer Liquid *282*
Nippon Ant Killer Powder *311*
Nippon Fly Killer Pads *236*
Nippon Fly Killer Pads *312*
Nippon Fly Killer Spray *328*
Nippon Fly Killer Spray *383*
Nippon Flying Insect Killer Tablets *228*
Nippon Killaquer Crawling Insect Killer *328*
Nippon Killaquer Crawling Insect Killer *383*
Nippon Ready For Use Ant And Crawling Insect Killer *311*
Nippon Ready For Use Fly Killer Spray *217*
Nippon Ready For Use Fly Killer Spray *339*
Nippon Ready-To-Use Ant And Crawling Insect Killer Spray *267*
Nippon Wasp Nest Destroyer Foam *328*
Nippon Wasp Nest Destroyer Foam *383*
Nippon Woodlice Killer Powder *311*
Nisa Fly And Wasp Killer *328*
Nisa Fly And Wasp Killer *383*
Nitromors Creosote *549*
Nitromors Mould Remover *420*
Nitromors Mould Remover *447*
Nitromors Timbrene Clear Environmental Formula *576*
Nitromors Timbrene Clear Environmental Formula *717*
Nitromors Timbrene Environmental Formula Wood Preservative *733*
Nitromors Timbrene Supreme Environmental Formula *574*
Nitromors Timbrene Supreme Environmental Formula *653*
Nitromors Timbrene Supreme Environmental Formula *718*
No Fleas On Me *311*
No Foul-Zo *163*
Noa-Noa Rame *27*
Noa-Noa Rame *94*
Noa-Noa Rame Black/Red *10*
Noah Gold *494*
Noah Gold *526*
Noah Gold *671*
Nomad Residex P *311*
Nordrift Antifouling *10*
Norimp 2000 Black *10*
Norshield 150 *301*
Norshield 150 *315*
Nu Wave A/F *39*
Nu Wave A/F *150*
Nu Wave A/F *178*
Nubex Emulsion Concentrate C (Low Odour) *551*

Nubex Emulsion Concentrate Cb (Low Odour) *557*
Nubex Emulsion Concentrate Cb (Low Odour) *716*
Nubex Wdr *439*
Nubex Wdr *440*
Nubex Wdr *594*
Nubex Wdr *596*
Nubex Woodworm All Purpose Cb *557*
Nubex Woodworm All Purpose Cb *716*
Nut Brown Creosote *549*
Nuvan Staykil *277*
Nuvan Staykil *294*
Nuvanol N 500 Sc *293*
Nylar 10 Ec *362*
Nylar 100 *362*
Nylar 4ew *362*
Nylar 650 Ulv *337*
Nylar 650 Ulv *365*
Oakmere Creosote Type 2 *549*
Odex Fly Spray *372*
Omega *633*
Omega Biocide *441*
Omega Biocide *600*
Omega Plus *652*
Omega Plus *719*
One-Shot Aircraft Aerosol Insect Control *335*
Orvec Aircraft Insecticide *335*
Os Color Wood Stain And Preservative *575*
Os Color Wood Stain And Preservative *657*
Os Color Wood Stain And Preservative *738*
Os Color Wood Stain And Protector *666*
Os Color Wr *575*
Os Color Wr *657*
Os Color Wr *738*
Osmose Abs33 *517*
Osmose Abs33 *619*
Osmose Abs33a *515*
Osmose Endcoat *714*
Osmose Endcoat Brown *600*
Osmose Endcoat Clear *600*
Osmose Endcoat Green *530*
Osmose Endcoat Green *603*
Osmose K33 C50 *468*
Osmose K33 C50 *522*
Osmose K33 C50 *535*
Osmose K33 C58.7 *468*
Osmose K33 C58.7 *522*
Osmose K33 C58.7 *535*
Osmose K33 C72 *468*
Osmose K33 C72 *522*
Osmose K33 C72 *535*
Osmose K55 *521*
Osmose K55 *539*
Outright Household Flea Spray *311*
Ovoline 275 Golden Creosote *524*

HSE
PRODUCT INDEX

P.J. Fungicidal Solution *420*
P.J. Fungicidal Solution *447*
Palace Base Coat Wood Preservative *714*
Palace Dual Purpose Fi 8 (Microemulsion) *643*
Palace Dual Purpose Fi 8 (Microemulsion) *675*
Palace Fungicidal Wash *419*
Palace Fungicidal Wash *453*
Palace Fws (Microemulsion) *441*
Palace Microfine 8 -Insecticide Concentrate *633*
Palace Microfine Fungicide Concentrate *714*
Palace Microfine Fungicide Insecticide
    (Concentrate) *557*
Palace Microfine Fungicide Insecticide
    (Concentrate) *716*
Palace Microfine Insecticide Concentrate *551*
Palace Mould Remover *439*
Palace Mould Remover *440*
Palace Timbertreat Ecology *655*
Palace Timbertreat Ecology *740*
Palace Woodworm Killer 8 (Microemulsion) *633*
Pale Green Wood Preservative *533*
Pale Green Wood Preservative *715*
Panacide M21 *434*
Panaclean 736 *434*
Pandrol Timbershield Rods *582*
Paneltone *709*
Paragon Formula 10 Cockroach Bait *245*
Paramos *419*
Patente Laxe *15*
Patente Laxe *108*
Path And Patio Cleaner *419*
Patriot Cik *257*
Patriot Flying And Crawling Insect Killer *347*
Patriot Flying And Crawling Insect Killer
    Aerosol *347*
Patriot Rtu *311*
Pc Insect Killer *236*
Pc Insect Killer *312*
Pc-D *451*
Pc-D Concentrate *451*
Pc-F/B *714*
Pc-H/3 *551*
Pc-H/4 *557*
Pc-H/4 *716*
Pc-K *505*
Pc-K *593*
Pc-Xj/2 *714*
Pcx-12/P *551*
Pcx-12/P Concentrate *551*
Pcx-122 *557*
Pcx-122 *716*
Pcx-122 Concentrate *557*
Pcx-122 Concentrate *716*
Pedigree And Whiskas Exelpet Household Flea
    Spray *301*
Pedigree And Whiskas Exelpet Household Flea
    Spray *315*
Pedigree Care And Whiskas Care Household Anti-
    Flea Spray *306*
Pedigree Care And Whiskas Care Household Anti-
    Flea Spray *356*
Pedigree Care Anti-Flea Carpet Powder *311*
Pedigree Care Bedding Spray *217*
Pedigree Care Bedding Spray *339*
Pedigree Care Household Anti Flea Spray *306*
Pedigree Care Household Anti Flea Spray *356*
Pedigree Care Household Spray *301*
Pedigree Care Household Spray *315*
Penetone Fly Killer *368*
Penetone Fly Killer *386*
Penguin Aqualine *123*
Penguin Non-Stop *23*
Penguin Non-Stop *65*
Penguin Non-Stop White *50*
Penguin Non-Stop White *66*
Penguin Racing *10*
Penguin Racing White *50*
Penguin Racing White *66*
Peripel *206*
Perma Aq Dual Purpose *652*
Perma Aq Dual Purpose *719*
Permabor Boracol 10rh *423*
Permabor Boracol 10rh *437*
Permabor Boracol 20 *582*
Permabor Boracol 40 Paste *582*
Permacide Masonry Fungicide *420*
Permacide Masonry Fungicide *447*
Permadex Masonry Fungicide Concentrate *439*
Permadex Masonry Fungicide Concentrate *440*
Permadip 9 Concentrate *582*
Permapaste Pb *652*
Permapaste Pb *719*
Permarock Fungicidal Wash *420*
Permarock Fungicidal Wash *447*
Permasect 0.5 Dust *311*
Permasect 10 Wp *311*
Permasect Powder *311*
Permethrin *311*
Permethrin Dusting Powder *311*
Permethrin F And I Concentrate *652*
Permethrin F And I Concentrate *719*
Permethrin Ph - 25wp *311*
Permethrin Wettable Powder *311*
Permethrin Ww Conc *633*
Permoglaze Micatex Fungicidal Treatment *439*
Permoglaze Micatex Fungicidal Treatment *440*
Permoglaze Micatex Fungicidal Treatment *594*
Permoglaze Micatex Fungicidal Treatment *596*
Permost 0.5% Dust Powder *311*
Permost 25 Wp *311*
Permost Uni *328*

Permost Uni *383*
Perycut Cockroach Mat *236*
Perycut Cockroach Mat *312*
Pesguard Ns 6/14 Oil *338*
Pesguard Ns 6/14 Oil *394*
Pesguard Oba F 7305 C *338*
Pesguard Oba F 7305 C *394*
Pesguard Oba F 7305 D *338*
Pesguard Oba F 7305 D *394*
Pesguard Oba F 7305 E *338*
Pesguard Oba F 7305 E *394*
Pesguard Oba F 7305 F *338*
Pesguard Oba F 7305 F *394*
Pesguard Oba F 7305 G *338*
Pesguard Oba F 7305 G *394*
Pesguard Oba F7305 B *338*
Pesguard Oba F7305 B *394*
Pesguard Ps 102 *217*
Pesguard Ps 102 *339*
Pesguard Ps 102a *217*
Pesguard Ps 102a *339*
Pesguard Ps 102b *217*
Pesguard Ps 102b *339*
Pesguard Ps 102c *217*
Pesguard Ps 102c *339*
Pesguard Ps 102d *217*
Pesguard Ps 102d *339*
Pesguard Wba F-2656 *338*
Pesguard Wba F-2656 *394*
Pesguard Wba F-2692 *338*
Pesguard Wba F-2692 *394*
Pesguard Wba F2714 *340*
Pesguard Wba F2714 *395*
Pestkill Household Spray *338*
Pestkill Household Spray *394*
Pestrin *311*
Pet Bedding And Household Flea Spray *338*
Pet Bedding And Household Flea Spray *394*
Pet Rescue Household Flea Spray *266*
Pet Rescue Household Flea Spray *314*
Pharaohkill *245*
Pharorid *298*
Pharorid S *299*
Phernal Brand Moth Balls *204*
Phillips Pestkill Karpet Killer *335*
Pif Paf Crawling Insect Killer *236*
Pif Paf Crawling Insect Killer *312*
Pif Paf Crawling Insect Killer Aerosol *241*
Pif Paf Crawling Insect Killer Aerosol *326*
Pif Paf Crawling Insect Killer Aerosol *384*
Pif Paf Crawling Insect Powder *311*
Pif Paf Flying Insect Killer *236*
Pif Paf Flying Insect Killer *312*
Pif Paf Flying Insect Killer Aerosol *236*
Pif Paf Flying Insect Killer Aerosol *312*
Pif Paf Insecticide *234*

Pif Paf Insecticide *278*
Pif Paf Insecticide *325*
Pif Paf Mosquito Mats *228*
Pif Paf Yellow Flying Insect Killer *237*
Pif Paf Yellow Flying Insect Killer *324*
Pilot *15*
Pilot *108*
Pilot Antifouling *10*
Pilt 80 Rfu *600*
Pilt Nf4 R.F.U. *600*
Pjg Boron Rods *486*
Pla Products Dry Rot Killer *451*
Pla Wood Preserver *733*
Pob Mould Treatment *420*
Pob Mould Treatment *447*
Polycell 3 In 1 Mould Killer *423*
Polycell 3 In 1 Mould Killer *437*
Polycell Mould Cleaner *439*
Polycell Mould Cleaner *440*
Polyphase Solvent Based Ready For Use *600*
Polyphase Water Based Concentrate *600*
Polyphase Water Based Ready For Use *600*
Portland Brand Ant Killer *223*
Powder Ant Killer *311*
Precor Ulv Ii *301*
Precor Ulv Ii *315*
Premier Ant Killer *223*
Premier Barrettine Green Wood Preserver *529*
Premier Barrettine New Universal Fluid D *574*
Premier Barrettine New Universal Fluid D *653*
Premier Barrettine New Universal Fluid D *718*
Premier Barrettine New Wood Preserver *576*
Premier Barrettine New Wood Preserver *717*
Premier Barrettine New Woodworm Killer *633*
Premier Pro-Tec (S) *576*
Premier Pro-Tec (S) *717*
Premier Pro-Tec(S) Environmental Formula Wood
    Preservative *733*
Premiere Fly And Wasp Killer *372*
Premium Grade Wood Treatment *509*
Premium Grade Wood Treatment *656*
Premium Grade Wood Treatment *735*
Premium Wood Treatment *446*
Premium Wood Treatment *457*
Premium Wood Treatment *655*
Premium Wood Treatment *740*
Preservative For Wood Green *529*
Preservative Gel *480*
Preservative Gel *584*
Prevent *347*
Preventol Of *659*
Prima *27*
Prima *94*
Pro Control Cik Super *267*
Pro Control Cik Superfast (P) *268*
Pro Control Cik Superfast (P) *355*

HSE
PRODUCT INDEX

Pro Mite Solvent Based Fabric Protector   *311*
Pro Mite Water Based Carpet Protector   *311*
Probor 10   *423*
Probor 10   *437*
Probor 10   *480*
Probor 10   *584*
Probor 20 Brushable Gel   *565*
Probor 20 Brushable Gel   *588*
Probor 50   *423*
Probor 50   *437*
Probor 50   *480*
Probor 50   *584*
Probor Powder   *582*
Procontrol Cik Super (P)   *267*
Procontrol Fik H20   *232*
Procontrol Fik H20   *244*
Procontrol Fik Super   *232*
Procontrol Fik Super   *244*
Professional   *10*
Professional Soft Antifouling   *27*
Professional Soft Antifouling   *94*
Profilan - Primer   *607*
Profilan - Primer   *635*
Propeller T.F.   *47*
Propeller T.F.   *114*
Protect And Shine Spray   *633*
Protect B Flea Killer Aerosol   *306*
Protect B Flea Killer Aerosol   *356*
Protek 9 Star L40   *754*
Protek 9 Star Wood Protection   *582*
Protek Double 9 Star Wood Protection   *582*
Protek Shedstar   *582*
Protek Wood Protection Fencegrade 9:1   *582*
Protek Wood Protection Shed Grade   *582*
Protek Woodstar   *582*
Protim 210   *618*
Protim 210   *727*
Protim 210 C   *618*
Protim 210 C   *727*
Protim 210 Cwr   *618*
Protim 210 Cwr   *727*
Protim 210 Wr   *618*
Protim 210 Wr   *727*
Protim 215   *618*
Protim 215   *727*
Protim 215 Pp   *725*
Protim 220   *658*
Protim 220   *743*
Protim 220 Cwr   *658*
Protim 220 Cwr   *743*
Protim 220 Wr   *658*
Protim 220 Wr   *743*
Protim 220c   *658*
Protim 220c   *743*
Protim 23 Wr   *618*
Protim 23 Wr   *727*

Protim 230   *654*
Protim 230   *724*
Protim 230 Wr   *654*
Protim 230 Wr   *724*
Protim 230l   *617*
Protim 230l   *723*
Protim 250   *607*
Protim 250   *635*
Protim 250 Wr   *607*
Protim 250 Wr   *635*
Protim 260 F   *714*
Protim 340 Wr   *643*
Protim 340 Wr   *675*
Protim 340e   *643*
Protim 340e   *675*
Protim 415   *646*
Protim 415   *682*
Protim 415   *707*
Protim 415 T Special Solvent   *646*
Protim 415 T Special Solvent   *682*
Protim 415 T Special Solvent   *707*
Protim 415t   *646*
Protim 415t   *682*
Protim 415t   *707*
Protim 415twr   *646*
Protim 415twr   *682*
Protim 415twr   *707*
Protim 418   *646*
Protim 418   *682*
Protim 418   *707*
Protim 418j   *611*
Protim 418j   *677*
Protim 418j   *698*
Protim 418v   *646*
Protim 418v   *682*
Protim 418v   *707*
Protim 418wr   *646*
Protim 418wr   *682*
Protim 418wr   *707*
Protim 600wr   *633*
Protim 80   *615*
Protim 80   *631*
Protim 80   *728*
Protim 80 C   *615*
Protim 80 C   *631*
Protim 80 C   *728*
Protim 80 Cwr   *615*
Protim 80 Cwr   *631*
Protim 80 Cwr   *728*
Protim 80 Wr   *615*
Protim 80 Wr   *631*
Protim 80 Wr   *728*
Protim 800   *463*
Protim 800   *637*
Protim 800 C   *463*
Protim 800 C   *637*

Protim 800 Cwr   *463*
Protim 800 Cwr   *637*
Protim 800 Wr   *463*
Protim 800 Wr   *637*
Protim 800p   *459*
Protim 800p   *464*
Protim 800p   *636*
Protim 90   *618*
Protim 90   *727*
Protim Aq   *633*
Protim Aquachem-Insecticidal Emulsion P   *633*
Protim B10   *582*
Protim B610   *591*
Protim B610   *680*
Protim B610   *705*
Protim Brown   *615*
Protim Brown   *631*
Protim Brown   *728*
Protim Brown Cdb   *463*
Protim Brown Cdb   *637*
Protim Cb   *652*
Protim Cb   *719*
Protim Cca Oxide 50   *468*
Protim Cca Oxide 50   *522*
Protim Cca Oxide 50   *535*
Protim Cca Oxide 58   *468*
Protim Cca Oxide 58   *522*
Protim Cca Oxide 58   *535*
Protim Cca Oxide 72   *468*
Protim Cca Oxide 72   *522*
Protim Cca Oxide 72   *535*
Protim Cca Salts Type 2   *469*
Protim Cca Salts Type 2   *545*
Protim Cca Salts Type 2   *685*
Protim Cdb   *652*
Protim Cdb   *719*
Protim Cedar   *579*
Protim Cedar   *726*
Protim E410   *590*
Protim E410   *648*
Protim E410   *681*
Protim E410   *704*
Protim E410(I)   *590*
Protim E410(I)   *648*
Protim E410(I)   *681*
Protim E410(I)   *704*
Protim E415   *590*
Protim E415   *648*
Protim E415   *681*
Protim E415   *704*
Protim E415 (I)   *590*
Protim E415 (I)   *648*
Protim E415 (I)   *681*
Protim E415 (I)   *704*
Protim E470   *502*
Protim E470   *563*

Protim E480   *582*
Protim Fdr 2125c   *725*
Protim Fdr 215c   *725*
Protim Fdr 220   *741*
Protim Fdr 230   *722*
Protim Fdr 250   *600*
Protim Fdr 800   *460*
Protim Fdr-H   *630*
Protim Fdr-H   *729*
Protim Fdr210   *725*
Protim Fdr215   *725*
Protim Fdr418   *676*
Protim Fdr418   *708*
Protim Fdr418j   *611*
Protim Fdr418j   *677*
Protim Fdr418j   *698*
Protim Fdr418v   *676*
Protim Fdr418v   *708*
Protim Fentex Europa 1 Rfu   *487*
Protim Fentex Europa 1 Rfu   *710*
Protim Fentex Europa I   *487*
Protim Fentex Europa I   *710*
Protim Fentex Green Concentrate   *592*
Protim Fentex Green Concentrate   *693*
Protim Fentex Green Rfu   *592*
Protim Fentex Green Rfu   *693*
Protim Fentex I   *489*
Protim Fentex I   *602*
Protim Fentex M   *592*
Protim Fentex M   *693*
Protim Fentex P   *488*
Protim Fentex P   *661*
Protim Fentex Phoenix   *744*
Protim Fentex Phoenix   *755*
Protim Fentex T   *744*
Protim Fentex T   *755*
Protim Fentex Tg   *744*
Protim Fentex Tg   *755*
Protim Fentex Tp   *744*
Protim Fentex Tp   *755*
Protim Fentex Tps   *744*
Protim Fentex Tps   *755*
Protim Fentex Tw   *744*
Protim Fentex Tw   *755*
Protim Fentex Twr Plus   *511*
Protim Fentex Twr Plus   *683*
Protim Fentex Wr   *692*
Protim Gc   *628*
Protim Green E   *532*
Protim Green E   *641*
Protim Green Wr   *529*
Protim Insecticidal Aq8   *633*
Protim Insecticidal Emulsion 8   *633*
Protim Insecticidal Emulsion P   *633*
Protim Joinery Lining 280   *659*
Protim Jp   *630*

Protim Jp *729*
Protim Jp 210 *725*
Protim Jp 220 *741*
Protim Jp 230 *722*
Protim Jp 800 *460*
Protim Kleen Ii *592*
Protim Kleen Ii *693*
Protim Panelguard *692*
Protim Paste *618*
Protim Paste *727*
Protim Paste 220 *658*
Protim Paste 220 *743*
Protim Paste 220 F *741*
Protim Paste 800 *463*
Protim Paste 800 *637*
Protim Paste 800 F *460*
Protim Paste P *633*
Protim Plug Compound *454*
Protim Plug Compound *692*
Protim R Clear *630*
Protim R Clear *729*
Protim Solignum D.O.T. *582*
Protim Solignum Softwood Basestain Cs *579*
Protim Solignum Softwood Basestain Cs *726*
Protim Stainguard *622*
Protim Stainguard *713*
Protim Universal Aq 250 *607*
Protim Universal Aq 250 *635*
Protim Wall Solution 250 *441*
Protim Wall Solution 250 *600*
Protim Wall Solution Ii *439*
Protim Wall Solution Ii *440*
Protim Wall Solution Ii Concentrate *439*
Protim Wall Solution Ii Concentrate *440*
Protim Woodworm Killer P *633*
Protim Wr 260 *652*
Protim Wr 260 *719*
Protrol *291*
Protrol 9% *292*
Provence Fly And Wasp Killer *372*
Pure Zap *257*
Purge *237*
Purge *324*
Purge Insect Destroyer *328*
Purge Insect Destroyer *383*
Py Kill 25 Plus *338*
Py Kill 25 Plus *394*
Pybuthrin 2/16 *347*
Pybuthrin 33 *347*
Pybuthrin 33 Bb *231*
Pybuthrin 33 Bb *243*
Pybuthrin Fly Killer *236*
Pybuthrin Fly Killer *312*
Pybuthrin Fly Spray. *236*
Pybuthrin Fly Spray. *312*
Pygo *335*

Pynamin Forte 40mg Mat *209*
Pynamin Forte Mat 120 *209*
Pynosect 10 *328*
Pynosect 10 *383*
Pynosect 2 *368*
Pynosect 2 *386*
Pynosect 4 *368*
Pynosect 4 *386*
Pynosect 6 *311*
Pynosect Extra Fog *328*
Pynosect Extra Fog *383*
Pynosect Pco *311*
Pynosect Pcp *311*
Pynosect Powder *311*
Pyrakill Flying Insect Spray *372*
Pyrasol C Rtu *257*
Pyrasol Cp *257*
Pyrbek *336*
Pyrbek *361*
Pyro-Black *582*
Pyro-Red *486*
Pyrolith 3505 Ready To Use *505*
Pyrolith 3505 Ready To Use *593*
Quantum Hygiene Fly Killer *368*
Quantum Hygiene Fly Killer *386*
Quicksilver Antifouling *47*
Quicksilver Antifouling *114*
R.I.P. Fleas *235*
R.I.P. Fleas *303*
R.I.P. Fleas *313*
Rabamarine A/F 2500 *39*
Rabamarine A/F 2500 *150*
Rabamarine A/F 2500 *178*
Racumin Contact Powder *407*
Racumin Rat Bait *407*
Raffaello 3 *28*
Raffaello 3 *83*
Raffaello 3 *134*
Raffaello Alloy *47*
Raffaello Alloy *114*
Raffaello Plus Tin Free *15*
Raffaello Plus Tin Free *108*
Raffaello Racing *55*
Raffaello Racing *84*
Raffaello Racing *135*
Raid Ant And Crawling Insect Killer *263*
Raid Ant And Crawling Insect Killer *376*
Raid Ant And Roach Killer *328*
Raid Ant And Roach Killer *383*
Raid Ant Bait *248*
Raid Ant Killer Powder *311*
Raid Cockroach Killer Formula 2 *261*
Raid Cockroach Killer Formula 2 *354*
Raid Dry Natural Flying Insect Killer *347*
Raid Flea Killer Plus *307*
Raid Flea Killer Plus *360*

Raid Fly And Wasp Killer 216
Raid Fly And Wasp Killer 389
Raid Fly Wasp And Plant Insect Killer 338
Raid Fly Wasp And Plant Insect Killer 394
Raid Flyguard 221
Raid Flying Insect Killer 372
Raid Mothproofer 206
Raid Outdoor Insectguard 214
Raid Outdoor Insectguard 334
Raid Residual Crawling Insect Killer 263
Raid Residual Crawling Insect Killer 376
Raid Wasp Nest Destroyer 252
Raid Wasp Nest Destroyer 375
Ranch Preservative 576
Ranch Preservative 717
Rapid Exit Fly Spray 368
Rapid Exit Fly Spray 386
Rat Eyre Rat Bait 404
Ratak 408
Ratak Cut Wheat 408
Ratak Wax Blocks 408
Ravax Af 10
Ready Kill 257
Red Can Fly Killer 372
Regency Garden Buildings Wood Preserver 568
Remecology Insecticide R8 633
Remecology Spirit Based Fungicide R5 600
Remecology Spirit Based Insecticide R6 633
Remecology Spirit Based K7 652
Remecology Spirit Based K7 719
Remtox Aq Fungicide R7 600
Remtox Aq Fungicide/Insecticide R9 607
Remtox Aq Fungicide/Insecticide R9 635
Remtox Borocol 10 Rh Masonry Biocide 416
Remtox Borocol 10 Rh Masonry Biocide 436
Remtox Borocol 20 Wood Preservative 582
Remtox Dry Rot F. W. S. 441
Remtox Dry Rot Paint 441
Remtox Dry Rot Paint 600
Remtox Dual Purpose F/1 K7 599
Remtox Dual Purpose F/1 K7 606
Remtox Dual Purpose F/1 K7 668
Remtox Dual Purpose Paste K9 607
Remtox Dual Purpose Paste K9 635
Remtox Fungicidal Wall Solution Rs 441
Remtox Fungicide Microemulsion M7 600
Remtox Fungicide Paste K6 600
Remtox Fungicide/Insecticide Microemulsion
  M9 607
Remtox Fungicide/Insecticide Microemulsion
  M9 635
Remtox Insecticide Microemulsion M8 633
Remtox Insecticide Paste R4 633
Remtox M9 Dual Purpose Super
  Concentrate 599

Remtox M9 Dual Purpose Super
  Concentrate 606
Remtox M9 Dual Purpose Super
  Concentrate 668
Remtox Micro Fungicidal Wall Solution Rs 441
Remtox Microactive Fungicide W7 600
Remtox Microactive Fws W6 441
Remtox Microactive Insecticide W8 633
Remtox Remwash Extra 419
Remtox Spirit Based F/I K7 607
Remtox Spirit Based F/I K7 635
Remtox-Silexine B40 Paste Wood
  Preservative 582
Remtox-Silexine Boron Rods 582
Remwash Masonry Steriliser 432
Rentex 467
Rentex 690
Rentex 691
Rentokil Ant And Insect Killer 223
Rentokil Ant And Insect Powder Professional 223
Rentokil Carbon Dioxide 246
Rentokil Chlorpyrifos Gel 248
Rentokil Classic Wax Polish 633
Rentokil Control Paste Sb 505
Rentokil Control Paste Sb 593
Rentokil Dry Rot And Wet Rot Treatment 456
Rentokil Dry Rot And Wet Rot Treatment 733
Rentokil Dry Rot Fluid (E) For Bonded
  Warehouses 425
Rentokil Dry Rot Fluid (E) For Bonded
  Warehouses 486
Rentokil Dry Rot Paste (D) 513
Rentokil Dry Rot Paste (D) 734
Rentokil Dual Purpose Fluid 655
Rentokil Dual Purpose Fluid 740
Rentokil Flying Insect Killer 228
Rentokil Houseplant Insect Killer 359
Rentokil Houseplant Insect Killer 367
Rentokil Iodofenphos Gel 293
Rentokil Mould Cure Spray 422
Rentokil Mould Cure Spray 430
Rentokil Mould Cure Spray 433
Rentokil Permethrin 2.5 Ec 633
Rentokil Rat Killer 401
Rentokil Total Wood Treatment 428
Rentokil Total Wood Treatment 431
Rentokil Total Wood Treatment 449
Rentokil Total Wood Treatment 501
Rentokil Total Wood Treatment 556
Rentokil Total Wood Treatment 672
Rentokil Wasp Killer 223
Rentokil Wasp Nest Destroyer 338
Rentokil Wasp Nest Destroyer 394
Rentokil Wood Preservative 655
Rentokil Wood Preservative 740
Rentokil Wood Preserver 486

HSE
PRODUCT INDEX

587

Rentokil Woodworm Killer   633
Rentokil Woodworm Treatment   633
Repel Fabric Spray   207
Residex   347
Residex P   311
Residual Powerkill   311
Resigen   237
Resigen   324
Reslin Premium   237
Reslin Premium   324
Reslin Super   237
Reslin Super   324
Restor-8-Dual Purpose Concentrate   652
Restor-8-Dual Purpose Concentrate   719
Restor-8-Insecticide Concentrate   633
Restor-Mfc   441
Restor-Mfc   600
Restor-Paste   652
Restor-Paste   719
Rhodaquat Rp50   419
Rhodaquat Rp50   474
Richard Burbidge Green End Coat   714
Rid Insect Powder   370
Rid Moss   419
Rid-Ant   257
Ridento Ready To Use Rat Bait   404
Rlt Bactdet   452
Rlt Clearmould Spray   432
Rlt Halophen   432
Rlt Halophen Ds   432
Roachbuster   245
Rodewod 50 Sl   470
Rodine   401
Roebuck Eyot Aircraft Disinsection Spray   335
Ronseal Low Odour Wood Preserver   733
Ronseal Wet Rot Prevention Tablets   582
Ronseal Wood Preservative Tablets   582
Ronseal's All-Purpose Wood Preserver   733
Ronseal's All-Purpose Wood Preserver
   Green   529
Ronseal's Universal Wood Treatment   655
Ronseal's Universal Wood Treatment   740
Roofcoat Tp15   185
Room Fogger   302
Room Fogger   327
Rout   404
Roxem D   226
Roxem D   309
Roxem D   318
Roxem D   349
Roxem D   424
Roxem D   442
Roxem D   445
Roxem D   450
Ruberoid Fungicidal Wash   439
Ruberoid Fungicidal Wash   440

Ruberoid Fungicidal Wash Solution   420
Ruberoid Fungicidal Wash Solution   447
Ruby Rat   404
Rwn Building Products Wood Preserver   568
S C Johnson Professional Raid Flying Insect
   Killer   338
S C Johnson Professional Raid Flying Insect
   Killer   394
S C Johnson Wax Raid Ant And Cockroach
   Killer   263
S C Johnson Wax Raid Ant And Cockroach
   Killer   376
S C Johnson Wax Raid Ant Killer Powder   267
S C Johnson Wax Raid Fly And Wasp Killer   338
S C Johnson Wax Raid Fly And Wasp Killer   394
Sactif Flying Insect Killer   328
Sactif Flying Insect Killer   383
Sadolin New Base   576
Sadolin New Base   717
Sadolin Quick Drying Wood Preserver   610
Sadolin Quick Drying Wood Preserver   667
Sadolin Shed And Fence Preserver   600
Sadolin Wb 100   582
Sadolin Wood Preserver   607
Sadolin Wood Preserver   635
Safe Kill Rtu   242
Safeclean Promite Solvent Protector   311
Safeguard Antiflame 4050 Wd Wood
   Preservative   582
Safeguard Boracol 10 Rh   423
Safeguard Boracol 10 Rh   437
Safeguard Boracol 20   582
Safeguard Boracol 40 Paste   582
Safeguard Deepwood Fungicide   714
Safeguard Deepwood I Insecticide Emulsion
   Concentrate   633
Safeguard Deepwood Ii Dual Purpose Emulsion
   Concentrate   652
Safeguard Deepwood Ii Dual Purpose Emulsion
   Concentrate   719
Safeguard Deepwood Iii Timber Treatment   652
Safeguard Deepwood Iii Timber Treatment   719
Safeguard Deepwood Iv Woodworm Killer   633
Safeguard Deepwood Paste   595
Safeguard Deepwood Paste   597
Safeguard Deepwood Paste   642
Safeguard Deepwood Paste   652
Safeguard Deepwood Paste   719
Safeguard Deepwood Surface Biocide
   Concentrate   439
Safeguard Deepwood Surface Biocide
   Concentrate   440
Safeguard Fungicidal Wall Solution   451
Safeguard Mould And Moss Killer   439
Safeguard Mould And Moss Killer   440
Safetray P   666

Safetray Sl *470*
Sainsbury's Crawling Insect And Ant Killer *236*
Sainsbury's Crawling Insect And Ant Killer *312*
Sainsbury's Fly And Wasp Killer *328*
Sainsbury's Fly And Wasp Killer *383*
Sainsbury's Fly, Wasp And Mosquito Killer *216*
Sainsbury's Fly, Wasp And Mosquito Killer *389*
Sakarat D (Whole Wheat) *408*
Sakarat Ready To Use (Cut Wheat Base) *413*
Sakarat Ready To Use (Whole Wheat) *413*
Sakarat X Ready To Use Warfarin Rat Bait *413*
Samsung Natural Insect Spray *347*
Sandhill Aqueous Fungicide - Insecticide *652*
Sandhill Aqueous Fungicide - Insecticide *719*
Sandhill Fungicide - Insecticide *652*
Sandhill Fungicide - Insecticide *719*
Sandhill Wood Preservative Clear *714*
Sandtex Fungicide *453*
Sanmex Ant & Crawling Insect Killer *328*
Sanmex Ant & Crawling Insect Killer *383*
Sanmex Fly And Wasp Killer *328*
Sanmex Fly And Wasp Killer *383*
Sanmex Supakil Insecticide *328*
Sanmex Supakil Insecticide *383*
Sc Johnson Flying Insect Killer *347*
Sc Johnson Raid Fly And Wasp Killer *372*
Sc Johnson Wax Raid Ant & Cockroach Killer With Natural Pyrethrum *317*
Sc Johnson Wax Raid Ant & Cockroach Killer With Natural Pyrethrum *358*
Scorpio Fly Spray For Flying Insects *328*
Scorpio Fly Spray For Flying Insects *383*
Scotwest Antifouling *10*
Screwfix Woodworm Killer *633*
Sea Grandprix 500 Tci *13*
Sea Grandprix 500 Tci *74*
Sea Grandprix 500 Tci *126*
Seaflo 10 *38*
Seaflo 10 *148*
Seaflo 10 *159*
Seaflo 10 *179*
Seaflo 15 *36*
Seaflo 15 *144*
Seaflo 15 *155*
Seajet 033 *27*
Seajet 033 *94*
Sealocrete Timber Master Wood Preserver *576*
Sealocrete Timber Master Wood Preserver *717*
Seaprince *10*
Seashield (Base) *23*
Seashield (Base) *65*
Seatech *15*
Seatech *108*
Seatech Antifouling *27*
Seatech Antifouling *94*
Seatech Antifouling White *54*
Seatech Antifouling White *95*
Seatender 10 *10*
Seatender 12 *29*
Seatender 12 *86*
Seatender 12 *129*
Seatender 15 *13*
Seatender 15 *74*
Seatender 15 *126*
Seatender 7 *10*
Sectacide 50 Ec *283*
Secto Ant And Crawling Insect Lacquer *270*
Secto Ant And Crawling Insect Spray *257*
Secto Ant And Crawling Insect Spray. *217*
Secto Ant And Crawling Insect Spray. *339*
Secto Ant Bait *223*
Secto Extra Strength Insect Killer *311*
Secto Flea Free Insecticidal Rug And Carpet Freshener *311*
Secto Fly Killer *328*
Secto Fly Killer *383*
Secto Fly Killer Living Room Size *274*
Secto Household Dust Mite Control. *335*
Secto Household Flea Spray *317*
Secto Household Flea Spray *358*
Secto Mini-Space Insect Killers *274*
Secto Moth Killer *272*
Secto Moth Killer *275*
Secto Slow Release Fly Killer Kitchen Size *274*
Selfpolishing Antifouling 2000 *50*
Selfpolishing Antifouling 2000 *66*
Selkil Crawling Insect Killer *267*
Sergeant's Dust Mite Patrol *335*
Sergeant's Dust Mite Patrol Injector And Spray *338*
Sergeant's Dust Mite Patrol Injector And Spray *394*
Sergeant's Dust Mite Patrol Spray *338*
Sergeant's Dust Mite Patrol Spray *394*
Sergeant's Household Spray *338*
Sergeant's Household Spray *394*
Sergeant's Insect Patrol *338*
Sergeant's Insect Patrol *394*
Sergeants Car Patrol *335*
Sergeants Carpet Patrol *335*
Sergeants Dust Mite Patrol *335*
Sergeants Rug Patrol *335*
Sewarin Extra *413*
Sewarin P *413*
Shearwater Racing Antifouling *10*
Shelltox Cockroach And Crawling Insect Killer 3 *263*
Shelltox Cockroach And Crawling Insect Killer 3 *376*
Shelltox Cockroach And Crawling Insect Killer 4 *263*

Shelltox Cockroach And Crawling Insect Killer 4   *376*
Shelltox Crawling Insect Killer Liquid Spray   *210*
Shelltox Crawling Insect Killer Liquid Spray   *265*
Shelltox Extra Flykiller   *280*
Shelltox Extra Flykiller   *378*
Shelltox Flying Insect Killer 2   *338*
Shelltox Flying Insect Killer 2   *394*
Shelltox Flying Insect Killer 3   *338*
Shelltox Flying Insect Killer 3   *394*
Shelltox Flykiller   *280*
Shelltox Flykiller   *378*
Shelltox Insect Killer   *328*
Shelltox Insect Killer   *383*
Shelltox Insect Killer 2   *328*
Shelltox Insect Killer 2   *383*
Shelltox Mat 1   *228*
Shelltox Mat 2   *209*
Sherley's Flea Busters   *245*
Sherley's Flego   *266*
Sherley's Flego   *314*
Sherley's Flego Trigger Pack   *266*
Sherley's Flego Trigger Pack   *314*
Sherley's Rug-De-Bug   *311*
Shiprite Sailing   *10*
Shiprite Speed   *15*
Shiprite Speed   *108*
Shiprite Traditional   *10*
Siege Ii   *257*
Sigma Alphagen 20   *17*
Sigma Alphagen 20   *73*
Sigma Fungicidal Solution   *419*
Sigma Pilot Ecol Antifouling   *8*
Sigma Pilot Ecol Antifouling   *43*
Sigma Pilot Ecol Antifouling   *176*
Sigmalife Impregnant   *568*
Sigmaplane Ecol Ha 120 Antifouling   *30*
Sigmaplane Ecol Ha 120 Antifouling   *85*
Sigmaplane Ecol Ha 120 Antifouling   *113*
Sigmaplane Ecol Ha 120 Antifouling   *125*
Sigmaplane Ha Antifouling   *39*
Sigmaplane Ha Antifouling   *150*
Sigmaplane Ha Antifouling   *178*
Sigmaplane Hb Antifouling   *39*
Sigmaplane Hb Antifouling   *150*
Sigmaplane Hb Antifouling   *178*
Sigmaplane Ta Antifouling   *39*
Sigmaplane Ta Antifouling   *150*
Sigmaplane Ta Antifouling   *178*
Signpost Creosote   *549*
Sikkens Wood Preserver   *580*
Sikkens Wood Preserver   *731*
Silexine Anticon   *191*
Silexine Anticon   *196*
Silexine Fungi-Chek Emulsion   *191*
Silexine Fungi-Chek Emulsion   *196*

Silexine Fungi-Chek Oil Matt Paint   *191*
Silexine Fungi-Chek Oil Matt Paint   *196*
Sinesto B   *466*
Sinesto B   *694*
Single-Shot Permethrin Spray   *311*
Skeetal Flowable Concentrate   *222*
Slaymor   *401*
Slaymor Bait Bags   *401*
Slaymor Mouse Bait   *401*
Slaymor Rat Blocks   *401*
Slippy Bottom   *27*
Slippy Bottom   *94*
Slipstream Antifouling   *23*
Slipstream Antifouling   *65*
Slow Release Fly Killer Controllable Cassette   *274*
Smite   *253*
Snowcem Algicide   *419*
Soft Antifouling   *27*
Soft Antifouling   *94*
Solignum Colourless   *655*
Solignum Colourless   *740*
Solignum Cut End   *652*
Solignum Cut End   *719*
Solignum Dark Brown   *524*
Solignum Dry Rot Killer   *439*
Solignum Dry Rot Killer   *440*
Solignum Fencing Fluid   *549*
Solignum Fungicide   *439*
Solignum Fungicide   *440*
Solignum Green   *532*
Solignum Green   *641*
Solignum Insecticidal Aq8   *633*
Solignum Remedial Concentrate   *652*
Solignum Remedial Concentrate   *719*
Solignum Remedial Pb   *652*
Solignum Remedial Pb   *719*
Solignum Universal   *652*
Solignum Universal   *658*
Solignum Universal   *719*
Solignum Universal   *743*
Solignum Wood Preservative Paste   *655*
Solignum Wood Preservative Paste   *740*
Solignum Woodworm Killer   *633*
Solignum Woodworm Killer Concentrate   *633*
Solo Blox   *400*
Solo Mouse Blox   *400*
Solo Rodenticide   *400*
Solo Super Blox   *400*
Solvex Rapid Mould Cleaner   *420*
Solvex Rapid Mould Cleaner   *447*
Sorex Brodifacoum Bait Blocks   *400*
Sorex Brodifacoum Rat And Mouse Bait   *400*
Sorex Brodifacoum Sewer Bait   *400*
Sorex Brodifacoum Sewer Bait Sachets   *400*
Sorex Cd Mouse Bait   *406*

Sorex Cd Mouse Bait 410
Sorex Fatal 405
Sorex Fly Spray Rtu 338
Sorex Fly Spray Rtu 394
Sorex Super Fly Spray 338
Sorex Super Fly Spray 394
Sorex Warfarin 250 Ppm Rat Bait 413
Sorex Warfarin 500 Ppm Rat Bait 413
Sorex Warfarin Sewer Bait 413
Sorex Wasp Nest Destroyer 368
Sorex Wasp Nest Destroyer 386
Sorexa Cd Concentrate 403
Sorexa Cd Concentrate 409
Sorexa Cd Mouse Killer 406
Sorexa Cd Mouse Killer 410
Sorexa Checkatube 400
Sorexa D 408
Sorexa D Mouse Killer 408
Sorexa Gel 408
Southdown Creosote 549
Sov Aq Micro I 633
Sovac F 600
Sovac F/I 607
Sovac F/I 635
Sovac F/I Wr 607
Sovac F/I Wr 635
Sovac Fwr 600
Sovac I 633
Sovaq Flx F/I 599
Sovaq Flx F/I 606
Sovaq Flx F/I 668
Sovaq Flx I 598
Sovaq Micro F 600
Sovaq Micro Fws 441
Sovaq Micro I 633
Sovereign Aq F/I 607
Sovereign Aq F/I 635
Sovereign Aq/Ft 714
Sovereign Aqf 600
Sovereign Aqueous Fungicide/Insecticide 2 652
Sovereign Aqueous Fungicide/Insecticide 2 719
Sovereign Aqueous Insecticide 2 633
Sovereign Coloured Wood Preservative 714
Sovereign Coloured Wood Preservatives 600
Sovereign Dentolite Solution Concentrate 419
Sovereign Dipcoat 714
Sovereign Green Timber Preservative/
  Dipcoat 530
Sovereign Green Timber Preservative/
  Dipcoat 603
Sovereign Insecticide/Fungicide 2 652
Sovereign Insecticide/Fungicide 2 719
Sovereign Masonry Sterilising Wash 432
Sovereign Sovaq Micro F/I 607
Sovereign Sovaq Micro F/I 635
Sovereign Sovaq Micro I 633

Sovereign Timber Preservative 714
Sovereign Timber Preservative Pale Green 533
Sovereign Timber Preservative Pale Green 715
Sovereign Timbor Rod 582
Sp.153 Sterilising Detergent Wash 434
Sp.154 Fungicidal And Bactericidal
  Treatment 434
Speedclean Antifouling 10
Spencer Stormcote Fungicidal Treatment 420
Spencer Stormcote Fungicidal Treatment 447
Spencer Wood Preservative Xp 730
Spira 'No-Bite' Mosquito Killer 228
Spira 'No-Bite' Outdoor Mosquito Coils 208
Spraydex Ant & Insect Killer 236
Spraydex Ant & Insect Killer 312
Spraydex Houseplant Spray 236
Spraydex Houseplant Spray 312
Spraydex Insect Killer 236
Spraydex Insect Killer 312
Sprint 57
Sprint 80
Sprint 92
Sprytech 1 633
Sprytech 2 652
Sprytech 2 719
Sprytech 3 441
Square Deal Deep Protection Wood
  Preserver 733
St Michael Flyspray 372
Standard Antifouling 10
Stanhope Biocide 25x 441
Stapro Insecticide 311
Staykil Household Flea Spray 266
Staykil Household Flea Spray 314
Sterashield 182
Sterashield 186
Sterashield 193
Stericide 198
Stericide Am 195
Stericide Am 197
Stericide Am 199
Stop Insect Killer 328
Stop Insect Killer 383
Storm 412
Storm Mouse Bait Blocks 412
Storm Sewer Bait 412
Storm Sewer Bait Sachets 412
Stradz Mosquito Killer 228
Sumilarv 10 Ec 362
Sumithion 20 Mc 283
Sumithion 20% Mc 283
Sumithion 40 Wp 283
Sumithrin 10 Sec 335
Sumithrin 10 Sec Carpet Treatment 335
Super Andy Man Wood Protection 582
Super Fly Spray 338

Super Fly Spray *394*
Super Raid Ii *338*
Super Raid Ii *394*
Super Raid Insect Killer Formula 2 *372*
Super Shelltox Crawling Insect Killer *263*
Super Shelltox Crawling Insect Killer *376*
Super Shelltox Flying Insect Killer *338*
Super Shelltox Flying Insect Killer *394*
Super Tropical Antifouling *10*
Super Tropical Extra Antifouling *10*
Superdrug Fly And Wasp Killer Pump Spray *347*
Superdrug Fly Killer *328*
Superdrug Fly Killer *383*
Superdrug Mosquito Killer *228*
Superdrug Slow Release Fly Killer *274*
Superdrug Small Space Fly/Moth Strip *274*
Supergrade Wood Preserver *614*
Supergrade Wood Preserver *629*
Supergrade Wood Preserver Black *614*
Supergrade Wood Preserver Black *629*
Supergrade Wood Preserver Brown *614*
Supergrade Wood Preserver Brown *629*
Supergrade Wood Preserver Clear *614*
Supergrade Wood Preserver Clear *629*
Supergrade Wood Preserver Green *529*
Superspeed *10*
Superyacht 900 *38*
Superyacht 900 *148*
Superyacht 900 *159*
Superyacht 900 *179*
Superyacht Antifouling *60*
Superyacht Antifouling *149*
Superyacht Antifouling *160*
Superyacht Antifouling *180*
Swak Natural *347*
Swat *368*
Swat *386*
Swat A *368*
Swat A *386*
Swat Gel *248*
Swot *368*
Swot *386*
Takata LII Antifouling *35*
Takata LII Antifouling *139*
Takata LII Antifouling Hi-Solid *35*
Takata LII Antifouling Hi-Solid *139*
Takata LII Antifouling Ls *35*
Takata LII Antifouling Ls *139*
Takata LII Antifouling Ls Hi-Solid *35*
Takata LII Antifouling Ls Hi-Solid *139*
Takata LII Antifouling No 2001 *35*
Takata LII Antifouling No 2001 *139*
Talon Rat And Mouse Bait (Cut Wheat) *400*
Talon Rat And Mouse Bait (Whole Wheat) *400*
Tanalith (3419) Cbc *492*
Tanalith (3419) Cbc *523*

Tanalith (3419) Cbc *538*
Tanalith 3302 *468*
Tanalith 3302 *522*
Tanalith 3302 *535*
Tanalith 3313 *468*
Tanalith 3313 *522*
Tanalith 3313 *535*
Tanalith 3357 *469*
Tanalith 3357 *545*
Tanalith 3357 *685*
Tanalith 3422 *497*
Tanalith 3422 *537*
Tanalith 3487 *493*
Tanalith 3487 *525*
Tanalith Cbc Paste 3402 *500*
Tanalith Cbc Paste 3402 *546*
Tanalith Cbc Paste 3402 *687*
Tanalith Cl (3354) *469*
Tanalith Cl (3354) *545*
Tanalith Cl (3354) *685*
Tanalith Cp 3353 *469*
Tanalith Cp 3353 *545*
Tanalith Cp 3353 *685*
Tanalith E (3485) *496*
Tanalith E (3485) *528*
Tanalith E (3485) *701*
Tanalith E (3492) *495*
Tanalith E (3492) *527*
Tanalith E (3492) *679*
Tanalith E (3492) *702*
Tanalith E (3494) *495*
Tanalith E (3494) *527*
Tanalith E (3494) *679*
Tanalith E (3494) *702*
Tanalith Oxide C3309 *468*
Tanalith Oxide C3309 *522*
Tanalith Oxide C3309 *535*
Tanalith Oxide C3310 *468*
Tanalith Oxide C3310 *522*
Tanalith Oxide C3310 *535*
Tanalith Oxide C3314 *468*
Tanalith Oxide C3314 *522*
Tanalith Oxide C3314 *535*
Tanamix 3743 *517*
Tanamix 3743 *619*
Target Flying Insect Killer *217*
Target Flying Insect Killer *339*
Teak Oil Extra *666*
Teamac Antifouling 'D' (C/258 Series) *27*
Teamac Antifouling 'D' (C/258 Series) *94*
Teamac Killa Copper *40*
Teamac Killa Copper *162*
Teamac Killa Copper Plus *6*
Teamac Killa Copper Plus *26*
Teamac Killa Copper Plus *70*
Teamac Killa Copper Plus *161*

Teamac Killa Copper Premium Antifouling (C/262/24) *27*
Teamac Killa Copper Premium Antifouling (C/262/24) *94*
Teamac Super Tropical *40*
Teamac Super Tropical *162*
Teamac Tropical Copper Antifouling (C/260/65) *10*
Teamac Wood Preservative *730*
Teamac Woodtec Green *529*
Tecca Ccb 1 *499*
Tecca Ccb 1 *543*
Tecca Ccb 1 *665*
Tecca Lq1 *468*
Tecca Lq1 *522*
Tecca Lq1 *535*
Tecca P2 *469*
Tecca P2 *545*
Tecca P2 *685*
Teepol Products Vapona Fly Killer *274*
Teknar Hp-D *222*
Teknar Hp-Fc *222*
Tenopa *219*
Tenopa *289*
Tent Treatment *311*
Terminator *338*
Terminator *394*
Terminator Fly And Wasp Killer *338*
Terminator Fly And Wasp Killer *394*
Terminator Slow Release Fly Killer *274*
Terminator Small Space Fly And Moth Killer *274*
Terminator Wasp & Insect Nest Destroyer *338*
Terminator Wasp & Insect Nest Destroyer *394*
Terminator Wasp And Insect Nest Destroyer *338*
Terminator Wasp And Insect Nest Destroyer *394*
Tetra Construction Fungicidal Wash *453*
Tetra Fungicidal Wash *453*
Texas Ant Gun *257*
Texas Creosote *549*
Texas Fungicidal Wash *419*
Tfa 10 *29*
Tfa 10 *86*
Tfa 10 *129*
Tfa 10 H *29*
Tfa 10 H *86*
Tfa 10 H *129*
Tfa 10 Hg *13*
Tfa 10 Hg *74*
Tfa 10 Hg *126*
Tfa 10 La *12*
Tfa 10 La *128*
Tfa 10g *13*
Tfa 10g *74*
Tfa 10g *126*
Tfa-10 La Q *32*
Tfa-10 La Q *77*

Tfa-10 La Q *91*
Thompson's 60 Second Cleaner *453*
Thompson's All Purpose Wood Preserver *733*
Thompson's All Purpose Wood Preserver Green *529*
Thompson's End Grain Preserver *494*
Thompson's End Grain Preserver *526*
Thompson's End Grain Preserver *671*
Thompson's Interior And Exterior Fungicidal Spray *439*
Thompson's Interior And Exterior Fungicidal Spray *440*
Thompson's Interior Mould Killer *439*
Thompson's Interior Mould Killer *440*
Thompson's Mould And Moss Killer *419*
Thompson's Patio And Paving Cleaner *419*
Thompson's Universal Wood Treatment *655*
Thompson's Universal Wood Treatment *740*
Tiger Cruising *10*
Tiger Cruising White *47*
Tiger Cruising White *114*
Tiger Cruising-A Antifouling *54*
Tiger Cruising-A Antifouling *95*
Tigerline *15*
Tigerline *108*
Tigerline Antifouling *47*
Tigerline Antifouling *114*
Tim-Bor *582*
Timber Fungicide *666*
Timber Master 6 In 1 Wood Treatment *574*
Timber Master 6 In 1 Wood Treatment *653*
Timber Master 6 In 1 Wood Treatment *718*
Timber Paste *652*
Timber Paste *719*
Timber Preservative *568*
Timber Preservative *733*
Timber Preservative Clear Tfp7 *578*
Timber Preservative Clear Tfp7 *721*
Timber Preservative Clear Tfp7 *739*
Timber Preservative Green *529*
Timber Preservative With Repellent *714*
Timber Treatment Plus *666*
Timber Water Repellent *600*
Timber-Gel Ready To Use Microemulsion Timber Fluid *608*
Timber-Gel Ready To Use Microemulsion Timber Fluid *644*
Timber-Gel Ready To Use Microemulsion Timber Fluid *669*
Timber-Treat Ecology *607*
Timber-Treat Ecology *635*
Timbercare Microporous Exterior Preservative *581*
Timbercare Microporous Exterior Preservative *737*
Timberdip *474*

HSE PRODUCT INDEX

Timberguard 1 Plus 7 Concentrate *465*
Timberlife Extra *568*
Timbertek *684*
Timbertek Conc *684*
Timbertex Pd *589*
Timbertex Pd *624*
Timbertex Pd *712*
Timbertex Pi *582*
Timbertex Pi/2 *684*
Timbertone T10 Preservative *621*
Timbor Paste *582*
Timbrene Green Wood Preserver *529*
Timbrene Supreme *574*
Timbrene Supreme *653*
Timbrene Supreme *718*
Timbrene Wood Preserver *733*
Timbrene Woodworm Killer *633*
Timeless Timber All Purpose End Coat And
    Preserver *714*
Titan Fga *27*
Titan Fga *94*
Titan Fga Antifouling *15*
Titan Fga Antifouling *108*
Titan Fga Antifouling White *47*
Titan Fga Antifouling White *114*
Titan Fga-C Antifouling *61*
Titan Fga-C Antifouling *171*
Titan Tin Free *14*
Titan Tin Free *137*
Titan Ultra *15*
Titan Ultra *108*
Tomcat 2 Blox *401*
Tomcat 2 Mouse Bait Station *401*
Tomcat 2 Rat And Mouse Blox *401*
Tomcat 2 Rat And Mouse Killer *401*
Tomcat 2 Rodenticide *401*
Tomcat 2 Super Blox *401*
Tomcat Blox *411*
Top Kill Insect Killer *286*
Top Kill Insect Killer *379*
Torkill Fungicidal Solution 'W' *439*
Torkill Fungicidal Solution 'W' *440*
Total Home Guard *302*
Total Home Guard *327*
Total Mouse Killer System *401*
Tourist Flying Insect Killer *344*
Townex Mouse Poison *399*
Townex Sachets *407*
Tox Exterminating Fly And Wasp Killer *286*
Tox Exterminating Fly And Wasp Killer *379*
Trade Ronseal Low Odour Wood Preserver *733*
Traditional Preservative Oil *567*
Traditional Preservative Oil *645*
Traditional Preservative Oil *674*
Transocean Cleanship Antifouling 2.90 *10*
Transocean Masterline Antifouling 2.84 *38*

Transocean Masterline Antifouling 2.84 *148*
Transocean Masterline Antifouling 2.84 *159*
Transocean Masterline Antifouling 2.84 *179*
Transocean Optima Antifouling 2.34 *27*
Transocean Optima Antifouling 2.34 *94*
Trappit Ant Powder *347*
Travel Masters Anti-Mosquito Coils *208*
Travel Masters No Bite Mosquito Killer
    Replacement Tablets *228*
Travis Perkins Creosote *549*
Travis Perkins Fungicidal Wash *419*
Tribor 10 *423*
Tribor 10 *437*
Tribor 20 *582*
Tribor Gel *582*
Triflex Sterilising Solution Concentrate *419*
Trilanco Farm Fly Spray *347*
Trilanco Fly Killer *368*
Trilanco Fly Killer *386*
Trilanco Fly Spray *231*
Trilanco Fly Spray *243*
Trilanco Original Fly Spray *347*
Trilanco Super Fly Spray *347*
Trilux *50*
Trilux *66*
Trimethrin 20s *633*
Trimethrin 2aq *633*
Trimethrin 3aq *633*
Trimethrin 3os *633*
Trimethrin Aq Plus *652*
Trimethrin Aq Plus *719*
Trimethrin Os Plus *652*
Trimethrin Os Plus *719*
Tripaste *607*
Tripaste *635*
Tripaste Pb *652*
Tripaste Pb *719*
Tripaste Pp *607*
Tripaste Pp *635*
Trisol 120 *666*
Trisol 22 *448*
Trisol 23 *441*
Tritec *633*
Tritec 120 *633*
Tritec 120 Plus *643*
Tritec 120 Plus *675*
Tritec 121 *633*
Tritec 121 Plus *643*
Tritec 121 Plus *675*
Tritec Plus *652*
Tritec Plus *719*
Triton 120 *633*
Triton Timber Rot Killer *666*
Triton Woodworm Killer *633*
Tropical Super Service Antifouling Paint *10*
Ts 10240 Antifouling Ada160 Series *10*

Turbair Beetle Killer   *284*
Turbair Beetle Killer   *323*
Turbair Beetle Killer   *366*
Tyrax   *238*
Tyrax   *319*
Tyrax   *350*
Ukc-2 Dry Rot Killer   *441*
Ukc-2 Dry Rot Killer Concentrate   *441*
Ukc-2 Dry Rot Killer Concentrate   *600*
Ukc-8 Woodworm Killer Concentrate   *633*
Ukc-8 Woodworm Killer Microemulsion
   Concentrate   *633*
Ultra Tech 2000 D   *643*
Ultra Tech 2000 D   *675*
Ultra Tech 2000 F   *666*
Ultra Tech 2000 I   *633*
Ultra Tech 2000m   *441*
Ultrabond Wood Preservative   *568*
Ultrabond Woodworm Killer   *576*
Ultrabond Woodworm Killer   *717*
Unisil S Silicone Waterproofing Solution   *439*
Unisil S Silicone Waterproofing Solution   *440*
Unitas Antifouling Paint Chocolate   *10*
Unitas Antifouling Paint Red   *10*
Unitas Antifouling Paint White   *45*
Unitas Fungicidal Wash-Exterior (Solvent
   Based)   *439*
Unitas Fungicidal Wash-Exterior (Solvent
   Based)   *440*
Unitas Fungicidal Wash-Interior (Water
   Based)   *439*
Unitas Fungicidal Wash-Interior (Water
   Based)   *440*
Universal   *463*
Universal   *637*
Universal Fluid (Grade Pb)   *652*
Universal Fluid (Grade Pb)   *719*
Universal P.I.   *608*
Universal P.I.   *644*
Universal P.I.   *669*
Universal Wood Preservative   *554*
Universal Wood Preservative   *572*
Universal Wood Preserver   *652*
Universal Wood Preserver   *719*
Vacsele 2611   *658*
Vacsele 2611   *743*
Vacsele Aqua 6106   *643*
Vacsele Aqua 6106   *675*
Vacsele P2312   *658*
Vacsele P2312   *743*
Vacsol (2:1 Conc) 2711/2712   *658*
Vacsol (2:1 Conc) 2711/2712   *743*
Vacsol 2234 J Conc   *725*
Vacsol 2243   *725*
Vacsol 2244   *725*
Vacsol 2622/2623 Wr   *654*

Vacsol 2622/2623 Wr   *724*
Vacsol 2625/2626 Wr 2:1 Concentrate   *654*
Vacsol 2625/2626 Wr 2:1 Concentrate   *724*
Vacsol 2627 Azure   *646*
Vacsol 2627 Azure   *682*
Vacsol 2627 Azure   *707*
Vacsol 2652/2653 Jwr   *722*
Vacsol 2709/2710   *658*
Vacsol 2709/2710   *743*
Vacsol 2713/2714   *654*
Vacsol 2713/2714   *724*
Vacsol 2718 2:1 Conc   *609*
Vacsol 2718 2:1 Conc   *647*
Vacsol 2718 2:1 Conc   *678*
Vacsol 2718 2:1 Conc   *697*
Vacsol 2719 Rtu   *609*
Vacsol 2719 Rtu   *647*
Vacsol 2719 Rtu   *678*
Vacsol 2719 Rtu   *697*
Vacsol 2746/2747 J   *722*
Vacsol Aqua 6101   *485*
Vacsol Aqua 6101   *699*
Vacsol Aqua 6105   *484*
Vacsol Aqua 6105   *650*
Vacsol Aqua 6105   *700*
Vacsol Aqua 6107   *609*
Vacsol Aqua 6107   *647*
Vacsol Aqua 6107   *678*
Vacsol Aqua 6107   *697*
Vacsol Aqua 6108   *646*
Vacsol Aqua 6108   *682*
Vacsol Aqua 6108   *707*
Vacsol Aqua 6108 Rtu   *646*
Vacsol Aqua 6108 Rtu   *682*
Vacsol Aqua 6108 Rtu   *707*
Vacsol Aqua 6109   *646*
Vacsol Aqua 6109   *682*
Vacsol Aqua 6109   *707*
Vacsol Azure 2720   *611*
Vacsol Azure 2720   *677*
Vacsol Azure 2720   *698*
Vacsol Azure 2721   *611*
Vacsol Azure 2721   *677*
Vacsol Azure 2721   *698*
Vacsol Azure 2722   *609*
Vacsol Azure 2722   *647*
Vacsol Azure 2722   *678*
Vacsol Azure 2722   *697*
Vacsol J (2:1 Conc) 2744/2745   *741*
Vacsol J 2742/2743   *741*
Vacsol J Rtu   *725*
Vacsol Jwr (2:1 Conc) 2642/2643   *741*
Vacsol Jwr 2640/2641   *741*
Vacsol Jwr Concentrate   *725*
Vacsol Jwr Rtu   *725*
Vacsol Mwr Concentrate 2203   *618*

HSE
PRODUCT INDEX

Vacsol Mwr Concentrate 2203   *727*
Vacsol Mwr Ready To Use 2204   *618*
Vacsol Mwr Ready To Use 2204   *727*
Vacsol P Ready To Use (2335)   *741*
Vacsol Wr (2:1 Conc) 2614/2615   *658*
Vacsol Wr (2:1 Conc) 2614/2615   *743*
Vacsol Wr 2612/2613   *658*
Vacsol Wr 2612/2613   *743*
Vacsol Wr Ready To Use 2116   *618*
Vacsol Wr Ready To Use 2116   *727*
Vallance Fungicidal Wash   *439*
Vallance Fungicidal Wash   *440*
Valspar Creosote   *549*
Vape Mat   *344*
Vapona Ant And Crawling Insect Killer   *236*
Vapona Ant And Crawling Insect Killer   *312*
Vapona Ant And Crawling Insect Killer
  Aerosol   *263*
Vapona Ant And Crawling Insect Killer
  Aerosol   *376*
Vapona Ant And Crawling Insect Killer
  Powder   *311*
Vapona Ant And Crawling Insect Spray   *210*
Vapona Ant And Crawling Insect Spray   *265*
Vapona Ant And Woodlice Killer Powder   *267*
Vapona Ant Bait   *248*
Vapona Ant Bait   *398*
Vapona Ant Chalk   *257*
Vapona Ant Trap   *398*
Vapona Anti Dust Mite   *335*
Vapona Antpen   *257*
Vapona Carpet And Household Flea Powder   *311*
Vapona Citronella Candle   *200*
Vapona Fly And Wasp Killer   *236*
Vapona Fly And Wasp Killer   *312*
Vapona Fly And Wasp Killer Spray   *280*
Vapona Fly And Wasp Killer Spray   *378*
Vapona Fly Killer   *274*
Vapona Fly Killer Window Sticker   *221*
Vapona Flypen   *257*
Vapona Green Arrow Fly And Wasp Killer   *347*
Vapona House And Plant Fly Aerosol   *347*
Vapona House And Plant Fly Spray   *338*
Vapona House And Plant Fly Spray   *394*
Vapona Max Concentrated Fly And Wasp
  Killer   *340*
Vapona Max Concentrated Fly And Wasp
  Killer   *395*
Vapona Micro-Tech   *248*
Vapona Mosquito Travel Plug   *344*
Vapona Moth Killer   *274*
Vapona Mothaks   *203*
Vapona Mothaks   *205*
Vapona Plug-In   *228*
Vapona Plug-In Flying Insect Killer   *209*
Vapona Professional Cockroach Killer   *274*

Vapona Small Space Fly Killer   *274*
Vapona Wasp And Fly Killer   *338*
Vapona Wasp And Fly Killer   *394*
Vapona Wasp And Fly Killer Spray   *338*
Vapona Wasp And Fly Killer Spray   *394*
Vapona Wasp Killer Aerosol   *240*
Vapona Wasp Killer Aerosol   *387*
Vc 17m   *1*
Vc 17m Tropicana   *5*
Vc 17m Tropicana   *107*
Vc 17m-Ep   *1*
Vc 17m-Hs   *3*
Vc 17m-Hs   *106*
Vc 17m-Hs   *166*
Vc Offshore   *15*
Vc Offshore   *47*
Vc Offshore   *108*
Vc Offshore   *114*
Vc Offshore Extra 100 Series   *41*
Vc Offshore Extra 100 Series   *170*
Vc Prop-O-Drev   *47*
Vc Prop-O-Drev   *114*
Vc17m   *4*
Vectobac 12as   *222*
Vet Kem Acclaim 2000   *302*
Vet Kem Acclaim 2000   *327*
Vet-Kem Acclaim   *307*
Vet-Kem Acclaim   *360*
Vet-Kem Pump Spray   *302*
Vet-Kem Pump Spray   *327*
Vetpet Crusade   *311*
Vetzyme Household Patrol   *322*
Vetzyme Household Patrol   *364*
Vetzyme Pestkill Household Spray   *338*
Vetzyme Pestkill Household Spray   *394*
Vetzyme Rug Patrol   *335*
Vf For Fleas   *245*
Vijurrax Spray   *271*
Vijurrax Spray   *276*
Viniline   *10*
Vinilstop 9926 Red.   *10*
Vinyl Antifouling 2000   *10*
Vitapet Flearid Household Spray   *338*
Vitapet Flearid Household Spray   *394*
Vitax Kontrol Ant Killer Powder   *311*
Vulcan C   *257*
W.S.P. Wood Treatment   *684*
Wahl Envoyage Mosquito Killer   *228*
Wall Fungicide   *441*
Warfarin Concentrate   *413*
Warfarin Ready Mixed Bait   *413*
Wasp Destroyer Foam   *338*
Wasp Destroyer Foam   *394*
Wasp Killer   *232*
Wasp Killer   *244*
Wasp Nest Destroyer   *223*

Wasp Nest Destroyer *368*
Wasp Nest Destroyer *386*
Wasp Nest Killer Professional *223*
Waspend *342*
Waspend *369*
Waspend *385*
Waspex *293*
Water Based Wood Preserver (Concentrate) *553*
Water Based Wood Preserver (Concentrate) *604*
Water Based Woodworm Fluid *633*
Water Based Woodworm Killer *551*
Water-Based Wood Preserver *553*
Water-Based Wood Preserver *604*
Waterbased Woodworm Killer 5x *551*
Waterline *15*
Waterline *108*
Waterline White *47*
Waterline White *114*
Waterways Antifouling *27*
Waterways Antifouling *94*
Weathershield Aquatech Preservative
     Basecoat *516*
Weathershield Aquatech Preservative
     Basecoat *586*
Weathershield Exterior Preservative
     Basecoat *577*
Weathershield Exterior Preservative
     Basecoat *720*
Weathershield Exterior Preservative
     Basecoat *732*
Weathershield Exterior Preservative Primer *462*
Weathershield Exterior Preservative Primer *569*
Weathershield Exterior Timber Preservative *534*
Weathershield Exterior Timber Preservative *736*
Weathershield Fungicidal Wash *439*
Weathershield Fungicidal Wash *440*
Weathershield Fungicidal Wash *594*
Weathershield Fungicidal Wash *596*
Weathershield Mulit-Surface Fungicidal
     Wash *423*
Weathershield Mulit-Surface Fungicidal
     Wash *437*
Weathershield Mulit-Surface Fungicidal
     Wash *480*
Weathershield Mulit-Surface Fungicidal
     Wash *584*
Weathershield Preservative Basecoat *462*
Weathershield Preservative Basecoat *569*
Weathershield Preservative Primer *577*
Weathershield Preservative Primer *720*
Weathershield Preservative Primer *732*
Weathershield Timber Preservative *460*
Westland Ant Killer Rtu *267*
Westland Kill-All Ant Killer Powder *223*
Whiskas Care Bedding Spray *217*
Whiskas Care Bedding Spray *339*

Whiskas Care Anti-Flea Carpet Powder *311*
Whiskas Care Bedding Pest Control Spray *217*
Whiskas Care Bedding Pest Control Spray *339*
Whiskas Care Household Anti Flea Spray *306*
Whiskas Care Household Anti Flea Spray *356*
Whiskas Care Household Spray *301*
Whiskas Care Household Spray *315*
White Tiger Cruising *122*
White Tiger Cruising *165*
Wickes All Purpose Wood Treatment *463*
Wickes All Purpose Wood Treatment *637*
Wickes Creosote *524*
Wickes Decorative Wood Preserver *462*
Wickes Decorative Wood Preserver *569*
Wickes Decorative Wood Preserver Rich
     Cedar *462*
Wickes Decorative Wood Preserver Rich
     Cedar *569*
Wickes Exterior Wood Preserver *460*
Wickes Fungicidal Wash *419*
Wickes Mouldkill Emulsion *182*
Wickes Mouldkill Emulsion *186*
Wickes Mouldkill Emulsion *193*
Wickes Wood Preserver Clear *460*
Wickes Woodworm Killer *633*
Wilko Ant Destroyer *223*
Wilko Ant Killer Lacquer *270*
Wilko Ant Killer Spray *257*
Wilko Creosote *524*
Wilko Exterior System Preservative Primer *581*
Wilko Exterior System Preservative Primer *737*
Wilko New Ant Killer Spray *267*
Wilko Woodlice Killer *223*
Willo-Zawb *335*
Willodorm *335*
Wintox *273*
Wintox *281*
Wintox *377*
Wocosen 100 SI *448*
Wocosen 100 SI *666*
Wocosen 100 SI-C *555*
Wocosen 100 SI-C *673*
Wocosen 12 OI *643*
Wocosen 12 OI *675*
Wocosen S *666*
Wocosen S-P *643*
Wocosen S-P *675*
Wolmanit Cx-P *494*
Wolmanit Cx-P *526*
Wolmanit Cx-P *671*
Wolsit Kd 20 *643*
Wolsit Kd 20 *675*
Wolsit Kd 20c *643*
Wolsit Kd 20c *675*
Wood Preservative *568*
Wood Preservative *730*

Wood Preservative Aa 155/00   *579*
Wood Preservative Aa 155/00   *726*
Wood Preservative Aa 155/01   *725*
Wood Preservative Aa 155/03   *579*
Wood Preservative Aa 155/03   *726*
Wood Preservative Aa155   *579*
Wood Preservative Aa155   *726*
Wood Preservative Clear   *655*
Wood Preservative Clear   *740*
Wood Preservative Type 3   *524*
Wood Preserver Green   *529*
Woodlice Killer   *223*
Woodman Creosote And Light Brown
   Creosote   *549*
Woodstain Aquarethane   *750*
Woodstain Aquarethane   *753*
Woodtreat   *614*
Woodtreat   *629*
Woodtreat 25   *607*
Woodtreat 25   *635*
Woodtreat Boron 40   *582*
Woodtreat Bp   *655*
Woodtreat Bp   *740*
Woodworm 8 Micro Emulsifiable
   Concentrate   *633*
Woodworm Fluid (B) Ec   *633*
Woodworm Fluid (B) For Bonded
   Warehouses   *633*
Woodworm Fluid B   *633*
Woodworm Fluid Fp   *633*
Woodworm Fluid Z Emulsion Concentrate   *633*
Woodworm Killer   *551*
Woodworm Killer   *633*
Woodworm Killer   *652*
Woodworm Killer   *719*
Woodworm Killer (Grade P)   *633*
Woodworm Killer And Rot Treatment   *417*
Woodworm Killer And Rot Treatment   *444*
Woodworm Killer And Rot Treatment   *472*
Woodworm Killer And Rot Treatment   *638*
Woodworm Killer P8   *633*
Woodworm Paste B   *633*
Woodworm Treatment Spray   *633*
Woolworth Mosquito Killer   *228*
Woolworths Ant Killer   *223*
Woolworths New Ant Killer Spray   *267*
Woolworths Woodlice Killer   *223*
X Moss   *418*
X Moss   *473*
X Thallo   *432*
X Thallo   *560*
X-Gnat Advanced Insect Repellent - Fabric
   Spray   *207*
X-Gnat M/C (Mosquito Control)   *317*
X-Gnat M/C (Mosquito Control)   *358*
Xm Anti-Fouling C2000 Cruising Self Eroding   *15*
Xm Anti-Fouling C2000 Cruising Self
   Eroding   *108*
Xm Anti-Fouling P4000 Hard   *15*
Xm Anti-Fouling P4000 Hard   *108*
Xm Antifouling Hs3000 High Performance Self
   Eroding   *15*
Xm Antifouling Hs3000 High Performance Self
   Eroding   *108*
Xyladecor Matt U 404   *471*
Xyladecor Matt U 404   *571*
Xylamon Brown U 101 C   *472*
Xylamon Brown U 101 C   *638*
Xylamon Curative U 152 G/H   *472*
Xylamon Curative U 152 G/H   *638*
Xylamon Primer Dipping Stain U 415   *471*
Xylamon Primer Dipping Stain U 415   *571*
Z-Stop Anti-Wasp Strip   *311*
Z.144 Fungicidal Wash   *420*
Z.144 Fungicidal Wash   *447*
Zappit   *328*
Zappit   *383*
Zappit Fly And Wasp Killer   *338*
Zappit Fly And Wasp Killer   *394*
Zodiac + Flea Spray   *302*
Zodiac + Flea Spray   *327*
Zodiac + Household Flea Spray   *307*
Zodiac + Household Flea Spray   *360*
Zodiac + Pump Spray   *302*
Zodiac + Pump Spray   *327*
Zodiac Fogger   *302*
Zodiac Fogger   *327*
Zodiac Household Flea Spray   *308*
Zodiac Household Flea Spray   *388*
Zodiac Maxi Household Flea Spray   *302*
Zodiac Maxi Household Flea Spray   *327*

10
HSE ACTIVE INGREDIENT INDEX

Acypetacs Copper *458*
Acypetacs Copper + Acypetacs Zinc +
 Permethrin *459*
Acypetacs Zinc *460*
Acypetacs Zinc (Equivalent To 3% Zinc Metal) +
 Dichlofluanid *461*
Acypetacs Zinc + Dichlofluanid *462*
Acypetacs Zinc + Permethrin *463*
Acypetacs Zinc + Permethrin + Acypetacs
 Copper *464*
Alkylaryltrimethyl Ammonium Chloride *414*
Alkylaryltrimethyl Ammonium Chloride *465*
Alkylaryltrimethyl Ammonium Chloride + Boric
 Acid *415*
Alkylaryltrimethyl Ammonium Chloride + Disodium
 Octaborate *416*
Alkyltrimethyl Ammonium Chloride + Sodium
 Tetraborate *466*
Allethrin *208*
d-Allethrin *209*
d-Allethrin + Cypermethrin *210*
d-Allethrin + Cypermethrin + Permethrin +
 Tetramethrin *211*
d-Allethrin + Cypermethrin + Tetramethrin *212*
d-Allethrin + Dichlorvos + Permethrin +
 Tetramethrin *213*
d-Allethrin + Permethrin *214*
d-Allethrin + Permethrin + Tetramethrin *215*
d-Allethrin + Tetramethrin *216*
d-Allethrin + d-Phenothrin *217*
Alphachloralose *399*
Alphacypermethrin *218*
Alphacypermethrin + Flufenoxuron *219*
Alphacypermethrin + Tetramethrin *220*
Ammonium Bifluoride + Sodium Dichromate +
 Sodium Fluoride *467*
Arsenic Pentoxide + Chromium Trioxide + Copper
 Oxide *468*
Arsenic Pentoxide + Copper Sulphate + Sodium
 Dichromate *469*
Azaconazole *470*
Azaconazole + Dichlofluanid *471*
Azaconazole + Permethrin *417*
Azaconazole + Permethrin *472*
Azamethiphos *221*
Bacillus Thuringiensis Var Israelensis *222*
Bendiocarb *223*
Bendiocarb + Pyrethrins *224*
Bendiocarb + Tetramethrin *225*
Benzalkonium Bromide *418*
Benzalkonium Bromide *473*
Benzalkonium Chloride *419*
Benzalkonium Chloride *474*
Benzalkonium Chloride + 2-Phenylphenol *420*
Benzalkonium Chloride + 2-Phenylphenol *475*
Benzalkonium Chloride + Boric Acid *421*

Benzalkonium Chloride + Boric Acid *476*
Benzalkonium Chloride + Boric Acid +
 Dialkyldimethyl Ammonium Chloride +
 Methylene Bis(Thiocyanate) *477*
Benzalkonium Chloride + Carbendazim +
 Dialkyldimethyl Ammonium Chloride *422*
Benzalkonium Chloride + Copper Oxide *478*
Benzalkonium Chloride + Dialkyldimethyl
 Ammonium Chloride *479*
Benzalkonium Chloride + Disodium
 Octaborate *423*
Benzalkonium Chloride + Disodium
 Octaborate *480*
Benzalkonium Chloride + Disodium Octaborate +
 3-Iodo-2-Propynyl-N-Butyl Carbamate *481*
Benzalkonium Chloride + Lindane *482*
Benzalkonium Chloride + Naphthalene +
 Permethrin + Pyrethrins *226*
Benzalkonium Chloride + Naphthalene +
 Permethrin + Pyrethrins *424*
Benzalkonium Chloride + Permethrin *483*
Benzalkonium Chloride + Permethrin +
 Tebuconazole *484*
Benzalkonium Chloride + Tebuconazole *485*
Benzyl Benzoate *227*
Bioallethrin *228*
S-Bioallethrin *229*
S-Bioallethrin + Bioallethrin *230*
Bioallethrin + Bioresmethrin *231*
S-Bioallethrin + Bioresmethrin *232*
S-Bioallethrin + Deltamethrin *233*
S-Bioallethrin + Dichlorvos + Permethrin *234*
Bioallethrin + Methoprene + Permethrin *235*
Bioallethrin + Permethrin *236*
S-Bioallethrin + Permethrin *237*
Bioallethrin + Permethrin + Pyrethrins *238*
Bioallethrin + S-Bioallethrin *239*
S-Bioallethrin + Tetramethrin *240*
S-Bioallethrin + Tetramethrin + Permethrin *241*
Bioresmethrin *242*
Bioresmethrin + Bioallethrin *243*
Bioresmethrin + S-Bioallethrin *244*
Boric Acid *245*
Boric Acid *425*
Boric Acid *486*
Boric Acid + 2-
 (Thiocyanomethylthio)Benzothiazole *487*
Boric Acid + 2-Phenylphenol *488*
Boric Acid + 3-Iodo-2-Propynyl-N-Butyl
 Carbamate *489*
Boric Acid + Alkylaryltrimethyl Ammonium
 Chloride *426*
Boric Acid + Benzalkonium Chloride *427*
Boric Acid + Benzalkonium Chloride *490*
Boric Acid + Chromium Acetate + Copper Sulphate
 + Sodium Dichromate *491*

Boric Acid + Chromium Trioxide + Copper Oxide *492*

Boric Acid + Copper Carbonate Hydroxide *493*

Boric Acid + Copper Carbonate Hydroxide + Propiconazole *494*

Boric Acid + Copper Carbonate Hydroxide + Propiconazole + Tebuconazole *495*

Boric Acid + Copper Carbonate Hydroxide + Tebuconazole *496*

Boric Acid + Copper Oxide *497*

Boric Acid + Copper Sulphate *498*

Boric Acid + Copper Sulphate + Potassium Dichromate *499*

Boric Acid + Copper Sulphate + Sodium Dichromate *500*

Boric Acid + Cypermethrin + Propiconazole *428*

Boric Acid + Cypermethrin + Propiconazole *501*

Boric Acid + Dialkyldimethyl Ammonium Chloride *502*

Boric Acid + Dialkyldimethyl Ammonium Chloride + Methylene Bis(Thiocyanate) + Benzalkonium Chloride *503*

Boric Acid + Disodium Octaborate *504*

Boric Acid + Disodium Tetraborate *429*

Boric Acid + Disodium Tetraborate *505*

Boric Acid + Methylene Bis(Thiocyanate) *506*

Boric Acid + Methylene Bis(Thiocyanate) + 2-(Thiocyanomethylthio)Benzothiazole *507*

Boric Acid + Permethrin *508*

Boric Acid + Permethrin + Zinc Octoate *509*

Boric Acid + Propiconazole *510*

Boric Acid + Sodium 2,4,6-Trichlorophenoxide *511*

Boric Acid + Sodium 2,4,6-Trichlorophenoxide *744*

Boric Acid + Sodium Tetraborate *512*

Boric Acid + Zinc Octoate *513*

Boric Oxide *514*

Brodifacoum *400*

Bromadiolone *401*

Calciferol *402*

Calciferol + Difenacoum *403*

Carbendazim *181*

Carbendazim *515*

Carbendazim + Dialkyldimethyl Ammonium Chloride + Benzalkonium Chloride *430*

Carbendazim + Dichlorophenyl Dimethylurea + 2-Octyl-2h-Isothiazol-3-One *745*

Carbendazim + Dichlorophenyl Dimethylurea + 2-Octyl-2h-Isothiazolin-3-One *182*

Carbendazim + Dichlorophenyl Dimethylurea + 2-Octyl-2h-Isothiazolin-3-One *746*

Carbendazim + Dichlorophenyl Dimethylurea + Dithio-2,2'-Bis(Benzmethylamide) + 2-Octyl-2h-Isothiazolin-3-One *183*

Carbendazim + Disodium Octaborate *516*

Carbon Dioxide *246*

Ceto-Stearyl Diethoxylate + Oleyl Monoethoxylate *247*

5-Chloro-2-Methyl-4-Isothiazolin-3-One + 2-Methyl-4-Isothiazolin-3-One *517*

5-Chloro-2-Methyl-4-Isothiazolin-3-One + Copper Sulphate + 2-Methyl-4-Isothiazolin-3-One *518*

Chlorophacinone *404*

Chlorpyrifos *248*

Chlorpyrifos + Cypermethrin *249*

Chlorpyrifos + Cypermethrin + Pyrethrins *250*

Chlorpyrifos + Pyrethrins *251*

Chlorpyrifos + Tetramethrin *252*

Chlorpyrifos-Methyl *253*

Chlorpyrifos-Methyl + Permethrin + Pyrethrins *254*

Cholecalciferol *405*

Cholecalciferol + Difenacoum *406*

Chromium Acetate + Copper Sulphate + Sodium Dichromate *519*

Chromium Acetate + Copper Sulphate + Sodium Dichromate + Boric Acid *520*

Chromium Trioxide + Copper Oxide *521*

Chromium Trioxide + Copper Oxide + Arsenic Pentoxide *522*

Chromium Trioxide + Copper Oxide + Boric Acid *523*

Citronella Oil *200*

Coal Tar Creosote *524*

Copper *1*

Copper + Dichlorophenyl Dimethylurea *2*

Copper + Zinc Pyrithione + 2-Methylthio-4-Tertiary-Butylamino-6-Cyclopropylamino-S-Triazine *3*

Copper Carbonate Hydroxide + Boric Acid *525*

Copper Carbonate Hydroxide + Propiconazole + Boric Acid *526*

Copper Carbonate Hydroxide + Propiconazole + Tebuconazole + Boric Acid *527*

Copper Carbonate Hydroxide + Tebuconazole + Boric Acid *528*

Copper Metal *4*

Copper Metal + 2-Methylthio-4-Tertiary-Butylamino-6-Cyclopropylamino-S-Triazine *5*

Copper Naphthenate *529*

Copper Naphthenate + 3-Iodo-2-Propynyl-N-Butyl Carbamate *530*

Copper Naphthenate + Cuprous Oxide + Dichlofluanid + Zinc Naphthenate *6*

Copper Naphthenate + Disodium Octaborate *531*

Copper Naphthenate + Permethrin *532*

Copper Naphthenate + Tri(Hexylene Glycol)Biborate *533*

Copper Naphthenate + Zinc Octoate *534*

Copper Oxide + Arsenic Pentoxide + Chromium Trioxide *535*

Copper Oxide + Benzalkonium Chloride *536*

Copper Oxide + Boric Acid 537
Copper Oxide + Boric Acid + Chromium
Trioxide 538
Copper Oxide + Chromium Trioxide 539
Copper Oxide + Dialkyldimethyl Ammonium
Chloride 540
Copper Resinate + Cuprous Oxide 7
Copper Resinate + Cuprous Oxide + Zineb 8
Copper Sulphate + 2,4,5,6-Tetrachloro
Isophthalonitrile 9
Copper Sulphate + 2-Methyl-4-Isothiazolin-3-One
+ 5-Chloro-2-Methyl-4-Isothiazolin-3-One 541
Copper Sulphate + Boric Acid 542
Copper Sulphate + Potassium Dichromate + Boric
Acid 543
Copper Sulphate + Sodium Dichromate 544
Copper Sulphate + Sodium Dichromate + Arsenic
Pentoxide 545
Copper Sulphate + Sodium Dichromate + Boric
Acid 546
Copper Sulphate + Sodium Dichromate + Boric
Acid + Chromium Acetate 547
Copper Sulphate + Sodium Dichromate +
Chromium Acetate 548
Coumatetralyl 407
Creosote 549
Cuprous Oxide 10
Cuprous Oxide + 2,3,5,6-Tetrachloro-4-(Methyl
Sulphonyl)Pyridine 11
Cuprous Oxide + 2,4,5,6-Tetrachloro
Isophthalonitrile 12
Cuprous Oxide + 2,4,5,6-Tetrachloro
Isophthalonitrile + 4,5-Dichloro-2-N-Octyl-4-
Isothiazolin-3-One 13
Cuprous Oxide + 2-
(Thiocyanomethylthio)Benzothiazole 14
Cuprous Oxide + 2-Methylthio-4-Tertiary-
Butylamino-6-Cyclopropylamino-S-Triazine 15
Cuprous Oxide + 2-Methylthio-4-Tertiary-
Butylamino-6-Cyclopropylamino-S-Triazine +
4,5-Dichloro-2-N-Octyl-4-Isothiazolin-3-
One 16
Cuprous Oxide + 4,5-Dichloro-2-N-Octyl-4-
Isothiazolin-3-One 17
Cuprous Oxide + Copper Resinate 18
Cuprous Oxide + Cuprous Sulphide 19
Cuprous Oxide + Cuprous Thiocyanate + 2,3,5,6-
Tetrachloro-4-(Methyl Sulphonyl)Pyridine + 2-
Methylthio-4-Tertiary-Butylamino-6-
Cyclopropylamino-S-Triazine 20
Cuprous Oxide + Cuprous Thiocyanate + 2-
Methylthio-4-Tertiary-Butylamino-6-
Cyclopropylamino-S-Triazine + 4,5-Dichloro-2-
N-Octyl-4-Isothiazolin-3-One 21

Cuprous Oxide + Cuprous Thiocyanate +
Dichlorophenyl Dimethylurea + Zineb + 2,3,5,6-
Tetrachloro-4-(Methyl Sulphonyl)Pyridine 22
Cuprous Oxide + Dichlofluanid 23
Cuprous Oxide + Dichlofluanid + 2-Methylthio-4-
Tertiary-Butylamino-6-Cyclopropylamino-S-
Triazine 24
Cuprous Oxide + Dichlofluanid + Dichlorophenyl
Dimethylurea + 2-Methylthio-4-Tertiary-
Butylamino-6-Cyclopropylamino-S-Triazine 25
Cuprous Oxide + Dichlofluanid + Zinc Naphthenate
+ Copper Naphthenate 26
Cuprous Oxide + Dichlorophenyl
Dimethylurea 27
Cuprous Oxide + Dichlorophenyl Dimethylurea +
2,3,5,6-Tetrachloro-4-(Methyl
Sulphonyl)Pyridine 28
Cuprous Oxide + Dichlorophenyl Dimethylurea +
2,4,5,6-Tetrachloro Isophthalonitrile 29
Cuprous Oxide + Dichlorophenyl Dimethylurea +
2,4,5,6-Tetrachloro Isophthalonitrile + 2-
Methylthio-4-Tertiary-Butylamino-6-
Cyclopropylamino-S-Triazine 30
Cuprous Oxide + Dichlorophenyl Dimethylurea + 2-
Methylthio-4-Tertiary-Butylamino-6-
Cyclopropylamino-S-Triazine 31
Cuprous Oxide + Dichlorophenyl Dimethylurea +
4,5-Dichloro-2-N-Octyl-4-Isothiazolin-3-
One 32
Cuprous Oxide + Dichlorophenyl Dimethylurea +
Tributyltin Methacrylate + Tributyltin Oxide 33
Cuprous Oxide + Tributyltin Acrylate 34
Cuprous Oxide + Tributyltin Methacrylate 35
Cuprous Oxide + Tributyltin Methacrylate +
Tributyltin Oxide 36
Cuprous Oxide + Tributyltin Methacrylate +
Tributyltin Oxide + 4,5-Dichloro-2-N-Octyl-4-
Isothiazolin-3-One 37
Cuprous Oxide + Tributyltin Methacrylate +
Tributyltin Oxide + Zineb 38
Cuprous Oxide + Tributyltin Methacrylate +
Zineb 39
Cuprous Oxide + Zinc Naphthenate 40
Cuprous Oxide + Zinc Pyrithione 41
Cuprous Oxide + Zineb 42
Cuprous Oxide + Zineb + Copper Resinate 43
Cuprous Sulphide + Cuprous Oxide 44
Cuprous Thiocyanate 45
Cuprous Thiocyanate + 2,3,5,6-Tetrachloro-4-
(Methyl Sulphonyl)Pyridine + 2-Methylthio-4-
Tertiary-Butylamino-6-Cyclopropylamino-S-
Triazine + Cuprous Oxide 46
Cuprous Thiocyanate + 2-Methylthio-4-Tertiary-
Butylamino-6-Cyclopropylamino-S-Triazine 47
Cuprous Thiocyanate + 2-Methylthio-4-Tertiary-
Butylamino-6-Cyclopropylamino-S-Triazine +

4,5-Dichloro-2-N-Octyl-4-Isothiazolin-3-One + Cuprous Oxide *48*

Cuprous Thiocyanate + 4,5-Dichloro-2-N-Octyl-4-Isothiazolin-3-One *49*

Cuprous Thiocyanate + Dichlofluanid *50*

Cuprous Thiocyanate + Dichlofluanid + 2-Methylthio-4-Tertiary-Butylamino-6-Cyclopropylamino-S-Triazine *51*

Cuprous Thiocyanate + Dichlofluanid + Dichlorophenyl Dimethylurea + 2-Methylthio-4-Tertiary-Butylamino-6-Cyclopropylamino-S-Triazine *52*

Cuprous Thiocyanate + Dichlofluanid + Zinc Pyrithione + 2-Methylthio-4-Tertiary-Butylamino-6-Cyclopropylamino-S-Triazine *53*

Cuprous Thiocyanate + Dichlorophenyl Dimethylurea *54*

Cuprous Thiocyanate + Dichlorophenyl Dimethylurea + 2,3,5,6-Tetrachloro-4-(Methyl Sulphonyl)Pyridine *55*

Cuprous Thiocyanate + Dichlorophenyl Dimethylurea + 2-Methylthio-4-Tertiary-Butylamino-6-Cyclopropylamino-S-Triazine *56*

Cuprous Thiocyanate + Dichlorophenyl Dimethylurea + 4,5-Dichloro-2-N-Octyl-4-Isothiazolin-3-One *57*

Cuprous Thiocyanate + Dichlorophenyl Dimethylurea + Zineb + 2,3,5,6-Tetrachloro-4-(Methyl Sulphonyl)Pyridine + Cuprous Oxide *58*

Cuprous Thiocyanate + Tributyltin Methacrylate + Tributyltin Oxide *59*

Cuprous Thiocyanate + Tributyltin Methacrylate + Tributyltin Oxide + Zineb *60*

Cuprous Thiocyanate + Zinc Pyrithione *61*

Cyfluthrin *550*

Cyfluthrin + Pyriproxyfen *255*

Cyfluthrin + Transfluthrin *256*

Cypermethrin *257*

Cypermethrin *551*

Cypermethrin + 2-Phenylphenol *552*

Cypermethrin + 3-Iodo-2-Propynyl-N-Butyl Carbamate *553*

Cypermethrin + Chlorpyrifos *258*

Cypermethrin + Dichlofluanid *554*

Cypermethrin + Methoprene *259*

Cypermethrin + Permethrin + Tetramethrin + D-Allethrin *260*

Cypermethrin + Propiconazole *555*

Cypermethrin + Propiconazole + Boric Acid *431*

Cypermethrin + Propiconazole + Boric Acid *556*

Cypermethrin + Pyrethrins *261*

Cypermethrin + Pyrethrins + Chlorpyrifos *262*

Cypermethrin + Tetramethrin *263*

Cypermethrin + Tetramethrin + d-Allethrin *264*

Cypermethrin + Tri(Hexylene Glycol)Biborate *557*

Cypermethrin + Zinc Versatate *558*

Cypermethrin + d-Allethrin *265*

Cyromazine + Permethrin *266*

Deltamethrin *201*

Deltamethrin *267*

Deltamethrin *559*

Deltamethrin + Pyrethrins *268*

Deltamethrin + S-Bioallethrin *269*

Dialkyldimethyl Ammonium Chloride *432*

Dialkyldimethyl Ammonium Chloride *560*

Dialkyldimethyl Ammonium Chloride + 3-Iodo-2-Propynyl-N-Butyl Carbamate *561*

Dialkyldimethyl Ammonium Chloride + Benzalkonium Chloride *562*

Dialkyldimethyl Ammonium Chloride + Benzalkonium Chloride + Carbendazim *433*

Dialkyldimethyl Ammonium Chloride + Boric Acid *563*

Dialkyldimethyl Ammonium Chloride + Copper Oxide *564*

Dialkyldimethyl Ammonium Chloride + Disodium Octaborate *565*

Dialkyldimethyl Ammonium Chloride + Methylene Bis(Thiocyanate) + Benzalkonium Chloride + Boric Acid *566*

Dialkyldimethyl Ammonium Chloride + Permethrin + Propiconazole *567*

Diazinon *270*

Diazinon + Dichlorvos *271*

Dichlofluanid *62*

Dichlofluanid *184*

Dichlofluanid *568*

Dichlofluanid + 2-Methylthio-4-Tertiary-Butylamino-6-Cyclopropylamino-S-Triazine + Cuprous Oxide *63*

Dichlofluanid + 2-Methylthio-4-Tertiary-Butylamino-6-Cyclopropylamino-S-Triazine + Cuprous Thiocyanate *64*

Dichlofluanid + Acypetacs Zinc *569*

Dichlofluanid + Acypetacs Zinc (Equivalent To 3% Zinc Metal) *570*

Dichlofluanid + Azaconazole *571*

Dichlofluanid + Cuprous Oxide *65*

Dichlofluanid + Cuprous Thiocyanate *66*

Dichlofluanid + Cypermethrin *572*

Dichlofluanid + Dichlorophenyl Dimethylurea + 2-Methylthio-4-Tertiary-Butylamino-6-Cyclopropylamino-S-Triazine + Cuprous Oxide *67*

Dichlofluanid + Dichlorophenyl Dimethylurea + 2-Methylthio-4-Tertiary-Butylamino-6-Cyclopropylamino-S-Triazine + Cuprous Thiocyanate *68*

Dichlofluanid + Dichlorophenyl Dimethylurea +
Zinc Pyrithione 69
Dichlofluanid + Permethrin + Tebuconazole 573
Dichlofluanid + Permethrin + Tri(Hexylene
Glycol)Biborate 574
Dichlofluanid + Permethrin + Zinc Octoate 575
Dichlofluanid + Tri(Hexylene Glycol)Biborate 576
Dichlofluanid + Tri(Hexylene Glycol)Biborate +
Zinc Naphthenate 577
Dichlofluanid + Tri(Hexylene Glycol)Biborate +
Zinc Octoate 578
Dichlofluanid + Tributyltin Oxide 579
Dichlofluanid + Zinc Naphthenate 580
Dichlofluanid + Zinc Naphthenate + Copper
Naphthenate + Cuprous Oxide 70
Dichlofluanid + Zinc Octoate 581
Dichlofluanid + Zinc Pyrithione + 2-Methylthio-4-
Tertiary-Butylamino-6-Cyclopropylamino-S-
Triazine + Cuprous Thiocyanate 71
4,5-Dichloro-2-N-Octyl-4-Isothiazolin-3-One + 2-
Methylthio-4-Tertiary-Butylamino-6-
Cyclopropylamino-S-Triazine 72
4,5-Dichloro-2-N-Octyl-4-Isothiazolin-3-One +
Cuprous Oxide 73
4,5-Dichloro-2-N-Octyl-4-Isothiazolin-3-One +
Cuprous Oxide + 2,4,5,6-Tetrachloro
Isophthalonitrile 74
4,5-Dichloro-2-N-Octyl-4-Isothiazolin-3-One +
Cuprous Oxide + 2-Methylthio-4-Tertiary-
Butylamino-6-Cyclopropylamino-S-Triazine 75
4,5-Dichloro-2-N-Octyl-4-Isothiazolin-3-One +
Cuprous Oxide + Cuprous Thiocyanate + 2-
Methylthio-4-Tertiary-Butylamino-6-
Cyclopropylamino-S-Triazine 76
4,5-Dichloro-2-N-Octyl-4-Isothiazolin-3-One +
Cuprous Oxide + Dichlorophenyl
Dimethylurea 77
4,5-Dichloro-2-N-Octyl-4-Isothiazolin-3-One +
Cuprous Oxide + Tributyltin Methacrylate +
Tributyltin Oxide 78
4,5-Dichloro-2-N-Octyl-4-Isothiazolin-3-One +
Cuprous Thiocyanate 79
4,5-Dichloro-2-N-Octyl-4-Isothiazolin-3-One +
Cuprous Thiocyanate + Dichlorophenyl
Dimethylurea 80
4,5-Dichloro-2-N-Octyl-4-Isothiazolin-3-One +
Tributyltin Methacrylate + Tributyltin Oxide 81
4,5-Dichloro-2-N-Octyl-4-Isothiazolin-3-One + Zinc
Pyrithione 82
1,4-Dichlorobenzene 202
1,4-Dichlorobenzene + Dichlorvos 272
1,4-Dichlorobenzene + Naphthalene 203
Dichlorophen 185
Dichlorophen 434
Dichlorophen + Dichlorvos + Tetramethrin 273

Dichlorophenyl Dimethylurea + 2,3,5,6-
Tetrachloro-4-(Methyl Sulphonyl)Pyridine +
Cuprous Oxide 83
Dichlorophenyl Dimethylurea + 2,3,5,6-
Tetrachloro-4-(Methyl Sulphonyl)Pyridine +
Cuprous Thiocyanate 84
Dichlorophenyl Dimethylurea + 2,4,5,6-Tetrachloro
Isophthalonitrile + 2-Methylthio-4-Tertiary
Butylamino-6-Cyclopropylamino-S-Triazine +
Cuprous Oxide 85
Dichlorophenyl Dimethylurea + 2,4,5,6-Tetrachloro
Isophthalonitrile + Cuprous Oxide 86
Dichlorophenyl Dimethylurea + 2-Methylthio-4-
Tertiary-Butylamino-6-Cyclopropylamino-S-
Triazine + Cuprous Oxide 87
Dichlorophenyl Dimethylurea + 2-Methylthio-4-
Tertiary-Butylamino-6-Cyclopropylamino-S-
Triazine + Cuprous Oxide + Dichlofluanid 88
Dichlorophenyl Dimethylurea + 2-Methylthio-4-
Tertiary-Butylamino-6-Cyclopropylamino-S-
Triazine + Cuprous Thiocyanate 89
Dichlorophenyl Dimethylurea + 2-Methylthio-4-
Tertiary-Butylamino-6-Cyclopropylamino-S-
Triazine + Cuprous Thiocyanate +
Dichlofluanid 90
Dichlorophenyl Dimethylurea + 2-Octyl-2h-
Isothiazol-3-One + Carbendazim 747
Dichlorophenyl Dimethylurea + 2-Octyl-2h-
Isothiazolin-3-One + Carbendazim 186
Dichlorophenyl Dimethylurea + 2-Octyl-2h-
Isothiazolin-3-One + Carbendazim 748
Dichlorophenyl Dimethylurea + 4,5-Dichloro-2-N-
Octyl-4-Isothiazolin-3-One + Cuprous
Oxide 91
Dichlorophenyl Dimethylurea + 4,5-Dichloro-2-N-
Octyl-4-Isothiazolin-3-One + Cuprous
Thiocyanate 92
Dichlorophenyl Dimethylurea + Copper 93
Dichlorophenyl Dimethylurea + Cuprous
Oxide 94
Dichlorophenyl Dimethylurea + Cuprous
Thiocyanate 95
Dichlorophenyl Dimethylurea + Dithio-2,2'-
Bis(Benzmethylamide) + 2-Octyl-2h-Isothiazolin-
3-One + Carbendazim 187
Dichlorophenyl Dimethylurea + Tributyltin
Methacrylate + Tributyltin Oxide + Cuprous
Oxide 96
Dichlorophenyl Dimethylurea + Zinc
Pyrithione 97
Dichlorophenyl Dimethylurea + Zinc Pyrithione + 2-
(Thiocyanomethylthio)Benzothiazole + 2-
Methylthio-4-Tertiary-Butylamino-6-
Cyclopropylamino-S-Triazine 98

Dichlorophenyl Dimethylurea + Zinc Pyrithione + 2-Methylthio-4-Tertiary-Butylamino-6-Cyclopropylamino-S-Triazine 99

Dichlorophenyl Dimethylurea + Zinc Pyrithione + Dichlofluanid 100

Dichlorophenyl Dimethylurea + Zineb + 2,3,5,6-Tetrachloro-4-(Methyl Sulphonyl)Pyridine + Cuprous Oxide + Cuprous Thiocyanate 101

Dichlorvos 274

Dichlorvos + 1,4-Dichlorobenzene 275

Dichlorvos + Diazinon 276

Dichlorvos + Iodofenphos 277

Dichlorvos + Permethrin + S-Bioallethrin 278

Dichlorvos + Permethrin + Tetramethrin + d-Allethrin 279

Dichlorvos + Tetramethrin 280

Dichlorvos + Tetramethrin + Dichlorophen 281

Difenacoum 408

Difenacoum + Calciferol 409

Difenacoum + Cholecalciferol 410

Diphacinone 411

Disodium Octaborate 435

Disodium Octaborate 582

Disodium Octaborate + 3-Iodo-2-Propynyl-N-Butyl Carbamate + Benzalkonium Chloride 583

Disodium Octaborate + Alkylaryltrimethyl Ammonium Chloride 436

Disodium Octaborate + Benzalkonium Chloride 437

Disodium Octaborate + Benzalkonium Chloride 584

Disodium Octaborate + Boric Acid 585

Disodium Octaborate + Carbendazim 586

Disodium Octaborate + Copper Naphthenate 587

Disodium Octaborate + Dialkyldimethyl Ammonium Chloride 588

Disodium Octaborate + Methylene Bis(Thiocyanate) + 2-(Thiocyanomethylthio)Benzothiazole 589

Disodium Octaborate + Permethrin + Propiconazole + Tebuconazole 590

Disodium Octaborate + Propiconazole + Tebuconazole 591

Disodium Octaborate + Sodium Pentachlorophenoxide 592

Disodium Tetraborate 282

Disodium Tetraborate + Boric Acid 438

Disodium Tetraborate + Boric Acid 593

Dithio-2,2'-Bis(Benzmethylamide) 188

Dithio-2,2'-Bis(Benzmethylamide) + 2-Octyl-2h-Isothiazolin-3-One + Carbendazim + Dichlorophenyl Dimethylurea 189

Dodecylamine Lactate + Dodecylamine Salicylate 439

Dodecylamine Lactate + Dodecylamine Salicylate 594

Dodecylamine Lactate + Dodecylamine Salicylate + Permethrin 595

Dodecylamine Salicylate + Dodecylamine Lactate 440

Dodecylamine Salicylate + Dodecylamine Lactate 596

Dodecylamine Salicylate + Permethrin + Dodecylamine Lactate 597

Fenitrothion 283

Fenitrothion + Permethrin + Resmethrin 284

Fenitrothion + Permethrin + Tetramethrin 285

Fenitrothion + Tetramethrin 286

Fipronil 287

Flocoumafen 412

Flufenoxuron 288

Flufenoxuron 598

Flufenoxuron + Alphacypermethrin 289

Flufenoxuron + Propiconazole + 3-Iodo-2-Propynyl-N-Butyl Carbamate 599

Hydramethylnon 290

Hydroprene 291

S-Hydroprene 292

3-Iodo-2-Propynyl-N-Butyl Carbamate 190

3-Iodo-2-Propynyl-N-Butyl Carbamate 441

3-Iodo-2-Propynyl-N-Butyl Carbamate 600

3-Iodo-2-Propynyl-N-Butyl Carbamate 749

3-Iodo-2-Propynyl-N-Butyl Carbamate + Benzalkonium Chloride + Disodium Octaborate 601

3-Iodo-2-Propynyl-N-Butyl Carbamate + Boric Acid 602

3-Iodo-2-Propynyl-N-Butyl Carbamate + Copper Naphthenate 603

3-Iodo-2-Propynyl-N-Butyl Carbamate + Cypermethrin 604

3-Iodo-2-Propynyl-N-Butyl Carbamate + Dialkyldimethyl Ammonium Chloride 605

3-Iodo-2-Propynyl-N-Butyl Carbamate + Flufenoxuron + Propiconazole 606

3-Iodo-2-Propynyl-N-Butyl Carbamate + Permethrin 607

3-Iodo-2-Propynyl-N-Butyl Carbamate + Permethrin 750

3-Iodo-2-Propynyl-N-Butyl Carbamate + Permethrin + Propiconazole 608

3-Iodo-2-Propynyl-N-Butyl Carbamate + Permethrin + Propiconazole + Tebuconazole 609

3-Iodo-2-Propynyl-N-Butyl Carbamate + Propiconazole 191

3-Iodo-2-Propynyl-N-Butyl Carbamate + Propiconazole 610

3-Iodo-2-Propynyl-N-Butyl Carbamate + Propiconazole + Tebuconazole 611

Iodofenphos 293

Iodofenphos + Dichlorvos 294

Lambda-Cyhalothrin 295
Lindane 296
Lindane 612
Lindane + Benzalkonium Chloride 613
Lindane + Pentachlorophenol 614
Lindane + Pentachlorophenol + Tributyltin
  Oxide 615
Lindane + Pentachlorophenyl Laurate 616
Lindane + Tetramethrin 297
Lindane + Tributyltin Naphthenate 617
Lindane + Tributyltin Oxide 618
Methoprene 298
S-Methoprene 299
Methoprene + Cypermethrin 300
Methoprene + Permethrin 301
S-Methoprene + Permethrin 302
Methoprene + Permethrin + Bioallethrin 303
Methoprene + Permethrin + Pyrethrins 304
Methoprene + Permethrin + Tetramethrin 305
Methoprene + Pyrethrins 306
S-Methoprene + Pyrethrins 307
S-Methoprene + Tetramethrin 308
2-Methyl-4-Isothiazolin-3-One + 5-Chloro-2-
  Methyl-4-Isothiazolin-3-One 619
2-Methyl-4-Isothiazolin-3-One + 5-Chloro-2-
  Methyl-4-Isothiazolin-3-One + Copper
  Sulphate 620
Methylene Bis(Thiocyanate) 621
Methylene Bis(Thiocyanate) + 2-
  (Thiocyanomethylthio)Benzothiazole 622
Methylene Bis(Thiocyanate) + 2-
  (Thiocyanomethylthio)Benzothiazole + Boric
  Acid 623
Methylene Bis(Thiocyanate) + 2-
  (Thiocyanomethylthio)Benzothiazole + Disodium
  Octaborate 624
Methylene Bis(Thiocyanate) + Benzalkonium
  Chloride + Boric Acid + Dialkyldimethyl
  Ammonium Chloride 625
Methylene Bis(Thiocyanate) + Boric Acid 626
2-Methylthio-4-Tertiary-Butylamino-6-
  Cyclopropylamino-S-Triazine + 2,3,5,6-
  Tetrachloro-4-(Methyl Sulphonyl)Pyridine 102
2-Methylthio-4-Tertiary-Butylamino-6-
  Cyclopropylamino-S-Triazine + 4,5-Dichloro-2-
  N-Octyl-4-Isothiazolin-3-One 103
2-Methylthio-4-Tertiary-Butylamino-6-
  Cyclopropylamino-S-Triazine + 4,5-Dichloro-2-
  N-Octyl-4-Isothiazolin-3-One + Cuprous
  Oxide 104
2-Methylthio-4-Tertiary-Butylamino-6-
  Cyclopropylamino-S-Triazine + 4,5-Dichloro-2-
  N-Octyl-4-Isothiazolin-3-One + Cuprous Oxide +
  Cuprous Thiocyanate 105

2-Methylthio-4-Tertiary-Butylamino-6-
  Cyclopropylamino-S-Triazine + Copper + Zinc
  Pyrithione 106
2-Methylthio-4-Tertiary-Butylamino-6-
  Cyclopropylamino-S-Triazine + Copper
  Metal 107
2-Methylthio-4-Tertiary-Butylamino-6-
  Cyclopropylamino-S-Triazine + Cuprous
  Oxide 108
2-Methylthio-4-Tertiary-Butylamino-6-
  Cyclopropylamino-S-Triazine + Cuprous Oxide +
  Cuprous Thiocyanate + 2,3,5,6-Tetrachloro-4-
  (Methyl Sulphonyl)Pyridine 109
2-Methylthio-4-Tertiary-Butylamino-6-
  Cyclopropylamino-S-Triazine + Cuprous Oxide +
  Dichlofluanid 110
2-Methylthio-4-Tertiary-Butylamino-6-
  Cyclopropylamino-S-Triazine + Cuprous Oxide +
  Dichlofluanid + Dichlorophenyl
  Dimethylurea 111
2-Methylthio-4-Tertiary-Butylamino-6-
  Cyclopropylamino-S-Triazine + Cuprous Oxide +
  Dichlorophenyl Dimethylurea 112
2-Methylthio-4-Tertiary-Butylamino-6-
  Cyclopropylamino-S-Triazine + Cuprous Oxide +
  Dichlorophenyl Dimethylurea + 2,4,5,6-
  Tetrachloro Isophthalonitrile 113
2-Methylthio-4-Tertiary-Butylamino-6-
  Cyclopropylamino-S-Triazine + Cuprous
  Thiocyanate 114
2-Methylthio-4-Tertiary-Butylamino-6-
  Cyclopropylamino-S-Triazine + Cuprous
  Thiocyanate + Dichlofluanid 115
2-Methylthio-4-Tertiary-Butylamino-6-
  Cyclopropylamino-S-Triazine + Cuprous
  Thiocyanate + Dichlofluanid + Dichlorophenyl
  Dimethylurea 116
2-Methylthio-4-Tertiary-Butylamino-6-
  Cyclopropylamino-S-Triazine + Cuprous
  Thiocyanate + Dichlofluanid + Zinc
  Pyrithione 117
2-Methylthio-4-Tertiary-Butylamino-6-
  Cyclopropylamino-S-Triazine + Cuprous
  Thiocyanate + Dichlorophenyl
  Dimethylurea 118
2-Methylthio-4-Tertiary-Butylamino-6-
  Cyclopropylamino-S-Triazine + Dichlorophenyl
  Dimethylurea + Zinc Pyrithione 119
2-Methylthio-4-Tertiary-Butylamino-6-
  Cyclopropylamino-S-Triazine + Dichlorophenyl
  Dimethylurea + Zinc Pyrithione + 2-
  (Thiocyanomethylthio)Benzothiazole 120
2-Methylthio-4-Tertiary-Butylamino-6-
  Cyclopropylamino-S-Triazine + Tributyltin
  Methacrylate + Tributyltin Oxide 121

2-Methylthio-4-Tertiary-Butylamino-6-Cyclopropylamino-S-Triazine + Zinc Pyrithione *122*
Naphthalene *204*
Naphthalene + 1,4-Dichlorobenzene *205*
Naphthalene + Permethrin + Pyrethrins + Benzalkonium Chloride *309*
Naphthalene + Permethrin + Pyrethrins + Benzalkonium Chloride *442*
No Recognised Active Ingredient *123*
2-Octyl-2h-Isothiazol-3-One + Carbendazim + Dichlorophenyl Dimethylurea *751*
2-Octyl-2h-Isothiazolin-3-One *192*
2-Octyl-2h-Isothiazolin-3-One *443*
2-Octyl-2h-Isothiazolin-3-One + Carbendazim + Dichlorophenyl Dimethylurea *193*
Octyl-2h-Isothiazolin-3-One + Carbendazim + Dichlorophenyl Dimethylurea *752*
Octyl-2h-Isothiazolin-3-One + Carbendazim + Dichlorophenyl Dimethylurea + Dithio-2,2'-Bis(Benzmethylamide) *194*
Oleyl Monoethoxylate + Ceto-Stearyl Diethoxylate *310*
Oxine-Copper *627*
Pentachlorophenol *628*
Pentachlorophenol + Lindane *629*
Pentachlorophenol + Tributyltin Oxide *630*
Pentachlorophenol + Tributyltin Oxide + Lindane *631*
Pentachlorophenyl Laurate + Lindane *632*
Permethrin *206*
Permethrin *311*
Permethrin *633*
Permethrin + 2-Phenylphenol *634*
Permethrin + 3-Iodo-2-Propynyl-N-Butyl Carbamate *635*
Permethrin + 3-Iodo-2-Propynyl-N-Butyl Carbamate *753*
Permethrin + Acypetacs Copper + Acypetacs Zinc *636*
Permethrin + Acypetacs Zinc *637*
Permethrin + Azaconazole *444*
Permethrin + Azaconazole *638*
Permethrin + Benzalkonium Chloride *639*
Permethrin + Bioallethrin *312*
Permethrin + Bioallethrin + Methoprene *313*
Permethrin + Boric Acid *640*
Permethrin + Copper Naphthenate *641*
Permethrin + Cyromazine *314*
Permethrin + Dodecylamine Lactate + Dodecylamine Salicylate *642*
Permethrin + Methoprene *315*
Permethrin + Propiconazole *316*
Permethrin + Propiconazole *643*
Permethrin + Propiconazole + 3-Iodo-2-Propynyl-N-Butyl Carbamate *644*

Permethrin + Propiconazole + Dialkyldimethyl Ammonium Chloride *645*
Permethrin + Propiconazole + Tebuconazole *646*
Permethrin + Propiconazole + Tebuconazole + 3-Iodo-2-Propynyl-N-Butyl Carbamate *647*
Permethrin + Propiconazole + Tebuconazole + Disodium Octaborate *648*
Permethrin + Pyrethrins *317*
Permethrin + Pyrethrins + 2,3,5,6-Tetrachloro-4-(Methyl Sulphonyl)Pyridine *195*
Permethrin + Pyrethrins + Benzalkonium Chloride + Naphthalene *318*
Permethrin + Pyrethrins + Benzalkonium Chloride + Naphthalene *445*
Permethrin + Pyrethrins + Bioallethrin *319*
Permethrin + Pyrethrins + Chlorpyrifos-Methyl *320*
Permethrin + Pyrethrins + Methoprene *321*
Permethrin + Pyriproxyfen *322*
Permethrin + Resmethrin + Fenitrothion *323*
Permethrin + S-Bioallethrin *324*
Permethrin + S-Bioallethrin + Dichlorvos *325*
Permethrin + S-Bioallethrin + Tetramethrin *326*
Permethrin + S-Methoprene *327*
Permethrin + Tebuconazole *649*
Permethrin + Tebuconazole + Benzalkonium Chloride *650*
Permethrin + Tebuconazole + Dichlofluanid *651*
Permethrin + Tetramethrin *328*
Permethrin + Tetramethrin + Fenitrothion *329*
Permethrin + Tetramethrin + Methoprene *330*
Permethrin + Tetramethrin + d-Allethrin *331*
Permethrin + Tetramethrin + d-Allethrin + Cypermethrin *332*
Permethrin + Tetramethrin + d-Allethrin + Dichlorvos *333*
Permethrin + Tri(Hexylene Glycol)Biborate *652*
Permethrin + Tri(Hexylene Glycol)Biborate + Dichlofluanid *653*
Permethrin + Tributyltin Naphthenate *654*
Permethrin + Zinc Octoate *446*
Permethrin + Zinc Octoate *655*
Permethrin + Zinc Octoate + Boric Acid *656*
Permethrin + Zinc Octoate + Dichlofluanid *657*
Permethrin + Zinc Versatate *658*
Permethrin + d-Allethrin *334*
Phenothrin *335*
d-Phenothrin + Pyrethrins *336*
d-Phenothrin + Pyriproxyfen *337*
d-Phenothrin + Tetramethrin *338*
d-Phenothrin + d-Allethrin *339*
d-Phenothrin + d-Tetramethrin *340*
2-Phenylphenol *659*
2-Phenylphenol + Benzalkonium Chloride *447*
2-Phenylphenol + Benzalkonium Chloride *660*
2-Phenylphenol + Boric Acid *661*

HSE ACTIVE INGREDIENT INDEX

2-Phenylphenol + Cypermethrin 662
2-Phenylphenol + Permethrin 663
Pirimiphos-Methyl 341
Pirimiphos-Methyl 664
Pirimiphos-Methyl + Resmethrin +
  Tetramethrin 342
Potassium Dichromate + Boric Acid + Copper
  Sulphate 665
Potassium Salts Of Fatty Acids 343
Prallethrin 344
Propiconazole 448
Propiconazole 666
Propiconazole + 3-Iodo-2-Propynyl-N-Butyl
  Carbamate 196
Propiconazole + 3-Iodo-2-Propynyl-N-Butyl
  Carbamate 667
Propiconazole + 3-Iodo-2-Propynyl-N-Butyl
  Carbamate + Flufenoxuron 668
Propiconazole + 3-Iodo-2-Propynyl-N-Butyl
  Carbamate + Permethrin 669
Propiconazole + Boric Acid 670
Propiconazole + Boric Acid + Copper Carbonate
  Hydroxide 671
Propiconazole + Boric Acid + Cypermethrin 449
Propiconazole + Boric Acid + Cypermethrin 672
Propiconazole + Cypermethrin 673
Propiconazole + Dialkyldimethyl Ammonium
  Chloride + Permethrin 674
Propiconazole + Permethrin 345
Propiconazole + Permethrin 675
Propiconazole + Tebuconazole 676
Propiconazole + Tebuconazole + 3-Iodo-2-
  Propynyl-N-Butyl Carbamate 677
Propiconazole + Tebuconazole + 3-Iodo-2-
  Propynyl-N-Butyl Carbamate + Permethrin 678
Propiconazole + Tebuconazole + Boric Acid +
  Copper Carbonate Hydroxide 679
Propiconazole + Tebuconazole + Disodium
  Octaborate 680
Propiconazole + Tebuconazole + Disodium
  Octaborate + Permethrin 681
Propiconazole + Tebuconazole + Permethrin 682
Propoxur 346
Pyrethrins 207
Pyrethrins 347
Pyrethrins + 2,3,5,6-Tetrachloro-4-(Methyl
  Sulphonyl)Pyridine + Permethrin 197
Pyrethrins + Bendiocarb 348
Pyrethrins + Benzalkonium Chloride +
  Naphthalene + Permethrin 349
Pyrethrins + Benzalkonium Chloride +
  Naphthalene + Permethrin 450
Pyrethrins + Bioallethrin + Permethrin 350
Pyrethrins + Chlorpyrifos 351
Pyrethrins + Chlorpyrifos + Cypermethrin 352

Pyrethrins + Chlorpyrifos-Methyl +
  Permethrin 353
Pyrethrins + Cypermethrin 354
Pyrethrins + Deltamethrin 355
Pyrethrins + Methoprene 356
Pyrethrins + Methoprene + Permethrin 357
Pyrethrins + Permethrin 358
Pyrethrins + Resmethrin 359
Pyrethrins + S-Methoprene 360
Pyrethrins + d-Phenothrin 361
Pyriproxyfen 362
Pyriproxyfen + Cyfluthrin 363
Pyriproxyfen + Permethrin 364
Pyriproxyfen + d-Phenothrin 365
Resmethrin + Fenitrothion + Permethrin 366
Resmethrin + Pyrethrins 367
Resmethrin + Tetramethrin 368
Resmethrin + Tetramethrin + Pirimiphos-
  Methyl 369
Silicon Dioxide(Amorphous Non Coated
  Precipitated) 370
Sodium 2,4,6-Trichlorophenoxide 754
Sodium 2,4,6-Trichlorophenoxide + Boric
  Acid 683
Sodium 2,4,6-Trichlorophenoxide + Boric
  Acid 755
Sodium 2-Phenylphenoxide 451
Sodium 2-Phenylphenoxide 684
Sodium Dichlorophen 452
Sodium Dichromate + Arsenic Pentoxide + Copper
  Sulphate 685
Sodium Dichromate + Boric Acid + Chromium
  Acetate + Copper Sulphate 686
Sodium Dichromate + Boric Acid + Copper
  Sulphate 687
Sodium Dichromate + Chromium Acetate + Copper
  Sulphate 688
Sodium Dichromate + Copper Sulphate 689
Sodium Dichromate + Sodium Fluoride +
  Ammonium Bifluoride 690
Sodium Fluoride + Ammonium Bifluoride + Sodium
  Dichromate 691
Sodium Hypochlorite 453
Sodium Pentachlorophenoxide 454
Sodium Pentachlorophenoxide 692
Sodium Pentachlorophenoxide + Disodium
  Octaborate 693
Sodium Tetraborate 371
Sodium Tetraborate + Alkyltrimethyl Ammonium
  Chloride 694
Sodium Tetraborate + Boric Acid 695
Tebuconazole 696
Tebuconazole + 3-Iodo-2-Propynyl-N-Butyl
  Carbamate + Permethrin + Propiconazole 697
Tebuconazole + 3-Iodo-2-Propynyl-N-Butyl
  Carbamate + Propiconazole 698

Tebuconazole + Benzalkonium Chloride 699
Tebuconazole + Benzalkonium Chloride + Permethrin 700
Tebuconazole + Boric Acid + Copper Carbonate Hydroxide 701
Tebuconazole + Boric Acid + Copper Carbonate Hydroxide + Propiconazole 702
Tebuconazole + Dichlofluanid + Permethrin 703
Tebuconazole + Disodium Octaborate + Permethrin + Propiconazole 704
Tebuconazole + Disodium Octaborate + Propiconazole 705
Tebuconazole + Permethrin 706
Tebuconazole + Permethrin + Propiconazole 707
Tebuconazole + Propiconazole 708
2,4,5,6-Tetrachloro Isophthalonitrile 124
2,4,5,6-Tetrachloro Isophthalonitrile + 2-Methylthio-4-Tertiary-Butylamino-6-Cyclopropylamino-S-Triazine + Cuprous Oxide + Dichlorophenyl Dimethylurea 125
2,4,5,6-Tetrachloro Isophthalonitrile + 4,5-Dichloro-2-N-Octyl-4-Isothiazolin-3-One + Cuprous Oxide 126
2,4,5,6-Tetrachloro Isophthalonitrile + Copper Sulphate 127
2,4,5,6-Tetrachloro Isophthalonitrile + Cuprous Oxide 128
2,4,5,6-Tetrachloro Isophthalonitrile + Cuprous Oxide + Dichlorophenyl Dimethylurea 129
2,3,5,6-Tetrachloro-4-(Methyl Sulphonyl)Pyridine 198
2,3,5,6-Tetrachloro-4-(Methyl Sulphonyl)Pyridine + 2-Methylthio-4-Tertiary-Butylamino-6-Cyclopropylamino-S-Triazine 130
2,3,5,6-Tetrachloro-4-(Methyl Sulphonyl)Pyridine + 2-Methylthio-4-Tertiary-Butylamino-6-Cyclopropylamino-S-Triazine + Cuprous Oxide + Cuprous Thiocyanate 131
2,3,5,6-Tetrachloro-4-(Methyl Sulphonyl)Pyridine + Cuprous Oxide 132
2,3,5,6-Tetrachloro-4-(Methyl Sulphonyl)Pyridine + Cuprous Oxide + Cuprous Thiocyanate + Dichlorophenyl Dimethylurea + Zineb 133
2,3,5,6-Tetrachloro-4-(Methyl Sulphonyl)Pyridine + Cuprous Oxide + Dichlorophenyl Dimethylurea 134
2,3,5,6-Tetrachloro-4-(Methyl Sulphonyl)Pyridine + Cuprous Thiocyanate + Dichlorophenyl Dimethylurea 135
2,3,5,6-Tetrachloro-4-(Methyl Sulphonyl)Pyridine + Permethrin + Pyrethrins 199
Tetramethrin 372
Tetramethrin + Alphacypermethrin 373
Tetramethrin + Bendiocarb 374
Tetramethrin + Chlorpyrifos 375
Tetramethrin + Cypermethrin 376

Tetramethrin + Dichlorophen + Dichlorvos 377
Tetramethrin + Dichlorvos 378
Tetramethrin + Fenitrothion 379
Tetramethrin + Fenitrothion + Permethrin 380
Tetramethrin + Lindane 381
Tetramethrin + Methoprene + Permethrin 382
Tetramethrin + Permethrin 383
Tetramethrin + Permethrin + S-Bioallethrin 384
Tetramethrin + Pirimiphos-Methyl + Resmethrin 385
Tetramethrin + Resmethrin 386
Tetramethrin + S-Bioallethrin 387
Tetramethrin + S-Methoprene 388
Tetramethrin + d-Allethrin 389
Tetramethrin + d-Allethrin + Cypermethrin 390
Tetramethrin + d-Allethrin + Cypermethrin + Permethrin 391
Tetramethrin + d-Allethrin + Dichlorvos + Permethrin 392
Tetramethrin + d-Allethrin + Permethrin 393
Tetramethrin + d-Phenothrin 394
d-Tetramethrin + d-Phenothrin 395
2-(Thiocyanomethylthio)Benzothiazole 709
2-(Thiocyanomethylthio)Benzothiazole + 2-Methylthio-4-Tertiary-Butylamino-6-Cyclopropylamino-S-Triazine + Dichlorophenyl Dimethylurea + Zinc Pyrithione 136
2-(Thiocyanomethylthio)Benzothiazole + Boric Acid 710
2-(Thiocyanomethylthio)Benzothiazole + Boric Acid + Methylene Bis(Thiocyanate) 711
2-(Thiocyanomethylthio)Benzothiazole + Cuprous Oxide 137
2-(Thiocyanomethylthio)Benzothiazole + Disodium Octaborate + Methylene Bis(Thiocyanate) 712
2-(Thiocyanomethylthio)Benzothiazole + Methylene Bis(Thiocyanate) 713
Thiophanate-Methyl 455
Transfluthrin 396
Transfluthrin + Cyfluthrin 397
Tri(Hexylene Glycol)Biborate 714
Tri(Hexylene Glycol)Biborate + Copper Naphthenate 715
Tri(Hexylene Glycol)Biborate + Cypermethrin 716
Tri(Hexylene Glycol)Biborate + Dichlofluanid 717
Tri(Hexylene Glycol)Biborate + Dichlofluanid + Permethrin 718
Tri(Hexylene Glycol)Biborate + Permethrin 719
Tri(Hexylene Glycol)Biborate + Zinc Naphthenate + Dichlofluanid 720
Tri(Hexylene Glycol)Biborate + Zinc Octoate + Dichlofluanid 721
Tributyltin Acrylate + Cuprous Oxide 138
Tributyltin Methacrylate + Cuprous Oxide 139
Tributyltin Methacrylate + Tributyltin Oxide 140

Tributyltin Methacrylate + Tributyltin Oxide + 2-Methylthio-4-Tertiary-Butylamino-6-Cyclopropylamino-S-Triazine *141*

Tributyltin Methacrylate + Tributyltin Oxide + 4,5-Dichloro-2-N-Octyl-4-Isothiazolin-3-One *142*

Tributyltin Methacrylate + Tributyltin Oxide + 4,5-Dichloro-2-N-Octyl-4-Isothiazolin-3-One + Cuprous Oxide *143*

Tributyltin Methacrylate + Tributyltin Oxide + Cuprous Oxide *144*

Tributyltin Methacrylate + Tributyltin Oxide + Cuprous Oxide + Dichlorophenyl Dimethylurea *145*

Tributyltin Methacrylate + Tributyltin Oxide + Cuprous Thiocyanate *146*

Tributyltin Methacrylate + Tributyltin Oxide + Zinc Pyrithione *147*

Tributyltin Methacrylate + Tributyltin Oxide + Zineb + Cuprous Oxide *148*

Tributyltin Methacrylate + Tributyltin Oxide + Zineb + Cuprous Thiocyanate *149*

Tributyltin Methacrylate + Zineb + Cuprous Oxide *150*

Tributyltin Naphthenate *722*

Tributyltin Naphthenate + Lindane *723*

Tributyltin Naphthenate + Permethrin *724*

Tributyltin Oxide *725*

Tributyltin Oxide + 2-Methylthio-4-Tertiary-Butylamino-6-Cyclopropylamino-S-Triazine + Tributyltin Methacrylate *151*

Tributyltin Oxide + 4,5-Dichloro-2-N-Octyl-4-Isothiazolin-3-One + Cuprous Oxide + Tributyltin Methacrylate *152*

Tributyltin Oxide + 4,5-Dichloro-2-N-Octyl-4-Isothiazolin-3-One + Tributyltin Methacrylate *153*

Tributyltin Oxide + Cuprous Oxide + Dichlorophenyl Dimethylurea + Tributyltin Methacrylate *154*

Tributyltin Oxide + Cuprous Oxide + Tributyltin Methacrylate *155*

Tributyltin Oxide + Cuprous Thiocyanate + Tributyltin Methacrylate *156*

Tributyltin Oxide + Dichlofluanid *726*

Tributyltin Oxide + Lindane *727*

Tributyltin Oxide + Lindane + Pentachlorophenol *728*

Tributyltin Oxide + Pentachlorophenol *729*

Tributyltin Oxide + Tributyltin Methacrylate *157*

Tributyltin Oxide + Zinc Pyrithione + Tributyltin Methacrylate *158*

Tributyltin Oxide + Zineb + Cuprous Oxide + Tributyltin Methacrylate *159*

Tributyltin Oxide + Zineb + Cuprous Thiocyanate + Tributyltin Methacrylate *160*

Trichlorphon *398*

Warfarin *413*

Zinc Naphthenate *730*

Zinc Naphthenate + Copper Naphthenate + Cuprous Oxide + Dichlofluanid *161*

Zinc Naphthenate + Cuprous Oxide *162*

Zinc Naphthenate + Dichlofluanid *731*

Zinc Naphthenate + Dichlofluanid + Tri(Hexylene Glycol)Biborate *732*

Zinc Octoate *456*

Zinc Octoate *733*

Zinc Octoate + Boric Acid *734*

Zinc Octoate + Boric Acid + Permethrin *735*

Zinc Octoate + Copper Naphthenate *736*

Zinc Octoate + Dichlofluanid *737*

Zinc Octoate + Dichlofluanid + Permethrin *738*

Zinc Octoate + Dichlofluanid + Tri(Hexylene Glycol)Biborate *739*

Zinc Octoate + Permethrin *457*

Zinc Octoate + Permethrin *740*

Zinc Pyrithione *163*

Zinc Pyrithione + 2-(Thiocyanomethylthio)Benzothiazole + 2-Methylthio-4-Tertiary-Butylamino-6-Cyclopropylamino-S-Triazine + Dichlorophenyl Dimethylurea *164*

Zinc Pyrithione + 2-Methylthio-4-Tertiary-Butylamino-6-Cyclopropylamino-S-Triazine *165*

Zinc Pyrithione + 2-Methylthio-4-Tertiary-Butylamino-6-Cyclopropylamino-S-Triazine + Copper *166*

Zinc Pyrithione + 2-Methylthio-4-Tertiary-Butylamino-6-Cyclopropylamino-S-Triazine + Cuprous Thiocyanate + Dichlofluanid *167*

Zinc Pyrithione + 2-Methylthio-4-Tertiary-Butylamino-6-Cyclopropylamino-S-Triazine + Dichlorophenyl Dimethylurea *168*

Zinc Pyrithione + 4,5-Dichloro-2-N-Octyl-4-Isothiazolin-3-One *169*

Zinc Pyrithione + Cuprous Oxide *170*

Zinc Pyrithione + Cuprous Thiocyanate *171*

Zinc Pyrithione + Dichlofluanid + Dichlorophenyl Dimethylurea *172*

Zinc Pyrithione + Dichlorophenyl Dimethylurea *173*

Zinc Pyrithione + Tributyltin Methacrylate + Tributyltin Oxide *174*

Zinc Versatate *741*

Zinc Versatate + Cypermethrin *742*

Zinc Versatate + Permethrin *743*

Zineb + 2,3,5,6-Tetrachloro-4-(Methyl Sulphonyl)Pyridine + Cuprous Oxide + Cuprous Thiocyanate + Dichlorophenyl Dimethylurea *175*

Zineb + Copper Resinate + Cuprous Oxide *176*

Zineb + Cuprous Oxide *177*

Zineb + Cuprous Oxide + Tributyltin
  Methacrylate *178*
Zineb + Cuprous Oxide + Tributyltin Methacrylate +
  Tributyltin Oxide *179*
Zineb + Cuprous Thiocyanate + Tributyltin
  Methacrylate + Tributyltin Oxide *180*